PALEOECOLOGY OF BERINGIA

Paleoecology of Beringia, Wenner–Gren Foundation for Anthropological Research
Symposium #81, Burg Wartenstein, Austria, June 8–17, 1979

PALEOECOLOGY OF BERINGIA

Edited by

DAVID M. HOPKINS
U.S. Geological Survey
Menlo Park, California

JOHN V. MATTHEWS, JR.
Geological Survey of Canada
Ottawa, Ontario, Canada

CHARLES E. SCHWEGER
Department of Anthropology
University of Alberta
Edmonton, Alberta, Canada

STEVEN B. YOUNG
Center for Northern Studies
Wolcott, Vermont

Technical Editor
Vincent Stanley
Pine, Colorado

1982

ACADEMIC PRESS
A Subsidiary of Harcourt Brace Jovanovich, Publishers
NEW YORK LONDON
PARIS SAN DIEGO SAN FRANCISCO SÃO PAULO SYDNEY TOKYO TORONTO

ACADEMIC PRESS, INC.
111 Fifth Avenue, New York, New York 10003

United Kingdom Edition published by
ACADEMIC PRESS, INC. (LONDON) LTD.
24/28 Oval Road, London NW1 7DX

Library of Congress Cataloging in Publication Data

Wenner-Gren Foundation for Anthropological Research.
Symposium (81st : Burg Wartenstein, Austria)
Paleoecology of Beringia.

Bibliography: p.
Includes indexes.
1. Paleoecology--Bering Land Bridge--Congresses.
I. Hopkins, David Moody, Date. II. Title.
QE720.W46 1982 560'.45'0916451 82-22621
ISBN 0-12-355860-3

PRINTED IN THE UNITED STATES OF AMERICA

82 83 84 85 9 8 7 6 5 4 3 2 1

CONTENTS

CONTRIBUTORS

Numbers in parentheses indicate the pages on which the authors' contributions begin.

Thomas A. Ager (75), *U.S. Geological Survey, 12201 Sunrise Valley Drive, Reston, Virginia 22092*

Roger G. Barry (195), *CIRES, Campus Box 449, University of Colorado, Boulder, Colorado 80302*

G. F. Baryshnikov (267), *Institute of Zoology, Academy of Science, Leningrad 199164, USSR*

Lawrence C. Bliss (241), *Department of Botany, University of Washington, Seattle, Washington 98195*

Jacques Cinq-Mars (353), *Archaeological Survey of Canada, National Museum of Man, Ottawa, Ontario K1A 0M8, Canada*

Les C. Cwynar (113), *Scarborough College, University of Toronto, 1265 Military Trail, West Hill, Ontario M1C 1A4, Canada*

Tzvi Gal-Chen (205), *Goddard Space Flight Center, Greenbelt, Maryland 20770*

R. E. Giterman (43), *Geological Institute, Academy of Science, Moscow 109017, USSR*

R. Dale Guthrie (307), *Institute of Arctic Biology, University of Alaska, Fairbanks, Alaska 99701*

C. Vance Haynes (383), *Department of Anthropology, University of Arizona, Tucson, Arizona 85721*

Dennis Hibbert (153), *North Seattle Community College, 9600 College Way North, Seattle, Washington 98103*

David M. Hopkins (3), *U.S. Geological Survey, 345 Middlefield Road, Menlo Park, California 94025*

Henryk Kubiak (281), *Institute of Systematic and Experimental Zoology, Polish Academy of Sciences, Slawkowska 17, 31-016, Krakow, Poland*

Stuart A. Marks (409), *603 Turnpike Road, Laurinburg, North Carolina 28352*

P. Jerome Martin (259), *Norwest Feed Research, 4216-121 Street, Edmonton, Alberta T6J 1Y8, Canada*

Paul S. Martin (399), *Department of Geosciences, University of Arizona, Tucson, Arizona 85721*

J. V. Matthews, Jr. (43, 151), *Geological Survey of Canada, 601 Booth Street, Ottawa, Ontario K1A 0E8, Canada*

Richard E. Morlan (353), *Archaeological Survey of Canada, National Museum of Man, Ottawa, Ontario K1A 0M8, Canada*

Hansjürgen Müller-Beck (329), *Institut für Urgeschichte, Universität Tübingen, D-7400 Tübingen 1, German Federal Republic*

Robert C. D. Olivier (291), *29 Alston Place, Colombo, #2, Sri Lanka*

R. E. Redmann (223), *Department of Plant Ecology, University of Saskatchewan, Saskatoon, Saskatchewan S7N 0W0, Canada*

James H. Richards (241), *Department of Range Science, College of Natural Resources, Utah State University, Logan, Utah 84322*

James C. Ritchie (113), *Scarborough College, University of Toronto, 1265 Military Trail, West Hill, Ontario M1C 1A4, Canada*

Charles E. Schweger (95), *Department of Anthropology, University of Alberta, Edmonton P6G 2H4, Canada*

A. V. Sher (43), *Paleontological Museum, Academy of Science, Lenin Prospekt 17, Moscow 117071, USSR*

S. V. Tomirdiaro (29), *Complex Scientific Research Institute (DVNTS), Academy of Science, Gorod Magadan 685010, USSR*

N. K. Vereshschagin (267), *Institute of Zoology, Academy of Science, Leningrad 199164, USSR*

Steven B. Young (179), *Center for Northern Studies, Wolcott, Vermont 05680*

Boris A. Yurtsev (157), *Komarov Botanical Institute, Ul. Popova 2, Leningrad 197022, USSR*

PREFACE

Paleoecology of Beringia is the product of a symposium organized by its editors, sponsored by the Wenner–Gren Foundation for Anthropological Research, and held at the foundation's conference center in Burg Wartenstein, Austria, June 8–17, 1979. It is, however, more than a conference volume. Many of the reports included here are markedly different from the manuscripts submitted at the conference, and all of the contributions were revised in light of the discussions and questions raised at Burg Wartenstein.

Some reports were written after the conference. Hopkins' "Aspects of the Paleogeography" represents his attempt to periodize the Wisconsin cold cycle for unglaciated Beringia. He became aware of the need for this work only at the conference itself, after noting the tendency of the discussions to treat the Wisconsin stage as a single, undifferentiated cold cycle and Ritchie's emphasis on the possible importance of a more productive mid-Wisconsin landscape. Hopkins' inventory and analysis of aeolian deposits, also new, was prepared in response to Barry's emphasis upon the value to paleoclimatic modelers of a broad spectrum of proxy data.

The post-conference contribution by Giterman, Sher, and Matthews compares Siberian and Alaskan paleoecological data and confirms the persistent differences in the East and West Beringian biotas inferred from modern plant records by both Yurtsev and Young.

Morlan and Cinq-Mars' inventory of the early archaeological record in East Beringia is also entirely new. Research published between 1979 and 1982 is discussed by many of the authors and in the synthesis, which was written by the editors in the spring of 1982.

The focus of this volume is on the paradox central to all studies of the unglaciated Arctic during the last Ice Age: that vertebrate fossils indicate that from 45,000 to 11,000 years BP an environment considerably more diverse and productive than the present one existed, whereas the botanical record, where it is not silent, supports a far more conservative appraisal of the region's ability to sustain any but the sparsest forms of plant and animal life.

Two intellectual approaches to the composition of Beringia's late Pleistocene environment have, in part, created the incompatible view of apparently depauperate vegetation and comparatively abundant mammal life. The first approach is largely historical and empirical and looks strictly to the botanic (largely fossil pollen) record for the presence or absence of taxa. The second approach is rooted in ecology and permits scientists to reconstruct the landscape in accordance with the assumed food requirements and climatic restrictions of a number of large, gregarious herbivores once present.

The editors, confronted with contradictory evidence and methodology in Beringian research, prepared a comprehensive guest list for the Burg Wartenstein conference. The discussions of the productivity paradox involved not only an international core group of scientists studying Beringia, but also many specialists with whom Beringian workers have had little previous communication. This second group included two climatologists experienced in climate modeling; two ecologists expert in the productivity of, respectively, modern tundra and grasslands; a range-management analyst specializing in the digestive physiology of ungulates; a student of the ecology and behavior of modern elephants; and an ethnologist who has worked with a contemporary aboriginal hunting tribe.

The diversity of the participants in many ways made the process of the conference as interesting as its contents. The interactions that took place contributed much to the arguments included in this volume's synthesis and to a rich multidisciplinary methodology for future work. The puzzle is not solved, and not many more pieces are in place, but it can be viewed in sharper focus than was possible before. Those who have prepared this volume hope that this is evident in the following reports and synthesis.

ACKNOWLEDGMENTS

Support for the Burg Wartenstein Conference and for the preparation of this book came from many sources. A planning grant from the Research Committee of the National Geographic Society provided an invaluable initial impetus. The National Geographic Society grant also defrayed part of our travel and publication costs. A grant from the National Science Foundation provided major support for travel by American participants in the Burg Wartenstein Conference and made the publication of this book possible. Above all, we are deeply indebted to the Wenner–Gren Foundation for assistance in planning the conference, for hospitality at Burg Wartenstein, and for funding the travel of several participants. Administrative support for both conference and book was provided by the Center for Northern Studies, Wolcott, Vermont. The editors were supported by their home institutions during organization of the conference and preparation of the book. For this our thanks go to the United States Geological Survey, the Geological Survey of Canada, the University of Alberta, and the Center for Northern Studies.

In addition to overseeing activities in support of the conference in New York and at Burg Wartenstein, Dr. Lita Osmundsen, Director of Research for the Wenner–Gren Foundation, provided us with much useful guidance in planning and structuring conference discussions and in planning the book.

Janice Young, while serving as chief administrative officer at the Center for Northern Studies, was a model of patience while providing administrative support and financial management in our chaotic early days.

Vincent Stanley, our text editor, played an exceptionally active role in the preparation of the book. As conference rapporteur and an active participant in our subsequent get-togethers, Vince proved to have a sharp eye for inconsistencies and unresolved contradictions. We owe him great thanks for his patient efforts in overseeing the flow of manuscripts, editing texts for style and consistency, overseeing preparation of galleys and camera-ready copy, and maintaining liaison with Academic Press.

Dorothy Vitaliano provided us with an excellent translation of Tomirdiaro's contribution, which had reached us in Russian. George Soleimani translated correspondence from Russian contributors. We owe great thanks to Dorothy Vitaliano, George Soleimani, and the anonymous translators of the Canadian Government Translation Service for providing us with English renditions of many Russian articles and books that form part of the intellectual underpinnings of this volume.

Robert Ackerman, Jacques Cinq-Mars, Richard Morlan, and John Rick assisted us by providing independent reviews of several contributions and by critiquing successive versions of the synthesis.

Mary Malloy assisted David Hopkins in maintaining communication with contributors and fellow editors, prepared much of the manuscript material, and performed a host of other services indispensable to the planning, organization, and preparation of the book. The late and much-mourned Nancy

Tamamian assisted us at an early stage with advice on book-preparation problems and editorial matters. Peggy Smith did much of the preliminary work on the indexing, verified most of the difficult items in the bibliography, and assisted in many other ways. Nancy Blair also helped us by verifying particularly difficult items in the bibliography. To these colleagues at the United States Geological Survey, we extend our sincere gratitude.

Michael Matthews of Ottawa, Ontario, developed the computer program that was used for organization of the indices. Michael Matthews, Linda Barton, and Nadine Ives spent many hours on the final preparation of the topical and taxonomic indexes.

Tau Rho Alpha enhanced the book by designing the logos that head the major sections. David L. Jones prepared the endpaper map, the location maps in Tomidiaro's and Young's contributions, and all of the illustrations in Hopkins' paper. Other linedrawings in the papers of Matthews, Giterman, Yurtsev, and Tomidiaro were prepared by Ed Belec. To all of these illustrators we extend our thanks.

Most of this volume was set in type and prepared for the camera by Richard Ellington.

Finally, we thank our contributors, who enriched our lives by their presence at Burg Wartenstein and who gave much thought and effort to their papers in this volume.

PALEOGEOGRAPHY

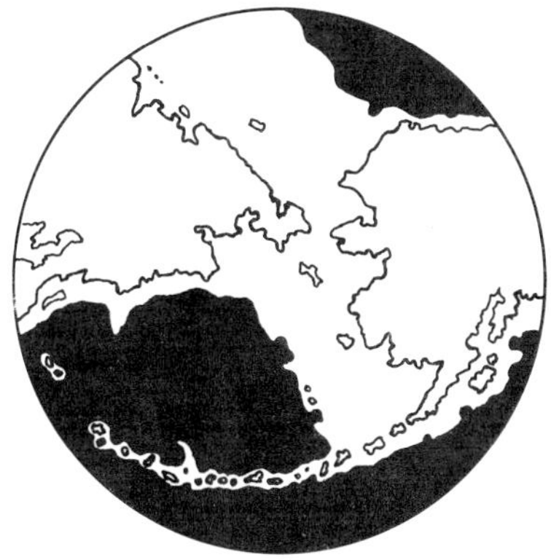

Geography to a large extent determines climate and, in combination with climate, provides the matrix within which ecosystems exist, function, and evolve. The geography of northern regions is dynamic rather than static, and paleogeographic changes underlie the drastic changes in Beringian ecosystems detailed in this book.

In the first of the two reports that follow, Hopkins briefly updates earlier discussions of Beringian paleogeography and attempts to periodize the late Pleistocene of Beringia in order to clarify the series of paleogeographic and paleoclimatic changes that have taken place there during the last 80,000 or 100,000 years; the scheme was developed after the Burg Wartenstein Conference and consequently was not available for use in most of the companion reports in this volume. The inventory of Beringian aeolian deposits that follows underlines the evident former climatic aridity of Beringia and reinforces the paleobotanical and palynological evidence, discussed later, for a predominance of sparse, low, open vegetation.

Tomirdiaro's companion paper treats the evolution of lowland landscapes in northeastern Siberia during late Quaternary time. The perennially frozen, often ice-rich, silty Quaternary sediments that mantle Siberian lowland areas seem surprisingly familiar to North American visitors. These deposits are quite similar to the frozen silt that blankets large areas in and near unglaciated lowlands in Alaska and Yukon Territory, but the origins proposed for the Siberian deposits by most Soviet authors are quite different from those suggested by North American workers. Soviet researchers generally term these deposits "lacustrine-alluvial," reflecting the general belief that they accumulated on river flood plains of enormous extent during Pleistocene alluvial cycles. Tomirdiaro departs from this traditional and still prevalent Soviet interpretation, recognizes the ultimate windblown origin of the widely distributed silt deposits, and then goes on to emphasize a concept important to later discussions in this book—the unity of late Pleistocene periglacial lowland landscapes and the wide distribution of sparse vegetation over a dry, cold region that once extended from western Europe to Beringia.

—David M. Hopkins

1

ASPECTS OF THE PALEOGEOGRAPHY OF BERINGIA DURING THE LATE PLEISTOCENE

David M. Hopkins

This paper reviews knowledge of Beringian glacial chronology, shoreline history, and paleoceanography.

The Happy, Boutellier, and Duvanny Yar intervals are named as subdivisions of the last (Wisconsin) cold cycle in Beringia. The Happy interval, which corresponds to the Itkillik Glaciation, was a cold interval that ended more than 60,000 and perhaps more than 80,000 years ago. The Boutellier interval was a long, complex interstadial interval that began 80,000 or 60,000 years ago and ended about 30,000 years ago. The Duvanny Yar interval was a cold, dry interval that corresponds to such modest glacial advances of late Wisconsin time as the Walker Lake Glaciation in the Brooks Range. An unnamed interval corresponding approximately to the birch pollen-assemblage zone and extending from 14,000 to 8500 years BP constitutes the transition to the climatic and landscape conditions that characterized most of the Holocene.

Sea level probably was low during the Happy interval. A return to approximately present sea level during deposition of the Flaxman Formation on the Beaufort Sea coast may correspond to a mesic, peat-forming episode that marked the beginning of the Boutellier interval. Sea level later fell to about −80 m but returned briefly to a level near −20 m about 42,000 years BP. Rising sea level seems to have submerged the Alaska-Siberia land connection as early as 15,500 years BP.

Many areas in Beringia show a record of moderate aeolian activity during the Boutellier interval, intense activity during the Duvanny Yar interval, diminishing activity during the time of the birch zone, and stabilization about 8500 years BP. During the Duvanny Yar interval, sand movement was active in winter as well as in summer. The paleowind data indicate winter circulation consisting of a counterclockwise gyre concentric to a low in the Gulf of Alaska and summer circulation consisting of southeasterly airflow from the Gulf of Alaska over coastal mountains to interior Beringia.

Key words: Happy interval, Boutellier interval, Duvanny Yar interval, birch pollen-assemblage zone, aeolian deposits, sea-level history, paleoclimates, paleowind directions

INTRODUCTION

In this report, I attempt to reconstruct a part of the physical stage upon which were played the climatic, biological, and anthropological events discussed elsewhere in this volume. To facilitate this discussion, I attempt to periodize the late Quaternary of unglaciated parts of Beringia and propose a series of intervals or stades based on type selections chosen from both sides of the Bering Strait. This report also includes the first

ISBN 0-12-355860-3

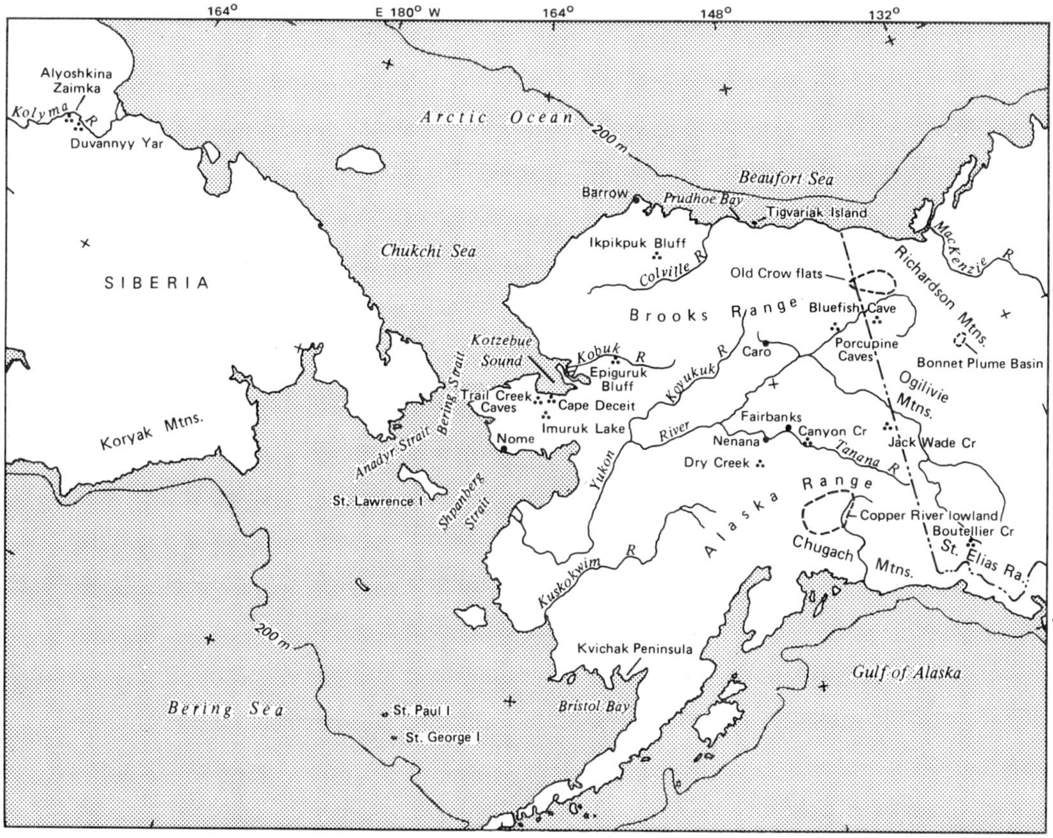

Figure 1. Location map of Beringia showing most of the localities mentioned in text

attempt in many years to inventory and interpret the widespread aeolian deposits—dunes, sand sheets, and loess—in Beringia. Sea-level history and the glacial sequence are covered only briefly. A more complete account of the late Cenozoic paleogeography of Beringia is given by Hopkins (1972, 1973); these accounts are partly updated for the latest Pleistocene by Hopkins (1979a). Chronology and correlation of glacial events in Alaska are re-evaluated by Porter et al. (in press), Hamilton and Thorson (in press), and Hamilton and Hopkins (in press).

NOMENCLATURE

The period of interest here is the interval between about 120,000 and 8000 years ago; it includes isotope stages 5 through 2 and part of stage 1 as defined in deep-sea cores (Shackleton and Opdyke, 1973; Ruddiman and McIntyre, 1981). Because Beringia (Fig. 1) lies removed from and almost midway between the western European and central North American regions that provide type sections for the most familiar subdivisions of the last cold cycle, and because there is continuing disagreement on the timing, duration, and correlation of climatic oscillations recorded in those two remote regions, clarity will be served if a series of provincial names, applicable throughout Beringia, can be established for the major phases of the Wisconsin or Weichsel climatic cycle (Fig. 2). Ideally, type localities for these provincial stades should be chosen in places where the interval of concern is represented by well-studied and well-dated sequences of fossiliferous, nonglacial, nonmarine sediment. The type sequences should display definable lower and upper boundaries beyond which sediment assignable to the next older and younger stades can be recognized. Chances of inter-

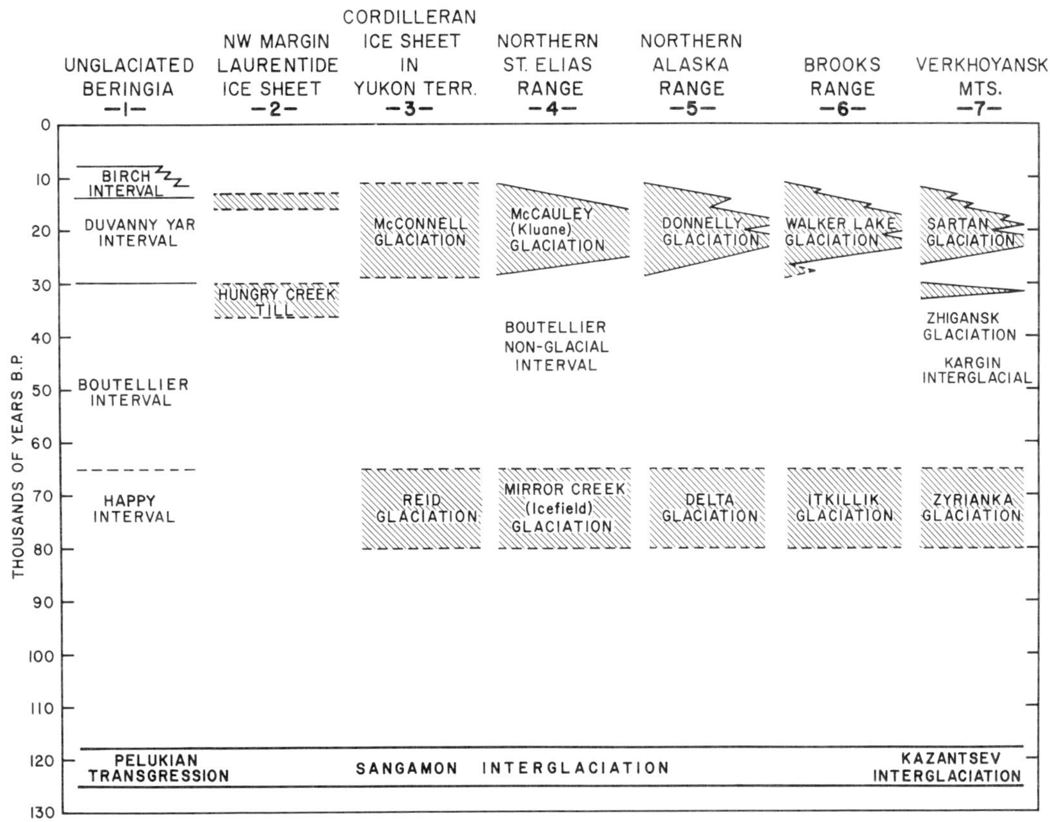

Figure 2. A comparison between time-stratigraphic nomenclature proposed for unglaciated regions (column 1) and the chronology and nomenclature of glacial sequences (shaded) in various parts of Beringia (cc. 2–7). Column 1 is based on data presented in this paper; (2) upon Hughes et al. (1981) and J. V. Matthews, Jr., personal communication; (3) after Bostock (1966) and Hughes et al. (1972); (4) after Denton and Stuiver (1967) and Denton (1974); (5) after Porter et al. (in press); (6) after Hamilton (in press); and (7) after Kind (1975). Solid boundaries are well controlled and dashed boundaries poorly controlled by radiometric dating.

national acceptance will be enhanced, it seems to me, if the type localities can be distributed throughout northwestern Canada, Alaska, and northeastern Siberia.

These ideals can be only partly realized at the present time. It has been vigorously suggested to me, in fact, that our knowledge of even the best-studied stratigraphic sections is still so incomplete that it is premature to attempt to erect a provincial Beringian time-stratigraphic nomenclature for the subdivisions of the last climatic cycle. However, these very uncertainties make the application within Beringia of a nomenclature developed for western Europe or eastern North America most inappropriate and potentially misleading. The need for a provincial nomenclature is *now,* while many aspects of the geochronology are still unclear, rather than several decades from now, when we may hope that all external and internal correlations will be well established and agreed upon and ages securely ascertained. With these observations in mind, I propose here an imperfect scheme in full realization that it will need modification and correction as our knowledge grows.

The Happy Interval

Two major glacial episodes are recorded during the period since the last (Sangamon or Kazantzev) interglacial interval in Beringia. Accumulating evidence indicates that the earlier of these two episodes was by far the larger and that it took place more than 50,000 years ago. The Zyrianka Glaciation of Kind (1975) in the Verkhoyansk Mountains of Yakutia;[1] the first phase of the late Pleistocene glaciation of Braitseva et al. (1968) in Kamchatka; the Itkillik Glaciation of Hamilton (in press) in the Brooks Range; the Salmon Lake Glaciation of Hopkins on the Seward Peninsula; the Healy Glaciation of Wahrhaftig; and the Delta Glaciation of Péwé in the Alaska Range (all in Péwé, et al., 1953); the Mirror Creek or Icefield Glaciation of Denton and Stuiver (1967) in the St. Elias Mountains of southern Yukon Territory; and probably the Reid advance of the Cordilleran ice sheet of Bostock (1966) and Hughes et al. (1972) farther east in southern Yukon Territory all seem to have taken place since the last interglacial but at a time beyond the reach of conventional radiocarbon dating. I name this earlier glacial and periglacial interval the "Happy interval," designating the type section as the upper part of the Gold Hill Loess at Eva Creek in the Fairbanks, Alaska, placer-mining district (Péwé, 1975b). More precisely, I designate the type section of the Happy interval as that part of the Gold Hill Loess designated as Eva unit 2-2 by Guthrie (1968b, Fig. 2). This interval takes its name from Happy Station, the nearest stopping point on the Alaska Railroad, 5.5 km to the northeast of the type locality on Eva Creek.

Although the Gold Hill Loess was thought by Péwé (1975b) to have been deposited during the Illinoian Glaciation, recent tephrochronologic studies (Briggs and Westgate, 1978; J. A. Westgate, written communication 1981) suggest that it consists, instead, of an early Wisconsin upper unit separated by a previously unrecognized unconformity from a much older, lower unit. The Ester Ash Bed, found near the base of the Gold Hill Loess, has a fission-track age of 0.55 million years, and thus is of middle Pleistocene age. This is probably the same ash bed that Guthrie (1968b) observed within the unit he designated Eva 2-1, near the base of what Péwé (1975b) later designated as the "Gold Hill Loess." The Dome Ash Bed, a tephra found near the top of the Gold Hill Loess at the type locality (Péwé, 1975b), is also present as angular redeposited lumps of ash at Canyon Creek, 85 km to the southeast (Weber, et al., 1981). Collagen extracted from bones associated with the redeposited ash at Canyon Creek yielded a radiocarbon age of about 40,000 years. Repeated uranium-series analyses of a horse phalange from the same horizons yielded concordant $^{238}U/^{230}Th$ and $^{235}U/^{231}Pa$ ages of 78,000 and 73,000 years, respectively (J. L. Bischoff, oral communication 1982). Thus, the age of the Dome Ash Bed remains in doubt, but it seems safe to say that it is older than 40,000 and possibly older than 80,000 years.

Small-mammal faunas from the type section of the Happy interval are described by Guthrie (1968b, coll. 3-7 and 25).

The glacial/periglacial episode that is the most distinctive feature of the Happy interval is recorded in the Old Crow Flats of northern Yukon Territory by ice-wedge pseudomorphs underlying disconformity A of Morlan (1980a) and by at least the upper part of the alluvium (Morlan's unit 3) that they intrude. The upper part of alluvium unit 3 is also host to the Old Crow tephra, a tephra unit chemically distinct from and even more widespread than the Dome Ash Bed. Radiocarbon and glass fission-track determinations on material from the type locality indicate that the Old Crow tephra is older than 56,000 but younger than 100,000 years; a uranium series age of 81,068 ± 7500 years was determined by J. L. Bischoff on mammoth bone collected on unconformity A one meter above the tephra (Westgate, et al., in press). The Old Crow tephra has also been found in primary loess exposed in roadcuts along the highway from Fairbanks to Nenana in central Alaska (T. D. Hamilton, oral communication 1980). In the Koyukuk River valley of northern Alaska, the Old Crow tephra has been found in loess, lacustrine sediment, and alluvial overbank deposits, and seems to have been deposited during the waning phase of the Itkillik Glaciation (T. D. Hamilton, oral communication 1981).

The primary loess constituting the upper part of unit 2 of the Deering Formation at Cape Deceit (Guthrie and Matthews, 1971; Matthews, 1974b) evidently was deposited during the Happy interval. Dunes of Happy age are probably represented by the crossbedded, ripple-marked sand, Schweger's unit A, exposed as a buried terrace-like feature in the upstream part of Epiguruk Bluff on the Kobuk River (Schweger, 1976). In Siberia, the Happy interval is probably represented in the bluff at

[1] Localities not shown in Figure 1 can be found on the endpaper map inside the cover of this volume.

Duvannyy Yar on the lower Kolyma River by the basal unit, horizon I of Sher et al. (1979) (also see Kaplina, et al., 1978; Giterman, et al., this volume). Horizon I consists of homogeneous silt, evidently loess, containing typical steppe-tundra pollen assemblages.

The Happy interval includes a time of extensive mountain glaciation and synchronous loess deposition in nearby lowlands. This glacial/periglacial interval seems most likely to correspond to isotope stage 4 of deep-sea cores, which represents a time of major lowering of global sea level due to ice accumulation on the continents between about 75,000 and 60,000 years ago. However, uranium-series age determinations of about 80,000 years for the Old Crow tephra and for the less securely dated Dome Ash Bed, both of which appear within primary loess sequences of Happy age, suggest that the glacial/periglacial episode recorded within the Happy interval may have coincided, instead, with one of the earlier and globally less significant eustatic low-sea-level episodes within isotope stage 5. Thus, the age limits of the Happy interval and its correlation beyond Beringia remain in doubt and await the accumulation of additional geochronological and stratigraphic data.

The Boutellier Interval

The Happy interval was separated from a late Wisconsin or Weichselian glacial/periglacial interval by a lengthy interval of intermediate climate and generally less extensive glaciation in Beringia. In northeastern Siberia, this interval has occasionally been termed the "Kargin Interstadial or Interglacial" (e.g., Kind, 1975) after a type locality near the mouth of the Ob' River in northwestern Siberia, some 3700 km west of the Bering Strait (Kind, 1967). For the foothills of the St. Elias Range, southwestern Yukon Territory, this interstadial interval has been named the nonglacial "Boutellier interval," on the basis of an interglacial soil and nonglacial alluvium containing a few detrital organic beds between underlying and overlying drift sheets of early and late Wisconsin age (Denton and Stuiver, 1967). I propose to apply the name "Boutellier interval" throughout Beringia. Radiocarbon dates from deposits of the Boutellier interval at the type locality range from 29,600 ± 460 (GSC-779) to more than 49,000 years (Y-148) (Denton, 1974). Pollen and mosses from peat lenses in Boutellier beds at and near the type locality are described by Schweger and Janssens (1980).

A better stratigraphic record of the Boutellier nonglacial interval is available in the Old Crow Flats of northern Yukon Territory (Morlan, 1980a). There, disconformity A, its underlying paleosol, and the overlying alluvial units 4 and 5 clearly postdate the Happy interval, and predate the late Wisconsin cold cycle, expressed locally by the glaciolacustrine clay of unit 6 (Morlan, 1980a, pp. 17–19, Fig. 2.2). Radiocarbon dates confirm that the base of this sequence is older than 40,000 and probably older than 50,000 years, and, as I noted above, a mammoth bone resting on unconformity A yielded a uranium-series date of 81,068 ± 7500 years (Westgate, et al., in press). The youngest radiocarbon dates in alluvial unit 5 average about 30,000 years, but these come from levels a meter or more below the lakebeds of unit 6 (J. V. Matthews, Jr., oral communication 1981). Furthermore, the proglacial lake in which unit 6 was deposited correlates with a maximum advance of Laurentide ice that may not have been exactly synchronous with advances of the Cordilleran ice sheet in Yukon Territory and montane glacier systems in Yukon and Alaska (O. L. Hughes, written communication 1981).

In Alaska, the Boutellier interval is represented by the Eva Formation and all but the uppermost part of the Goldstream Formation of the Fairbanks district (Péwé, 1975b; cf. Matthews, 1970); by peat 5 and the lower part of unit 2 of the Deering Formation at Cape Deceit (Guthrie and Matthews, 1971; Matthews, 1974b); and by organic silt and sand and interbedded peat of unit B at Epiguruk Bluff (Schweger, 1976, this volume).

The Boutellier interval is represented at Duvannyy Yar in northeastern Siberia by horizon II of Sher et al. (1979; cf. Giterman, et al., this volume), a sequence of *alas* (thaw-lake) deposits containing pollen and plant remains of forest-tundra vegetation and yielding radiocarbon dates of more than 45,000 years (MGU-578) at the base and 36,900 years (MGU-469) somewhat above the base. Lozhkin (1975) and Kaplina (1979) describe numerous other northeastern Siberian sections in which the interstadial here named the Boutellier interval is represented by peaty silt or *alas* deposits containing pollen spectra indicative of the presence of a shrub- or forest-tundra cover and yielding radiocarbon dates ranging from more than 44,000 to about 24,000 years BP.

A complex fine structure of glacial advances and retreats and of oscillations of climatic and vegetational boundaries can be discerned within mid-Wisconsin time in central North America and mid-Weichsel time in western Europe (e.g., Terasmae and Dreimanis, 1976; Grootes, 1977; Woillard, 1978a). Kind (1972, 1975) and Lozhkin (1975) summarized evidence for comparable climatic and vegetational oscillations within this interval in north-central and northeastern Siberia, and Bombin (1980) suggested that several climatic fluctuations are recorded in the Old Crow Flats within beds assigned here to the Boutellier interval.

In both Alaska and the Yukon, the Boutellier interval was evidently ushered in by a period when the climate was milder and more mesic than at any other time between the last interglacial and the beginning of the Holocene (Matthews, 1980). The episode is marked by soil formation and evidence of deep thawing, ice-wedge collapse, and accumulation locally of autochthonous peat beneath and on unconformity A in the Old Crow Flats (Morlan, 1980a), by forest beds in the peaty Eva Formation of the Fairbanks district (Péwé, 1975b), by peat 5 in the Cape Deceit section (Matthews, 1974b), and probably by the pond deposits at the basal 16-m level in Schweger's (1976) Epiguruk II pollen profile. Another occurrence, not yet studied in detail, is represented at Caro Bluff on the Chandalar River; there, the interstadial at the beginning of the Boutellier interval is recorded by bryophytic peat containing rooted spruce stumps more than 41,000 years old that rests on outwash of the Itkillik Glaciation (Hamilton, 1979a).

The long pollen cores from Imuruk Lake, Seward Peninsula (Colinvaux, 1964; Shackleton, 1979; Colinvaux and Noltimier, in press) appear to extend through beds representing most of the Wisconsin time. A tephra encountered 5 m deep might represent either the Old Crow tephra or the Dome Ash Bed. If so, then Colinvaux's (1964) pollen zones F through I would seem to have been deposited during the Boutellier interval, and at least two intra-Boutellier episodes of relatively mild, mesic climates are recorded, centered at about 75,000 and 45,000 years BP. No other Beringian pollen cores seem to have reached levels representing these warm episodes, but several cores penetrate beds representing the later parts of the Boutellier interval (e.g., Rampton, 1971; Anderson, 1981; Cwynar, 1982; Linda Brubaker, written communication 1980). Schweger's (1976) pollen profile through the Epiguruk Bluff may provide a record of the complete duration of the Boutellier interval. Pollen sequences from the later part of the interval have been described from several river and coastal sections scattered across the northeastern Siberian maritime lowland between the Kolyma and Lena Rivers (Kaplina, 1979). All these studies show pollen concentrations in beds of late Boutellier age to be low and taxonomic diversity to be limited. Trees and shrubs and such mesic indicators as Ericales and *Sphagnum* are present in low abundance or may be absent. Pollen records combined with macrofossil evidence cited by Hopkins, Smith, and Matthews (1981) suggest that, although spruce, larch, birch, and cottonwood or aspen trees may have extended to near their present geographic limits as late as 35,000 years ago, they were certainly much sparser than at present. These mesic tree species probably were confined to groves along streambanks and to open, parklike stands on moist and sheltered slopes. Though richer and more diverse than the steppe-tundra species that characterize the Duvanny Yar interval, discussed below, pollen spectra from all but the earliest part of the Boutellier interval nevertheless suggest a cold and arid landscape.

Interestingly, the earliest evidence of a possible human presence in Siberia (Mochanov, 1977) and northwestern Canada (Morlan, 1980a; Morlan and Cinq-Mars, this volume) dates from the Boutellier interval. In Alaska, the Jack Wade Creek fauna, with its butchered caribou skulls, can be assigned with confidence to this time (Porter, 1979; Porter and Hopkins, in press). Most of the northeastern Siberian finds of soft-parts remains of mammoth, horse, and bison also date from this interval (Vereshchagin and Baryshnikov, this volume). Thorson et al. (1981) have reported proboscidian remains in interstadial beds of Boutellier age at Tyone Creek in the Copper River lowland of southern Alaska. The large-mammal fauna from Canyon Creek dates from the Boutellier interval, although J. L. Bischoff's preliminary uranium-series date on a horse phalange indicates that it may be considerably older than the 40,000-year radiocarbon dates based on bone collagen and reported by Weber et al. (1981). Small mammals from sediments laid down in the Fairbanks area during the Boutellier interval are described by Guthrie (1968b, coll. 9, 10, 12, 14, 18, 27, 29).

The Boutellier interval certainly includes most of isotope stage 3 of deep-sea cores. If the 80,000-year uranium-series age determination for mammoth bone on unconformity A in the Old Crow Flats is correct, then isotope stages 4 and 5a are also included, isotope stage 5a being represented by unconformity A and its associated peat in the Old Crow Flats, by the Eva Formation in the Fairbanks district, and by correlative early Wisconsin peat beds elsewhere in Beringia.

Duvanny Yar Interval

Isotope stage 2 of deep-sea cores corresponds in a general way to the Sartan Glaciation of northern Siberia (Kind, 1967) and the Walker Lake or Itkillik II glacial advance in the Brooks Range of northern Alaska (Hamilton, in press; Hamilton and Hopkins, in press; Porter, et al., in press) and to an extremely dry, probably cold, periglacial interval in the unglaciated parts of Beringia. As a distinguishable time-stratigraphic unit on land, this interval terminates with evidence of rapid glacial retreat, changing hydrologic regime, and vegetation changes as discussed below. For this most recent periglacial interval in Beringia, I propose the name "Duvanny Yar interval."[2] The type section is horizon III of Sher at al. (1979) in the well-studied section at Duvannyy Yar on the lower Kolyma River. Beds in the type section consist of silt almost devoid of organic remains that Sher et al. (1979) believe to consist of overbank floodplain deposits but which Tomirdiaro (1980) and I believe to consist mostly of primary loess. No radiocarbon dates are available from within horizon III at Duvannyy Yar, although radiocarbon dating of other northeastern Siberian sections suggests that intensive loess deposition was initiated about 25,000 or 24,000 years ago (Lozhkin, 1975). *Alas* deposits 9200 years old that occupy a thermokarst depression incised into horizon III (Sher, et al., 1979) provide a minimum age for the end of the Duvanny Yar interval. However, pollen studies indicate that deposition of horizon III terminated at the time that birch pollen began to increase, an event that took place 14,000 to 13,000 years ago throughout Beringia (Giterman, et al., this volume; Ager, this volume).

The Duvanny Yar interval is represented at Epiguruk Bluff on the Kobuk River by sandy alluvium and aeolian sand (Schweger, 1976; T. D. Hamilton, oral communication 1981). On the Arctic Coastal Plain of Alaska, large dunes were active during the Duvanny Yar interval (Carter, 1981a, 1981b), and many other Alaskan dune complexes are at least partly of Duvanny Yar age, as is some part of the extensive loess mantle (Péwé, 1968, 1975a, 1975b). Loess of Duvanny Yar age is clearly identified in a large-diameter core from Isabella Creek basin near Fairbanks (Matthews, 1974a) and in the exposed section above the fauna of Boutellier age at Jack Wade Creek (Porter, 1979). In northern Yukon Territory, the Duvanny Yar interval is represented by lakebeds in the lowland basins of the upper Porcupine River drainage (Hughes, et al., 1981).

The Duvanny Yar interval corresponds to the herb zone displayed in long Beringian pollen sequences (e.g., Colinvaux, 1964, 1981; Rampton, 1971; Shackleton, 1979; Kaplina, 1979; Anderson, 1981; Ager, this volume; Ritchie and Cwynar, this volume). A notable feature of herb-zone pollen cores is that many of them display evidence that the lakes from which they were recovered were shallower or even intermittently dry during the Duvanny Yar interval (e.g., Colinvaux, 1981; Cwynar, 1982). The most recent studies have shown, however, that the cold, dry, herb-zone climate of the Duvanny Yar interval was interrupted briefly, about 22,000 to 19,000 years ago, by a slightly more mesic interval (Anderson, 1981; Cwynar, 1982; Hamilton, in press; Giterman, et al., this volume).

Studies of macrofossils indicate clearly that spruce became extinct in eastern Beringia during the Duvanny Yar interval (Hopkins, Smith, and Matthews, 1981). Cottonwood and possibly aspen may have persisted in a refugial area near Bering Strait, judging from the present distribution of relict populations and early appearances of *Populus* macrofossils and pollen in western and northern Alaska (Hopkins, Smith, and Matthews, 1981; Ager, this volume). A somewhat weaker case can be made for persistence of alder in this same refugium (Anderson, 1981).

Surprisingly, deposits of Duvanny Yar age are

[2] Spelling of the name of the interval is simplified from the standard transliteration of the name of the bluff which, as indicated, is Duvannyy Yar.

either missing or poorly represented in the exposures of predominantly silty sequences constituting the Deering Formation at Cape Deceit and the Goldstream Formation in the Fairbanks area. Lithology, radiocarbon dating, and paleontological and palynological studies all suggest that there is no primary loess of Duvanny Yar age exposed within the upper part of unit 2 of the Deering Formation nor within most exposures of the Goldstream Formation (Guthrie, 1968b; Guthrie and Matthews, 1971; Matthews, 1974a, 1974b; Péwé, 1975a, 1975b; Sellman, 1967, 1972). In practically all these exposures, redeposited loess laid down during the time of the birch zone rest unconformably upon redeposited loess laid down during the Boutellier interval. Interestingly enough, a similar situation is recorded at Mamontova Gora in Yakutia at the opposite end of Beringia (Péwé, et al., 1977) (location on endpaper map) and a similar disconformity separates organic silt (probably redeposited loess) of Boutellier age and overlying Holocene peat in the Klondike district, Yukon Territory (Dyck and Fyles, 1964, comment on GSC-88; O. L. Hughes, written communication 1981). The apparent presence of silt of Duvanny Yar age in the Isabella Basin core (Matthews, 1974a) indicates that in the Fairbanks district, at least, we are seeing a record of erosion and removal of deposits of Duvanny Yar age during the time of the birch zone rather than a record of nondeposition during the Duvanny Yar interval. The wide occurrence of this hiatus suggests that it records a Beringia-wide change in landscape processes at the end of the Duvanny Yar interval.

Among the few in situ faunas of Duvanny Yar age that have been described are the scattered bones from horizon III at Duvanny Yar and the larger assemblage from nearby Alyoshkina Zaimka (Sher, et al., 1979) and a bone accumulation in cave 1 of the Porcupine Caves (E. J. Dixon, written communication 1978, 1979). Occupation of Dyukhtai Cave on the Aldan River at the western edge of Beringia seems to have begun as early as 16,000 or 15,000 years ago, but aside from this, no evidence of a human presence during the Duvanny Yar interval has yet been found in Beringia.

Birch Zone

The Pleistocene ends in Beringia with an interval of rapid environmental change ushered in by an increase in the abundance of dwarf birch shrubs 14,000 to 12,000 years ago and completed by the initiation of widespread peat accumulation 10,000 to 8000 years ago. This interval is clearly defined in pollen sequences by the rapid rise to dominance of birch pollen; as a pollen zone, it ends with the appearance, at different times in different places, of abundant spruce and alder pollen, although these additions to the pollen flora do not necessarily coincide with the time of initiation of peat accumulation. Deposits dating from this interval are recognized in outcrop in many parts of Beringia, but none has been studied in sufficient detail to constitute a satisfactory type section. For the moment, I will simply term this late Pleistocene and early Holocene interval the time of the birch zone.

At Cape Deceit, the birch zone is represented by platy (colluvial?) organic silt within the upper part of unit 2 of the Deering Formation (Matthews, 1974b). Elsewhere on the northern Seward Peninsula and around the shores of Kotzebue Sound, the birch zone is represented by exceptionally woody thaw-lake deposits that contain remains of extralimital trees (balsam poplar, aspen, or tree birch), remains of aquatic plants outside their present limits, beaver-chewed wood, and aquatic mollusks that do not live in the region today (McCulloch and Hopkins, 1966; Spiker, et al., 1978; Hopkins, Smith, and Matthews, 1981). These deposits have been interpreted as indicating an early Holocene climate warmer than that of today (McCulloch and Hopkins, 1966). However, many of the distinctive features of these late Wisconsin and early Holocene thaw-lake deposits might be produced by a snow cover thicker and fluffier than the wind-packed snow that covers the region during present-day winters; such a snow cover would result in thinner pond ice and persistence through the winter of a greater volume of unfrozen water beneath the ice.

In the Fairbanks district, the birch zone is expressed by redeposited loess rich in organic debris, constituting in some places the uppermost part of the Goldstream Formation and in others the Ready Bullion Formation of Péwé (1975b). A permafrost research tunnel penetrating these latest Pleistocene and earliest Holocene deposits displays irregular folded gravel strata intercalated in the base of a sequence of organic-silt beds deposited 14,000 to 11,000 years ago. These rest unconformably on silt (loess?) deposited at least 30,000 years ago (Sellmann, 1967, 1972). The loess-covered slopes of the Fairbanks district and many other loess areas of central Alaska are furrowed by inactive, healed gullies leading to fans that coalesce to form an apron on lower valley walls (Péwé, 1955, pl. 3; 1975b). The exposures in the Fairbanks permafrost tunnel suggest that these gullies and fans were formed during the time of the birch zone and that solifluction and mud flowage as well as rill wash played a role in their development. Because gully and fan development is inactive at the present time, it would seem that the birch zone records an interval when the hydrologic regime differed from that of the present as well as from the preceding Duvanny Yar interval.

On the western Arctic Coastal Plain, the birch interval is represented by the earliest late Wisconsin thaw-lake deposits recognized thus far in the region and by river terraces that record a reestablishment of through-flowing drainage in areas where drainage had previously been defeated by active dunes (Carter, 1981b, L. D. Carter, oral communication 1980). Leaves of balsam poplar, a tree no longer to be found on the western Arctic coastal plain, occur in detrital peat lenses in 13,500- to 9500-year-old river terrace deposits (Hopkins, Smith, and Matthews, 1981). Paleosols and intercalated zones of buried willows in aeolian sand indicate that dune activity diminished and was intermittently interrupted by periods of at least local sand stabilization. Here again the record indicates a change in the moisture regime and the appearance of a climate more mesic than that of the Duvanny Yar interval, although the moisture regime must have been different from the present one, because regional peat accumulation did not begin in northern Alaska until about 8500 years ago (Carter, 1981b; D. M. Hopkins, unpublished data, 1977–81).

Aside from pollen cores (e.g., Rampton, 1971; Cwynar, 1982), deposits dating from the time of the birch zone are poorly known in Yukon Territory. Bison-bearing gravel incised into sediments of the proglacial lake in the Old Crow Flats (unit 7a of Morlan, 1980a) dates from the time of the birch zone and provides evidence that the lake had been drained but that streams had not yet incised appreciably into the abandoned lake plain by 12,500 years ago.

Several large in situ faunas, mostly archaeological assemblages, date from the time of the birch zone. Notable among these assemblages are the material from Dry Creek in central Alaska (R. D. Guthrie, written communication 1980, 1981), all except the lowest levels at Bluefish Cave in the Porcupine River drainage (Cinq-Mars, 1979; Morlan and Cinq-Mars, this volume), and much of the material from the Trail Creek Caves on Seward Peninsula (Larsen, 1968). The small group of *Bison* preserved in the gully fill in the Old Crow Flats evidently does not involve human activity, but the Berelekh site in northeastern Siberia, which contains remains of many more than 100 individual mammoths as well as other faunal material (Vereshchagin, 1977) almost surely is somehow an archaeological or at least anthropogenic assemblage (Mochanov, 1977).

In summary, the stratigraphic record of the birch zone provides unequivocal evidence of rapid, drastic climatic and environmental change. Widespread evidence of newly vigorous trunk streams in northern Alaska and gullying and solifluction on slopes in central Alaska indicate that precipitation must have increased. The evidence of extension of the range of beaver and other aquatic fauna preserved in the thaw-lake deposits of western Alaska is consistent with an increase in winter snow cover. The evidence of continued dune activity in northern Alaska and probably elsewhere indicates that if winters featured a good snow cover, then summers must have been dry. Also, the rapid, almost catastrophic wastage and retreat of mountain glaciers at the time of the initial expansion of dwarf birch indicates that if winters were snowy, then summers must have been warm. Evidently the birch zone records a time of increased winter snowfall and of warm dry summers, at least in central, western, and northern Alaska.

SEA-LEVEL HISTORY

Estimates of worldwide sea-level history during the past 130,000 years are based on the oxygen-isotope record in Caribbean deep-sea cores (Fig. 3a) (Emiliani, 1955; Shackleton and Opdyke, 1973) and on uplifted coral reefs on tropical islands (Fig. 3b) (Chapell, 1974; Bloom, et al., 1974). These curves provide standards, albeit dubious (Fairbanks, 1980), for comparison with the sea-level record in Beringia (Fig. 3c).

The oxygen-isotope and coral-reef records agree

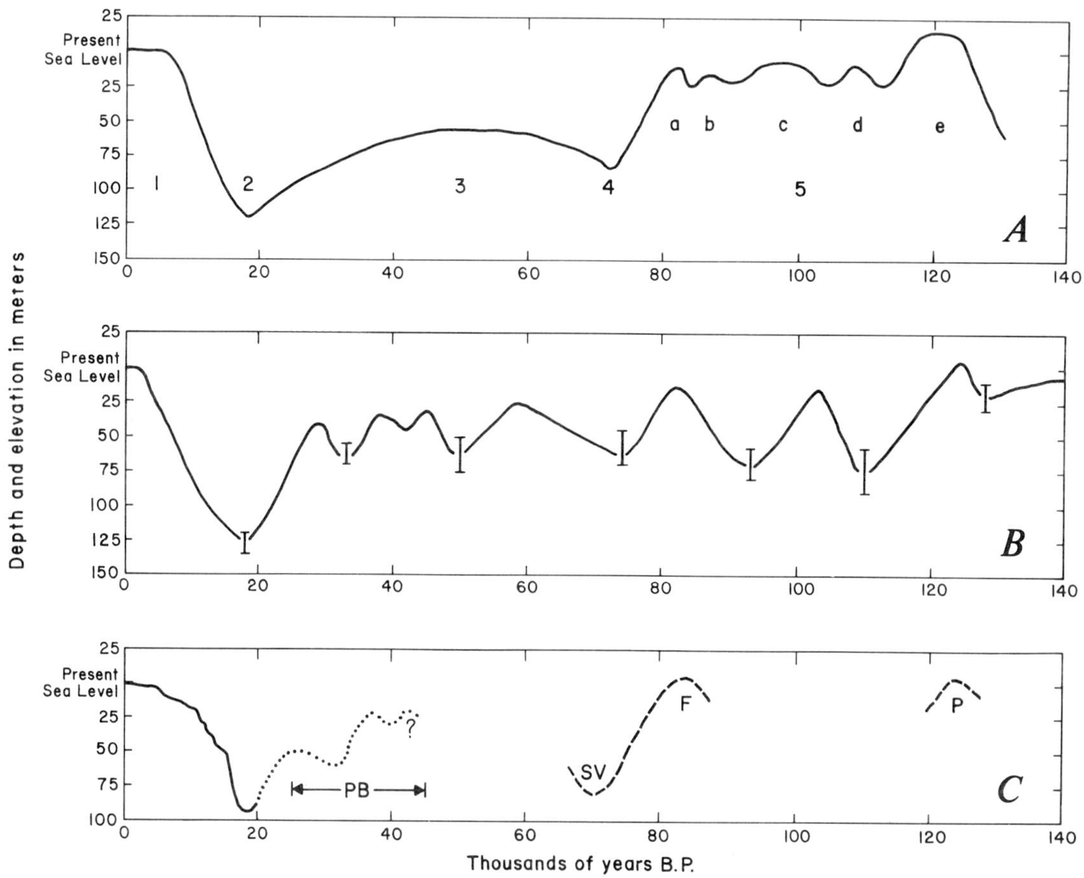

Figure 3. Worldwide sea-level histories postulated on the basis of (a) oxygen-isotope data from benthic foraminifers in a Caribbean deep-sea core (numbered and lettered subdivisions of curve are oxygen-isotope stages (Shackleton and Opdyke, 1973) and (b) raised coral reefs in New Guinea (Bloom, et al., 1974) with (c) sea-level history interpretations for Beringia. Portion of Beringian curve younger than 20,000 years after Hopkins (1973) and McManus et al. (in press); a sea-level curve applicable to the Mackenzie Bight area of eastern Beaufort Sea (Forbes, 1980) is not presented here. *PB* portion is speculative and based on interpretation of radiocarbon-dated stratigraphy of offshore boreholes and onshore gravel pits at Prudhoe Bay (Hartz, et al., 1979; Hopkins, Robinson, and Buckley, 1981). Older part of Beringian curve is not yet dated by local evidence. *SV* represents regression recorded by valleys carved across Beaufort Sea shelf more than 32,000 years ago (Hopkins, et al., 1979). *F* represents Flaxman transgression, recorded by glaciomarine deposits present at altitudes below 5 m on the coast of Beaufort Sea (Hopkins, 1979b) but absent from coasts of Chukchi and Bering Seas. *P* represents Pelukian transgression, recorded by shoreline features at an altitude of about 7 m along much of coasts of Bering, Chukchi, and Beaufort Seas (Hopkins, 1967c). Pelukian transgression dates from the last interglacial and no doubt corresponds to isotope stage 5e

in indicating pronounced oscillations of sea level during isotope stage 5 after the peak of the last interglaciation and before the extreme reduction in sea level of isotope stage 4. Ruddiman and McIntyre (1979) show that these oscillations of 115,000 to 75,000 years ago reflect abortive glaciations—the development and then destruction of icecaps at high latitudes. A rise in sea level during this interval to a relative position several meters above the present shoreline of Beaufort Sea is recorded in northern Alaska by the Flaxman Formation, a glaciomarine unit containing abundant striated boulders foreign to Alaska and derived from Precambrian and early Paleozoic terranes

(Hopkins, 1979b; L. D. Carter and D. M. Hopkins, unpublished data, 1977–81). The Flaxman Formation is unconformable on sediment of the Pelukian Transgression, and is older than a drastic reduction in sea level, which in turn was earlier than 40,000 years BP. Beach ridges as much as 7 m above sea level at Point Barrow (Sellmann and Brown, 1973) probably were deposited during the Flaxman transgression, despite the fact that they have yielded some radiocarbon dates as young as 25,000 years. The Flaxman Formation is evidently the product of an aborted glacial event that involved the expansion and then the destruction of icecaps on the continental shelf at high latitudes during some part of isotope stage 5. Reports of erratic boulders on the Canadian coast of the Beaufort Sea (J-S. Vincent, oral communication 1977) and on the Siberian coast of the Chukchi Sea indicate that the Flaxman transgression reached levels above present sea level in many places on the Arctic coast. However, deposits of the Flaxman Formation are absent and there is no evidence of an intra-Wisconsin high-sea-level episode along the Alaska coasts of the Chukchi and Bering Seas between Point Barrow and Bristol Bay.

The shores of the Kvichak Peninsula, southwestern Alaska, preserve a record of glaciomarine sedimentation that indicates Bristol Bay was flooded at some time before 33,000 years BP, at a time earlier than a mid-Wisconsin interstadial and when glaciers were at their maximum Wisconsin extent.[3] My age assignments for early Wisconsin glacial advances in Beringia are colored, perhaps unduly, by the 80,000-year uranium-series dates on fossil bones from the Old Crow Flats and Canyon Creek. If these are accepted as correct age estimates, however, then the maximum Wisconsin glaciation in Beringia must be placed at earlier than 80,000 years BP. On this assumption, the deposition of the glaciomarine beds of Kvichak Peninsula may be synchronous with the Flaxman transgression in northern Alaska.

The oxygen-isotope record indicates that enough water was stored in growing glaciers during isotope stage 4 to cause sea level to drop about 85 m, reaching a minimum 60,000 or 70,000 years ago (Fig. 3a) (in the coral-reef sequence, Fig. 3b, the estimated sea level drop is more modest). Although most of Beringia lacks a recognized record of sea level during what is evidently the early part of the Boutellier interval, offshore drilling on the Beaufort Sea shelf near Prudhoe Bay indicates that valleys were excavated to a depth of nearly 80 m below sea level at some time after deposition of the Flaxman Formation and before 42,000 years ago (Hopkins, et al., 1979). It is reasonable to assume that this erosion took place during the low-sea-level episode represented by isotope stage 4. Probably the Alaskan shelves, including the Bering land bridge, were very broadly exposed about 70,000 years ago, and sea level lay 80 m below its present position in the north and perhaps 70 m below its present position in the south.

The oxygen-isotope record indicates a recovery of sea level during isotope stage 3 and a broad peak at −55 m about 50,000 years ago (Fig. 3a). The coral-reef record, however, records a series of sharply defined sea-level peaks that decline from −20 m about 60,000 years BP to −45 m about 28,000 years BP (Fig. 3b). In western Siberia, the Kargin Interstade is recorded partly by raised beaches "radiocarbon dated but only in a wide range from 36–38 to 45 thousand years" (Arkhipov, 1979; detailed by Arkhipov, et al., 1977). Though isostatically uplifted above their original positions, these beaches do record a period when sea level must have recovered to within a few tens of meters of its present position.

Relative sea-level data from within Beringia during this interval is very sparse. A submerged and buried delta that prograded to the north into an east-west-trending depression south of Nome records an ancient local relative sea level at −20 m.[4] The submerged delta was once thought to be of mid-Wisconsin age (Hopkins, 1973), but the presence of more than 5 m of covering sediment suggests that it may be much older.

Some inkling of the position of the ancient shoreline on the Beaufort Sea shelf near Prudhoe

[3] An Illinoian or next-to-last glacial age was suggested for the glaciomarine deposits of Kvichak Peninsula by Hopkins (1973), but pollen analyses by R. E. Giterman (summarized by Ager, this volume) later showed that an organic horizon in loess everlying the glaciomarine beds is interstadial rather than interglacial.

[4] The submerged and buried delta south of Nome was first observed by D. Moore (1964) and has since been confirmed on many high-resolution seismic reflection profiles by C. H. Nelson (U.S. Geological Survey, unpublished data 1970–75).

Bay is provided by paleosols, thin layers of buried turf, and discontinuous zones of detrital peat interbedded with outwash gravel in mainland gravel pits and offshore boreholes dated 43,000 to 26,000 BP (Hopkins, et al., 1979). The gravel fills valleys that had previously been excavated during times of rising sea level. The peaty deposits record brief cessations of gravel deposition during the Boutellier interval, probably during peaks in the series of partial recoveries of sea level 60,000 to 28,000 years ago. That sea level lay somewhere near the level of the peaty accumulations is indicated by the recovery of a marine amphipod, possibly *Onisimus affinis* (C. Holmquist, in Hopkins and Hartz, 1978a), in nonmarine detrital peat 13 m below sea level in borehole PB-7 near the mouth of Prudhoe Bay; the peat is dated as 42,800 ± 1440 years old (USGS-249). A submerged and buried shoreline feature at −33 ± 2 m detected by James Craig (U.S. Geological Survey, unpublished data 1981) on high-resolution seismic-reflection records from Harrison Bay, north of the Colville River delta, may also represent this high stand about 42,000 years ago. Pre-Holocene beach gravel encountered at 20 m depth interbedded with gravel valley fill in borehole HLA-15 near Tigvariak Island (Hartz, et al., 1979) also may be equivalent.

The deep-sea oxygen-isotope record (Shackleton and Opdyke, 1973) is interpreted as indicating that global sea level fell to −120 m during isotope stage 2 (Fig. 3a), but the position of the ancient shoreline on any given segment of the continental margin differs as a result of local differences in tectonic history and local isostatic effects (Clark, et al., 1978; Bloom, 1980). Data on sea level in Beringia during isotope stage 2 are fairly abundant (Fig. 3c). Sea level fell to a minimum position of −90 m on the southern Bering Shelf (Knebel and Creager, 1973). On the Beaufort Sea shelf, the shoreline features marking minimal late Wisconsin sea level now lie at −97 m in the west but deepen to −115 m near the Alaskan-Canadian border (Dinter, in press).

A recent reevaluation of sea-level and shoreline history, based mostly on cores from the Chukchi Sea (McManus, et al., in press) indicates that as early as 15,500 years ago, the Bering Strait was flooded and the Bering land bridge reduced to a narrow isthmus at the present site of the Anadyr Strait. Final separation of Siberia from Alaska was accomplished by flooding of this isthmus about 14,400 years ago. St. Lawrence Island remained broadly attached to the Alaskan mainland, and Alaska was then separated from Siberia by a shallow and sinuous seaway extending from the southern Bering Sea through the Anadyr and Bering Straits northward across the partly flooded continental shelf of the Chukchi Sea to the Arctic Ocean (Hopkins, 1979a, Fig. 1.7a). By 12,000 years BP, sea level had risen to −30 m, the Shpanberg Strait was flooded, and St. Lawrence Island was isolated from the Alaskan mainland (Hopkins, 1979a, Fig. 1.7b). Areas of exposed continental shelf were shrinking rapidly, and by 10,000 years ago, sea level stood near −20 m, so that only the shallowest of the modern embayments in the coast of western Alaska were left exposed (Hopkins, 1979a, Fig. 1.7c). On the Beaufort Sea shelf, flooding by rising sea level made shallow estuaries of the paleovalleys of major streams. The present offshore barrier islands of Beaufort Sea are evidently gravel residue formed by the erosion of peninsulas and headlands between the ancient valleys (Hopkins and Hartz, 1978b).

Until recently, discussions of the history of sea ice in the seas adjoining Beringia have been purely speculative. New studies of fossil diatom and foraminifer assemblages by Sancetta (1981, in press) and McManus and colleagues (in press) now provide better insight into the paleoceanography and sea-ice history of the southeastern Bering and Chukchi Seas. Sancetta concludes that winter sea ice was present over the submerged outer shelf and abyssal basin of Bering Sea—areas now ice free at all seasons—throughout Wisconsin time. Annual ice cover was especially prolonged and extensive during isotope stages 4 and 2 (75,000–60,000 and 25,000–14,000 years BP). Marine productivity was high during isotope stages 5a and 3 but low during stages 4 and 2.

When sea level was at a minimum, the arctic continental shelves were almost completely emergent, and the zone of shallow summer-warmed open water was very narrow. It seems likely that sea ice lay at or near the shore throughout the year, during the interval 25,000 to 17,000 BP, and food resources available to man must have been minimal. Cores from the Chukchi Sea shelf indicate that the marine embayment that formed there when sea level began to rise 17,000 years ago remained ice covered throughout most of the year.

A warming climate and an influx of Bering Sea water increased salinities and temperatures about 14,000 or 13,000 years ago and lengthened the summer open-water season (McManus, et al., in press).

EXTENT OF GLACIATION

Because several reviews of the Wisconsin glacial chronology of Alaska and the Yukon are pending (Hughes, et al., 1982; Hamilton, in press; Hamilton and Hopkins, in press; Hamilton and Thorson, in press; Porter, et al., in press) and because, on the other hand, I have seen no recent reports on the glacial history of northeastern Siberia, I give here only a cursory treatment.

The Cordilleran ice sheet in Alaska was, according to Hamilton and Thorson (in press) "a complex body of ice streams, intermontane icefields, and piedmont glaciers that coalesced with ice caps on parts of the continental shelf south of the Alaska Peninsula and the Aleutian Islands." Much of southern Alaska supports large glaciers today, and Hamilton and Thorson (in press) suggest that some areas may have remained sufficiently glaciated during the Boutellier interval that earlier and later ice advances are not distinguishable. They note that sufficient ice seems to have persisted in the Chugach Mountains, for example, to obstruct the canyon of the Copper River and thus to maintain a proglacial lake in the interior Copper River basin throughout most, if not all, of the Wisconsin. Fluctuations in the extent of the Cordilleran icecap are indicated by moraines in north-draining valleys that record extensive outlet and piedmont glaciers of early Wisconsin age and somewhat less extensive glaciers of late Wisconsin age. The early Wisconsin glacier system on Kamchatka also was evidently substantially larger than the late Wisconsin system (Braitseva, et al., 1968).

The more northern highlands (e.g., the Ogilvie Mountains of northern Yukon Territory, the Alaska Range and the Brooks Range of Alaska, and the Verkhoyansk Mountains in Yakutia) were extensively glaciated during the early Wisconsin Happy interval but much less intensely ice covered during the late Wisconsin Duvanny Yar interval. Glaciers evidently were not much larger than at present during much of the intervening Boutellier interval, but a significant advance, the Zhigansk Glaciation, is recorded about 32,000 years BP in the Verkhoyansk Mountains (Kind, 1975). In the Brooks Range, a glacial advance older than 24,000 years that Hamilton (in press) includes in the Walker Lake Glaciation may be correlative.

Recent restudy of Pleistocene glaciation limits in Alaska (Hamilton, 1981; Hamilton and Thorson, in press; Porter, et al., in press) illuminate moisture distribution and air-mass movements during the Duvanny Yar interval. Lower glaciation limits along broad valley reentrants confirm the importance of topography in guiding penetration of moist air masses into major mountain ranges. A minimal lowering of glaciation limits by only 100 m in the eastern Brooks Range testifies to the great aridity there during Duvanny Yar time. In the southwestern Brooks Range, on the other hand, lowering of the glaciation limit by some 300 m indicates that the southern Bering Sea was the major source of precipitation (Hamilton, 1981).

The Laurentide ice sheet seems to have had a somewhat different history. Although there is no clear evidence of an early Wisconsin advance older than 40,000 years BP along the Laurentide ice-sheet boundary in eastern Beringia, the Hungry Creek till in the Bonnet Plume basin at the southern end of the Richardson Mountains records an early advance of Laurentide ice between 36,900 and 16,000 years BP (Hughes, et al., 1981; J. V. Matthews, Jr., written communication 1981). Possibly this advance is responsible for the moraines of the Buckland Glaciation (Rampton, in press), which can be traced continuously northwestward along the east flanks of the Richardson and Mackenzie Mountains to the Beaufort Sea coast. A later readvance of the Laurentide ice sheet is estimated to have reached its maximum position very late, near the beginning of the time of the birch zone, 14,000 to 13,500 years ago (Rampton, 1979).

AEOLIAN DEPOSITS

Deposits of windblown sand and silt of late Pleistocene age are widespread in Beringia. Although these deposits are potentially important sources of paleoclimatic and paleoecological information, their usefulness is limited because they are far from completely inventoried and because, with a few exceptions, their ages remain poorly known.

Still less is known about former seasons of deposition. Complementary to the large dunes and loess tracts are areas in which ventifacts and sand wedges are present that provide a record of deflation wherever they are found. It is fairly common to find combinations of deflationary features beneath a cover of windblown sand or silt or dune sand mantled by loess. At Canyon Creek on the north side of the Tanana River Valley in central Alaska, Weber et al. (1981) have reported a complete sequence consisting of a basal pavement of ventifacts and associated sand wedges, overlain by fine dune sand, overlain in turn by loess. In the vicinity of Cape Blossom and on the northern Seward Peninsula in western Alaska, and probably elsewhere, there are sequences of thaw-lake deposits whose sediment consists of fine sand and silt blown in by the wind (D. M. Hopkins and R. F. Rowland, unpublished data 1970).

Distribution

Accumulations of aeolian sand and silt which are widespread in the unglaciated parts of Alaska (Black, 1951), are concentrated in and near lowland areas and on the slopes and in nearby tributary valleys of the larger streams draining glaciated areas (Fig. 4). Although most of these loess tracts have been recognized for many years (Péwé, 1968), recognition of dune tracts has progressed much more slowly. Absence of mineral resources and consequent sparsity of roads, excavations, and settlements within the major sand seas; modification of original dune morphology by the formation of lakes in interdune areas; presence of a surficial blanket of loess several meters thick; and presence of a thick forest cover are all factors that have retarded recognition or led to misinterpretation of large dune tracts in various parts of Alaska.

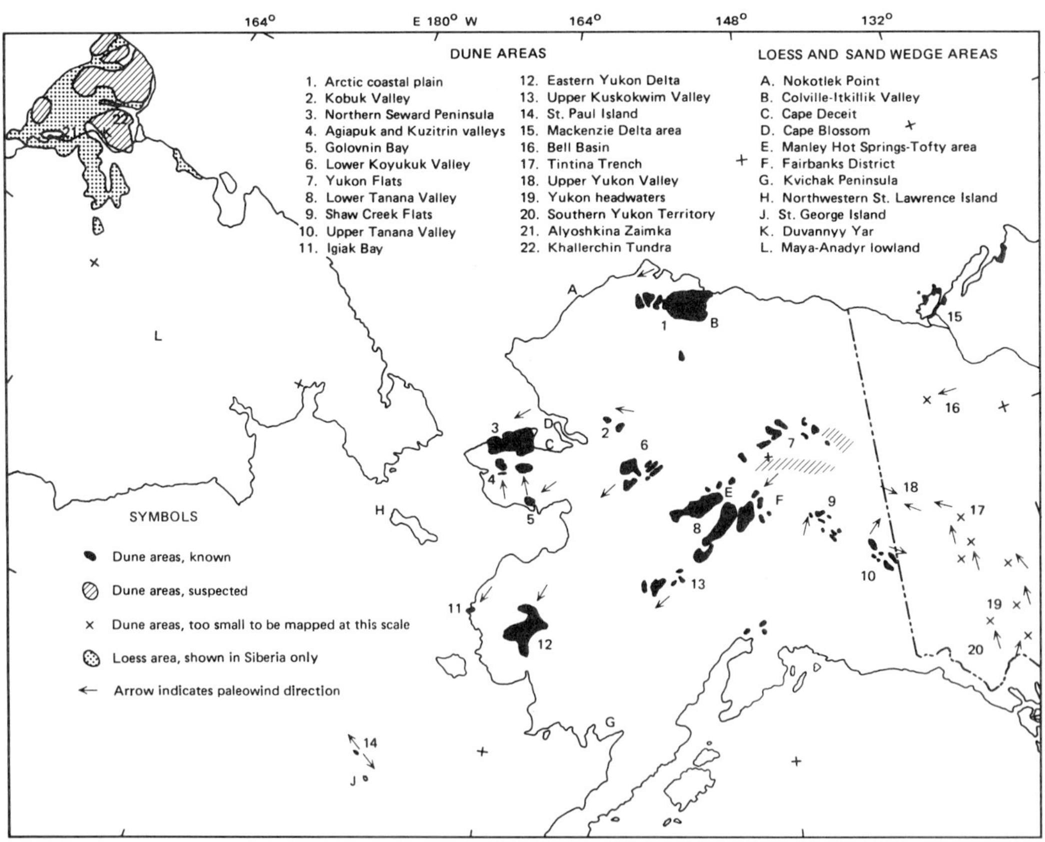

Figure 4. Pleistocene dunes and loess areas showing indicated wind vectors. Data sources given in text

Notable are large sand seas covering 7000 km^2 on the Alaskan Arctic Coastal Plain (Carter, 1981a) and several thousand square kilometers in the lowland extending southward from the lower Tanana River west of Nenana (Collins, 1958; Reed, 1961; Péwé, et al., 1966; Chapman, et al., 1975a, 1975b). Smaller sand seas cover the coastal plain of northern Seward Peninsula (D. M. Hopkins, unpublished data 1970), and linear tracts of dune areas are extensive along the south side of some of the ancient courses of the Yukon River in the delta area (Dupré, 1978). Williams (1962) mapped several large dune tracts at the northwest margin of the Yukon Flats. Other large areas of low hillocks interspersed with irregular lakes on the south, east, and northeast margins of the Yukon Flats are also probably dune tracts, although Williams (1962) mapped them as loess,[5] and they certainly do bear a continuous loess blanket several meters thick.

Other notable dune tracts occur near the north flank of the Alaska Range in the upper Kuskokwin drainage (Fernald, 1960), in and near the Shaw Creek Flats in the middle of the Tanana River valley (Péwé, 1965, 1975a; Weber, et al., 1978), in the upper Tanana Valley (Fernald, 1965), in the lowland along the Koyukuk River at the confluence with the Yukon River (Weber and Péwé, 1970), and scattered about the central and southern Seward Peninsula (D. M. Hopkins, unpublished data 1968, 1973). The Kobuk River valley contains ancient dune tracts as well as a small active sand sea, the Great Kobuk Dunes (Fernald, 1964). In short, dunes are to be found in or near most of the lowland and broad-valley segments of streams that drain formerly glaciated areas, as well as in some of the coastal lowlands of northwestern Alaska.

Loess and redeposited loess are also widespread in unglaciated areas of Alaska. In addition to the loess areas recorded by Péwé (1963, 1975a, Fig. 17) and Williams et al. (1977), loess deposits at least 10 m thick have recently been discovered in the narrow north-trending lowland between the Colville and Itkillik River valleys in northern Alaska (L. D. Carter, oral communication 1975) and on the Kvichak Peninsula of southwestern Alaska (D.M. Hopkins, unpublished data 1971). Other unrecognized loess areas no doubt await discovery.

Ancient sand wedges (frost-crack structures that were filled with blowing sand rather than ice) have been reported in the Shaw Creek Flats, on the campus of the University of Alaska in the Fairbanks district, in roadcuts along the highway leading from Fairbanks to Nenana, in a roadcut in the village of Manley Hot Springs, and in a series of kilometer-long prospecting trenches on a south-facing hillside in the Tofty mining district about 20 km north of Manley Hot Springs (Péwé, 1965, 1975b; D. M. Hopkins, unpublished data 1955–61, 1974). The sand wedges of central Alaska generally intrude weathered bedrock, commonly show deformation by hillside creep subsequent to formation, and lie buried beneath a mantle of loess. Indeed, sand wedges may be widespread beneath the forest carpet in the uplands of central Alaska.

Pre-Holocene ice wedges are rare, and ancient sand wedges are almost universally distributed along the Beaufort Sea coast and inland on the Arctic Coastal Plain of northern Alaska. Composite sand-and-ice wedges are exposed at Nokotlek Point and in a few other places along the northern Chukchi Sea coast (D. M. Hopkins, L. D. Carter, and J. K. Brigham, unpublished data 1980–81).

Although dune tracts and loess deposits in Yukon Territory are incompletely inventoried, it is clear that their distribution is much more restricted than in Alaska, perhaps because relief in the unglaciated areas is more uniformly rugged and because many of the tectonic lowlands were occupied by proglacial lakes. Recognized dune tracts are limited mostly to small areas adjoining the sandy shoreline deposits of proglacial lakes, such as glacial Lake Champagne and unnamed proglacial lakes in the areas of modern Lake Dezadeash and Lake Bennett in southern Yukon (Fig. 4, loc. 20), to narrow belts on outwash terraces in the more open parts of the upper Yukon River valley, and to sandy lacustrine sediments in the Bell Basin (Kindle, 1953; Wheeler, 1961; Campbell, 1967; O. L. Hughes and V. N. Rampton, written communication 1980). Ventifacts, sand wedges, and truncations of buried soil profiles testify to repeated

[5] During an overflight in July 1980, I was surprised to recognize a large tract of heavily forested dunes interspersed with crescentic and teardrop-shaped ponds in the silt belt on the south side of the Yukon Flats. During Williams' foot traverses in the heavily forested areas marginal to the Yukon Flats in 1948 and 1949, he found only silt in soil pits and in the rare, low, natural exposures. Participants in a recent expedition to collect pollen cores also found only silt in natural exposures on the lakeshores (Mary Edwards, written communication 1981).

cycles of deflation in open terrain in and near the Tintina Trench in central Yukon Territory (Hughes, et al., 1972; Foscolos, et al., 1977), although the final deposition site of the deflated sediment has yet to be recognized.

Aeolian sand may be much more widely distributed than has been hitherto realized in low-lying areas along the Beaufort Sea coast of northwestern Canada. A widespread buried sheet of oxidized, nearly stone-free, unfossiliferous brown sand, as much as 15 m thick, is exposed on Garry, Richardson, and other high-standing islands at the front of the Mackenzie River delta and on the shores of the Tuktoyaktuk Peninsula and Liverpool Bay farther to the east (Mackay, 1963; Fyles, et al., 1972). Although this sand commonly has been interpreted as fluvial or deltaic, an aeolian origin is strongly suggested by grain-size analyses (Rampton and Bouchard, 1975, App. A) and the presence of steep, planar crossbedding (Mackay, 1963, Fig. 2). More than one episode of aeolian activity is evidently recorded because the brown sand is buried beneath outwash gravel. However, vegetated paraboloid dunes are also recognized at the surface in some areas (Mackay, 1963, Figs. 12, 13).

Kolpakov (1978) reported widely distributed ventifacts of early, middle, and late Pleistocene age in the Yana and Vilyui valleys and interpreted these as evidence of widespread deflation deserts on the plains and intermontane valleys of northwestern Yakutia, just west of Beringia.

Large tracts of windblown silt are certainly present in northeastern Siberia. They go largely unrecognized because of the prevailing view among Soviet geographers and Quaternary geologists that extensive silt deposits in and near large arctic and subarctic rivers represent "lacustrine-alluvial deposits" (e.g., Baranova and Biske, 1964; Sher, et al., 1979). A convincing case for the loessial origin of northeastern Siberian *yedoma* deposits is made by Tomirdiaro (this volume) and confirmed by Péwé et al. (1977). Tomirdiaro's map of *yedoma* areas in the Yana-Indigirka-Kolyma region (1980, Fig. 8) provides a partial inventory of northeastern Siberian loess tracts (Fig. 4), and other tracts are probably present in such places as the Maya-Anadyr lowland farther east.

Although no large dune tracts are mapped in northeastern Siberia, they are undoubtedly present. Participants in the Kolyma River excursion of the 14th Pacific Science Congress were shown excellent exposures of dune sand at Alyoshkina Zaimka, interpreted, however, as "interbedded channel and floodplain deposits" (Sher, et al., 1979). The dune sand at Alyoshkina Zaimka is permeated by rootlets and contains several levels of ice, sand, and composite wedges (Sher, et al., 1979, Fig. 22). The unabraded, articulated skeleton of a bison was exposed and excavated in the presence of the excursionists; it was inclined at an angle of about 15° and evidently lay on the slip face of a dune. Alyoshkina Zaimka lies opposite to and apparently provides a section through the Khallerchin Tundra, a sandy, lake-dotted lowland tract more than 100 km wide that extends northward to the Arctic coast. Vegetated dunes have been recognized on the surface of this lowland tract (A. V. Lozhkin, oral communication 1979). Sher et al. (1979, p. 94) regard the Khallerchin Tundra as an alluvial terrace "formed at the very end of the Late Pleistocene." Tomirdiaro (1980, Fig. 8) interpreted the Khallerchin Tundra as a lowland formed by thermokarst processes during the Holocene but this seems at odds with the structure and sandy composition of the sediments. In view of the sandy soils, the presence of vegetated dunes on surfaces away from the river, and the composition of the Alyoshkina Zaimka section on the opposite bank of the Kolyma River, it seems likely that the Khallerchin Tundra is a dune tract formed on the surface of a braided floodplain of the Kolyma River during the Duvanny Yar interval. The loess in the Duvannyy Yar type section nearby must represent a distal facies of finer windblown sediment carried in suspension from the same source. Possibly other broad tracts in the northeast Siberian maritime lowland that Tomirdiaro (1980) interprets as "Holocene thermokarst lake-*alas* lowlands" are also large dune tracts; this interpretation is adopted in Figure 4.

The exposed continental shelves adjoining Beringia were important sources of windblown silt and sand during intervals of low sea level. Locally, the shelves themselves seem to have been covered by extensive sand seas.[6] Tomirdiaro (1980) empha-

[6] The discussion of aeolian sediments on the Bering and Chukchi continental shelves is based mostly on my unpublished fieldwork between Cape Espenberg and Shishmaref on northern Seward Peninsula (1966, 1970, 1974), on northeastern St. Lawrence Island (1968), on St. Paul

sized the significance of the high salt content in frozen loess in coastal regions of the northeastern Siberian maritime lowland, and I suspect that at least part of the sand in the extensive sand sea on the western Arctic Coastal Plain of Alaska was also derived from the continental shelf. The stabilized dune field that covers the coastal plain of the northern Seward Peninsula is clearly the onshore remnant of a larger sand sheet that once extended far beyond the present shore. Derivation of the sand from the formerly exposed Chukchi shelf is indicated by the salty taste of the sand and by the presence in coarse-sieve residues of weathered foraminifers and bits of marine mollusk shells. Texturally identical sand blankets the sea bottom for tens of kilometers northward on the shallow floor of the Chukchi Sea (McManus and Creager, 1965).

Low-lying parts of St. Paul Island are covered with paraboloid and seif dunes, now stabilized, vegetated, and protected by a hard-pan soil.[7] The upper and lower parts of the sand sheet are composed of basaltic sand evidently derived from beaches, now submerged, that existed when sea level was no more than a few tens of meters lower than at present; the central several meters, however, consist of quartz-microcline-mica sand derived from sources on a formerly exposed continental shelf remote from the present island. The sandy texture of the bottom sediment beneath adjoining waters suggests that the emergent dunes may be only a small part of a sand sea that once occupied a much larger area in the vicinity of the Pribilof Islands.

Thin loess is present away from the sand sheet on St. Paul Island and also on parts of St. Lawrence and St. George Islands. The mineralogy of the loess on the Pribilof Islands clearly indicates that it was derived from continental-shelf sources rather than from the basaltic rocks that constitute the islands.

Island (1963, 1965, 1971, 1975), and on St. George Island (1963, 1965), as well as on my unpublished mineralogical and textural studies of sand beds in Colinvaux's (1981) 14-m pollen core from Cagaloq Lake, St. Paul Island.

[7] Active paraboloid dunes are also present adjacent to sandy beaches on St. Paul Island.

Age of the Aeolian Features

Evidence of aeolian activity as ancient as 1.8 million years has been recognized within Beringia. Several aeolian complexes now buried beneath younger deposits are known to have formed during the Happy interval. Activity at many of the Beringian dune tracts was initiated before the end of the Boutellier interval, reached peak intensity during the Duvanny Yar interval, and continued episodically during the time of the birch zone. A few tracts were reactivated from time to time during the later Holocene and remain locally active at the present time.

The earliest evidence of aeolian redistribution of sediments in a cold climate comes from St. George Island, at the southern edge of the land bridge. There, layers of cemented frost rubble in a matrix of baked loess are interspersed in a sequence of basaltic lava flows slightly older than 1.8 million years, but lava sequences older than 2.0 million years do not display evidence of frost action or loess sedimentation (Cox, et al., 1966; D. M. Hopkins and Th. Einarsson, unpublished data 1965). Just west of Beringia, early evidence of dry environments, thin vegetation, and aeolian redistribution of sediments is provided by the widespread sheets of ventifacts in the Yana and Vilyui River basins of northern Siberia. There, ventifact-bearing beds overlie sediment containing an early Quaternary spore-pollen flora and underlie gravel containing a Tiraspol mammal fauna (Kolpakov, 1978), which indicates that the ventifacts were formed at some time within the broad interval 2.0–0.5 million years ago.

The dune sands at Alyoshkina Zaimka evidently accumulated during the latter part of the Duvanny Yar interval, judging from radiocarbon dates of 15,000 ± 200 and 14,980 ± 100 years (MAG-468 and MAG-470) on organic material from underlying peaty silt alluvium (Sher, et al., 1979). The geomorphology suggests that the entire Khallerchin Tundra sand tract to the north dates from the Duvanny Yar interval.

Formation of the dunes on St. Paul Island occupied a lengthy time span that probably began during the Boutellier interval and that continues to the present time in areas adjoining the modern sandy beaches. However, study of material from the Cagaloq Lake pollen core (Colinvaux, 1981) indicates that deposition of the exotic quartz-feldspar-mica sand was contemporaneous with deposition of herb-zone pollen and thus that it was blown in off the continental shelf during the Duvanny Yar interval.

At least three stages of dune formation are associated with a succession of abandoned courses of the Yukon River in its delta area (Dupré, 1978; W. R. Dupré, oral communication 1981). The youngest, a dune complex south of Igiak Bay, is placed within the time of the birch zone by bracketing radiocarbon dates of 12,550 ± 110 years (Tx-4047) on wood in underlying alluvium and 5070 ± 60 years (USGS-215) on basal overlying peat and by association of the dune tract with an ancient river course leading to a now-submerged delta determined to be about 12,000 years old (Knebel and Creager, 1973). Older dune tracts lying to the east and southeast may be of Duvanny Yar or possibly of Boutellier age.

The sand sea that covers the northern coastal plain of the Seward Peninsula west of Cape Espenberg is mostly and perhaps entirely of Wisconsin age, since Pelukian (Sangamon) marine sediments and alluvium are exposed beneath the sand in several places. Radiocarbon dates on organic debris that collapsed into thaw lakes later filled with windblown sediments, on grass(?) roots sieved from dune sand, and on organic litter buried beneath tephra layers indicate that active sand movement began at some time beyond the range of conventional radiocarbon dating and continued during the Boutellier and Duvanny Yar intervals. Basal dates on peat beds and organic-rich thaw-lake deposits indicate that sand movement ended prior to 11,500 BP (Reeburgh and Young, 1976; Spiker, et al., 1978).

Aeolian activity in the Kobuk River valley has evidently persisted throughout much of Wisconsin time. Exposures at Epiguruk Bluff (Schweger, this volume) suggest the presence there of at least 20 m of aeolian sand of Happy age preserved beneath a draping blanket of organic silt and sand that accumulated during the Boutellier interval. The geometry of this organic sand-silt mantle suggests that it consists largely of windblown sediment deposited slowly on an undulating vegetated landscape that included, at times, some marshes and small ponds. The younger sediment exposed in Epiguruk Bluff consists of pebble-free river alluvium of Duvanny Yar age that seems to form the surface on which the widespread surficial dunes, now mostly stabilized, are situated. The alluvial aggradation of the Kobuk Valley during Duvanny Yar time may have been a response to increased river load due to reactivation of a preexisting sand sea (T. D. Hamilton and Gail Ashley, oral communication 1981). These observations suggest that the Kobuk Valley sand sea was again highly active during the Duvanny Yar interval. Activity has continued on a diminishing scale to the present. The Great Kobuk Dunes are a vigorously active remnant of a late Pleistocene sand sea and afford an opportunity to study the interaction of blowing sand, bonding hoarfrost, drifting snow, and frost cracking within a still-active Arctic sand sea.

The field of large dunes on the western Arctic Coastal Plain has been activated and stabilized several times (Carter, 1981a). Although the time of inception of aeolian activity is uncertain, radiocarbon dates indicate that a major phase of dune building began shortly after 36,000 years ago and continued through the latter part of the Boutellier interval and throughout the Duvanny Yar interval. Paleosols and zones of rooted willows suggest two episodes of temporary and at least local stabilization of sand movement about 13,500 years BP and between 11,800 and 10,800 years BP. Shortly after 10,800 years BP, the regimen of sand movement changed, and a sand sheet several meters thick was formed, blanketing the large linear dunes formed earlier and extending beyond them. Although encroaching vegetation rapidly stabilized the sand sheet between 8500 and 8000 years ago, small paraboloid dunes, now vegetated, record another episode of reactivation some time during the last 3500 years (Carter, 1981b).

Two sets of dune tracts are distinguished by Fernald (1960) near the front of the Alaska Range in the upper Kuskokwim River drainage. The fresher dunes are on an outwash terrace of the Farewell Glaciation of Duvanny Yar age and the more subdued, older dunes are on higher outwash of the Selatna Glaciation of Happy age.

The aeolian sequence in the Shaw Creek Flats north of the Tanana River opposite the mouth of the Delta River is thought by Péwé (1965, 1975a) to consist of ventifacts and dunes of Delta (=Happy) age overlain by loess of presumed Donnelly (=Duvanny Yar) age. A pollen core from Quartz Lake, a few tens of kilometers to the east, however, records a transition from sand to silt at the 10,000- or 11,000-year level (T. A. Ager, written communication 1981). At Canyon Creek on the north bank of the Tanana River, a few tens of kilometers west of the Shaw Creek Flats, ventifacts and sand wedges of Duvanny Yar age are overlain by aeolian sand that began to accumulate during the time of the birch zone and continued to be deposited for some time after 9600 years BP (Weber, et al., 1981). Loess covering the dunes at Canyon Creek is of middle and late Holocene age.

Mixtures of aeolian and alluvial sand may have accumulated in the upper Tanana River valley throughout much of the Wisconsin, but the conspicuous large dune of the area accumulated mostly during the time of the birch zone. Carter stated that at Riverside Bluff (not shown in Fig. 4), "a thin pond deposit immediately below the lowest eolian bed and above alluvial deposits dates 12,230 ± 120 B.P. (USGS-1037), and a pond deposit in the middle of the eolian section dates 11,880 ± 180 B.P. (11,704). Elsewhere in the upper Tanana Valley, dates that Fernald (1965) relates to eolian sand are 12,400 ± 450 (W-1212), 11,250 (I-302), and 8,200 (W-1206)." (L. D. Carter, written communication 1981).

In other Alaskan dune tracts, data are insufficient to estimate ages. The late age of the ventifact-aeolian sand-loess sequence at Canyon Creek discredits the significance of a loess blanket as an age criterion. Ventifacts and sand wedges simply record episodes of deflation. Aeolian sand moves by traction, and movement cannot persist if a cover of shrubs or trees is present because shrubby or forest vegetation increases friction and reduces wind shear near the ground. Loess is moved in suspension, but mostly within a few hundred meters of the ground; the presence of forest vegetation reduces wind shear at low levels and thus functions to impede transport of windblown dust. Therefore, the ventifacts-aeolian sand-loess sequence may reflect little more than a decline in wind effectiveness related to successional changes in the vegetation cover in response to a shift toward a more moist climatic regime.

Most, if not all, of the small dune tracts in central and southern Yukon Territory were formed during the Duvanny Yar interval. Sand forming local dune tracts that lie north of the St. Elias Mountains (Fig. 4, Loc. 20) are derived from the sandy beaches of proglacial lakes confined by the Cordilleran ice sheet during the McCauley Glaciation; the other dune tracts rest on or are derived from glacial outwash of equivalent age (O. L. Hughes, written communication 1980). Relation to drift sheets and paleosols of various ages show that the sand wedges and ventifact pavements of the Tintina Trench, however, record a long history of intermittent deflation and sand movement. The older features evidently date from the Reid Glaciation, and the younger from the McConnell Glaciation (Hughes, et al., 1972; Foscolos, et al., 1977; O. L. Hughes, written communication 1980).

The sand deposits that I believe to be windblown on lowlands along the Beaufort Sea coast in northwestern Canada probably have a history of dune formation and reactivation similar in complexity to the history established in northern Alaska, but because the aeolian origin of the older deposits has hardly been recognized, this history can not yet be elucidated. The principal deposit, a sheet of brown sand up to 50 m thick (the deltaic sand of Fyles et al., 1972), overlies marine deposits probably of latest interglacial (Pelukian) age but is mantled by glaciofluvial sand and gravel that is beyond the reach of radiocarbon dating. Thus, a major episode of dune-sand accumulation seems to have taken place during the Happy interval. However, Mackay (1963, Figs. 12. 13) shows oriented lakes in a sandy terrain on the Tuktoyaktuk Peninsula in which interlake ridges are crowned by vegetated and stabilized paraboloid dunes—a terrain so similar to the northern Alaskan sand sea of Carter (1981a, 1981b) that one is tempted to suggest that this area, some 100 km east of the Mackenzie River delta, may also display large linear dunes principally of Duvanny Yar age and that the paraboloid dunes may record late Holocene reactivation.

Much remains to be learned about the ages of loess sequences. Substantial thicknesses of loess deposited during the Duvanny Yar interval are exposed along the rivers and coastal bluffs of the northeastern Siberian maritime lowland, but few, if any, exceed the 15-m thickness of horizon III at Duvannyy Yar itself (Lozhkin, 1975; Kaplina, 1979; Sher, et al., 1979; Tomirdiaro, 1980). Loess evidently of Happy age is also present in many places in northeastern Siberia, although exposed thicknesses are small, and most of this older loess evidently lies below present stream grade. The intervening organic-rich deposits, representing the

Boutellier interval, are typically 2 or 3 m thick and may consist partly of loess intermixed with organic accumulations.

In western Alaska, loess of Happy age at Cape Deceit is less than 5 m thick, and primary loess of Duvanny Yar age is evidently not preserved there at all (Matthews, 1974b). At Cape Blossom, a moraine of the Nome River Glaciation (Illinoian) is covered by several meters of pre-Sangamon loess, a forest bed of Sangamon age, and then about 10 m of loess of undifferentiated Wisconsin age. The loess sequence on the Kvichak Peninsula, southwestern Alaska, accumulated intermittently throughout most of the latest glacial cycle. At one site on the west shore of the peninsula, the following section was observed (D. M. Hopkins, unpublished data 1971).[8]

1.0 m	Peat (Holocene)
0.5 m	Loess (birch zone)
0.5 m	Peat
6.0 m	Loess (Duvanny Yar interval)
1.5 m	Peaty silt (loess rained on a well-vegetated surface?) (Boutellier interval)
1.0–2.0 m	Loess (late Happy interval)
Thickness unknown	Glaciomarine drift (early Wisconsin)

To summarize, the stratigraphic evidence and distribution data in Alaska, where aeolian sand bodies have been most adequately mapped and studied, indicate that at least some of the recognized dune tracts were the sites of intense activity during the Happy interval. The Boutellier interval was a time of diminished sand movement, and some tracts may have been largely stabilized. The Duvanny Yar interval was a time of maximal activity of dune tracts throughout Beringia. Sand movement diminished during the time of the birch zone and largely ended about 8500 years BP.

The scanty data available thus far indicate that loess preserved on the Beringian landscape today includes deposits as old as the next-to-latest (Illinoian) glacial cycle. During the Wisconsin cycle, aeolian silt was laid down in two major pulses, of which the earlier (Happy interval) was the more important. Along many of the major rivers of Beringia, the Boutellier interval saw continued slow deposition of windblown silt on vegetated landscapes, resulting in accumulations of organic silt with intercalated peat and pond deposits. A second but lesser pulse of loess deposition took place during the Duvanny Yar interval. In places where sources are available and wind fetch sufficient, loess deposition continues at a slow pace to the present time.

Evidence of Climate and Season of Activity

Until very recently, little attention has been given to evidence of the season of dune activity or to collateral paleoclimatic evidence involving the aeolian features of Beringia, but well-focused studies promise to provide considerable new information. Pond sediment interbedded with aeolian deposits, for example, provides evidence of a former fluctuating water table that occasionally rose above the level of interdune hollows. Rootlets in aeolian sand or silt indicate the presence of sparse vegetation during accumulation. Deposition of mixtures of sand and snow on slip faces or advance of dunes over persisting snowbanks should be recorded by collapse structures and penecontemporaneous slump features disturbing steep crossbedding. Contortion of bedding due to cryoturbation probably indicates saturation during either freeze-up or spring thaw.

Features related to frost cracking are especially informative. Ice wedges and sand wedges result from cracking due to contraction during extreme cooling of frozen ground. Ice wedges should result if the contraction cracks become filled with hoarfrost, snow, or snow-melt water, and sand wedges form if the cracks are filled by blowing sand. A thick snow cover damps temperature fluctuations in the soil and thus impedes frost cracking. Ancient ice wedges thus provide evidence of severe winters and a thin snow cover, probably wind packed. Sand wedges indicate that the ground was bare and sand movement active at the time that frost cracks were open in late winter or early spring.

In the one sand sea that is still active in Beringia, sand movement responsible for the gross dune form takes place during winter. In early summer, the dune field consists of simple transverse and barchanoid forms with west-facing slip faces, reflecting shaping by down-valley winds out of the east-northeast. By early autumn, however, each dune has a small liplike ridge along its crest just above the slip face, reflecting modification of

[8] Ager (this volume) reports on pollen from this site.

the major dune form by strong fair-weather winds blowing cross-valley out of the northwest (T. D. Hamilton, oral communication 1981).

The sand sheet on St. Paul Island conveys considerable information concerning climate and seasonality. The distribution of discrete sand layers in the Cagaloq Lake core (Colinvaux, 1981) indicates that intense sand movement into the crater was somewhat episodic, although the aeolian activity that formed the dunes in more open, low-lying areas may have been more constant. Wedge structures are present but rare in the sand sheet. Many natural sections in coastal bluffs display cryoturbated bedding and large basalt blocks scattered in sand. Large basalt blocks are also found resting on the surface of thick aeolian sand a few hundred meters from the base of one of the cinder cones. Evidently there was a substantial winter snow cover, sufficient to inhibit frost cracking and provide enough melt water to saturate the sand, support severe frost-heaving, and promote lateral gliding of boulders over very gentle slopes during spring. These observations suggest that sand movement took place on St. Paul Island mainly during the summer. In the low coastal exposures, common permeation of the sand by roots indicates that a sparse grass cover was present in interdune areas; this in turn may indicate summer precipitation. Counterbalancing these indications of a humid climate is the evidence from Colinvaux's (1981) pollen core, which indicates that the climate was dry enough to cause the presently perennial Cagaloq Lake to become intermittent during deposition of the sediment of the herb zone, that is, during the Duvanny Yar interval.

In the Alyoshkina Zaimka section in northeastern Siberia, lenses of silt interbedded in the sand probably represent deposits of interdune ponds and suggest that the summer water table fluctuated and sometimes lay above the surface of the interdune hollows. The all-pervading rootlets and the presence of a bison skeleton suggest the presence of a thin grass cover, and indicate at least some summer rainfall. The diverse wedge types indicate that sand movement was active during winter but that blowing snow was also often present. My impression is that the Alyoshkina Zaimka dunes were active throughout the year.

A moderately mesic situation is indicated for the northern Seward Peninsula sand sea. The eastern part of the complex consists of deposits of thaw lakes that existed during the Duvanny Yar and earlier intervals and that became filled with fine aeolian sand. The western part consists of true dunes interspersed with sandy interdune pond deposits. Short rootlets permeating one dune were dated as 28,900 ± 1,000 years (W-2804) (Spiker, et al., 1978; D. M. Hopkins, unpublished data 1971). The climate of the northern Seward Peninsula during the Duvanny Yar and Boutellier intervals and probably the Happy interval was dry enough and the vegetation thin enough to permit active deflation of marine sand on the exposed continental shelf, although moist enough to permit the existence of thaw lakes and interdune ponds. The presence of ice-wedge pseudomorphs in many exposures indicates that blowing snow was present but that the snow cover was thin. Sand movement to create the northern Seward Peninsula sand sea probably took place mostly during winter.

Sand wedges evidently of Duvanny Yar age occur in areas of the Alaskan Arctic Coastal Plain with a thin or discontinuous regional sand cover but have not been observed in the dune ridges of the major sand sea (L. D. Carter and D. M. Hopkins, unpublished data 1981). This suggests that the dune sand was dry and incoherent in winter and thus prevented the formation of contraction cracks. However, thin pond deposits have been observed in the few good exposures of interdune trough deposits within the sand sea. Summer rainfall was evidently insufficient to saturate the dune ridges but sufficient to form small, probably ephemeral interdune ponds. Good exposures of stratification within the dune ridges show no evidence of slumping or collapse such as might have resulted if snow had been mixed with or buried by blowing sand. The overall impression is that during Duvanny Yar time, the area of the Arctic Coastal Plain was very dry, with little snowfall and little summer rain. Carter (1981b) has proposed that the dunes were active either entirely during the summer or year around, with an absence of snow cover. The presence of sand wedges almost universally outside the area of the sand sea indicates that sand movement was certainly active during late winter and spring, when frost cracks would have been open.

The widespread sand wedges in central Yukon Territory (Hughes, et al., 1972; Foscolos, et al., 1977) indicate that near the margin of the Cordilleran ice sheet snow cover was thin, bare ground widespread, and sand movement active during winter, during both the Happy and the Duvanny Yar intervals. Similar inferences can be drawn from the presence of sand wedges of Duvanny Yar age at Canyon Creek in central Alaska (Weber, et al., 1981). The widespread sand wedges in other parts of the Yukon-Tanana uplands of central Alaska (Péwé, 1975a, 1975b) also provide evidence of winter sand movement, although perhaps at a more ancient time.

In summary, features of the aeolian sand accumulations suggest a dry climate and discontinuous vegetation throughout Beringia during the Duvanny Yar interval and only slightly more mesic conditions during the preceding Boutellier interval. Although winter snow cover was substantial in the area of the Pribilof Islands, the summer climate was dry enough to permit sand to blow and to lower or even dry out crater lakes. A patchy, discontinuous snow cover is indicated for the Kolyma River lowland and the northern Seward Peninsula, where enough winter moisture was present to permit the formation of ice and composite sand-ice wedges. Most frost cracks were infiltrated by sand rather than snow along the Beaufort Sea coast and in the Tintina Trench; these regions must have been near-deserts.

These inferences merely scratch the surface. Much more detailed paleoclimatic information can be gleaned by careful and detailed study of Beringian aeolian deposits and features.

Ancient Wind Directions

Paleowind directions can be interpreted from the morphology of ancient dunes in Beringia. At first glance, a plot of wind vectors indicated by the dunes (Fig. 4) seems a maelstrom of inconsistencies and contradictions. The apparent confusion would have been even greater had I plotted the distribution of ancient loess deposits because in several areas windblown silt was deposited in positions seemingly upwind from areas of sand deposition. In the lower Tanana Valley, for example, the enormous sand sea shaped by northeasterly winds lies south of the river; north of the river, thick loess mantles a belt extending tens of kilometers into the Yukon-Tanana upland (e.g., near Fig. 4, E, F). Obviously, the aeolian deposits of Beringia were not all deposited simultaneously. The question is whether the time separation was seasonal or whether these paleowind indicators are of radically different ages.

Most of the Beringian dune areas seem to have been active during the Duvanny Yar interval, and I think it probable that most of the paleowind indicators in Figure 4 reflect air circulation during that interval (exceptions are the dunes on St. Paul Island, on the Yukon delta south of Igiak Bay, and in the middle and upper Tanana Valley, as discussed later). If my assumption is correct, then the confusion of paleowind indicators must be largely due to differences in season of activity, although local orographic effects have also evidently played a role.

The existence during the Duvanny Yar interval of a large winter gyre, centered roughly over the Gulf of Alaska and influenced in form and position by the presence of the continental ice sheets to the east, is suggested by evidence of southeasterly winter winds in the Tintina Trench, east-northeasterly winds on the Alaskan Arctic Coastal Plain, and northeasterly winter winds on the northern Seward Peninsula. Dune areas that are consistent with this pattern and that probably were formed by winter circulation over Beringia during the Duvanny Yar interval are widespread in central and western Alaska and include the southwest-trending dunes in the lower Tanana Valley, dunes in the upper Kuskokwim River valley,[9] most of the dunes on the Yukon delta, dunes in the lower Koyukuk River Valley, dunes south of the mouth of the Fish River and Golovnin Bay, southern Seward Peninsula, and dunes in the Kobuk River valley.

The southerly and southeasterly wind vectors

[9] In an earlier report (Hopkins, 1972), I followed Fernald (1960) in stating that ancient dunes in the upper Kuskokwim valley were deposited by katabatic winds from a southerly quadrant. F. R. Collins (written communication 1980) points out, however, that "although strong southwesterlies certainly do accompany summer storms, and southeast winds whistle through Rainy Pass . . . the dunes are southwest of the Kuskokwim River and its tributaries from the Alaska Range which are the most immediate source of sand. From Fernald's 1960 photograph the dunes resemble, in their general shape and orientation, the central Alaskan dunes—perhaps the sand was moved mostly by northeast winds, but the northwestern steepening and convexity (Fernald, 1960, p. 246) may have been superimposed by Rainy Pass winds." This interpretation is adopted in Figure 4.

scattered from central Seward Peninsula through the Tanana River valley to southern Yukon Territory (Fig. 4) probably reflect the southeasterly summer air flow over Beringia postulated for the Duvanny Yar interval by Barry (this volume). In addition, the dichotomy of loess north of the lower Tanana River and dunes to the south built by northeasterly winds may be a dichotomy between strong southerly or southeasterly summer winds and strong northeasterly winter winds. Interestingly enough, nearly all these examples of northward or northeastward transport of sand and dust during the Duvanny Yar interval lie in the lee of mountain masses to the south; this suggests that they were created in places where the southeasterly summer airflow was accelerated and dried by passage over topographic highs, as postulated by Streten (1974).

Contradictions remain, some easily explained, some not. The opposing wind vectors west of Dawson (Fig. 4, loc. 18) represent small dune tracts on opposite sides of the Yukon River, where the river crosses the orographic funnel of the Tintina Trench. A southwesterly wind vector near Tok (Fig. 4, loc. 10) represents stabilized dunes built by katabatic winds from the Wrangell Mountains icecap that funneled northward through the Tok River Valley in the Alaska Range. The sand wedges of Duvanny Yar age at Canyon Creek, which lie on a rocky bluff north of the Tanana River, must also represent a record of orographic winds, draining episodically during winter, from the high, glacier-covered segment of the Alaska Range that lies to the south.

Northeastern Siberia remains a blank on our paleocirculation map. In an area where the existence of sand dunes and loess deposits is barely acknowledged, evidence for ancient wind vectors remains likewise unrecognized.

Circulation during the time of the birch zone cannot be reconstructed so confidently. Southerly orographic winds at an unknown season easily account for the early Holocene dune at Canyon Creek and for the Shaw Creek dunes, which were formed at least partly during the time of the birch zone. The northeasterly wind vector indicated by the dunes south of Igiak Bay probably reflects winter circulation, although no detailed studies of seasonality have been undertaken. More anomalous is the northwesterly wind vector at an unknown season indicated by the form of dunes in the upper Tanana Valley for which radiocarbon dates clearly indicate a birch-zone age.

The large seif dune at St. Paul Island took its present form during the time of the birch zone and was evidently formed by alternating northwesterly and southeasterly winds. The high influx of spruce, alder, and birch in Cagaloq Lake during the time that the seif dune was most recently active seems to indicate that mid-summer airflow was from the southeast, as Barry (this volume)[10] postulates for a slightly earlier time. If the summer winds had come from the northwest, pollen introduced by long-distance transport should have included pine and little or no spruce because stone pine (*Pinus pumila*) is the abundant gymnospermous pollen producer and spruce is entirely absent in the northeastern Siberia regions that lie northwest of St. Paul Island. The opposing northwesterly winds that contributed to the construction of the seif dune must have been an autumn phenomenon, active at a time when *Pinus pumila* was no longer shedding pollen but prior to heavy snow accumulation. The scarcity of frost-crack features and the stratigraphic evidence of wet ground during spring make it unlikely that the seif dune was active during winter or spring, and its orientation, in any case, is inconsistent with any probable winter circulation pattern.

In summary, this highly speculative review suggests that, during the Duvanny Yar interval, winter circulation in eastern Beringia was dominated by a huge counterclockwise gyre centered on a low approximately over the Gulf of Alaska. The winds, as one might expect, were very dry. Little snowfall penetrated into Yukon Territory, central and western Alaska, and the Kolyma lowland; even less reached the Arctic coast north and northeast of the Brooks Range. The summer pattern was evidently dominated by a southeasterly air flow whose moister winds generally moved sand or loess only where accelerated and dried by passage over mountain masses.

Circulation during the time of the birch zone is less firmly established; however, the northwesterly winds indicated by dunes in the upper Tanana Valley and the apparent shaping of the large seif

[10] In earlier reports (Hopkins, 1970, 1972) I misinterpreted the dune morphology on St. Paul Island and concluded that the dunes recorded northerly or northeasterly summer winds during the time of the birch zone.

dune on St. Paul Island by northwesterly winds in autumn may signal an important change in the circulation pattern during at least part of the year. Could these paleowind indicators point to wide expanses of open water in the Beaufort Sea and adjacent parts of the Arctic Ocean during the time of the birch period?

A SYNTHESIS

In the past, scarcity of chronological and stratigraphic information has placed severe constraints upon our ability to reconstruct the paleogeographic and environmental history of Beringia. Our thinking has had to be framed in terms of simple vignettes that portray an interglacial Beringia that existed about 120,000 years ago and a full-glacial Beringia that existed about 18,000 years ago (Hopkins, 1972, 1979a).

Field studies during the past few years permit us to think now in much more fluid terms. This report has attempted to periodize the latest cold cycle in Beringia, to briefly review current knowledge of Beringian glacial chronology, shoreline history, and paleoceanography and, in more detail, to examine the possibility that the aeolian sediment of Beringia can be used to reconstruct atmospheric circulation patterns during at least the later part of the latest cold cycle.

A major advance in our understanding has been the recognition and clear demonstration that alpine and Cordilleran glaciers of Alaska advanced to their maximum Wisconsin extent more than 40,000 years ago and that late Wisconsin alpine glaciation was of very limited extent. This dichotomy has been recognized for many years in Siberia and northwestern Canada but was doubted until recently by most Alaskan Quaternary geologists, including myself. The precise age of this extensive early Wisconsin glaciation, however, remains unsolved. A determination as to whether it coincided with isotope stage 4 or, instead, correlates with one of the insolation maxima within isotope stage 5 must await accumulation of more geochronological, tephrochronological, and geomagnetic data.

Problems of correlation limit our ability to reconstruct in any detail the landscape, vegetation, and climate during the Happy and the early part of the Boutellier intervals. Beginning about 40,000 years ago, however, the picture becomes clearer. Sea level then was not far below its present position. The land bridge probably was reduced to a fairly narrow isthmus that at times may have been flooded. The coastal mountains remained heavily glaciated, but alpine glaciers elsewhere were probably not much larger than at present. Laurentide ice may have advanced to a maximum Wisconsin position, filling the Mackenzie Valley as early as 32,000 years ago—a time when alpine glaciers also probably underwent a brief expansion.

Stratigraphic sequences of Boutellier age commonly show evidence of incipient dune activity or of slow loess fall upon herbaceous or ericaceous vegetation. The vegetation during the latter part of the Boutellier interval thus seems to have been less continuous and less shrubby than at present but denser than during the ensuing Duvanny Yar interval—a presumption supported by comparison of pollen spectra from the Boutellier and Duvanny Yar intervals. The landscape was diversified by the presence of some conifers, tree birches, cottonwoods, and perhaps aspen, possibly forming sparse woodlands on sheltered (shady?) slopes.

That most of the large-mammal carcasses found in Alaska and northeastern Siberia date either from 14,000 to 10,000 or from 45,000 to 30,000 years ago suggests that similar conditions favorable for burial and preservation prevailed during both the time of the birch zone and the latter part of the Boutellier interval. This analogy suggests that after 45,000 years BP, the Boutellier interval featured snowy winters, heavy spring runoff, and very dry summers. The presence of ice in the coastal mountains, the persistence nearby of the Laurentide ice sheet, and the continued existence of large areas of exposed continental shelf during the Boutellier interval resulted in Beringian climates drier and more continental than at present but more mesic and less extremely continental than during the Duvanny Yar interval.

The Duvanny Yar interval was a time of extremely dry continental climates in Beringia. The land bridge was maximally exposed; the Cordilleran ice sheet was large and extensive, as were glaciers in the coastal mountains of northeastern Siberia and Kamchatka; and the Laurentide ice sheet stood near its maximum northwestern limits. Alpine glaciers, on the other hand, were smaller than they had been during the Happy interval, especially in the more northern ranges.

Global circulation was such that summer air-mass movement in Beringia was northward or northwestward rather than northeastward as at present. This dominant northward and northwestward transport meant that moist air masses from the southern Bering Sea rarely penetrated into central and northern Alaska and northwestern Canada but moved northward, instead, to nourish glaciers in the highlands on either side of the present-day Bering Strait and perhaps also in northeastern Siberia. The westward lowering of late Wisconsin glaciation limits in the Brooks Range and the predominance of more icy frost-wedge structures of Duvanny Yar age in western Alaska and the Kolyma lowland, still farther west, in comparison to the predominance of ancient sand wedges along the present-day Beaufort Sea coast, evidently reflect this east-west moisture gradient.

The presence of large active dunes, steppe-tundra vegetation, and ephemeral ponds in basins that now contain perennial lakes on the Pribilof Islands indicates a surprisingly dry climate at the south shore of the land bridge. These observations gain significance in the light of Sancetta's recent studies (1981, in press) showing that the inflow of Pacific water was drastically curtailed and that an extensive winter sea-ice cover was present on southern Bering Sea during the Duvanny Yar interval.

The driest and most hostile Beringian environments seem to have been on the Arctic lowlands adjoining the Beaufort Sea coast. Large areas seem to have been sandy desert—an alternation of active, mobile dune complexes and of sandy deflation areas. Large streams draining the glaciated mountains or carrying melt water from the Laurentide ice sheet maintained flow, but smaller streams heading in foothills were defeated by encroaching dunes on the coastal plains.

The diversified topography around the present-day Bering Strait, on the other hand, may have been the site of one of the most mesic areas in Beringia. The more frequent passage of northward-moving moist air masses may have made this area the refugium from which cottonwood, aspen, and possibly alder dispersed during the rapid climatic changes that ushered in the time of the birch zone.

Despite evidence of great aridity, there are enough well-dated fossil remains of Duvanny Yar age to establish firmly the presence of mammoth, horse, bison, caribou, and saiga. Whether or not human hunters were also present remains an open question for all except the westernmost areas of Beringia.

The end of the Duvanny Yar interval was marked by rapid climatic and environmental changes. Abrupt warming is suggested by catastrophic retreat of glaciers, but there must have been a drastic change in the moisture regime as well. Throughgoing streams were reestablished as early as 13,500 years BP through the northern Alaska sand sea. The gullying of loess-covered slopes and the frequent burial of mammal carcasses in the resulting fans of organic silt indicate that the altered hydrologic regime extended to central Alaska as well. Water levels in small lake basins rose, and thermokarst lakes became a prominent feature of the landscape throughout the permafrost regions of Alaska and Yukon Territory and probably throughout Beringia. As more moisture became available, the plant cover thickened, and aeolian-sand movement diminished in intensity. Shrub tundra dominated by dwarf birch appeared at the present-day Alaskan coast of Bering Sea and spread eastward and northward so rapidly, according to Anderson's (1981) isopoll diagrams, that the base of the birch-pollen zone in eastern Beringia can be regarded as practically isochronous. Ericaceous pollen displays a similar, though less dramatic record. These patterns evidently reflect the release and proliferation of shrub species that were already widely distributed but very inconspicuous in Beringian vegetation. An interesting characteristic of Beringian large-mammal faunas 13,000 to 11,000 years old is the appearance of wapiti (*Cervus*), perhaps reflecting the proliferation of more shrubby but still open and generally treeless vegetation.

Cottonwood and perhaps aspen became prominent in Beringian vegetation early in the time of the birch zone. Meanwhile, spruce slowly dispersed northwestward in eastern Beringia, and larch and stone pine may have had a similar history in northeastern Siberia. During the time of the birch zone, these tree species seem to have formed open woodlands, and closed forests did not become a major element in Beringian vegetation until well after 8500 years ago.

Significant for climatic reconstruction is the fact that in situ peat accumulation was not initiated until several millenia after the beginning of the birch zone; in northern Alaska, where the inception of peat deposition has been best studied, in situ accumulation did not begin until about 8500 years ago.

L. D. Carter (1981b) suggests that the principal climatic change in northern Alaska 13,500 years BP was an increase in winter snowfall and a consequent intensification of the spring snow-melt flood. The epoch of rapid in situ peat accumulation, he suggests, represents waterlogging of soils associated with the inception of the present-day wet summer climate. Much of the paleoclimatic evidence from other parts of Beringia seems explicable in terms of an increase in snowfall about 13,500 years ago and a delay of inception of summer precipitation until around 8500 years ago.

Important factors in the changing climates of the time of the birch zone may have been the occurrence of a summer insolation maximum intensely felt at the latitude of Beringia (Kutzbach, 1981), the flooding of the Bering Sea shelf, the destruction of shelf-based icecaps off southern Alaska and the Aleutians and the persisting presence of remnants of the Laurentide ice sheet in Canada to the east. The result was better penetration of snow-bearing air masses during winter into what had been the extremely continental interior of eastern and probably also western Beringia but a blockage of moist air penetration during summer.

Ice cover on the Arctic Ocean basin may have been less extensive than at present during the time of the birch zone, and large open-water areas may have been present during late summer and autumn. This might be the explanation, for example, for the sketchy evidence of northwesterly autumn winds in the Pribilof Islands and the upper Tanana Valley, the presence of cottonwood groves on the Arctic Coastal Plain (a region in which cottonwood no longer grows), the evidence in the same region for dry summers but a heavy winter snow pack, and the record for very late readvances of valley glaciers in the Brooks Range and of the Mackenzie Valley ice stream of the Cordilleran ice sheet.

EPILOG

This is a maiden effort, and interpretations are put forward in full expectation that they will be extensively modified, improved, and corrected by future work. I hope that this report will arouse controversy and stimulate discussion and research that can lead to a more soundly based reconstruction of the changing climates of Beringia during late Quaternary time.

ACKNOWLEDGMENTS

Almost everyone I work with and talk to has contributed to the preparation of this report. I am especially grateful to Tom Ager, Pat Anderson, Dave Carter, Florence Collins, Jim Craig, Tom Hamilton, Owen Hughes, Ross Mackay, John Matthews, Bob Nelson, Vern Rampton, Constance Sancetta, Charley Schweger, Robert Thorson, and Florence Weber for unselfishly sharing unpublished data with me. Perceptive, sometimes scathing, reviews of early drafts of the manuscript were provided by T. A. Ager, L. D. Carter, T. D. Hamilton, J. V. Matthews, R. M. Thorson, and J. R. Williams. Peggy Smith assisted in research and compilation and Mary Malloy in preparation of the manuscript. David M. Jones patiently worked with me through many revisions of the illustrations. Above all, I thank Lita Osmundsen and the Wenner-Gren Foundation for providing the setting, and my fellow participants in the "Paleoecology of the Arctic Steppe-Mammoth Biome" symposium for providing the stimulation that brought this report into being.

2

EVOLUTION OF LOWLAND LANDSCAPES IN NORTHEASTERN ASIA DURING LATE QUATERNARY TIME[1]

S. V. Tomirdiaro

Interfluve areas on the lowland plains of northeastern Siberia are a mosaic of small plateaus and hillocks (*yedoma*) separated or indented by depressions and basins (*alasy*). *Alasy* are thermokarst depressions formed by thawing of the ice-rich sediments of the *yedoma*. The *yedoma* sediments have customarily been interpreted as the products of late Pleistocene floodplain aggradation and accumulation in floodplain lakes. However, N. A. Shilo postulated as early as 1964 that the *yedoma* deposits must consist of perenially frozen loess of late Pleistocene age. The absence of channel sand and gravel, the absence of remains of fresh-water fauna, the obscure stratification, and the presence of syngenetic ice wedges all argue strongly against either a lacustrine or a floodplain origin for the *yedoma* sediments. The overwhelming predominance of silt in the very uniform, well-sorted *yedoma* sediment, the blanketing distribution as a cover on alluvial terraces of diverse ages and on rough bedrock relief, and the common presence of permeating vertical rootlets of herbaceous plants demonstrate that the *yedoma* loam is a windblown sediment. Lack of woody remains, presence of pollen spectra of the steppe-tundra type, and presence of a substantial carbonate and salt content indicate that the *yedoma* loam accumulated in a climate drier than the present one.

The *alasy* are a Holocene phenomenon. Presence in them of thaw lakes and lake sediments and thick peat layers testifies to a contrast between a relatively mild, humid, maritime Holocene climate and the much drier, more rigorously continental climate of the late Pleistocene. The evidence of a former cold, dry continental climate must mean that the late Pleistocene Arctic Ocean remained frozen even in summer. The abrupt shift to a much moister, more maritime Holocene climate resulted from the break-up of the perennial sea-ice cover when rising sea level admitted relatively warm Atlantic water into the Arctic Basin. Thermokarst lakes developed on the loess-ice plain ahead of the encroaching sea, causing a general irreversible reorganization and lowering of the late Pleistocene landscape.

Loess-ice plains, of which remnants persist as *yedoma* in northeastern Siberia, once

[1]This paper originally appeared in the *Izvestiya* of the All-Union Geographic Society (Tomirdiaro, 1975), and was translated for this volume by Dorothy Vitaliano. Because much of Dr. Tomirdiaro's argument and interpretation represents a controversial view in the Soviet Union, it is useful to point the reader to his more recent work, which provides much new and concrete data relevant to the paper published here. Useful photographs and extensive supporting data on sedimentary structures, granulometry, mineralogy, geochemistry, palynology, and geochronology may be found in Tomirdiaro (1978 and 1980, see especially pp. 5–73 of the 1980 paper).

ISBN 0-12-355860-3

stretched from western Europe across Siberia to Alaska. Holocene "alasification" has been a general process in unglaciated lowlands in the middle and northern latitudes, one that has been completed with the complete destruction of permafrost in eastern Europe and western Siberia but which is still in progress in northeastern Siberia and Alaska. The contrast between the present-day marshy and lake-dotted lowland landscapes of northeastern Siberia and the dry, continental loess plains of eastern Europe is a strictly Holocene phenomenon. *(Abstract by D.M.H.)*

Key words: *alas,* Arctic Ocean ice cover, ice wedges, loess, periglacial landscapes, thermokarst, *yedoma*

The phenomenon of active cryogenic evolution of landscapes was first discovered and studied by I.P. Gerasimov. He noted that before the end of the Pleistocene there was a uniform periglacial landscape in the central Yakutian lowland, outwardly resembling supra-floodplain cold steppe and forest-steppe, and that its destruction was brought about by the Holocene climatic change and by consequent development of thermokarst processes. According to Gerasimov (1952), "The geomorphologic evolution of the landscape consisted of gradual degradation of ancient ground ice, their self-burial beneath mantles of thawing *melkozem* cover, and development of thaw lakes which were gradually transformed into landscapes of *alas* type"[2] (Gerasimov, 1952). Although Gerasimov came to these conclusions only in reference to the plains of central Yakutia, I am convinced as a result of many years' work in northern Yakutia and Chukotka that they should also be the basis of paleogeographic formulations for all of the arctic and subarctic plains of Northeast Asia (Fig. 1).

Until recently, Siberian plains were thought to be regions of late Pleistocene alluvial or lacustrine-alluvial deposition. If this were true the development of a multitude of lakes and *alasy* would have taken place, not only in the Holocene, but also in the late Pleistocene, the epoch in which the original ice deposits were formed (Rozenbaum, 1971).

N.A. Shilo (1964) was the first to come out against the idea of an alluvial origin for the ice-saturated mantle of late Pleistocene age on the plains of Northeast Asia. This mantle, often called by the Yakuti name '*yedoma,*'[3] consists of thick syngenetic ice wedges, separated by polygonal blocks of perennially frozen loess. The thicknesses of the wedges and intervening loess blocks vary considerably, sometimes within the same section, because of differences in the rates of sedimentation in the various layers of the deposits. In spite of these differences, the ground ice in all varieties of *yedoma* is composed exclusively of syngenetic polygonal ice, and without exception dates from the middle and late Pleistocene (Baranova, et al., 1968; Boyarskaya, et al., 1971; Geology of the USSR, 1970; Dosotovalov and Kudryavtsev, 1967; Konishchev, 1970; Lavrushin, 1963; Popov, 1965; Sher, 1971).

Shilo subjected to convincing criticism all versions of the alluvial concept of formation of the *yedoma* on the plains of Northeast Asia such as those advanced by N.N. Romanovskiy, Yu. A. Lavrushin, V.E. Vtyurin, and others. He exposed the fundamental fallacy, inherent in the concept of alluvial origin, of the idea that a sudden change in direction of tectonic movements occurred throughout Northeast Asia at the end of the late Pleistocene. Only a massive tectonic shift could explain the cessation at that time of the formation of alluvial plains and the deep recent downcutting (35-45 m) that has formed confined river valleys. Furthermore, we know that *yedoma* was present over a broad geographical area, including the Aldan River valley in southern Yakutia (Boyarskaya, et al., 1971), central and northern Yakutia (Baranova, et al., 1968), West and East Chukotka

[2]*Alas* (pl. *alasy*) is a Yakuti word designating a grassy, treeless meadow occupying a flat-floored thermokarst depression in central Yakutia that, in a moister climate, would contain a thermokarst lake. The word is also applied in Soviet geomorphic literature to the shrub-covered or sometimes lightly forested depressions marking the basins of drained thermokarst lakes in more northern parts of Siberia—*Eds.*

[3]The Yakuti word *yedoma* refers to hillocks and small, steep-sided plateaus which geologists have shown to be underlain by ice-rich sediment in the great alluvial lowlands of Yakutia—*Eds.*

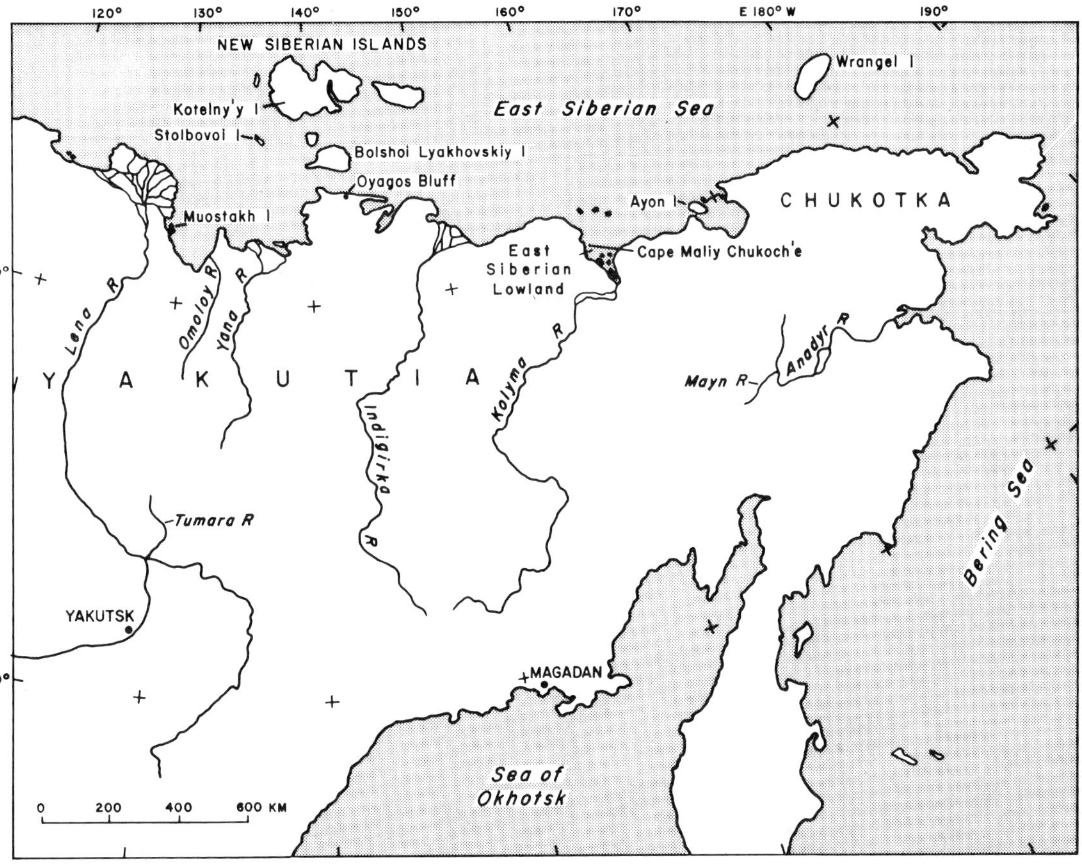

Figure 1. Map of northeastern Siberia showing places mentioned in the text

(Tomirdiaro and Orlova, 1970), and even Alaska (Péwé, 1968). It is hard to imagine that a complete reversal in tectonic movements could have taken place at the same time in areas so extremely diverse in geologic character, simultaneously causing prolonged subsidence to give way to abrupt uplift. According to Shilo (1971), "The presence of very large ice wedges outside the floodplain exactly indicates the existence of more favorable conditions for their formation in places far from the floodplains, places which, by the way, are often erroneously attributed to river terraces."

The absence of channel facies from all known *yedoma* exposures is further evidence against an alluvial origin (Rozenbaum, 1971). In fact, the observed thick ground-ice masses are incompatible with the existence of the thawed ground that necessarily underlies channels of water courses (Popov, 1965; Rozenbaum, 1971), suggesting, again, that there were never vast regions of "alluvial accumulation" involving channel activity.

Shilo's work convinced many investigators to abandon the alluvial concept, but others, still bound to the idea of subaqueous lithogenesis, proposed several variants of a lacustrine-alluvial origin for the *yedoma* (Konishchev, 1970; Lavrushin, 1963; Rozenbaum, 1971). However, our investigations of the lacustrine deposits beneath a group of lakes on the Anadyr tundra have shown that these deposits are always well sorted and stratified. They may even assume the characteristics of alternating channel and floodplain facies in that associated organic material consists solely of intercalations of detrital peat. This distinguishes lake deposits from the homogenous sediments of the *yedoma* (Tomirdiaro, 1972). The presence of vertical roots of herbaceous plants, often penetrating the whole thickness of the *yedoma,* rules out the possibility of subaqueous formation of these deposits.

Our investigations of the thermal regime beneath the lakes on the Anadyr tundra (Tomirdiaro, 1972) also demonstrate the utter impossibility of ground ice being preserved beneath any fresh-water body. As V. A. Kudryavtsev shows (Dosotovalov and Kudryavtsev, 1967), "under the conditions of thermokarst lake formation, the thermokarst, once formed, will progress until the ground ice is completely melted." Our investigations of year-round heat balance under the Anadyr tundra lakes show that heat accumulates constantly in far-northern lakes because the process of spring thawing of the ice is far more intensive than the fall/winter freeze-up. During the thaw, no insulating layer is formed above the melting boundary because melt water drains beneath the ice; during freeze-up, the thickening ice itself becomes an insulator (Tomirdiaro, 1972).

It turns out that the abundance of thermokarst lakes is a result of a feedback process involving the growth of ice wedges. Upturned ground at the margins of ice wedges forms closed polygonal basins which hold shallow polygon ponds. Thermokarst lakes grow rapidly in the face of the continuous thermoabrasional retreat of their ice-saturated banks. This rapid growth is followed inevitably by breaching and discharge into the drainage network and then by the "alasification" of the tundra. Once the wedges reach a critical size, they generate new thermokarst lakes (Tomirdiaro, 1966). According to our investigations, the thermokarst cycle is completed within 1100 to 1300 years (Tomirdiaro, 1972).

Comparison of the structure and composition of the *yedoma* deposits with those of thermokarst lakes rules out the possibility that the *yedoma* could have formed in lake basins per se. It is impossible that any kind of lake landscape could have existed on these plains during the formation of the *yedoma.* Polycyclic lake-thermokarst would have destroyed the growing ice, preventing it from growing sufficiently, as is happening now on all the young plains, including the Holocene arctic maritime plains and high floodplain areas in transition to the terrace stage (Rozenbaum, 1971). It is not by chance that the Holocene has produced no *yedomas,* even though 10,00C years is sufficient time for very thick ground-ice layers to have formed. *Alasy* and lake plains of late Pleistocene age are also unknown.

Even in the present arctic tundra zone, most *alasy* are covered by thick peat bogs containing numerous wood macroremains, indicating that they were formed at the time of the Holocene climatic optimum (Baranova and Biske, 1964; Sher, 1971); this has been confirmed by absolute age determinations. But the formation of lakes and *alasy* apparently began during the very early Holocene, as is indicated by the absolute dating of the older *alasy* of the Berelekh River at 8670± years BP (Baranova and Biske, 1964). On the whole, the vast mantles of lake and *alas* deposits that developed on the Yakutia plains belong to the Holocene (Geology of the USSR, 1970). Yu. P. Baranova and S.F. Biske (1964) first showed that "as a result of the development of thermokarst the surface of the late Quaternary plains of the East Siberian lowland underwent essential changes during the Recent epoch. Due to the melting of large masses of underground ice, the vast *alas* plains were formed 20-30 m lower than the original plain . . . The *alas* deposits date from the end of the late Quaternary epoch to Recent time."

Remnants of the *yedoma* have been preserved in the maritime lake-*alas* lowland in the form of many isolated hills and plateaus rising 20 to 30 m above the countless lakes and bogs surrounding them. *Yedoma* is preserved best near river valleys incised into the plain: valley incision tends to drain thermokarst lakes formed nearby (Tomirdiaro, 1970b). The *yedoma* is preserved best of all deep in the Asian continent south of the boundary between humid and semiarid climatic zones that cuts the East Siberian lowland (Kalesnik, 1947).

The evidence for the incompatibility of the presence of *yedoma* and the development of the marshy "lacustrified" landscapes that typify the Holocene arctic plains persuades us that the *yedoma* was part of a very cold, dry landscape, already recognized on other grounds as characteristic of the late Pleistocene (Velichko, 1973). This is confirmed by the spore-pollen diagrams obtained by Z.V. Orlova on samples from the *yedomas* that we investigated in northern Yakutia (Tomirdiaro, et al., 1972).

The discovery of the Maya-Anadyr *yedoma* on Chukotka in 1968, the first such find in the Asiatic part of the Pacific Ocean basin, enables us to unite into one region the so-called loess-ice *yedoma* plains of Yakutia and Chukotka and analagous deposits in Alaska (Tomirdiaro, 1970a). This discovery has also made it possible to suggest an aeolian-cryogenic origin for these deposits (Tomirdiaro and Orlova, 1970; Tomirdiaro, 1972)—an idea now fully confirmed by more recent field data (Tomirdiaro, et al., 1972) and incorporated in the study of the region in a number of new special works (Velichko, 1973; Kolpakov, 1973).

Yedoma has never been found to replace Pleistocene loess accumulations in any of the regions of the world where permafrost completely melted during the Holocene. (In West Siberia, the top of the permafrost melted deeply during the Holocene optimum. Although shallow permafrost has now been reestablished there to form a two-level frozen layer, the *yedoma* has been destroyed.) If one understands this, many features of the *yedoma* that formerly seemed extremely puzzling became easily explainable, for example:

(1) The universal restriction to the cold, dry glacial phases of the late Pleistocene (Velichko, 1973; Kolpakov, 1973; Popov, 1965).

(2) The superimposition of *yedoma* mantles of the same age on rocks and topographic forms of very different origin and age (Fig. 2), including river terraces of sharply different age and level in mountainous country (Boyarskaya, et al., 1971).

(3) The blanketing nature of occurrence, covering broad interfluves and divides with a mantle that even climbs the slopes of the foothills 80 m or more above the present valleys (Fig. 2) (Baranova, et al., 1968, p. 67).

(4) The astonishing uniformity, noted by all the investigators, of the granulometric and mineralogic composition, the complete dominance of the silt fraction (0.05–0.01 mm) in all *yedoma* sections that have been studied, and also the complete unity of facies typical of them (Kolpakov, 1973; Popov, 1965; Tomirdiaro, et al., 1972; Sher, 1971).

(5) The absence of remains of fish and other river or lake fauna of the Late Paleolithic, or of the so-called loess assemblage (Velichko, 1973; Vereshchagin, 1972).

(6) The absence of wood macroremains in spite of

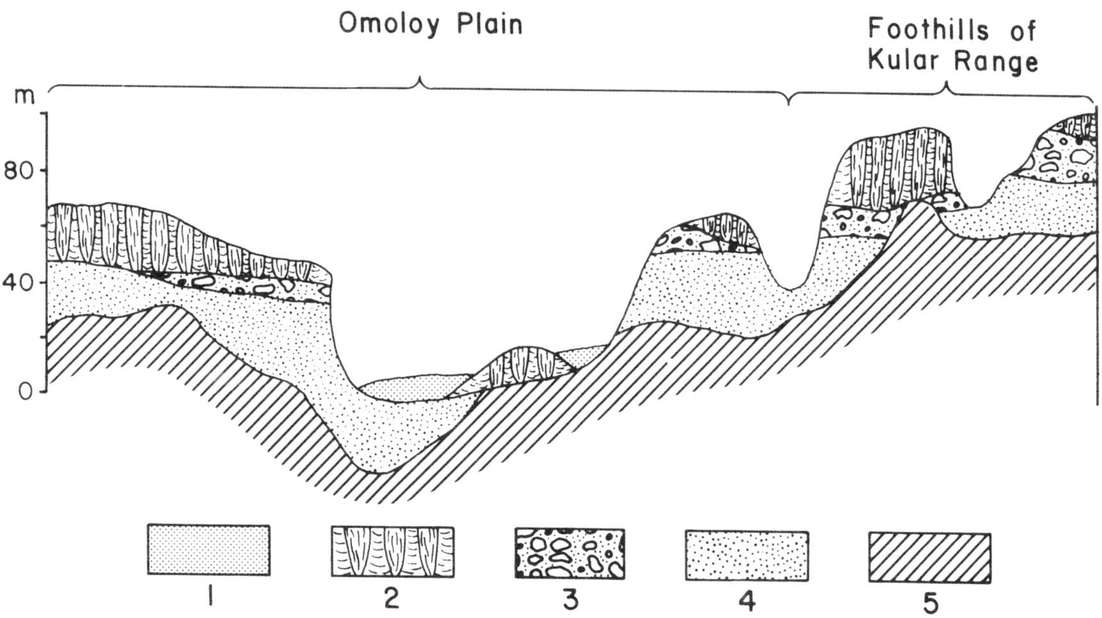

Figure 2. Diagrammatic cross-section through the Omoloy River Basin (after Baranova, et al., 1968). (1) Alluvium of Holocene age; (2) loess of late Pleistocene age with syngenetic ice wedges; (3) sandy gravel of Pleistocene age; (4) sand of Neogene age; and (5) bedrock

the preservation in situ of rootlets of herbaceous vegetation solidly penetrating the *yedoma* (Popov, 1965; Sher, 1071) and the abundance of herbaceous plant pollen, particularly grasses and *Artemisia*, in *yedoma* sections. The pollen spectra make it possible to distinguish a single steppe-tundra paleolandscape for all the *yedomas* from the Arctic islands to the Aldan basin (Tomirdiaro, 1972).

(7) The substantial carbonate content and salinity of *yedoma* silts, which are higher in salinity than present soil and lacustrine-alluvial deposits (Tomirdiaro, et al., 1972).

(8) Loess-type microaggregation, with interparticle and biogenic porosity discernible in thin section (Morozova, 1971), and also the general "freshness" and only slight transformation of the original permafrost loess material by diagenetic processes. We are dealing with material "which under suitable conditions (i.e., upon thawing out and becoming subject to loessification processes) can be transformed into loess in the classic sense" (Morozova, 1971, p. 104).

(9) The high porosity of the blocks of ground *(baydzbarakbas)* formed when the *yedoma* thaws subaerially, and the ability of this deeply thawed sediment, when dried out in the summertime, to maintain a vertical face and support columnar-jointed segments up to 2 or 3 m high (Tomirdiaro, 1972). Thawing produces a cohesive sediment that differs from the so-called "cold loess" of western Eurasia in the preservation of almost undecomposed organic remains.

(10) The ability of the *yedoma* to drape the old underlying relief.

Over the whole vast north-south section from the islands of the New Siberian Archipelago to Mamontova Gora in the Aldan Depression, the *yedoma* shows no sign of zonal differences. Not one of the major features of the *yedoma* enumerated above varies, not even the spore-pollen spectra. Neither do the ice wedges vary zonally in width, nor even in the size of their polygonal grid, which averages 10–12 m in polygons that we investigated in the Oyagos Bluff and Cape Maliy Chukoch'e *yedomas* on the north coast of the East Siberian Lowland, in the Vorontsov Bluff *yedoma* near Ozhgino settlement in the southwestern part of that lowland, and in the *yedoma* on the terraces of the Aldan River at Mamontova Gora in southern Yakutia. The relative widths of ice wedges and loess columns may vary, but the variation is independent of latitude. In some cases ice wedges clearly predominate in the section and in others, loess blocks; however, the same polygonal grid diameter is always maintained. Thus, B.I. Vtyurin (1961) reports the development of thick syngenetic ice wedges up to 8 m wide enclosing loess columns up to 3 m in diameter in such a relatively southern area as the Tumara River basin. Yet, even there, the same 11 m diameter of the polygonal frost-crack grid is found.

Under present conditions, the whole spectrum of natural zones is displayed from the north coast of Yakutia to the Aldan depression, from typical arctic tundra to deciduous and even pine forests mixed with prairie steppes (Karavayev and Skryabin, 1971). According to the investigations of N. N. Romanovskiy and V. A. Shaposhnikova (1971), the present ice wedges and Holocene ice wedges in general vary zonally in the size of their frost-crack polygons. The physics of the frost-cracking process dictates that this phenomenon of zoning is inevitable under present conditions, since the frost-crack polygon size is a direct function of the mean annual temperature and annual ground temperature amplitude; these are zonal values. Consequently, the uniformity in size and diameter of *yedoma* ice-wedge polygons indicates unequivocally that during the late Pleistocene the mean annual temperature and, especially, the annual temperature amplitude varied very little within the vast area in question.

The uniform development of cold-steppe and tundra-steppe periglacial landscapes in south and central Yakutia during the late Pleistocene epoch has been pointed out in numerous special investigations (Karavayev and Skryabin, 1971; Velichko, 1973; Garutt, et al., 1970). The development of analogous landscapes in northern Yakutia in the zone presently occupied by typical arctic tundra is confirmed by the abundance of mammoth, horse, bison, and even saiga bones in the loess-ice strata of these areas (Vereshchagin, 1972; Tomirdiaro, 1972; Sher, 1971). Further confirmation is offered by the first spore-pollen diagrams obtained from loess samples from the coastal bluffs of the arctic *yedoma* (Tomirdiaro and Orlova, 1970; Tomirdiaro, et al., 1972).

In the maritime lowland of northern Yakutia, a

sharp difference can be observed between polygonal grid size in the remnants of the late Pleistocene *yedoma* and in the surrounding *alas* plains. The diameter of the recent frost-crack polygonal grid in the Holocene *alasy* averages about 20 m, i.e., almost twice the size of the grid of Pleistocene wedges found in the nearby *yedoma*. From the standpoint of thermophysics and soil mechanics this is important and unequivocal evidence of an abrupt change in climate that took place in this region during the Pleistocene-Holocene transition. The acutely continental, frost climate of the late Pleistocene, which permitted the formation of a close frost-crack grid (Popov, 1965, p. 284), gave way to the fairly mild, maritime climate of the Holocene, permitting the formation of a polygonal grid half as dense. The change from an acutely continental to a maritime climate can be explained only by exposure of the East Arctic Ocean, which during the Pleistocene had been frozen even in the summer, in contrast to the prevalence of seasonally open water during the Holocene. The abrupt humidification of the climate during the Holocene brought about the development of peat bogs in the whole maritime lowland and other former steppe-plains in Yakutia.

Special thermophysical calculations have shown that the sea-ice covering the Arctic Ocean for many centuries during the late Pleistocene may have been hundreds of meters thick (Tomirdiaro, 1970a). As was the case during previous interglacials, the sills separating the Atlantic and Arctic basins were deepened during the Holocene due to the general glacioisostatic lowering of the European continental block. This sharply increased the Atlantic-Arctic water exchange and led not only to melting of the sea-ice cover in the Arctic, but also to the formation of an ice-free regime in the basin, i.e., to the formation of the Holocene climatic optimum (Tomirdiaro, 1970a). That powerful warming and amelioration of the climate, the abrupt weakening of its continental nature, and especially the overall increase in mean annual precipitation led to the flare-up of millions of thermokarst lakes throughout the loess-ice plains. At the same time, the sharp increase in humidity in the Arctic and the cessation of loess deposition led to the supplanting of steppe vegetation by a cover of mosses and lichens.

In this respect, the preservation until our time of small islands of relict grassy steppes in the circumpolar and polar regions is very telling. Islands of steppe landscapes can be found in the continental part of the Yana, Indigirka, and Kolyma basins (Karavayev and Skryabin, 1971), and according to the investigations of B.A. Yurtsev, steppe-tundra is developed even on Wrangel Island, where a local continental climate results in relative abundance of sunny days. At the same time, there are no steppe communities on the warmer but more humid mainland coast; exceptionally swampy tundras are developed there.

In the continental subarctic to the south, "light coniferous taiga[4] encroached on the vast older treeless prairie-steppe areas" (Gerasimov, 1952), initiating the Holocene transformation of the original periglacial landscape. The abruptly warming sea began to thaw and destroy the loess-ice plain on what then became the continental shelf. This was accomplished not by inundation, but by thermal attrition of ice-permeated bluffs and their undercutting and collapse into the sea, a process observed today in the last remnants of the plain represented by Moustakh, Stolbovoy, and Lyakovskiy Islands in the west, Kotel'nyy Island in the north, and Ayon Island in the east. Two other loess-ice islands, Vasil'yevskiy and Semenovskiy, were destroyed in just this manner during the present century, a fate shared by thousands of other islands into which the thawing loess-ice plain disintegrated. The puzzling Sannikov, Andreyev, and other vanished lands north of Siberia are also probably examples of islands destroyed by thermal attrition on the continental shelf.

Thermokarst lakes developed on the loess-ice plain ahead of the encroaching sea. That is why the greatest "alasification" of the plain and maximum development of lakes are now observed in the arctic-tundra zone where the climate is excessively humid and why the greatest preservation of loess-ice *yedoma* occurs in the continental subarctic zone where a less humid or semiarid climate prevails (Tomirdiaro, 1970b). As I have already mentioned, thermokarst processes cause both a general and irreversible reorganization of the old landscape and, in the tundra zone, an 1100-to-1300-year polycyclic succession of lake and *alas* landscapes.

---

[4]Soviet geographers distinguish between a "light coniferous taiga" dominated by various species of the deciduous larch and a "dark coniferous taiga" dominated by spruce and other evergreen conifers—*Eds.*

In conclusion, it is appropriate to compare the main features of the abrupt transformation of the landscape that occurred in Northeast Asia during the Pleistocene-Holocene transition with the natural processes that took place at the same time in the northwestern part of the Eurasian continent.

According to the latest theory of natural processes during the Pleistocene (Velichko, 1973), the familiar landscape zones of present-day western Eurasian temperate and cold belts developed only during the warm interglacial epochs, of which the Holocene is an example. During the cold glacial epochs, on the other hand, a colossal landscape superzone, the so-called periglacial steppe-tundra, formed throughout this same region, from the Polar Urals to the Black Sea. Neither the *yedoma* that we have examined, nor the European loess horizons of different ages, nor even the late Pleistocene cold loesses can be divided into north to south sectors that might reflect our present-day zonal structure. For example, the loess of western Eurasia often contains traces of ice wedges that thawed early in the Holocene (Velichko, 1973).

Velichko explains the origin of the late Pleistocene cryoxerotic steppe-tundra superzone by the formation at that time of thick oceanic ice over an area that included the northern Atlantic Ocean; this sharply reduced evaporation and advective heat exchange and correspondingly dried and cooled air masses arriving on the European continent. As a result, during the late Pleistocene vast cold, dry steppe-tundra with typical cryoxerophytic and halophytic vegetation (Velichko, 1973) prevailed even in what is now highly humid western Europe (Velichko, 1973); and it supported the characteristic xerophytic fauna of the mammoth assemblage (Vereshchagin, 1972). In fact, the extent of terrestrial glaciation was restricted in Europe during the late Pleistocene precisely because of the extreme dryness and unusually continental climate produced by the ice cover on the oceans. As we have seen, the natural setting in the northeast Eurasian continent was similar during the same time period.

Because global atmospheric transfer is westward, the amount of precipitation falling on the continents during the Pleistocene must have decreased sharply from west to east. Hence, in the face of the acute cryoxerotization of late Pleistocene landscapes of Europe, still greater desiccation and cooling of Northeast Asia is to be expected a priori. Furthermore, the development of powerful aeolian processes and active accumulation of "cold" aeolian loess at that time in Europe implies even greater development of these processes in the northeast.

Although similar natural processes operated in western and northeastern Eurasia, there seem to have been some differences—ordinary loess was formed in the west, and loess-ice *yedoma* mantles in the northeast. But as I have indicated earlier (Tomirdiaro, 1973), this difference merely reflects the Holocene melting of permafrost in Europe and resultant further diagenesis of the European loess. Continuous polygonal ice-wedge pseudomorphs are frequently found in the Late Würmian aeolian loess of western Europe as well as in the Valdai loess of the Russian plains (Velichko, 1973). Inasmuch as the most intensive accumulation of loess coincided here with the most rigorous climatic minimum of the late Pleistocene, and since ice wedges were also formed during this very cold epoch (Velichko, 1973), a special syngenetic frost layer was formed which might be called the loess-ice layer. In other words, it can be assumed that the original late Pleistocene loess-ice layers in Europe were not essentially different from those comprising the northeastern Siberian *yedoma.*

It becomes obvious, then, that there is no discordance between the processes involved in what is now generally recognized as the aeolian accumulation of loess in Europe and the processes of cryolithogenesis and formation of ice wedges. On the contrary, these processes were intimately related to one another during the late Pleistocene. Thus, ice-wedge pseudomorphs and other signs of cryolithogenesis are found in silty deposits on the Russian Plain that display such distinctive indications of an aeolian origin as incorporation of exotic minerals and of volcanic ash from the Caucasus (Remizov and Kovalev, 1965). The development of ground ice took place here in an environment that on the whole was very dry and cold throughout the Pleistocene (Dosotovalov and Kudryavtsev, 1967). Evidently, formation of underground ice was brought about by very low ground temperatures and a consequent rapid freezing of the snowmelt water that penetrated the open frost cracks in spring.

In addition, a considerable volume of water may have penetrated and frozen as hoarfrost in cracks and pores in the low-temperature loess layer directly from the atmosphere. It is no accident that investigations of the structure and composition of ice wedges in the Anadyr tundra show that even in the epigenetic ice wedges found there, a substantial percentage of the volume of bubbles "is due to xenogenic air, which indicates that hoarfrost plays a large part in the formation of the wedges" (Gasanov, 1969). But the Anadyr tundra is situated in the excessively humid climate of the maritime zone. In the more rigorous continental climate of the late Pleistocene, hoarfrost must have played a much greater role in the formation of ground ice both in eastern Europe and, as I have noted (Tomirdiaro, 1972), in East Siberian *yedoma.*

It appears, then, that similar landscape processes operated throughout middle- and high-latitude regions of the Eurasian continent during late Pleistocene time. The contrast between the present day marshy and lake-dotted landscapes of "lacustrine-alluvial plains" of northeastern Siberia and the dry continental loess plains of eastern Europe is a Holocene development. Recent thermokarst transformation of the landscape occurs most actively and completely in lowlands still mantled by the late Pleistocene loess-ice complex. However, the intensity and dynamics of formation of landscapes of the lake and *alas* types distinguished by Gerasimov (1952) depend not only on the ice content of the substrate but also on the climate. There are many more lakes and *alasy* in the maritime tundra zone and fewer in the zone of continental dry climate, although it is in the more continental region that ice-saturated loess mantles are best preserved.

ANCIENT VEGETATION–THE FOSSIL RECORD

Reconstructing the ecology of ice-age Beringia must begin at the first trophic level, the producing level, that is, the vegetation. Not until this has been accomplished can a fully satisfactory reconstruction be achieved and eventually modeled. To be sure, certain inferences can be drawn from the morphology and dietary requirements of the fossil mammals, but these must be validated by comparison with independent lines of paleobotanical evidence. One must proceed from description of the vegetation to consideration of the consumers—herbivores and carnivores both large and small—and eventually to the role of the human occupants of Beringia.

The pioneering Alaskan palynological research of D. A. Livingstone in the 1950s has led ultimately to a present-day wealth of rapidly accumulating data and interpretation from sites throughout Beringia. Indeed, the reviews presented here will soon be outdistanced by results of ongoing research of numerous colleagues. The central objective has been to establish the nature of the late Pleistocene vegetation cover, utilizing fossil pollen as well as macroremains, and involving increasingly sophisticated techniques and interpretations. The central question has been whether any living analogs could be found for the vegetation represented by late Pleistocene "steppe-tundra" or "herb-zone" pollen assemblages. Because similar fossil pollen spectra dominated by *Artemisia*,

Cyperaceae, and Gramineae have been discovered outside Beringia in Scotland, France, Germany, and Idaho, it is now possible to draw on a very wide range of interpretive experience and insight in dealing with this question. Indeed, living analogs seem to have been found, as we shall learn in a later section.

At a broader and perhaps more philosophical level, the quest for appropriate modern analogs involves the question of whether modern vegetation is an individualistic, stochastic, open system or a more closed system composed of discernible, predictable units. At stake is how to proceed in reconstructing the past vegetation in any region and, more specifically, what we may anticipate as the late Pleistocene vegetation of Beringia.

In the following pages, Giterman, Sher, and Matthews explore the time of first emergence of hypoarctic tundra and then of more xeric treeless arctic vegetation, using a Siberian palynological, macrofossil, and insect record that extends several million years back in time and an Alaskan record that covers parts of the last two million years. Critical pollen data from Siberia is here recalculated and presented in a form more familiar to western palynologists.

Late Pleistocene pollen records from newly studied regions of southwestern Alaska, presented by Ager, are critical for our understanding of the vegetation of the land bridge itself. Giterman, Sher, and Matthews emphasize east-west differences in the late Pleistocene vegetation of Beringia, but Ager emphasizes that the range of variability within and among plant communities must have been much less pronounced in land-bridge times than in present-day Beringia.

Palynological studies in Soviet Western Beringia are based almost entirely upon pollen spectra extracted from alluvium, loess, detrital organic accumulations, and in situ peat collected from exposed sections. Most palynological studies in Alaska and the Yukon are based on sediment cores from lakes. However, palynological studies of exposed sections have important uses. Schweger explores the potential and significance of pollen records from dated geological sections in Alaska and uses them to develop a model for Beringian lowland vegetation.

Especially carefully and critically studied pollen sequences from the northern Yukon are employed by Ritchie and Cwynar to challenge the interpretation that Pleistocene herb-zone spectra represent highly productive steppe-tundra or, indeed, anything different from the more sparsely vegetated biotopes (fell fields, polar desert) that can be found in arctic regions today. Their insistence on data-based interpretations has been of great value in correcting more optimistic and exuberant appraisals of the Pleistocene Beringian vegetation.

The section ends with Matthews' contribution, which considers the implications of plant, insect, and mammal macroremains for the contrasting interpretations of the Beringian pollen record. He provides an updated version of the steppe-tundra hypothesis in response to Ritchie and Cwynar's challenge.

Beringia, at its greatest extent, was an enormous region. Even today, the emergent eastern and western segments are huge and remain largely unresearched. While the new information presented here is a very significant contribution when added to the scattering of paleoecological data points already on the map of Beringia, the vast blank areas remaining dramatize how much more we must come to know.

—*Charles E. Schweger*

3

COMPARISON OF THE DEVELOPMENT OF TUNDRA-STEPPE ENVIRONMENTS IN WEST AND EAST BERINGIA: POLLEN AND MACROFOSSIL EVIDENCE FROM KEY SECTIONS

R. E. Giterman, A. V. Sher, and J. V. Matthews, Jr.

Pollen and macrofossils from several important exposures in the lower Kolyma basin of eastern Siberia and one in western Alaska record development of tundra-steppe in Beringia.

The Krestovka exposure near the Kolyma River contains sediments dating from late Pliocene to late Pleistocene. Its oldest unit, the Begunov Suite, formed in a forested environment. The overlying Kutuyakh beds are rich in mammalian fossils that, in concert with paleomagnetic data, suggest a late Pliocene age for most of the unit. The Kutuyakh interval was one of increase in abundance of herbs, but also of high percentages of *Artemisia, Selaginella sibirica,* and other xeric types indicative of steppe-like biocenoses that appear first in the overlying beds of the late Pliocene (early Pleistocene of North American usage) Olyor Suite.

A long time gap separates the Olyor sediments from the late Pleistocene and Holocene units capping the Krestovka exposure. Some details on the environment of this period are supplied by pollen spectra from the Utkinsky Kamen section.

At Krestovka, late Pleistocene spectra contain stone pine and other shrubs, as well as abundant herbs and *Selaginella sibirica.* This trend is best displayed in the thick Sartan (late Wisconsin) sequence of "ice complex" sediments at the Duvannyy Yar exposure, another important site on the Kolyma River. The Duvannyy Yar exposure also reveals that a period of enhanced abundance of shrubs (presumably due to milder climate) occurred during the Sartan. Fossils from unit II at Duvannyy Yar illustrate the difference between Sartan and Karginsk environments in eastern Siberia.

The Cape Deceit section in Alaska contains sediments ranging in age (in North American nomenclature) from at least early Pleistocene to late Pleistocene. The oldest unit, Cape Deceit Formation, is intermediate in age between the Olyor and Kutuyakh Suites in Siberia and represents tundra different from that of the present, but that contains few of the xeric (steppe?) indicators seen in the late Pleistocene unit (Deering Formation) or in other Wisconsin pollen sequences from East Beringia.

Fossils from various levels of the Deering Formation suggest existence of a tundra-steppe vegetation mosaic in East Beringia; however, the peak in development of steppe-like biocenoses may have occurred during the early Wisconsin and not the late Wisconsin as in eastern Siberia. There are other differences between the Siberian and Alaska-Yukon assemblages (e.g., *Selaginella sibirica* percentages, various groups of Coleoptera, large

ISBN 0-12-355860-3

mammals) that caution against the conclusion that Beringia was characterized by a monolithic steppe biome during the late Pleistocene.

Key words: tundra-steppe, Krestovka, Olyor, Duvannyy Yar, *Chrysobyrrhulus, Selaginella sibirica,* Sartan, Cape Deceit, fossil Coleoptera

INTRODUCTION

Paleoecological data obtained during the last few decades from various sites in northeastern Siberia (West Beringia) and Alaska and Yukon Territory in northwestern North America (East Beringia) seem to show that a dominant characteristic of the vegetation of the cold epochs of the Pleistocene, especially the latest Pleistocene (Sartan and late Wisconsin), was the occurrence of extensive tracts of tundra-steppe.[1] These tracts formed due to a combination of cold climate and extreme continentality, which was enhanced to different degrees in various parts of Beringia by imposition of a large land bridge in place of the present-day Bering and Chukchi Seas. As a vegetation type of regional extent, the Pleistocene tundra-steppe has no modern analogs. Nevertheless, some existing plant communities in alpine and relict-steppe areas do preserve species of plants and animals that formerly characterized tundra-steppe communities.

V. P. and M. P. Grichuk (1960) were among the first workers to note the association of tundra and steppe elements in assemblages representing the periglacial vegetation of European U.S.S.R. In that region the tundra-steppe associations were seen as a special type of vegetation associated with the second, cryoxerotic phase of glaciation, at the time when climate was both cold and dry and glaciers were in a state of recession. We feel that the epithet "periglacial" is no longer appropriate for the vegetation of the cold, dry epochs of the northeastern U.S.S.R. or of East Beringia because most of that region was not "peri" glacial in the literal sense (Péwé, 1975a). Proximity to glaciers was not an important factor in the development of regional vegetation of Beringia. If anything, the relationship of tundra-steppe and glaciation is an inverse one. For example, in West Beringia, tundra-steppe seems to have been best developed during the latest Pleistocene when the glaciated areas were less extensive than during earlier cold epochs (Baranova, et al., 1968).

Even though tundra-steppe seems to have peaked during the latest Pleistocene, it is wrong to assume that tundra-steppe associations first appeared at that time. The existing paleoecological record shows that steppe-like associations qualitatively different from present tundra existed at various times during the Pleistocene and late Pliocene and that the late Pleistocene represents only the culmination of a prolonged trend to ever drier and more continental climates.[2]

The purpose of this report is to document that developmental sequence. We are concerned primarily with palynological data, mostly that from several key sections in the lower Kolyma basin of northeastern Siberia. However, significant macrofossil evidence is now available and we will attempt to collate it with the pollen data. In addition, the pollen and macrofossil record of the Cape Deceit site in Alaska is discussed.

---

[1] In addition to "tundra-steppe," the steppe-like biocenoses of Pleistocene Beringia have been called "periglacial tundra-steppes," "steppe-tundra," "arctic steppe," and "mammoth steppe" (cf. Giterman, 1963; Giterman and Golubeva, 1965; Giterman, et al., 1968; Matthews, 1976a; Hibbert, this volume; Guthrie, this volume).

[2] A common source of confusion in attempting to relate evidence from East and West Beringia is that the established concepts for the duration of the Pleistocene are different in North America and the USSR. According to the stratigraphic scheme used in the Soviet Union (MSK code, 1964), the start of the Pleistocene was at about 0.7 million years BP. The Upper Pliocene includes two stages: the Akchagyl and Apsheron. In Alekseev et al. (1979) the base of the former equals the base of the Pliocene Astian stage of North American workers (3.5–3.3 million years ago), and the base of the Apsheron equals the base of the Calabrian, which marks the Gilså paleomagnetic event and the start of the Pleistocene (1.8–1.6 million years) for North American workers. The Pleistocene according to the North American system includes the interval between approximately 1.8–1.6 million years (the Olduvai paleomagnetic event) and 10,000 years BP (the start of the Holocene). In neither system does the start of the Pleistocene mark the start of northern-hemisphere continental glaciation, which began as early as 3 million years ago, during the late Tertiary.

Some of the information in this report has not been published in either Russian or English and much of it is new to North American audiences or is portrayed in a different style than in the original publications (e.g., Sher, et al., 1979). The Cape Deceit record, too, is interpreted differently than in the original report (Matthews, 1974b).

R. E. Giterman, the senior author, is responsible for all the pollen data from the eastern Siberian sections and preparation of the initial version of the manuscript. The geological and paleontological information on the key sections of Krestovka, Bol. Chukoch'ya and Utkinsky Kamen is the responsibility of the second author, A. V. Sher. T. N. Kaplina (PNIIS, Gosstroy, U.S.S.R.) is responsible for stratigraphic information in the Duvannyy Yar section. Sher and Kaplina collected the pollen samples at the Krestovka and Duvannyy Yar sections respectively. Micromammalian remains have been studied by V. S. Zazhigin; fossil insects by S. V. Kiselyov; and magnetic samples by E. I. Virina. The third author, J. V. Matthews, Jr., is responsible for the updated section on the Cape Deceit site, for the preparation of the diagrams, and final editing.

METHODS

Though they are based on the same raw data, the pollen diagrams in this paper differ from those presented elsewhere (Kaplina, et al., 1978; Sher, et al., 1979; Giterman, in press). Bar diagrams rather than integrated plots are used because the latter can be misleading when applied to alluvial sequences. The diagrams are all based on the same sum: total pollen and spores, and they are all constructed from primary counts. Ericales pollen, which in the USSR is usually grouped with herb pollen, is included with tree and shrub pollen in this paper. Several pollen studies of lake cores in East Beringia (e.g. Cwynar, 1980) have shown that Ericales pollen fluctuations tend to be in phase with those of shrub-birch and arboreal types; hence, grouping Ericales with the herbs tends to dampen the paleoenvironmental signal of the combined tree-shrub curve.

Many of the sections discussed here comprise discrete sedimentary units separated by disconformities. It makes little sense to attempt to zone the pollen diagrams without taking account of stratigraphy because the time interval between samples within a zone that includes a disconformity may span tens of thousands of years. Consequently, in this paper the stratigraphic unit is the first-order subdivision for comparison and discussion of the pollen and macrofossil sequences.

GEOGRAPHICAL AND VEGETATION SETTING

West Beringia

The eastern Siberian paleontological data discussed here were derived from several key sections along rivers in the Kolyma lowland, the eastern part of the vast maritime lowland that extends from the North Anyuy Range westward to the northern part of the Verkhoyansk Range (Fig. 1). It takes its name from the Kolyma River, which drains a large area of the northeastern U.S.S.R. and flows north through the lowland, entering the Arctic Ocean in a large delta. Within the lowland the Kolyma River is joined by eastern tributaries such as the Maly Anyuy, Bol'shoy Anyuy, and Omolon, which drain the Anyuy Mountains and Yukagir upland. The Bol'shaya Chukoch'ya (and its tributary the Olyor), Bol'shaya Kon'kovaya, Bol'shaya Kuropatoch'yah, and Alazeya are small rivers west of the Kolyma that tend to parallel its course and drain directly into the Arctic Ocean.

The basement rocks of the lowland are folded and faulted, creating an assortment of blocks, some of which must have been uplifted as recently as the Pleistocene epoch, judging from evidence at the Krestovka exposure (discussed below). For the most part, however, the lowland is underlain by frozen, ice-rich sediments known locally as the "*yedoma* sequence." The high ice content of *yedoma* or "ice-complex" sediments gives rise to numerous *alas* or thermokarst depressions, creating a region with surfaces at multiple levels. A prominent surface, not of thermokarst origin, is the "Khallerchin Tundra," located on the lower reaches of the Kolyma 15–20 meters above the present floodplain but well below the top of the *yedoma* surface. The origin of both the Khallerchin Tundra and the *yedoma* are topics of controversy. Some workers believe *yedoma* to be floodplain deposits but Tomirdiaro and others (including some western geologists) have interpreted them as primary and

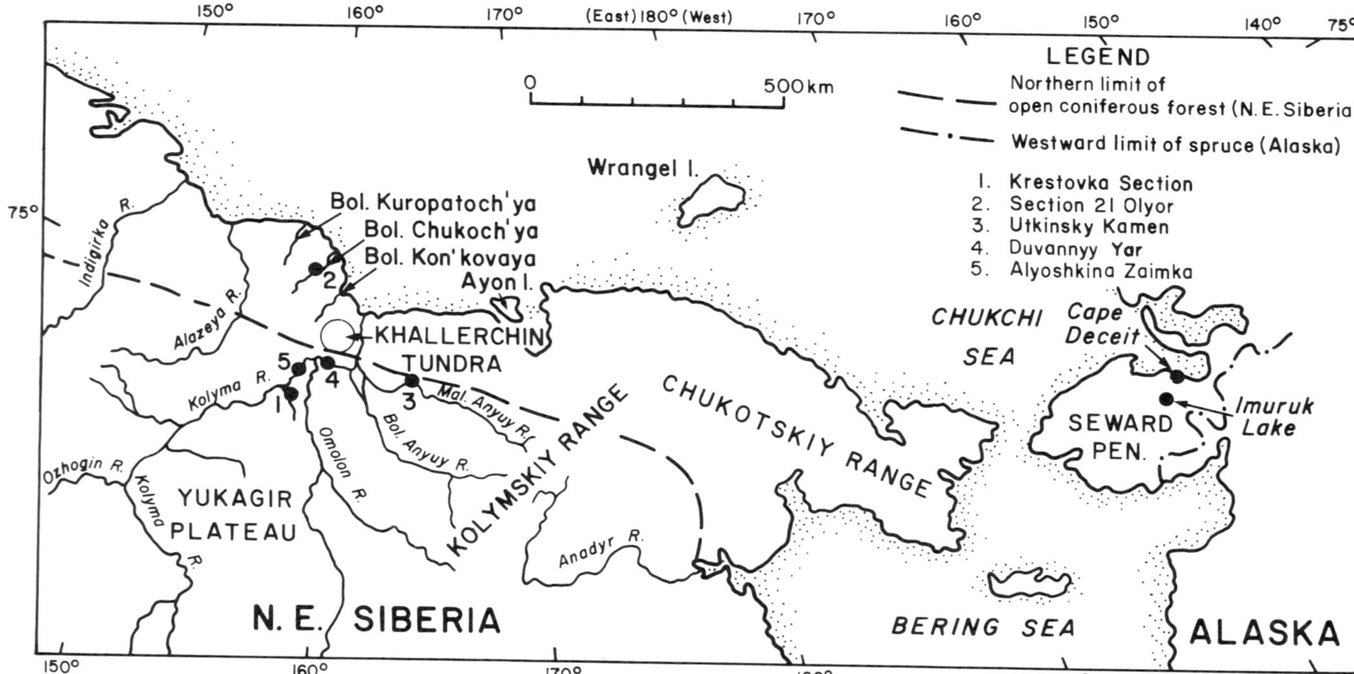

Figure 1. Map showing localities mentioned in the text. Northern limit of spruce in Siberia lies outside of region shown. Limit of spruce in Alaska represents the limit of coniferous forest

secondary loess (Sher, et al., 1979; Tomirdiaro, this volume). Hopkins (this volume) posits an aeolian origin for both the *yedoma* silt and the sand underlying the Khallerchin Tundra.

The history of geological exploration of the Kolyma basin adds perspective to the important record yielded by some of the key exposures there. Sher (1974) and Sher et al. (1979) have provided insight on the evolution of thinking about the Pleistocene environments of the region. Western readers should note that the maritime lowland and offshore islands have been the site of some of the most famous "frozen carcass" finds (Vereshchagin and Baryshnikov, this volume).

Several latitudinal vegetation zones occur within the Kolyma lowland. A narrow band along the coast of the Arctic Ocean is occupied by the "arctic tundra" zone of Yurtsev (1966, 1974b), which is probably equivalent to Young's zone 3 tundra (Young, 1971). Farther south is a zone of "typical" tundra characterized by various shrub-moss-lichen-dominated communities. Permafrost is present at shallow depths and is indirectly responsible, through the vagaries of microrelief and drainage, for complex microscale-mesoscale vegetation patterns. South of the "typical" tundra zone are the larch forest-tundra and northern larch-taiga zones. Drainage in those areas is also poor, causing formation of extensive moors dominated by mosses and herbs.

Several altitudinal vegetation zones are evident in the mountainous region on the right bank of the Kolyma River. The lower forested belt contains communities dominated by larch (*Larix dahurica*) and shrubs such as the Japanese stone pine (*Pinus pumila*), *Betula* sect. *Nanae,* and *Salix.* Above this are open larch woodlands and higher still a shrub belt containing shrub alder (*Alnaster fruticosus*).[3] Except for the narrow zone of arctic tundra near the coast, all of these zones, both lowland and montane, fall within the hypoarctic botanical-geographical region of Yurtsev (1966, 1974b) (equivalent to the tundra zone 4 of Young, 1971).

[3]According to the latest revision of North American species of Alder (Furlow, 1979), *Alnaster fruticosus* (Rupr.) Ledb. = *Alnus viridus* ssp. *crispa* (Ait.) Turril. Thus, despite the different generic designation, the *Alnaster* of eastern Siberian pollen diagrams is the same as what would be identified as "*Alnus*" sp. or "*Alnus crispa* type" by North American workers—*Eds.*

Eastern Siberia is dotted with tracts of relict-steppe vegetation, some of them quite large (Yurtsev, this volume). Many are defined floristically rather than in terms of physiognomy or community structure. Nevertheless, it is clear that most such sites could be subsumed under the most general definition of steppe. Yurtsev (this volume) details some of the floristic attributes of these relict steppe areas, and in the process documents some distributional disjunctions that appear to require a former continuity of East and West Beringian steppe habitat. In another work, Berman et al. (1979) have discussed the relationship of East and West Beringian steppe insect faunas.

Surface pollen spectra from the vegetation zones mentioned above and from intrazonal steppe regions have been studied by A. P. Vas'kovsky (1957) and more recently were summarized by Kaplina (1979). Yurtsev (this volume) also refers to surface samples from certain relict steppe areas in the far-eastern U.S.S.R.

In the tundra regions, surface spectra contain an average of 14% tree and shrub pollen (excluding Ericales).[4] No *Larix* pollen is recorded, and pollen of the shrubs *Pinus pumila, Betula* sect. *Nanae, Alnaster,* and Ericales usually occur in only small amounts. Gramineae and Cyperaceae occur in the herb portion of the spectra; percentages of the latter often exceed those of the former. Stony-tundra samples contain rather high percentages of *Artemisia* (up to 28%) and some Chenopodiaceae (3% maximum), which are similar to samples from certain fell-field areas in East Beringia (Cwynar, 1980; Ritchie and Cwynar, this volume).

The surface spectra of the eastern Siberian lowland larch forests (which lack *Pinus pumila*) contain an average of 32% arboreal-shrub pollen, with *Larix* accounting for as much as 15%; *Betula* sect. *Nanae,* 33%; and *Pinus pumila,* 22%. In certain samples Ericales pollen may have relatively high values. Bryales are the dominant spores.

Tree pollen accounts for up to 57% of the pollen rain in the montane larch forests. *Betula* percentages are high, and *Larix* averages 8%. Ericales is usually present and Gramineae and Cyperaceae

[4]These and following figures from eastern Siberian surface samples are based on sums of individual pollen categories, e.g., tree and shrubs, and herbs. Surface pollen results quoted for Alaska-Yukon (East Beringia) are based on a sum of total pollen and spores.

comprise the bulk of the nonarboreal pollen. Significantly there is practically no pollen of *Artemisia* or Chenopodiaceae. The spores are mostly of *Lycopodium*, Polypodiaceae, *Sphagnum*, and *Selaginella sibirica*.

Tree and shrub pollen, especially of *Pinus pumila*, dominate surface spectra from the upper limit of the montane forest zone. Missing from all forest-zone spectra is spruce (*Picea*) pollen, which is not surprising because the nearest outlying spruce groves are nearly 1000 km south of the lower Kolyma region.

Spores or pollen of *Selaginella sibirica*, Chenopodiaceae, and *Artemisia* are abundant in samples from steppe communities. Most such sites are edaphic, occurring mostly on southern mountain slopes in the upper and middle reaches of the Yana, Indigirka, and Kolyma Rivers. Some sites yield surface spectra containing an abundance of allochthonous *Pinus pumila* pollen. Others are dominated by pollen of herbs seldom recorded in Pleistocene pollen spectra (Yurtsev, this volume).

East Beringia

Cape Deceit in western Alaska is the main East Beringian site discussed here (cf. Guthrie and Matthews, 1971; Matthews, 1974b). The type section is couched low against the cape, a prominent metalimestone outcrop. Although this formidable bedrock barrier has undoubtedly directly sheltered the Quaternary sediments of the exposure, other factors such as tectonism may have protected sediments from inundation and erosion during Pleistocene marine transgressions (Hopkins, 1967c).

Cape Deceit lies in a region of continuous permafrost. Mean annual temperatures range from -4° to -8°C. Regional vegetation is shrub tundra representing Young's tundra zone 4 (Young, 1971). Calcareous regosols at the top of the Cape support *Dryas*-fell field vegetation. Silty soil of the uplands near the Cape contains communities dominated by *Eriophorum* and *Carex* as well as numerous heaths (Ericales) and shrub birch (*Betula* sect. *Nanae*). Sedges (chiefly *Carex*), grasses, and mosses are prevalent on the more moist lowland sites. Alders, willows, and shrub birches of higher stature grow on sunny slopes as well as on the floodplains of the larger streams in the area.

Treeline is located approximately 75 km east of Cape Deceit, yet outlying groves of coniferous trees are found as close as 40 km away. White spruce (*Picea glauca*) is the main treeline species in this area, but groves of balsam poplar are sometimes present beyond the spruce limit. In contrast to eastern Siberia, larch is a rare element of the forests of Alaska and the Yukon.

The local pollen rain at Cape Deceit contains relatively high percentages of *Betula* (up to 35%) and *Alnus* (≈ 10%), sedge (30–60+%), and Ericales pollen (≈ 10–35%) but little spruce (usually less than 10%) and even less *Artemisia*. The pollen rain is typical of that in other lowland sites within the hypoarctic tundra zone. In most instances such areas yield surface spectra containing less than 5% *Artemisia*, very low percentages of spruce, and significant percentages of birch, alder, and Cyperaceae, the latter usually having values higher than those for Gramineae (Nelson, 1979; Matthews, this volume).

KOLYMA BASIN SITES

Krestovka Section

Seventy-five kilometers above the mouth of the Omolon River and the village of Kolymskoye, the right bank of the Kolyma River intersects a high terrace (60–70 m) adjacent to the slopes of the Yukagir Plateau. The terrace consists entirely of unconsolidated late Cenozoic sediments that are exposed along several tributaries of the Kolyma River. Some of these terraces reveal sedimentary units not seen at exposures elsewhere in the Kolyma lowland.

One of the most important of these is the Krestovka section, on the Krestovka River approximately 2 km above its juncture with the Kolyma. Sher, the first to map the section, defined the sandy gravel at its base as the Begunov Suite (Sher, 1975).[5] More detailed studies involving pollen and other macrofossils followed (Sher, et al., 1977). The most recent description of the section includes paleomagnetic analyses (Sher, et al., 1979), and differs in several other important ways from the 1977 compilation.

Five major units are exposed at the Krestovka section (Fig. 2). The Begunov Suite at the base varies from 4 to 6 m in thickness and consists

[5]The term "suite" is approximately equivalent to the "formation" of North American geological literature.

mostly of ferruginous, coarsely bedded sandy gravel with lenses of sand and organic debris. Near its top the sand and gravel gives way to 1–1.5 m of thin-bedded sand and blue-black silt containing wood. Angular clasts of local lithology are abundant in the gravel; hence the gravel and associated sand probably represent the local alluvium of a small, relatively high-energy stream.

The upper iron-stained sand beds of the Begunov Suite contain wedge-shaped cracks. Some are small seams of sand, 0.5 to 1 m in height and 4 cm thick, truncated by overlying beds, and extending downward to form narrow vertical to subvertical ferruginous streaks. Although they are probably not ice-wedge pseudomorphs, the vertical features may represent some sort of soil-frost phenomena.

Unit II, the Kutuyakh Suite, consists of 5–10 m of thin-bedded sand, silt and peat interrupted by well-defined and morphologically varied ice-wedge pseudomorphs. The top of the unit is an erosional unconformity.

Kutuyakh Beds exhibit considerable lateral facies changes. The base of the thickest exposure is a sequence of compact thin-bedded sand, locally clay-rich, containing lenses of fine gravel and coalified wood. The sand is overlain first by gray silt and gray-brown peaty sand and then by a peat 0.4–0.8 m thick that is broken laterally by ice-wedge pseudomorphs filled with blue-gray lacustrine silt from the overlying sequence. The uppermost Kutuyakh sediments consist of bluish gray fine sand and gray loam with iron concretions. An abundance of organic material and a high degree of compaction characterize all but the uppermost sand sequences. The lower sediments are alluvial and the peat is autochthonous, having formed probably in a floodplain depression that was later filled by organic-rich alluvium and lacustrine silt.

Some Kutuyakh ice-wedge pseudomorphs may have formed due to local rather than regional climate; nevertheless, the existence of the pseudomorphs, and their large size, shows conclusively that at least part of the Kutuyakh Suite was deposited under an arctic climate in which permafrost was actively aggrading.

The third major unit at the Krestovka section comprises sediments of the Olyor Suite, which outcrop at many other sites within the Kolyma lowland. It is distinguished by its unique assemblage of large and small mammals (see discussion of Olyor type section, below).

The Olyor beds at Krestovka range from 7–12 m in thickness, and are bounded above and below by unconformities. According to the evolutionary stage of the fossils, the hiatus between the Krestovka and the underlying Kutuyakh sediments is a major one. Two Olyor subdivisions are evident (Fig. 2). The lower one, unit IIIa, consists mostly of thin-bedded alluvial sand with intercalations of sandy silt and prominent foreset beds (0.5–1.5 m in thickness) of ferruginous sandy gravel. The latter are particularly rich in vertebrate fossils. Unit IIIb is composed mostly of faintly to massively bedded gray sandy silt with interbeds of sand. Peat lenses occur in the upper part of the sequence near the erosional contact with the next higher unit.

The presence in the Olyor Suite at Krestovka of ice-wedge pseudomorphs of varying style testifies to a complex sequence of permafrost aggradation and degradation; however, since the sediments containing the pseudomorphs are largely alluvial, the fluctuations do not necessarily imply a similarly complex record of regional freeze-thaw events.

Unit IV at Krestovka consists of bedded alluvial sand and silt lying unconformably on the Olyor beds, extending 6–15 m upward to nearly the top of the exposure.[6] The sand contains pockets of vegetation detritus, zones of cemented silt peds, and inactive ice wedges.

Unit V (not shown in Fig. 2) consists of the *alas* sediments. They occur sporadically near the top of the section as the fillings of thermokarst depressions formed during partial thawing of the upper part of unit IV. In sharp contrast to unit IV, unit V consists of compact, massive, iron-stained silt. The upper part of the *alas* sequence in some parts of the exposure contains bedded peaty sand and silt with abundant mollusks, tree-size wood (some in growth position), and a zone of ice-wedge pseudomorphs.

There are no ^{14}C dates from the Krestovka section, and indeed, as the following data show, much of the section is probably well beyond the ^{14}C range. Dating is based instead on plant macrofossils, the stage of evolution of the fossil mammals, and the paleomagnetic data.

Mammalian fossils, both large and small, are abundant in some parts of the exposure; however,

---

[6]Some workers, including Hopkins, consider these sediments to be aeolian in origin—*Eds.*

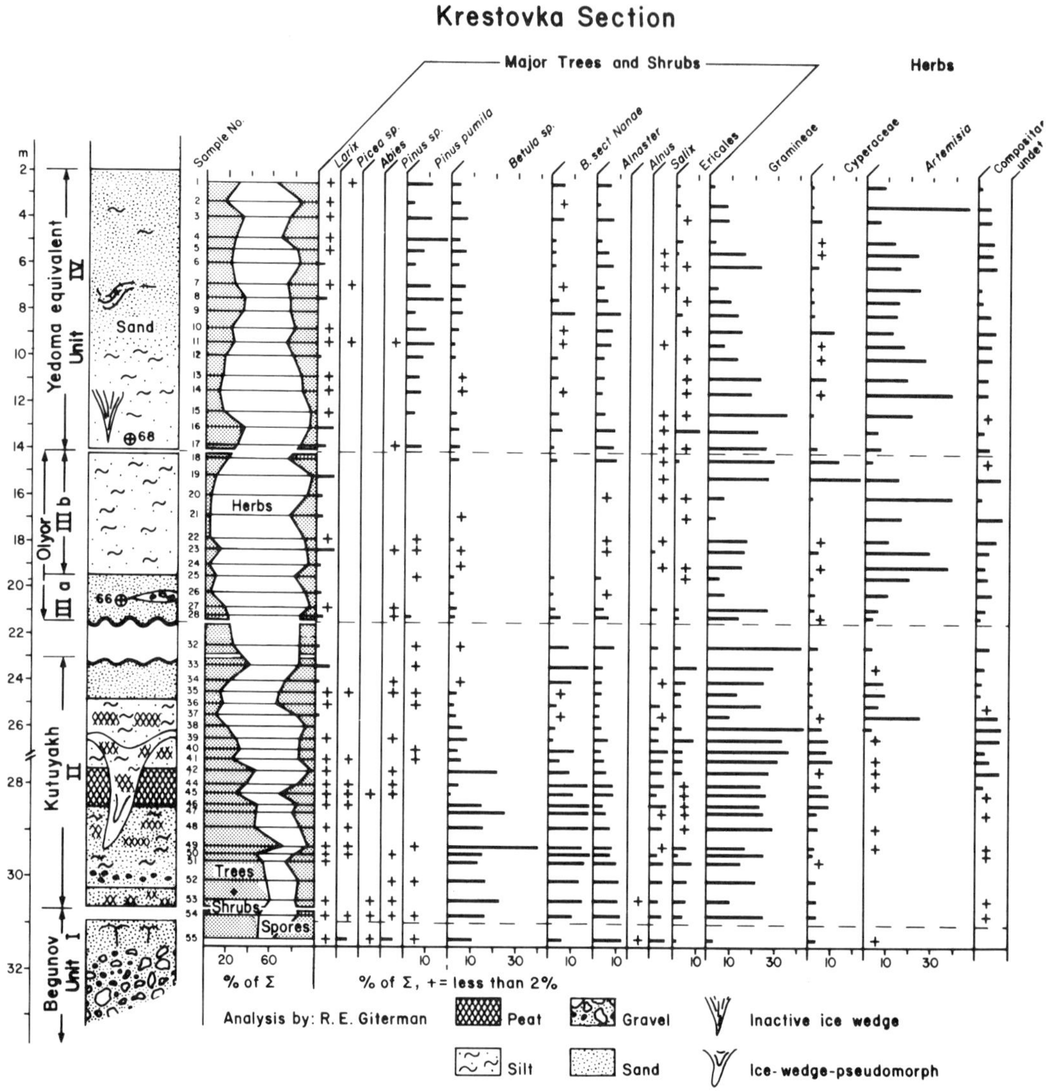

Figure 2. Composite pollen diagram for the Krestovka section. Note horizontal- and vertical-scale changes between summary diagram and the detailed diagram. Percentages are based on a sum of all pollen and spores. Pollen analyses by R. E. Giterman; pollen samples collected by A. V. Sher

to date, none have been found in the Begunov sediments. The Begunov sediments do contain an abundance of larch cones. These cones, studied by P. I. Dorofeev and E. G. Bobrov (R. E. Giterman, personal communication), apparently belong to three extinct species, two of which are closely related to the extant northeastern Asian species *Larix olgensis* Henry and *L. daburica* Turcz. The third group of cones is tentatively referred to as *L. omoloica* Dorof., an extinct form that first appears in the late Miocene Khapchan flora from the Omoloy River. *L. omoloica* shares characters with the extant North American species *L. occidentalis* Nutt. and the two Siberian species, *L. sibirica* Lebed. and *L. gmelina* Rupr. The Begunov gravel also contains a few spruce cones, a significant fact in the light of the present distribution of spruce in northeastern Siberia. These cones are morphologically intermediate between several extant species and the extinct species *Picea wolloso-*

Krestovka Section (cont.)

Og = Onagraceae Um = Umbelliferae Px = *Phlox* Dp = Dipsaceae Cl = *Claytonia*

Cp = Caprifoliaceae Pl = *Polemonium* Sx = *Saxifraga* Lb = *Linnea borealis*

Lg = Leguminosae R = Rubiaceae My = *Myrica* La = Labiatae

wiczii. Several other Begunov plants (*Potamogeton jacuticus* Dorof., *P. omoloicus* Dorof., *Epipremnum crassum* C. and M. Reid, *Suida* cf. *jacutica* Dorof., and *Rubus* ex. gr. *tenuicarpus* Dorof.) are either found in the Khapchan flora or in other Neogene contexts, or resemble taxa found in the Omoloy flora and the late Miocene Mamontova Gora flora. Such comparisons have led some workers to suggest that the Begunov and Khapchan floras are of equivalent age. But the Begunov flora is much less diverse taxonomically; it includes none of the Khapchan "exotic" broadleaved trees and thermophilous herbs; and the several pine species seen in the Khapchan are replaced by a single pine and a complex of larches in the Begunov. These facts suggest that the Begunov Suite is slightly younger than the Khapchan, a conclusion that also seems consistent with the differences between Khapchan and Begunov pollen spectra (Baranova, et al., 1968; Kartashova, 1974). Accordingly, the maximum possible age of the Begunov suite is late Miocene. Its minimum age is established by the late Pliocene age for the overlying Kutuyakh Suite. Because the Begunov flora is much more like the Khapchan than the Kutuyakh, the age of the Begunov suite is probably early to middle Pliocene (i.e., 4.0–3.5 million years BP, or late Miocene-early Pliocene in North American usage).[7] Only a few Begunov samples have yielded paleomagnetic data. All have normal polarity.

The Kutuyakh beds are rich in small-mammal remains. The evolutionary stage of some of them suggests correlation with late Pliocene assemblages from Podpousk, Lebyazhiye, and Liventskovka, of western Siberia, or with the middle Villifranchian Khaprov complex of the European U.S.S.R.

At another locality slightly upstream from the Krestovka section is a sequence of sediments that may bridge the time gap between the Kutuyakh and Olyor units at the main section. Silt at this upstream exposure contains a species of the rodent *Phenacomys,* which is more primitive than the *Phenacomys (Pliomys) deeringensis* from Cape Deceit, and teeth of other rodents that are apparently more advanced than their Kutuyakh counterparts. Thus the Kutuyakh sequence must be older than the Cape Deceit Formation (see below).

Plant macrofossils from the Kutuyakh Suite contribute little information on its probable age; we know that samples at some levels do contain: (1) a few fossils of the Neogene plant *Epipremnum crassum;* (2) a few fossils of extinct or at least distinctly nonarctic aquatics such as *Nuphar pumila* and *Brasenia;* and (3) spruce macroremains.

The upper part of the Kutuyakh sequence is paleomagnetically reversed (E. I. Virina, in Sher, et al., 1979; Virina, 1979). In view of the mammalian evidence, the sediments were probably deposited in the early part of the Matuyama Epoch. Sediments lower in the section, and occurring stratigraphically below mammalian fossils comparable to those of the Khaparov complex of eastern Europe, have normal polarities, which probably means that the Kutuyakh beds include the Matuyama-Gauss boundary (E. I. Virina, in Sher, et al., 1979). A thin, positive polarity zone exposed at one section is thought to be the Gilsà or Olduvai event, and if so the Kutuyakh Suite is definitely late Pliocene (possibly as old as 2.5 million years).

Although large-mammal fossils are rather rare in the lower part of the Olyor Suite at Krestovka (unit IIIa) and even more so in the upper part (IIIb), those present do suggest a relationship to the fauna at the Olyor type section (see below). Small-mammal fossils and paleomagnetic data provide a much more definite statement on the age of the unit. The lower subdivision (IIIa) has a reversed polarity but the upper one (IIIb) is normal. Because the small-mammal fossils from the lower unit predate the Tiraspol (contrary to earlier correlations of the Olyor suite—Sher, 1974) and are most closely related to the Taman faunal complex (upper Apsheron), the paleomagnetic reversal within horizon IIIa must be the Matuyama-Bruhnes of around 700,000 years BP. The small mammals in IIIb imply deposition during the early part of the Bruhnes epoch.

The Olyor Suite at Krestovka is significant on two accounts. First, the upper unit is new, and not exposed at the Olyor type section. Its mammalian fauna is much more similar to Tiraspol faunas than is the fauna from the lower unit. Secondly, the Krestovka exposure shows that the Olyor Suite, even as defined at the type section, probably encompasses a considerable period of time because its faunas mark several stages in the evolution of certain small mammal lineages (Sher, et al., 1979).

The base of unit IV contains fossils of much more advanced small mammals. This sequence is thus probably of late Pleistocene age and, if so, correlates with the Yedoma Suite of icy silt and sand found at other Kolyma-lowland exposures (Sher, et al., 1979). Unit IV sediments have normal polarity.

Figure 2 is a pollen diagram from the Krestovka section. Although Sher et al. (1979) also discussed the palynology of the section, their presentation was somewhat more generalized than ours, which is based on new, unpublished data. The sequence shown in Figure 2 is a composite drawn from study of sections 7, 11, and 12 at the exposure (Sher et al., 1979, Fig. 2).

The Begunov Suite is represented by a single sample. Tree and shrub pollen is abundant as might be expected by the taxonomic content of the macrofossil assemblages. It contains higher percen-

[7]The Begunov Suite probably correlates with the earlier part of the late Miocene-early Pliocene Clamgulchian stage in Alaska—*Eds.*

tages of spruce (some of it *Picea* sect. *Omorica*) and pine than any other sample at the section. The pollen spectrum includes many other types, mostly arborescent forms, which represent either extinct species or extant species now occurring far south of the site. Among the shrubs are traces of *Pinus pumila*, *Alnaster* (10+%), and a relatively low percentage of *Betula* sect. *Nanae*. Herb percentages are also low. An especially interesting feature is the low value for grass and sedge (both less than 5%) and the near-absence of xerophytes such as *Artemisia* and Chenopodiaceae. *Sphagnum* and Filicales account for most of the spores.

The basic pollen trend within the Kutuyakh beds is the decline of arboreal pollen and increase in herbs. At the base of the unit arboreal percentages approach 60%; at the top they are less than 20%. Many of the Kutuyakh tree and shrub pollen types are the same as those from the Begunov, and spruce, which is present only in trace amounts, makes its last consistent appearance. *Pinus pumila* is also represented but, like spruce, only in trace amounts. The pollen type identified as "*Betula* sp." declines from mean values of between 10% and 20% (one peak at 40%) in the middle part of the unit to less than 5% at the top. Except for a rise of percentages in the upper two samples (33, 34) the curve for *Betula* sect. *Nanae* is similar.

The increase of herb values toward the top of the unit is due primarily to a rise of grass percentages. Grass percentages exceed those of sedge throughout the sequence. Xerophytes are poorly represented; however, percentages of *Artemisia* climb noticeably in the upper part of the unit. This trend is not accompanied by any significant increase in *Selaginella sibirica* as is the case at other levels of the exposure. Uppermost Kutuyakh beds show a slight rise in arboreal percentages, particularly of *Larix*. *Larix* macrofossils also occur in that part of the section (Sher, et al., 1979).

At the 28m level peat is responsible for the rise of Bryales values, and moss fragments dominate the macrofossil assemblage. Except for two seeds of *Epipremnum crassum*, of uncertain provenance, the peat does not contain exotic elements. It has also yielded some extinct species (e.g., several types of *Potamogeton*, *Nympheae* sp.) but these are the only features distinguishing the Kutuyakh macroflora from that of samples higher in the section. Insect fragments from unit II include species of poorly drained tundra sites, such as *Blethsia catenaria* and *Pterostichus costatus*, in addition to others that feed on larch and/or *Pinus pumila*.

Though not illustrated in Figure 2, pollen samples from an upstream exposure that is thought to bridge the gap between the Kutuyakh and Olyor Suites show as much as 30% tree and shrub types, including *Betula*, *Alnus*, and *Larix*. *Larix* and several species of common water plants, including an extinct pond weed and an unusual species of *Brasenia*, are also found among the plant macrofossils from that section, but the exposure's greatest dissimilarity to the Kutuyakh beds is its insect fauna. In addition to a diverse assemblage of tundra and forest-tundra beetles, a number of taxa are also present that have clear steppe affinities, including abundant fossils of the pill beetle *Chrysobyrrhulus rutilans*, which is thought to indicate cryoxerotic conditions (S. Kiselyov, in Sher, et al., 1979).

In summary, the paleoecological data from the Kutuyakh unit at Krestovka portray an environment that initially supported few trees, and then became colder and more treeless before reverting once again to open larch forest. Although the thaw of the ice wedges in the Kutuyakh beds may have been the result of local disturbance, pollen and macrofossils from the upper part of the sequence do seem to indicate a slight regional warming of the climate. Spruce and several other types of water plants probably last grew in the Kolyma lowland during Kutuyakh time. The first appearance of the steppe and cryoxerotic species that typify the later Pleistocene of the Kolyma lowland occurs in sediments thought to be intermediate in age between the Kutuyakh and Olyor Suites. To date, however, these beds have not yielded remains of true xerophytic plants.

The Krestovka version of the Olyor sequence as shown in Figure 2 includes the upper part of the Olyor units IIIa and IIIb. Tree and shrub percentages decline from approximately 40% near the base to 5% (the lowest in the section) at the middle before climbing to somewhat higher values near the top. Herb-pollen percentages account for nearly 80% of the total pollen and spores in some of the middle samples, and Gramineae percentages are markedly higher than those of sedge except near the top of IIIb, where both are about equally represented. Spore percentages fluctuate between 10% and 20% over much of the unit; in contrast to samples from the Kutuyakh and Begunov suites, *Selaginella sibirica* is the dominant type.

Artemisia, Chenopodiaceae, and Caryophyllaceae abound in the middle part of the Olyor unit. Some of the peaks, e.g., of Chenopodiaceae in the lower part of the section, may be related to the local prevalence of sandy substrates; however, it is clear that regional vegetation was mostly herbaceous because even the percentages of *Betula* sect. *Nanae, Salix*, and *Ericales* drop to trace levels in this part of the sequence. On the other hand, *Larix* percentages are relatively high in these samples, and macrofossils reinforce the pollen evidence. The regional vegetation, though essentially treeless, was not entirely so.

Plant macrofossils from unit IIIb include sedges, green mosses, *Larix*, and even a few species of Nymphaceae and Ranunculaceae that presently grow south of the Kolyma lowland. Insect fossils from the same level consist of a variety of taiga and wet-tundra taxa, in contrast to unit IIIa, which contains abundant fossils of steppe weevils, *Chrysobyrrhulus*, and leaf beetles that currently reside in meadow-steppe areas (Kiselyov, in Sher, et al., 1979).

Although the insect fossils from unit IIIa indicate a dry steppe-like environment, the majority of the plant macroremains represent taxa normally found around the margins of or in small ponds. Kiselyov (in Sher, et al., 1979), noting that this paradox is typical of a number of Siberian macrofossil samples, concluded that the plants portray proximal floodplain sites, while the insects and pollen represent the regional upland environment. East Beringian macrofossil samples show similar disparities.

A long gap in time, representing perhaps much of the middle Pleistocene, separates Olyor and unit IV sediments at Krestovka. Unit IV starts with treelike shrub percentages nearly as high as those of the Olyor unit. Macrofossils of *Larix* and such tree-associated insects as the carpenter ant (*Camponotus*) and the weevil (*Pissodes*) occur in a sample near the base. Aquatic plants are well represented, but none of them refer to species now found in the Kolyma lowland. Once again, the insect assemblage (dominated by *Chrysobyrrhulus*) and plant macroremains seem to represent different types of environments.

Higher in unit IV the pollen spectra take on a distinctive character. The percentage of trees and shrubs remains relatively high (20–30%) but the only tree likely to have been present, and sporadically at that, is *Larix*. Stone pine (*Pinus pumila*), a shrub, seems to have been more abundant at this time than at any other, and apparently was interspersed with a few shrub birch and alders. *Artemisia* percentages reach their highest value in unit IV. Chenopodiaceae is present in only trace amounts, in marked contrast to some samples from unit III. An arid climate with discontinuous plant cover is indicated by the high percentages of *Selaginella sibirica*—the highest seen in the entire Krestovka sequence.

The palynological characteristics of the *alas* sequence are not shown in Figure 2, but it is noteworthy that a peat in the upper part of the *alas* sequence contains *Betula* fruits similar to those of the arboreal species *B. dahurica* as well as insects normally found in taiga regions. However, the insect *Aelia frigida*, a steppe species, also occurs as a rare element in unit V.

In summary, the paleoecological data from the Krestovka Section provide an interesting though punctuated picture of the development of the vegetation of the Kolyma lowland. Several stages are evident. The earliest, Begunov, records forests containing an assortment of *Larix*, tree birch, *Picea*, and arboreal *Pinus* species. Hypoarctic elements such as *Alnaster* are present even at this earliest time level. The Kutuyakh Suite marks development of essentially treeless communities and permafrost. Kutuyakh sediment contains evidence of incipient xeric communities, but these tendencies do not become pronounced until deposition of unit III, the Olyor Suite, which yields the first firmly dated cryoxerotic insect assemblages. Unit IV, representing a climate that on the basis of sedimentary structures seems to be the most severe in the sequence, yielded spectra less dominated by herbs than those from the Olyor Suite and that contained significant amounts of stone-pine pollen. Both units III and IV display spectra with high percentages of *Artemisia* and *Selaginella sibirica*, a significant departure from the *alas* sequence, which presumably represents vegetation and climate similar to the present.

Olyor Type Section

Sediments containing the typical Olyor vertebrate assemblage (Sher, 1974) occur not only at the Krestovka section but also on the right bank of the Kolyma River, on the Kamenushika River (a tributary of the Kolyma), and along the Alazeya, Bolshoy Khomus-Yuryakh, Sededoma, Ozhogin and Bol'shaya Chukoch'ya Rivers.

The type section is Sher's (1974) section 21, on the right bank of the Bol'shaya Chukoch'ya River, 7.7 km above the mouth of the Mastaakh-Yuryvye River (Fig. 3). There the river impinges on an *alas* basin that has a floor 5–10 m below the *yedoma* plain. As a result, the base of the section is not obscured by slumped and thawed silt of the Yedoma Suite as at most exposures. The upper 7 m of section 21 is thought to be of middle to late Quaternary age and consists of fine, locally silty cross-bedded sand, with isolated pockets of woody debris. The contact with the underlying unit is a disconformity having as much as 1 m relief.

The lower 21 meters of the type section contain sediment of the Olyor Suite. They include silt, sandy silt, silty sand, and peat beds up to 1 m thick. The best-developed peats apparently formed in the middle of low-centered ice-wedge polygons (Katz, et al., 1970; Sher, 1974). Ice-wedge pseudomorphs are present, some of them evidently having formed after thawing under water.

The evolutionary development of the fauna from the Olyor Suite suggests that section 21's age is similar to that of unit III at Krestovka. The entire Olyor type section is paleomagnetically reversed, while at Krestovka only the lower subunit is reversed. Thus the type section is not an exact equivalent of the Olyor sequence at Krestovka. Nevertheless, the type section undoubtedly dates from the end of the Matuyama Epoch and is older than 700,000 years. This makes the Olyor type beds late Pliocene (Apsheron) in age under the Soviet system (or early Pleistocene in North American usage—Butzer and Isaac, 1975).

The Olyor-section pollen sequence is rather uniform and similar to that of the Olyor level at Krestovka. A peak of shrub pollen and Bryales spores occurs in sample 9 but this is probably due to overrepresentation of local pollen from plants that grew near the site. In general, trees and shrubs average approximately 30% with peaks to 50%; herbs account for 25–60%; and spores about 20%. The occasional grains of *Picea* and *Pinus* in some samples are probably allochthonous. *Larix* is also a minor element, although *Larix* wood occurs in the section.

Of the herbs, grass pollen is usually more abundant than sedge. *Artemisia* percentages do not exceed 10% and generally are less than 5%; some of the other Compositae pollen types attain relatively high values. An important feature of the herb spectra is their taxonomic diversity. It is notable as well that *Selaginella sibirica* is present but does not dominate the spore spectra.

Like insect assemblages from Olyor beds at Krestovka, most of those from the type section and other exposures of the Olyor Suite along the Bol'shaya Chukoch'ya River are dominated by the pill beetle *Chrysobyrrhulus rutilans* (Table 1). Assemblages also include a number of taxa typical of dry tundra, a few steppe species, and even (in some samples) a small number of taiga forms (S. Kiselyov, in Sher, et al., 1979; Kiselyov, 1979, and personal communication).

Katz et al. (1970) have studied collections of plant macroremains from several levels at section 21. The peat zones consist mostly of Bryales fragments. Samples from other parts of the section contain abundant *Potamogeton, Menyanthes,* and *Carex.* If the fossils from these assemblages were grouped in the same manner as has been done for several East Beringian samples (Hughes, et al., 1981; Matthews, this volume), the aquatic *Potamogeton-Carex* group would dominate (80–90%). The insect assemblages, in sharp contrast, contain hardly any aquatic or hygrophilous species (Table 1), once again illustrating a divergence of the insect and plant macrofossil evidence.

Most plants in the Olyor macrofossil assemblages can be found within the present hypoarctic zone. The few that appear to be limited to more southern sites should not be given too much emphasis, at least not until the distribution of aquatic plants in northeastern Siberia is better known (Katz, et al., 1970).

Utkinsky Kamen Section

The type locality for the middle Pleistocene in the Kolyma lowland is the Utkinsky Kamen section on the right bank of the Maly Anyuy River, approximately 28 km from the village of Anyuisk (Fig. 1). The section lies near the juncture of the Anyuisk lowland and the piedmont of the North Anyuisk ridge (Sher, 1974). The Utkinsk beds in the lower 5 m of the sequence are assigned to the

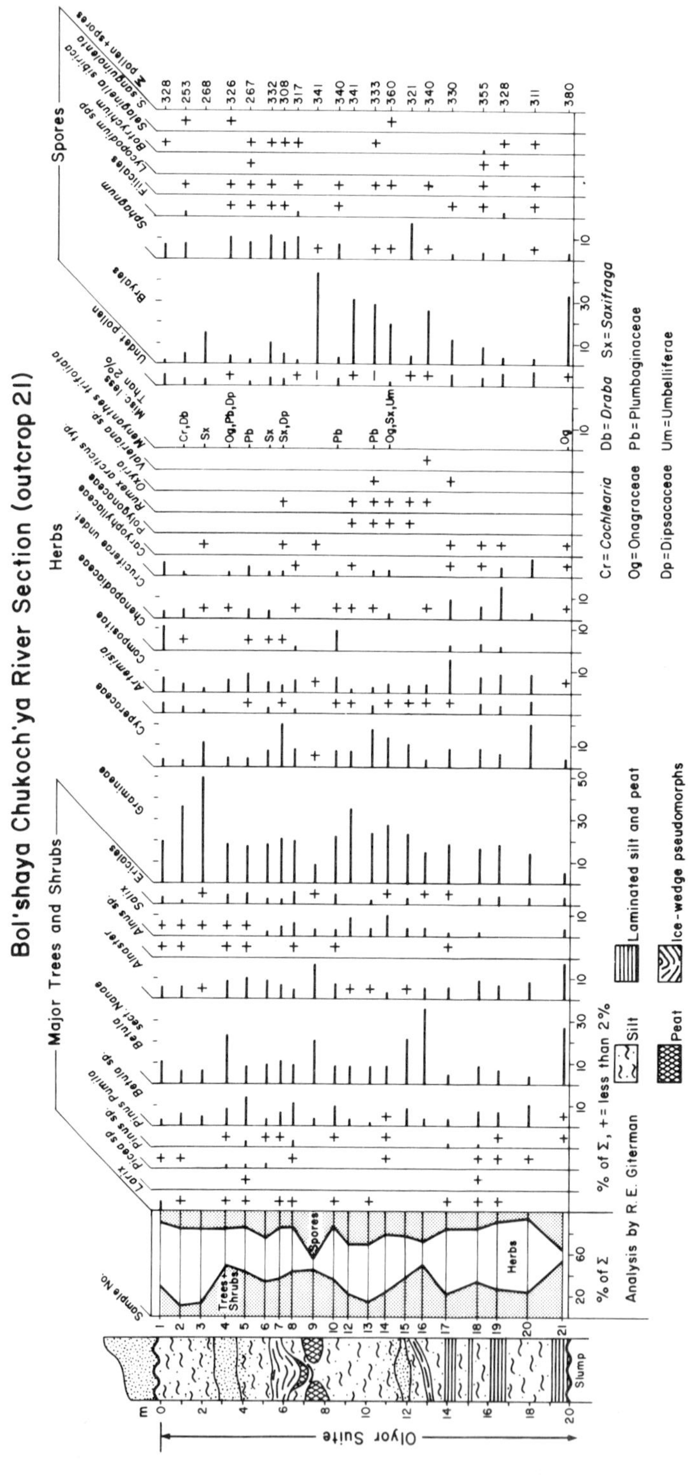

Figure 3. Pollen sequence from the type section of the Olyor Suite (section 21). Note horizontal-scale change between summary diagram and the detailed diagram. Percentages are based on a sum of all pollen and spores. Pollen analyses by R. E. Giterman

Table 1
Comparison of Siberian and East Beringian Insect Assemblages

% OF INSECT GROUPS IN KOLYMA LOWLAND ASSEMBLAGES (upper headings) / % OF INSECT GROUPS IN EAST BERINGIAN ASSEMBLAGES (lower headings)

Sample	No.	Tundra Meso. / Cryobius	Tundra Xeric	— / Lepido-Morychus	Chryso-byrr. / (Lepido-Morychus)	Steppe Artem.	Steppe Meadow	Taiga / Form-icid.	Larch Pine	— / Misc. Phytoph.	Others	— / Tachi-nus	— / Sil-phid.	— / Aphod.	Aquat. / Hygro. Aquat.	Sample	No.
Dy13-16	*55*	*18*	*2*		*7*				*6*		*31*				*37*		
		10		2				6		5			+	+	76	CRH44	204
		12		19	(2)			28		6		+			28	R.B.	89
Alsb101	*415*	*1*	*4*		*37*	*29*	*21*				*7*		*+*		*1*		
Alsb102	*807*	*4*	*9*		*37*	*9*	*32*				*7*		*+*	*+*	*1*		
		10		68	(16)					9		9		4	+	S-6	1538
Dy7-8	*122*	*7*	*30*		*27*	*2*	*22*		*3*		*8*			*5*	*3*		
Dy9-11	*111*	*17*	*28*		*39*	*4*	*2*		*3*		*6*	*+*			*2*		
Dy4-6	*84*	*7*	*6*		*78*	*1*	*6*				*2*		*+*	*+*			
		31		65	(+)			+		+		+		9	2	Eva3C	378
K-68	*410*	*7*	*9*		*58*	*7*	*13*	*+*	*+*		*4*	*+*	*+*	*+*	*+*		
Dy 3	*83*	*11*	*18*		*50*	*8*	*1*				*4*	*+*		*2*	*6*		
		61		16						6		17			+	S-5	379
		51		24	(+)			+		4					18	Eva1A	297
		13		70	(10)					7		+	+	3	8	CRH32	72
		6		71						15		7			+	S-1	241
		16		23*						15		45			+	S-10-67	727
K-67	*103*	*15*	*19*		*6*		*5*	*2*			*26*	*2*	*4*	*2*	*19*		
K-66	*512*	*10*	*13*		*49*	*9*	*2*				*15*	*+*	*+*		*3*		
		52		17	(+)			5		4		+	+	+	13	JVM-5	528

Read Kolyma-lowland Table from top down, using samples listed on left margin.
Read East Beringian Table from bottom up, using samples listed on right margin.

*Lacks *Lepidophorus* as do all Kolyma Lowland samples.

Number in columns = % of "No." which is the minimum number of individuals.

Number in parentheses = % of *Morychus* in Lepidophorus-Morychus group (e.g., CRH32 contains 70% Lepido-Morychus group of which 10% are *Morychus*.

Upper set of samples (separated by dashed horizontal line): Holocene; next lower set: Sartan (Duvannyy Yar interval); next lower set: Karginsk (mid-Wisconsin); next lower set: Deering interval; next lower set: early Pleistocene; lowest sample: late Miocene.

middle Pleistocene because they have yielded an early form of *Mammuthus primigenius*, a large form of *Equus*, and a large form of *Alces*. A typical late Pleistocene mammalian fauna characterizes the upper 7.5 m of the exposure (Fig. 4) (Sher, 1974).

Except in the two lowest samples (Fig. 4) the Utkinsky spore-pollen diagram contains relatively more herb pollen than is typical of samples from the Olyor Suite (Fig. 3). The percentage of spores is also lower in the Utkinsk beds. Trace amounts of *Larix, Picea, Alnus,* and *Pinus* sp. occur among the tree and shrub pollen but all of these are probably allochthonous. Pollen of shrubs such as *Alnaster, Salix,* and *Pinus pumila* amount to less than 10%, and the percentages of *Betula* sect. *Nanae* exceed this value in only a few samples.

The herb component of the diagram is dominated by grass and *Artemisia*. Grass-sedge ratios are generally high, and *Artemisia* values peak at between 20% and 40%. The spectra also contain a diversity of other herbs, including Compositae (excl. *Artemisia*), Chenopodiaceae, Caryophyllaceae (*Cerastium* type), *Armeria arctica, Polemonium,* Ranunculaceae, and consistent but low percentages of *Sanguisorba*. Bryales dominates the spore fraction. *Selaginella sibirica,* while represented in most samples, never accounts for more than 2% of the total pollen and spores.

These characteristics suggest that the Utkinsk sequence represents a type of tundra-steppe vegetation, but one in which *Selaginella sibirica* was less important than later during the Pleistocene or than at some of the present-day relict steppe sites in eastern Siberia. Since the Utkinsky beds have less *Selaginella,* they may record vegetation having a relatively higher percentage of herb cover. Of all the pollen spectra from the Kolyma lowland those from Utkinsky Kamen are the closest counterparts of the so called "steppe-tundra" spectra at East Beringian sites (Matthews, this volume).

Insect fossils have been recovered from various levels of the Utkinsky Kamen section (Kiselyov, 1976). The lower horizons are dominated by tundra species and contain a few taiga forms, but very few steppe taxa. In the middle horizons, tundra and steppe species dominate, and taiga species are rare.

The two lowest samples shown in Figure 4 probably represent a period of expanded forest-tundra vegetation. Tree and shrub pollen is prominent and taiga forms such as *Larix, Pinus, Alnaster,* and *Salix* are present. *Betula* sect. *Nanae* and Ericales are well represented and herb-pollen percentages are low; both *Artemisia* and Gramineae remain below 12%.

No plant macrofossils have been reported from the Utkinsky type section, but beds at other sections presumed to be equivalent to the lower part of the Utkinsky sequence contain fossils of *Larix dahurica, Calla palustris,* and several other plants that no longer grow this far north (Sher, 1974). These assemblages are also dominated by what Matthews (this volume) terms the *Carex-Potamogeton* group of fossils.

The Upper Pleistocene portion of the Utkinsky Kamen section, like the Utkinsky beds, contains abundant herb pollen, although the percentage of Gramineae pollen is lower and *Artemisia* peaks are higher (approaching 70% in several samples). Pollen of *Larix, Pinus* sp. (cf. *P. silvestris*), *Pinus pumila,* tree and shrub birch, *Alnaster,* and Ericales round out the tree and shrub fraction of the spectra; all have low frequencies. Spores are also rare; however, in contrast to the Utkinsky beds, a few of these late Pleistocene samples contain in excess of 2% *Selaginella sibirica*. The *Selaginella* percentages are, however, lower than in some of the other Upper Pleistocene samples from the Kolyma lowland.

Duvannyy Yar Section

The Duvannyy Yar section is located along a 9-km stretch of the right bank of the Kolyma River, 35–45 km downstream from the mouth of the Omolon River (Figs. 1, 5). The exposure reveals a suite of sediments underlying the Omolon-Anyuy *yedoma* that has a surface 70–100 m above sea level and *alasy* up to 40 m in depth.

The Duvannyy Yar section is an unusually thick exposure of the so called "ice complex" or Yedoma Suite; in addition it allows a rare view of sediment underlying the ice complex (Fig. 5). All units are of late Pleistocene age but record a complex series of sedimentary and climatic events. For example, ice-wedge pseudomorphs in the massive blue-gray silts of unit I, nearest water level, indicate deposition under a climatic regime cold enough for permafrost. Unit II, composed mainly of woody, peaty sediments, probably represents thaw-lake debris (Hopkins, this volume). Some of

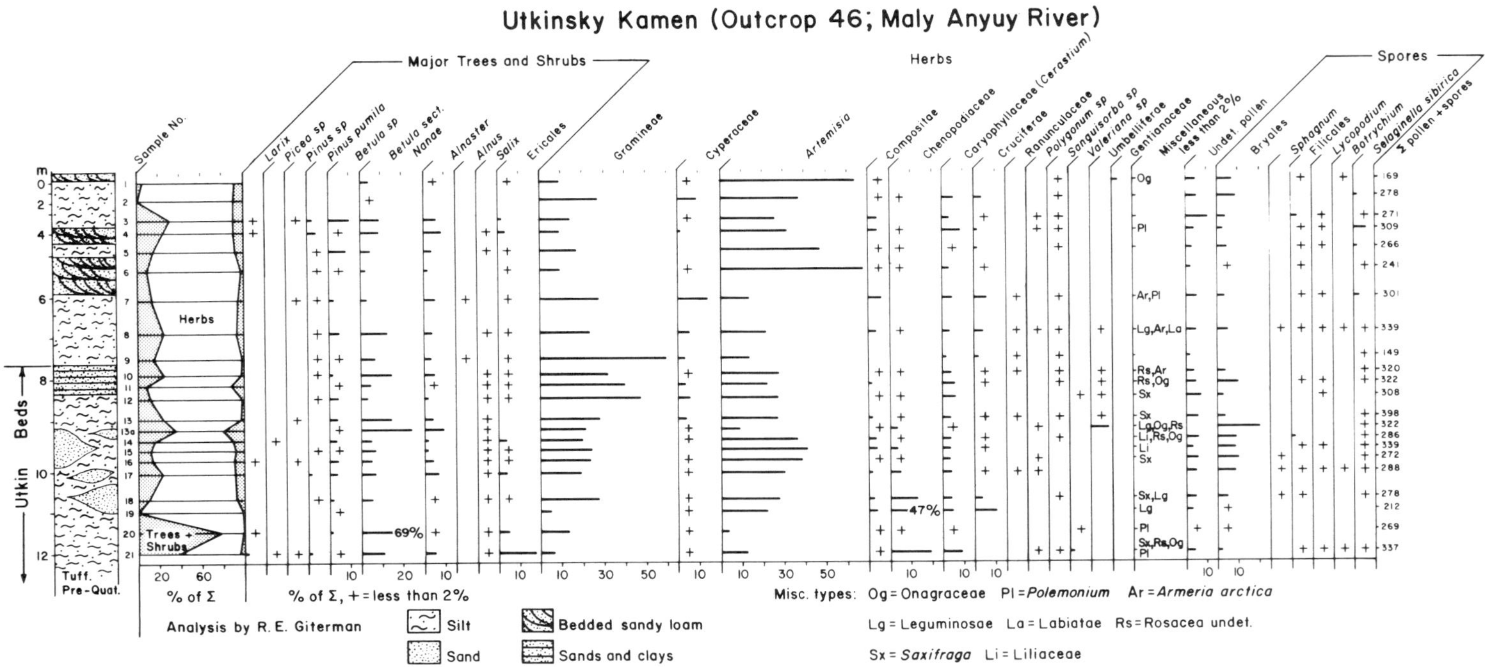

Figure 4. Pollen diagram from the Utkinsky Kamen Section. The Utkinsky beds encompass only the lower part of the section (as indicated). Note horizontal- and vertical-scale changes between summary diagram and the detailed diagram. Percentages are based on a sum of all pollen and spores. Pollen analyses by R. E. Giterman

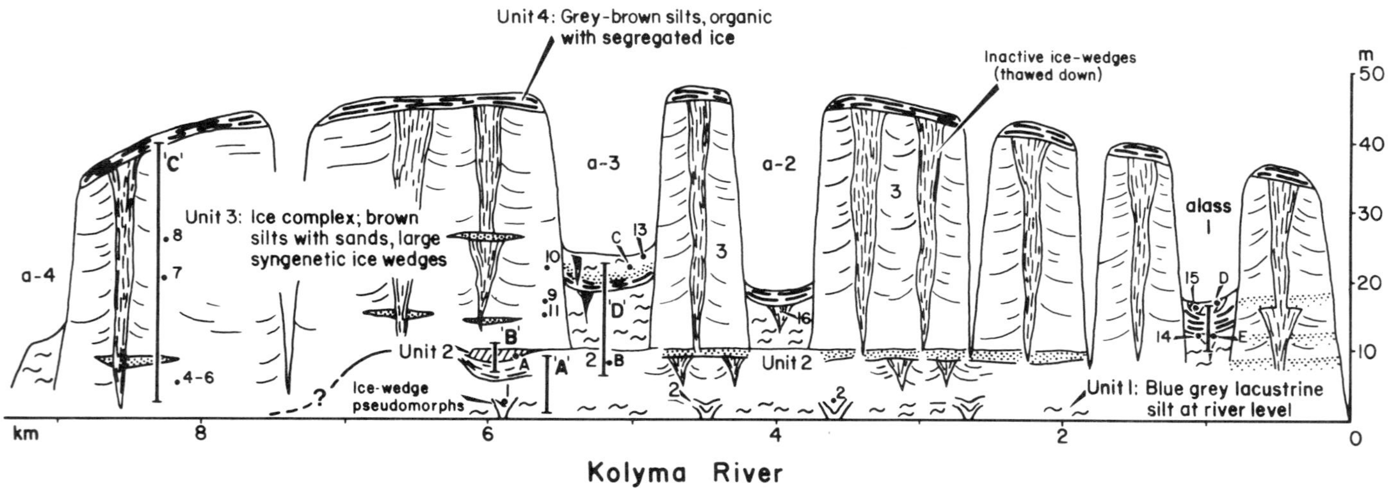

Figure 5. Generalized view of the Duvannyy Yar section (modified from Sher, et al., 1979). Numbers refer to fossil insect samples studied by S. Kiselyov. Capital letters indicate ^{14}C dates: A = 36,900 ± 500 (MGU-469); B = 37,600 ± 1100 (MGU-468); C = 5400 ± 110 (MGU-470); D = 3955 ± 80 (MAG-161); E = 6300 (MAG-156). Vertical bars indicate position of the pollen sequences: 'A' = unit 1 of Fig. 6; 'B' = unit II of Fig. 6; 'C' = unit III of Fig. 6. Unit III is the "ice complex" or "*yedoma* unit" mentioned in the text

its sediments fill the pseudomorphs of unit I, thus indicating that thaw of the original ice wedges was due to inundation by water rather than regional warming. Although ice wedges do occur in unit II, they are inactive and their tops have been partially thawed down. The base of unit II has been ^{14}C dated at around 40,000 years, or during the Karginsk interglacial ("Boutellier interval" of Hopkins, this volume).

The bulk of the exposure is made up of the oversaturated, ice-rich silt, sand, and loam of the Yedoma Suite. Radiocarbon dates from Duvannyy Yar and other exposures in the Kolyma lowland (Lozhkin, 1979) indicate a Sartan (late Wisconsin) age. Alluvial sand and occasional fine gravel characterize the base of the *yedoma* sequence. Gray-brown silt and finely bedded organic silt, with zones of in situ grass roots, occur through the rest of the unit.

Several generations of ice wedges occur at the base of the *yedoma* unit. The lower one consists of wedges 3–4 m wide and spaced at intervals of 9–10 m. A second one has veins 0.15–0.4 meters in thickness and 3-m spacing; they are smaller than those now forming in the area. Higher still, ice wedges are larger and once again comparable to those forming at the present surface, except that unlike the existing polygons they include a subsidiary net of ice veins that represent very small polygons. The scale of this subsidiary set is similar to present-day polygons formed where snow is blown clear and the mean ground temperature not higher than −9°C. At some levels of unit III the spacing of the primary ice wedges is as low as 2–2.3 m, or smaller than contemporary polygons in the Kolyma lowland.

The veneer of gray-brown silt that makes up unit IV is generally 0.7 to 1.2 m thick but sometimes reaches 3–4 m in depressions over partially thawed ice wedges. The Duvannyy Yar section includes exposures of several *alas* sequences (Fig. 5). Their sediments are dated as between 9200 and 5400 years and contain small ice wedges as well as pseudomorphs that formed as a result of local inundation. Fillings of the pseudomorphs have abundant wood, mostly larch and tree birch.

Figure 6 is a composite pollen diagram of the Duvannyy Yar section. The few spectra from unit I suggest a cold, treeless climate. Although herb pollen is abundant and Gramineae is the dominant element, the assemblage also includes significant quantities of *Artemisia*, other Compositae, and Caryophyllaceae. Values for Chenopodiaceae are less than 20%, except in sample 1, which probably represents local overrepresentation. *Betula* sect. *Nanae* and *Alnaster* percentages each average less than 5%; and only one sample contains a trace of *Pinus pumila* pollen. Spore percentages, mostly of *Selaginella sibirica*, range as high as 40+%, averaging about 20%.

Tree and shrub pollen are abundant in the peaty sediments of unit II. Dominant are *Pinus pumila* (especially in sample 12), *Betula* sect. *Nanae, Alnaster*, and Ericales. Of the herbs, Cyperaceae is more abundant than Gramineae and peaks at values near 30% in the middle part of the section, probably due partly to local influence. *Selaginella sibirica* is rare, and most spores are of Bryales. In other publications (e.g. Kaplina, et al., 1978), three pollen phases are recognized in unit II. These are not evident in Figure 6 because Ericales is grouped there with the trees and shrubs. In any event, the phases probably only mark changing local vegetation during progressive paludification.

Plant macrofossils from the pseudomorphs intruded into unit I from unit II are the type that normally occur on the floodplain of a large river within the northern taiga or forest-tundra zone. The insect assemblages are dominated by *Chrysobyrrhulus* and other taxa indicative of xeric or steppe biotopes. Macrofossils (wood and cones) show that *Larix dahurica* grew near the site at the start of deposition of unit II, although only traces of *Larix* occur in the pollen diagram. This is expected because *Larix* pollen is often severely underrepresented in terms of the actual abundance of the plant. However, it is peculiar that the nearby presence of shrub birch, a prolific pollen producer, is indicated by macrofossil evidence, although only small quantities of pollen occur.

The pollen sequence from unit III contains four phases (Figure 6). The first, consisting of the three basal samples, has tree and shrub percentages as high as 20%, including significant quantities of *Pinus pumila, Betula* sect. *Nanae,* and *Alnaster.* Grass is abundant and outranks sedge. Several other herbs are present in significant amounts but some of these peaks (e.g., 53% Caryophyllaceae at the very base) undoubtedly reflect local vegetation. *Selaginella sibirica*, the dominant spore taxon, accounts for about 10% of the total. Phase 2

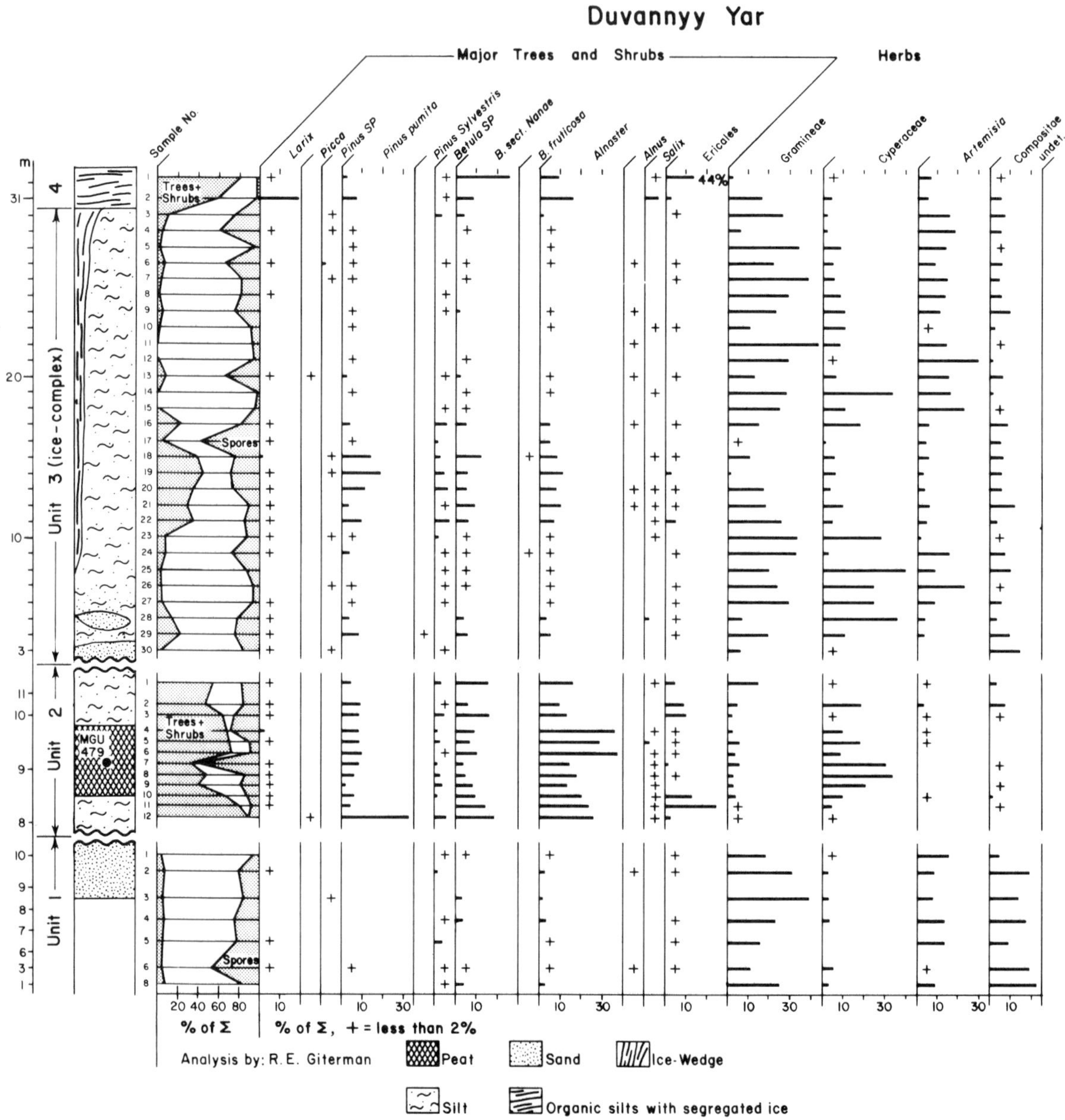

Figure 6. Pollen diagram from the Duvannyy Yar section. Note horizontal-scale changes between summary diagram and the detailed diagram; also change in vertical scale in units I and II. Percentages are based on a sum of all pollen and spores. Pollen analyses by R. E. Giterman

(samples 23–27) is characterized by lower arboreal percentages. Gramineae and Cyperaceae are abundant and percentages of *Artemisia* are rather high (≈ 25% in one sample). Spores are rare, of those present most are *Selaginella sibirica.* In phase 3 (samples 16–22) tree and shrub pollen once again exceed 20%; spores are abundant, with *Selaginella sibirica* peaking at over 40% in one sample; and the tree and shrub category is dominated by *Pinus pumila, Betula,* and *Alnaster. Artemisia,* other Compositae, and Caryophyllaceae have values approaching 5–10% but grass registers a major de-

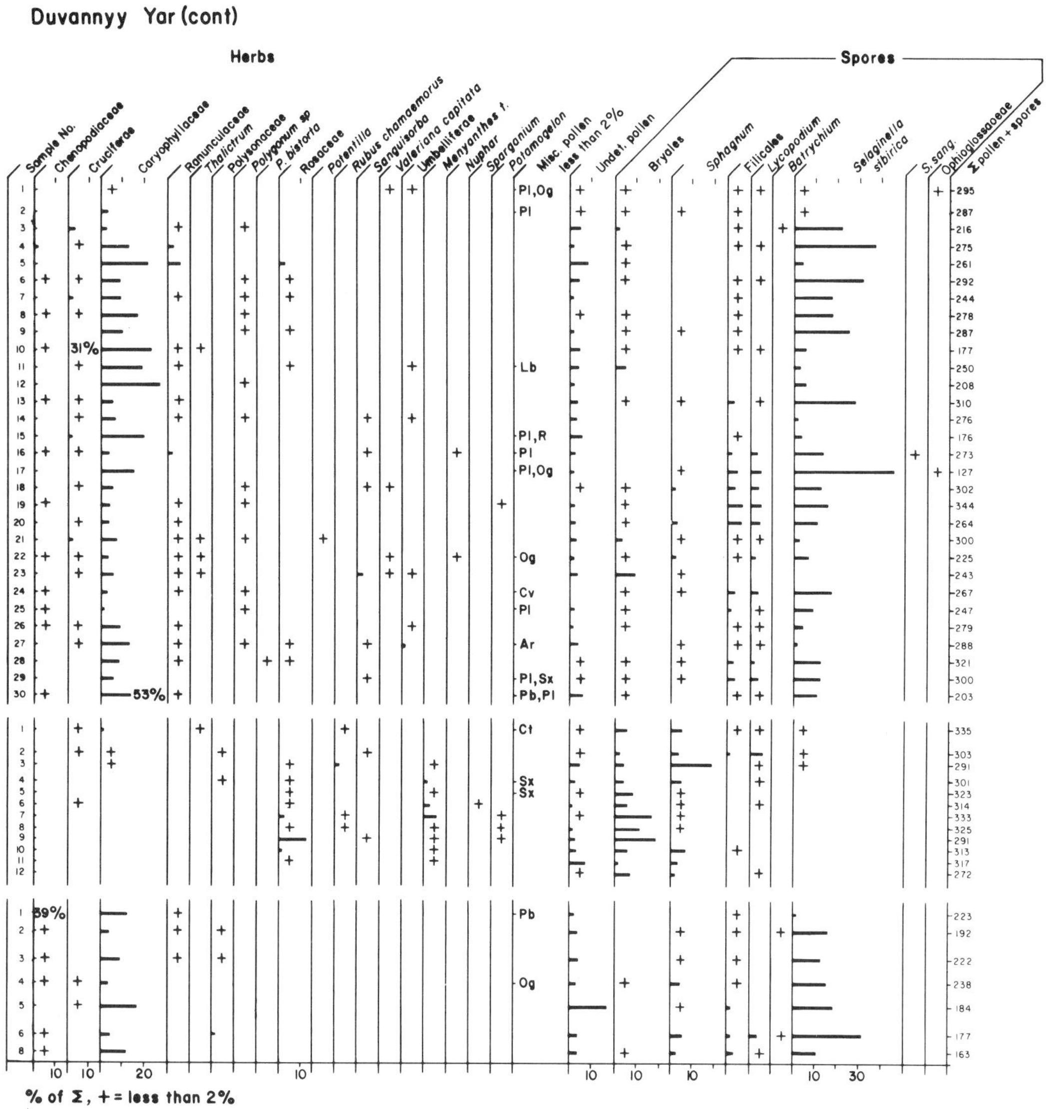

Pl = *Polemonium* Lb = Labiatae Cv = Convolvulaceae Sx = *Saxifraga* Og = Onagraceae R = Rubiaceae
Ar = *Armeria* Pb = Plumbaginaceae Ct = *Cerastium*

cline. Phase 4 (samples 3–15) represents a return of herb dominance (≈ 80%). Grass averages about 25% and easily outranks sedge; *Artemisia* ranges from 10–20% (30% in one sample); the Caryophyllaceae attain some of the highest consistent percentages seen in any section; and there is a diversity of other herb types. Spores average 20% and virtually all of them are *Selaginella sibirica.* It is unlikely that these pollen phases recognized within unit III reflect mere local vegetation change, because they involve fluctuations of several pollen types and are not correlated with any pronounced lithologic change. Thus, Duvannyy Yar, like the other exposures in northeastern Siberia (Kaplina, 1979) suggests that the Sartan (or late Wisconsin in North America) contains at least one phase of climatic amelioration.

Kiselyov has studied several insect assemblages

from three levels of unit III (Kaplina, et al., 1978; Sher, et al., 1979) (see Fig. 5). The lower assemblages (Dy 4–6, Table 1) contain abundant *Chrysobyrrhulus,* several other steppe Coleoptera, and specimens of the ground beetle subgenus *Cryobius* (most of whose species reside in mesic tundra). Another sample (Dy 9–11, Table 1), from the levels of the exposure at which tree and shrub pollen rise, contains a relatively large number of specimens representing xeric-adapted species yet shows a lower percentage of *Chrysobyrrhulus* and more mesic-tundra Coleoptera. Higher in the ice-complex sequence the percentage of *Cryobius* and other mesic-tundra taxa drops and steppe species once more become abundant, except that, unlike the lowest assemblages, most of the steppe species from the upper part of the unit live in meadow-steppe biotopes (compare Dy 4–6 and Dy 7–8 in Table 1).

Although few of the vertebrate fossils from the Duvannyy Yar section have been found in situ, it is highly probable that most come from various levels of unit III. The vertebrate assemblage supports the conclusion drawn from pollen and insect data because it includes an abundance of ungulates (e.g., horse and bison) that feed primarily on graminoid species and that are physiologically adapted for coexistence in areas with a mosaic of dry and mesic communities (cf. Guthrie, this volume). In addition, fossils of the other typical members of the late Pleistocene mammoth fauna *Mammuthus primigenius, Rangifer, Cervus elaphus, Ovibos, Coelodonta, Canis lupus, Alopex lagopus,* and *Panthera spelaea*) also occur at Duvannyy Yar.

These several threads of evidence suggest formation of the ice-complex sediments in a cold, dry climate. Similar conditions are implied by pollen, insects, and vertebrates from the Alyoshkin Suite, which includes the sediment that underlies the surface of the Khallerchin Tundra (Fig. 1). Originally, the Alyoshkin Suite was defined by Sher (1974) as a sequence of floodplain sands inset into the *yedoma* surface. Instead, it may only represent an aeolian sand facies of the *yedoma* silt (Hopkins, this volume). Dates of 14,980 ± 100 (MAG-470) and 15,000 ± 200 (MAG-468) have been obtained on organics at the Alyoshkina Zaimka section (Sher, et al., 1979). The section also contains bones of the saiga antelope, *Mammuthus, Equus, Bison, Rangifer,* and others, plus an abundance of insect fossils from xeric or steppe species (samples Alsh. 101, Table 1). This particular characteristic is most pronounced in the uppermost sample (Alsh. 101), in which the abundant fossils of petrophylic weevils suggest enhancement of *Artemisia*-herb-type steppe at the expense of tundra and meadow-steppe. The Alyoshkina Zaimka pollen sequence (Sher, et al., 1979, Fig. 23) shows a decline of shrub birch in the basal part followed by increase in the percentage of herbs, particularly grass, *Artemisia,* other Compositae, and Cruciferae. Most of the spores are *Selaginella sibirica.*

The Pleistocene pollen and macrofossil assemblages at Duvannyy Yar and Alyoshkina Zaimka stand in sharp contrast to the pollen spectra from Holocene units (unit IV and the *alas* sections, Figs. 7, 8). Note, for example, in the upper two samples of Figure 6 the sudden increase in tree and shrub percentages, expressed mostly by *Larix,* dwarf birch, and Ericales. Such a shift must mark inception of a warmer, more moist climate similar to that of the present. A peak of *Larix, Alnaster,* and *Pinus pumila* in sample 2 of Figure 6 may even portray a slightly warmer climate than at present because such high values neither occur in surface samples from Duvannyy Yar nor in surface spectra from other Kolyma-lowland sites (Kaplina, 1979).

A more complete Holocene record is to be found in the various *alas* sections (Figs. 7 and 8). The basal 2 meters of the *alas* 1 section (Fig. 7) and the basal 10 meters of the *alas* 3 section undoubtedly (Fig. 8) represent a compressed and truncated version of units II and III. Holocene sediments equivalent to unit IV begin in the 5-m level with an abrupt rise of tree and shrub percentages. Pollen spectra from both *alasy* are seriously influenced by local vegetation; nevertheless, the decline of *Alnaster* percentages in the upper part of both sections probably marks regional climatic deterioration following a Holocene optimum.

Insect fossils from the *alas* deposits also contrast with those from the upper part of the ice complex (Dy 13–16, Table 1) (Sher, et al., 1979). *Chrysobyrrhulus* is almost gone. In its place are species of ground beetles, rove beetles, and weevils more typical of mesic-tundra sites. Aquatic and near-shore species are also abundant, and as might be expected, the assemblages include several taxa associated with trees or shrubs, but, curiously, no bark beetles (Scolytidae). Although local biotopes may also be biasing the insect assemblage, it is clear that the *alas* insects represent a radically different regional environment than do the fossils from either the ice complex at Duvannyy Yar or the Alyoshkina exposure.

The most notable feature of the Duvannyy Yar section is its thick sequence of ice-complex sediments. Hopkins (this volume) has designated it as the stratotype for the Duvannyy Yar interval, a period of time during the last major world glaciation (Sartan or late Wisconsin) when the climate of all Beringia, and especially of the Kolyma lowland, was extremely cold and dry. Grasses, other herbs, and *Selaginella sibirica* must have been the main elements of vegetation at this time. The Duvannyy Yar interval was also a time when the insect fauna of northeastern Siberia contained the greatest number of xeric and steppe species. Furthermore, it is assumed to have been a time when a diverse assemblage of large mammalian herbivores lived in northeastern Siberia. Man was probably also a member of that mammalian community.

EAST BERINGIA

No single site nor any complex of sites in East Beringia displays as complete a picture of Pleistocene environmental change as do the various sections in the lower Kolyma basin. The closest analog, in terms of the duration of the record and availability of different types of fossils, is the Cape Deceit exposure in western Alaska (Fig. 1).

Cape Deceit

The stratigraphy and paleontology of the Cape Deceit exposure are detailed elsewhere (Matthews, 1974b) and need only be discussed briefly here. However, certain revisions require mention, and there is a need to present the Cape Deceit data in a format comparable to that used for the Kolyma-lowland sections. One revision involves Matthews' original interpretation of the age of the sequence. It consists of three formations bounded by disconformities. The middle unit, Inamachuk Formation, is poorly exposed and of uncertain age. It probably does not deserve its status as a major stratigraphic unit and will not be discussed further here. As a result the discussion and Figure 9 concern comparison of only the oldest and youngest parts of the Cape Deceit section.

The Cape Deceit Formation is of Pleistocene or

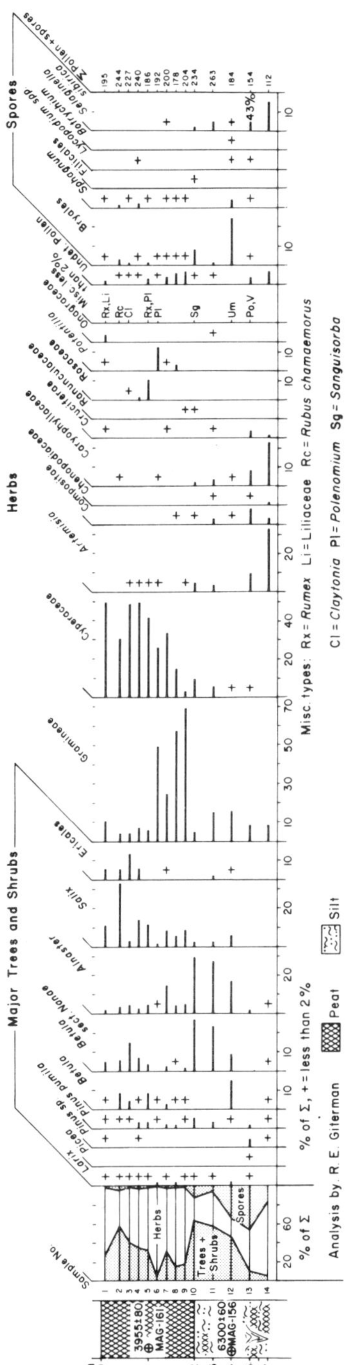

Figure 7. Pollen diagram for Duvannyy Yar *alas* 1. Note horizontal-scale change between summary diagram and the detailed diagram. Percentages are based on a sum of all pollen and spores. Pollen analyses by R. E. Giterman

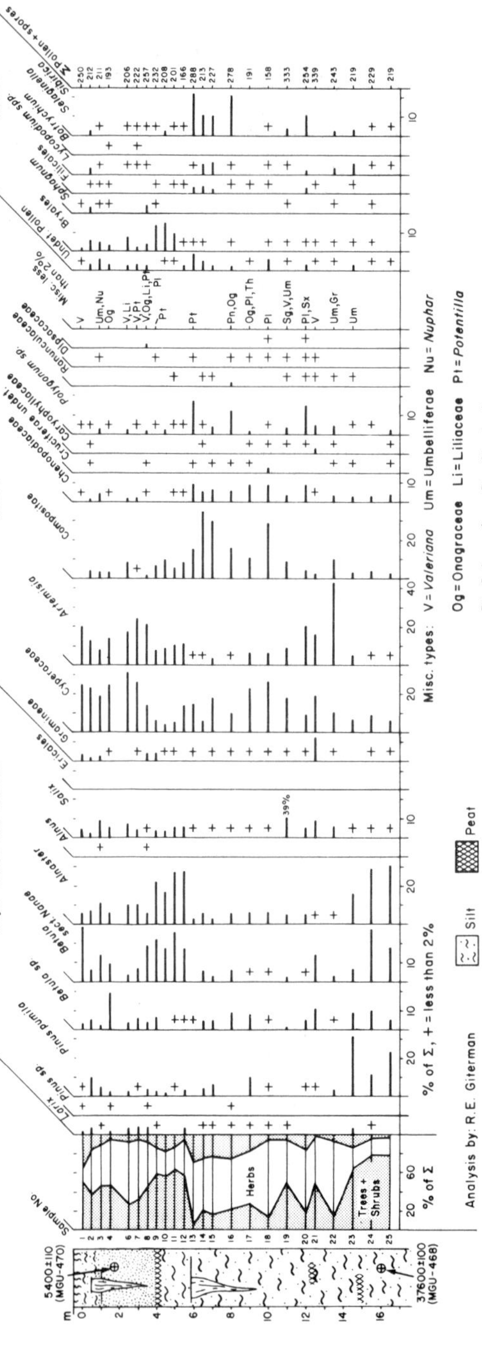

Figure 8. Pollen diagram for Duvannyy Yar *alas* 3. Note horizontal-scale change between summary diagram and the detailed diagram. Percentages are based on a sum of all pollen and spores. Pollen analyses by R. E. Giterman

Pliocene age.[8] It is dated by its content of small-mammal fossils. In 1974 the best estimate was that it was at least 700,000 years old and possibly equivalent to the Olyor Suite. Recent studies (Sher, et al., 1979) have shown, however, that the microtine species in the Cape Deceit local fauna are intermediate in age between those of the Olyor and Kutuyakh Suites, which means that the Cape Deceit Formation is considerably older than 700,000 years, possibly as old as 1.8–2.5 million years, or Pliocene in age (Repenning, 1980). Repenning (1980) points out that the entire Cape Deceit local fauna is a "... Siberian fauna found in Beringia on the wrong side of Bering Strait." This implies that the Cape Deceit Formation might have been deposited at a time when forest barriers blocked Asiatic species from moving any further into North America than western Alaska (Repenning, 1980). Though it is clear that part of the Cape Deceit Formation was deposited when climate was too cold for trees, we know as well that deposition was twice punctuated by peat formation and the growth of trees at the site. Even though the mammalian fossils are from sediments deposited in a tundra environment, Cape Deceit may have been only slightly beyond the limit of trees, as it is at present, and if so (contrary to Matthews, 1974b), the Cape Deceit Formation might represent a single interglacial or interstadial. This could explain some of the features of the pollen spectra. For example: sedge dominates grass, *Betula* percentages are relatively high, and *Alnus* (macrofossil evidence) was present locally. Although the insects from the Cape Deceit Formation represent treeless biotopes, further details are obscured because the assemblages are probably biased by local taxa like *Tachinus* (Table 1).

The middle unit shown in Figure 9 is actually the lower part of the Deering Formation—the upper unit at the site. The peat at its base was deposited when spruce forest grew as far west as the site. This peat (peat 4 in Matthews, 1974b) was thought to predate the last interglacial but we now suspect that it is younger, certainly no older than the last interglacial.[9] Peat 5, deposited after peat 4 (Fig. 9), is at least as old as mid-Wisconsin (^{14}C age: >39,900). Hopkins (this volume) considers it to represent in part isotope stage 4 of the marine record.

According to both pollen and macrofossil evidence, deposition of the lower part of the Deering Formation began when treeline was west of its current position. Sea level was intermittently higher than at present and the land bridge was undoubtedly flooded. Next, tundra returned to the site but differed from present tundra in its dearth of alders and Ericales and abundance of grass and sedge (about equal amounts). *Artemisia* pollen, though consistently present, accounts for less than 10% of pollen and spores in most samples from unit 1. Insects from the same level suggest a scantily vegetated, dry local environment differing in some major but undefined way from that existing during deposition of the much older Cape Deceit Formation. At the top of unit 1 there is a distinctive horizon of dessication cracks that probably formed in an arid, scantily vegetated environment.

Near the middle of the Deering Formation, the *Alnus* curve climbs precipitously to values exceeding 40%. Deposition slowed and allowed the formation of an autochthonous peat (peat 5) that was subsequently disturbed by solifluction. The visible

[8]Unless otherwise noted, ages of Cape Deceit units are cited according to the North American concept of the Pleistocene and Pliocene. See footnote 2 for details.

[9]The primary basis for the original conclusion was that the peat lay stratigraphically below sediments containing fossils of the lemming (*Dicrostonyx henseli*). *D. henseli* had been thought of as the evolutionary antecedent to the extant collared lemming (*D. torquatus*). Because *D. torquatus* fossils had been found in sediments of early Wisconsin Age in Alaska (Guthrie, 1968b), Matthews assumed that any unit containing *D. henseli* must predate the Wisconsin. But it is now well known that collared lemmings with the "primitive" *henseli*-type dentition occur among living populations (e.g., Banks Island) (J. Guilday, personal communication to R. E. Morlan 1979). The frequency of occurrence of the *henseli* morph in a fossil assemblage still has historical and evolutionary significance, but at least 100 specimens are required for an assessment. The sample from Unit I, Deering Formation, contained only a few *Dicrostonyx* fossils. Furthermore, we also now know: (1) that *henseli*-type fossils occur in sediments of early mid-Wisconsin age (specifically those correlative with part of the Boutellier interval—Hopkins, this volume); and (2) that the presumed early Wisconsin records of *D. torquatus* listed by Guthrie (1968b) are probably no older than mid-Wisconsin. Thus, there is no longer any compelling evidence for assigning the "*henseli* zone" at the base of the Deering Formation an age older than early Wisconsin.

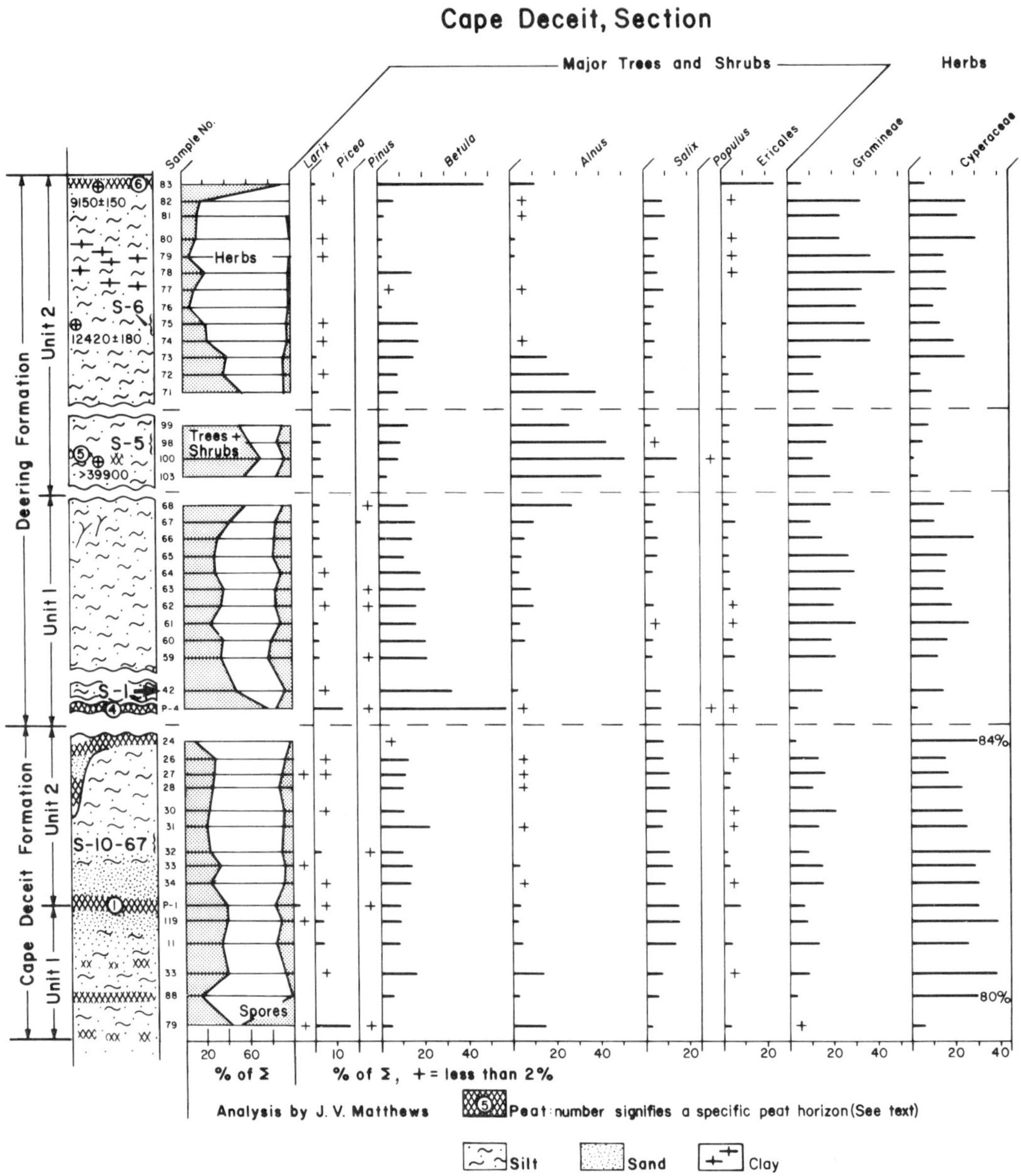

Figure 9. Composite pollen diagram of the Cape Deceit locality in western Alaska. Note horizontal-scale change between summary diagram and the detailed diagram. Percentages are based on a sum of all pollen and spores. Pollen analyses by J. V. Matthews, Jr. Note that number associated with peat horizons is the designation used for the peat in the text and in Matthews (1974b). See Matthews (1974b) for actual position of the samples plotted here. Height of plotted composite section is approximately 9–10 m

Cape Deceit, Section (cont.)

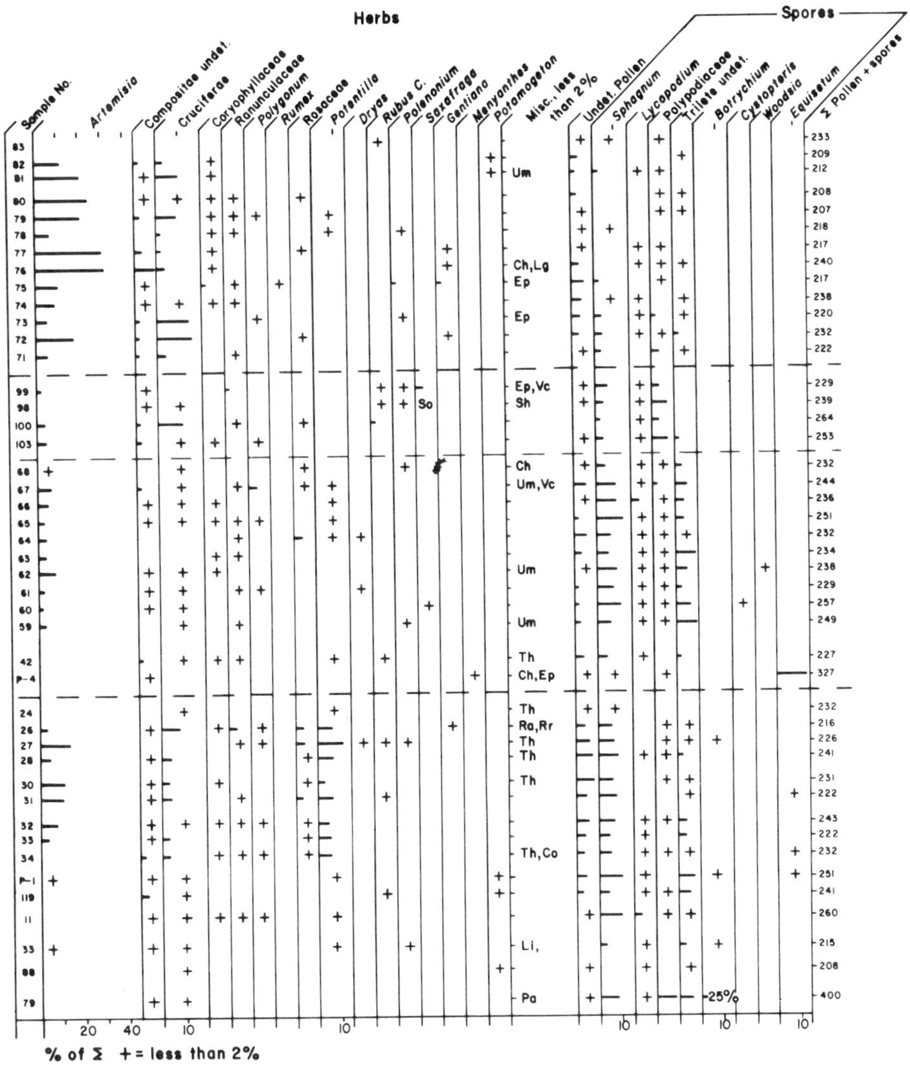

% of Σ + = less than 2%

Um = Umbelliferae Ch = Cheopodiacea - Amaranthaceae Lg = Leguminosae

Ep = *Epilobium* Vc = *Valeriana capitata* Sh = *Shepherdia canadensis*

Th = *Thalictrum* Ra = *Rosa acicularis* Rr = *Rubus arcticus* Co = *Cornus*

Li = Liliaceae Pa = *Parnassia* So = *Saxifraga oppisitifolia*

evidence of cryoturbation plus the pollen, plant macrofossils, and insects from samples S-5 and S-4 (the peat) imply mesic tundra and relatively moist regional climate (Matthews, 1974b) (Table 1), similar to that of the present.

The upper part of the Deering Formation undoubtedly contains one or more unrecognized unconformities, especially if it encompasses, as proposed here, more than 39,000 years. Nevertheless, a few of the samples display lower tree and shrub values than in any other part of the section. Herbs dominate, especially *Artemisia*, but since most of the pollen samples from the upper part of the Deering Formation postdate 12,400 years, they obviously cannot represent the peak of late Wisconsin climatic severity.

Betula percentages increase abruptly in the uppermost peat, dated at approximately 9150 years, but a few samaras of dwarf birch from the underlying clayey silt (sample S-6) show that they grew near the site as early as 12,400 years BP. Other plant macrofossils from the same 12,400-year-old pond sediments are about equally divided between members of the *Carex-Potamogeton* group (shoreline or aquatic sites) and the *Potentilla-Chenopodium* group (upland and dry sites) (Matthews, this volume). The insect taxa, like those from Siberian samples, are virtually all typical of dry sites (e.g. *Morychus,*[10] *Lepidophorus,* and others) (see Sample S-6 in Table 1).

A comparison of the S-6 insect assemblage with one representing a Holocene tundra pond (S-6 and CRH-44 in Table 1) reveals the marked difference between late Pleistocene shrub-tundra fauna and that of present tundra. The deviations are especially marked for the Hygrophilous and *Morychus-Lepidophorus* groups as defined in Matthews (this volume) and Hughes et al. (1981).

The Cape Deceit sequence is the only one now known from East Beringia that contains fossiliferous terrestrial sediments of both early and late Pleistocene age. If our revised correlation of the Cape Deceit Formation is correct it is clear that the appearance of coastal tundra in East Beringia predates the Pleistocene as defined in both North American and Soviet systems. Cape Deceit also shows that both its early and late Pleistocene versions of tundra differed from most existing tundra biotopes. On the other hand, because the Cape Deceit Formation may represent an interglacial, there now appears less justification for suggesting that the differences between Cape Deceit and Deering Formation faunas and floras represent progressive development of steppe-like tundra during successive cold epochs of the Pleistocene (cf. Matthews, 1974b, in press).

Better evidence of the state of late Pleistocene regional vegetation in the Cape Deceit-Seward Peninsula area is provided by the several cores from Imuruk Lake (Colinvaux, 1964; Shackleton, 1979) (Fig. 1). Like the Cape Deceit data they show that late Pleistocene tundra was more grassy and xeric than at present; Shackleton's core also suggests that the maximum period of aridity occurred during the early Wisconsin, probably at the same time that the unit with dessication cracks was deposited at Cape Deceit.

[10] The fossils identified as *Morychus* at Alaska-Yukon sites are very close to those identified as *Chrysobyrrhulus* in Siberia. They may represent different species, but both are at least congeneric (P. Johnson, personal communication 1979).

DISCUSSION

Comparison of Pollen Spectra from East and West Beringia

A number of minor elements shown in the Kolyma basin pollen diagrams have special ecological implications. They may be ranked according to the following major groups:

(1) Taxa of the arctic tundra: *Rumex arcticus* (moist tundras, sandy seashores), *Cochlearia arctica* (seashores in Arctic tundras), *Armeria arctica* (dry lichen and rock debris tundra—also in steppe communities on sandy slopes—Yurtsev, 1972b).

(2) Taxa commonly found in arctic tundra and alpine mountain belts: *Oxyria digyna, Draba* spp., *Saxifraga* sp., *Lycopodium alpinum, Lycopodium appressum,* and *Lycopodium pungens.*

(3) Boreal taxa: *Selaginella sibirica* (among unshaded rocks, hillside waste and debris slopes in forest and tundra), *Botrychium lunaria* (humid larch forests), *Sanguisorba* sp. (meadows, borders of forests, river banks, margins of peat bogs), *Valeriana capitata* (forests, tundras, bogs, steppe meadows), and *Polemonium boreale* (moss and lichen tundras, moist debris slopes in forests).

(4) Arcto-steppe species: *Cerastium maximum.*

(5) Steppe species: *Eurotia ceratoides.*

(6) Saline habitat taxa: *Suaeda corniculata.*

Schweger (this volume) presents a list of the minor pollen types and their ecological implications from various sites in East Beringia, and Cwynar (1980) has emphasized the significance of some of the identifiable minor elements at Hanging Lake in northern Yukon Territory.

There are some important distinctions between East and West Beringian pollen spectra. Saline and steppe species have yet to be found in East Beringian pollen diagrams (Ritchie and Cwynar, this volume), although such taxa are represented in macrofossil assemblages (Matthews, this volume).

Potentilla, an important element in spectra from the Cape Deceit Formation,[11] is at best a minor element in Kolyma-lowland samples.

Another contrast between East and West Beringian pollen samples involves *Selaginella sibirica*. In the Kolyma lowland, peaks of *S. sibirica* coincide with times of maximum cold and drought. The species is especially characteristic of the ice complex at Duvannyy Yar, and is also well represented in some surface samples from relict-steppe sites (Yurtsev, this volume). *S. sibirica* grows today in Eastern Beringia and was undoubtedly also present there during the Pleistocene, but it is never represented at more than trace levels in late Pleistocene pollen spectra. Enough different types of East Beringian sites have been studied to show that this deficiency is not simply the result of collecting or counting biases. Furthermore, the Siberian record suggests that *S. sibirica* was abundant at lowland sites, meaning that if it was ever common in Alaska-Yukon its spores should appear in alluvial pollen spectra such as those discussed by Schweger (this volume). The fact that they do not appear shows that there were real differences between full-glacial vegetation of East and West Beringia. The prevalence of *S. sibirica* in eastern Siberian sections may signal a greater degree of regional aridity than ever existed in Alaska-Yukon, but there may be other explanations as well. One possibility is that expansion of *S. sibirica* in East Beringia during late Wisconsin time was checked by competition from one or more species not native to Siberia. Whatever the explanation, the rarity of *S. sibirica* in Alaska-Yukon pollen sequences remains one of the most perplexing differences between East and West Beringian records. *Pinus pumila*, stone pine, is an example of another departure. Despite its ability to withstand severely cold and dry climate, the species apparently did not cross into East Beringia via the land bridge because pine pollen never reaches more than trace levels in Alaska-Yukon full-glacial pollen spectra.

The Kolyma samples also share many characteristics with East Beringian pollen spectra. Late Pleistocene sequences from both areas nearly lack tree and shrub pollen. Both areas also show high percentages of *Artemisia*. The species of *Artemisia* cannot be ascertained at the present time; therefore, we should not assume that the high percentages imply equivalent environments in both regions. For example, Ritchie and Cwynar (this volume) argue that high *Artemisia* values in the Yukon may be due to the fell-field species *A. arctica* and consequently represent nothing more than an expansion of fell-field environments. On the other hand, Yurtsev (this volume) believes that some of the eastern Siberian *Artemisia* pollen might be from the steppe species *A. frigida*. Regardless of which species the pollen represent, the prevalence of *Artemisia* in many late Wisconsin-Sartan spectra implies the occurrence of substrates generally drier than in most existing tundra areas as well as deeper seasonal thawing (Matthews, this volume). It is noteworthy that some of the eastern Siberian relict-steppe sites that contain abundant *Selaginella sibirica* and *Artemisia frigida* yield surface spectra closely comparable to eastern Siberian tundra-steppe fossil assemblages (Yurtsev, this volume). By the same token, Cwynar (1980) presents an *Artemisia*-dominated surface spectrum from Yukon Territory that comes from a plant community that is more similar to polar-desert fell field than to the type of steppe communities postulated above.

Insect Assemblages

Fossil insects have become an important source of paleoecological information in the Kolyma basin, due largely to the pioneering research of S. Kiselyov (i.e., 1976, 1979). Sufficient data now exist to allow some tentative east-west comparisons. Table 1 is a first attempt to do this. In it the various faunas are shown in terms of the percentage of individuals belonging to designated "ecological groups." The groups used for eastern Siberian assemblages are not the same as those used in Alaska-Yukon (compare Sher et al., 1979, with Matthews, this volume, and Hughes, et al., 1981). One objective of the table is a reconciliation of the two rankings (see columns at top and bottom). Table 1 reveals certain distinctions between the former fauna of East and West Beringia. For example, fossils of the rove beetle *Tachinus* are relatively abundant in some East Beringian samples but comprise only a small fraction of the "others" category in the samples from the Kolyma basin.

[11]The species is unknown but the abundant *Potentilla* macrofossils represent neither *P. plaustris* (=*Comarum palustris)* nor any of the shrubby species. *Potentilla* achenes are also abundant in other Cape Deceit samples that strangely yield much less *Potentilla* pollen.

Another difference concerns *Chrysobyrrhulus:* if it and *Morychus* are congeneric as indicated above, then Siberian samples show a much earlier onset of high *Chrysobyrrhulus-Morychus* percentages than Alaska-Yukon samples. This distinction may reflect real differences in the sequence of development of xeric environments in the two regions. However, we must also be aware that most samples from East Beringia come from nonalluvial sediments, or if alluvial, from rivers and floodplains much smaller than those of the Kolyma region. Consequently, the differences may reflect contrasting local, rather than regional, environments.

The weevil *Lepidophorus* poses a particularly thorny dilemma in comparing east and west insect assemblages. Because of its xeric habitat requirements, *Lepidophorus* seems the logical taxon to have achieved a Holarctic distribution during the late Pleistocene. It was present at Cape Deceit 12,400 years ago and its fossils occur in practically all East Beringian faunas; yet it is absent from all of the Kolyma assemblages and is but a rare element of the Chukotkan fauna today (S. Kiselyov, personal communication 1978).

The best generalization to be drawn from Table 1 is that it supports the indication of plant macrofossils and pollen that full-glacial environments on both sides of the land bridge were both dry and cold. In the Kolyma basin the onset of this trend occurred early in the Pleistocene and culminates in samples only 15–18,000 years old. In East Beringia (Alaska-Yukon) dry regional climate seems more of a late Pleistocene phenomenon.

Sample S-5 from Cape Deceit comes closest to representing conditions of existing mesic tundra. In contrast sample S-1, representing the early Wisconsin Deering interval (Table 1, Fig. 9), portrays an environment that may have been even drier and more scantily vegetated than during late Wisconsin time. Shackleton's (1979) evidence from Imuruk core V shows a similar pattern.

Table 1 also includes a late Tertiary sample that is thought to represent a very early phase in development of northern-hemisphere lowland tundra (Matthews, 1976b). Although it differs from Holocene shrub-tundra and open taiga assemblages (loc. 44, R.B., Table 1) in its content of extinct species, the relative abundance of some of the various taxon groups is similar. It contrasts sharply with most of the late Pleistocene insect assemblages shown in Table 1.

Tundra-Steppe in Northeastern Siberia

Analyses of pollen, insects, and plant macrofossils from various sections in northeastern Siberia show that during certain periods the landscape was characterized in part by grassy vegetation having no present-day regional analogs. Apparently this steppe-like vegetation was most prevalent during the coldest phases of the Pleistocene—a time when climate was also drastically continental. Winters would have been cold, but with little snow, and summers warm, dry, and sunny. These tundra-steppe conditions seem to have been best developed during the latest cold epoch of the Pleistocene and are probably responsible for the many steppe species that grow in dry edaphic sites of northeastern Siberia today (Karavaev and Skryabin, 1971; Yurtsev, 1972b).

Not only was tundra-steppe vegetation different physiognomically from modern tundra but it deviated floristically as well. The flora included a mixture of mesic species (*Rumex arcticus, Cochlearia arctica, Oxyria digyna, Valeriana capitata, Potentilla* spp., *Polygonum* spp., *Sanguisorba* sp., and *Thalictrum*) and xeric or saline types (*Eurotia ceratoides, Armeria arctica, Selaginella sibirica,* and *Suaeda corniculata).* Other taxa that played a considerable role in the tundra-steppe communities were various species of the families Caryophyllaceae, Cruciferae, Compositae and Ranunculaceae. Together with the grasses and sedges they formed steppe-like communities that were part of a larger mosaic of tundra-steppe biotopes. The most steppe-like areas of this mosaic probably provided seasonal range for the diverse grazing ungulate community of the late Pleistocene, and evidently supported the steppe mammals and insects that had moved north from Mongolia. We believe the best name for this mosaic is "tundra-steppe," a term that conveys the range of vegetational diversity of the environment yet is not linked in any direct way to glacial phenomena.

The best development of tundra-steppe seems to have coincided with fully glacial regional climates and drastically lower world sea level. Consequently, the two primary reasons for formation of tundra-steppe in northeastern Asia were the enhancement of the strong high-pressure region over the Arctic Basin caused by exposure of the continental shelf during periods of lower sea level and the coexistence of a perennially frozen Arctic Ocean

(Velichko, 1975). Although permafrost was widely distributed during the tenure of tundra-steppe environments, seasonal thawing was probably deeper and the terrain generally better drained than most areas of existing tundra.

Tundra-Steppe in East Beringia

Pollen and macrofossil evidence from East Beringia also indicate a tundra-steppe mosaic. Its best development occurred during the late and/or early Wisconsin. Several incongruities of the Siberian and Alaska-Yukon record caution us against assuming too quickly that uniform tundra-steppe ever stretched across all of Beringia. For example, such mammals as the woolly rhinoceros never entered the new world and neither *Camelops* (camel) nor *Taxidea* (badger) crossed into Siberia, yet all inhabited Beringia during the late Pleistocene. Certain plants and steppe-adapted insects also failed to cross the land bridge (Berman, et al., 1979). And, finally, some plant species, such as *Selaginella sibirica,* occurred across all of Beringia but responded differently to late Pleistocene climates on either side of the land bridge. These differences suggest that tundra-steppe in Siberia was not identical to tundra-steppe in Alaska-Yukon.

History of Tundra-Steppe

A still meager fossil record shows that the inception of lowland tundra, structurally similar to extant zone 4 tundra, occurred during the late Tertiary. Conditions foreshadowing those of the late Pleistocene tundra-steppe were developing in the Kolyma basin well before 700,000 years ago. This trend probably started during the early Quaternary with local areas of dry tundra, such as that found at existing edaphic sites which expanded by Sartan time to form large tracts of steppe-like vegetation. We cannot be so certain that the same sequence occurred in Alaska-Yukon, because the peak in steppe-like communities there seems to have been in early, rather than the late, Wisconsin. It is nevertheless clear that the late Pleistocene environments of both East and West Beringia were both very cold and dry.

The driving force behind the development of tundra-steppe on both sides of Beringia was probably climatic continentality brought on by exposure of the Bering land bridge and the Arctic shelf (Barry, this volume). The culmination of the trend to ever drier, colder climates during the Pleistocene is what A. A. Velichko (1973) has aptly termed the "climatic minimum of the Pleistocene." However "minimum" implies neither low diversity nor simple ecosystem structure, because as several other authors in this volume show, the tundra-steppe environment of the late Pleistocene was if anything richer in species and more productive than existing tundra.

ACKNOWLEDGMENTS

We are indebted to the Wenner-Gren Foundation for their invitation to attend the workshop for which the initial version of this paper was prepared. Thanks are extended to Olga Kukal, Linda Barton, and Carole Matthews for their editorial help and to Ed Belec for drafting the pollen diagrams.

4

VEGETATIONAL HISTORY OF WESTERN ALASKA DURING THE WISCONSIN GLACIAL INTERVAL AND THE HOLOCENE

Thomas A. Ager

Lacustrine sediment cores from the Yukon Delta and St. Michael Island in Norton Sound, western Alaska, provide pollen records spanning about 24,000 and 16,000 years, respectively. Three pollen zones of regional significance have been recognized in these cores: a basal Herb Zone, an intermediate *Betula* Zone, and an uppermost *Alnus* Zone. These pollen assemblage zones are similar in both composition and sequence to those described from northern Alaska by D.A. Livingstone, but the vegetation changed much earlier in western Alaska. The Herb Zone pollen assemblages represent an herbaceous tundra vegetation in which Gramineae, Cyperaceae, *Salix, Artemisia,* and a variety of forbs were important elements. This herb-dominated vegetation covered most of the Beringian landscape during the Wisconsin glacial interval until about 14,000 years ago, when shrubs began to increase in some areas. The subsequent *Betula* Zone pollen assemblages record the expansion of a mesic shrub tundra vegetation including dwarf birch in the Yukon Delta region between about 14,000 and 6000 to 7000 years ago. This vegetation of *Betula nana*, ericales, *Salix*, and numerous herbs suggests a shift to a less continental climate characterized by increased moisture and probably warmer summer temperatures. The *Alnus* Zone records the rapid dispersal and establishment of *Alnus* shrubs within the mesic shrub tundra vegetation of the Yukon Delta beginning sometime between 7000 and 6000 years ago. *Picea* invaded the eastern fringes of this region about 5500 years ago, establishing riparian forests in the interior along the Yukon and Kuskokwim Rivers and some of their tributaries.

Pollen data from coastal exposures of perennially frozen loess and peat in the Bristol Bay region of southwestern Alaska provide an outline of vegetation history that roughly parallels that from the Yukon Delta region 400 km to the north. The Bristol Bay data, however, extend the record farther back in time beyond the limit of radiocarbon dating. Pollen spectra from the lowermost organic-matter-bearing units of the exposures are interpreted to represent a slightly mesic herb-shrub tundra that probably covered southwestern Alaska during the middle Wisconsin interstadial. During the late Wisconsin glacial interval, the climate became increasingly arid, and the vegetation of the Bristol Bay region shifted to a predominantly herbaceous tundra that included a few shrubs.

Key words: vegetation, climate, history, Quaternary, palynology, Yukon Delta, Norton Sound, Bristol Bay, Alaska

ISBN 0-12-355860-3

INTRODUCTION

This paper summarizes unpublished and previously published palynological research on sediments of Quaternary age in western Alaska. The geographic focus of the paper is primarily upon the Yukon Delta-Norton Sound area (Fig. 1, Fig. 2) of western Alaska, where I have been conducting research since 1976 in cooperation with William Dupré, University of Houston.[1] Additional unpublished data have been provided by D.M. Hopkins and R.E. Giterman[2] for pollen-bearing sediments from the coast of Bristol Bay (Fig. 1, loc. 7). This paper then summarizes previously reported palynological data from western Alaska in the context of the new evidence presented in this paper. From that summary I have excluded those areas of western Alaska north of Seward Peninsula.

The region of the Yukon-Kuskokwim coastal lowland (Fig. 1, Fig. 2) has received little attention from Quaternary researchers until recently (e.g., Dupré and Thompson, 1979; Dupré, 1977a, 1977b). Most of the previous geologic work done in the region consists of reconnaissance geologic mapping (e.g., Beikman, 1974; Coonrad, 1957). This area lies far beyond the maximum limits of Quaternary glaciations (Fig. 1), and was linked repeatedly with the Bering land bridge during low stands of sea level (Hopkins, 1967c, 1972, 1973, 1979a). The region's extensive deposits of Quaternary sediments of fluvial, deltaic, lacustrine, aeolian, and marine origin preserve a wealth of paleoenvironmental information.

METHODS

Cores discussed in this paper were obtained by means of a modified Livingstone Piston Sampler from an anchored floating platform (Ager, 1975). Cores were subsampled in the laboratory, usually at 10 cm intervals; the sediment subsamples were processed for pollen analysis by methods modified from those described by Faegri and Iversen (1975). Pollen identifications were made by comparison with reference material from Alaska and the Yukon Territory in the collections of the U.S. Geological Survey Palynology Laboratory at Reston, Virginia. At least 300 pollen grains were counted for most samples.

REGIONAL SETTING

Fluvial and deltaic sediments of the lower Yukon and Kuskokwim Rivers intermingle to form a vast coastal lowland of more than 75,000 square kilometers in western Alaska (Fig. 1, Fig. 2). The area shown in Figure 2 includes several uplands that are composed of Upper Cenozoic basalts or Cretaceous sedimentary and volcanic rocks (Beikman, 1974).

The lowlands form a predominantly wet landscape pitted by thousands of lakes and ponds and laced by a complex network of meandering rivers and sloughs. Most of the lakes are of thermokarst origin, but some are oxbow lakes and elongate abandoned sloughs. Relatively few lakes in the region are likely sites from which to obtain continuous, long-term, reliable pollen and sediment records. The areas having the greatest potential for yielding long pollen records are lakes in the basaltic uplands of the Ingakslugwat Hills (Fig. 1, loc. 1), St. Michael Island (Fig. 1, loc. 2) in Norton Sound, and Nunivak Island (Fig. 2). No sites on Nunivak Island have yet been investigated, but several crater lakes appear to be suitable. One lake in the Ingakslugwat Hills and two lakes and a small pond on St. Michael Island were cored in 1976,

---

[1]Fieldwork for the research presented in this paper from the Yukon Delta and St. Michael Island in Norton Sound was conducted in 1976 and 1978 as part of a regional investigation of modern coastal processes and Quaternary geological history of the Yukon Delta led by William R. Dupré, Principal Investigator, University of Houston. That project was funded in part by the U.S. Bureau of Land Management through an interagency agreement with the U.S. National Oceanic and Atmospheric Administration (NOAA); the project was managed by NOAA's Outer Continental Shelf Environmental Assessment Program (OCSEAP) Office. Additional support was provided by the U.S. Geological Survey. Radiocarbon dates were provided by Meyer Rubin and Elliott Spiker of the U.S. Geological Survey, Reston, Va.

[2]Unpublished field notes, and preliminary notes for a manuscript, along with unpublished pollen data from the Bristol Bay coastal sections, were generously provided by D.M. Hopkins, U.S. Geological Survey, Menlo Park, Calif., and R.E. Giterman, Geological Institute, Academy of Sciences, Moscow, U.S.S.R.

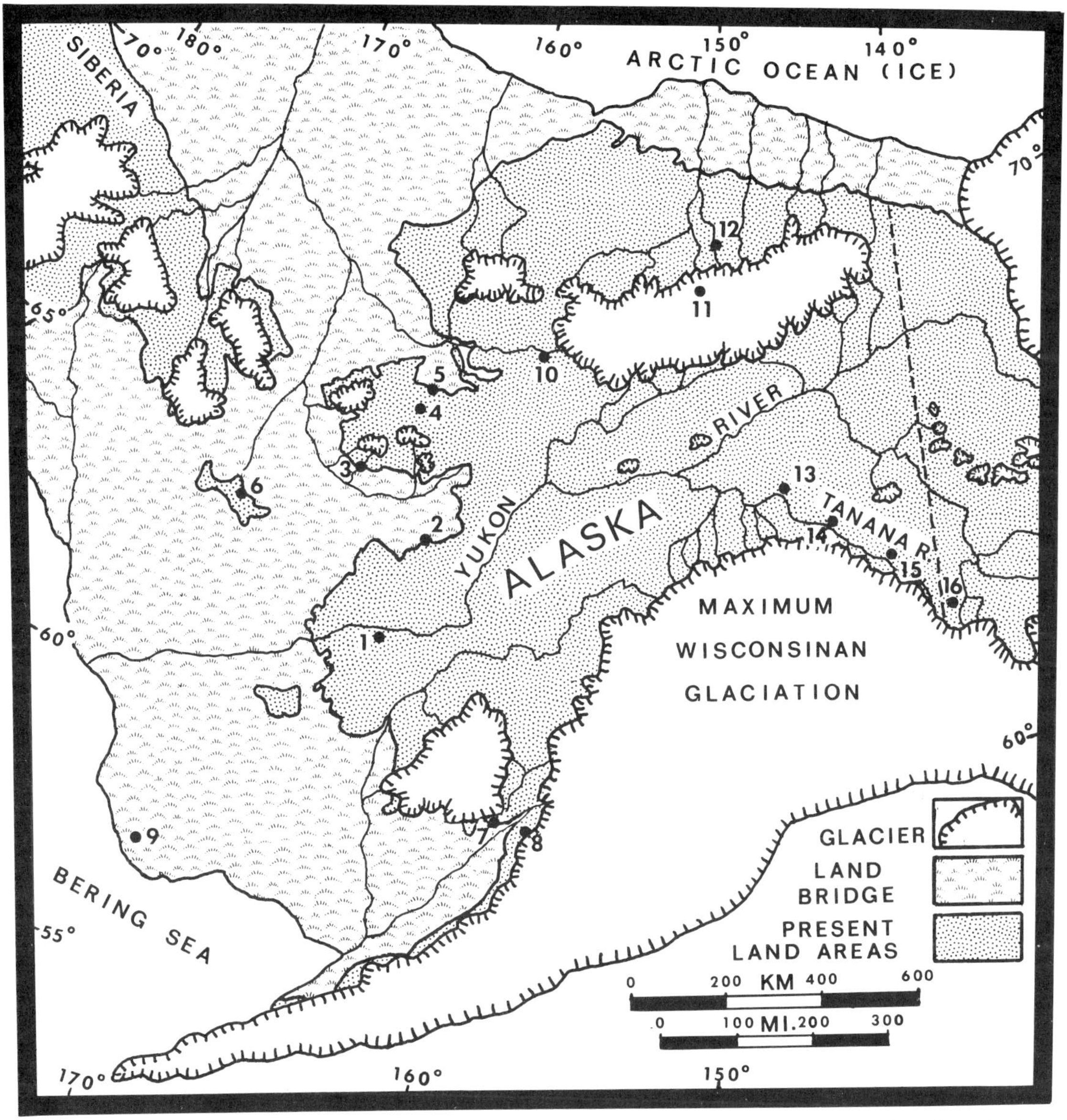

Figure 1. Map of Alaska and the Bering land bridge during the maximum extent of glaciation during Wisconsin time (modified from Hopkins, 1979; Péwé, 1975; and Coulter, et al., 1965). Numbers on the map correspond to sites from which palynological data are available and which are discussed or referred to in this paper. Aleutian sites discussed in the text are outside the map area. 1) Ingakslugwat Hills, Yukon Delta (this paper); 2) St. Michael Island, Norton Sound (this paper); 3) Nome, Seward Peninsula (Hopkins, et al., 1960); 4) Imuruk Lake, Seward Peninsula (Colinvaux, 1964, 1967a; Colbaugh, 1968); 5) Cape Deceit, Seward Peninsula (Matthews, 1974b); 6) Flora Lake, St. Lawrence Island (Colinvaux, 1967c); 7) Bristol Bay (this paper; D. M. Hopkins, R. E. Giterman, and others, unpublished data); 8) Naknek River, Bristol Bay-Alaska Peninsula (Heuser, 1963); 9) Pribilof Islands, Bering Sea (Colinvaux, 1967a, 1967b; Parrish, 1979); 10) Onion Portage area, northwest Alaska (Schweger, 1975; Schweger and Habgood, 1976; Schweger, this volume); 11) Chandler Lake, northern Alaska (Livingstone, 1955); 12) Umiat, northern Alaska (Livingstone, 1957); 13) Isabella Basin, Fairbanks area, Tanana Valley (Matthews, 1974a); 14) Birch Lake, Tanana Valley (Ager, 1975); 15) Lake George, Tanana Valley (Ager, 1975); 16) Antifreeze Pond, southwest Yukon Territory (Rampton, 1971)

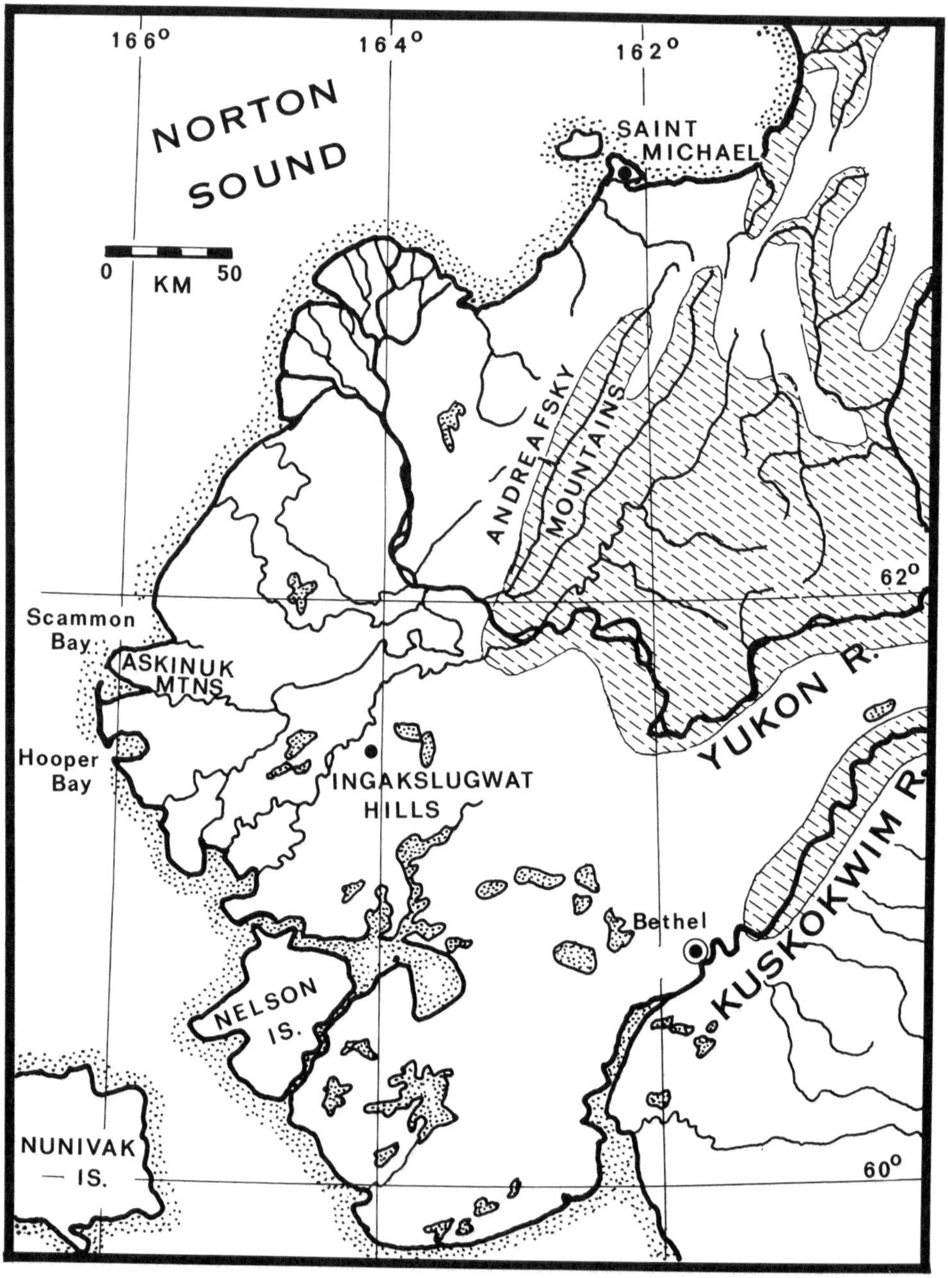

Figure 2. Index map of Yukon-Kuskokwim deltaic lowland in western Alaska. Shaded areas mark the approximate distribution of spruce trees in this region, based upon examination of Landsat imagery and visual reconnaissance during the 1976, 1978, and 1979 field seasons. Alder shrubs have a patch distribution over most of this region, reaching the vicinity of the coast near Nelson Island, Scammon Bay, and St. Michael Island. Poplar (*Populus balsamifera*) extends beyond the limit of spruce, primarily along the lower Yukon River and rivers draining the Andreafsky Mountains

1978, and 1979. These lakes are within volcanic craters of Quaternary age (Hoare and Condon, 1971a, 1971b). The terrain in these basaltic uplands is characterized by volcanic features such as frost-shattered basalt flows, cinder cones, thick pyroclastic deposits, and craters of several types. St. Michael Island has several deep, circular to elongate maars. A radiocarbon-dated sediment core suggests that at least one of these maars has been occupied by a lake for more than 39,000 years (W-4379). Craters and other types of depressions in the basalt fields of the Ingakslugwat Hills enclose many lakes that are probably older than 20,000 years.

CLIMATE

The climate of the Yukon-Kuskokwim coastal lowland is transitional between maritime and continental (Johnson and Hartman, 1971). Mean annual temperature ranges from about -1°C on Nunivak Island to about -3.5°C on St. Michael Island. Mean July temperature is about 9.1°C on Nunivak Island, 10°C on St. Michael Island, and 12°C at Bethel. Mean annual precipitation is about 450 mm for the central Yukon Delta. The region shown in Figure 2 lies within the zone of discontinuous permafrost (Péwé, 1975a). Permafrost features such as thermokarst lakes, palsas, and high-center ice-wedge polygons are common. Many of the ice wedges formed during Holocene time and appear to be actively growing at present (W.R. Dupré, personal communication 1979).

FLORA AND VEGETATION

Botanical exploration of this region has been confined mostly to the coastal areas and along the main channels of the lower Yukon and Kuskokwim rivers. The published floristic data suggest that the regional vascular flora includes about 450 species (Hultén, 1968). Only a few reconnaissance floristic and vegetation descriptions have been published for this region (e.g., Hultén, 1962, 1966; Ager and Ager, 1980).

Nevertheless, the regional vegetation can usefully be divided into five broadly defined types, modeled in part upon those of Viereck and Little (1972). These types are wet tundra, mesic tundra, dry tundra, deciduous scrub-forest, and spruce forest.

Wet tundra vegetation covers tens of thousands of square kilometers of the Yukon-Kuskokwim lowlands. Wet-tundra plant communities are quite variable in composition and complexity. The simplest assemblages of wetland plants are found in coastal lowlands, whereas the freshwater bog communities are more complex. Sedges (Cyperaceae), grasses (Gramineae), and a variety of aquatic and semi-aquatic plants are the most common elements of the lowland wet tundra, including such taxa as *Carex*, *Eriophorum*, *Luzula*, *Potentilla*, *Hippuris*, *Ranunculus*, and prostrate species of *Salix* (e.g., *S. fuscescens*). *Sphagnum* and other mosses are important elements of inland wet tundra vegetation but are less common in coastal settings of the deltaic lowlands.

Mesic tundra covers a considerably smaller area of the region than does wet tundra; it is present primarily on the lower to middle slopes of uplands such as Ingakslugwat Hills, Nelson Island, the Scammon Bay area, and the Andreafsky Mountains (Fig. 2). The upland areas offer a greater variety of habitats than do the wet lowlands; this diversity of habitats is reflected in the relative richness of the flora of the mesic tundra. Protected hollows, stream valleys, and gullies support thickets of willows such as *Salix alaxensis* and *S. pulchra*; thickets of *Alnus crispa* are not as common, but they are present in many localities in such habitats.

On interfluves, low woody shrubs such as *Betula nana*, *Salix* spp., *Ledum palustre* ssp. *decumbens*, *Empetrum nigrum*, *Vaccinium uliginosum*, *V. vitis-idaea*, *Arctostaphylos alpina*, and scattered patches of *Spiraea beauverdiana*, are important elements of the ground cover. *Cornus suecica* and *Rubus chamaemorus* are also common.

Common non woody plants of the mesic tundra in this region include many species of *Carex*; several species of *Eriophorum*; grasses such as *Arctagrostis latifolia*, *Calamagrostis canadensis*, *Arctophila fulva*, and *Hierochloe alpina*; and herbs such as *Valeriana capitata*, *Claytonia* sp., *Iris setosa*, *Arnica lessingii*, *Pedicularis* spp., *Polygonum viviparum*, *Petasites* spp., *Rumex arcticus*, *Angelica lucida*, *Saxifraga* spp., and many others. *Equisetum* spp. and several species of ferns also are found in the mesic tundra of this region. The ferns tend to be restricted to sheltered habitats such as gullies and crevices in rocky outcrops. *Sphagnum* spp. and other mosses and lichens, particularly *Cladonia* spp. are important elements of the ground cover.

Dry tundra vegetation grows primarily on the upper slopes and ridgetops of the more rugged uplands (e.g., Nelson Island, Askinuk Mountains, Andreafsky Mountains). These areas are generally mantled by sparsely vegetated fields of frost-shattered rubble overlying bedrock. Crustose lichens are abundant on boulder fields and bedrock surfaces of these areas. Common vascular plants include taxa such as *Empetrum nigrum*, *Dryas octopetala*, *Minuartia*, *Papaver*, and various Saxifragaceae.

Deciduous scrub-forest is distributed primarily as riparian forest or scrub-forest along the Yukon River. These corridors of trees and large shrubs extend beyond the limits of spruce to within a few kilometers of the Bering Sea coast along the northernmost active distributary of the Yukon River delta; they are also present along rivers in the Andreafsky Mountains. This vegetation type includes *Populus balsamifera; Alnus crispa; Salix glauca, S. alaxensis* and other willows; the low shrubs *Myrica gale* and *Betula nana*; various Gramineae and Cyperaceae; Umbelliferae such as *Heracleum lanatum*; numerous other herbs; and *Equisetum*.

Spruce forest includes riparian forests along low terraces and levees of the Yukon River east of about long. 163° W. (Fig. 2) and open stands of spruce in several of the valleys of the Andreafsky Mountains. No sites have yet beeen investigated palynologically within areas where spruce (*Picea*) trees grow today in this region (Fig. 2), but pollen of *Picea* is dispersed by wind far beyond the spruce tree limit in the delta (T.A. Ager, unpublished data). Tree and shrub taxa found in the spruce forest include *Picea glauca, Populus balsamifera, Alnus Crispa, Salix alaxensis, S. glauca, S. pulchra*, and other willows. *Populus tremuloides* and *Betula papyrifera* grow on well-drained sites only in the eastern half of the area of Figure 2.

INGAKSLUGWAT HILLS POLLEN RECORD

The hilly volcanic terrain of the Ingakslugwat Hills (Fig. 2, Fig. 3) encloses lake-filled craters and depressions, many of which form ideal sediment traps from which to obtain long pollen records. One such crater-lake, Tungak Lake (Fig. 3), was cored from a raft during the 1976 and 1978 summer field seasons.

The surrounding vegetation is predominantly shrubby mesic tundra, but the various habitats provided by the hilly volcanic landscape support pockets of wet tundra in scattered depressions and many patches of dry tundra vegetation on rocky ridgetops and steep slopes of basalt and scoria. Hollows and ravines provide habitat for thickets of *Alnus crispa* and *Salix* (e.g., *S. alaxensis, S. pulchra, S. glauca*). The Ingakslugwat Hills are about 50 km beyond the edge of *Populus balsamifera* and *Picea glauca* forests near the Yukon River (Fig. 2). One small stand of young *Populus balsamifera*, presumably a clone, was found 1 km east of Tungak Lake, however, so small stands are likely to be widely distributed in the central Delta.

One of the four sediment cores (Core 76-3) obtained at Tungak Lake has been analyzed for pollen content, and the preliminary results are presented in a pollen percentage diagram (Fig. 4). Radiocarbon-dated core segments provide the basis for an absolute chronology of vegetation changes in the central Yukon Delta during the late Wisconsin glacial interval and Holocene time. No radiocarbon date is available from the base of the core, but extrapolation from other dates higher in the core suggests that the core may span about 24,000 years. One section in the lower part of the core (502–433 cm) was discarded when it was found to have been seriously contaminated owing

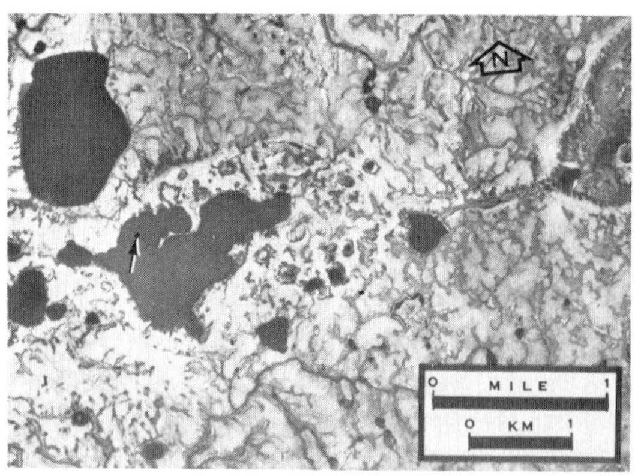

Figure 3. Air photo of part of the Ingakslugwat Hills, Yukon Delta. Arrow points to the coring site in Tungak Lake. The lake occupies part of a large basaltic crater complex of Quaternary age

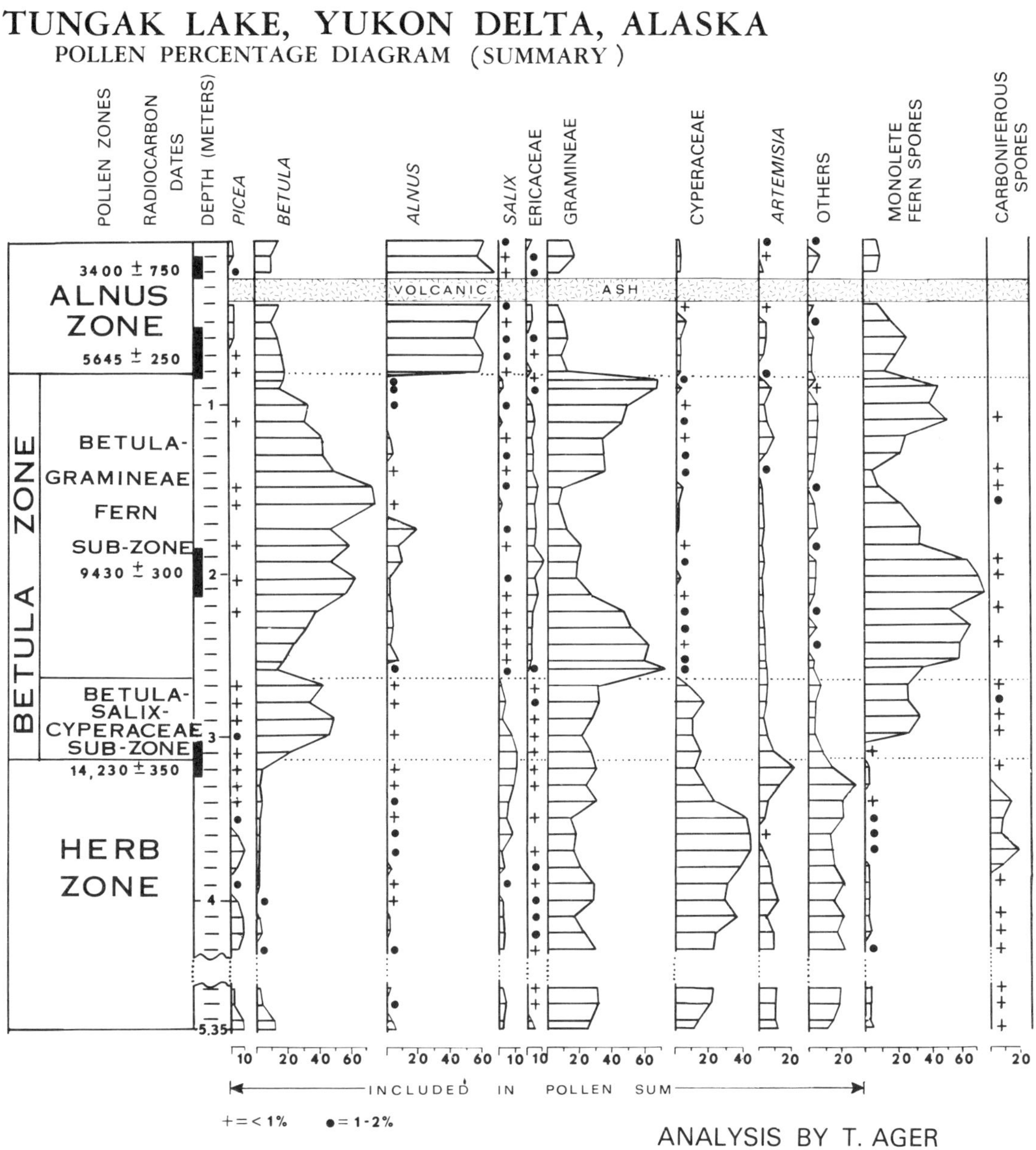

Figure 4. Pollen percentage diagram (Summary) for Core 76-3 from Tungak Lake, Ingakslugwat Hills, Yukon Delta

to piston slippage during the coring operation. Preliminary examination of a second core from the same lake suggests that the missing section probably contains pollen assemblages quite similar to those in adjacent core samples within the Herb Zone.

The pollen percentage diagram for Tungak Lake (Fig. 4) is divided into three pollen assemblage zones of regional significance, plus several subzones that appear to reflect local vegetational changes. These assemblage zones are similar to those described from northern Alaska at Chandler Lake (Fig. 1, loc. 11) and Umiat Fig. 1, loc. 12) by D.A. Livingstone (1955, 1957).

Ingakslugwat Hills Herb Zone

The basal Herb Zone is characterized by only small amounts of tree and shrub pollen and relatively large amounts of herb pollen, particularly that of Cyperaceae, Gramineae, *Salix*, and *Artemisia*. Minor elements of the Herb Zone pollen spectra include such taxa as *Thalictrum* and other Ranunculaceae, *Rumex*, Cruciferae, Caryophyllaceae, *Plantago*, *Potentilla*, Compositae (Tubuliflorae, Liguliflorae), *Polemonium*, and others. Spores of ferns and *Sphagnum* are present in small amounts in the Herb Zone; their occurrence suggests that some mesic to wet habitats persisted in the Ingakslugwat Hills during the late Wisconsin glacial interval. Persistence of some mesic habitats is also suggested by small amounts of pollen of Ericaceae and *Betula* (probably the dwarf birch shrub, *Betula nana*) in Herb Zone samples from this locality. *Picea* pollen is rare or absent in upper Herb Zone samples, but in the lower part of the zone, *Picea* percentages sometimes rise to several percent of the pollen sum. Calculation of pollen influx rates from a second core from Tungak Lake may show that the actual amount of *Picea* pollen entering the lake was much lower than the percentages imply, as was seen in Zone 1 samples from Birch Lake (Fig. 1, loc. 14) in interior Alaska (Ager, 1975). Some or possibly all of the *Picea* pollen in Herb Zone samples may be recycled late Cenozoic pollen that entered the lake as part of the loess rain during late Wisconsin time. This interpretation is supported by the presence of as much as several percent recycled spores of Carboniferous age in Herb Zone samples in this lake (Fig. 4).

I cannot, however, entirely rule out the possibility that some or all the Herb Zone *Picea* pollen was derived from isolated groves of spruce persisting somewhere in southwestern Alaska during late Wisconsin time as suggested by *Picea* influx data from the Pribilof Islands (Colinvaux, 1967b; Fig. 1, loc. 9, this report).

Ingakslugwat Hills *Betula* Zone

The *Betula* Zone of the Tungak Lake Core (Fig. 4) spans core interval 312 to 82 cm. The *Betula* Zone from this locality may be divided into two subzones. These subzones are tentative and may be modified or abandoned upon examination of additional cores from this lake. Comparison of the Tungak Lake core with the Puyuk Lake core from St. Michael Island (this paper) suggests that the Tungak Lake pollen subzones within the *Betula* Zone may be of local rather than regional significance.

The *Betula* Zone is characterized by relatively large amounts of *Betula*, Gramineae, and Cyperaceae pollen and fern spores. Within the *Betula* Zone are several substantial fluctuations in the percentages of these major constituents of the pollen and spore spectra.

The *Betula-Salix*-Cyperaceae Subzone spans core interval 312 to 256 cm. Radiocarbon dates suggest that this core segment was deposited during the time interval from about 14,000 to 12,000 radiocarbon years ago. The pollen assemblages in this subzone differ from the underlying Herb Zone assemblages primarily in that *Betula* percentages are substantially higher (generally 20% to 50% vs. about 2% to 3% in the Herb Zone), and fern spores are much more abundant. The most abundant pollen types include *Betula*, *Salix*, Gramineae, Cyperaceae, and *Artemisia*; monolete fern spores are also abundant. Minor constituents include scattered grains of Compositae (Tubuliflorae), *Rumex*, Cruciferae, *Epilobium*, and *Polemonium*.

The *Betula*-Gramineae-Fern Subzone spans core interval 265 to 82 cm. Radiocarbon-dated core segments suggest that this part of the core was deposited between about 12,000 and 6000 years ago. This pollen subzone is characterized by relatively high percentages of pollen of *Betula* and Gramineae, and monolete fern spores, and small but fairly consistent amounts of Ericaeae, Cyperaceae, and *Artemisia* pollen. In some samples *Alnus* percentages rise to the 5% to 15% range. Ericaceae percentages are generally somewhat higher than those in the underlying *Betula-Salix*-Cyperaceae Subzone, and percentages of *Salix* and Cyperaceae pollen are conspicously lower. Ericaceae tend to be greatly underrepresented in modern and fossil pollen records, particularly in lacustrine sediments. Therefore, the increase in Ericaceae pollen percentages probably reflects an actual increase in the abundance of ericaceous plants in the basin. The ericales most likely to have contributed substantially to this pollen increase include *Ledum palustre* ssp. *decumbens*, *Vaccinium vitis-idaea*, *V. uliginosum*, and *Empetrum nigrum*. This rise in Ericaceae percentages implies an increase in mesic habitats within the drainage basin and perhaps throughout much of the region during this time interval. Substantial fluctuations of percentages of several constituents occur within this subzone and may be a basis for further subdivisions when additional cores are examined. Minor constituents of the *Betula*-Gramineae-Fern Subzone include Umbelliferae, Ranunculaceae, *Epilobium*, and Compositae (Tubuliflorae).

Ingakslugwat Hills *Alnus* Zone

The *Alnus* Zone from Tungak Lake (Fig. 4) spans core interval 82 to 0 cm, which was deposited during approximately the past 6000 years. *Alnus* Zone pollen spectra are characterized by a high relative abundance (50% to 60%) of *Alnus* pollen, high percentages of fern spores, and lesser amounts (generally more than 10% each) of pollen of *Betula* and Gramineae. The apparent decrease in *Betula* and Gramineae may be more apparent than real, because the substantial increase in *Alnus* pollen results in the decline of relative frequencies of other taxa. These lower percentages, therefore, do not necessarily imply a significant change in the importance of *Betula* and Gramineae in the local vegetation.

The *Alnus* rise apparently was very abrupt, but less so than the pollen diagram implies. The boundary between the *Alnus* Zone and the underlying *Betula* Zone coincides with the boundary between two core sections. About 10 cm of sediment may be missing between these sections owing to coring problems associated with penetration of the thick volcanic ash layer. *Alnus* populations appear to have expanded very rapidly in many areas of Alaska during the time interval of about 6000–5500 years ago (e.g., Hopkins, Smith, and Matthews, 1981; Schweger, this volume; T. A. Ager, unpublished data).

Pollen of Cyperaceae, *Artemisia, Salix,* and Ericaceae are present in amounts of a few percent each. Several percent of *Picea* pollen occur in most samples in the upper 50 cm of the core. This increase reflects the westward spread of spruce forests, probably initially as riparian forests, along the Yukon and Kuskokwim Rivers and their tributaries during middle Holocene time. *Picea* probably reached the eastern part of this region (Fig. 2) about 5500 years BP; it now grows within about 60 km of the Ingakslugwat Hills (Fig. 2).

Minor elements of the *Alnus* Zone pollen spectra from Tungak Lake include *Myrica, Potentilla,* and *Epilobium*, and scattered grains of Cruciferae, *Polemonium*, Umbelliferae, and other herbaceous plants.

ST. MICHAEL ISLAND POLLEN RECORD

St. Michael Island is in Norton Sound, northeast of the modern distributary system of the Yukon River delta (Fig. 1, loc. 1; Fig. 2). The island is underlain by basalt flows and pyroclastic deposits of probable Quaternary age (Hoare and Condon, 1971b). The terrain consists of subdued rounded hills; the island's highest hills are formed by volcanic cones. Most of the island surface is mantled by peat deposits, but some scattered outcrops of lichen-encrusted basalt boulders are present. Several maars on the island contain freshwater lakes. Coastal erosion has breached several probable maars, forming arcuate embayments in the coastline. Some of these embayments are nearly filled by sediments and support brackish-water marshes. Two maar lakes were cored during the 1978 and 1979 field seasons. During winter coring operations in 1979 a 15 m core having a basal rediocarbon age of >39,000 years BP (W-4379) was obtained from a maar lake on the southeast side of the island (Lake VI or Zagoskin Lake, Fig. 5). Pollen and diatom analysis of that core is in progress.

Another maar lake (Lake III or Puyuk Lake, Fig. 5) was cored in 1978; the results of pollen analysis of that core are presented in Figure 6. Two radiocarbon dates from the core provide a preliminary chronology. Core segment 150–130 cm yielded a date of 7630 ± 70 years BP (W-4506); core segment 344–319 cm is dated as 15,610 ± 220 years BP (W-4503). The thick volcanic ash layer in the upper part of the core is probably from the same eruption that deposited similar ash in Tungak Lake sometime between 5645 and 3400 radiocarbon years BP (Fig. 4).

St. Michael Island Herb Zone

The pollen diagram for Puyuk Lake (St. Michael Island Core 78-2) (Fig. 6) is divided into three pollen zones that are quite similar to the three

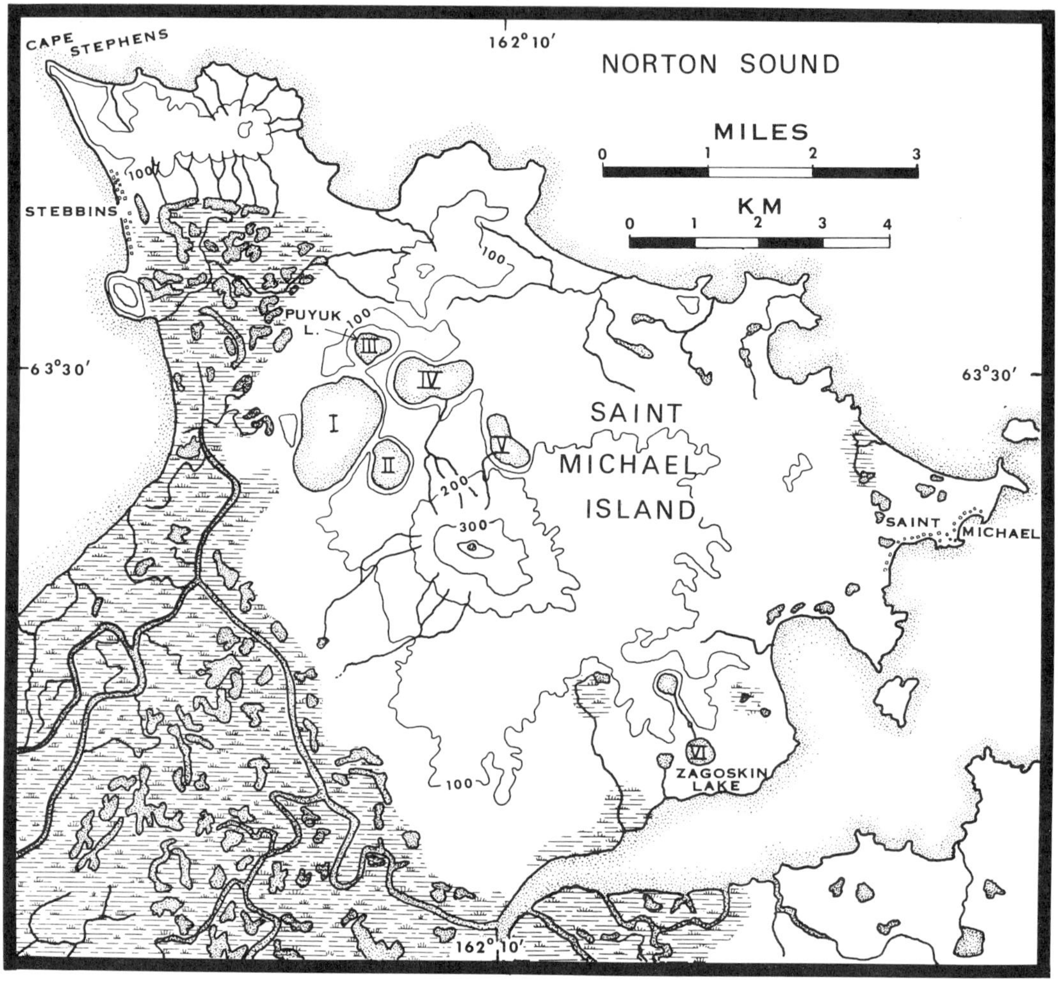

Figure 5. Map of St. Michael Island, Norton Sound, Alaska. The island is composed of basalt flows and pyroclastic deposits of Quaternary age. Many of the island's upland lakes occupy maar craters. Lakes III (Puyuk Lake) and VI (Zagoskin Lake) were cored in 1978 and 1979. Contours in feet. Map modified from U.S. Geological Survey topographic map of St. Michael (B-1) and (C-1) quadrangles, scale 1:63,360

major zones described from Tungak Lake (Ingakslugwat Hills, Core 76-3). The basal Herb Zones from St. Michael Island and the Ingakslugwat Hills share a common predominance of pollen percentages of Gramineae, Cyperaceae, *Salix*, and *Artemisia*. However, the Herb Zones from these two sites do show some differences that are worthy of mention. In making such comparisons, it is important to keep in mind that the Tungak Lake Core Herb Zone spans a time interval of approximately 24,000 to 14,000 years BP, whereas the Puyuk Lake Herb Zone record spans only about 16,000 to 14,000 years BP. Analysis of the 15 m core (Zagoskin Lake) from St. Michael should provide a

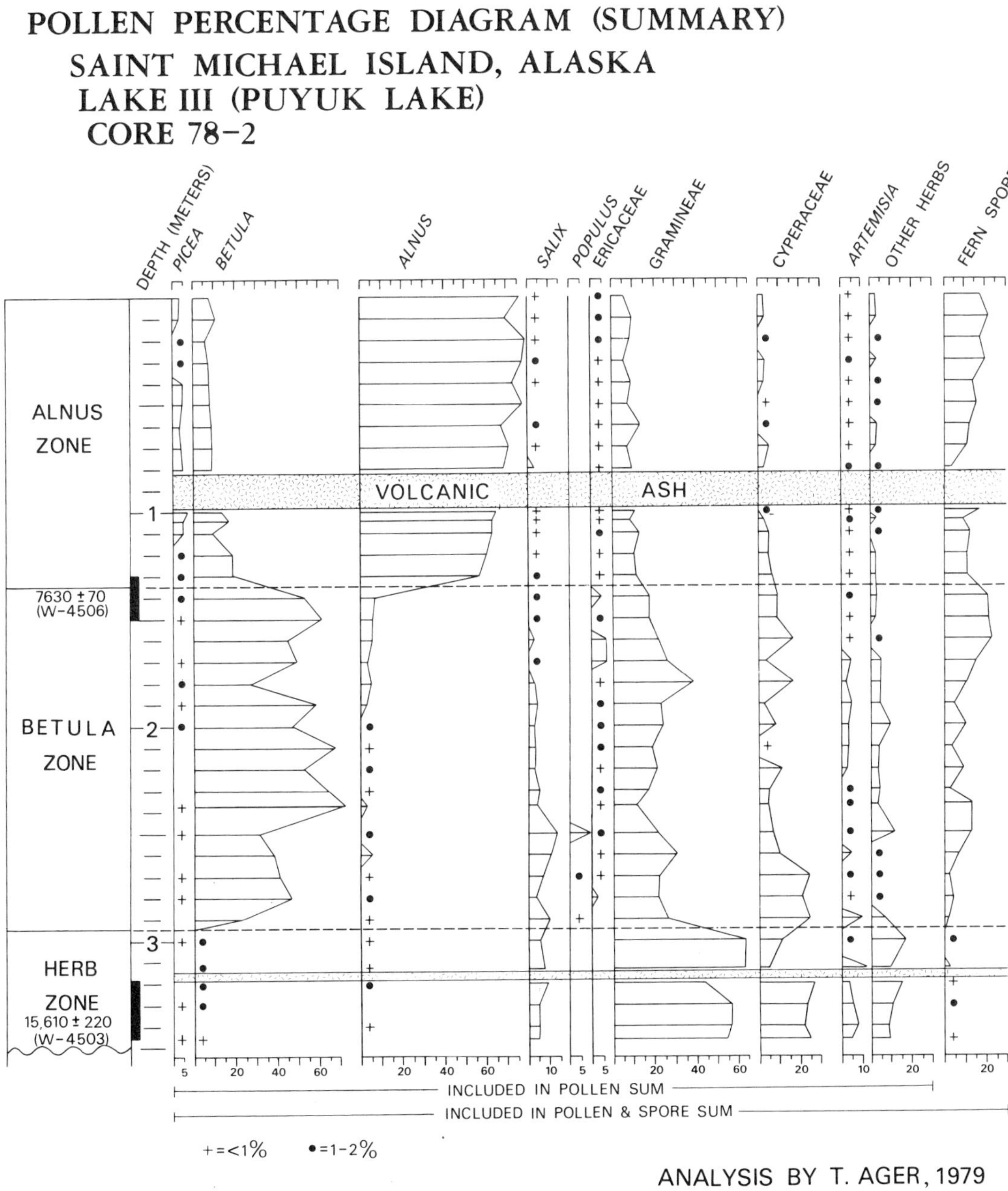

Figure 6. Pollen percentage diagram (Summary) for Core 78-2 from Lake III (Puyuk Lake), St. Michael Island, Norton Sound, Alaska

better basis for comparison of Herb Zone characteristics in the St. Michael and Ingakslugwat Hills areas.

The Puyuk Lake Herb Zone displays only trace amounts of *Betula* pollen whereas the Tungak Lake record shows several percent *Betula* pollen in most Herb Zone samples. In spite of low percentages of *Betula* pollen in the former, evidence exists that at least a few *Betula* shrubs grew near the lake; a single bractlet of *Betula* was found in Puyuk Lake core interval 313 to 306 cm (identified by J. V. Matthews, Jr., written communication Dec. 1979).[3] The bractlet was not an entire specimen, and the species of *Betula* that produced it could not be identified. Present-day distribution of *Betula* species in western Alaska suggests that it was probably the dwarf birch *Betula nana.*

No trace of Ericaceae pollen has yet been found in the Puyuk Lake Herb Zone samples, whereas small amounts of that pollen type occur in most samples from the Herb Zone from Tungak Lake. In all Herb Zone samples from Puyuk Lake, percentages of Gramineae are higher than percentages of Cyperaceae, whereas in some samples from the Tungak Lake Herb Zone, Cyperaceae percentages exceed those of Gramineae. Percentages of *Artemisia* pollen from St. Michael tend to be slightly lower than those in Herb Zone samples from Tungak Lake. *Sphagnum* spores are rare in the Puyuk Lake Herb Zone, but contribute several percent to the pollen-spore spectra from Tungak Lake.

St. Michael Island *Betula* Zone

Core interval 295 to 135 cm is assigned to the *Betula* Zone. It contains fewer major fluctuations in dominant pollen and spore types than do the *Betula* Zone samples from Tungak Lake in the Ingakslugwat Hills, and consequently it has not been divided into subzones. The *Betula* Zone from St. Michael differs from that from Ingakslugwat Hills in that Cyperaceae pollen percentages are much higher and percentages of *Alnus*, Ericaceae, and *Artemisia* pollen and fern spores are generally lower. Several samples from the St. Michael Island *Betula* Zone contain *Populus* pollen, whereas no *Populus* pollen has yet been identified in the Ingakslugwat Hills record.

St. Michael Island *Alnus* Zone

The core interval 135 to 0 cm in St. Michael Island Core 78-2 is assigned to the *Alnus* Zone, which closely resembles the *Alnus* Zone described for Tungak Lake. Only minor differences exist between the pollen spectra from these two localities; several percent of *Picea* pollen occur in most *Alnus* Zone samples from core 78-2, as they do in *Alnus* Zone samples of the Tungak Lake record. The pollen record suggests that the *Alnus* rise occurred in Norton Sound at least several hundred years prior to the appearance of *Picea* in the eastern part of the region of Figure 2.

VEGETATIONAL INTERPRETATIONS

Pollen data from Puyuk and Tungak Lakes provide the basis for attempting the first interpretations of vegetation history of the Yukon Delta-Norton Sound regions during approximately the past 24,000 years. The Herb Zone pollen assemblages from these two sites are interpreted to reflect full-glacial environmental conditions when sea level was eustatically lowered about 100 m to expose the vast expanse of the Bering land bridge. The pollen assemblages of the Herb Zone suggest an herb-dominated tundra (or tundra-like) vegetation in which Gramineae, Cyperaceae, *Salix*, *Artemisia*, and a variety of forbs were important elements. The vegetation may have been discontinuous, as suggested by the presence of pollen of "pioneer" type plants such as *Plantago,* Cruciferae, Caryophyllaceae, and various Compositae. Small amounts of *Betula* and Ericaceae pollen suggest that mesic habitats suitable for shrubs were much less common than such habitats are today.

The reduction of shrub taxa, the presence of pioneer plants that can tolerate bare, unstable mineral soil, and the relative abundance of grass, sage, and sedge pollen, along with an absence of pollen of spruce and alder point to an environment considerably drier than today's environment. This increased aridity during full-glacial times is to be expected because of the shift in moisture sources far to the south (owing to the exposure of the Bering land bridge), probably southward shifts in

---

[3]Additional plant macrofossils identified by J. V. Matthews, Jr., Geological Survey of Canada, from the Herb Zone core segments from St. Michael Island (Puyuk Lake) include achenes of *Ranunculus* sp., and *R. trichophyllus* type, and oogonia of *Chara* or *Nitella.*

Insect fossils from these Herb Zone samples were also identified by Matthews and include three genera of Coleoptera: *Colymbetes* cf. *C. sculptilis* Harris (predaceous diving beetle), *Tachyporus* cf. *T. rulomus* Blackw. (rove beetle), and *Aphodius* sp. (scarab beetle). The *Colymbetes* fossils are the most abundant insect remains in these samples, and these beetles may have been consumed by fish during winter, when *Colymbetes* are one of the few beetles that remain active under the lake ice cover (J.V. Matthews, Jr., written communication Dec. 1979).

storm tracks, and perhaps cooler sea surface temperatures that would reduce evaporation in the North Pacific (Hopkins, 1972; Hopkins, Smith, and Matthews, 1981; Barry, this volume; Gal-Chen, this volume). The Herb Zone pollen assemblages from these two lakes rather closely resemble herb-dominated full-glacial pollen assemblages such as Zone J at Imuruk Lake (Fig. 1, loc. 4; Colinvaux, 1964; Colbaugh, 1968), 240 km north of St. Michael Island, and 470 km north of Ingakslugwat Hills in the Yukon Delta. Colinvaux (1964, p. 308) interpreted the Zone J assemblages to represent a cold-climate, grassy tundra vegetation.

Considerable controversy still exists as to whether the full-glacial landscape of Beringia was treeless or whether some scattered groves of trees survived (Hopkins, 1972; Hopkins, Smith, and Matthews, 1981). Pollen evidence suggests that even trees and large shrubs that produce large quantities of pollen under most conditions (e.g., spruce or alder) are poorly represented in full-glacial sediments. The small amounts of such pollen that do occur can perhaps be explained as reworked pollen from interglacial or interstadial sediments, as mentioned previously, or as very long-distance-transported pollen accumulating slowly in places where the local vegetation produces far too little pollen to mask its presence. Some samples, however, contain more arboreal pollen than can be easily explained in this manner. Clearly, pollen analysis alone cannot adequately resolve this question, and the available macrofossil evidence is scanty. Thus far, no radiocarbon-dated tree-size wood of full-glacial age (about 25,000 to 12,000 years ago) has been found from Alaska (Hopkins, Smith, and Matthews, 1981). Even alluvial sediments of full-glacial age have not yet yielded tree-size wood in Alaska. Riparian forests in Alaska today commonly extend into otherwise treeless areas. Riparian habitats would seem to have been some of the most likely sites where trees such as spruce and poplar might have persisted during the late Wisconsin glacial interval. Therefore, the absence of wood from alluvial sediments supports the interpretation that trees were very rare in Beringia, or perhaps absent entirely at that time. In spite of the lack of evidence, I believe that *Populus balsamifera* and *P. tremuloides* were probably present in Beringia throughout the last glacial interval. I suspect that they were not abundant but survived in small scattered groves, perhaps in wind-sheltered sites on south-facing slopes. The appearance of *Populus* pollen in the St. Michael Island core from Puyuk Lake between about 14,000 and 11,000 years ago suggests that *Populus* was probably also present in the region prior to 14,000 years ago and that these scattered trees probably rapidly expanded their range when climatic conditions became more favorable for them, perhaps as early as about 14,000 years ago, and certainly before 11,000 years ago.

The evidence for the persistence of *Picea, Larix, Betula papyrifera,* and *Alnus* remains weak. At the moment, most of the evidence suggests that *Picea* became extinct in Alaska during late Wisconsin time (Hopkins, Smith, and Matthews, 1981). The modern distribution of *Larix* in Alaska suggests that it may have survived in the unglaciated interior during the Wisconsin (Viereck and Little, 1972). The history of *Alnus* distribution is less clear. Some fluctuations in *Alnus* pollen percentages within the *Betula* Zone of Tungak Lake, and less consistently in the Puyuk Lake core, suggest that some *Alnus* may have been present in western Alaska, perhaps as early as about 12,000 years ago. If so, then a few *Alnus* shrubs may have persisted in scattered sites in Beringia throughout the full-glacial interval.

High percentages (16%-32%) of *Alnus* pollen also occur in 10,000-year-old samples from a peat core from northwest Yukon Delta (T.A. Ager and W.R. Dupré, unpublished data). The abrupt rise in *Alnus* percentages 6000-7000 years ago in the Tungak Lake and Puyuk Lake cores suggests that an environmental threshold was reached that permitted the scattered alder shrubs to begin a rapid population expansion and perhaps an increase in pollen production.

Sea level rose to approximately its present position during the interval of 10,000-6000 years ago; this eastward transgression by the sea certainly altered moisture conditions in the Yukon-Kuskokwim lowland, perhaps favoring the spread of alders.

As indicated by the increase in *Betula* pollen about 14,000 years ago, the landscape of much of Beringia underwent a shift from predominantly herbaceous tundra vegetation to a mesic shrub tundra in which Ericaceae, *Betula nana* and *B. glandulosa* (in the interior), *Salix* spp., *Eriophorum* spp. and mosses became major elements. This transition appears to have taken place during an interval of about 2000 years over most of

Alaska. Ferns, which had been very restricted by the arid full-glacial climate, quickly spread into the expanded mesic habitats in sheltered sites in ravines, hollows, and rocky crevices.

The final vegetation change of regional significance recorded in these two pollen records is the westward expansion of spruce forest into the eastern parts of the Yukon-Kuskokwim coastal lowland and adjacent uplands (Fig. 2). This event took place 6000–5000 years ago, apparently following a gradual westward spread into Alaska from northwest Canada as suggested by radiocarbon dates and other evidence (Hopkins, Smith, and Matthews, 1981). *Picea* probably spread initially along the rivers of interior Alaska, forming riparian forests. The pattern of present-day distribution of spruce in the region is shown in Figure 2.

The present edge of spruce forests appears to coincide roughly with the position of the 10°C July isotherm (Hopkins, 1959b; Young, 1971). The available pollen data in this region are from sites several tens of kilometers beyond the edge of the spruce forest (Fig. 2). It is not yet possible to document minor shifts in the position of the edge of the spruce forest since middle Holocene time in western Alaska.

BRISTOL BAY

Coastal exposures of Quaternary sediments on the Kvichak Peninsula in the Bristol Bay region of southwestern Alaska (Fig. 1, loc. 7) were studied in 1972 by D. M. Hopkins. Perennially frozen loess and peat deposits from several outcrops were subsampled for pollen and macrofossil analysis and radiocarbon dating. Pollen samples were analyzed by R.E. Giterman, Geological Institute, Academy of Sciences, Moscow. A detailed account of the Quaternary geology and paleoecology of these deposits has not yet been published, but a summary of the data, illustrated by a pollen diagram from one of the studied sections (Fig. 7), is presented here.

This outcrop, Hopkins' Station 72/17, is within Nushagak Bay Quadrangle at lat. 58° 37′ N, long. 158° 14′ W. The sediments at this and many other sites along this coastline consist of glacial till overlain by glaciofluvial sand, loess, and peat. Ice-wedge pseudomorphs and cryoturbation structures are evident in some parts of these sections (Fig. 7). Radiocarbon dates from this locality and nearby sites suggest that the exposed pollen-bearing loess and peat span the time from the present to beyond the limit of radiocarbon dating.

The pollen data from these localities provide a good outline of vegetation changes in this region for more than 33,000 years, as indicated by radiocarbon dates (Fig. 7). The pollen sampling intervals are irregular and commonly widely spaced, however, and therefore the reconstructed history of vegetation changes here is not as detailed as the history reconstructed from the lacustrine sediment cores from the Yukon Delta region.

The Bristol Bay pollen data imply a vegetation history that roughly parallels the history seen in the Yukon Delta, although the chronology appears to differ somewhat. In addition, the pollen spectra from parts of the sections beyond the limits of radiocarbon dating provide insight into regional vegetation during early and/or middle Wisconsin time. These older sediments consist of loess interbedded with peat lenses and beds; they yield pollen spectra that somewhat resemble Herb Zone pollen spectra from Yukon Delta, except that some Bristol Bay samples appear to contain more pollen of shrub taxa (*Betula,* Ericaceae) and more *Sphagnum* spores. This apparent abundance of shrubs and *Sphagnum* suggests that at least local, and perhaps regional, conditions were somewhat mesic and perhaps slightly warmer than conditions during late Wisconsin time. Local changes in drainage conditions may account for some of the increase in moisture. These slightly shrubbier herbaceous "tundra" pollen spectra may reflect either early Wisconsin stadial or middle Wisconsin interstadial climatic conditions, but comparisons between these herb-shrub spectra and the middle Wisconsin pollen spectra from Imuruk Lake (Fig. 1; Colbaugh, 1968; Colinvaux, 1964) lead me to conclude that the Bristol Bay samples dated as >33,000 years BP probably represent middle Wisconsin interstadial conditions. The few sites from which middle Wisconsin interstadial pollen data have been studied suggest that vegetation changes associated with that climatic event (or events) were minor in western Alaska, while vegetation underwent somewhat more dramatic changes in other parts of Alaska and Yukon Territory (Colbaugh, 1968; Rampton, 1971; Matthews, 1974a; Schweger, this volume).

Pollen spectra from loess overlying the herb-

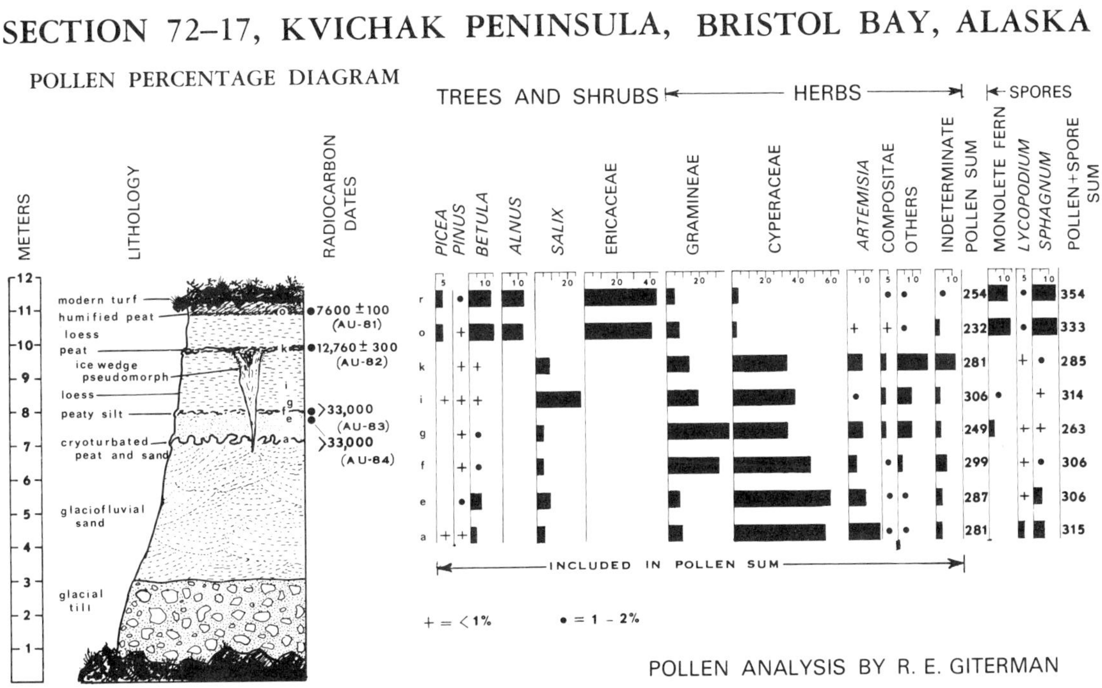

Figure 7. Pollen percentage diagram from coastal exposure 72/17 from Kvichak Peninsula, Bristol Bay region, southwestern Alaska

shrub assemblages are quite similar to pollen spectra from the Herb Zone samples from Yukon Delta region. These Herb Zone-type pollen spectra occur in the section (Fig. 7) between peaty beds dated as >33,000 years BP (AU-83) and 7600 ±100 years BP (Au-81). *Salix*, Cyperaceae, Gramineae, and assorted herbs, including *Artemisia,* dominate the pollen spectra, but the percentages are quite variable from sample to sample. This variation is a reflection of overrepresentation of pollen types from plant taxa that happened to be growing at that particular spot at the time the sampled sediment was deposited.

By about 12,760 ± 300 years ago, peat began to accumulate at this locality (Fig. 7). The pollen data do not give any other clear indication of vegetation change at that time. No *Betula* Zone is apparent. The pollen profile from section 72/17 (Fig. 7) gives the impression of an apparent simultaneous rise in *Betula, Alnus,* and even *Picea* percentages about 7600 years BP, but this impression is probably an effect of the wide sampling interval; pollen samples from another nearby section (72/42) record an undated *Betula* rise that preceded the *Alnus* rise. The *Betula* rise probably took place sometime between about 12,700 and 7600 years ago in this region. This chronology does not coincide precisely with that from the Yukon Delta. However, most dates for the *Betula* rise in western Alaska do fall at various times between 14,000 and 12,000 years ago. Apparently the expansion of dwarf *Betula* and other elements of low arctic shrub tundra was not simultaneous throughout western Alaska. Instead, *Betula* spread first into the most favorable new habitats from preexisting local seed sources. Eventually these expanding patches of shrub tundra vegetation merged to form the nearly continuous ground cover observed in the low arctic tundra of western Alaska today.

Holocene pollen profiles studied by Heusser (1963) indicate that *Betula* and *Alnus* were well-established in the Naknek River drainage on Alaska Peninsula, south of Bristol Bay (Fig. 1, loc. 8), by at least 5600 years ago. *Picea* pollen appears in trace amounts in these profiles, except in the uppermost samples in which several percent *Picea* occur. This suggests that spruce migrated into the Naknek River region in late Holocene time, probably within the past 2000 years.

OTHER POLLEN RECORDS FROM WESTERN ALASKA

Since the publication of Paul Colinvaux's review of pollen evidence from arctic Alaska (Colinvaux, 1967a) several additional palynological studies have been completed in western Alaska. In addition to the data presented in this paper from the Yukon Delta and Bristol Bay regions, important new information was reported from Seward Peninsula by Colbaugh (1968) and Matthews (1974b). Other studies in western Alaska published since the last review include several from the Aleutian Islands (Everett, 1971; Heusser, 1973, 1978) and the Pribilofs (Parrish, 1979). Below is a brief review of these recent studies and a discussion of the earlier work in light of the latest evidence.

The longest pollen records from Alaska yet studied are from the Seward Peninsula. Lacustrine sediment cores from Imuruk Lake (Fig. 1, loc. 4) permitted reconstruction of an extraordinarily long-term history of vegetation and climate of Seward Peninsula (Colinvaux, 1964, 1967a; Colbaugh, 1968); that unique record apparently spans three interglaciations and two complete glacial cycles. However, the pre-Wisconsin vegetational history from the Imuruk Lake pollen sequence will not be discussed here because the Yukon Delta-Norton Sound records are not old enough to permit comparisons.

The Imuruk Lake pollen zones that span the Wisconsin glacial interval (Zones J-K) and Holocent time (Zone L of Colinvaux, 1964; Zones L-M of Colbaugh, 1968) can be compared with the new data from adjacent regions, however. Zone J_3 from Imuruk Lake is very similar to the Herb Zone from Tungak Lake and Puyuk Lake cores. Despite the differences in altitude (about 320 m) and latitude (as much as 450 km) between the Imuruk Lake area and Yukon Delta sites, the full-glacial herbaceous vegetation of western Alaska was apparently rather homogeneous. Site to site variations in vegetation reflected local moisture and substrate conditions within a generally arid environment; most of the landscape was covered by herbaceous tundra-like vegetation. Radiocarbon dates for the transition from Zone J to Zone K (the *Betula* Zone) range from 13,000 to 12,000 years in different cores from Imuruk Lake. One of the dates (13,250±700) suggests, however, that the J-K transition was somewhat earlier, because the entire dated core interval is above the J-K boundary.

Zone K (the *Betula* Zone) marks the top of the pollen record in one of the Imuruk Lake cores (Core II, Colbaugh, 1968); apparently the last 10,000 years of pollen-bearing sediments are missing from part of the lake basin. Other cores from Imuruk Lake include a partial Holocene pollen record, interpreted by Colinvaux (1964) and Colbaugh (1968) to represent an early Holocene vegetation that produced significant amounts of *Alnus* and *Picea* pollen (Zone L of Colinvaux, 1964). Radiocarbon dates from core segments suggested that the K-L Zone boundary is between 10,700 and 8600 years BP (Colbaugh, 1968). However, data from peat exposures composing the "intermediate terrace" along the Imuruk Lake shore suggest that at least part of the early Holocene pollen record is compositionally similar to Zone K, the *Betula* Zone (Colinvaux, 1964).

The new pollen data from western Alaska (Fig. 1, loc. 1, 2, 5; Schweger, 1975, and this volume; Schweger and Habgood, 1976) leads me to conclude that the Holocene pollen record from the Imuruk Lake cores is unreliable. Lake-level changes that took place in early Holocene time (Colinvaux, 1964) apparently caused truncation of the early to perhaps middle Holocene sediments in Imuruk Lake. The thin veneer of Holocene sediments preserved in some parts of the lake basin contains significant amounts of pollen of *Alnus* and *Picea*. Those sediments probably have been extensively reworked by wave action and currents in the shallow lake, and perhaps by burrowing organisms (Black, 1966; Hopkins, 1972). Therefore, no complete and reliable Holocene pollen record is yet available for the central Seward Peninsula. Ongoing research in the Imuruk Lake area by Paul Colinvaux and William Patterson may fill this gap in the record with a pollen record from peat beds exposed in the vicinity of the lake (W. Patterson, personal communication 1980).

Cape Deceit

The impressive coastal exposures of unconsolidated sediments at Cape Deceit near Deering, Alaska (Fig. 1, loc. 5; Matthews, 1974b) include sediments of late Quaternary age (upper part of Unit 2, Deering Formation). Pollen and plant macrofossils and insect fossils from these deposits indicate that a Wisconsin "steppe-tundra" vegetation persisted in the area of northeast Seward Peninsula until about 13,000 to 12,000 years ago. Pollen spectra of early Holocene age from Peat 6 (within Unit 2 of the Deering Formation) are comparable to *Betula* Zone pollen assemblages described from the Yukon Delta region (this paper). No detailed Holocene pollen record is available from Cape Deceit, but surface samples from the area contain abundant *Alnus* pollen and are comparable to *Alnus* Zone samples from Yukon Delta.

Nome

The pollen record from Coastal Plain exposures at Nome on Seward Peninsula (Fig. 1, loc. 3) includes three pollen spectra of the Sangamon Interglaciation and a composite section spanning approximately the past 13,000 years (Hopkins, et al., 1960). These pollen data were previously discussed by Colinvaux (1967a); I offer only a few additional comments. The Nome data suggest that the transition from an Herb Zone to a Birch Zone apparently took place sometime between 13,000 and 10,000 years ago, later than on the opposite side of Norton Sound at St. Michael Island (this paper). In addition, the alder record from Nome shows no abrupt rise in alder pollen percentages at any time during the past 13,000 years. Alder pollen percentages rise gradually at about the midpoint of the pollen profile, but no radiocarbon date exists for that event, although it probably took place sometime in middle Holocene time.

The low percentages of *Alnus* pollen in the middle to late Holocene record from Nome is of interest because alder is common on southwestern Seward Peninsula today (Hultén, 1968), and ordinarily where alder occurs it tends to be heavily overrepresented in the pollen rain. The low percentages of *Alnus* pollen at Nome may be due to predominant winds blowing onshore from the Bering Sea during the flowering season. It is also possible that present-day climate on southwestern Seward Peninsula suppresses pollen production.

St. Lawrence Island

The St. Lawrence Island pollen data from Flora Lake (Fig. 1, loc. 6) of Colinvaux (1967a, 1967c), provide an outline of vegetational history, primarily during the Wisconsin glacial interval and the Holocene. Some pollen spectra are interpreted as interglacial, probably Sangamon. The late Quaternary part of the pollen diagram is divided into three pollen zones that are quite similar to those defined for the Yukon Delta region. A grass-sedge zone interpreted to represent a full-glacial herb zone is quite similar to that from the Yukon Delta, except that *Betula* pollen percentages are higher on St. Lawrence Island. This St. Lawrence Island "herb zone" is overlain by a *Betula* Zone. Unfortunately, no radiocarbon date exists for samples near the boundary between the herb and *Betula* zones. The Flora Lake *Betula* Zone sample is quite similar to the *Betula* Zone samples from the Yukon Delta region, except that the abundant fern spores seen on the mainland are rare in the post-Sangamon record from Flora Lake.

The upper *Alnus-Picea* Zone from Flora Lake is quite similar to the *Alnus* Zone described from the Yukon Delta. The single radiocarbon date from Flora Lake is near the base of this zone and suggests that the *Alnus-Picea* Zone spans about the past 6000 years, in close agreement with the age of the base of the *Alnus* Zone in the Yukon Delta. *Picea* percentages rise after the *Alnus* pollen rise, about 5500 years ago, and this too agrees with the timing of the *Picea* rise in the delta region. No *Alnus* or *Picea* grows on St. Lawrence Island, and the high pollen percentages of those taxa in many Flora Lake samples must be accounted for entirely by long-distance wind transport, probably from mainland Alaska. Shifting wind patterns during pollination seasons probably account for the erratic variations in percentages of *Alnus* and *Picea* pollen in the Flora Lake record.

Pribilofs

The Pribilof Islands in the southern Bering Sea occupied the southern edge of the Bering land bridge during the Wisconsin glacial interval (Fig. 1). Pollen samples from lacustrine sediment cores and peats from the Pribilofs were analyzed by Paul Colinvaux and Linda Parrish and some of the results have been reported (Colinvaux, 1967a, 1967b; Parrish, 1979), but no detailed account has yet been published of the Wisconsin pollen data from the 14 m core from a crater lake on St. Paul Island, except for a brief summary of *Picea* and a few other pollen types in these sediments (Colinvaux, 1967b).

Pollen data from the Pribilofs (Colinvaux, 1967b) are often cited as providing support for the idea that spruce persisted in southern Beringia during late Wisconsin time. A more detailed published account of the Pribilofs pollen data is needed to permit a complete evaluation of the interpretations of the history of spruce in southern Beringia. If some spruce trees did indeed persist there during Wisconsin time, they probably were drowned during rapid flooding of the land bridge that began about 14,000 years ago (Hopkins, 1972). The late appearance of spruce in western Alaska (in middle Holocene time) argues against spruce from the land bridge contributing to reforestation of the adjacent mainland.

The Holocene records from the Pribilofs (Colinvaux, 1967a, 1967b; Parrish, 1979) are of interest because they provide histories of a rich herbaceous tundra vegetation far removed from sources of pollen of birch, alder, and spruce. These pollen types are consistently represented in Holocene pollen spectra from the Pribilofs, further attesting to the efficiency of long-distance transport by wind. No *Betula* Zone or *Alnus* Zone is in evidence in the Pribilofs pollen records. The pollen data from St. Paul Island (Colinvaux, 1967a) suggest no vegetation change in that area of the Pribilofs during the past 9500 years. Pollen profiles from St. George Island, however, imply that the vegetation changed slightly there in middle and late Holocene time, and that those changes may be related to climate changes during Hypsithermal and Neoglaciation intervals (Parrish, 1979).

Aleutians

Only a few pollen records have been reported from the Aleutians (e.g., Heusser, 1973, 1978; Everett, 1971), and those few are from peat profiles of Holocene age. A preliminary search of radiocarbon dates from the Aleutian Islands turned up no published samples older than 10,255 years. Most of the islands were extensively glaciated during the Quaternary, and the precise time of late Wisconsin deglaciation is uncertain. The radiocarbon dates on basal peats from the Aleutians suggest, however, that the moist, stormy maritime climate that typifies the Aleutian chain today may have come into being as recently as about 10,000 years ago. The pre-Holocene glacial climate of the Aleutians may not have been conducive to peat formation in unglaciated areas and in areas where ice retreated earlier than 10,000 years ago. These interpretations are highly speculative, but if they are correct they suggest that the North Pacific was colder and drier during Wisconsin time than during the Holocene and that even islands in large expanses of high-latitude open ocean had little ameliorating maritime influence on their climates. These interpretations agree with those of Hopkins, Smith, and Matthews (1981), Barry (this volume), and Gal-Chen (this volume) that suggest lower evaporation in the North Pacific due to colder sea temperatures and that suggest a southward shift in storm tracks from their present pathways. These interpretations also suggest that the speculations about the southern edge of the Bering land bridge acting as a refugium for mesic vegetation comparable to modern low-arctic tundra (e.g., Hopkins, 1972) are probably invalid. Dry, sparse herbaceous "tundra" may have extended to the very southern edge of the Bering land bridge. The Aleutian Island records suggest that, at least in some localities, vegetation changes that took place during middle Holocene time may have been related to the Hypsithermal interval (Heusser, 1973, 1978). The Aleutian pollen records also demonstrate the extraordinary capacity of the atmosphere to transport trace amounts of spruce, alder, and other pollen types vast distances from the mainlands bordering the North Pacific and Bering Sea.

COMMENTS ON THE BERINGIAN "ARCTIC STEPPE"

Our current interpretations of full-glacial (Wisconsin) vegetation of Beringia are based on an admittedly thin network of sites spread over a vast territory. Large regions of Beringia (e.g., the Kuskokwim River drainage basin) remain to be studied by pollen analysts or other Quaternary paleoecologists. These "unknown" regions may eventually yield data that will alter our thinking on the nature of Beringian vegetation, but the available data suggest that no radically different Wisconsin pollen spectra will be found there. The pollen records now available display an extraordinary uniformity of pollen spectra in central and eastern Beringia, over great distances, hundreds of meters of altitude, and many degrees of latitude (e.g., Ager, 1975, Fig. 1, loc. 14, 15; Fig. 1, loc 1, 2; Rampton, 1971, Fig. 1, loc. 16; Matthews, 1974a, Fig. 1, loc. 13, 1974b, Fig. 1, loc. 5; Schweger, this volume; Colinvaux, 1964, Fig. 1, loc. 4; Ritchie and Cwynar, this volume). These pollen spectra suggest that herbaceous tundra communities were the overwhelmingly predominant vegetation type. The topographic complexity and environmental variability of Beringia logically leads to the idea that vegetation communities also varied compositionally and structurally from place to place, reflecting local conditions and perhaps forming a "mosaic" of plant assemblages. Yet the pollen data suggest that the range of variability within and between plant communities during the Wisconsin was clearly much less dramatic than is present variability (e.g., Viereck and Little, 1972).

The Wisconsin vegetation "mosaic" appears to have been composed of predominantly dry tundra communities in which some mesic and wet tundra communities occupied limited habitats. Trees and shrubs appear to have been very rare, and some species probably were eliminated from Beringia.

The limitations of routine pollen analytical methods of herb-dominated pollen spectra leave considerable latitude for argument and interpretation of the precise nature of these herbaceous communities. These pollen spectra may represent vegetation types having modern analogs (e.g., polar desert, alpine herb tundra) or they may represent vegetation types having no precise modern analog (e.g., "arctic steppe" or "steppe-tundra"). I believe that to reject or accept either point of view totally would be premature. Research may yet show that both types of vegetation existed in Beringia, perhaps even producing similar pollen spectra. This controversy may not be resolvable until additional macrofossil studies permit species-level reconstructions of Wisconsin vegetation assemblages, or until scanning-electron-microscopic investigations of pollen permit more precise identifications, particularly of dominant elements of the Herb Zone pollen spectra such as Gramineae, Cyperaceae, and *Artemisia.*

5

LATE PLEISTOCENE VEGETATION OF EASTERN BERINGIA: POLLEN ANALYSIS OF DATED ALLUVIUM

Charles E. Schweger

Dated geological sections, exposed along river valleys of eastern Beringia (Alaska-Yukon), have yielded late Pleistocene geological and paleoecological records. Pollen analyses of five of these sections are presented as relative percentage diagrams. Epiguruk, Kobuk River, Alaska, provides pollen records representing lowland and upland vegetation. These records dominated by Cyperaceae, Gramineae, and *Artemisia* indicate a birch-shrub tundra during the middle Wisconsin, Itkillik/Walker Lake nonglacial interval. Three sections in the Koyukuk River valley, Alaska, have also yielded pollen profiles. Section EIII represents part of either the early Wisconsin Itkillik Glaciation or the middle Wisconsin nonglacial interval. Sections EII and EIV date from 29,000 to 13,000 years BP, clearly spanning the late Pleistocene Walker Lake Glaciation. The three pollen records are dominated by Cyperaceae, include lesser amounts of Gramineae and *Artemisia,* and display a large variety of herbs. These records indicate a sedge-dominated tundra vegetation with local *Salix* stands. The minor elements strongly suggest meadow-type vegetation. It is concluded that these pollen records represent a variety of floodplain communities. They compare favorably to modern floodplain surface samples from Banks Island, arctic Canada; the comparison indicates the extent to which alluvial pollen records represent the local pollen component.

Beringian vegetation was a complex mosaic reflecting a number of biotic and abiotic factors, water and elevation being the two most important influences. Pollen records suggest a vegetation continuum from lowland-mesic and wet-meadow floodplain vegetation to fell field at higher elevations, extending upslope to polar-desert vegetation. The Pleistocene herbifauna probably relied heavily upon the higher production values of floodplain vegetation.

Key words: Beringia, Alaska, Yukon, pollen analyses, Pleistocene, late Wisconsin, paleoecology, alluvium, vegetation, tundra, meadow

INTRODUCTION

Rivers of eastern Beringia, such as the Kobuk, Old Crow, Stewart, and Porcupine, are entrenched in Pleistocene valley-fill sediment along parts of their courses. Cutbank exposures display excellent stratigraphic sections that can be related to geomorphic features such as moraines or outwash terraces. Typically, alluvial sections portray a wide variety of sedimentary environments, including foreset beds representing point-bar deposits, topset beds from overbank sedimentation, channel cut-and-fill sequences, and fine-grained lacustrine

ISBN 0-12-355860-3

deposits resulting from infilling of oxbow lakes or floodplain ponds. Features witnessed in alluvial sections that provide evidence for interpretation include buried soil horizons, ice-wedge pseudomorphs, beaver dams, autochthonous peat, detrital wood zones, and even buried shrubs and trees in growth position. Fossil pollen and spores, as well as botanical macroremains, can be recovered from the sediments, and in addition to the unique stratigraphic features, provide a rich record for paleoenvironmental reconstruction. Pollen analysis of alluvium has already provided important paleoecological records in a number of localities, including the American Southwest (Schoenwetter, 1966; Mehringer, et al., 1967), the lower Mississippi River valley (Delcourt, et al., 1980), and Egypt (Mehringer, et al., 1979).

Fossil pollen records derived from valley-fill alluvium reflect the wide variety of local environments—terrace soils, subaerial peat lands, fens, lakes, willow groves, and so forth—as well as more regional and long-distance pollen sources. Unfortunately, rapid changes in facies and rates of sediment deposition mean that pollen-influx measurements are not possible or certainly questionable. Also, in many alluvial facies, especially channel-fill or floodplain deposits, it is possible that some of the fossil pollen may have been derived from preexisting sediments. The problem of pollen redeposition is, however, not exclusive to alluvium, but may be a factor in bogs and lakes as well. The resulting mixed assemblages are easy to identify when the contaminating pollen is derived from Tertiary or older sediment sources. Recognition is more difficult when the source material is of Quaternary age because it is not evident what constitutes the rebedded pollen. Unfortunately, there has been only limited study of rebedded pollen in alluvial sediments (Cushing, 1964; Grichuk, 1967; Schweger, 1976; Delcourt and Delcourt, 1980). Careful and detailed stratigraphic work is needed to isolate and collect the most suitable facies. Autochthonous peats are preferable to detrital organic sediments, lacustrine silts and clays to channel sand and silt, and paleosol surfaces to point-bar alluvium. Nested or matched sampling is also desirable and can provide bulk samples for screening of macroremains (seeds, mosses, insects) that can then be compared directly with the fossil pollen assemblages.

Various reconstructions of Pleistocene Beringia's climate and vegetation have emphasized the arid nature of the climate and its limiting effects on the vegetation (Hopkins, 1972; Ager, 1975; Matthews, 1976a; Schweger, 1976). If precipitation values were low, distribution of surface waters may have had an even more significant influence on the vegetation mosaic than it does at present. Drainage lines, valley bottoms, and floodplains should have supported a vegetation that was more lush and more diverse than that of the uplands. It might therefore be expected that pollen records derived from valley-bottom alluvial sections provide a record representative of the most productive vegetation in eastern Beringia. This paper will present pollen records derived from five dated alluvium sections of eastern Beringia. The pollen records will be correlated directly to the glacial history of northern Alaska and provide the basis for proposing a vegetation continuum for eastern Beringia.

METHODS

Stratigraphic sections along the Kobuk and Koyukuk Rivers in Alaska were described and collected during the summers of 1969 and 1973, respectively. Locations of these sections and others discussed in this paper are shown in Figure 1. Research on the Kobuk River sections was done as part of the Onion Portage Archaeological Project, Brown University (Anderson, 1980); that on the Koyukuk River was done in conjunction with the field-mapping and stratigraphic studies of T. D. Hamilton of the U.S. Geological Survey. Research in the Old Crow, Porcupine, Stewart, and Silver Creek River valleys, Yukon Territory, was conducted from 1975 to 1980 as part of the Yukon Refugium Project, a project of the Museum of Man (Archaeological Survey of Canada) involving scientists from the University of Alberta and the Geological Survey of Canada.

Pollen samples were collected from various exposures and from a variety of sediment types in conjunction with detailed stratigraphic descriptions. Laboratory processing utilized a heavy-liquid technique (Schweger, 1976). All processed residue was mounted in glycerine and counted with a Leitz microscope; identifications were aided by the pollen-reference collection of the Laboratory of Paleoenvironmental Studies, Anthropology Department, University of Alberta.

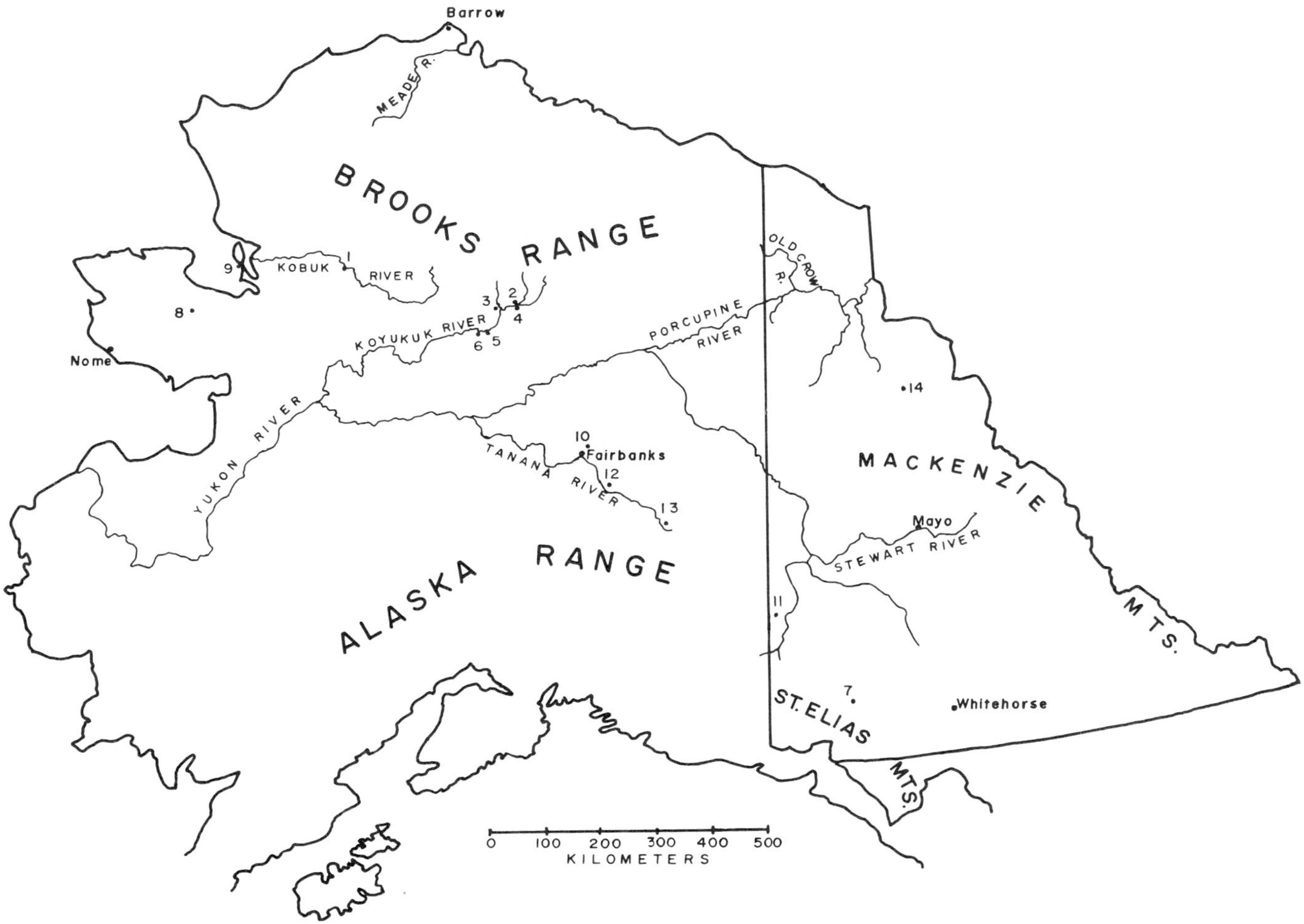

Figure 1. Map of Alaska and Yukon showing locations and sites discussed in the text. Numbered locations refer to the following: 1) Epiguruk. 2) Koyukuk exposure EII. 3) Koyukuk exposure EIV. 4) Koyukuk exposure EIII. 5) Koyukuk exposure EXI. 6) Koyukuk exposure EXII. 7) Silver Creek exposures. 8) Imuruk Lake. 9) Baldwin Peninsula. 10) Isabella Basin. 11) Antifreeze Pond. 12) Birch Lake. 13) Lake George. 14) Hungry Creek exposure

The number of grains counted varied between 200 and 250. The basic pollen sum includes identifiable pollen grains. Spores and indeterminate pollen were not included in the sum but their percentages were calculated to the same percentage base. In some samples pollen was very abundant, but in others was so sparse that as many as four slides had to be counted in order to reach the 200-grain sum, and many samples were entirely sterile. Pollen-percentage values were calculated and presented as histogram pollen diagrams. Those pollen taxa that generally occurred at frequencies of 2% or less are shown on the diagram as abbreviations. A summary of the minor taxa and a key to the abbreviations is presented in Table 1 and and their ecological significance is given in Table 2.

Geochronology was provided by radiocarbon dating of selected organic materials collected in stratigraphic context. Table 3 provides a list of the dates cited in the text.

Table 1
Pollen Taxa Generally Occurring at 2% or Less (Identifications Based Upon Pollen Reference Collection)

Abbreviation	Taxa	Epiguruk IIa	Epiguruk IIb	Koyukuk II	Koyukuk III	Koyukuk IV
Abi	*Abies*	—	—	X	X	X
Ace	*Acer*	—	—	X	—	—
Ane	*Anemone*	—	—	X	—	X
Ast	*Astragalus*	X	X	—	—	—
Cal	*Callitriche*	—	—	X	X	X
Car	Caryophyllaceae	—	—	X	X	X
Cas	*Castilleja*	—	—	—	X	X
Che	Chenopodiaceae/ *Amaranthus*	X	X	X	X	X
Comp Lig	Compositae-Liguliflorae	—	X	X	X	X
Corb	*Cornus*	—	X	X	—	—
Cory	*Corylus*	—	—	X	—	—
Dry	*Dryas*	X	X	—	—	—
Epi	*Epilobium*	X	—	X	—	—
Eri	Ericaceae	—	X	X	X	X
Erio	*Eriogonum*	—	—	X	—	X
Gen	*Gentiana*	—	X	X	—	—
Ger	*Geranium*	—	—	X	—	—
Goo	*Goodyera repens*	—	X	—	—	—
Hip	*Hippuris*	—	—	—	—	X
Lar	*Larix*	—	—	X	—	—
Leg	Leguminosae	—	—	—	X	X
Lil	Liliaceae	—	—	X	—	X
Ment	*Mentha*	—	—	X	—	—
Myr	*Myriophyllum*	—	—	—	—	X
Nup	*Nuphar*	—	—	—	—	X
Ost	*Ostrya/Carpinius*	—	—	—	—	X
Oxy	*Oxyria*	—	X	—	—	X
Pap	*Papaver*	—	X	X	—	—
Par	*Parnassia palustris*	—	X	—	—	—
Ped	*Pedicularis*	—	—	—	—	X
Phl	*Phlox sibirica*	—	—	X	X	X
Pic	*Picea*	X	—	—	—	—
Pin	*Pinus*	X	—	—	—	—
Plan	*Plantago*	—	—	X	—	X
Pol bis	*Polygonum bistorta*	X	—	X	X	—
Pol con	*Polygonum convolvulus*	X	—	X	—	—
Pol lap	*Polygonum lapathifolium*	—	X	X	—	X
Pol per	*Polygonum persicaria*	X	—	—	—	—
Pole	*Polemonium*	X	—	—	X	X
Pop	*Populus*	—	—	X	—	—
Pot	*Potamogeton/Triglochin*	X	X	X	X	X
Pote	*Potentilla*	—	—	—	—	X
Pte	*Pterocarya*	—	—	—	—	X
Que	*Quercus*	—	—	—	—	X
Ran	*Ranunculus*	—	—	X	X	X
Ros	Rosaceae	—	—	X	X	X
Rub	*Rubus*	X	—	—	—	—
Rub cha	*Rubus chamaemorus*	—	—	X	—	—
Sag	*Sagittaria*	—	—	X	—	—
San	*Sanguisorba*	X	X	—	—	—
Sax hir	*Saxifraga hirculus*	—	X	—	—	X
Sax niv	*Saxifraga nivalis*	X	X	X	X	X
Sax oppo	*Saxifraga oppositifolia*	X	—	X	X	—
Sax ste	*Saxifraga stellaris*	X	X	X	X	X
Spa	*Sparganium*	—	—	X	—	—
Sta	*Stachys*	X	—	—	—	—
Tax	*Taxus*	—	—	X	—	—
Tha	*Thalictrum*	—	X	X	X	X
Tof	*Tofieldia*	—	—	—	—	X
Tsu	*Tsuga*	—	—	X	—	X
Typ	*Typha angustifolia*	—	X	—	—	—
Ulm	*Ulmus*	—	—	X	—	X
Umb	Umbelliferae	X	—	—	—	—
Val	*Valeriana*	X	X	X	X	X
Vic	*Vicia*	—	—	—	—	—

SAMPLE SITES

Epiguruk, Kobuk Valley (67°5′N, 158°10′W)

This locality (Fig. 1, loc. 1), named after its Inuit place name, is a north-facing cutbank exposure, 40 m high and 2.5 km long, along the Kobuk River, approximately 15 m downstream from the village of Ambler. The Epiguruk exposure (Fig. 2) is subdivided into five sedimentary units described in Table 4; eleven radiocarbon dates provide geochronological control (Table 3). Detailed pollen analysis has been completed on four profiles: IIa and IIb will be discussed here; profiles I and II as well as further details of Epiguruk stratigraphy have been presented elsewhere (Schweger, 1976). Correlations between Epiguruk and the Brooks Range glacial history follow Hamilton (in press). Unit A sand is assigned to the early Wisconsin Itkillik Glaciation, and unit 8 represents part of the middle Wisconsin nonglacial interval. The overlying unit C, dated between 28,000 and 16,000 BP, represents outwash alluvium of the late Wisconsin Walker Lake Glaciation.

Units A and C are sterile of pollen; unit B is polleniferous and two profiles were collected for analysis (Fig. 2). Epiguruk IIa (Fig. 3) is a 4.5 m

Table 2
Habitat Characterization of Minor Pollen Taxa

Taxa	Hultén's Habitat Description
Anemone	Dry slopes, heaths, meadows, dry forest
Astragalus	Dry stony slopes, meadows, gravel bars, streambanks
Callitriche	Shallow water
Castilleja	Meadows, stream beds, open areas, dry slopes
Cornus	Woods, marshes, stream edges
Dryas	Gravel bars, heaths, wet forests
Epilobium	Meadows, forests, river edges, wet to moist areas
Gentiana	Meadows, bogs, dry slopes
Geranium	Forest meadows, moist meadows, woods
Goodyera repens	Mossy woods
Hippuris	Shallow water, creeks
Mentha	Wet places, river banks
Myriophyllum	Shallow water, sluggish streams
Nuphar	Ponds, slow streams
Oxyria	Wet places, snow beds in mountains and tundra
Papaver	Sandy gravelly soil
Parnassia palustris	Wet meadows, heaths, thickets
Pedicularis	Dry to wet meadows, rocky slopes, stony tundra
Phlox sibirica	Dry mountain slopes
Plantago	Wet places, salt marshes, open waste areas, saline soil
Polygonum bistorta	Meadows, heaths, tundra bogs
Polygonum convolvulus	No data
Polygonum lapathifolium	No data
Polygonum perisicaria	No data
Polemonium	Wet meadows, stream edges, dry sandy to rocky areas
Potamogeton/Triglochin	Lakes, ponds, streams/wet meadows, marshes
Potentilla	Dry open rocky to sandy areas, stream edges
Ranunculus	Wet meadows, ponds, moist places, moist forest
Rubus	Wet meadows, woods, thickets, river edges
Rubus chamaemorus	Peat bogs
Sagittaria	Wet shores, shallow water
Sanguisorba	Moist places, bogs, snow beds, meadows
Saxifraga hirculus	Bogs, meadows
Saxifraga nivalis	Dry slopes in mountains
Saxifraga oppositifolia	Calcareous rocks
Saxifraga stellaris	Wet rocky slopes, snow beds, along creeks
Sparganium	Deep to shallow water, along creeks
Stachys	Fields, clearings, moist woods
Thalictrum	Alpine meadows, meadows, woods, thickets
Tofieldia	Stony tundra, wet to dry places, heaths, marshes, shores
Typha	Shallow water, marshes
Valeriana	Moist places
Vicia	Forests, thickets, stream edges, meadows

section through lacustrine and peaty sediments. Epiguruk IIb (Fig. 4), 450 m further upstream and 20 m higher in the exposure, is a 1.8-m section of organic silt and sand. These two pollen profiles are believed to represent lacustrine-fen and upland environments for the middle Wisconsin and are dated older than 38,000 BP.

Epiguruk IIa shows a pollen flora dominated by Cyperaceae (20% to 80%), Gramineae (5% to 20%, and over 60% in the basal sample), *Artemisia* (10%) and other herbaceous types. *Betula* and *Salix* are the only shrubby taxa that are significantly abundant. Ranunculaceae (*Ranunculus* and *Thalictrum* types) are the most frequent herbaceous taxa; others reach frequencies of less than 5%.

The Epiguruk IIb pollen diagram is also dominated by Cyperaceae (45% to 80%) and Gramineae (5% to 20%), but *Betula* and *Salix* are important because each rises to 15% frequency. *Artemisia* is of only minor importance and remains below 5%, as do the other herbaceous elements. The higher frequency of indeterminate pollen is an indicator of the generally poorer degree of pollen preservation in this section.

It is evident from the stratigraphy (Fig. 2) that pollen sequence IIa represents a lowland, pond-fen facies, and that sequence IIb represents a nearby upland. As the two are derived from the same middle organic unit it is assumed that they are nearly contemporaneous. Aside from the significantly lower percentages of *Artemisia* in IIb, the two pollen diagrams are quite similar. That *Artemisia* varies inversely with Cyperaceae in IIa may indicate the local significance of Cyperaceae and the extralocal or regional significance of *Artemisia* in the vegetation (Janssen, 1966). If so, fluctuations in the local source of Cyperaceae, due perhaps to changes in pond-water levels, would have a strong effect on *Artemisia* frequency. *Artemisia* frequencies would have been greatest in sediment deposited when water levels were higher and the Cyperaceae pollen sources more distant from the collection section. Cyperaceae was no doubt always important on the nearby slopes, as represented by the diagram of Epiguruk IIb (Fig. 4); as a result *Artemisia* frequencies remain low throughout the profile.

No doubt the central Kobuk Valley featured treeless vegetation that supported only scattered dwarf birch and willow shrubs during the nonglacial interval represented by Epiguruk IIa and IIb. Cyperaceae must have been the dominant ground cover, but grasses and the varied herbaceous elements indicate variety in the vegetation cover. The steppe-like element in the vegetation may be the result of the well-drained sandy substrate that is so widespread in the central Kobuk Valley (Young and Racine, 1977).

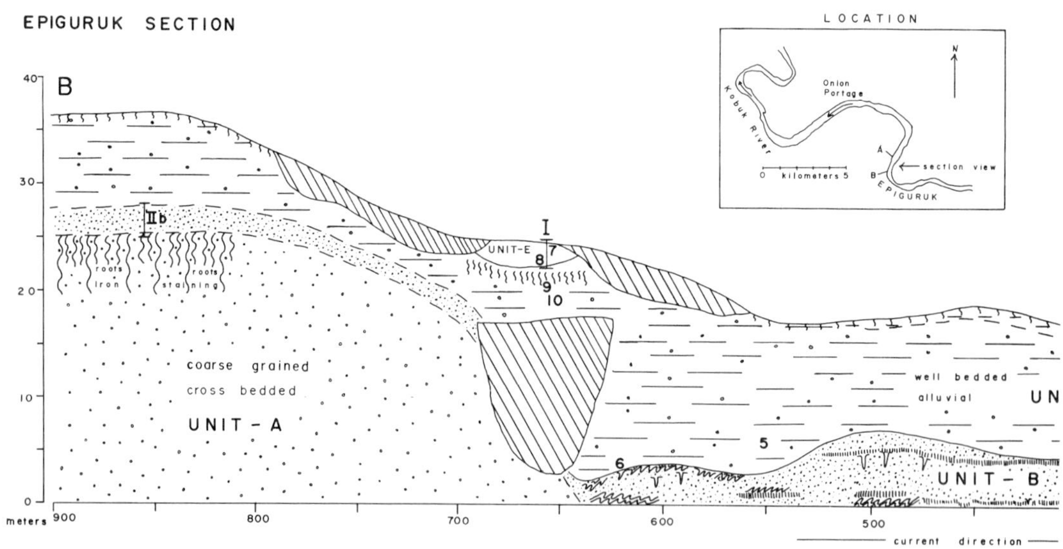

Figure 2. Stratigraphic cross-section of Epiguruk exposure, Kobuk River, showing

Koyukuk River Exposures

A number of important stratigraphic sections are to be found along the upper and middle portions of the Koyukuk River and its tributary, the John River. Both rivers drain southward from the Brooks Range and through this region traverse end moraines of Wisconsin age. Consequently, the cutbank river exposures display sequences of till and outwash as well as fine-grained alluvium. Detailed surficial mapping by Hamilton (1979, in press) has significantly modified the southern Brooks Range stratigraphy outlined by Hamilton and Porter (1975).

Late Pleistocene glaciation is divided into an Itkillik advance of early Wisconsin age and a Walker Lake advance of late Wisconsin age. Radiocarbon dates place the Itkillik/Walker Lake nonglacial interval as beginning before 53,000 years BP. The Walker Lake advance is clearly dated from 25,000 (T. D. Hamilton, personal communication 1981) to 11,500 years BP, including a minor readvance dated 13,000 to 12,500 years BP. (Hamilton 1980, in press; Hamilton, et al., 1980).

Pollen analyses were carried out on five Koyukuk sections (Fig. 1, loc. 2, 3, 4, 5, 6); three of these (loc. 2, 3, 4) have been completed and are discussed here. A complete stratigraphic description of these sections, EII, EIII, and EIV, is provided by Hamilton (1980, Fig. 2, 3, and 4, respectively; in press).

Exposure EII (67°2′N, 151°5′W), (Hamilton, in press, Fig. 4, KY-2)

In this south-facing 25-m-high bluff, sand and gravel at the base is overlain by a glacial till. Weathering of the till below a stone line indicates an unconformity, which is overlain by bedded sand. The sand becomes finer-grained and eventually silty near the top. The till is assigned to the Itkillik Glaciation (early Wisconsin), and the alluvial sand and silt is correlated with the late Wisconsin Walker Lake Glaciation. The base of the sand is dated as 23,500 ± 380 years BP, and the bottom and top of the 11-m section of silty sand, from which fossil pollen was collected, is dated as 15,455 ± 130 and 13,160 ± 170 years BP, respectively (Fig. 5).

Although three pollen-sterile zones were encountered in the analyzed section and some samples had nearly 40% to 50% indeterminate pollen, the pollen was generally abundant and well preserved. The pollen assemblages are dominated by Cyperaceae (30% to 86%), Gramineae (2% to 17%), and *Artemisia* (8% maximum). Cruciferae pollen reaches 18% and an unusual maximum of 50% in one sample (see Giterman, et al., this volume, Fig. 6, Cruciferae). In addition to Tubuliflorae pollen, a wide variety of minor herbaceous elements

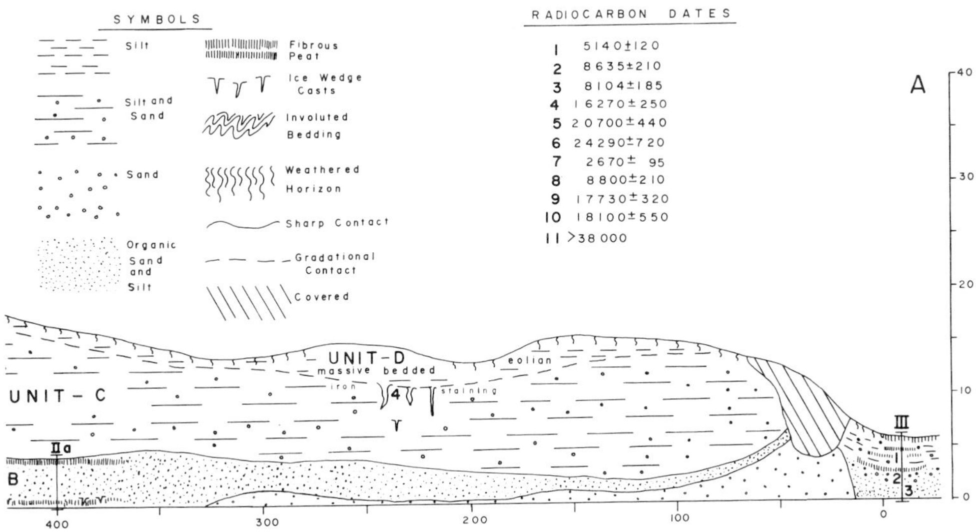

location of radiocarbon dates and pollen profiles. Date II is from pollen profile IIb

were discovered (Table 1). Although pollen from arboreal sources is infrequent, *Picea* pollen does rise to 10% at 5.25 m. *Salix* is the most important shrub element in the pollen flora (up to 48%). *Betula* and *Alnus* pollen frequencies never rise above 10%.

Exposure EIV (66°57′N, 151°45′W), (Hamilton, in press, Fig. 5, KY-4)

This 27-m-high exposure, located on the west bank of the John River, has been subdivided into seven units. Unit 1, a basal gravel, is overlain by a 4-m-thick clay bed, unit 2. Units 3, 4, and 5, totaling 10 m in thickness, consist of silt and sand, silt, and silt with sand, respectively. The middle of unit 3 is dated as 29,000 ± 700 years BP; the unit 4/5 contact is dated as 19,700 ± 360 years BP; and the top of unit 5 is dated as 17,420 ± 180 years BP. Unit 6, 1.5 m of sand, is capped by 5.5 m of loess that forms unit 7 and tops the section. This entire sequence may represent an initial proglacial-lake phase grading upward to the deposition of fine-

Table 3
Radiocarbon Dates Associated with Alluvial Pollen Sections. Epiguruk and Koyukuk-John River Dates
(Described in further detail in Schweger [1976] and Hamilton in press, respectively)

Location	Lot No.	Sample	Date
Epiguruk			
1	GX-1443	Allochthonous peat frags.	5,140 ± 120
2	GX-1442	Angiosperm wood frags.	8,635 ± 210
3	GX-1441	Angiosperm wood frags.	8,105 ± 185
4	GX-4778	Angiosperm wood frags.	16,270 ± 250
5	I-4779	Angiosperm wood frags.	20,700 ± 440
6	GX-1446	Angiosperm wood frags.	24,290 ± 720
7	GX-1444	Autochthonous peat	2,670 ± 95
8	GX-1445	Twigs	8,800 ± 210
9	I-4777	Twigs	17,730 ± 320
10	I-4776	Twigs	18,100 ± 550
11	GX-1447	Silty peat	>40,000
Koyukuk			
Exposure II	SI-1877	Unidentified wood	13,160 ± 170
	SI-1876	Willow wood	15,455 ± 130
	SI-1875	Sedge and grass	23,500 ± 380
Exposure EIV	SI-1882	Unidentified wood	17,420 ± 180
	SI-1881	Willow wood	19,700 ± 360
	SI-1880	Willow wood	29,000 ± 700
Exposure EIII	SI-1879	Bryophytes and wood frags.	>40,000

Table 4
Description of the Epiguruk Sedimentary Units Shown in Figure 2

Unit A:	Coarse-grained sand, gray to buff, crossbedded, with current ripple marks. Between 50 and 350 m, upper contact is largely unconformable. Upstream from 650 m contact is transitional over 1.5 m. Between 800 to 900 m it is an iron-stained, weathering zone with rootlets. Unit is interpreted as an alluvial sand; it may be aeolian in part.
Unit B:	Highly variable deposit, dark-brown organic silt and sand, peat, and gray lacustrine silt. Prominent fibrous autochthonous peats, between 350 to 650 m, interbedded with organic sand and lacustrine marly silts containing mollusca. Small ice-wedge pseudomorphs associated with peats and near 600 m bedding strongly involuted. Upstream from 600 m unit becomes organic silty sand. Upper contact a marked unconformity over most of section. One radiocarbon date of greater than 38,000 years BP (Table III) taken from middle of pollen profile IIb (Fig. 2). Unit is interpreted as an accretionary soil dropping in elevation into a lacustrine-fen basin.
Unit C:	Fine-grained sand and silt, tan to gray. Uniform in character over exposure. Bedding units, one to three cm thick, display ripple marks, clay partings, and small cut-and-fill structures. Mostly sterile of organics. Local ice-wedge pseudomorphs located near 250 m. Over most of exposure upper contact grades rapidly into overlying unit. Five radiocarbon dates (Table III and Fig. 2) place unit age between 24,300 and and 16,300 BP. This unit is interpreted as an alluvial floodplain deposit.
Unit D:	Fine-grained sand and silt, tan, massive, discontinuous. This unit is interpreted as an aeolian deposit.
Unit E:	Silt, dark gray, thinly laminated with fine-grained detrital organics grading upward into fibrous peat and then into modern forest soil. Two radiocarbon dates (Table III and Fig. 2) place age between 8800 BP and 2700 BP. This unit is interpreted as a local lacustrine basin grading upward to a small fen.

grained alluvial outwash and eventually loess and may be associated with the Walker Lake Glaciation of the John Valley.

Pollen analysis (Fig. 6) was carried out on a 9-m section representing units 3, 4, and 5. Unfortunately, a 2.5-meter interval of unit 4 was slumped and refrozen into the permafrost and therefore was not collected. Four pollen-sterile zones were identified. Indeterminate pollen ranges from 10% to 30% in the polleniferous samples. Cyperaceae (42% to 74%) is the most frequent pollen taxon, Gramineae ranges from 6% to 25%, and *Artemisia* is generally less than 8%. Of the shrubs, only *Salix* frequencies rise above 10%, reaching a maximum abundance of 16%; *Picea* pollen reaches only 5%. Pollen from Tubuliflorae reaches 10%. Cruciferae reaches 5%, and there is a rich assemblage of herbaceous taxa with frequencies of 2% or less (Table 1).

Exposure EIII (67°1′N, 151°8′W), (Hamilton, 1980b, Fig. 3, Exposure 3)

This 50-m high, north-facing exposure is located on the Koyukuk River. The lower 20 m is bedrock overlain by nearly 20 m of coarse gravel and sand; a loess unit of variable thickness tops the section. At the midpoint of the section, 25 m above the river and within the gravel unit, is a 2–3 m interval of gray to brown organic silt. Peat and

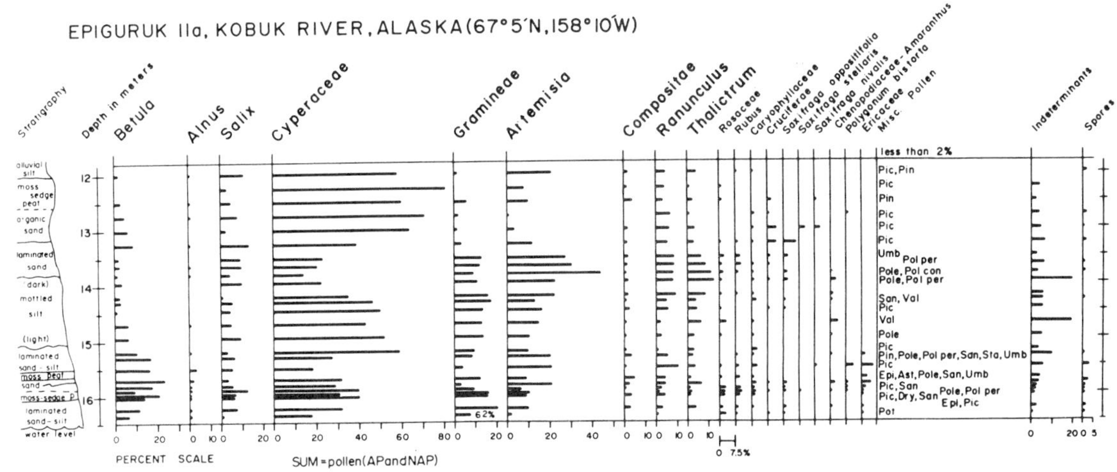

Figure 3. Epiguruk IIa relative percent pollen diagram

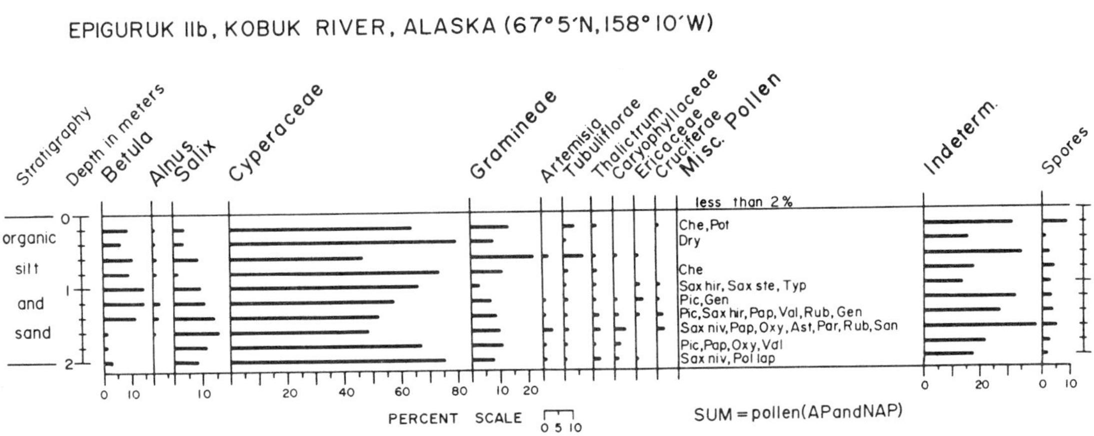

Figure 4. Epiguruk IIb relative percent pollen diagram

wood fragments from this unit are dated as greater than 40,000 years BP, and the fine-grained sediments yield a pollen record (Fig. 7). T. D. Hamilton (personal communication 1980) suspects that these sediments represent an interstadial within the Itkillik Glaciation; however, they could also represent part of the middle Wisconsin interstadial.

The frequency of indeterminate pollen reaches a maximum of nearly 45%. Cyperaceae dominates the pollen assemblages (88% maximum), Gramineae ranges from 5% to 15%, and *Artemisia* varies, reaching a maximum of 7%. Tubuliflorae and Cruciferae reach 13% and 8%, respectively. *Salix* dominates the shrub-pollen frequencies, rising in one sample to 52%. Arboreal elements such as *Pinus* and *Picea* are very rare and the suite of minor herbaceous elements and aquatics is somewhat restricted (Table 1).

These three Koyukuk records, derived largely from alluvial sediments of an aggrading floodplain, are remarkably similar even though they, according to their ^{14}C dates and stratigraphy, probably represent markedly different regional climates. Cyperaceae completely dominates the pollen assemblages; Gramineae rarely rises above 20%, and is generally no more than 10%. *Artemisia* fluctuates below 10%, but when combined with pollen of Tubuliflorae and Liguliflorae, lends the Compositae group considerable importance.

Arboreal pollen is rare; *Picea* rises to 10% in only one instance (exposure II), and is generally well below 5%. *Betula* percentages are mostly under 10%, and *Alnus* is even less abundant; of the shrubs only *Salix* is abundant. It is possible that some, or all, of the arboreal pollen may represent redeposition from preexisting sources upstream. The herbaceous pollen forms are varied: *Callitriche*, Caryophyllaceae, Chenopodiaceae/*Amaranthus*, Compositae-Liguliflorae, Ericaceae, *Phlox sibirica*, Ranunculaceae (*Ranunculus* type and *Thalictrum*), Rosaceae, *Saxifraga nivalis*, *Saxifraga stellaris*, and *Valeriana* are found in all three Koyukuk pollen diagrams (Table 1).

Indeterminate pollen is frequent, ranging from 5% to as high as 50%. These grains were either broken or corroded to the extent that confident identification could not be made. This portion of

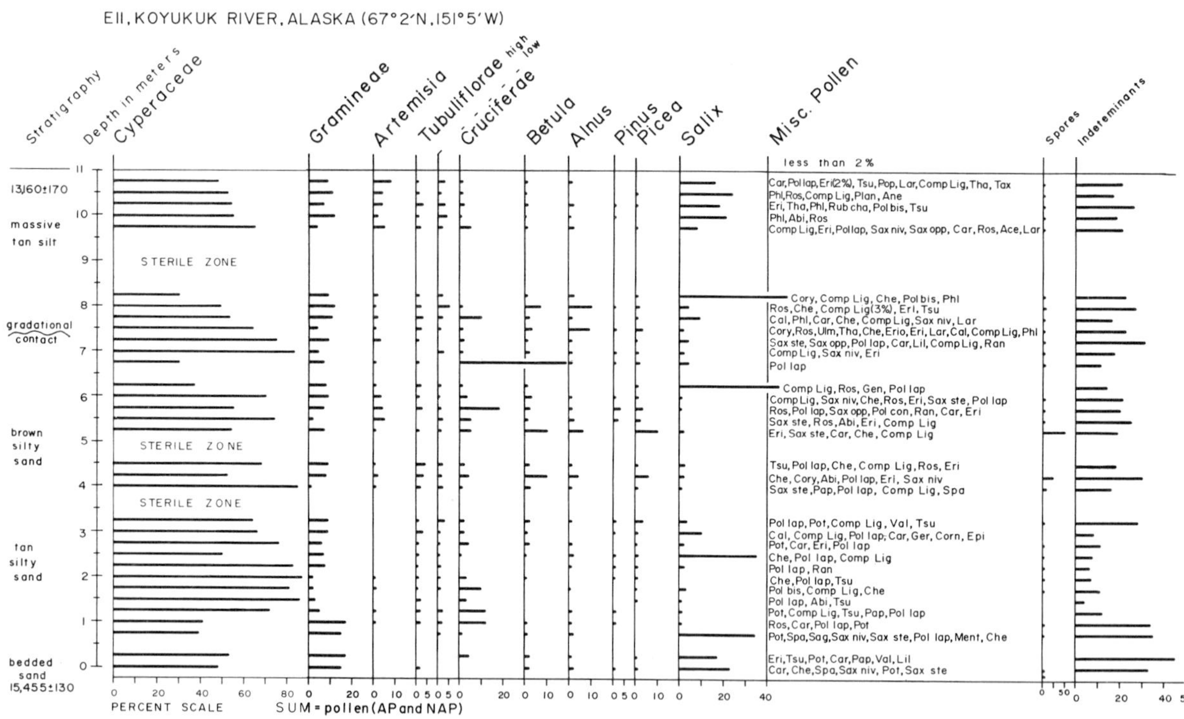

Figure 5. Koyukuk EII relative percent pollen diagram

the assemblage may represent reworked pollen indicating that recycling may be largely a destructive process. Sections EII and EIV have several sterile zones that occur mostly in sandy silt or sand. Because the frequency of indeterminate pollen does not rise immediately above or below them, sterile zones are probably the result of hydrological factors or dilution by rapid sedimentation rather than of post-depositional destruction of the pollen.

These pollen profiles taken together record a treeless tundra vegetation having a much reduced shrub component. Willow shrubs were certainly locally abundant on the aggrading floodplain of the Koyukuk River. The very high frequencies of Cyperaceae pollen imply a sedge-rich vegetation. Sedge fens, pioneer sedge seres, and sedge-rich meadows must have dominated the floodplain vegetation. Sedge was no doubt also important in the upland vegetation, but figures much less prominently in lacustrine pollen cores of the same age (e.g., Colinvaux, 1965; Ager, this volume).

Erosion, deposition, and frost action, as well as

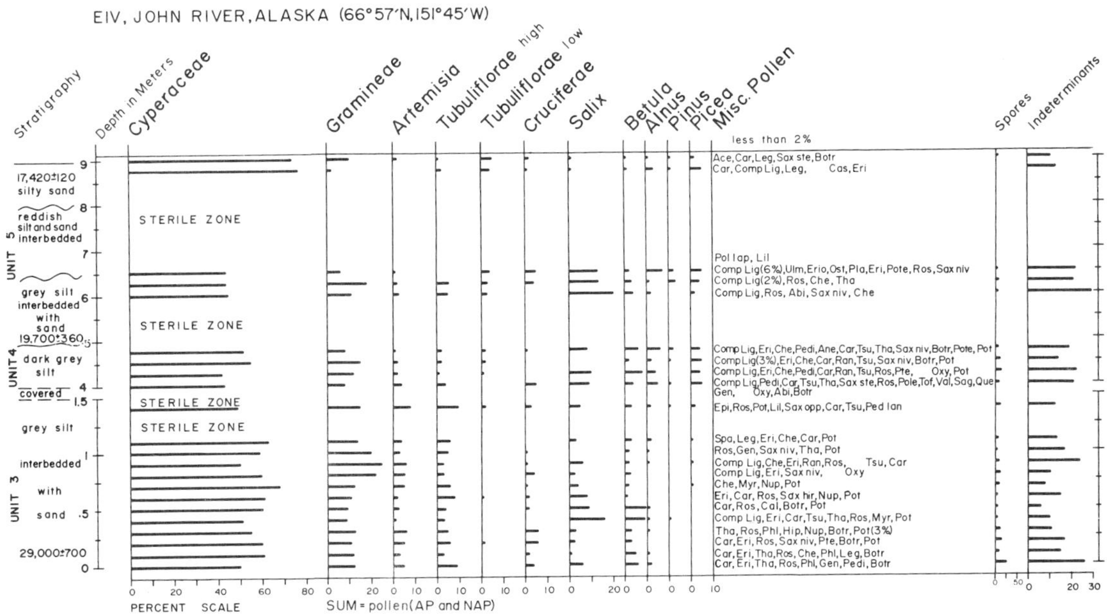

Figure 6. Koyukuk EIV relative percent pollen diagram

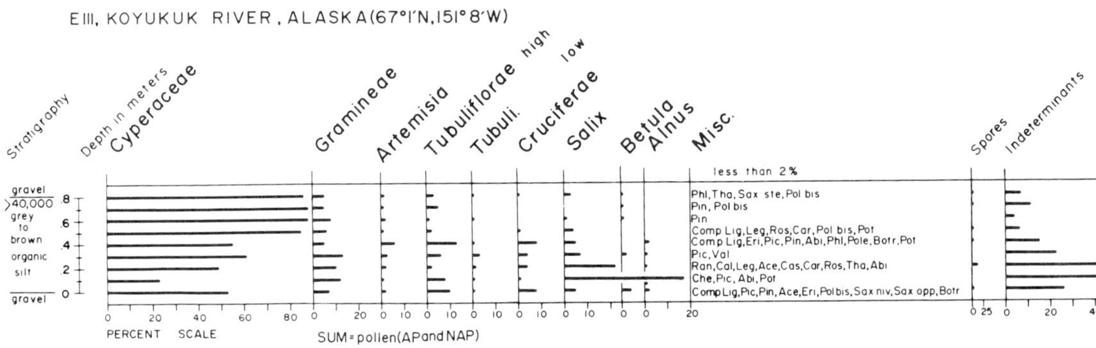

Figure 7. Koyukuk EIII relative percent pollen diagram

the generally more mesic conditions along the Koyukuk River, must have created a variety of habitats suitable for the numerous herbaceous types evidenced in the pollen record. There are general similarities among the taxa discovered here, and the species lists compiled from wet-sedge meadows and floodplain communities in northern Alaska (Spetzman, 1959; Johnson, et al., 1966), and even Devon Island (Muc, 1977). However, mosses most frequently share dominance with sedge in the modern communities. In contrast, the fossil pollen records presented here show very low spore frequencies; it seems that bryophytes were not as important in these unique late Pleistocene communities as they are at the present time.

The Koyukuk Valley pollen records represent time periods ranging from more than 40,000 years ago to as recently as 13,000 years BP, yet the pollen assemblages indicative of glacial-interstadial oscillations are not evident, nor is the birch period, despite its prominence in latest Pleistocene segments of lacustrine diagrams. Figure 8 illustrates how these five pollen records relate to the glacial record of the south-central Brooks Range. Pollen records from Koyukuk exposures II and IV represent glacial conditions; whereas the records from the Epiguruk bluff on the Kobuk River are from some portion of the long nonglacial interval between the Itkillik and Walker Lake Glaciations. Koyukuk exposure III may also represent part of this nonglacial interval or possibly an interstadial within the Itkillik Glaciation

Yukon Territory Alluvial Exposures

Field and laboratory research by the Yukon Refugium Project has led to pollen analysis of several alluvial sections in Yukon Territory, mostly in the unglaciated regions, the easternmost provinces of Beringia. Pollen studies that follow the pioneering work of Lichti-Federovich (1973, 1974) are being conducted on the numerous and formidable Pleistocene sections of the Old Crow Flats (Fig. 1). Other work concentrates on stratigraphic sections along the middle and upper portions of the Porcupine River and its tributary, the Bluefish River (Fig. 1). Because these promising studies are still in progress they will not be discussed here.

Farther south, the Hungry Creek section of the Bonnet Plume basin (Fig. 1, loc. 14) has yielded an important pollen record that, together with seed and insect fossils, indicates a well-developed boreal forest environment 35,000 years ago, just prior to the late Wisconsin advance of the Laurentide ice sheet over the site (Hughes, et al., 1981). In central Yukon Territory, the Stewart River valley (Fig. 1) concentrated valley glaciers from the Mackenzie Range. Downstream from Mayo are the end-moraine positions of the Reid and McConnell Glaciations (Hughes, et al., 1972). Here, as in the Koyukuk Valley, the river has cut through alluvial-glacial valley fill; organic deposits intercalated with outwash and till have been collected for fossil pollen studies. Although this work is not yet completed, preliminary results indicate significant environmental changes from forest to tundra that

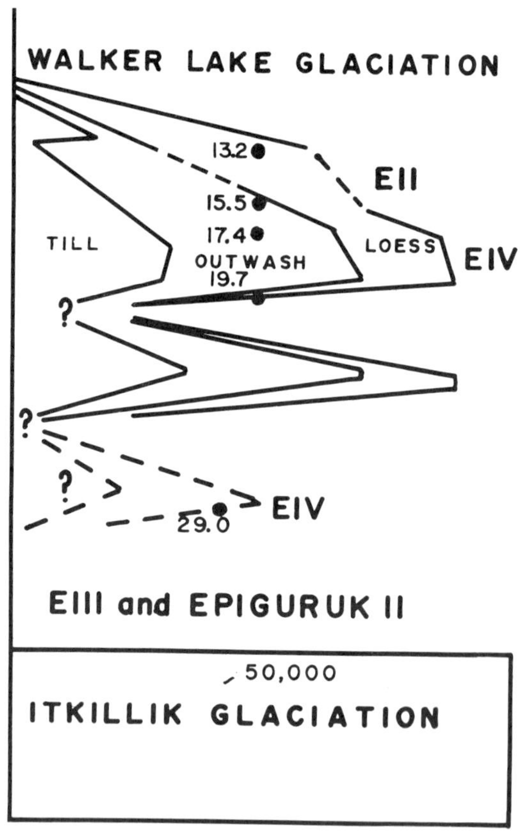

Figure 8. Diagram of south-central Brooks Range glacial history (modified from Hamilton, in press). Radiocarbon dates shown are from indicated pollen profiles

may be related to the glacial history of the Stewart River valley.

Alluvial sediments exposed along Silver Creek, St. Elias Mountains (Fig. 1, loc. 7), dated from 37,700 ± 1300/1500 to 29,600 ± 460 years BP, represent the Boutellier nonglacial interval (Denton and Stuiver, 1967; Hopkins, this volume). Pollen and bryophyte remains indicate a treeless alpine meadow vegetation that can be related to an altitudinally controlled middle Wisconsin vegetation pattern featuring shrub tundra, and spruce forest-tundra at lower elevations (Schweger and Janssens, 1980).

DISCUSSION

Minor Pollen Types

Over 60 minor pollen taxa were identified from the pollen assemblages of the five northern Alaskan sections presented here (see Table 1). Of these *Abies, Acer, Corylus, Ostrya/Carpinus, Pterocarya, Quercus, Taxus, Tsuga,* and *Ulmus* are mostly rare single-grain occurrences, and are rejected as unlikely representatives of the late Pleistocene flora of Beringia. These taxa may have been derived from Tertiary deposits and then redeposited in late Pleistocene alluvium. This possibility must be considered because late Tertiary source rocks have been found in a limited area of the Koyukuk Valley (Hamilton, 1979b). However, pollen grains of the above are not stained darkly as is so often the case with older palynomorphs.

Chenopodiaceae/*Amaranthus, Potamogeton/Triglochin, Saxifraga nivalis, Saxifraga stellaris,* and *Valeriana* occur as minor components in all five pollen sections, whereas Ericaceae and *Thalictrum* are found in four of the five sections. These taxa must have been widespread, and certainly locally abundant (see Matthews, this volume).

In order to make use of the information that minor taxa can lend to paleoecological reconstruction, we consulted Hultén's *Flora of Alaska and Neighboring Territories (1968)* for habitat descriptions of each identified genus and species. They are summarized and listed in Table 2. Because most genera display a considerable range of habitat preference this approach is definitely limited; nevertheless the results are interesting and useful.

Ponds, shallow water, or wet-soil habitats are represented by *Callitriche, Hippuris, Myriophyllum, Nuphar, Potamogeton/Triglochin, Sagittaria, Sparganium,* and *Typha.* Forests and thickets are frequently mentioned by Hultén, but with the exception of *Goodyera repens,* only in combination with meadows, stream edges, or open fields. Bogs and heaths are also common, and excepting *Rubus chamaemorus,* in combination with meadows. Dry tundra, dry slopes, and gravelly and sandy substrates are commonly mentioned along with river edges, snow beds, and wet places. The most frequent reference by far is to meadows, particularly moist or wet meadows. Only *Saxifraga nivalis* and *Saxifraga oppositifolia* refer exclusively to dry, rocky habitats.

At first, it is tempting to conclude that the minor pollen elements represent an open, mesic meadow vegetation. However, there are sufficient dry-site entries to suggest dry, open, well-drained, or coarse-grained habitats as well. An aggrading floodplain, besides featuring numerous shallow ponds and wetlands would also have open, well-drained, and even dry sand-and-gravel bars and dunes. Levees, low terraces, point bars, stream edges, gravel bars, cutoff and oxbow ponds, wet flats and moist slopes all indicate the variety of environments to be found. It is also noteworthy that taxa indicative of acid, organic substrate are not important. This lends support, as do the earlier observations that moss spores are rare and that Ericaceae pollen rarely exceeds 5% (Fig. 3), to the conclusion that late Pleistocene meadows were different than at present because of their dominance by bryophytes and peaty substances.

Modern Pollen-Rain Comparisons

Modern pollen-rain surface samples (Table 5) from Banks Island, arctic Canada, provide a revealing comparison to fossil pollen assemblages presented here. Moss polsters were collected along a 1-km transect from a well-drained hill site to the wet Thomson River floodplain, 30 m below. Sample 218-E5 represents a 60% vegetation cover developed on a regolithic substrate, dominated by *Oxytropis* sp., *Polemonium boreale,* and *Potentilla nivea,* as well as grasses. Sample 218-E10 represents an *Eriophorum* tussock vegetation on the upper floodplain having 100% cover that includes *Dryas integrifolia, Salix* sp., *Oxytropis* sp., grasses, and bryophytes. The wet lower floodplain, adjacent to

Table 5
Modern Pollen Surface Samples (Moss Polsters) from Thomson River, Banks Island, arctic Canada (73°39′8″N, 120°0′20″W)
Percentages Are Based Upon Pollen Sum

Taxa	Dry Hill Top 218-E5	Upper Flood-plain 218-E10	Wet Flood-plain 218-E11	Wet Flood-plain 218-E12
Picea	0.2	1.2	0.5	0.7
Pinus	0.2			1.0
Alnus	1.3	3.5	1.9	1.5
Betula	0.7	0.8	0.2	1.0
Salix	4.0	17.3	15	17.2
Cyperaceae	5.1	18.9	70.9	53
Gramineae	4.4	15.7	8.7	19.4
Artemisia	0.7	0.8	0.5	1.2
Rosaceae	1.5	28	1.5	2.4
Potentilla type	14.3	3.1	0.2	
Dryas type	0.4	4.7		
Caryophyllaceae	2.9	0.8	0.5	1.5
Cruciferae	1.3	0.4		0.7
Tubuliflorae	0.2	0.4		0.2
Leguminosae		2.8		
Oxytropis type	53.2	1.6		
Polemonium	8.8			
Ranunculaceae	0.2			
Chenopod./*Amaranthus*	0.2			
Taraxacum	0.2			
Polygonum bistorta type				0.2
Pollen Sum	453	254	412	413
Indeterminate Pbllen	16.8	15.7	10.4	1.2
Trilete spores	1.1	1.8		
Bryophyte spores		24.8	65.8	158
Pre-Quaternary spores	2.4	2.8	0.5	1.7

the river, is represented by samples 218-E11 and 218-E12; *Eriophorum* tussock vegetation dominates, as do grasses, *Dryas integrifolia*, *Salix* sp., Caryophyllaceae, and mosses. The floodplain here, including these two sample sites, is a heavily used musk-oxen pasture covered with feces.

Pollen spectra derived from these surface samples point out the treeless and largely shrubless nature of Banks Island vegetation; only *Salix* (4% to 17%) is abundant. Hilltop sample 218-E5 is dominated by the local pollen types *Oxytropis* (53%), *Potentilla* type (14%), and *Polemonium* (9%), whereas Cyperaceae (5%) and Gramineae (4%) frequencies are low. However, in sample 218-E10 *Salix* jumps to 17%; it stays high in the floodplain samples, reflecting its local importance in the vegetation. Rosaceae reaches a maximum of 28%, probably reflecting the significance of *Dryas*, although only 5% of the pollen could be confidently identified as *Dryas*. Cyperaceae (19%) and Gramineae (16%) frequencies also rise. In the remaining floodplain samples (218-E11 and 218-E12), grass continues to be important (9% and 19%, respectively), and Cyperaceae dominate, reaching 71% and 53%, respectively.

These samples clearly indicate the tremendous importance of the local vegetation and its influence as the local component of the pollen rain. The rapid and often very sizable changes in alluvial fossil pollen frequencies reflect in a similar manner the influence of local vegetation. For example, Cruciferae and *Salix* were obviously components in the local Koyukuk floodplain vegetation (Fig. 5). In general, the fossil pollen records bear closest resemblance to surface samples 218-E11 and 218-E12 from the floodplain of the Thomson River, Banks Island—to the point that the latter records also contain relatively high frequencies of indeterminate pollen and redeposited pre-Quaternary spores. This comparison cannot, however, be extended to the bryophyte spores, which are very abundant in the modern surface samples but nearly absent in the fossil spectra. Although these comparisons support the conclusion that the alluvial fossil pollen records of Beringia represent local treeless-floodplain vegetation dominated by Cyperaceae, they also indicate that the Wisconsin-age communities could not have been identical to the modern wet-floodplain vegetation in which mosses frequently dominate. It may be important to remember that the Pleistocene floodplains were aggrading and that there may not have been a period of stability sufficiently long for moss-dominated vegetation to develop (Komarkova and Webber, 1980).

Temporal Relationships

It is generally believed that the Sangamon Interglaciation witnessed development of boreal forest-type vegetation throughout most of eastern Beringia (Colinvaux, 1967a; Hopkins, 1972; Lichti-Federovich, 1974; Matthews, 1970, 1974b; Nelson,

1979). That spruce treeline may have advanced beyond present limits suggests a climate warmer than at present. However, the woefully small number of sites and lack of geochronological control (Westgate, et al., 1978) indicates just how poor our knowledge is of the last interglacial.

There is more pollen evidence correlated to or dated from the middle Wisconsin interval. In northern Yukon Territory the pollen stratigraphy indicates a grass-herbaceous or grass-birch-herbaceous tundra replaced by a spruce forest-tundra at the height of the middle Wisconsin (Lichti-Federovich, 1973, 1974). However, dating and correlation of these units is not complete at this time (Morlan and Matthews, 1978). The Hungry Creek locality provides clear evidence of a middle Wisconsin boreal forest dated as 35,000 years BP (Hughes, et al., 1981). For Seward Peninsula (Fig. 1, loc. 8) the middle Wisconsin (Imuruk Lake, zone I), is dated from 55,000 to 35,000 years BP, based on expansion of boreal forest into the regions (Shackleton, 1979). A pollen sample dated as nearly 27,000 years old from Kotzebue Sound (Fig. 1, loc. 9), indicates a birch shrub-tundra vegetation (Hopkins, et al., 1976). In the Fairbanks area (Fig. 1, loc. 10), Matthews (1974a) found evidence of a taiga-like vegetation occurring between 35,000 and 32,000 years BP. Rampton has reported (1971) that, farther eastward into southern Yukon (Fig. 1, loc. 11) after 31,000 and before 27,000 years BP, fell-field or sedge-grass tundra was replaced by birch shrub-tundra vegetation. And pollen recovered from sediments in the St. Elias Mountains dated as 38,000 to 30,000 years BP (Fig. 1, loc. 7) indicates open-tundra vegetation (Schweger and Janssens, 1980).

These data, and the pollen records presented above from the Kobuk and Koyukuk River valley sections, indicate that the middle Wisconsin was an interval of complex climate. A climate initially warmer, and/or, more mesic than that of the present allowed the spread of boreal forest westward onto the Seward Peninsula. A climatic reversal around 35,000 years BP saw boreal environments replaced by tundra. However, a delineation of eastern Beringia's interstadial and stadial units made solely on the basis of pollen assemblages would seem to be potentially misleading.

During late Wisconsin full-glacial conditions, the more numerous, and better dated, pollen records indicate a treeless, and largely shrubless, open vegetation (Livingstone, 1955; Colinvaux, 1964; Rampton, 1971; Matthews, 1974a; Ager, 1975, this volume; Cwynar and Ritchie, 1980; Ritchie and Cwynar, this volume). This cold, probably xeric, vegetation was replaced by a birch shrub-tundra around 14,000 years BP in central Alaska (Ager, 1975); after 13,000 years BP in the Koyukuk Valley (Fig. 5); after 11,000 years BP in the Kobuk Valley (Schweger, 1976); and 9000–8000 years BP on the north slope of Alaska (Nelson, 1979). This time-transgressive vegetation change seems to have been the result of a rapid climatic shift that saw warmer temperatures, and perhaps more precipitation. It marks the end of the Pleistocene and the beginning of the late glacial/Holocene in eastern Beringia.

Wisconsin Vegetation

It is unreasonable to assume that the late Pleistocene vegetation of Beringia was a homogeneous cover easily characterized by terms such as steppe, grassland, herbaceous meadow, or even tundra. Vegetation always exists in intimate relationship with substrate, topography. and climate (Batzli, 1980). The Pleistocene vegetation of eastern Beringia was probably a complex mosaic, no doubt comprising some communities markedly different in composition from modern plant communities. Environmental gradients and vegetation continua were just as much part of the Wisconsin-age landscape as they are at present (Schweger and Janssens, 1980).

The modern plant-herbivore relationships in the IBP experimental area near Atkasook on the Meade River, Alaskan Arctic Coastal Plain (Batzli, 1980), may serve as a useful comparison, perhaps even an analog, for the Wisconsin-age vegetation represented by the alluvial pollen records presented here (Batzli, 1980). Large-scale vegetation mapping has resulted in recognition of 21 physiognomic and floristic units. These range from desert scrub on floodplain sand dunes, through snow-patch, lowland, aquatic, and upland communities. The total effect is a complex pattern controlled primarily by moisture and permafrost, in which snow-cover duration and disturbance by river and wind erosion strongly affect the local vegetation (Komarkova and Webber, 1980). Microtopography and succession, as well as grazing and manuring by animals, have been emphasized as significant forces affecting vegetation (Peterson and Billings, 1980; McKendrick, et al., 1980). Net primary production is higher for wet sites and drops off for dry sites, whereas the highest production is that of the streambank scrub of various species of *Salix* (Komarkova and Webber, 1980).

During the Wisconsin, valleys of eastern Beringia would have represented a relatively favorable environment. Even though aridity was pervasive (Barry, this volume), surface waters would have provided local mesic conditions. Across large areas of aggrading floodplain, permafrost would have been deep or absent, and disturbances would have maintained early successional stages. Valley-bottom vegetation assured abundant fodder, and valleys provided shelter for animals, which like those of today must have had to endure long and cold winters.

If arboreal vegetation was present in Beringia it seems reasonable to assume that it would have occurred along the river valleys. *Populus balsamifera,* the cottonwood, is presently found in restricted stands along some river valleys of the Alaskan north slope (Murray, 1980a), and on spruce trails westward along the Noatak and Kobuk Rivers. The pollen evidence presented here does not indicate the presence of riparian *Populus* forest, let alone isolated trees; however, it is well known that *Populus* pollen is generally not well preserved, and hence the negative pollen evidence is not too meaningful. Only in one pollen sample do spruce percentages rise to 10% (Fig. 5, Koyukuk exposure II, 5.25 m), thereby placing spruce frequency within the range of pollen surface samples from forest-tundra. Radiocarbon dating places this peak at around 14,000 years BP, the time of the *Betula* pollen increase in the Tanana Valley (Ager, 1975) and ice recession in the central Brooks Range (Hamilton and Porter, 1975).

Palynological data from northwestern North America clearly indicate that spruce reappeared in Beringia at the end of a long migration from refugia south of the continental ice sheet (Hopkins, Smith, and Matthews, 1981). Thus, even if spruce was present in Beringia during the glacials, it was not the stock from those relict populations that eventually recolonized the region during the Holocene. In the absence of convincing fossil evidence for the presence of spruce in Beringia, the most critical question centers on the potential spruce may have for long term asexual reproduction: that is, whether spruce, most likely *Picea mariana,* could have survived as clones reproducing asexually, and producing pollen only when unusually good climatic periods were experienced (Hopkins, Smith, and Matthews, 1981). Although the 10% *Picea* pollen from Exposure II hints at this, the possibility of redeposition of spruce pollen from older sources cannot be eliminated.

The pollen frequencies of the shrub taxa *Alnus* and *Betula* are likewise equivocal, although it is probable that *Salix* was present in the valleys of Beringia. Occasionally very high *Salix* pollen frequencies, as seen in the Koyukuk River exposures, are good evidence that *Salix* shrubs were present locally on the floodplain. *Alnus* pollen frequencies are so low that it is very unlikely that alder was present. Regional pollen studies also indicate a postglacial reinvasion of Beringia by alder and not the spread of local indigenous populations. *Betula* pollen frequencies present an even more curious problem because they suggest that *Betula,* if it was even present, was not abundant in the Koyukuk Valley. *Betula* was certainly present in the Kobuk Valley during the time in which Epiguruk unit 2 was deposited.

Alluvial pollen records show the presence of more xeric elements, mainly *Artemisia,* Gramineae, and Compositae, in addition to minor amounts of *Phlox sibirica, Dryas, Potentilla,* and *Saxifraga nivalis.* Cyperaceae and a variety of herbaceous taxa indicate wet-meadow vegetation. Both vegetation types may have been present on the floodplains, although the present-day xeric vegetation along the Meade River floodplain is of a very different composition, in which *Artemisia* is not prominent.

The xeric or steppe pollen types are better represented in records derived from sites with surrounding uplands, although it should be borne in mind that "upland" is a relative distinction. The presence of xeric types in the alluvial pollen records may indicate a xeric upland vegetation partially "masked" by the more mesic local vegetation. On this basis one could propose two broadly defined vegetation types, an upland xeric tundra dominated by sedges, *Artemisia,* and grasses, as well as numerous, but less abundant, herbaceous meadow (or tundra) with frequent willow shrubs.

These relationships may be illustrated by a hypothetical cross-section, as shown in Figure 9, ascending upslope from a valley floodplain to a series of higher terraces, to upland flats or pediments, to foothills, and finally to mountains. Just such a topographic profile characterizes the Kobuk and Koyukuk Valleys, as well as many other valleys in Alaska and Yukon Territory. Although Wisconsin glacial climates were colder than at present, and life zones would have followed the glacial snowlines downslope, there is no reason to assume that the plant communities moved downslope as discrete units. Instead, species reactions to the unique environmental conditions in Beringia would have been "individualistic." The result led to new mixtures, or communities, not necessarily resembling those of today.

It is proposed here, and by others in this volume (e.g., Ager, Ritchie and Cwynar, Yurtsev, and Young) that Beringia could not have had a uniform vegetation cover that can be easily summarized as steppe-tundra or arctic steppe (Matthews, 1976a). Instead, the vegetation of eastern Beringia displayed a complex array of plant assemblages reflecting the complexity of the abiotic and biotic environments. To be more specific, on the basis of pollen evidence presented here and elsewhere, the vegetation displayed a zonation related to two of the most important factors in present-day arctic vegetation, surface water and elevation. During the late Wisconsin, valley bottoms, representing the most mesic situation, were characterized by sedge-*Salix* meadows that had a variety of open and aquatic habitats. Because of the region's active, ever-changing channels, sites of recent sedimentation, areas of deflation, and loess accumulation, it is unlikely that the vegetation cover was complete. Along the present Meade River, for example, slightly more than 1% of the area has been mapped as bare ground, 2% as active sand dunes, and nearly 4% as stabilized sand dune (Komarkova and Webber, 1980). During glacials, rivers draining glacial terrain would have been subject to significant diurnal and seasonal changes in discharge; this together with higher sediment loads would have resulted in even more areas of bare ground.

Farther upslope on higher terraces and flats, tundra meadow and arctic grassland vegetation is indicated. Still higher on pediments and into foothills, a discontinuous, xeric, cold-adapted, fell-field tundra has been reconstructed (Cwynar and Ritchie, 1980; Ritchie and Cwynar, this volume). The mountains may have been largely unvegetated, bearing, perhaps, a plant cover similar to polar desert of the modern high Arctic.

This proposed moisture-elevation gradient implies rapid changes in vegetation composition, physiognomy, and primary production in most parts of eastern Beringia. Primary production in the mesic sedge meadows and willow-scrub stands would have been very high in comparison to other parts of the continuum (Muc, 1977; Komarkova and Webber, 1980). The high productivity of lowland mesic vegetation must have been important to the great variety of Wisconsin-age herbivores of

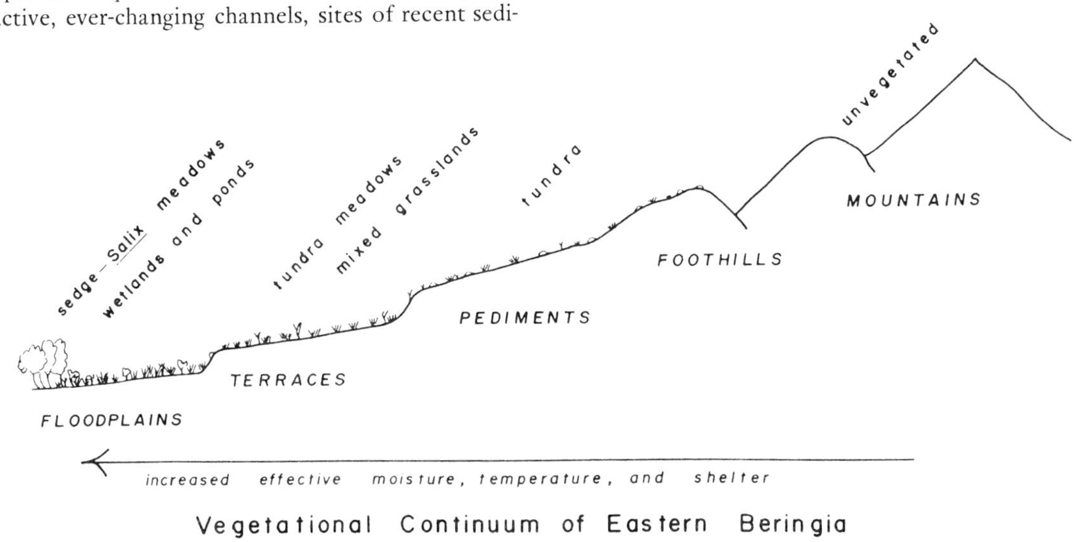

Figure 9. Proposed late Wisconsin vegetation continuum from lowlands to mountain for eastern Beringia

Beringia. Valley-bottom meadow and willow-scrub vegetation would have been intensively grazed in the way that similar meadows are now grazed by musk oxen on Banks Island (Tener, 1965). Because of the very process of herbivority, the meadows no doubt would have been manured by droppings and decaying carcasses, and production and nutrient values thus would have been all the higher. The relationship between herbivores and their diet is complex and leads to important influences on the vegetation (Redmann, this volume). Thus, any attempt to reconstruct ice-age vegetation exclusively on the basis of modern vegetation patterns, which exist without the influence of the extinct Pleistocene fauna, may be naive.

Paleoecologists have come to recognize that "the selection of the study site determines the level of detail that can be resolved in the reconstruction" (Jacobson and Bradshaw, 1981). And although there is a need to standardize the choice for "reference sites" (Berglund, 1979), it is also recognized that no single study site is without limitations and problems. Therefore, "selecting combinations of sites with different properties within a study area should permit detail in the reconstruction of vegetational change that adds a new dimension to paleoecological studies" (Berglund, 1979). Pollen analysis of alluvial sections, while not without their limitations and problems, may be used in combinations with other sites, particularly lakes and bogs of various sizes, to more effectively reconstruct the vegetational history of Beringia.

ACKNOWLEDGMENTS

Dr. Thelma Habgood, Department of Anthropology, University of Alberta, assisted me in preparing and counting the pollen samples. Tom Andrews, also of the Department of Anthropology, collected the Banks Island surface samples. Dr. Robert Stuckenrath, Smithsonian Institution, provided radiocarbon dates. Dr. Thomas Hamilton, U.S. Geological Survey, was a valued field companion, and generously shared his experience and interpretations. To these individuals I express my greatest appreciation. David Hopkins, John Matthews, Jr., and Thomas Hamilton critically reviewed this paper and made numerous valuable contributions.

6

THE LATE QUATERNARY VEGETATION OF THE NORTH YUKON

James C. Ritchie and Les C. Cwynar

The arctic-steppe concept is tested by scrutiny of two detailed late Pleistocene herb-zone pollen records from Hanging Lake (tundra) and Lateral Pond (woodland) in northern Yukon Territory and by consideration of other published pollen sequences in eastern Beringia. In particular, we examine the idea that Alaska and northern Yukon, when joined to Siberia by the Bering land bridge, "constituted the eastern end of a huge grassland biome, unequaled today in size and character."

Herb-zone pollen spectra consist largely of taxa whose modern affinities are arctic-alpine. Contrary to expectations of the arctic-steppe concept, herb-zone spectra do not show diversity of pollen taxa higher than in the modern pollen rain. Pollen influx rates are as low or lower than during the Holocene at the same sites and an order of magnitude lower than in modern central Canadian prairies. Influx rates for *Artemisia,* as well as for most of the other prominent herbs and grasses, are about the same in the herb zone as in younger pond sediments, although herb-zone taxa form much smaller percentages of the total pollen rain after birch, alder, and spruce appear in younger sediments. Pleistocene pollen spectra from western Europe are interpreted as representing open, discontinuous tundra or polar desert severely affected by frost action. The eastern Beringian herb zone also probably records discontinuous tundra vegetation floristically and physiognomically similar to communities found in eastern Beringia today on xeric, regosolic sites above the montane or beyond the arctic treeline. The communities represented by the herb zones are not extinct but are restricted in distribution now due to competition by species adapted to the more leached and humic Holocene soils and species that have immigrated or expanded in response to climatic amelioration.

We conclude that the "arctic steppe biome" never existed in eastern Beringia during the late Quaternary. The large and diverse ungulate populations probably were present during Pleistocene interstadials more than 30,000 years ago—intervals for which we lack adequate palynological records—rather than during the time of the herb zone, 30,000 to 14,000 years ago. It is doubtful that herb-zone vegetation could support a large and diverse population. Termination of the herb zone by the dwarf brich "rise" about 14,000 years ago signals a climatic amelioration (warmer and wetter). It is unlikely that the spread of this more productive vegetation was a factor in late Pleistocene mammalian extinctions.

Key words: arctic steppe, herb-pollen zone, birch-pollen rise, *Artemisia,* Hanging Lake, Lateral Pond

ISBN 0-12-355860-3

INTRODUCTION

This paper examines the arctic-steppe concept from a botanical perspective, derived from both late Quaternary pollen stratigraphic data and modern plant ecological and floristic information. These findings are based primarily on investigations in the northern Yukon (Fig. 1). The area forms the eastern extremity of the Beringian sector of the Holarctic (sensu Yurtsev, 1974b, Fig. 14). It is transected by the western limits of Laurentide Ice (Fig. 1), and it marks the eastern limit of such Beringian floristic elements as *Selaginella sibirica, Plantago canescens,* and *Salix phlebophylla.* The tundra vegetation of many montane and upland sites in this region bears a closer floristic resemblance to communities in Alaska, and to a lesser extent, eastern Siberia, than it does to tundras of continental subarctic Canada.

It would not be appropriate in this context to present and discuss the detailed pollen stratigraphy of the individual sites from which data have been recovered and on which most of our ideas are based.

Instead, a general summary will be made, illustrated by two representative biostratigraphic sequences, and we will examine the relevance of these findings to the question of the existence and nature of an arctic-steppe biome. This problem will also be examined in the light of interpretations of comparable assemblages from various other parts of the northern hemisphere reported by other workers, and by reference to modern floristic and vegetational configurations.

THE NORTHERN YUKON POLLEN STRATIGRAPHIC RECORD

The Late Pleistocene and Holocene

Two northern Yukon lake sites, one in the boreal forest and the other in tundra, have yielded detailed pollen stratigraphic information for the past 33,000 years. Summary percentage and influx pollen diagrams are shown here, as Figures 2 and 3. The tundra site, referred to informally as Hanging Lake and reported in detail elsewhere by Cwynar (1981), is 35 km northeast of the Old Crow Flats lowland and consists of a small (60 ha) lake, 9.5 m deep, perched in a local bedrock feature with no inlet streams and a single small outlet. It is surrounded by cottongrass tussock tundra on poorly to moderately drained surfaces and various open dwarf shrub-lichen tundras on upland mineral substrata. Radiocarbon-age determinations for 21 levels and very close-interval samples with large pollen sums have made possible a thorough analysis of the patterns of pollen influx over the 33,000 years of continuous primary sedimentation (Fig. 2). The second site is in the southern Richardson Mountains (Fig. 1) and is referred to informally as Lateral Pond. It is small (3 ha) and shallow (3.4 m),

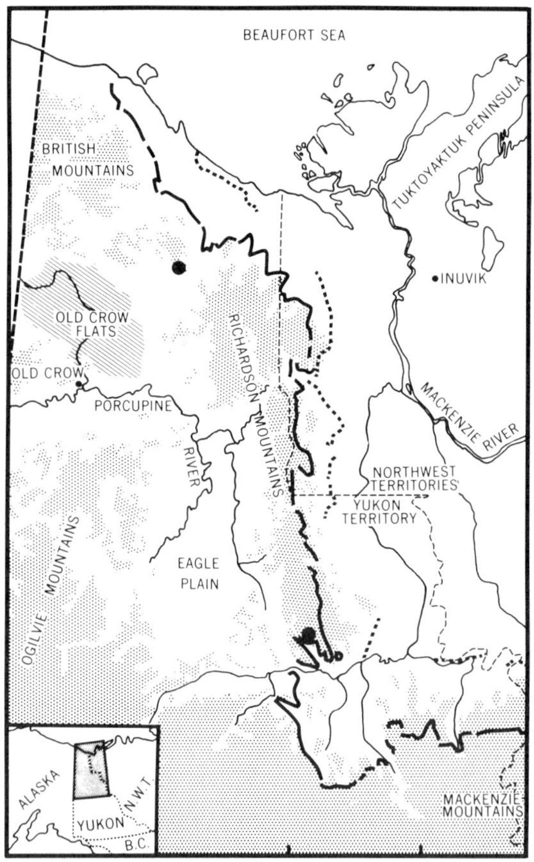

Figure 1. A sketch map of northern Yukon and adjacent Northwest Territories. The solid dots mark the locations of Hanging Lake (northern) and Lateral Pond. The heavy line indicates the maximum (westernmost) extent of Wisconsin glaciers

occurring in the spruce woodland zone at an elevation of roughly 600 m above sea level and 200 m below modern treeline. A 2.4 m core of primary sediment yielded a pollen sequence spanning roughly the past 16,000 years. Figure 3 presents a summary diagram showing only the main pollen types; the detailed results will be published elsewhere (Ritchie, 1982).

We will discuss here only the pollen stratigraphy of the late Pleistocene portions of these sections. The Holocene changes shown are not pertinent to the theme of this book.

Both percentage diagrams (Fig. 2 and 3) are broadly similar to the patterns recorded elsewhere in eastern Beringia (Livingstone, 1955; Colinvaux, 1964; Ager, 1975, among others). There is a basal herb zone with high relative values for *Artemisia*, grass and sedge-type pollen; willow percentages peak near the upper limit of the herb zones, and there is a marked increase in dwarf birch at roughly 14,000 BP. However, the influx diagrams illustrate that the percentage values give an erroneous impression of the vegetation. Influx values of *Artemisia* and grass between 30,000 and 14,000 BP are very low and were no higher during that period than during the Holocene. Only three of the important herb pollen types (Cruciferae, Chenopodiaceae and *Plantago*) show higher values for influx during the herb zones than subsequently. It is of further significance to the present discussion that the herb zone pollen assemblages (Cwynar, 1981; Ritchie, 1982) include a large number of taxa (not shown in the summary diagrams of Figs. 2 and 3) whose modern affinities are mostly arctic-alpine. These include *Aconitum, Bupleurum triradiatum*, Caryophyllaceae, *Dryas, Astragalus, Oxytropis, Hedysarum, Androsace, Lesquerella arctica, Pedicularis, Phlox, Polemonium, Polygonum alaskanum, P. viviparum, Oxyria, Rumex, Saxifraga hieracifolia, S. tricuspidata, Saussurea, Selaginella sibirica, Thalictrum,* and *Lagotis glauca.* It should be noted also that at the Hanging Lake site the total number of pollen taxa and their diversity have slightly lower values before 14,000 BP than after (Fig. 2), indicating that the flora then was less diverse than in Holocene and modern times.

Early and Middle Wisconsin Pollen Stratigraphy

One of the critical gaps in our knowledge of the vegetation/environmental history of eastern Berin-

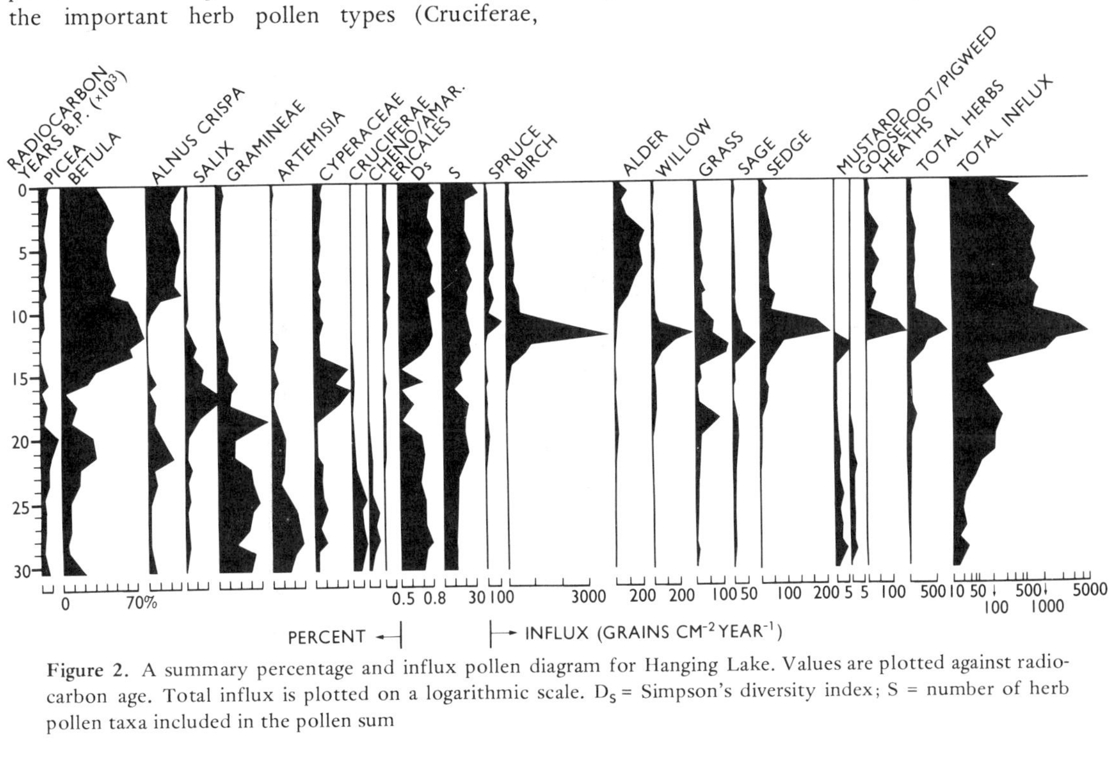

Figure 2. A summary percentage and influx pollen diagram for Hanging Lake. Values are plotted against radiocarbon age. Total influx is plotted on a logarithmic scale. D_S = Simpson's diversity index; S = number of herb pollen taxa included in the pollen sum

Lateral Pond, Yukon Territory
Relative pollen frequency

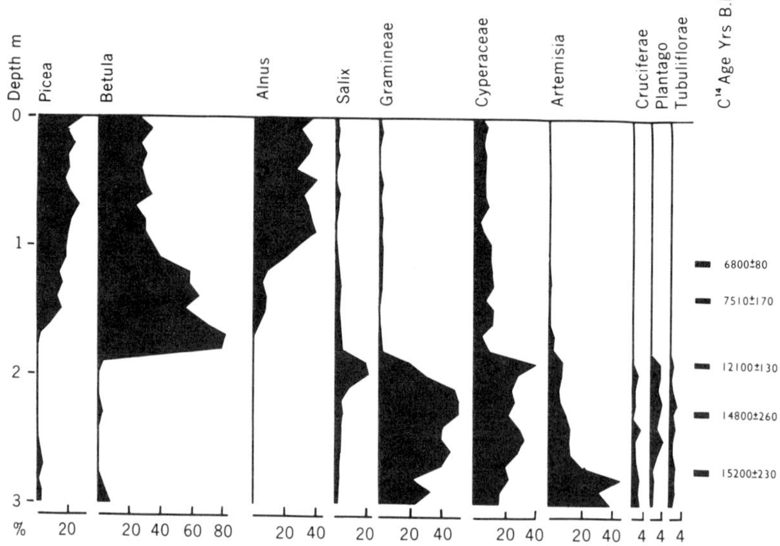

Lateral Pond, Yukon Territory

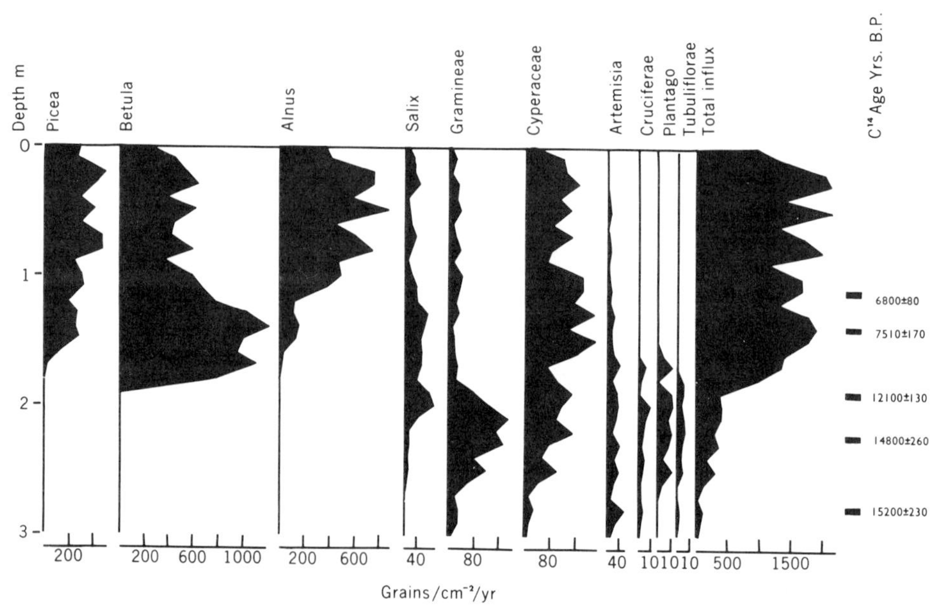

Figure 3. Pollen percentages (a) and influx values (b) for the main pollen types, plotted against depth in the sediment at 10 cm intervals. The ^{14}C values are corrected ages, determined at the Radiocarbon Laboratory, Geological Survey of Canada. The laboratory numbers from youngest to oldest are: GSC 2854, GSC 2852, GSC 2808, GSC 2785 and GSC 2758

gia is the time span preceding the latest stadial event. It is essential that a repository of primary sediment, extending back from the present continuously for 30,000 to 60,000 years be located, sampled effectively (using coring equipment that can penetrate the resistant clay-silt-sand sediments likely to have been deposited during stadial intervals), and analyzed critically in a laboratory with the capacity to identify the approximately 100 pollen types likely to be encountered in such material.

We are restricted so far in the northern Yukon to pollen data from reworked sediments of doubtful lithostratigraphic and chronostratigraphic provenance. The most definite statement that can be made at present, from the data published by Lichti-Federovich (1973, 1974) and Rampton (1979), is that several distinct pollen assemblages have been recorded, apparently representing boreal forest, forest-tundra and tundra vegetation types, but at present there is no reliable information to set them in a temporal framework. Further, there is no adequate evidence to demonstrate that any particular assemblage of vertebrates was associated in time and space with any particular pollen assemblage type, because of the stratigraphic uncertainties of both types of data (Schweger and Habgood, 1976).

These are the several pollen stratigraphic facts known to date, presented in summary form. Current and future investigations will undoubtedly replace, refine, and modify the information.

INTERPRETATION OF THE EVIDENCE

Studies of modern pollen representation in the Northwest (e.g., Ritchie, 1977; and unpublished work by Cwynar and by Ritchie) show that many important tundra elements are unrepresented or grossly underrepresented in pollen spectra (*Dryas*, *Potentilla*, *Saxifraga*, *Phlox*, *Salix*, *Ranunculus*, Leguminosae, Caryophyllaceae, Ericaceae and others) while others (*Alnus* and *Betula glandulosa*) are highly overrepresented. These are not new discoveries, yet their significance for interpretation is often ignored. The general consequence has been an oversimplified, erroneous interpretation of those pollen zones dominated by nonarboreal pollen. It is known that a large proportion of the many tundra vegetation types recorded from Alaska (Murray and Batten, 1977) and from northern Yukon (Hettinger, et al., 1973) are dominated by cryptogams and by vascular plants with low pollen production. The identification of these from fossil pollen assemblages will require a high level of interpretive skill, reinforced by a thorough knowledge of modern floristics and plant ecology. Such approaches to tundra paleoecology are more advanced elsewhere (e.g., Greenland, by Fredskild –1973; and Funder–1978).

With that cautionary preface we will turn to some of the central paleoecological questions of Beringia amenable to solution by pollen stratigraphic interpretation, using the data presented above from the northern Yukon.

(1) Does the dwarf birch rise at around 14,000 BP signify a major change in vegetation, and was it climatically induced?

Our interpretation of the pollen data is that many subarctic species, which had been absent or greatly restricted during the previous 10,000 year period, spread rapidly; that *Betula glandulosa* was one of these and that it increased to occupy approximately 30% of upland surfaces on noncalcareous substrata. One of us has shown elsewhere that dwarf birch is overrepresented in pollen spectra from lake sediments by a factor of five (Ritchie, 1977, p. 414), so pollen frequencies of 60% to 80% do not indicate a continuous birch-dominated landscape. Furthermore, dwarf birch today is rare or absent from tundras on the summit ridges of mountains with limestone bedrock. The influx values for the birch rise (Figs. 2 and 3) suggest a closed vegetation, with extensive ericad, herb, willow, and although not registered in the sediments, moss and lichen components. A crucial fact, emerging from the increasing number of pollen influx diagrams, is that the abundance of the herb pollen types remains constant or increases during the entire period encompassed by the herb pollen zones and the dwarf birch zone. In other words, percentage diagrams can give a spurious indication when nonarboreal pollen types predominate. The arctic-montane herb element which dominates the herb zones probably changed little in total amount or cover, being confined by competition from the vigorous chamaephytes to unstable or regosolic substrata, as they are in the modern landscape.

Today, *Betula glandulosa* shows its maximum modern development in low arctic and subarctic regions. Ager's (1975) interpretation of the dwarf birch pollen zone boundary in eastern Alaska in terms of "a sudden climatic change to warmer and moister conditions" appears to apply as well in northern Yukon.

From these interpretations it follows that, at least quantitatively, the transition from the herb zones to the dwarf birch zone at around 14,000 BP signaled the development of a more productive, stable tundra complex, very similar to what occurs in the northern Yukon today on noncalcareous surfaces above treeline. Further, it is our conclusion that any correlation of reduced numbers and diversity of vertebrates with this environmental change should certainly not be interpreted as a cause and effect relationship.

It should be noted in passing that one of us interpreted a similar dwarf birch zone in the Mackenzie delta region in terms of differential migration of *Betula* without any paleoclimatic inferences because it occurred at that site at roughly 11,000 BP (Ritchie, 1977). A parallel interpretation is reported by Funder (1978) from eastern Greenland, in which he suggests that the *Betula* rise at 8000 BP has no climatic significance but should be interpreted as a slow dispersal effect.

(2) What was the composition and structure of the vegetation represented by the herb pollen zones, and what can be adduced about the environment? Was it an arctic steppe? Was it relatively productive, and adequate to support a complex vertebrate ecosystem?

It might be useful before attempting to reconstruct the vegetation of these zones to point out that the northern Yukon and adjacent terrain to the east and west consists of a mosaic of ecological regions with quite varied patterns of landforms, soils, drainage, and associated vegetation. There is every probability that this physical variability was present also in the late Pleistocene. The modern ecological regions, distinguished by terrain characteristics and prevailing vegetation, are shown for eastern Beringia in Figure 4, based on maps and reports by Oswald and Senyk (1978) and Ritchie (1980). We recognized that some of these physical features would have been different 15,000 to 25,000 years ago—e.g., the coastal plain extended somewhat farther north, large proglacial lakes occupied lowland basins (Fig. 4, region 3), and glaciers covered region 6 and adjacent areas to the north and east (Fig. 4). But in most general terms the unglaciated, western part of the area of Figure 4 has three major landform types—mountain massifs (the Mackenzie, Ogilvie, Richardson, and British Mountains) occupy approximately 50% of the land surfaces; rolling upland plains, formed from residual materials and pediments (the Eagle Plain, Porcupine uplands, Peel River uplands and Yukon slope pediments) amount to 40%; and 10% is occupied by low, flat, poorly drained sediments with high water tables (Yukon coastal lowlands, Old Crow, Bluefish and Bell lowlands). Except for the coastal plain, much of the lowland area named was covered by lakes with water levels up to 335 m during the late Wisconsin. That these surfaces are widely different ecologically is reflected directly today by their contrasting vegetation—tundra and forest-tundra on the mountain ridges and upper slopes; tundra in the north and woodlands in the south on the rolling plains; and sedge-cottongrass-moss mires in the lowlands. Within each major type, particularly in the mountains, there is great diversity of plant community related to bedrock type, aspect, stability and slope. Any regional paleoecological reconstruction that does not take account of such variation will have limited value. We propose to approach answers to these questions by marshaling three categories of information—the pollen stratigraphic record for the area (northern Yukon) and time period (33,000 to 14,000 BP); a comparison of interpretations offered by others for similar pollen assemblage zones; and an examination of the nature—floristic and vegetational—of modern examples of "arctic steppe" and of modern equivalents or analogs of the vegetation suggested by the herb pollen zones.

1) The pollen stratigraphic record, regardless of the interpretation put on it, shows three clear and important facts. First, pollen concentration/influx values are low for all pollen zones—10,000/ml and about 100/cm^2/yr. Second, there is no pronounced diminution in the concentration/influx values for these forb/herb/dwarf shrub pollen types throughout the late Pleistocene and early Holocene record. This fact has probably been overlooked as other diagrams have been percentage diagrams, and the pollen sum used has remained constant throughout a particular section with the result that where dwarf birch and alder predominate the herb types

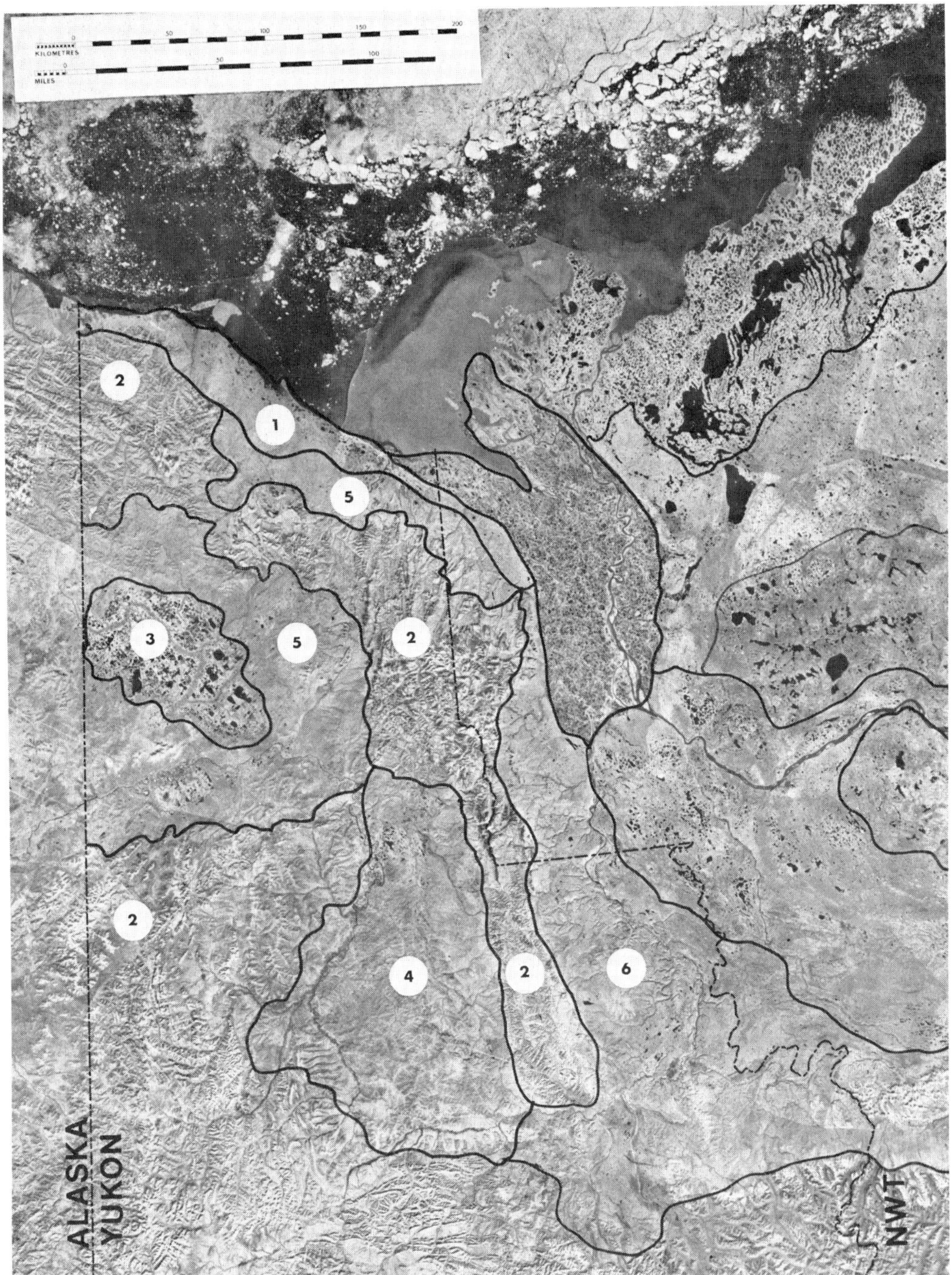

Figure 4. A satellite photograph showing the approximate limits of the ecological regions of the northern Yukon; lowlands (1 and 3), upland plains (4 and 6), and mountains (2)

have been poorly documented. (It is necessary to count not less than 1000 grains per level in Holocene samples to secure even an approximate registration of nonarboreal pollen types.) Our data suggest that the representation of these elements in the modern and late Holocene vegetation is not significantly less than it was during the herb-dominated period. It should be noted of course that precise estimates of this effect will vary with the ecological region being studied. And third, the identifiable taxa represented in the herb zones are predominantly perennial herbs characteristic of modern montane and high-arctic regions (sensu Yurtsev, Tolmatchev, and Rebristaya, 1978). The closest correspondence with modern pollen influx and floristic composition is found in mid-to-high arctic vegetation, in those open montane tundra communities on upper slopes and ridges in northwest North America, and on exposed, xeric sites with regosolic substrata in the low arctic of (e.g.) northern Yukon and adjacent Alaska. We suggest that the *zonal* vegetation (using the mesic site concept described by Richard—1976—and referred to in more detail below) consisted of discontinuous, open vegetation made up of combinations of the following common taxa (among others), all of which have relatively wide ecological amplitudes:

Grasses
Alopecurus alpinus
Deschampsia arctica
Festuca spp.
Poa spp.
Hierochloë alpina

Shrubs
Salix polaris
S. arctica
S. phlebophylla
S. reticulata
S. pulchra

Herbs
Saxifraga spp.
Papaver spp.
Ranunculus spp.
Draba spp.
Cerastium alpinum
Stellaria edwardsii
Cochlearia arctica
Lesquerella arctica
Oxyria digyna
Potentilla spp.
Dryas spp.
Polemonium spp.
Oxytropis spp.

Cyperaceae
Carex spp.
Eriophorum spp.

2) Similar pollen assemblages from elsewhere have been described by many workers from a large number of sites. We will cite only a few, but they are fully representative.

The highly significant discoveries from the Grande Pile site in northern France (Woillard, 1978a, 1978b) provide an example of the interpretation put on *Artemisia*-herb-*Salix* pollen zones by European paleoecologists. She identifies nine stadial episodes in the latest 65,000 years by these pollen zones, interpreted as representing open, discontinuous tundra or polar desert, and reflecting very cold, harsh climates correlative with a glacial climatic regime.

The full and late glacial vegetation of Scotland has been studied intensively in the last decade, in the light of radiocarbon, pollen, and other biostratigraphic and physical data from about 50 sites. It is instructive to note in passing that even this large number of sites, studied by modern techniques and with a high level of palynological expertise involving six to eight laboratories, has not been sufficient to solve all of the paleoecological problems of this relatively small area (Scotland $\approx$ 25,000 km^2; northern Yukon $\approx$ 100,000 km^2). Both the influx and percentage pollen data reported by Pennington (in Gray and Lowe, 1977) are very similar to those recorded in the northern Yukon and Alaska for stratigraphically equivalent (stadial-interstadial) episodes. Pennington (Gray and Lowe, 1977) interprets these data, supplemented by information on diatom stratigraphy and sediment geochemistry, as follows:

> The characteristic association of high *Artemisia* percentages with pollen of families which include many herbs of open ground (Caryophyllaceae, Compositae, Cruciferae) indicates a breakup at these sites of the continuous vegetation cover, for which the most obvious explanation would be an increase in frost disturbance, during freeze-thaw cycles, of ground not protected by snow cover.

Hyvärinen (1975) draws similar conclusions from more or less identical pollen data from sites in Finnish Lappland and his interpretation is: "an open, discontinuous vegetation composed of dominantly herbaceous communities, with some willow, but without, for example, ericads or much dwarf birch, i.e., of communities of bare mineral ground and disturbed soils."

Bell (1969) provides a perceptive analysis of the apparent paleoecological contradictions arising from the admixture of northern and southern "steppe" elements in full-glacial sediments from southern Britain. She concludes that the climate was generally cold, dry and continental, the last characteristic involving local high summer temperatures. She writes: (p. 917), "A large annual temperature range leading to active freeze-thaw processes would be consistent with the open character of the vegetation as well as with the southern species." She proceeds to evaluate the so-called steppe element in these fossil assemblages, and points out that: (p. 919) "It is immediately clear that the British taxa on this list are not members of steppe communities as such in their present habitats, and this same situation may apply to their presence in the full-glacial floras. As now, they may merely have grown on open, dry substrates, or in some cases on saline ones discussed later. *Artemisia* is a genus which may be quoted as a weed, xerophyte or halophyte, and therefore can only give generalized ecological information. Without knowing the species involved, it cannot be used to indicate a floristic resemblance to the Russian Steppe." She further concludes that the halophyte element in these assemblages (which of course has never been recorded from any site in eastern Beringia) was "very probably associated with saline soils formed over permafrost, resembling the situation in Yakutia today," a point that has been made fully by Giterman (1975) in discussing late Pleistocene assemblages from northeastern Siberia.

Fredskild (1973) interprets percentage and influx data from sites in Greenland which differ from ours only in the absence of *Artemisia* as follows: "To begin with, there was an open pioneer vegetation of widely distributed arctic plants, which included plants such as *Saxifraga oppositifolia* type, *S. nivalis* type, *S. caespitosa* type, *Erigeron* type, *Carex* and other Cyperaceae, *Poa, Puccinellia,* as well as other Gramineae, *Armeria, Koenigia, Oxyria, Cerastium cerastoides, Silene acaulis, Minuartia* type, *Huperzia,* and *Sagina.*"

It is important to note that so far no taxon has been recorded in any late Pleistocene deposit from Alaska-Yukon which can be ascribed to either a southern steppe element (as for example *Ephedra, Gypsophila, Helianthemum* pollen, common in European full- and late-glacial deposits) or to the tundra-steppes mapped by Yurtsev, Tolmatchev, and Rebristaya (1978, Fig. 2) in western Greenland and in the Kolyma and Indigirka River watersheds of northern Chukotka.

3) Modern vegetation pollen analogs and pseudo-analogs can be useful. In a recent illuminating treatise Richard (1977) has set out the procedures used by many botanist-paleoecologists in reconstructing past vegetation from pollen data. We subscribe to his basic premise that the first step in paleoecology is to reconstruct the vegetation. Others believe that one can translate fossil pollen stratigraphic data directly into paleoclimatic conclusions without the troublesome intermediate step of vegetation reconstruction (e.g., Webb and Bryson, 1972). The pivotal step, according to Richard (p. 97 ff) is to identify the zonal vegetation type reflected by the pollen spectra. He writes (p. 97): "Many authors emphasize the influence of topography on the type of vegetation distribution within a territory and, consequently, its bearing on the pollen spectrum produced by this vegetational cover. The notion of climax domain allows this factor to be integrated in a coherent whole." On page 99 he continues, "The advantage of this concept in the interpretation of pollen diagrams lies in the fact that it is an entity having climate as the paramount controlling factor, provided all the plants that could potentially participate in it are available. If it proved possible to characterize the various existing climax domains by well-defined contemporaneous pollen spectra, it would then be possible to trace the vegetational evolution through pollen diagrams and to express the results in terms of replacements of climax domains. The amount of pollen shed by the plant community (or communities) corresponding to the climatic climax (or the climax complex) will depend on the abundance of this (or these) community (or communities) on the territory and, hence, on the type of landscape (topography) controlling the distribution of the mesic sites." (Literal translation from French by Pierre Richard, at editors' request.) This is done either by a comparison with modern pollen influx data, and/or by adducing from the pollen data a reconstruction compatible with the maximum internal ecological concordance of the data.

We have abstracted the pertinent unpublished influx data from an earlier study by Ritchie and Lichti-Federovich (1965) of the modern pollen spectra of the northern grasslands of central Canada. These results (Table 1) are not directly comparable with influx values from lake sediments. However, they are the only available data from modern northern grasslands. They suggest that the annual influx of such ecosystems is 2000 grains/cm^2/year, more than an order of magnitude greater than values recorded from the northern Yukon herb zone sediments. It is also important to note that these modern prairie landscapes are heavily cultivated so that roughly 50% of the landscape produces little pollen and that the arboreal elements are partly from planted species of spruce and pine.

Is there a zonal type (or types) of vegetation in modern northwestern American (and/or northeastern Siberian) tundra landscapes which matches reasonably the pollen assemblages represented collectively or individually by the herb pollen zones? We suggest that there are several communities, described from Alaska and northern Yukon, which fit with the palynological data. One, perhaps the most important, is the widespread, open, discontinuous vegetation occupying shale, sandstone and granite and other noncalcareous residual regosolic soils on upper montane slopes and ridges, and xeric surfaces of the rolling plains region of the northern Yukon (Fig. 5). The vegetation composition is illustrated by the following particular *relevée* (Table 2), which is representative of a common regional vegetation of many upland tundra situations in Alaska-northern Yukon. (cf. Murray and Batten, 1977; Hettinger, et al., 1973). Cwynar (1981) presents pollen spectra from moss polsters in these upland fell field communities that resemble closely the herb zone assemblages.

Artemisia arctica is the most widespread of the eight species of the genus found in modern Alaska-northern Yukon tundras. It is characteristically associated on xeric uplands with *Salix* (*phlebophylla, glauca, arctica*), grasses (*Festuca*) and a wide variety of perennial herbs. The other species of *Artemisia* never, or rarely, occur as members of

Table 1

Pollen Influx Values for Five Sites in the Canadian Grasslands

(A detailed account of methods etc. as well as complete percentage data, is available in J. C. Ritchie and S. Lichti-Federovich, *Pollen et Spores,* 7:64–87, 1965)

Main Pollen Types	Sites: Pierson, Man. 101°15′/49°11′	Reston, Man. 101°05′/49°32′	Boissevain 100°03′/49°14′	Goodlands 100°36′/49°06′	Fort Shilo* 99°37′/49°48′	
Picea	12	18	2	14	24	
Pinus	106	74	101	100	694	
Populus	116	228	72	120	142	
Acer negundo	438	196	52	14	11	
Other AP	84	464†	368†	196	131	
Total AP	756	980	595	444	1008	pollen grains/cm^2/year collected in open dish samples during 1961–63. Mean values for the 3 year period are given.
Gramineae cereale	320	278	709	261	254	
Gramineae other	142	282	174	220	94	
Cheno-Amar	260	140	136	309	44	
Artemisia	190	92	70	162	470	
Compositae other	187	52	107	192	132	
Leguminosae	32	88	18	58	34	
Cruciferae	19	11	11	6	—	
Cyperaceae	20	26	20	12	16	
Other NAP	56	28	95	75	20	
Total NAP	1226	997	1340	1295	1054	
Total pollen	1982	1977	1935	1739	2062	

*Extensive plantations of pine
†Extensive riverine, elm, ash woodlands

a zonal vegetation type; they are locally common, sometimes locally dominant, in azonal, fragmented communities. *A. glomerata* is of local occurrence, found as a community dominant only on south-facing slopes with *Saussurea, Draba,* and *Arabis* in parts of the Brooks Range, Alaska (Murray and Batten, 1977). *A. tilesii* is locally common on unstable alluvial habitats and solifluction terraces. *A. unalaskensis* is confined to alpine meadows of the Aleutian-Pribilof Island chains. *A. borealis* is rare in eastern Beringia, never occurring as a community dominant. Elsewhere it is locally common, and it is of interest that it is the codominant of certain herb communities, referred to above, described by Bocher (1954) from south-facing slopes in western Greenland. He describes them (p. 184 ff) as "low arctic continental xeromorphic grasslands or steppes." However, this species is equally common in high-arctic polar desert types in Eurasia, e.g., in Novaya Zemlya (Alexandrova, 1977), where it occurs in open vegetation with *Saxifraga oppositifolia, S. hirculus, Luzula arctica* and *Deschampsia alpina.* Throughout the Yukon and adjacent Alaska and Alberta, *A. frigida* is locally dominant in communities on south-facing riverbank sites where steep bluffs of silty loam soils have been exposed by downcutting by major rivers. These sites are common in the Porcupine and Old Crow watersheds (northern Yukon), but it is significant that in the northern Yukon, and probably elsewhere, *A. frigida* is confined to these habitats as a local community dominant. It is totally absent from the Eagle Plain and South Richardson Mountains area for example. Young (1976) has suggested that these local *A. frigida-Calamagrostis* facies may be the survivors of the Pleistocene arctic steppe. We suggest that, at least as far as the period 30,000 to 14,000 BP is concerned, this is an improbable notion for several reasons.

1) These local communities lack most of the perennial herbs characteristic of the pollen assemblages of the herb zone and the only similarity is the generic one for the two taxa, *Artemisia* and *Gramineae.*

2) The modern physiognomic and floristic equivalent of such a zonal "steppe" would be the true steppes of south-central Siberia or the prairie

Figure 5. A sorted shale fell field with discontinuous vegetation at Hanging Lake

Table 2

The Floristic Composition of a Representative Tundra Community, Recorded on a Shale-sandstone Summit Ridge, Elevation 640 m asl, in the Doll Creek Area, Northern Yukon 66°04′N and 135°50′W

Vasculares	Musci
Abundant:	*Pogonatum dentatum*
Ledum decumbens	*Rhacomitrium lanuginosum*
Salix phlebophylla	*Polytricastrum alpinum*
Arctostaphylos alpina	*Schistidium gracilis*
Frequent:	*Dicranum groenlandicum*
Artemisia arctica	*Aulacomnium turgidum*
Betula glandulosa	*Ditrichum flexicaule*
Saxifraga tricuspidata	
Hierochloë alpina	**Lichenes**
Trisetum spicatum	
Dryas octopetala	*Dactylina arctica*
Empetrum nigrum	*Cladonia macrophylla*
Salix glauca	*Cetraria richardsonii*
S. arctica	*Icmadophila ericetorum*
Occasional:	*Asahinea chrysantha*
Pedicularis capitata	*Cornicularia divergens*
Minuartia arctica	*Cetraria islandica*
Saussurea angustifolia	*C. nivalis*
Lycopodium selago	
Festuca ovina ssp. *alaskana*	
Papaver macounii	
Smelowskia calycina	
Rhododendron lapponicum	
Salaginella sibirica	
Lagotis glauca	
Juncus triglumis	
Carex bigelowii	

region of the northern plains of North America. The pollen influx values for the latter zone (Ritchie and Lichti-Federovich, 1965) are 10 to 20 times the values determined so far from the fossil assemblages (Table 1).

3) The modern south-facing river bluff surfaces are characterized by a local environment with very high net radiation and high plant and soil temperatures. A reconstruction which would cover extensive level or rolling surfaces with this vegetation would require zonal paleoclimates well beyond any mean values conceivable for stadial conditions at high latitudes (CLIMAP, 1976).

4) These sites have fine grained silty soils with a deep active layer permitting the deep-rooting *A. frigida* and *Calamagrostis* to establish. Such a soil environment, reconstructed zonally, would imply a thermal regime totally incompatible with the regional paleoclimates suggested by all other evidence for the same period.

Additional floristic and structural modern equivalents of these reconstructions have been described recently in detail from the upper slopes of Sheep Mountain, southwest Yukon, by Hoefs et al. (1975), such as their *Oxyrio* (*digynae*)-*Artemisietum arcticae-tilesii* association, the *Artemisio* (*arcticae*)-*Salico* (*reticulatae*) *Festucetum altaicae*, as well as tundras with abundant *Artemesia alaskanum* and *A. hyperboreum* (=*A. furcata*) on screes and colluvial regosols. On the other hand, their *Artemisia frigida* grasslands occur at low elevations and have associated species (Hoefs, et al., 1975, Table XVI, p. 161) of southern, prairie affinity completely absent from *A. frigida* communities in northern Yukon and from the three pollen zones at Hanging Lake and Lateral Pond. We suggest, however, that species of *Artemisia* other than *A. arctica* might well have been locally important in these late Pleistocene Beringian communities, and in particular we point to those species that are ecologically very similar to *A. arctica*—i.e., adapted to exposed, regosolic sites. Possible species are *A. borealis,* which occurs on talus slopes locally in northern Alaska (Johnson, et al., 1966); *A. globularia; A. glomerata;* and *A. furcata,* all of which are locally common in *Dryas* fell fields, frost scar habitats, talus slopes and *Dryas* stone-stripe habitats on the north slope of Alaska (Johnson, et al., 1966; Murray and Batten, 1977).

Currently we are engaged in a scanning electron microscope survey of northern hemisphere arctic-subarctic *Artemisia* pollen, with the ultimate aim of identifying the fossil grains from the herb zone with whatever species or species group can be differentiated by pollen morphology. Meantime, we suggest that the most probable interpretation of the herb pollen zones in northern Yukon (and elsewhere) is in terms of an open, discontinuous tundra, floristically and physiognomically similar to the communities found commonly today in northern Yukon and Alaska on xeric, regosolic sites, beyond both the montane and arctic treelines, and which are very similar to the community described by the list in Table 2. The *Artemisia* species involved was in our opinion probably *A. arctica,* the ecological vicariad of *A. norvegica,* which has been identified as the species represented in equivalent stadial pollen zones in northwestern Europe.

Furthermore, we suggest that the mixed ecological assemblage of heliophilous, chionophilous, and calciphilous elements at the Doll Creek sites, southern Richardson Mountains (Ritchie, 1982), can be interpreted as a mosaic of snow patch communities (*Oxyria, Ranunculus pygmaeus, Salix reticulata, Cassiope*), and open communities of apophytes adapted to regosolic, unstable soils. These communities are not extinct. They have persisted to the present day, but their areas and habitat ranges have been constricted by a great variety of physical and biotic factors in the local environment: by competition from species adapted to developed soils with humus accumulation and nutrient leaching (ericads, juniper); by competition from species whose expansion has been stimulated by climatic amelioration (e.g., *Betula glandulosa,* ericads); and by shading and other microclimatic and edaphic influences induced by the spreading spruce, poplar, and large woodlands. Our current thrust is to accumulate pollen and macrofossil data from the herb pollen zones (and others) in sufficient detail and with the necessary close-interval time control to work out quantitatively the dynamics of the vegetation in terms of chronologies of invasion, stability, and replacement, along the lines suggested recently by Watts (1973).

One of us (Ritchie) has developed a simple working hypothesis to account for the pollen zonation at Lateral Pond, in the form of a preliminary conceptual framework (Fig. 6) that is compatible with the fossil facts.

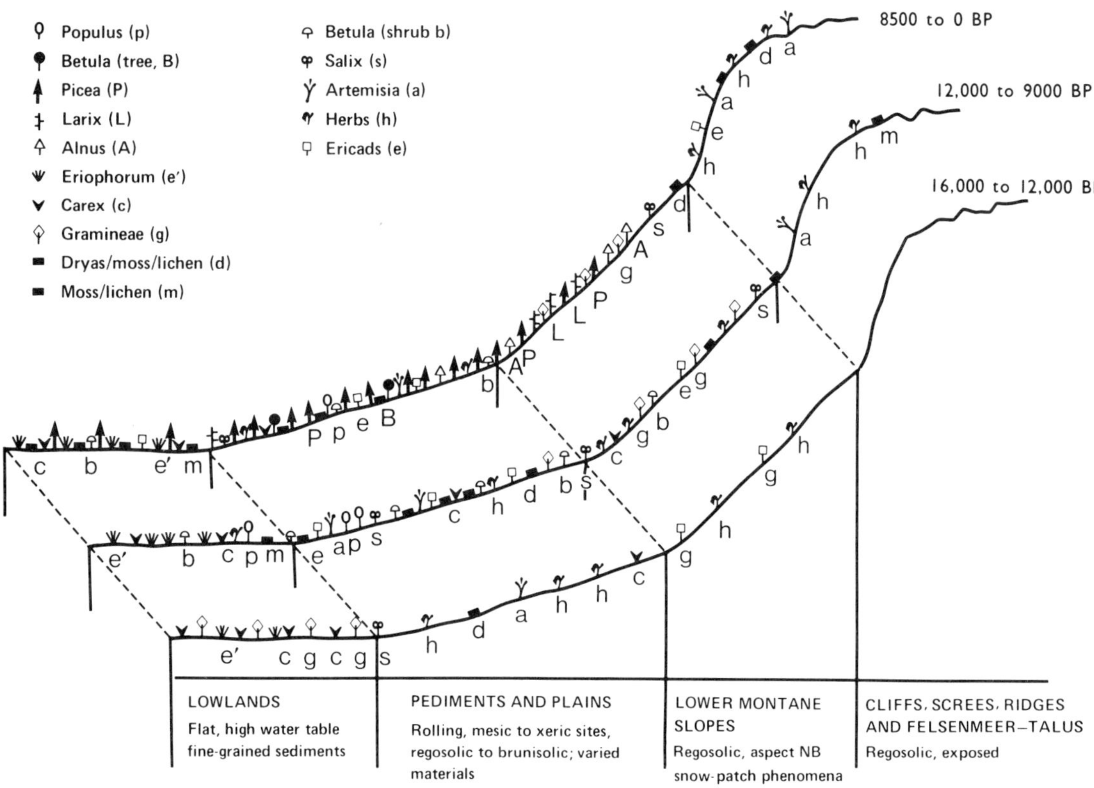

Figure 6. A highly schematic diagram summarizing the physiognomy, approximate floristic composition and habitat ranges of the vegetation reconstructed for the three main periods of the late Pleistocene and Holocene, as interpreted from pollen diagrams of sites in the Doll Creek valley of the northern Yukon. The central thesis is that the open herb tundra of the first period becomes progressively restricted to higher and/or more xeric sites but retains its floristic and physiognomic characteristics in the process

1) The communities that formed the pre-14,000 BP tundras, i.e., the herb zone tundras, persisted throughout the late Pleistocene and Holocene but became progressively confined to montane ridges, local scarps and screes by the increasing populations of shrubs and trees.

2) The rise of birch pollen curves at 14,000 BP signals an amelioration of climate favoring the development of closed herb-shrub tundra on all lower slopes and rolling plains and an extension of mire-forming communities (fens, bogs) on lowland sites. There is increasing evidence that this was a global paleoclimatic event. These tundras would have been the most productive in terms of net primary production and biomass of the entire period from 30,000 BP to present. A paleoclimate as warm as present is suggested. This is in agreement with the independent evidence from ostracod analysis by Delorme et al. (1977), taking account of the pertinent caveat added by MacKay (1978). The herb-tundra communities became confined to the competition-free habitats on montane ridges and xeric-regosolic sites of the plains.

3) The spruce and tree birch, and later alder, zones give in the percentage diagrams an exaggerated conception of real vegetation change. It is suggested that the changes between 11,000 and 6500 BP involved little modification of montane ridge and upper-slope habitats, while the closed heath-birch tundras of mesic and lower-slope sites were invaded by spruce, local poplar and scattered birch, converting them into modern northern boreal woodland and no doubt producing differential responses in the population dynamics of the various low shrub and herb elements. The invasion of alder in the mid-Holocene is interpreted as a minor physiognomic change in the general structure of the woodland vegetation.

A recent reconstruction of late Pleistocene and Holocene vegetation for Eurasia, based on numerical transfer function techniques (Avenarius, et al., 1978, Fig. 4), gives a general indication of the extent of tundra types in western Beringia for 18,000 BP and for a mid-Holocene episode. It is significant that the mean July temperature reconstructed for 18,000 BP is 0°C. However, it would be premature to draw firm conclusions from these general reconstructions. So far no site in western (Siberian) Beringia has been reported with primary, dated lacustrine sediment from late Pleistocene time.

Finally, we note two significant facts which also argue strongly against the primary proposition of the arctic-steppe biome concept that, "during the late Pleistocene these regions (Alaska and northern Yukon), when joined with central and eastern Asia by the Bering Land Bridge, constituted the eastern end of a huge grassland biome, unequalled today in size or character" (Matthews, 1976). Such a late Pleistocene Beringian continuity, lasting for several thousand years, is incompatible with the fact that the modern vegetation of the land masses on adjacent sides of the Bering Strait displays more differences than similarities between the two land areas (cf. among many authors, the descriptive accounts and maps in Lavrenko and Sochava, 1956; in Alexandrova, 1977; in Viereck and Little, 1972). And secondly, while amphi-Beringian floristic elements do exist ($\approx$ 100 species) (Hultén 1968), the total vascular plant complexes of the two land areas are well-differentiated taxonomically and floristically and it is not surprising, therefore, that floristic phytogeographers (Yurtsev, 1974b) and taxonomists (Johnson and Packer, 1967) have concluded that the significant floristic connections occurred, probably several times, during late Tertiary and early Pleistocene times.

As a corollary of these conclusions we suggest that the "arctic-steppe biome" never existed in the northern Yukon (nor in Alaska) during the most recent stadial-interstadial cycle (25,000 to 0 BP), but this position does not pose a serious paleoecological impasse since, as it happens, there is so far no evidence from that period of large numbers of either species or individuals of herbivorous vertebrates in the far northwest of the continent.

ACKNOWLEDGMENTS

Research for this paper was supported by Northern Yukon Research Programme, Contribution No. 19, University of Toronto; and by Grant A 6320, National Science and Engineering Research Council of Canada. The senior author's contribution was prepared during the tenure of a Senior Killam Scholarship of the Canada Council.

7

EAST BERINGIA DURING LATE WISCONSIN TIME: A REVIEW OF THE BIOTIC EVIDENCE

J. V. Matthews, Jr.

The nature of the late Pleistocene environment of Beringia is a topic of current debate. Some have concluded that it was a large arctic-steppe biome which supported an ungulate fauna as diverse as that of certain temperate grasslands. Others argue that the late Wisconsin vegetation of much of East Beringia was similar to present arctic and alpine fell field. They claim that the evidence for either large numbers or diversity of ungulates during late Wisconsin time is deficient or nonexistent.

Although comparisons with the fauna of the African plains are inappropriate, several lines of evidence do show that the number of large-mammal species existing together in East Beringia during the late Pleistocene was several times larger than that of the existing tundra fauna. Prominent were *Mammuthus primigenius, Bison priscus, Equus,* several types of musk ox, and *Rangifer.* In spite of the apparent low pollen production in late Pleistocene plant communities, sufficient forage must have existed to support this unique ungulate community.

Herb-dominated pollen spectra representing the height of the late Wisconsin have attributes, such as high percentages of *Artemisia* and dominance of grass over sedge, that set them apart from the majority of lowland tundra surface samples; consequently, these herb-zone fossil spectra are thought to indicate an environment different from that of present tundra.

Macrofossils of plants and insects add dimension to this picture. They show that late Pleistocene treeless environments contained species and probably whole communities not found in East Beringia today. Many of the plant and insect fossils imply prevalence of steppe-like conditions.

When all evidence pertaining to Beringian environments is collated, the most probable reconstruction is of a treeless region composed of a mosaic of communities, among which were large tracts that can only be termed steppe-like. It is probable that both alpine fell field and some of the rare steppe sites in present day Alaska-Yukon harbor species which formerly occupied these steppe areas.

The late Wisconsin has been viewed as the last time when presently disjunct steppe species in the U.S.S.R. and East Beringia had continuous ranges. But if the land bridge itself acted to promote increased continentality in Beringia, then the exchange of steppe taxa may have occurred during the mid-Wisconsin or even earlier. Certainly, it is becoming clearer that the land bridge acted as a filter for certain steppe organisms during late Wisconsin time.

Key words: East Beringia, Alaska, Yukon, steppe, tundra, Carabidae, *Lepidophorus, Morychus, Chenopodium, Potentilla,* woolly mammoth, mammoth fauna, herb zone, late Wisconsin

ISBN 0-12-355860-3

INTRODUCTION

For English-speaking audiences the term Beringia first became well known in 1937 when the botanist Eric Hultén used it as an appellation for the Alaskan-Siberian land bridge that he postulated in order to explain the distribution of certain groups of plant species (cf. Hultén, 1968). Numerous neontologists and paleontologists accepted the existence of such an Asian-North American land connection, but the concept was not given strict form and temporal scope until Hopkins' (1959a) well-known 1959 publication. Since that date our concept of Beringia has changed, and many authors now use it, as I do here, to mark the entire region between the Kolyma River in eastern Siberia and the Mackenzie River in the Canadian Northwest Territories, including intervening continental shelf areas shallower than approximately 200 m.[1]

Regardless of how Beringia is conceived, it was, even in its most restricted form, an area that provided the primary late Cenozoic terrestrial link between the eastern and western hemispheres (Matthews, 1979a). Hence the history of Beringian environments has always been and will continue to be a prominent issue among paleontologists and biogeographers. At least five international meetings have dealt with Beringia. At two of those I presented papers (one an abstract) portraying the environment of late Pleistocene Beringia as a vast cold grassland—an "Arctic-steppe biome" (Matthews, 1976a, in press)—which was qualitatively and quantitatively different from extant tundra. Arctic steppe rather than arctic tundra was proposed as the environment in which Late Paleolithic man lived when he first penetrated the western hemisphere via Alaska and the Yukon. Following the lead set by others, I further stated that arctic steppe as a biome was presently extinct, having become so due to the combined destabilizing influences of warming climate and hunting by man.

These conclusions and generalizations have drawn criticism (Schweger and Habgood, 1976; Ritchie, 1977, 1980; Cwynar and Ritchie, 1980; Cwynar, 1980; Ritchie and Cwynar, this volume), while others, in my opinion, have too quickly accepted them. There is a real need for an overall review of the salient evidence relating to the arctic-steppe question. Such a review is the objective of this paper. In it I will focus on the environment of East Beringia (Alaska-Yukon) during late Wisconsin time, the period between approximately 25,000 and 10,000 years ago, when world climates were as cold (and/or as arid) as at any other time during the Pleistocene (Alekseyev, 1978; Giterman, et al., this volume). In doing so I propose to address a number of the issues also discussed by Ritchie and Cwynar in their contribution to this book, namely: (1) special aspects of the late Pleistocene pollen record, (2) the reality and reliability of the vertebrate fossil record, (3) other macrofossil and stratigraphic evidence, and (4) possible existing relicts or analogs.

When all facts are considered, there still appears to be a valid case for the existence in late Pleistocene East Beringia of an environment which differed from present-day tundra, most probably by containing within a mosaic of communities some large tracts that were physiognomically steppe-like. As a name for this environment the terms steppe-tundra or tundra-steppe seem appropriate. The term "arctic steppe" or references to an "arctic-steppe biome" (Matthews, 1976a) are misleading and should be abandoned.

POLLEN RECORD IN EAST BERINGIA

Paleoecological studies in Alaska and the Yukon Territory (Ager, Schweger, Ritchie and Cwynar, this volume) provide the data base for a general summary of late Pleistocene vegetational change in East Beringia. The sequence outlined below, based mainly on palynology, is intended as a framework for the later discussion of specific aspects of the late Wisconsin record.

Synopsis of Late Pleistocene Vegetation Change

The environment of the last interglacial in East Beringia is poorly known. Data from sites in western Alaska (Colinvaux, 1967a; Matthews, 1974b) and several localities on the northern Yukon coast (Rampton, in press) suggest that large areas of East Beringia were forested and that spruce and other

[1]B. A. Yurtsev's "Megaberingia" (1974) encompasses an even larger area, extending well west of the Kolyma River in Asia, and includes the Aleutian Archipelago and much of northwestern North America (cf. Yurtsev, this volume).

taiga species occurred north and west of their existing range during the Sangamon Interglaciation (Matthews, 1974b; Rampton, in press; D. M. Hopkins, Cape Espenberg data, personal communication 1975).

Early and middle Wisconsin environments are also poorly known. However, sites in northern Yukon Territory and a recently studied core from Imuruk Lake contain evidence of an early and mid-Wisconsin interstadial complex during part of which climate may have been warmer than present (Shackleton, 1979; Matthews, 1980). Spruce forests probably were more closed than they are today and contained shrub species that presently do not grow as far north (Matthews, unpublished). The period of deep thawing that is recorded at many silt exposures near Fairbanks, Alaska (Péwé, 1975a), occurred, in my opinion, during this interstadial rather than during Sangamon time (Matthews, 1979b, 1980).

According to pollen and macrofossil records from the Yukon and central Alaska (Rampton, 1971; Lichti-Federovich, 1973, 1974; Matthews, 1974a; Hughes, et al., 1981) spruce forest existed intermittently in the central part of East Beringia until 30,000 years ago. Significantly, some of the plant species growing in Alaska and the Yukon during the mid-Wisconsin are not members of the existing flora (Matthews, 1974a; Hughes, et. al., 1981).

East Beringia apparently lacked forests during the interval between 25,000 and about 10,000 years ago (Hopkins, Smith and Matthews, 1981). Until about 13,000 years ago shrub birch, alder, and many of the ericaceous shrubs characteristic of modern tundra were also rare. Shrub birch began to invade most areas about 14,000 years ago (Ager, 1975, this volume; Ritchie, 1977, 1980). Alder pollen percentages show a significant rise somewhat later, usually following the rise of the spruce curve at about 10,000 years BP. Most Alaska-Yukon sites, whether lakes from which cores have been raised or alluvial sections, show that prior to 14,000 years ago the treeless vegetation of East Beringia was dominated by grass and/or sedges and contained a diversity of herbs, among them *Plantago*, Cruciferae, Chenopodiaceae, and *Artemisia*. The high percentage of *Artemisia* pollen in many herb-zone spectra has been considered a sign that late Pleistocene "tundra" differed from modern tundra (Matthews, 1974b, 1976a). The abundance of grass pollen is interpreted by some as an indication that late Pleistocene vegetation in East Beringia was steppe-like.

Proponents of this steppe-tundra hypothesis have focused on the fact that modern tundra does not yield surface pollen spectra similar to those of the herb zone. They conclude that high *Artemisia* percentages imply better-drained substrates and possibly even deeper seasonal thawing than occurs at existing lowland tundra sites. Drawing on pollen and other lines of evidence, I suggested in 1976 that Beringia once formed an "arctic" (for its locus) "steppe" (for its vegetation) biome, which no longer exists.

A different viewpoint on late Wisconsin environments is detailed by Ritchie and Cwynar (this volume). Their reading of pollen evidence from the herb zone (especially the low yearly influx) rules out steppe-tundra and arctic steppe in favor of expanded fell-field vegetation similar to what exists in present-day montane and high-arctic regions. Only valley bottoms would have been dominated by sedges and grasses (Ritchie, 1977, 1980; Cwynar and Ritchie, 1980; Cwynar, 1980; Ritchie and Cwynar, this volume). They have also, by extended pollen counts and critical identification of many herb-pollen types, sought the floristic affinities of the herb zone. To them the pollen spectra are more clearly arctic-alpine than cold or temperate steppe. Because the influx of herb types does not drop at 14,000 years when birch percentages begin their rise, that transition is seen to mark a rather minor invasion of shrub birch under conditions of ameliorating climate rather than the abrupt change from steppe to shrub tundra postulated by others. They conclude that the warmer, more moist climate allowed birch-heath shrub tundra to occupy the more mesic sites, forcing herb tundra into marginal and xeric montane sites where it exists today as a relict of the vegetation that once characterized much of East Beringia.

These opinions represent a new and important view on the environmental history of Beringia and warrant detailed examination.[2]

---

[2]Ritchie and Cwynar are concerned almost exclusively with pollen evidence, whereas the reconstruction attempted in this report and elsewhere (Matthews, 1976a; Schweger, this volume) is based on the combined evidence from vertebrates, insect and plant macrofossils, stratigraphy, and pollen.

Herb-Zone Characteristics

Table 1 shows that amounts of *Artemisia* greater than 5% are found now in less than 12% of surface samples from tundra or forest-tundra sites in Beringia. Traces of *Artemisia* pollen occur in slightly more than half the samples. Many of the samples listed in the table are polsters, and thus are particularly subject to local overrepresentation of *Artemisia* (e.g., the polster for the 18% Tangle Lake sample was near flowering *A. tilesii*). A more accurate representation of the status of *Artemisia* in modern tundra would result if samples were restricted to surface muds from lakes. Unfortunately, the number of such samples listed in Table 1 is small, and for many the type of sediment yielding the surface spectrum is not indicated in the literature. Nevertheless, a scan of Table 1 (particularly the series from Ritchie, 1974) suggests that tundra surface samples from lakes rarely contain more than 5% *Artemisia.* In general then, despite occasional exceptions, the percentage of *Artemisia* pollen is low in tundra and forest-tundra surface samples from East Beringia. Note how Ritchie's modern samples contrast with herb-zone spectra (Table 2), most of which contain upwards of 10% and some as much as 40% *Artemisia* pollen. Ritchie and Cwynar (this volume) and Ritchie (1977) quite correctly point out the error of specifying former vegetation on the basis of high *Artemisia* percentages because the genus has a wide ecological amplitude and because individual species can not yet be distinguished on the basis of pollen alone. But regardless of which species the pollen represent, it is clear that during the late Pleistocene, in both Alaska-Yukon and eastern Siberia (Giterman, et al., this volume) *Artemisia* was more widespread than at present. Since nearly all *Artemisia* species grow where substrates are well drained and plant cover is low, high *Artemisia* percentages probably represent a reduction of the extent of the poorly drained, sedge-moss tundra typical of many lowland areas today. Most peaks of *Artemisia* occur at a time when Beringia was severely cold, essentially treeless and characterized by intense aeolian activity (Hopkins, this volume). This then is the physical backdrop for the high *Artemisia* percentages and other peculiarities of the herb-zone assemblages.

Some sites, like Hanging Lake in northern Yukon Territory, record no significant drop in *Artemisia* influx above the herb zone (Cwynar, 1981), while others of late Wisconsin age show abnormally low percentages. Hanging Lake is in the uplands of the Yukon and has probably been a suitable site for some species of *Artemisia* throughout the Pleistocene. Influx of *Artemisia* does decline above the herb zone in cores from lowland sites (Table 2). The fact that some samples show rather low percentages is evidence of the mosaic character of late Pleistocene vegetation (Schweger, 1976, this volume).

Tables 1 and 2 also illustrate another way in which tundra-surface pollen samples contrast with herb-zone spectra. In tundra samples, the percentage of sedge is often higher than that of grass, while the opposite condition exists in most herb-zone spectra. Also, herb-zone spectra usually contain a greater diversity of herb-pollen types and significant amounts of Chenopodiaceae and *Plantago.*

The low pollen influx (50–150 cm^2/yr) recorded by most herb-zone sequences is thought to rule out any possibility of a grassland or steppe environment because contemporary grasslands have influx values ten times higher (Ritchie, 1977; Cwynar and Ritche, 1980). Values of less than 100 gr/cm^2/yr suggest "a very sparse vegetation cover" (Cwynar and Ritchie, 1980, p. 1377; cf Birks, 1973). I advocate caution in assuming that pollen influx bears a direct and simple relationship to plant cover. The pollen rain at a modern site may vary greatly from year to year (Ritchie, 1977), and hence several years of data are required for reliable estimates of contemporary influx. In most cases influx estimates have been based on only one or two years of sampling. Also, few have attempted to relate pollen influx at a site to local vegetation, but where this has been done (Ritchie, 1977), site-related factors are shown to significantly influence the pollen influx (e.g., Maher, 1972; Hyvärinen, 1976).

Low plant cover is undoubtedly the prime reason why polar desert sites show low influx values (Ritchie and Lichti-Federovich, 1967), but the tendency of all tundra areas to exhibit low pollen production (150–800 gr/cm^2/yr) may be due in part to the propensity for asexual or intermittent reproduction among plants growing at high latitudes (Mosquin, 1966; Billings, 1974).

Table 1
Special Features of Surface Samples from Arctic and Subarctic Sites in North America

Region	Tundra Samples	Forest-Tundra	Artemisia (trace)[1]		Artemisia > 5%		Sedge % > Grass %		Type of Sample
			Tundra	F-tundra	Tundra	F-tundra	Tundra	F-tundra	
S. Yukon Territory (Birks, 1977)	10		10		1		8		Polsters
	6		6				6		Surface mud
S. Yukon Territory (Rampton, 1971)	3		No data		No data		3		Pond
	2		No data		No data				Polster
N. Yukon (C. E. Schweger, pers. comm. 1979)		2		2				2	Pond mud
N. Yukon (Cwynar, 1980)[4]	3		1		2				Polster
			2				2		Tauber traps[5]
		5			4			5	Tauber traps[5]
N. & W. Alaska (Nelson, 1979)	43		18		6		28		Polster and lake mud
Tangle Lks., Alaska (Ager & Sims, 1981)		9		8		1 (18%)		2	Soil and polsters
N.W. District of Mackenzie (N.W.T.) (Ritchie, 1974)	22		20		1		21		Lake/pond
	14		9				4		Polster
		8		8				7	Lake/pond
		21			4			11	Polster
M. Lake (Ritchie, 1977)		1				1		1	Lake mud
		2		2				2	Open dish[2]
N. Central N. Am. (Lichti-Federovich and Ritchie, 1968)	14		14				9		Surface mud
Arctic-subarctic Canada (Ritchie and Lichti-Fedrovich, 1967)	6		5				5		Open dish
		8		8				3	Open dish
N.E. North America (Davis, 1967)	22?		1				22		Polster and lake mud
N. Quebec (Bartley and Matthews, 1969)	1		1				1		Polster[3]
N. Quebec (Bartley, 1967)	18		14				10		Polsters and lichens
Baffin Island (Andrews, et al., 1979)	6		3				4		Moss polsters
Totals and %	170	56	(61%)	(50%)	(10.5%)	(7.1%)	(73.5%)	(55.4%)	

[1] In most cases less than 1%, but a few up to 4–5%
[2] Open dish samples taken in two successive years at same site
[3] 0–2 cm depth in *Sphagnum* cushion (Table 2, p. 57, Bartley and Matthews, 1969). Ten samples were taken from the same polster. All have the same characteristics—≈ 0.75% *Artemisia* and the sedge/grass ratio ≈ 3
[4] Includes data from Fig. 9 and Table 6 (Cwynar, 1980)
[5] Average of 2 or more traps from the same site

Table 2
Characteristics of Herb-Zone Spectra from East Beringian Lacustrine Sequences[1]

Site	Artemisia %	Grass/Sedge Ratio	Influx Data	Remarks
Hanging Lake, N. Yukon (Cwynar, 1980)	10–40% 18/39 samples ⩾20%	2–4, most sedge <10%	very low: <50–100 gr/cm²/yr	Zones HL 1A, HL 1B: Relatively high % Cruciferae and *Chenopodium* types. *Salix* less than 10%.
Lateral Pond (Ritchie and Cwynar, this vol.; Cwynar and Ritchie, 1980)	30–40% prior to 15,200 years	>1; ≈2 for most samples	≈200 gr/cm²/yr at 15,000 years	Influx of *Artemisia,* Cruciferae, *Plantago,* Tubuliflorae, declines slightly above herb zone; grass influx shows significant decline; *Salix* % relatively low
M. Lake, Inuvik area, N.W.T. (Ritchie, 1977)	10–35% (3 samples)	>1<2; 5 in one sample	150–1200 gr/cm²/yr	Zone 1: (Upper boundary of Zone 1 is shown higher in influx diagram): rise of *Betula* and end of herb zone is relatively late. Influx of *Artemisia* and grass drops after 11,300 BP; *Salix* >20%; very few samples represent true herb zone
Birch Lake II, Alaska (Ager, 1975)	All >10%; up to 30–50%	>2 in 5/6 samples	≈300 in lowest samples (excl. *Myriophyllum*)	Zone 1: influx of *Artemisia* declines at top of herb zone; grass influx remains constant; sedge influx rises in Zone 2 (*Birch* zone); *Plantago* % higher in herb zone than remainder; *Salix* 3–15%
Lake George, Alaska (Ager, 1975)	20–40%	>1, ≈2 in most samples	none	Zone 1: % *Plantago,* Tubuliflorae, Chenopodiaceae higher than in rest of core but no more than ≈5%; *Salix* 5–10%
Antifreeze Pond, S. Yukon (Rampton, 1971)	All <5%	6/10 samples 1 or >1	none	Zone 3a: Cruciferae and *Plantago* major peak (<10%); lowest *Spagnum* percentages. Possible thaw lake sequence
Imuruk Lake Core I (Colinvaux, 1964)	Zone J: 5–10+% Zone H: 20+% Zone G: 10–20%, ≈15%	Zone J: >1 Zone H: ≈1 Zone G: >1	none	Herb type spectra appear at several levels in sequence. Characteristic of type 0 and I assemblages
Imuruk Lake Core V (Shackleton, 1979)	Zone H: 15/21 >10%; up to 25% Zone J: 3–20%; most <10%	Zone H: >1 Zone J: >1 in most samples	none	Zones H and J: Zone H may represent early Wisconsin; J, late Wisconsin. *Thalictrum* peak in H; *Sphagnum* % low in both zones; *Salix* up to 10% in H, variable and up to 15% in J
Chandler Lake, Alaska (Livingstone, 1955)	5–15%, but some much less than 5%	<1; <0.2 near base	none	Zone I: no dates, but occurs below rise of *Betula* curve; *Salix* up to 20%
Tungak Lake, Yukon delta (Ager, this volume)	Variable, 5/16 >10%; most 5–10%; peak ≈20%	>1 in 7/16 samples; some <0.5	none	"Herb zone," below 14,230 BP date and rise of *Betula*; 5–10% *Salix*

[1]Percentage values and ratios derived by inspection of published diagrams, not from raw counts

Herb-Zone Floristics

By performing long counts and devoting much effort to critical generic and specific identifications of many minor pollen types, Ritchie and Cwynar in their individual and joint studies have done much to hone to a cutting edge what some have called a "blunt paleoecological instrument" (Colinvaux, 1967a). But it is important that conclusions not exceed the resolving capacity of modern palynology for representing past vegetation and floras. For example, in seeking to find the closest existing zonal vegetation type to that represented by herb-zone spectra, Ritchie and Cwynar (this volume) provide a list of genera and species that may have constituted such communities. Some of the genera listed (such as *Ranunculus, Potentilla,* and *Draba*) contain many species, most of which can not as yet be distinguished on the basis of their pollen. Moreover, of those that can, some (e.g., the crucifer *Lesquerella arctica*) are represented by only a few grains in widely separated samples (Cwynar, 1980, Fig. 11). In my opinion, reconstructions relying on such data must be viewed as very tentative. I show in a later section that the floristic character of the herb-zone spectra do not, at present levels of pollen identification, allow us to discriminate steppe from fell field.

Modern Fell-Field Spectra

Ritchie and Cwynar (this volume, p. 122) state that moss polsters from upland fell-field communities yield spectra that "resemble closely the herb-zone assemblages." They point out that the pollen data from the herb zone are perhaps most conformable with the widespread, open, discontinuous vegetation occupying shale, sandstone, and granite, and other noncalcareous residual, regosolic soils on upper montane slopes, ridges, and xeric surfaces of the rolling-plains region of the northern Yukon (p. 122). Table 3 portrays the surface pollen spectra from several such sites and compares them to spectra of the herb zone. The one from Old Crow mountain is most similar, especially when the presence of *Betula, Alnus,* and *Picea* is discounted, but the pattern is not as evident in other fell-field samples. Moreover, in making such comparisons it must be remembered that herb-zone fossil spectra do not all come from upland sites having regosolic soils. Many originate in lowland areas that were subject to intense aeolian activity and possessed deep alluvial or loessic soils (Hopkins, this volume). According to macrofossil data such sites supported herb-dominated vegetation somewhat different from the lowland vegetation postulated by Ritchie and Cwynar, but the differences are not of the type that are obvious from the pollen data alone.

Table 3 also includes the surface spectrum from a relict-steppe site on Wrangel Island. Both Yurtsev (this volume) and Murray (1981) claim a similarity to the herb-zone spectra, but there is an important difference—the high percentage of *Selaginella sibirica.* Siberian tundra-steppe pollen spectra differ in the same way from herb-zone spectra in Alaska-Yukon (Giterman, et al., this volume). Ritchie and Cwynar also note that another difference between East and West Beringian samples is the absence of pollen of halophytes in the latter. In fact, some halophytes do occur in the East Beringian macrofossil record, but they are species which as yet cannot be distinguished by pollen alone.

PLANT AND INSECT MACROFOSSILS

Macrofossils of plants, such as fruits and seeds, provide valuable additional insight to the past vegetation of Beringia. Because many "seeds" can be identified to genus and species, they offer a better source of floristic information than does pollen evidence. In addition, plant-macrofossil assemblages tend to reflect the local vegetation more than pollen does; consequently, accurate regional extrapolations depend on the repeated occurrence of similar assemblages.

Insect fossils, chiefly those of beetles (Coleoptera), have also begun to contribute new and detailed information on Beringian environments (Matthews, 1968, 1974a, 1974b, 1975; Kiselyov, 1979; Sher, et al., 1979). Identifiable fragments of Coleoptera and other insects occur in almost all organic sediments, especially in northern regions where permafrost enhances preservation. Fossil insect assemblages are usually more diverse taxonomically than plant-macrofossil floras, and the greater mobility of insects compared to plants makes them more sensitive to rapid fluctuations of climate.

Figure 1 presents a comparison of macrofossil assemblages from several Holocene and late Pleistocene sites in East Beringia. For comparison it

Table 3
Comparison of Herb-Zone Fossil Spectra with Fell-Field Surface Samples

Pollen Spore Type	SURFACE ALASKA-YUKON Circle #4 (Matthews, 1970)	Loc. A (Rampton, 1971)	Old Crow Mt.[1] (Cwynar, 1980)	Hanging Lk.[2] (Cwynar, 1980)	SIBERIA Wrangel I.[4] (Yurtsev, this vol.)	FOSSIL HERB-ZONE SPECTRA Hanging Lk. (Cwynar, 1980)	Birch Lake II (Ager, 1975)	Tungak Lake Yukon Delta (Ager, this vol))
Arboreal and Shrub								
Pinus					+			
Picea	5%	36%	10.6%	15%		2–15%	<2–6%	<10%
Betula	40%	11%	23.9%	25%	+	5–20%	<10%	most < 5%
Alnus	28%	29%	21.4%	38%	+	2–20%	< 5%	most < 2%
Salix	5%	7%	16.5%				3–17%	5–12%
Herbs								
Grass	<3%	10%	14.8%	0.5%	15%	10–30% ($\mu \approx 15\%$)	18–50%	15–30%
Sedge	10%	10%	1%	10%	5%	<10%	3–25%	10–40% (most > 20%)
Artemisia	>3%		8.4%	7.5%	50%	10–40% ($\mu \approx 15\%$)	>10% (up to 50%)	>5–20%
Other herbs					30%	≈20% (excl. aquat.)	<10%	
Spores								
Sphagnum						+	+	
Selaginella sibirica			0.6%		97%[3]	+		

+ indicates trace

[1]Average of values shown for three polster samples in Table 6, Cwynar (1980)
[2]From Fig. 10, Cwynar (1980)
[3]Percent of all pollen and spores. Values for herbs = % of sum excluding spores
[4]Relict steppe site according to Yurtsev, this volume

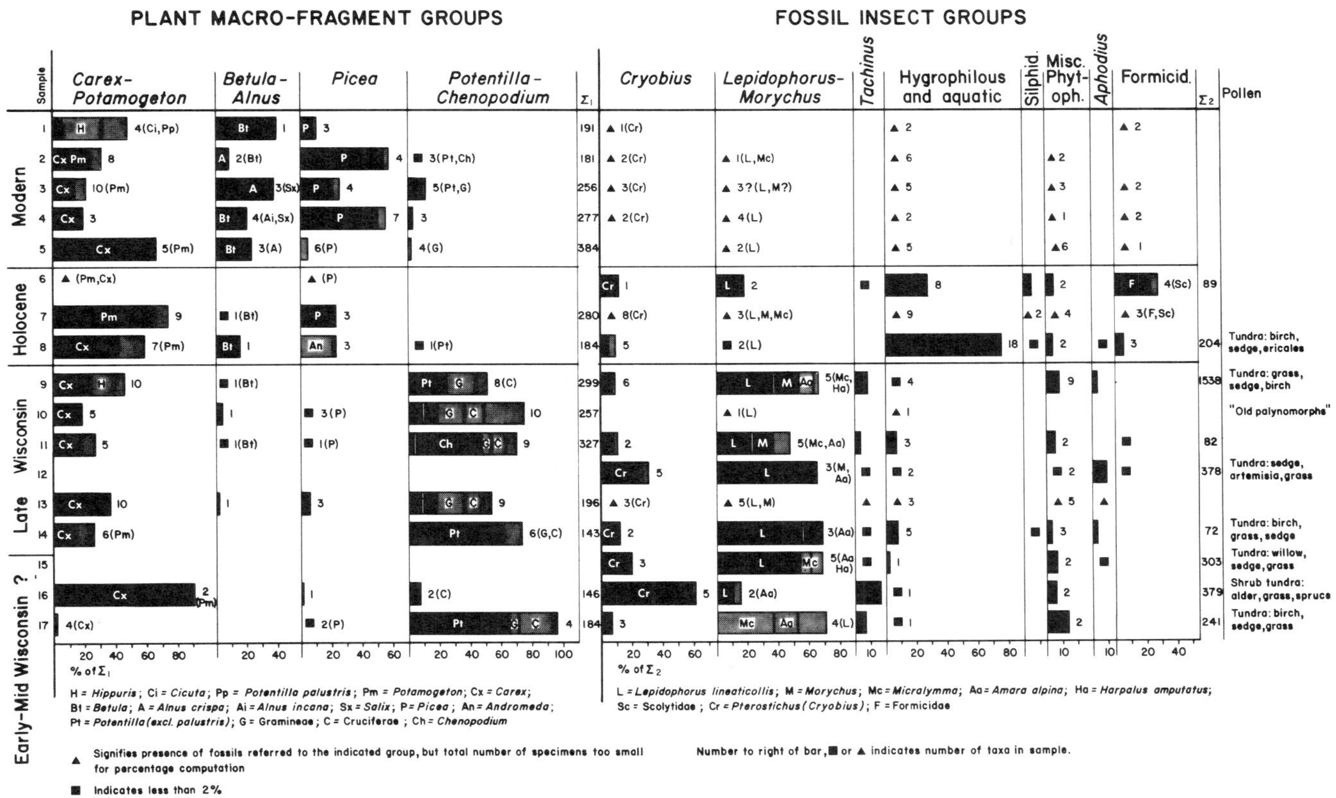

Figure 1. Percentage diagram comparing relative abundance of plant and insect macrofossils in late Wisconsin, Holocene, and modern detritus assemblages from various sites in East Beringia. Note the high percentages of POTENTILLA-CHENOPODIUM and LEPIDOPHORUS-MORYCHUS groups in the late Wisconsin samples. Black portion of bar indicates percentage of taxon or taxa for which the group is named (where black bar is in two parts, the first portion indicates % of first-named taxon of group: the remainder, the second taxon—see lettering on some of the bars). Number following bar indicates number of taxa in the assemblage. Pollen notes include previous conclusion as to the vegetational significance of the samples and the three most abundant pollen types. SITES: (1) Old Crow basin (Yukon Territory), 170 ± 50 yrs. BP (GSC-2389). (2) CRH-11, Old Crow basin (Y.T.), sample MRA 7-23-78-8, modern river alluvium. (3) Old Crow basin, sample MRA 7-31-77-1, organic alluvium from section exposed on modern floodplain. (4) HH 75-9, Bell basin (Y.T.), sample from MRA 7-17-77-2, modern river bar detritus. (5) Blackstone River (Y.T.), river detritus at treeline. (6) Ready Bullion Fm. (Fairbanks, Alaska), early Holocene. (7) CRH-32, Old Crow basin, 8100 ± 160 yrs. BP (GSC-1243). (8) CRH-44, Old Crow basin (Y.T.), 8460 ± 120 yrs. BP (GSC-2605). (9) Deering Fm., S-6, Cape Deceit (Alaska), 12,400 ± 180 yrs. BP (I-4781) (Matthews, 1974). (10) CRH-11, Old Crow basin, sample MRA 7-23-78-17, 13,000–12,000 yrs. BP (on basis of stratigtaphic position). (11) HH 75-9, Bell basin, 13,500 ± 310 yrs. BP (GSC-2553). (12) Gold Stream Fm. (Fairbanks, Alaska), 24,400 ± 650 yrs. BP (I-2116) (Matthews, 1968). (13) CRH-11, Old Crow basin, ca. 30,000 yrs. BP. (14) CRH-32, Old Crow basin, 31,300 ± 640 yrs. BP (GSC-1191). (15) Goldstream Fm. (Matthews, 1968). (16) Deering Fm., Peat 5, greater than 39,900 yrs. BP (Matthews, 1974b); probably represents an early Wisconsin interstadial (see text and Hopkins, this volume). (17) Deering Fm., S-1 (Matthews, 1974b), early Wisconsin?

also includes assemblages derived from samples of modern river detritus and soil in the northern Yukon. To facilitate comparison, the assemblages have been partitioned according to the relative abundance of taxa in groups defined by habitat or niche.[3] The time interval encompassed by each of the Pleistocene and Holocene samples varies, but may be prudently assumed to be at least on the order of decades and centuries. The Pleistocene and Holocene samples shown in Figure 1 are ranked according to their known or estimated age.[4]

Plant Assemblages

All of the modern samples except sample 5, which is from an alpine-tundra site, represent forest or forest-tundra conditions. Sample 1 represents an upland forest-tundra environment and sample 3 is the buried floodplain detritus from the same site.

Glaser (1978) has shown that macrofossil content of samples from the same community is subject to a taphonomic bias which is a function of the mode of deposition. Many of the samples in Figure 1 have similar depositional histories so that this type of bias does not fully account for their differences.[5]

The Wisconsin-age plant spectra in Figure 1 differ from their Holocene and modern counterparts chiefly in their high percentages of the *Potentilla-Chenopodium* group and low percentages of fossils in the *Picea* and *Betula-Alnus* groups. The dominant species of *Potentilla* is not known, but *P. palustris, P. anserina, P. hookeri,* and the species with sculptured or hirsute achenes are ruled out. Seeds of the *Chenopodium* type occur in significant numbers in five of the samples and are particularly abundant in sample 11. *Chenopodium*

---

[3]Following are some of the constituents of the various groups (see also Hughes, et al., 1981, Appendix B): (a) CAREX-POTAMOGETON grp.: Aquatic and near-aquatic plants; usually several different species of *Potamogeton* and *Carex.* (b) BETULA-ALNUS grp.: Mesic shrubs; shrub birch, both species of *Alnus* now occurring in East Beringia, and *Salix*; would include tree birch and *Populus* if these fossils occurred. (c) PICEA grp.: Spruce (largely represented by needles), *Larix,* and the ericaceous shrubs normally found at poorly drained black-spruce bogs. (d) POTENTILLA-CHENOPODIUM grp.: *Potentilla*; several types of *Chenopodium* seeds; Caryophyllaceae seeds (mostly *Stellaria* type); Cruciferae (except *Rorippa islandica*); undetermined Gramineae; *Androsace septentrionalis,* and other upland or xeric plants (e.g. *Linumn perenne* in sample 14). (e) CRYOBIUS grp.: includes the facultative and obligate tundra species of the ground-beetle subgenus *Cryobius* and others that occur at mesic tundra sites. (f) LEPIDOPHORUS-MORYCHUS grp.: *Lepidophorus lineaticollis,* a common weevil of dry sites in East Beringia; *Morychus,* a rarely collected pill beetle (Byrrhidae) that today lives along dry sandy river margins; and carabids such as *Amara alpina* and *Harpalus amputatus,* the former a common beetle of mesic to xeric tundra, the latter a prairie species reaching its northern limit in grassy steppe sites in central Alaska and the southern Yukon. (g) TACHINUS grp.: *Tachinus apterus* or *T. brevipennis,* both of which probably imply substrata rich in detrital organic matter, as, for example, some shorelines. (h) HYGROPHILOUS-AQUATIC grp.: Includes numerous aquatic taxa that occur at the poorly drained, vegetated shores of lakes or bogs. (i) SILPHID grp.: Carrion beetles. (j) MISC. PHYTOPH. grp.: Phytophagous species, including most weevils and chrysomelid beetles. (k) APHODIUS grp.: Dung- or mammal-burrow inhabitants. (l) FORMICID grp.: All ants (Formicidae), bark beetles (Scolytidae), and other taxa ordinarily restricted to forested sites, either open or closed.

[4]Samples 11, 12, and 13 cover the period between approximately 25,000 and 13,000 years ago. Sample 9 is 12,400 years old, and the three Holocene samples cluster around 8000 years BP. Sample 8 predates the invasion of spruce into the region of the site while sample 7, from approximately the same local area, predates the invasion of alder. Sample 14 is dated at 31,300 years BP. The ages of samples 15, 16, and 18 are not precisely known except that 15 is definitely older than 24,000 years while 16 and 17 are older than 39,900 years. Sample 15 could be as old as the early Wisconsin; but recent evidence suggests an age younger than 80,000 years (Matthews, 1979b; Hopkins, this volume). Sample 16 was originally thought to be of Sangamon age (Matthews, 1974b), but in view of Shackleton's (1979) recent findings, now seems more likely to represent an early Wisconsin interstadial (Giterman, et al., this volume). If so, then sample 17, from the same exposure but stratigraphically lower, is of earliest Wisconsin rather than Illinoian age (in contrast with Matthews, 1974b).

[5]Samples 3, 7, 10, 11, and 13 all come from ripple-bedded, sandy to silty alluvium containing laminae of organic detritus. In sample 2 the detrital organic horizons were interbedded with cross-stratified sand of an existing river bar, and samples 4 and 5 are from modern river bar detritus which had not yet been buried. Samples 9, 14, and 16 come from organic silt of colluvial and lacustrine origin. Sample 17 is also from silt, but one that is rather inorganic. Both samples 1 and 8 are from peat that formed at well-vegetated sites away from an active floodplain.

Table 4
Percentage Composition of Recent Plant Macrofossil Assemblages from McKinley National Park, Alaska[1,3]

Plant Macrofossil Groups including some of the major constituents (see footnote 3 in text and Fig. 1)	COMMUNITY TYPES							
	Forest-Tundra		Dryas-Heath			Dryas Fell Field		
	A[2]	B	A	B	C	A	B	C
Carex-Potamogeton grp.	14[3]	17.4	23	47.8	30.3	12.8	45	61.5
Carex spp.	(14)	(7.8)	(22.5)	(47.8)	(30.3)	(12.8)	(45)	(61.5)
Potamogeton spp.		(2.4)						
Hippuris		(1.6)						
Betula-Alnus grp.	67.1	47.2	54.4	29.5		54.7	21.5	
Betula (shrub)	(64)	(45.9)	(48.8)	(8.5)				
Salix spp.	(3.1)		(5.5)	(21)		(54.7)	(21.5)	
Picea grp.	31.4	32	10.3	9.8	3.9	0.8	3.9	
Picea	(31.4)	(30.7)						
Ericales		(1.3)	(10.3)	(9.8)	(3.9)	(0.8)	(3.9)	
Potentilla-Chenopodium grp.		3.5	12.3	16	29	27.9	29.4	38
Potentilla spp.		(2.5)	(0.2)	(0.6)	(1.1)			
Gramineae		(0.8)	(8)	(3.9)	(2)	(1.5)	(3.9)	(11.5)
Dryas spp.			(3.2)	(3.2)	(6.8)	(1.5)	(3.9)	(13.4)
Caryophyllaceae			(1.6)		(8.1)	(16.6)	(1.9)	(3.8)
Total macrofossils[4]	372	116	120	153	617	133	51	52

[1]Adapted from data presented in Tables 2, 3, 4 of Glaser, 1978
[2]A = snowfield assemblages; B = lake assemblages; C = shoreline floatsom windrows
[3]Number is % of fossils in indicated group, but excludes leaves (except conifer needles), and identified fragments other than seeds, fruits, and capsules. Where taxon is represented in Glaser (1978) by more than one type of fossil, the most abundant was tabulated or the one most often encountered in macrofossil assemblages. Number in parentheses is the % of the indicated taxon
[4]Indicated figure is sum of average abundance based on several samples. Other totals: sum of tabulated macrofossils in all samples studied (see Tables 2, 3, 4—Glaser, 1978)

may be relatively abundant because most samples represent a floodplain environment, but if this were the only cause, we should expect to see more *Chenopodium* fossils on the modern floodplain samples. Most of the spectra also contain significant quantities of unidentified Gramineae and Cruciferae.

Sample 16 differs from the rest of the Pleistocene samples. Its pollen spectrum suggests a mesic shrub-tundra environment, rather than the herb-rich tundra indicated by the pollen spectra of samples 7 and 9 from the same exposure. Both the latter have macrofossil spectra dominated by the *Potentilla-Chenopodium* group. In contrast, sample 16, with its abundance of *Carex*, is more similar to the macrofossil spectra from modern forest-tundra sites.

It could be argued that the differences between the Wisconsin and the Holocene or modern samples are due largely to the low percentages of the *Picea* and *Betula-Alnus* groups in the former.[6] To check this, macrofossil assemblages from existing tundra sites are required for comparison. Glaser (1978) has studied such samples in McKinley Park, Alaska, but because his spectra contain abundant leaves, they must be recast before attempting comparison with those in Figure 1. When these new totals are tabulated and partitioned according to the groupings listed in Figure 1, the spectra from forest-tundra in Mount McKinley National Park appear

[6]A few of the Wisconsin samples do contain one or two spruce needles (e.g., 11), but these fossils are thought to be rebedded since they are rare and are neither accompanied by other macroremains of spruce nor by bark beetles (Scolytidae).

similar to those from forest-tundra in the northern Yukon (compare Fig. 1 and Table 4). The latter contain higher percentages of spruce and alder (none in the McKinley Park samples) and lower percentages of dwarf birch, but these differences are easily reconciled by the fact that the samples in Figure 1 come from sediments deposited close to plant communities containing abundant spruce and alder.

The most significant deviation between Table 4 and Figure 1 is in the macrofossil assemblages from *Dryas* heath and *Dryas* fell field. Even though the total number of fossils in some of these samples is low, the major features of the spectra are clear: *Dryas* heath and *Dryas* fell field samples only yield up to 38% fossils of the *Potentilla-Chenopodium* group, and very few of the fossils actually refer to *Chenopodium* or *Potentilla*. *Dryas* seeds account for up to 13% of macrofossils in the *Chenopodium-Potentilla* group at McKinley Park, whereas *Dryas* is rare or absent in the East Beringia Wisconsin samples. The several ways in which the Wisconsin samples in Figure 1 differ from the McKinley Park fell-field assemblages suggest that the lowland environment of East Beringia during the late Wisconsin was neither a vegetational equivalent to present *Dryas* fell field and *Dryas* heath nor was it similar to that of existing sedge-dominated lowlands. The flora undoubtedly contained plants characteristic of modern alpine fell-field sites but, according to macrofossils, in combination with other taxa typical of open, well-drained, lowland areas. The autecology of some of these species (e.g., *Ranunculus cymbalaria, Carex maritima*) suggests alkaline substrates.

Bryophyte fossils from a number of the late Wisconsin sites (J. Janssens, personal communication) suggest a pattern similar to that indicated by the other plant macrofossils, i.e., late Wisconsin bryophyte assemblages appear to have no analog in the present environment, and many of them imply "rich-fen" conditions with a greater concentration of cations than is typical of existing forest-tundra and tundra fens (J. Janssens, personal communication 1980).

Fossil Insects

The fossil insect data (Fig. 1) reinforce the picture provided by plant macrofossils. Holocene samples contain an abundance and great diversity of species that live near or in water (Hygrophilous-Aquatic group), bark beetles that feed on conifers, and ants (Formicid group), but almost no *Tachinus* and low percentages of the *Lepidophorus-Morychus* group. Both the latter and the *Cryobius* group are well represented in late Wisconsin samples. No quantitative data from modern insect assemblages is shown in the table; nevertheless, the degree of divergence of the Pleistocene samples is clearly indicated by the autecology of the dominant species.

The carabid species, *Amara alpina*, is a tundra beetle of rather dry sites. Its presence in conjunction with the large number of species and individuals of the *Cryobius* group is good evidence that the late Wisconsin samples represent some type of treeless environment. The abundance of the weevil (*Lepidophorus lineaticollis*) and the pill beetle (*Morychus*) in many of the same samples indicates scanty vegetation and dry substrates. One typical grassland carabid, *Harpalus amputatus*, apparently is not found in present tundra areas, but was an occasional member of late Wisconsin assemblages. Some of the Wisconsin spectra contain 3% to 10% fossils of the dung beetle *Aphodius*, a characteristic which may signify abundance of ungulates (Coope, et al., 1961).

Fossil insect assemblages from the *yedoma* or "ice complex" sediments of eastern Siberia and Chukotka resemble those shown in Figure 1 (Kiselyov, 1979; Sher, et al., 1979; Giterman, et al., this volume). However, *Morychus* is replaced by closely related species *Chrysobyrrhulus rutilans*, and the eastern Siberian samples lack *Lepidophorus*. They contain several species that are currently typical of the Mongolian steppe, among them weevils of the subfamily Cleoninae (Sher, et al., 1979; Giterman, et al., this volume).[7]

[7]Kiselyov (in Sher, et al., 1979) considers *Chrysobyrrhulus* to be the real identity of the *Morychus* specimens in East Beringia and especially the *Morychus* fossils that I noted (Matthews, 1974c) in some samples from the Kolyma lowland. Until such time as the members of the byyrhid subfamily to which these two genera belong are revised, this question must remain in abeyance. I continue to refer to the Alaska-Yukon specimens as *Morychus* in anticipation that, when revised, *Morychus* and *Chrysobyrrhulus* will be shown to be congeneric (P. Johnson, personal communication 1979). Recent comparison of the Alaska-Yukon and Siberian fossils does show that the specimens from opposite sides of Beringia probably represent different species. Fossils of Cleoninae are a minor

THE VERTEBRATE FOSSIL RECORD

Pollen and macrofossil studies from northern Europe and central and northern Asia show that the parts of those regions that were unglaciated and not occupied by large melt-water lakes (Grosswald, 1980) supported treeless vegetation during the latest epoch of cold climate (Frenzel, 1968b; Grichuk, 1971; Klein, 1973; Kraev, 1975; Velichko, 1975). This panorama of interconnected treeless biotopes is presumed to have extended across the then-emergent Bering land bridge and into unglaciated East Beringia (Matthews, 1976a). As a consequence, northern-hemisphere mammalian faunas were less provincial and taxonomically more complex than at the present. Woolly mammoth (*Mammuthus primigenius*), steppe bison (*Bison priscus*), horse, and reindeer or caribou (*Rangifer*), the principal components of what Soviet paleontologists call the "mammoth fauna" or "Upper Paleolithic fauna," occur at site after site eastward from western Europe through the Ukraine and Siberia to Alaska and the Yukon. Inexplicably, certain taxa, such as the woolly rhinoceros and *Symbos* (a type of musk ox), failed to cross the Bering land bridge even though their fossils occur at sites clustered on one or the other side, but these represent minor exceptions to an otherwise hemispheric pattern.

Just as pronounced as the east-west exchange was a north-south intermingling of mammals on each side of the land bridge. In Siberia, where no taiga barrier intervened between tundra and steppe, tundra mammals such as arctic fox (*Alopex*), reindeer, musk ox (*Ovibos*), and collared lemming (*Dicrostonyx*) extended their range far south, while at the same time saiga antelope, steppe vole (*Lagurus*), and yak moved north (Vangengeim, 1975). In North America a different style of glaciation and physiographic barriers caused north-south exchange to be impeded. Even so, typical northern forms such as caribou, musk ox, collared lemming, and arctic fox invaded the mid-continent, and southern steppe species, like the American badger (*Taxidea taxus*) and steppe ferret (*Mustela nigripes*), entered Alaska-Yukon (Anderson, 1977; Guthrie, this volume).

Thus late Pleistocene Beringian faunas contain a mixture of taxa, including both arctic forms living in tundra today and others whose presumed or actual life habits seem inimical to a tundra existence (Vereshchagin and Baryshnikov, this volume). Guthrie (1968a), noting the high diversity of the ungulate fauna and its dominance by grazers, has proposed a grassland environment (see Guthrie this volume for an elaboration of the argument).

Beringian Local Faunas: Biocenoses or Mixed Assemblages?

Ritchie and Cwynar (this volume) view the pollen data as implying a Pleistocene environment for East Beringia much different from the one suggested by Guthrie (cf. Hamilton, 1981). Their reconstruction seems to call for regional vegetation too scant to support a diverse ungulate community. However, in their opinion, this does not pose a serious problem because "there is so far no evidence from that period (0–25,000 years BP) of large numbers of either species or individuals of herbivorous vertebrates" (Ritchie and Cwynar, this volume, p. 126). The critical question is whether the abundant mammalian fossils found at numerous late Pleistocene exposures in Alaska and Siberia represent biocenoses, i.e., mammalian communities. Ritchie and Cwynar doubt that they do, and refer to similar statements on the matter (e.g., Schweger and Habgood, 1976). Because other contributors to this book assume that the mammoth fauna is a biocenosis, the question merits closer scrutiny.

Bones typifying the members of the Upper Paleolithic fauna are found repeatedly at exposures in Siberia, Alaska, and the Yukon. In Siberia, most originate from the ice-rich silts of the "*yedoma* complex" (Sher, 1974; Tomirdiaro, this volume). Similar frozen, fetid silt, called "mucks," is the source of large numbers of bones in East Beringia. Only rarely are the bones found in situ. Instead, they usually occur out of context at the base of exposures, and may come from any stratigraphic level at the site. This is the crux of the problem; although such assemblages can acceptably be termed "local faunas" in the parlance of vertebrate paleontologists (Tedford, 1970) and do generally represent faunas of a restricted time interval, they

element of East Beringian assemblages, but as is the case for *Morychus*, the group must be revised before the fossils can be referred to species.

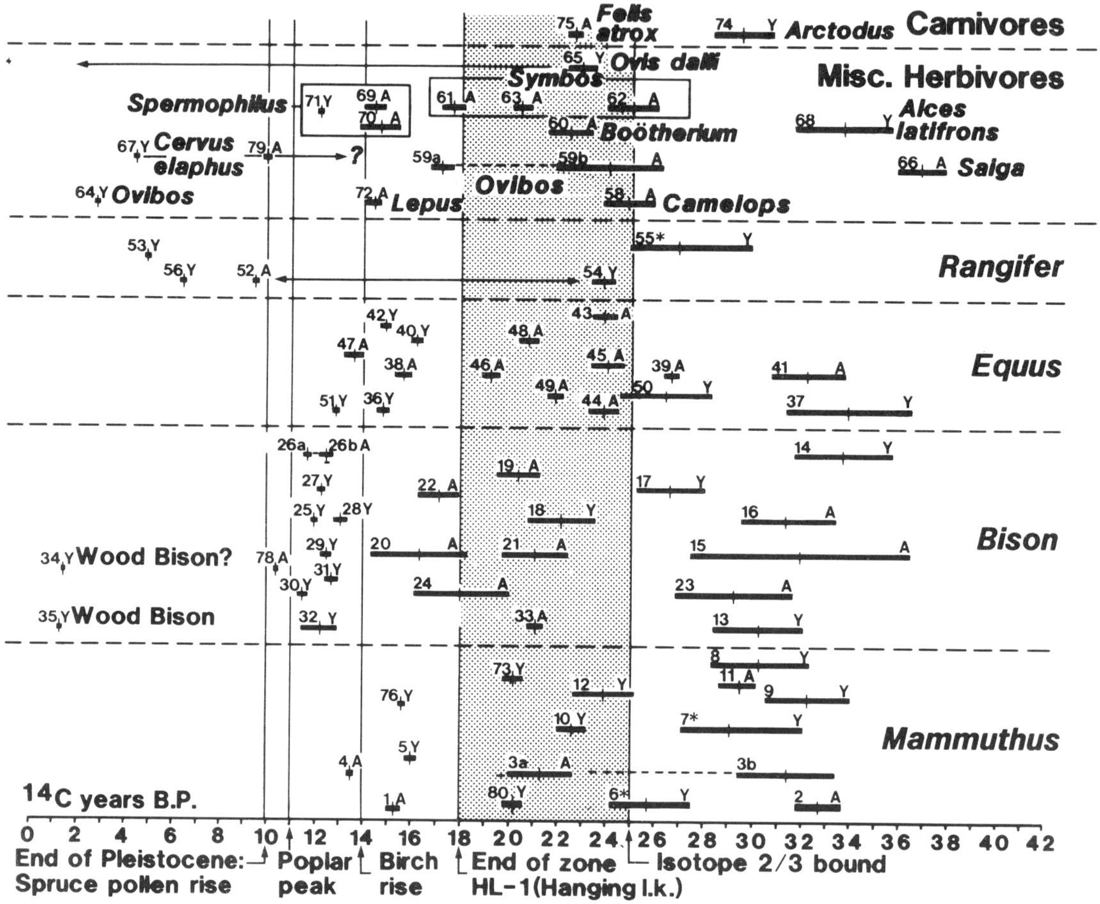

Figure 2. Finite C-14 dates on mammalian tissues (mostly bone collagen) from Alaska and Yukon Territory. Bar represents 2-sigma range (4 sigma for GSC dates). Star indicates bone-apatite dates. "A" signifies Alaskan sample; "Y" signifies Yukon Territory sample. Dashed line connects dates obtained from tissues of the same individual; arrow for *Cervus elaphus* indicates that it is expected at an earlier time (J. Cinq-Mars, pers. comm. 1980—Bluefish Cave data). (1) SI-453 (Péwé, 1975b); (2) ST-1632 (Péwé, 1975b); (3a) L-601 (Péwé, 1975b); (3b) ST-1721 (Péwé, 1975b—both samples from the same specimen, a carcass; (4) QL-1365 (N. Ten Brink, pers. comm. 1980); (5) GSC-1893 (Harington, 1977); (6) GX-1568 (Harington, 1977); (7) GX-1567 (Harington, 1977); (8) I-3576 (Harington, 1977); (9) I-4226 (Harington, 1977); (10) I-3573 (Harington, 1977); (11) DIC-1570 (J. Dixon, pers. comm. 1980); (12) I-11050 (Harington, 1980b); (13) I-3571 (Harington, 1977); (14) I-4224 (Harington, 1977); (15) SI-843 (Péwé, 1975a); (16) SI-1721 (Péwé, 1975b); (17) QU-781 (J. Cinq-Mars, pers. comm. 1979); (18) I-3570 (Harington, 1977); (19) SI-837 (Péwé, 1975b); (20) M-38 (Péwé, 1975b); (21) SI-839 (Péwé, 1975a); (22) SI-838 (Péwé, 1975b); (23) SI-842 (Péwé, 1975b); (24) SI-841 (Péwé, 1975a); (25) I-7765 (Harington, 1977); (26a) ST-1631 (Péwé, 1975b); (26b) SI-290 (Péwé, 1975b); (27) I-7764 (Harington, 1977); (28) K-1372 (Larsen, 1968); (29) I-3574 (Harington, 1977); (30) QU-780 (J. Cinq-Mars, pers. comm. 1979); (31) QU-782 (J. Cinq-Mars, pers. comm. 1979); (32) QU-783 (J. Cinq-Mars, pers. comm. 1979); (33) DIC-1333 (J. Dixon, pers. comm. 1980); (34) I-11051 (C. R. Harington, pers. comm. 1980); (35) I-5404 (Harington, 1980b); (36) I-3569 (Harington, 1977); (37) I-4222 (Harington, 1977); (38) K-1210 (Larsen, 1968); (39) SI-355 (Péwé, 1975a); (40) I-9271 (R. D. Guthrie, pers. comm. 1979); (41) I-9275 (R. D. Guthrie, pers. comm. 1979); (42) I-9316 (R. D. Guthrie, pers. comm. 1979); (43) I-9318 (R. D. Guthrie, pers. comm. 1979); (44) I-9321 (R. D. Guthrie, pers. comm. 1979); (45) I-9322 (R. D. Guthrie, pers. comm. 1979); (46) I-9371 (R. D. Guthrie, pers. comm. 1979); (47) I-9422 (R. D. Guthrie, pers. comm. 1979); (48) I-9274 (R. D. Guthrie, pers. comm. 1979); (49) DIC-1334 (J. Dixon, pers. comm. 1980); (50) QU-784 (J. Cinq-Mars, pers. comm. 1979); (51) GSC-2881 (Cinq-Mars, 1979); (52) K-1583 (Tauber, 1973); (53) I-8642 (Harington, 1977); (54)

do not necessarily represent biocenoses. Fossiliferous exposures in East Beringia often contain a sequence of sediments that collectively represents several different types of environments. In view of this, can it be shown that East Beringian local faunas of the mammoth type represent biocenoses? And was this faunal complex present as a biocenosis 18,000 years ago, during the most severe part of the late Wisconsin cold cycle?

Radiocarbon dates on mammalian tissues provide one possible means of answering this question. Figure 2 shows a plot of dates on bones and other tissues from Alaska-Yukon sites. Serious reservations have been expressed about the validity of radiocarbon dates on bone (Hassan and Ortner, 1977; C. V. Haynes, personal communication 1979), especially those based on bone apatite. Three apatite dates are included in Figure 2 and are flagged by asterisks. Most of the dates are on bone collagen or other tissues that had been frozen until shortly before collection and as a result are not as likely to have suffered the type of organic exchange described by Hassan and Ortner (1977). Furthermore, if there were serious general problems with the dates shown in Figure 2, their ages should deviate from those on associated organic materials. In fact, several of the dates shown there, e.g., 11 and 14, are in good agreement with dates on closely associated organics other than bone (N. Ten-Brink, personal communication 1980; J. Dixon, personal communication 1980). Bone dates from the Old Crow basin also reflect the hiatus of mammalian occupation caused by existence there of a large melt-water lake during the late Wisconsin (Morlan, 1980a; Morlan and Cinq-Mars, this volume).

A peculiarity of Figure 2 is the paucity of dates on caribou (*Rangifer*). Caribou remains are common in most local faunas, but are seldom dated, probably because they are seen as less interesting than those of extinct animals. According to Figure 2, caribou lived in the Yukon between about 24,000 and 9000 years ago. Caribou bones also occur at Bluefish Cave in association with a horse bone dated at 13,000 years BP (Cinq-Mars, 1979). It seems most likely, therefore, that this most adaptable of ungulates, which survives today even in polar desert, was a member of the East Beringian fauna during the height of the late Wisconsin.

In summary, Figure 2 suggests that the East Beringian large-mammal fauna included a minimum of eight ungulate species during the interval from 25,000–18,000 years ago. In addition to bison, horse, caribou, and mammoth—the traditional four members of the mammoth fauna—there were two types of musk ox (*Ovibos* and *Symbos* or *Böotherium*), sheep (*Ovis*), and possibly even camel. The lion, *Felis atrox* (=*Panthera atrox*), was also a resident of East Beringia during this time period. Soviet workers consider the coldest millenia of the Sartan glaciation (= late Wisconsin) to have occurred between 18,000 and 16,000 years ago (Tseitlin, 1979), and some believe that the periglacial climate in northeastern Siberia at that time was too cold to permit even the survival of man (Tseitlin, 1979). The dates spanning this interval in Figure 2 imply an East Beringian fauna containing mammoth, bison, horse, *Symbos*, possibly *Ovibos*, and caribou—a community more than twice as diverse as that of modern tundra.

Admittedly, the dates plotted in Figure 2 cover a large geographical area and some have large statistical errors. Local faunas from archaeological sites portray the mammalian community of smaller areas and usually over much shorter time intervals. Adequately dated late Pleistocene archaeological sites are rare in East Beringia; however, one site, Bluefish Cave near the Old Crow basin, does reveal the presence at 15,000 to 13,000 years BP of a diverse community of ungulates (including the typical mammoth complex) in association with pollen spectra of the herb-zone type (Cinq-Mars,

I-8580 (Harington, 1977); (55) GX-1640 (Harington, 1977); (56) I-4221 (Harington, 1977); (57) QL-1178 (Porter, 1979); (58) I-2117 (Harington, 1977); (59a) SI-454 (Péwé, 1975b); (59b) SI-455 (from same specimen as 59a, Péwé, 1975a); (60) SI-292 (Péwé, 1975b); (61) SI-851 (Péwé, 1975b); (62) SI-850 (Péwé, 1975a); (63) I-10649 (Harington, 1980a); (64) I-3568 (Harington, 1980b); (65) I-4225 (Harington, 1977); (66) GSC-3050 (Harington, 1981); (67) I-4225 (Harington, 1977); (68) I-4229 (Harington, 1977); (69) W-2703 (nest materials, Péwé, 1975b); (70) GX-0250 (nest materials, Péwé, 1975b); (71) GSC-2641 (nest materials); (72) feces (Porter, 1979); (73) I-10971 (Harington, 1980b); (74) I-11037 (Harington, 1980b); (75) SI-456 (Péwé, 1975a); (76) GSC- 3503 (J. Cinq-Mars, pers. comm. 1980); (77) SI-454 (Péwé, 1975b); (78) I-8582 (Harington, 1980a); (79) I-9998 (Harington, 1980a); (80) I-10971 (Harington, 1980b, and pers. comm. 1981)

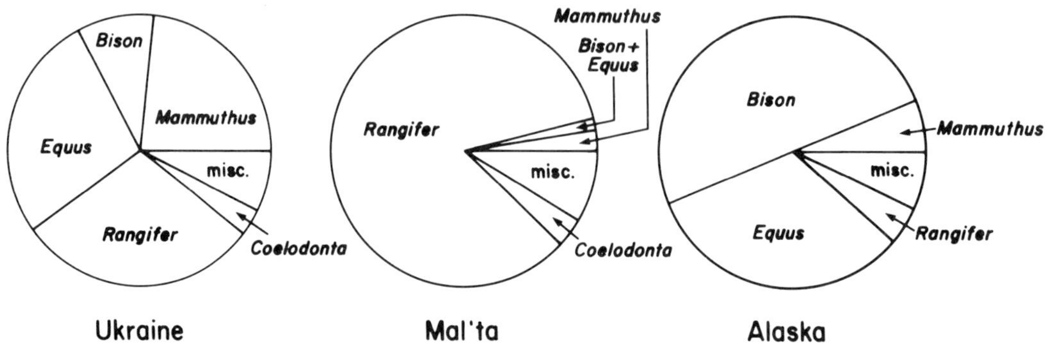

Figure 3. Comparison of the relative abundance of members of the mammoth fauna at various sites in Eurasia and East Beringia. Ukrainian diagram portrays data from 23 local faunas containing *Mammuthus, Bison,* and *Equus* (based on Pidoplichko, 1956). The Mal'ta archaeological site (data from Ermolova, 1978) is on the Angara River in Siberia, and has a C-14 date of 14,750 ± 120 yrs. BP, but Tseitlin (1979) considers the true age of the assemblage to be closer to 24,000 years. Miscellaneous portion of Mal'ta diagram includes mostly large and small carnivores, which are not tabulated in the other two diagrams. Alaska diagram portrays pooled percentages of ungulates from the four Fairbanks local faunas listed in Guthrie (1968a). Since the Mal'ta and some of the Ukrainian assemblages obviously reflect an archaeological bias, the three diagrams probably express the extremes of regional variation in the composition of the mammoth fauna

1979; Morlan and Cinq-Mars, this volume). Archaeological sites marking the Sartan (or late Wisconsin) interval are more numerous in Siberia (Mochanov, 1978a; Tseitlin, 1979). Most of them show clearly that the mammoth fauna was a biocenosis, and despite local variations (Fig. 3), temporary lacunae, or the selective aspects of human hunting, essentially the same fauna that characterized the entire unglaciated territory of the Holarctic region during the late Pleistocene. Soviet paleontologists are so convinced of the biocenotic nature of the mammoth fauna that they consider the question to be unworthy of the attention devoted to it here (A. V. Sher, personal communication 1979). Where numerous faunas have been studied, as for example in the Ukraine (Fig. 4) and eastern Siberia, the repeated association of the basic mambers of the mammoth fauna is spectacular.

Conclusions concerning the vegetation of East Beringia during the late Wisconsin must take account of the evidence of greater ungulate diversity. It need not indicate (as suggested in Matthews, 1976a) that the primary productivity of the late Wisconsin tundra was as high as in present regions having a comparable number of ungulates (Redmann, this volume). However, it is unlikely that the late Pleistocene ungulate community could have survived under the bryophyte and cushion plant dominated, low-productivity regime of existing polar desert and montaine fell field (Bliss and Richards, this volume) or in existing lowland-shrub and sedge-dominated tundra. In other words, the mammalian fossil record, despite its imperfections, imposes definite constraints on final arguments concerning the character of late Pleistocene environments in Beringia.

Indicator Species

Some mammals have been considered to be indicators of steppe. Saiga antelope is the best known example (Bannikov, et al., 1967; Sher, 1968). Its food habits also argue for existence of at least an integrated mosaic of steppe biotopes between Asia and East Beringia. Until recently there was no clear indication of when saiga lived in East Beringia, but a recent radiocarbon date, the first on saiga (Fig. 2, 66), shows that it was a member of the fauna of the north slope of Alaska during mid-Wisconsin time. Dates on associated organic material in eastern Siberia show saiga as a member of the mammoth fauna at 15,000 years BP (see Sher, et al., 1979 on the Alyoshkin Suite). Guthrie (this volume) and Harington (1981) assume that saiga was a member of the late Wisconsin fauna of East Beringia. In my opinion more East Beringian dates on saiga remains or on associated organics must be in hand before this assumption can be considered valid. Guthrie (this volume) also includes the steppe ferret and the badger in the late Wisconsin local faunas of Alaska (Anderson, 1977; Harington, 1977), but it must be pointed out that neither of these taxa have been dated—directly or indirectly.

Because of the well-publicized finds of frozen carcasses, the woolly mammoth (*Mammuthus primigenius*) has been the subject of more speculation

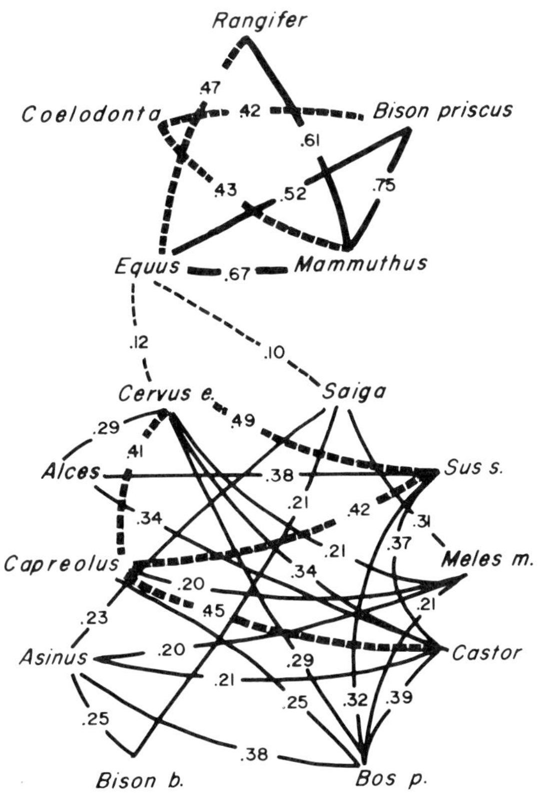

Figure 4. Degree of association (Jaccard's coefficient) of some mammalian taxa in Ukrainian fossil assemblages (based on a matrix constructed using 139 local faunas listed in Pidoplichko, 1956). Heavy line: 0.6–1.0; heavy dashed line: 0.4–0.6; light solid line: 0.2–0.4; and light dashed line (shown only for *Equus, Cervus,* and *Saiga*): less than 0.2. Note that the taxa fall into two clusters which are only weakly associated. The upper cluster represents the basic elements of the mammoth fauna. The weak link between the mammoth-fauna cluster and *Saiga* is supported by evidence from other sites (Muller-Beck, this vol.)

and comment than perhaps any of the other large Pleistocene mammals (Shilo, et al., 1977; Kubiak, this volume; Vereshchagin and Baryshnikov, this volume). Caution is required in interpreting the environmental significance of mammoth carcasses because it is often difficult to determine if the enclosing sediments and their fossils accurately represent the environment in which the mammoth formerly lived. There is also a danger in assuming that *Mammuthus primigenius* necessarily signals steppe-tundra because some of the best-preserved carcasses are of Karginsk interstadial age.[8] In spite of these reservations, I believe the evidence for existence of woolly mammoths in East Beringia during late Wisconsin time is irrefutable.

Figure 2 also shows dates for several small mammals. Remains of small mammals are abundant at archaeological sites in Beringia (Cinq-Mars, 1979; Tseitlin, 1979), and a number of specimens have been recovered from the late Pleistocene "mucks" in central Alaska and the Yukon (Guthrie, 1968b; Harington, 1977; Morlan, 1980a and unpublished). These finds show that the meadow vole (*Microtus gregalis*) lived in the lowlands of late Wisconsin Alaska at a time when the Coleoptera fauna was dominated by species of the *Cryobius* and *Lepidophorus-Morychus* groups, and the pollen by herb types (including 10–15% *Artemisia*). Apparently the distribution of *Microtus gregalis* in late Pleistocene East Beringia was tied to the same environmental conditions responsible for herb-type pollen spectra (Matthews, 1970). In Asia, *M. gregalis* is considered to be a representative of the Dahuro-Mongolian steppe and forest-steppe rodent complex (Kostenko, 1979).

[8]For example some of the carcasses or fragments thereof were obviously redeposited prior to final burial. The carcasses of animals that died as a result of becoming mired definitely intrude older sediments (see Harington, 1978, p. 71, for a modern example). Ukraintseva (1979) notes that the best finds are of animals that died either during the Karginsk interstadial or during the latest Pleistocene. She speculates that thawing of permafrost at these times is partly responsible for the miring of the animals and preservation of their corpses; consequently, their stomach contents may not represent optimum forage. For example, the *Larix* fossils in the stomachs of the Shadrin and Beresovka mammoths might represent marginal browse taken at a time of population and habitat decline (Ukraintseva, 1979); and if so, such stomach contents could lead to erroneous conclusions concerning the habitat requirements of woolly mammoths.

MODERN ANALOGS AND DISJUNCT DISTRIBUTIONS

In their contributions to this volume, as well as elsewhere, B. A. Yurtsev and S. B. Young argue that the present disjunct distributions of steppe plants in Alaska and Siberia imply existence of an extensive mosaic of steppe communities in Beringia during the last cold phase of the Pleistocene.

Ritchie and Cwynar (this volume) disagree. They point out, as does Kozhevnikov (1979) that the disjunct steppe species in the floras of Siberia and Alaska-Yukon may well be due to early Pleistocene, even late Tertiary, exchange (L. Cwynar, personal communication 1981).

Among the plants treated by Hultén (1968) which occur north of the Alaska Range and/or in any part of the Yukon Territory some 595 species (or 56.6% of the tabulated species) also occur in the part of Siberia lying east of the Lena River and north of Kamchatka (see Yurtsev, Fig. 3, this volume). Fifty amphi-Beringian species are adapted to xerophytic steppe conditions (Yurtsev, this volume). Many of them occur today in East Beringia on the warm, deeply thawed, south-facing slopes of central Alaska and the southern Yukon (Young, 1976).

Ritchie and Cwynar (this volume) doubt that these modern steppe sites provide valid analogs to East Beringian vegetation because: (1) the local environment of the sites is incompatible with the "zonal paleoclimates" predicted for 18,000 BP by CLIMAP (1976) and other climate models; (2) extrapolation of the deep active layer and silty soil of the present steppe sites to all of Beringia "would imply a thermal regime totally incompatible with regional paleoclimates suggested by all other evidence for the same period" (30,000–14,000 years BP); and (3) the steppe sites lack most of the perennial herbs characteristic of herb-zone pollen assemblages.

The first objection refers to climate-modeling experiments, which unfortunately are not yet capable of rendering reasonable estimates for an area the size of Beringia (Barry, Gal-Chen, this volume). Although the thermal regime of existing steppe sites in Alaska-Yukon does differ from that of most tundra sites, it does not follow that the more widespread occurrence of dry, well-drained soils indicated by the fossil evidence precludes occurrence of permafrost, ice wedges, altiplanation terraces, lower snowlines or any of the other features that suggest a periglacial climate for late Wisconsin Beringia. Significantly, some of the nonbiological evidence also implies drier substrates, increased regional aridity, increased aeolian activity, and enhanced continentality—all factors which would favor the expansion of steppe communities today and undoubtedly did so during the Pleistocene.

Ritchie and Cwynar's third point implies that pollen data can be used for floristic comparisons. Table 5 shows that the floristic composition of herb-zone spectra does not, at the present levels of pollen identification, justify such an assumption. It would be difficult, for example, to distinguish the herb-zone pollen flora at Hanging Lake (Cwynar, 1980) from the pollen flora of a lowland steppe site like some of those found in the southern Yukon. We know that such steppe sites feature some of the same substrate and microclimate characteristics presumed to have been prevalent over large areas of East Beringia during the late Pleistocene (Hopkins, Barry, this volume); this may be the reason for the assumption that the disjunct steppe plants found at those sites last had continuous Beringian ranges during late Pleistocene time. But even if it is valid this conclusion does not require that the late Pleistocene be the only time when such steppe plants might have had trans-Beringian distributions.

Insects

Some of the relict steppe areas in the Yukon and Siberia also contain insects having disjunct Beringian distributions. *Phimodera*, a largely Asiatic genus of desert-steppe Heteroptera, has recently been collected at Sheep Mountain in the Yukon on dry slopes which harbor approximately 20 other types of steppe insects (Scudder and Cannings, 1979) and a number of examples of disjunct steppe plants (Neilson, 1968; Douglas, 1974; Hoefs, et al., 1975). Recently specimens representing the North American steppe genus *Crophius* (Lygaeidae: Heteroptera) were collected in a relict steppe area near the Indigirka River in Siberia (Vinokurov, 1975). As is the case with plants, such disjunctions do not require that a continuity of steppe habitat occurred during the late Wisconsin. In fact, certain features of the insect faunas on either side of

Beringia could be interpreted to mean that the late Pleistocene land bridge was an effective barrier for some xeric-adapted insects. For example, none of the Mongolian and Dahurian Carabidae that presently live in eastern Siberia occur in Alaska-Yukon (Berman, et al., 1979). Similarly, species that are typical of the few steppe-like sites in the Yukon (e.g., *Harpalus amputatus, Cymindis unicolor, Bembidion obtusangulum*) are missing from the present eastern Siberian fauna. Such deficiencies extend to the fossil assemblages as well. The most perplexing example involves the weevil *Lepidophorus lineaticollis*. Although it is common in Alaska-Yukon late Pleistocene fossil assemblages and is typical of such xeric sites as fell field and river shorelines, *L. lineaticollis* is only a rare element of the present Chukotkan fauna and has yet to be found in any of the eastern Siberian fossil assemblages (S. V. Kiselyov, personal communication 1979). Riparian weevils are one of the groups that moved from Siberia into East Beringia during the Pleistocene (Table 6). Why *Lepidophorus* was not a member of a counter-exchange is not known.

Table 6 seems to contain several incompatibilities. According to the data presented there, the distribution of carabid beetles indicates a land bridge vegetated by forest-tundra; the Curculionidae (weevils), one rich in floodplain communities; and the Heteroptera, one with a diverse environment, ranging from tundra to semidesert. It is possible that these internal contradictions are due to gaps in our knowledge of the present insect faunas of Alaska-Yukon and Siberia. Alternatively, Table 6 may be another indication of the mosaic character of the late Pleistocene environment (Giterman, et al., Schweger, Yurtsev, this volume). However, because of the longevity of insect species, the apparent inconsistencies could also reflect the fact that the present fauna on either side of the land bridge represents an amalgam of exchanges, not all of which occurred during the late Pleistocene. What is clear from the table is that the land bridge during the last cold interval posed definite barriers to some steppe-adapted species.

Mammals

The search for modern analogs of the mammoth fauna has led some of us as far afield as the plains of Africa (Guthrie, 1968a; Matthews, 1976a) and to scenarios that attempted to relate greater diversity of the ungulate community to much higher levels of primary productivity (Matthews, 1976a, 1979b). Redmann (this volume) shows the fallacy of such extrapolations. Nevertheless, the evidence for a greater diversity of large mammals is clear. Some authors have also attributed the presence of mammals such as ground squirrels, badgers, and the saiga antelope in East Beringia to the existence of drier substrates and deeper seasonal thawing. While this may be so, the evidence that all of these mammals existed together in East Beringia at any single time, much less the late Wisconsin, is still highly circumstantial.

Many members of the mammoth fauna were grazers. Schweger and Martin (1976) suggested that Beringian ruminants and nonruminants may have partitioned the forage resource by forming a grazing succession as African ungulates do (Bell, 1971). But in areas with topographic diversity, spatial separation may be a better strategy for coexistence of several ungulates (Hudson, 1976). We will likely never know much about how Beringian ungulates interacted, but it is important to note that the low local relief of the land bridge could have precluded spatial separation of species—this might explain why it was a barrier for some of them. It should also be noted that loess deposition and runoff in areas near the major streams in interior Alaska created large tracts of loess gullies (Péwé, 1975a). Although these are subtle features today, during the late Pleistocene they were well developed and, as such, must have enhanced the fine-grained diversity of plant communities in interior Alaska. The net effect might well have been a more favorable milieu for a complex ungulate community. This contingency needs emphasis because it is an example of an environmental factor which may have been of great importance to a portion of the biota, yet is barely revealed by the regional pollen rain, or for that matter, by any other type of fossil evidence.

SUMMARY AND DISCUSSION

Two assumptions color my interpretation of the biotic evidence pertaining to late Wisconsin Beringia. The first is that a clear picture of Beringian paleoenvironments will emerge only after consideration of all of the available evidence, biotic and

Table 5

Beringian Pollen Assemblages as Floristic Indicators: Comparison of Floras and Pollen Types at Four Southern Yukon Steppe Communities with Pollen Flora and Suggested Flora of Hanging Lake Herb-Zone

S. Yukon Steppe Flora (Pollen type in bold face)[1]	Hanging Lake Zone HL1A-1B Pollen Types (presumed flora in brackets)[4]
	Dryopteris[3]
Equisetum[2] ←→	**Equisetum**
E. arvense	
E. pratense	
	Lycopodium
	L. selago
	L. annotinum
	L. complanatum
	Selaginella sibirica
	Picea
	Pinus
Juniperus	
J. communis	
Gramineae ←→	**Gramineae**
Agropyron caninum	(*Alopecurus alpinus*)[4]
A. Yukonense	(*Deschampsia arctica*)
Calamagrostis sp.	
C. neglecta	
C. purpurascens	
**Danthonia intermedia*	
Festuca ovina	(*Festuca* spp.—7)[4]
F. rubra	
Hierochloe odorata	(*Hierochloe alpina*)
Poa glauca	(*Poa* spp.—18)
Phleum alpinum	
Trisetum spicatum	

S. Yukon Steppe Flora (Pollen type in bold face)[1]	Hanging Lake Zone HL1A-1B Pollen Types (presumed flora in brackets)[4]
Cyperaceae ←→	**Cyperaceae**
Carex supina	(*Carex* spp.—123)
C. sabulosa	(*Eriophorum* spp.—8)
	Betula
	Alnus crispa
	A. incana
	Salix
	(*S. arctica*)
	(*S. phlebophylla*)
	(*S. polaris*)
	(*S. pulchra*)
	(*S. reticulata*)
	Oxyria (*O. Digyna*)
	Rumex
	Polygonum alaskanum
	P. bistorta/viviparum
	Caryophyllaceae
	(*Cerastium alpinum*)
	(*Stellaria edwardsii*)
Ranunculus/Anemone ←→	**Ranunculus/Anemone**
Anemone multifida	(*Ranunculus* spp.—21)
A. nutalliana	
Thalictrum ←→	**Thalictrum**
T. occidentale	
Delphinium	
D. glaucum	

[1] Combined flora of four lowland herb-rich communities in Alsek River region of southwestern Yukon (Douglas, 1974). Taxa occur in the following communities of Douglas' Table 1: *Artemisia frigida–Poa glauca, Carex sabulosa, Calamagrostis,* and *Artemisia alaskana.* All four communities occupy lowland sites, have scattered vegetation cover and are developed on well-drained alluvial or aeolian substrate. Asterisk indicates species now found only in the southern Yukon. All others occur in northern parts of Alaska-Yukon, many in tundra regions

[2] Bold face name in S. Yukon column indicates how flora (lower type face entries) would be represented in a pollen diagram (assuming all taxa produced pollen and no allochthonous pollen)

[3] Bold face indicates major pollen types in Hanging Lake zones HL1A-1B. Lower type face taxa are those whose pollen occurs in three or fewer nonconsecutive samples at low frequency (usually <1%) (according to Fig. 11, Cwynar, 1980)

S. Yukon Steppe Flora (Pollen type in bold face)[1]		Hanging Lake Zone HL1A-1B Pollen Types (presumed flora in brackets)[4]
Cruciferae	←→	**Cruciferae**
Arabis holboellii		*Lesquerella arctica* (+)
		(*Cochlearia arctica*)
		(*Draba* spp.—31)
Saxifraga tricuspidata	←→	*Saxifraga tricuspidata*
		S. hieracifolia
		S. cernua
		(*Saxifraga* spp.—25)
Potentilla	←→	**Potentilla**
P. anserina		(*Potentilla* spp.—21)
**P. diversifolia*		
P. pennsylvanica		
**Chamaerhodos erecta*		
Dryas	←→	**Dryas**
		(*Dryas* spp.—3)
Rosa	←→	**Rosa**
R. acicularis		*Rubus chamaemorus*
Leguminosae	←→	**Leguminosae**
**Lupinus kuschei*		
		Astragalus
Hedysarum	←→	**Hedysarum**
H. boreale		
Oxytropis	←→	**Oxytropis**
O. campestris		(*Oxytropis* spp.—17)
Shepherdia canadensis	←→	*Shepherdia canadensis*
Epilobium	←→	**Epilobium**
E. angustifolium		
E. latifolium		
		Bupleurum triradiatum
		Cnidium
		Cassiope

S. Yukon Steppe Flora (Pollen type in bold face)[1]		Hanging Lake Zone HL1A-1B Pollen Types (presumed flora in brackets)[4]
		Ledum
		Vaccinium
		Androsace
Polemonium pulcherimum	←→	*Polemonium*
		(*Polemonium* spp.—3)
		Phlox
Mertensia		
M. paniculata		
		Pedicularis lanata
Scrophulariaceae		
Penstemon procerus		
		Plantago canescens
Linum perenne		
Campanula		
C. rotundifolia		
Artemisia	←→	**Artemisia**
A. arctica		
A. alaskana		
A. frigida		
A. tilesii		
Tubuliforae	←→	**Tubuliflorae**
Achillea millefolium		
Antennaria rosea		
Arnica chamissonis		
Erigeron caespitosus		
E. compositus		
Solidago spathulata		
S. multiradiata		
		Saussurea
		Ligulflorae

[4] Taxa in brackets are ones suggested by Ritchie and Cwynar (this volume) as probable elements of herb-zone zonal flora during the late Pleistocene. Number following genera is the number of species within the current Alaska-Yukon flora (excluding regions south of Alaska Range and southeast coastal areas) (based on Hultén, 1968)
Comments: (1) The potential pollen flora of the S. Yukon steppe communities is similar to the pollen flora of major types at Hanging Lake. (2) Pollen flora from southern Yukon sites would poorly represent actual flora. (3) *Picea, Betula, Alnus crispa,* and *Alnus incana* are major pollen types of Hanging Lake herb-zone, but their pollen is probably allochthonous. *Lesquerella arctica,* considered to be present in former herb-zone vegetation on basis of pollen evidence is represented in Hanging Lake herb-zone (Fig. 11, Cwynar, 1980) in only one sample. (4) Spruce, birch, and alder occur in Alsek region and would undoubtedly be well represented in actual surface samples from the four herb communities.

Table 6
Distribution of Certain Insect Groups in Beringia[1]

(N.E. Siberia (West Beringia)	North America
CARABIDAE (GROUND-BEETLES)	
167 spp. ——→	[2]65 spp. (40%[3])
incl. 14 (8%) steppe species ——→ |	incl. 23 (14%) tundra
79 spp. (39%) ←——	**201** spp.[4] (Alaska-Yukon)
incl. 58 (29%) tundra	incl. 105 (52%) tundra
| ←——	Steppe sp., e.g., *Cymindis*, many *Harpalus*, many *Amara*, some *Bembidion*
CURCULIONIDAE (WEEVILS)	
105 spp. ——→	23 spp. (22%) (riparian complex)
incl. 11 (10%) taiga ——→ |	
14 (13%) steppe ——→ |	
? ←——	*Lepidophorus lineaticollis*[5]
HETEROPTERA (BUGS)	
315 spp. (incl. yakutian fauna) ——→	66 spp. (21%)
incl. 89 (28%) steppe - - - - →	incl. 2 steppe
	12 meadow-steppe

[1]Except as indicated, adapted from Berman et al., 1979
[2]Arrows indicate number of shared species, *not* direction of migration. Arrows for groups that do not occur on other continent are shown ending at median dividing line
[3]Percent of bold-faced number for all figures in parentheses
[4]Based on Lindroth's monograph on carabidae of Alaska and Canada (Lindroth, 1969). Includes species of north and central Alaska-Yukon (excluding species occurring only south of Alaska Range, Aleutians, Alaska Peninsula) as well as south interior Yukon and a few N.W.T. species recently found in the Yukon or expected there. Tundra species are those whose present distribution includes some tundra localities. Number of species with Holarctic distribution is higher than Berman et al. figure (79 vs. 65) because Alaska-Yukon data includes species found in Europe
[5]Occurs in Chukotka but not further west (S. Kiselyov, pers. comm. 1979)

abiotic. Lake cores undoubtedly provide long, uninterrupted sequences, but alluvial sections should not be discounted because in addition to pollen, they yield abundant macrofossils and better represent the environment of areas that were likely to have been most critical for the survival of large mammals and man. Vertebrate fossils define the character of that portion of the fauna most sensitive to changes in primary productivity and vegetation. The second assumption is that past plant and animal communities should not be expected to have exact modern counterparts (Watts, 1977; Wright, 1977). This means that "no analog" assemblages, whether of pollen, plant and insect macrofossils, or vertebrates, cannot automatically be discounted as due to rebedding or other types of taphonomic bias.

Any attempt to interpret the biotic record from Beringia must take account of the ways in which the physical environment differed from that of the present. Late Wisconsin time in East Beringia saw lower snowlines, growth of ice wedges and sand wedges and enhanced aeolian activity (Péwé, 1975a; Foscolos, et al., 1977; Hopkins, this volume). All are indicative of climates colder and drier than present. The Alaska-Yukon rivers that had glaciers at their headwaters developed wide, active floodplains during the Pleistocene. These and exposed coastal shelves were the source of the loess and sand that was deposited and redeposited over large areas of Alaska (Péwé, 1975a). Large basins like the Yukon Flats became huge alluvial plains crisscrossed by the anastomosing channels of numerous braided streams and surrounded by loess-covered slopes with a rillwork of steep sided gullies. Silt eroded from the gullies formed organic-rich fans that today yield most of the vertebrate remains in Alaska.

The Yukon differed from Alaska in that most of its lowland areas, except for the coastal plain, were inundated by melt-water lakes. Unlike Alaska, much of the glaciated part of the Yukon is a plateau. It received little loess fall during the late Pleistocene, but there is ample evidence of other types of aeolian activity (Foscolos, et al., 1977).

The Bering land bridge, in existence throughout Wisconsin time, became at times a connection of continental proportions. Its existence undoubtedly affected the climate of East Beringia, most probably, through enhancement of continentality (Hopkins, Barry, this volume).

These are some of the facts which must be accommodated in a paleoenvironmental synthesis. As well, it must be noted that tundra vegetation is sensitive to slight changes in drainage, nutrient cycling, and depth and tenure of snow cover (Haag, 1974; Bunnel, et al., 1975). Thus in arctic regions subtle changes of physical environmental inputs may telescope into substantial vegetation changes.

The Collated Biotic Scene

Pollen evidence from late Wisconsin sites strongly implies that much of East Beringia lacked forests during the interval between 24,000 and 10,000 years BP (Hopkins, Smith, and Matthews, 1981). Poplars probably persisted in isolated groves, as they do in certain otherwise treeless areas today (Murray, 1980a). Spruce may not have disappeared entirely from East Beringia, but there were obviously no coherent tracts of spruce forest, even of the riparian type, until after 10,000 years BP. Vegetation was dominated by herbs (and forbs) and alder, dwarf birch, and ericaceous shrubs were rare. These are the minimal conclusions to which the majority of Beringian paleoecologists would subscribe.

Agreement ends when attempts are made to relate the herb-dominated pollen spectra characterizing that interval to present environments and surface samples. A major point of difference is interpretation of high *Artemisia* and grass percentages. The species represented by these and many of the other herb-pollen types cannot be determined from the pollen alone; nevertheless, I believe it is clear that high percentages of *Artemisia* and other herbs distinguish the herb-zone spectra from the majority of tundra surface spectra, and portray a tundra vegetation different from the poorly drained type so widely distributed in East Beringia today. Pollen influx values range from less than 50 gr/cm^2/yr to nearly 500. Such low values certainly rule out a monolithic "arctic steppe" such as I proposed earlier (Matthews, 1976a), but they are still compatible with an environment containing a vegetational mosaic that included extensive tracts of steppe. Ritchie and Cwynar (this volume) discuss one element of the mosaic—upland fell field; and Schweger (this volume) another—sandy floodplains. Ager's Birch Lake core presumably represents the regional vegetation of a lowland region. As indicated above its herb-zone spectra do not differ significantly from herb-zone spectra at upland sites, but this reflects more the inadequacies of pollen data than any real similarity of regional vegetation. Macrofossil studies of upland sites have not yet been conducted, but those representing lowlands show that the vegetation there was more different from existing tundra than the pollen indicates. For example, samples from alluvial sections in the northern Yukon and Alaska contain abundant *Chenopodium* and occasional fossils of *Androsace septentrionalis*, *Corispermum hyssopifolium*, *Linum perenne*, and *Carex maritima*, all taxa that in West European late glacial assemblages are considered to be typical of physiognomic steppe (Bell, 1969). One 16,000-year-old northern Yukon sample contains fossils of submerged aquatics, emergent aquatics, acid-bog species, plus others that prefer slightly alkaline pioneer habitat.[9] The sample is associated with

[9] *Sparganium* sp., *Najas flexilis*, *Potamogeton* spp., *Hierochloë pauciflora*, other Gramineae, *Carex aquatilis*, *Carex maritima*, *Carex* spp., *Eriophorum* sp., *Kobresia myosuroides*, *Scirpus validus*, *Calla palustris*, *Chenopodium*, *Corispermum hyssopifolium*, *Melandrium apetalum*, *Stellaria* spp., *Ranunculus cymbalaria*, *Papaver*, *Rorippa islandica*, other Cruciferae, *Potentilla palustris*, *Potentilla*

insects of the *Cryobius*, *Lepidophorus-Morychus*, and *Aphodius* groups. No one would claim that all these fossils represent a single plant community, but they do signal a type of regional lowland vegetation that is not matched in present-day northern Yukon.

According to the vertebrate record, the late Wisconsin supported a more diverse community of ungulates than presently exists in tundra or boreal forest. Many of the species characterizing the ungulate group lived on through the severest phases of this stadial, becoming extinct only as climate once again ameliorated 12,000 to 10,000 years ago. I believe it unlikely that this diverse complex could have survived for long in an area as sparsely vegetated as present montane fell field or polar desert. But it is also wrong to assume that the level of diversity calls for grassland or steppe similar to that of the North American Great Plains or the steppe of central Africa. The mammalian evidence, like that from other sources seems best accommodated by a reconstruction that postulates an environment consisting of a mosaic of plant communities, some of which were drier and more steppe-like than current treeless areas of East Beringia.

The Land Bridge Variable

Emergence of the Bering land bridge is often thought of as being "in phase" with the periods of coldest world climate, but in reality, because of the mediating effect of local tectonism, existence of the land bridge is only partly dependent on eustatic fluctuations (Nelson, et al., 1974). The land bridge may have been breached for much of the early Pleistocene, even during periods of cold climate. On the other hand, it was probably in existence throughout all of Wisconsin time, including interstadials and stadials alike (Hopkins, et al., 1974; Hopkins, this volume). The land bridge must be considered as a semiindependent variable in the reconstruction of Beringian environments. It no doubt enhanced continentality of Eastern Beringia and may have been largely responsible, through enhancement of summer aridity, for disappearance of forests in East Beringia. The forests of eastern Yakutia exist under an extremely continental climate, but they are composed of different species and genera than East Beringian forests. Indeed, I suspect that if the climate of East Beringia were to become only as continental as that of present-day Yakutia, reproduction of spruce would be seriously hindered, if not precluded (Matthews, 1979b).

If the land bridge greatly enhanced the continentality of East Beringia, then it would have done so at all times of its existence, regardless of the state of world climate. This could explain why some middle and early Wisconsin interstadial assemblages from the Yukon have a distinctly interglacial appearance (Matthews, 1980; Hughes, et al., 1981). It is even possible that some of the steppe plants discussed above and by Yurtsev and Young in this volume became disjunct as a result of late Wisconsin rather than Holocene climate. In other words, the mid-Wisconsin, not the late Wisconsin, may have been the critical period for intermingling of eastern and western steppe faunas and floras.

ACKNOWLEDGMENTS

I am indebted to the Wenner-Gren Foundation, Lita Osmundsen and the gracious Burg-Wartensten staff for the opportunity to participate in this volume. I also thank my coeditors (D. M. Hopkins, C. E. Schweger, and S. B. Young) and a number of other valued colleagues (among them: J. C. Ritchie, L. C. Cwynar, R. D. Guthrie, C. R. Harington, R. E. Morlan, O. L. Hughes, R. Nelson, B. A. Yurtsev, D. Murray, A. V. Sher, R. E. Giterman, T. L. Péwé, S. Kiselyov, and M. Wood). Not one of them, I suspect, would subscribe to all of the conclusions presented here; nevertheless, their dissenting comments, probing questions, discussions, and generously offered unpublished data have contributed significantly to sharpening my perception of the Beringian environment of 180 centuries ago.

Thanks are also extended to Carole Matthews, Mike Matthews, Sylvia MacFabe, Linda Barton, Ed Belec, and Vince Stanley for their technical and editorial help in the preparation of this manuscript.

sp., *Rubus idaeus*, *Empetrum nigrum*, *Hippuris vulgaris*, *Myriophyllum exalbescens*, *Cicuta* sp., *Bupleurum triradiatum*, *Andromeda polifolia*, *Vaccinium vitis-ideae*, *Androsace septentrionalis*, *Taraxacum*, and even a few needles of spruce.

THE STEPPE-TUNDRA CONCEPT AND ITS APPLICATION IN BERINGIA

At Burg Wartenstein, the competing concepts of highly productive arctic steppe and minimally productive polar desert were hotly debated as alternative reconstructions of the late Pleistocene vegetation cover of Beringia. For most of us, the picture that eventually emerged involved a mosaic of vegetation types differing from present-day Beringian vegetation in its near absence of forests and woodlands, a much more restricted role for shrublands and marshes, and a predominance of very extensive steppe-tundra biotopes characterized by discontinuous herbaceous vegetation in which xerophytes were prominent.

The steppe-tundra concept is not new. Although reduced evaporation from land surfaces induced by the lowered temperatures of Pleistocene glacial intervals was once commonly thought to mean that glacial intervals were both cold and wet and that, conversely, interglacials were relatively dry as well as warm, some paleoclimatic evidence did not fit comfortably into this equation. In eastern Europe, especially, there has long been an awareness that glacial intervals either coincided with or included periods of relative drought. In the first paper in this section, Hibbert traces the roots of this awareness, the resulting development of the steppe-tundra concept, and the diffusion of the concept into the thinking and writing of North American students of the Quaternary history of Beringia.

Yurtsev sharpens our conception of tundra-steppe and underlines the distinction between tundra-steppe and true steppe in his description of relict patches of xeric vegetation that persist today on favorable sites in the taiga and tundra regions of northeastern Siberia. (Physiognomically similar vegetation has been discovered during the last decade, as Young reports, on steep, south-facing river bluffs in eastern Alaska. Though still largely undescribed, even larger and more extensive areas of *Artemisia*-grass steppe are present on sunny slopes within the boreal forests of adjoining southwestern Yukon Territory.) Yurtsev describes the floristics of the relict steppe and tundra-steppe communities in southeastern Siberia and discusses their significance in relation to former land-bridge biotopes. Like Giterman, Sher, and Matthews in the previous section, he emphasizes evidence of east-west differences in late Pleistocene Beringian floras.

Drawing upon knowledge of phytogeography, autecology, and synecology generated largely by his own extensive fieldwork in Alaska and discussing the paleobotanical and palynological data and inferences concerning vegetation-ungulate interactions discussed elsewhere in this volume, Young completes our conceptual discussion with an attempted reconstruction of ancient vegetation in eastern Beringia—a reconstruction that Yurtsev approaches for western Beringia only by implication. Young's focus is principally upon the time interval that Hopkins styles the Duvanny Yar interval (the cold, dry interval between 30,000 and 14,000 years ago), but he also discusses the apparently catastrophic transformation of Beringian steppe-tundra during the interval between 14,000 and 8500 years ago represented in pollen sequences by the birch-pollen assemblage zone.

—David M. Hopkins

8

HISTORY OF THE STEPPE-TUNDRA CONCEPT

Dennis Hibbert

The study of loess sheets, mammal assemblages, and pollen spectra from deposits dating to the last continental glaciation in northern Eurasia led workers in the late 19th and early 20th centuries to the conclusion that climate over much of the region had been extremely continental during times of glacial maxima. Soviet palynologists and vertebrate paleontologists working in western Beringia and later their North American counterparts in eastern Beringia developed the concept of a former extensive grassy vegetation characterized by the presence of both tundra and steppe elements. This vegetation has no exact modern analogue, and was named tundra-steppe by Soviet workers (usually steppe-tundra in English). More recent work suggests that Beringia contained a variety of vegetation types in a strongly continental climate.

Key word: Steppe-tundra

By the early 20th century scientists recognized that the climate that had existed across northern Eurasia at the time of Pleistocene glacial maxima was not the same as today's Arctic climate, nor was northern Eurasia then covered by mesophytic tundra. The idea that a simple southward expansion of typical hypoarctic tundra had occurred could not account for evidence of a widespread mammalian fauna containing elephants and antelope, nor could it account for the wide extent of loess deposition and dune fields that pointed to arid conditions.

Nehring (1890) found that vertebrate faunas dating from the time of late Pleistocene loess deposition in central Europe had a steppe-like character. By 1895 he was talking about an inferred "steppe climate with an arctic tinge to it" (quoted in Vangengeim, 1977, p. 130). A. Ya. Tugarinov (1929) suggested that the observed mixture of arctic and steppe elements in the late Pleistocene vertebrate faunas found in northern Eurasia represents real life-assemblages and not a mixed assemblage of fossil remains of species of different ages that had never actually lived together. He thought that these seemingly incompatible species could coexist because climatic and ecological conditions had been satisfactory to species associated with both northern and southern latitudes. Tugarinov pictured an "open landscape, noticeably xeric, rather cold, with little winter precipitation." (Tugarinov, 1929, p. 669, quoted in Vangengeim, 1977, p. 132). This description fits quite well the picture of the late Pleistocene landscape that had emerged by the 1940s in the study of European loess deposits and from research conducted by palynologists, particularly in the USSR.

Büdel (1949) reviewed work done in a variety of disciplines on the question of the climate and vegetation of Europe in the later Pleistocene. He noted that extensive loess deposition and steppe-life vegetation go together, judging from deposits in Europe, and (referring to V. P. Grichuk, 1946) that a common signature for this steppe vegetation in the pollen record consists of a combination of low percentages of tree pollen and high percentages of pollen from grasses and *Artemisia* (sagebrush or wormwood) in areas that are not steppe today. Büdel postulated that a dry climate zone had existed in Europe during the late Pleistocene glacial

ISBN 0-12-355860-3

maximum and that this zone contained the areas of extensive loess deposition, as well as two broad vegetation belts: *Lösstundra* in northwest Europe and, to the south, *Löss-steppe* that stretched eastward into central Asia. The forest belt that today separates the treeless Arctic from the steppes did not then exist; Büdel said that "*Löss-tundra* . . . was as much steppe as tundra" (ibid. p. 133), and that it does not exist in today's Arctic except perhaps as small remnants (ibid. p. 135). Büdel also pointed out (1949, p. 138) that a glacial cycle involves a shift in climate from a period of relatively cold and moist conditions that allows formation of ice sheets, to a cold and much drier climate that lasts through the time of maximum extent of the ice sheets. V. V. Reverdatto (1940, p. 58) had earlier defined a similar two-stage climate succession, naming the cold, dry stage the "cryoxerotic".

Soviet botanists and paleobotanists have found widespread evidence of a "periglacial" vegetation that had characteristics of tundra, bog, woodland, and steppe, and that had no necessary connection with glaciers. The area occupied by this vegetation if projected into Asia would lie within the limits of Büdel's *Löss-steppe* (Büdel, 1949, Fig. 6). Reverdatto suggested that steppe and arcto-alpine taxa existing in parts of Siberia that are today partly forested are remnants of "a more cold-loving vegetation adapted . . . to a harsher continental climate" than today's (Reverdatto, 1960, p. 116, quoted in Giterman, 1968, p. 196). He named the unique vegetation that he postulated for the periglacial zone "bog-steppe" (Reverdatto, 1949, p. 56). This vegetation occupied a landscape which supported xerophytes, hydrophiles, and alpine species, along with scattered larch. Far to the west, the pollen spectra again portray a vegetation present at the time of the last glacial maximum that has no exact modern analog. V. P. Grichuk (1951) noted abundant *Artemisia* (previously identified as *Salix*) in a tundra-like association in the European USSR, and K. K. Markov, discussing the work of V. P. Grichuk, M. P. Grichuk, and others, described a vegetation from the "periglacial region" near Leningrad that cannot be called tundra, forest, or steppe, but that contained elements of all three (Markov, 1956 p. 146). He called this vegetation cold wooded- or forest-steppe.

Palynologists working in eastern Siberia during the 1950s and 1960s produced numerous data on the Pleistocene vegetational history of the region, and the picture that began to emerge was one of increasing dryness and continentality of climate throughout the course of the Pleistocene. Grichuk and Grichuk (1960) attempted to reconstruct the vegetation of the Russian Plain and western Siberia during glacial maxima by assuming that vegetation zones similar to today's were shifted as a direct result of the climatic influence of continental ice sheets. However, later work by Vangengeim and Ravskii (1965) suggested that the situation in eastern Siberia could not be explained by this mechanism. Rather, the controlling factor seems to have been the extreme continentality of climate at times when the Bering land bridge was exposed (Hopkins, 1967b).

Eh. I. Ravskii (1961) brought together observations on glacial geology and permafrost in northeastern Siberia, results of palynological studies and the work of vertebrate paleontologists, and formulated a theory of periglacial climate and landscapes that links the treeless, cold, dry landscape of the pollen record with the "upper-paleolithic" (later also called "periglacial") mammoth-and-horse fauna. The flora and fauna were shown to belong to periods of intense glacial activity in parts of northeastern Asia. Vangengeim and Ravskii (1965) maintained that zones of climate and vegetation did not shift southward in eastern Siberia at times of glacial maxima, as was the case in Europe. Life zones were instead restructured as a result of the harsh continental climate in the region. Taiga disappeared entirely. Cold and dry landscapes of various types (periglacial cold steppe, arid steppe, semidesert) extended from the southern edge of the ice sheets in the north to the borders of temperate forests far to the south, and the way lay open for the mixing of tundra and steppe elements that gave rise to the steppe-tundra associations and to the "periglacial" fauna in the second, cold and dry, phase of each continental glaciation.

The Soviet model of a cold, xeric environment in western Beringia during the cold, dry phase of Pleistocene glacial periods was presented to a North American audience in papers by Giterman and Golubeva, Flerow, Vangengeim, and Kind delivered at the VII INQUA Congress in 1965 in Boulder, Colorado and published under the editorship of D. M. Hopkins (1967a). P. A. Colinvaux (1964) had come to similar conclusions with regard to the environment at Imuruk Lake on the

Seward Peninsula in western Alaska (eastern Beringia), and R. D. Guthrie, discussing the former more northerly range of the wapiti, had concluded in a paper submitted in early 1965 and published in 1966 that the vegetation of much of interior Alaska during the last glaciation had contained a higher percentage of grass and grasslike plants than is found in tundra. He called this environment "steppe-tundra", and this may be the earliest use of this term in English. In the same paper, Guthrie referred to the grassy flora that gave way to modern heath tundra and spruce forest as inhabiting a "tundra-steppe environment"; this may be the first time *this* term was used in English. Guthrie later suggested (1968a) that interior Alaska had been a cold grassland, judging from assemblages of large grazing mammals—horse, bison, and mammoth—found there. Indeed, Livingstone (1955) had noted that late Pleistocene pollen spectra from the north slope of the Brooks range suggest a treeless, tundra-like environment that contained abundant *Artemisia,* and that modern tundra at Point Barrow does not provide an exact analog to these spectra.[1]

Tundra-steppe—frequently "steppe-tundra" in English—is applied generally to an association with a codominance of steppe and arcto-alpine species (Yurtsev, 1972b, p. 476). The vegetation described is one that was like neither grassland nor tundra today, but which was a cold, dry grassland containing abundant elements that now grow in the tundra zone, along with abundant xerophytes such as *Artemisia*, chenopods, and (in Siberia) *Selaginella sibirica.* During the cold-dry stage of glacial climate, tundra itself was of more restricted extent, and drier, than now.

It was proposed (Giterman and Golubeva, 1967; Golubeva, 1968) that steppe-tundra occurred during the cold-dry phase of the Zyrianka glaciation because that time represented the extreme development of the trend to increasing continentality characteristic of Pleistocene cold periods. Although earlier glaciations had produced vegetation represented by pollen spectra showing tundra with an admixture of steppe elements (see Colinvaux, 1967a, and Golubeva, 1968, for examples from the Illinoian/Samarov), the strong representation of xerophytes held to signal steppe-tundra was at first not widely perceived in pollen records of these earlier cold cycles. Continued pollen work, however, coupled with a perhaps somewhat broader point of view, has led to the mention of steppe-tundra from cold phases as old as early Pleistocene (Giterman, 1975, 1976; Giterman, Sher and Matthews, this volume) in western Beringia.

Since 1967, pollen spectra that may record steppe-tundra in eastern Beringia have been reported (Matthews, 1974b; Ager, 1975). The search is also under way for modern analogues or remnants of steppe-tundra (Young, 1976). B. A. Yurtsev's extensive work in northeastern Siberia (1972a) has brought to light numerous areas of relict steppe vegetation in regions now far to the north of modern steppes, and "has led to a reappraisal of the significance of similar patches on south- and west-facing slopes in easternmost Alaska and, more extensively, in the Yukon" (D. M. Hopkins, personal communication 1979).

In 1968 (published 1972b) Yurtsev addressed the question whether Pleistocene steppe-tundra spectra could result from a mosaic of communities. Yurtsev stressed that all gradations between steppe and tundra are found in the tundra zone today and concluded that increased aridity at times of Pleistocene glacial maxima would have led to an increase in grassland associations but that the landscape producing the steppe-tundra spectra remained essentially tundra (ibid. p. 479). However, in the USSR in the 1960s vertebrate paleontologists studying the late Pleistocene mammalian fauna of Beringia found grazers strongly represented in the form of mammoth, woolly rhinoceros (a form that originated in arid central Asia—Vangengeim, 1977, p. 135), horse, and the distinctive saiga or steppe antelope. Sher (1971) pointed out that saiga prefers cold and dry conditions, with thin winter snow cover; but as late Pleistocene saiga fossils are found from northwest Europe to Alaska and

---

[1]The distinctive vegetation called *tundra-steppe* by Giterman and Golubeva (1967, p. 236ff; also see Golubeva, 1968, p. 191, 218ff) was described from pollen spectra attributed to the cold-dry phase of the Zyrianka glaciation in Siberia, which correlates with the early Weichsel or Wisconsin glaciation. The term *tundra-steppe* that occurs in Grichuk and Grichuk (1960) in Tables 1 and 2 and in Figure 13 should not be thought the same as the later usage, as it is applied to the phase of maximum cold and not that of maximum dryness in the postulated sequence: cool and moist—maximum cold, moist—cold and dry (M. P. Grichuk, 1960).

beyond, a widely distributed landscape more like steppe than tundra is suggested. J. V. Matthews (1974a) noted that late Pleistocene insect and megafossils support the idea of cold, dry steppe-tundra or arctic grassland. B. Frenzel (1968, Fig. 5) had used the term "arctic steppe" (for the German *Kältesteppe*) to describe a belt of treeless landscape stretching across northern Eurasia from Britain to the Bering Strait at the height of the Weichsel glaciation, and Matthews (1976) proposed that the name be adopted for what he suggested was a biome now extinct. Schweger and Habgood (1976) sounded a note of caution similar to Yurtsev's (1972b), suggesting that arctic steppe, if it existed, was mosaic in its character and not an uninterrupted association. They noted that pollen evidence from eastern Beringia does not always indicate former steppe-tundra where it might be expected and that mammalian fossils in the region lack stratigraphic control and are therefore shaky support for the arctic steppe hypothesis.

The Burg Wartenstein conference on "The Paleoecology of the Arctic Steppe-Mammoth Biome" was called as a forum for an exchange of views, with the aim to establish an accepted base for further research. The papers in the present volume reflect the growing awareness that the environmental character of Beringia during Pleistocene cold periods was probably no less complex than the environmental pattern of the region today.

9

RELICS OF THE XEROPHYTE VEGETATION OF BERINGIA IN NORTHEASTERN ASIA

Boris A. Yurtsev

Large and small tracts of steppe and tundra-steppe vegetation are scattered widely in the more continental parts of northeastern Siberia. Soils in these steppe and tundra-steppe areas share common features, including base saturation, lime-accumulation horizons, and a humic horizon involving saturation with plant roots rather than accumulation of surficial peat. A study of the character and distribution of these arctic steppe tracts provides insight into the former character of Beringian vegetation during late Pleistocene cold intervals when glaciers were expanded and sea level was at its lowest.

Steppe vegetation covers large areas in the Yana and Indigirka drainage basins. In the most arid and continental part of this region, larch forests are restricted to north slopes and especially moist sites with other exposures. Steppe vegetation commonly replaces burned-over larch woodlands, suggesting that climatic conditions may have been more optimal for woodlands at some past time in the Holocene.

Steppe patches diminish in size and abundance eastward, but nevertheless, steppe and tundra-steppe vegetation are found on many sunny, well-drained sites in the Kolyma and Anadyr River basins and in an arctic belt extending eastward through the Anyui Mountains and northern Chukotka. The surprising presence of islets of steppe vegetation on favorable sites amidst the hypoarctic tundra of northern Chukotka may reflect the absence of larch and stone pine, which otherwise might be expected to colonize these drier, warmer sites. Several endemic species of very local distribution and several spectacular range disjunctions suggest that the small steppe patches of Chukotka may have expanded recently from even more restricted areas during an earlier, more mesic part of the Holocene.

Especially significant are large tracts of tundra-steppe on Wrangel Island, a large island on the continental shelf well north of the northeast Siberian mainland. The interior and the southwestern parts of the island are covered largely by a mosaic of tundra-steppe and cryophytic meadow steppe with low willow thickets in the valleys. A recent pollen spectrum from one of the steppe-tundra tracts resembles the herb zone spectra encountered in pollen sequences that extend back to late Pleistocene levels. The vegetation of Wrangel Island may be the closest living analog to the sort of vegetation that clothed northern Beringia during Pleistocene time (abstract by D. M. Hopkins).

Key words: Disjunct plant distribution, endemic plants, steppe, tundra steppe, northeastern Siberia, Chukotka, Wrangel Island

ISBN 0-12-355860-3

INTRODUCTION

Study of the modern distribution and ecology of Beringian xerophytes and cryophytes has an important place alongside paleontologic and geologic studies in attempts to reconstruct the vegetation and landscape of Beringia during the cold-arid intervals of the Pleistocene. Of special paleoecological significance is information on the northern distribution of steppe plants ecologically alien to taiga and tundra environments.

The relict steppes in northeastern Yakutia[1] were first described by V. A. Sheludyakova (1938, 1957) and M. I. Yarovoi (1939). Similar steppes in Chukotka were described by the author (Yurtsev, 1968). A number of more recent works dealing with the steppe vegetation of northeastern Asia are listed by Yurtsev (1978).

The present paper outlines the distribution and ecology of relict steppe assemblages in northeastern Asia, with particular respect to the nature and distribution of cryoxeric landscapes in Pleistocene Beringia.

STEPPE VEGETATION AND ITS CRYOPHYTIC COUNTERPARTS

Steppe vegetation, in the ecological-physiognomic interpretation accepted here (Lavrenko, 1940; Yurtsev, 1978), comprises the associations of microthermous and hemicryophilous (that is, typical of temperate and cool temperate zones, respectively), xerophilous and mesoxerophilous perennial herbs. These plants inhabit areas with continental climates, are frost and drought resistant, and are normally confined to approximately neutral or slightly basic calcium-saturated soils. Characteristic of steppe phytocoenoses in climax and subclimax situations are discontinuity of the above-ground herbaceous vegetation, but high density of roots below ground and prevalence of graminoids, especially tufted grasses.

In both zonal and extrazonal steppe vegetation, three major subdivisions are usually recognized (Lavrenko, 1940; Yurtsev, 1978): (1) true tufted grass steppes; (2) meadow-steppes, dominated or at least codominated by mesoxerophytes, with mesophytes sometimes present and with a relatively continuous ground cover; and (3) petrophyte steppes in which subshrublets and petrophilous herbaceous dicots play an important role and in which the above-ground vegetation is sparse and discontinuous.

Each of the above units can be further subdivided into major climatic variants, depending primarily upon the thermic zone in which the dominant and associated species typically occur.

Microthermous plants are a basic component in every vegetation subtype. They predominate among steppe associations in the ground cover within the northern taiga belt in northeastern Yakutia.

Hemicyrophyte steppes are typical of the continental sectors of the Chukotkan tundra. They are commonly dominated by such species as *Carex supina* ssp. *spaniocarpa* and *Calamagrostis purpurascens.* They are also common in southwestern Greenland and in the alpine and subalpine zones of the higher mountains in areas of continental climate south of the tundra.

Another situation exists in which there is codominance of microthermous and/or hemicryophilous plants and cryophytes (arctic or alpine species characteristic of cold, treeless areas). This situation occurs on tundra-steppes, where the cryophilous component commonly includes dwarf shrubs such as *Dryas* species and occasionally also semiprostrate Ericaceae. It also occurs in purely herbaceous assemblages such as *Carex obtusata-C. rupestris* xeric meadows and *Kobresia*-dominated areas rich in steppe plants. These situations have been referred to by Lavrenko (1940) as cryophyte steppes, although they might more properly be called cryophyte-microthermous or microthermous cryophyte steppes, depending on which thermic species group dominates the community.

A final variant would be associations of herbaceous cryxerophytes that are characteristic of arctic and alpine regions but which have the general aspect of steppe vegetation. I consider these to be a separate type of vegetation, herbaceous-cryoxerophyte. It is best represented in montane landscapes on substrates rich in calcium carbonate, exemplified by dry meadows dominated by *Carex rupestris, C. nardina* s. l., and *C. glacialis.*

[1]Yakutia, though not named on the endpaper map, is the region that lies west of the dash-dotted line extending parallel to the lower course of the Kolyma River. Chukotka includes the Chuktotsk Peninsula and the Anadyr Mountains to the west—(*Eds.*).

RELICT VEGETATION IN THE ASIAN SECTOR OF BERINGIA

To reconstruct the composition of the late Pleistocene vegetation of Asian Beringia one must consider the relicts of this vegetation in the modern plant cover. The following section discusses these relicts on a geographical basis.

Northeastern Yakutia

The inner portion of the Verkoyansk-Kolyma Mountain region, particularly the Yana-Oimyakon Plateau and neighboring mountainous areas, contains the greatest concentration of relict steppe vegetation to be found north of latitude 60° N between the Alaska-Canada border and the eastern Taimyr Peninsula. In earlier publications (Yurtsev, 1972a, 1974b), the areas within the northern taiga landscapes of the Yana, Kolyma, and Indigirka drainages, where steppe phytocoenoses commonly occur, were outlined schematically within a single contour. This obviously does not accurately reflect the true situation. The areas rich in steppe vegetation, mapped schematically in Figure 1, are, in fact, separated by extensive areas in which steppe bluffs are lacking.

The steppe vegetation of northeastern Yakutia is composed mainly of 142 species of xerophytes, mesoxerophytes, and petroxerophytes (Table 1). In addition, a number of xeromesophytes are represented in the meadow-steppe communities. Approximately one quarter of the 142 species mentioned above also occur in xerophytic vegetation in western North America, and at least as many more are represented by closely related variant species. In Table 1 it can be seen that

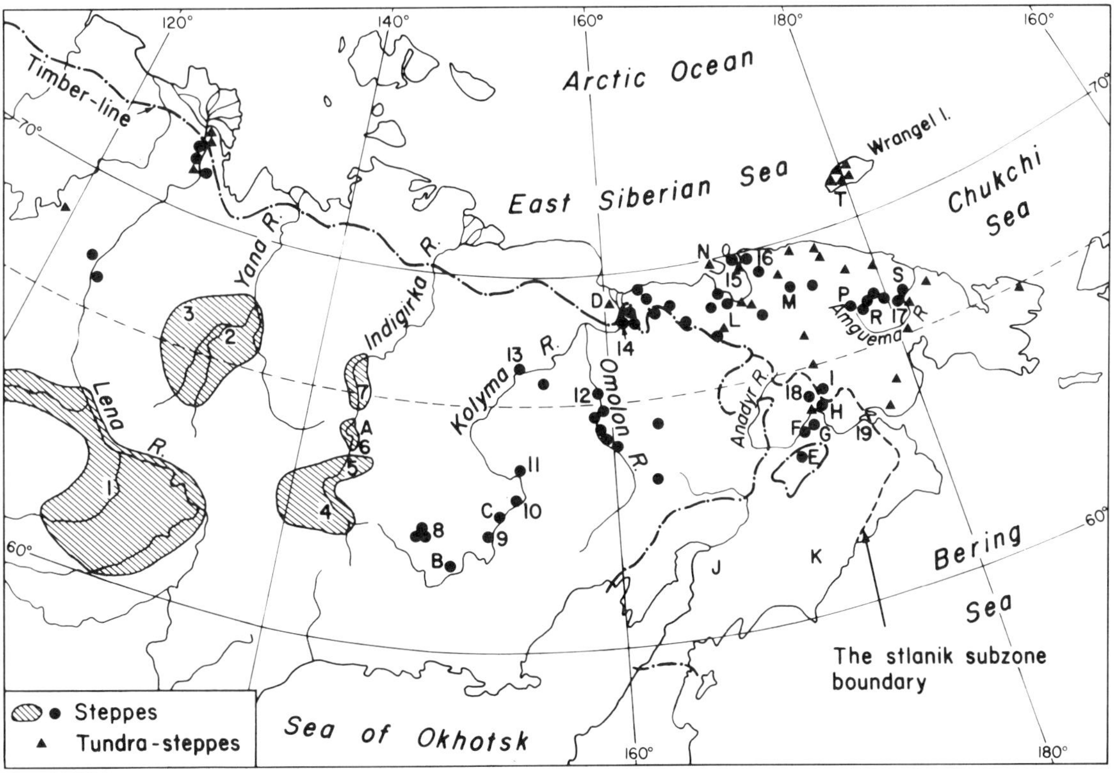

Figure 1. Distribution of steppe and tundra-steppe in northeastern Siberia. Steppe localities are indicated by cross-hatched areas and solid dots and tundra-steppe localities by crosses. Dash-dot line is the forest limit as defined by larch trees, and the dashed line is outer limit of *stlaniks* (shrublands and thickets of stone pine, birch, and poplar) in easternmost Siberia. Numbered localities are the weather stations listed on Table 2. Lettered localities are discussed in the text

Table 1
Distribution of Northeast Asian and Amphi-Beringian Taxa among Ecological Subdivisions of the Northeastern Yakutian Xerophyte Complex

Number of Species and Subspecies

	Total Xerophyte Complex		Asian Endemic		Amphi-Beringian		
Ecological Subdivisions	Number	Percent	Endemic to NE Asia	Endemic to NE Yakutia	Represented by American Vicariant	Present In North America	Total
Microthermous	103	72.6	25(24.3)[1]	14(13.6)	32(31.1)	21(20.4)	52(50.5)
Xerophytes	28	19.7	6(22.2)	5(18.5)	10(35.7)	3(11.1)	12(42.9)
Mesoxerophytes	55	38.7	11(19.6)	2 (3.5)	18(32.7)	17(30.3)	35(63.6)
Xeropetrophytes	20	14.1	8(40)	7(35)	4(20)	1 (5)	5(25)
Hemicryophilous	39	27.4	18(46.1)	8(20.5)	3 (7.7)	14(35.8)	17(43.5)
Xerophytes	16	11.3	7(43.7)	4(25)	1(63)	7(43.7)	8(50)
Mesoxerophytes	9	6.3	3(33.3)	2(22.2)	2(22.2)	4(44.4)	6(66.6)
Xeropetrophytes	14	9.8	8(57.1)	2(14.3)	0(0)	3(21.4)	3(21.4)
Total	142	100	43(30.3)	22(15.5)	35(24.6)	35(24.6)	69(48.6)

[1]Numbers in brackets are percentages

northeastern Yakutia is the center of genesis of montane xerophytes in northern latitudes. The majority of the endemic taxa are xeropetrophytes, inhabitants of montane and alpine steppes. The relatively few microthermous mesophytes occur in meadow steppes, xeric meadows, and open xeric woodlands.

The proportions of endemic species, amphi-Beringian species, and species that are one of a pair of vicariants differ markedly from one ecological group to another (Table 1). These proportional differences are instructive to one attempting to recreate past distribution patterns, migration routes, and environmental conditions. It can be seen that the proportion of endemics in the hemicryophyte mesoxerophytic group is twice as high as among microthermous steppe species.

Over half of the microthermous species are amphi-Beringian or are represented in North America by vicariant taxa; this is also true of nearly half of the hemicryophilous species. However, the proportion of amphi-Beringian to vicariant species is much different in the two groups. Less than half (21 of 52 species) of the microthermous species are amphi-Beringian, whereas 14 of the 17 species of hemicryophilous plants are known to occur on both continents. Moreover, there is a marked difference in the characteristic range of the other two habitat subgroups. Only about half the species of each xerophytic subgroup is amphi-Beringian or vicariant, while the proportion drops to one quarter or less in the xeropetrophytes.

These figures suggest that trans-Beringian migrations were most common among species adapted to colonization of the exposed continental shelf, and that the xeropetrophytes were less capable of this colonization than were the meadow-steppe plants. This is of considerable interest in view of the assumption that the more recent exchanges occurred during cold, arid intervals.

The composition of the steppe flora varies somewhat between the Yana and Indigirka drainages. A number of species, including *Stipa krylovii* (a eurixerophilous Daurian-Mongolian species which has spread southward to the desert-steppe of Mongolia), *Psathyrostachys juncea*, *Serratula marginata*, and *Polygonum gracile*, are known from the Yana but not the Indigirka drainage, probably because the Yana drainage is separated from the Lena-Aldan basin by lower and less numerous mountain ridges with lower and wider passes. In the Indigirka, but not the Yana, on the other hand, we find *Koeleria cristata,* a holarctic steppe species which occurs there as an isolated population. In addition, we find Daurian-Mongolian montane-steppe plants such as *Stellaria cherleriae* and *Chamaerhodos grandiflora,* and also endemic xeropetrophytes such as *Saxifraga multiflora, Arenaria tschuktschorum,* and *Artemisia kruhsiana* ssp. *multisecta,* plants that also occur in the Kolyma drainage. The petrophilous-steppe plant *Potentilla tollii* is endemic to the Yana, and some of the xerophilous and mesoxerophilous species are represented in the Yana and Indigirka areas by different races. It would seem that differences between the steppe assemblages of the Yana and Indigirka did not disappear completely even during periods of xerophyte expansion, which presumably can be explained by the limited montane glaciation in the area.

The interrelationship in our area of extrazonal steppe and the zonal northern taiga vegetation appears to be typical of moderately warm phases of interglacials. In steppe-rich regions, the steppe areas are mainly found on steep, south-facing slopes of the higher terraces and the mountainsides (Fig. 1). The steppe areas range from less than one hundred to several hundred meters in vertical extent, and they may extend as many as several kilometers horizontally (Yurtsev, et al., 1975). Different variants of steppe vegetation alternate within the steppe massifs. Most common are tufted-grass steppes dominated by *Festuca lenensis, Poa botryoides, Helictotrichon krylovii,* and *Agropyron jacutorum.* Locally there may be a considerable admixture of *Carex duriuscula* [=*C. stenophylla* s.l.—*Eds.*]; less commonly, *C. pediformis*; the rhizomatous *Agropyron karawaewii*; and various forbs. Also important are various petrophyte-steppe and meadow-steppe associations.

Petrophyte steppes occupy the steepest, often heavily eroded, parts of the slope. The common

Figure 2. Forest-steppe landscape on southerly exposures on the right bank of the Indigirka near Tebulyakh village. Forest of *Larix cajanderi* covers slight depressions and the bottoms of ravines. In the center are secondary steppe phytocoenoses with sparse shrubs of *Salix bebbiana* growing on the site of burned forest. The granitic Porozhnyi Mountains are seen in the background

dominants and codominants include *Potentilla tollii* (in the Yana), *Orostachys spinosa, Stellaria jacutica, Artemisia bargusinensis, Thalictrum foetidum,* other forbs, and some loose-tufted or rhizomatous grasses such as *Calamagrostis purpurascens, Agropyron jacutorum,* and *A. karawaewii.*

Meadow steppes, on the other hand, are characteristic of woodland margins, lowermost slopes and depressions within the slopes, and other habitats comparatively rich in moisture (Fig. 3). Meadow steppes differ from the above-mentioned steppe communities in that they show more complete vegetation cover, a larger proportion of mesoxerophilous and xeromesophilous dicots, and, frequently, the presence of shrubs such as *Rosa acicularis* and *Salix bebbiana.* Isolated individual larch (*Larix gmelinii* ssp. *cajanderi*) and aspen trees are sometimes found in these situations. Among the common dominant and codominant species are *Carex pediformis, C. obtusata, C. supina* ssp. *spaniocarpa, Poa botryoides, Potentilla arenosa,* and *Aster alpinus.* Mesoxerophilous mosses such as *Rhytidium rugosum, Thuidium abietinum,* and *Tortula ruralis* may be important in the ground layer.

There is normally no reproduction of larch in steppe phytocoenoses, due to the dry soils, frequent fires, and competition of the seedlings with steppe grasses (Fig. 4). However, the borders of the steppe areas are commonly larch woodlands with a sparse understory of *Rosa acicularis* and *Salix bebbiana* and with a herbaceous layer composed mainly of meadow-steppe species. A dwarf shrub layer of *Vaccinum vitis-idaea* ssp. *minus* may also be present, as may a ground layer dominated by the moss *Rhytidium rugosum.* These woodlands belong to the forest-steppe extrazonal subdivisions of the landscape. They contrast strongly with the zonal, northern taiga which consists of three basic types of vegetation: (1) larch woodlands commonly with an understory of *Alnus fruticosa, Betula middendorffii* or *B. exilis* [=*B. nana* ssp. *exilis* of Alaska–*Eds.*], or *Pinus pumila;* a dwarf shrub layer of Ericaceae such as *Ledum decum-*

Figure 3. A steppe area in the basin of an ephemeral lake on a high left-bank terrace of the Indigirka Valley upstream from the mouth of the Inyali River. The floor of the depression is occupied by a mixed meadow dominated by the steppe sedge, *Carex duriuscula,* the mesophytic grass, *Hierochloe odorata* s.l., and the semiaquatic *Polygonum amphibium.* The shoreline is seen as a white strip where soda and gypsum accumulate. Surrounding the lake is *Carex duriuscula* steppe on large-scale polygon relief, replaced upslope by *Koeleria cristata* steppe and by *Helictotrichon krylovii* steppe along the borders of the larch groves (photo by A. A. Korobkov)

bens and *Vaccinium* species, or of *Empetrum nigrum* s. l., and a ground layer that may include a variety of mosses, including *Sphagnum* as well as fruticose lichens; (2) thickets of semiprostrate shrubs such as *Pinus pumila, Alnus fruticosa,* and *Betula middendorffii* (*stlaniks*); and (3) tundra-like vegetation on low terraces in broad valleys and in intermontane depressions. This vegetation includes thickets of dwarf birch and willow, tussock areas dominated by *Eriophorum vaginatum* and *Carex lugens,* and lowland wet meadows of *Carex stans* [=*C. aquatilis* ssp. *stans* of Alaska—*Eds.*] and *Eriophorum angustifolium.*

The climatic conditions of the steppe-rich locations can be understood by referring to the data from the Sakkyryr, Oimyakon, Ust-Nera, and Ust-Moma weather stations (*3, 4, 5, 7,* Table 2 and Fig. 1). Note that the lowest precipitation in generally dry Yakutia is in the upper reaches of the Yana drainage. In this area the role of steppe vegetation is particularly significant; it extensively occupies even moderately steep slopes and hilltops. *Stipa krylovii, Agropyron cristatum,* and *Artemisia frigida* are known from these locations.

In the intermontane depression of the middle reaches of the Indigirka River, near the mouth of the Inyali River and Tebulyakh Village (*A,* Fig. 1), steppe vegetation dominates the lower elevations (Fig. 5). Steppe occupies not only south-facing slopes, but also those that face east or west, extending up the slopes and over the surfaces of low and high terraces to heights of 100 to 200 m. Throughout an area extending some 30 km along the river and from 1 to 4 km wide across the valley, montane taiga is replaced by forest steppe similar to that of the Khangai area of northern Mongolia (Korotkov, 1976). Larch woodlands "hide" on slopes with a northern exposure, at the base of slopes with other exposures, and in depressions and drainage lines. Except on north-facing slopes, the only hypoarctic dwarf shrub is *Vaccinium vitis-idaea.* The ground layer is dominated by *Rhytidium rugosum,* and the herbaceous layer also has a meadow-steppe aspect.

Figure 4. Secondary steppe dominated by forbs, *Helictotrichon krylovii,* and *Rhytidium rugosum* on the southeast-facing slope of the gully seen in the center of Figure 5. Fruiting *Pulsatilla multifida* is very abundant. Among the graminoids, *H. krylovii, Poa botryoides, P. stepposa,* and *Carex obtusata* are abundant. Burnt trunks and stumps of larch as well as a live branch with cones are seen; in the background are shrubs of *Salix bebbiana*

The distribution of vegetation on the large frost polygons found on terraces and above lakeshores is particularly interesting. Gallery stands of larch and tall shrubs form in the deep troughs between the polygons, but shallower troughs and the centers of the polygons are vegetated by a variety of steppe and meadow-steppe associations.

Larch reproduction is absent not only in the steppe areas, but also along the southern edge of the forest. Many woodlands fail to regenerate after fires; this is also true of the pseudo-taiga larch forests of the Khangai area (Korotkov, 1976). Instead they are replaced by secondary ecosystems with steppe vegetation and soils (Fig. 6). Even without fire the soils and associated biota of these woodlands are closer to those of steppe than to those of northern taiga.

A characteristic type of steppe meadow and

Table 2
Climatic Data for Weather Stations In or Near Relict Steppe Areas in Northeastern Siberia
(after Klyukin, 1960; Mozolevskaya, 1963)

Locality on Figure 1	Station Name	Mean Temperatures (°)					Mean Relative Humidity July, 1 P.M. (%)	Precipitation, mm			Mean of Decade Maxima Snow Depth, cm	Mean Duration of Frost-free Period, Days
		Annual	July (24 hours)	July (at 1 P.M.)	Annual Amplitude of Mean Monthly Temperatures	Sum of Mean Daily Temperatures Above 10°		Annual	July	Cold Months (Oct.–April) (% of Annual)		
1	2	3	4	5	6	7	8	9	10	11	12	13
1.	Yakutsk	−10	18.7	–	61.9	1565	46	192	37	27.1	31	123
2.	Verkhoyansk	−15.7	15.2	19.2	64.1	1084	49	154	35	27.3	26	69
3.	Sakkyryr	−14.8	12.8	17.4	53.7	——	——	170.1	44.5	——	30	64
4.	Oimyakon	−16.6	14.5	19.4	64.6	969	50	194	39	27.8	32	30
5.	Ust-Nera	−15.6	15.3	19.6	64.2	1016	48	224	51	21.4	36	53
6.	Predporozhnaya	−14.6	15.7	——	——	1207	——	163	38	20.2	26	——
7.	Ust-Moma	−15.0	15.0	18.9	61.9	1088	56	190	44	30	27	50
8.	Berelekh	−13.7	13.5	18.4	54.9	768	52	258	56	25.2	39	30
9.	Seimchan	−11.9	15.5	20.0	55.0	1106	52	286	43	48.2	——	55
10.	Balygychan	−11.9	15.4	19.2	55.0	1185	——	300	52	41.7	——	——
11.	Korkodon	−12.2	14.9	19.1	54.6	1028	56	283	51	44.2	63	57
12.	Ust-Oloi	−12.5	13.7	17.6	51.9	816	——	245	46	42.9	63	63
13.	Srednekolymsk	−12.6	13.6	16.2	51.8	——	57	195	34	45.1	63	78
14.	Kresty (Cherskiy)	−11.6	12.3	14.8	46.5	——	60	169	25	51.5	——	81
15.	Pevek	−10.4	7.5	9.0	34.6	——	75	136	24	40.4	——	——
16.	Krasnoarmeiskiy	−10.6	9.5	12.8	36.0	121	——	256	35	56.6	36	——
17.	Perevalnaya[1]	−11.5	9.8	13.0	38.8	(154)	62	(422)	(61)	(57.1)	(62)	49
18.	Mukhomornoye[2]	−9.9	13.0	16.0	42.4	(717)	(61)	256	53	39.1	73	70
19.	Anadyr	−7.7	10.5	12.1	33.2	372	——	312	41	57.7	25	81

[1]Data in brackets are for the neighboring station, Amguema
[2]Data in brackets are for the neighboring station, Snezhnoye

associated halophytes is found in the bottom of summer-dry ravines and along the shores of contracting lakes (Fig. 3). Here soda and sulphates accumulate and a salt crust covers the soil by the middle of the summer. In some places the sod is disturbed by horses seeking the salt, and a sparse vegetation of facultative halophytes such as *Puccinella hauptiana, Hordeum jubatum,* and *Chenopodium prostratum* develops (Fig. 3). There is no analog to this situation in the typical northern taiga landscape. The vertical extent of steppe vegetation in the area is also unusual. Even in the alpine zone there are petrophyte variants of microthermous-cryophyte steppes, dominated by *Selaginella sibirica* and *Festuca auriculata.* These occur on flat summits, slopes with southern exposure, and shallow saddles (Fig. 6).

The steppe nature of the Inyali-Tebulyakh region is also reflected in its lichen flora (Bredkina and Makarova, 1978). In addition to the widespread multiregional xerophytic lichens, there are four predominantly steppe species typical of semi-deserts as well. These are *Cornicularia steppae, Diploschistes steppicus, Acarospora schleicheri,* and the particularly abundant *Parmelia vagans.* There are also 7 desert-steppe lichens, including *Dermatocarpon desertorum, Endocarpon subfoliaceum,* and *Psora asiae-centralis.* Nine of the 19

Figure 5. Steppe vegetation in the lower reaches of the Ystan-Yuryakh River on the right bank of the Indigirka River valley. Shrub thickets of *Alnus fruticosa* and *Salix boganidensis* line the floodplain along the river in the foreground. The first terrace displays large-scale polygon relief dominated by *Carex duriuscula* steppe with an admixture of *Koehleria cristata* and *Artemisia pubescens.* The second and higher terraces, which slope gently southward, are dominated by *Koehleria* steppe and, in shallow depressions, by *Helictotrichon krylovii* steppe. Larch groves are confined to depressions and are replaced locally there by secondary steppe with shrubs of *Salix bebbiana.* The floor of the gully is occupied by *Carex duriuscula* steppe and secondary halyophytic associations that include *Puccinellia hauptiana, Chenopodium prostratum,* and *Taraxacum dissectum.* The ephemeral watercourse bears a strip of xeromexophytic meadow vegetation. Fruiting of *Pulsatilla multifida* is responsible for the white aspect of the steppe areas

species recorded by Looman (1964) for the lichen association Parmelietum chlorochroae (=*P. vagentis*) from the prairie provinces of Canada are known to occur here.

Among the variants of steppe in the more or less flat areas, microthermous associations are most common. Examples are the small-tufted Koelerietum cristatae and Helictotrichetum krylovii and the low sedge Caricetum duriusculae. *Koeleria* steppes occupy the driest locations. They are poor in vascular species, including on the average only 10 species. In their structure they resemble the southern dry (zonal) steppes. The spaces between tufts are occupied by various epigeic lichens, with *Parmelia vagans* dominant.

Helictotrichon steppes are characteristic of less dry areas and sometimes intergrade with meadow steppes. *Carex duriuscula* steppes occur on slopes, in depressions on the terrace surfaces, around lakes, and on the bottoms of wide ravines. Large-tufted steppe is prevalent on eroded south-facing slopes; common species are *Agropyron cristatum, A. jacutorum,* and sometimes *Artemisia frigida, Chamaerhodos grandiflora, Stellaria cherleriae,* and *Oxytropis scheludjakoviae* are also confined to montane steppes. In alpine and subalpine steppes the characteristic dominant species is *Festuca auriculata.* However *Helictotrichon krylovii, Carex pediformis, C. supina* ssp. *spaniocarpa*, and *Selaginella sibirica* may share in dominance.

Recent pollen samples from the Inyali-Tebulyakh area were collected by A. A. Korobkov and studied

Figure 6. Montane forest steppe on the right bank of the Indigirka River near the mouth of the Ystan-Yuryakh. In the foreground is secondary *Koeleria*-forb steppe with shrubs of *Salix bebbiana* and a ground layer of meadow-steppe herbs and the moss *Rhytidium rugosum,* growing on the site of a burned-over larch grove. This site can be seen in the background of Figure 5. The Indigirka River, out of sight at the foot of the extensive bluff at the right is at an altitude of about 400 m and the conical mountain summit at the left is nearly 1000 m in altitude. The forest-steppe belt extends to an altitude of about 700 m, but larch forest is limited to north-facing slopes, depressions on other slopes, and the surface of the 130-m terrace, shown in the middle of the view. In the subalpine belt at higher altitudes, steppes are confined to south-facing slopes and the higher saddles, other sites being occupied by shrub thickets of *Alnus fruticosa* and *Betula middendorffii,* larch woodlands with a dwarf shrub-moss substrate, herb-*Dryas* tundra steppe, forb-sedge-*Kobresia* cryophyte steppe, herb-dwarf shrub-moss tundra, *Dryas* fell fields, rocky cliffs, and stony barrens

by T. D. Davidovich (personal communication 1978). A sample collected in alpine forb-fescue-*Selaginella* steppe at an altitude of about 850 m produced a spore-pollen spectrum dominated by *Selaginella* spores (80% of the total). Other cryptogams make up 7%. Arboreal pollens comprise 4% of the total spectrum and consist of *Pinus pumila* (64% of the arboreal fraction), *Alnus fruticosa* (10%), *Betula* ex sect. *Nanae* (8%), and Larix (11%), all introduced by long-distance wind transport. Herb-dwarf shrub pollen comprised 9% of the total spectrum; of this component Caryophyllaceae made up 57%, *Artemisia* 12%, other Compositae 25%, *Polygonum* and Chenopodiaceae each 4%, and Cyperaceae and Ericaceae each 2%. The chenopodiacious pollen is probably *Chenopodium prostratum*, the only chenopod recorded in the area; it reaches an altitudinal limit some 400 m below the collection site. Recent pollen samples from steppes at lower altitude within the Indigirka forest-steppe belt consist of only 1% to 5% spores. Arboreal pollen comprises 10% to 40% of the total spore-pollen spectrum and includes *Pinus pumila* (39% to 46%), *Larix* (8% to 44%), *Alnus fruticosa* (10% to 31%), and *Betula* ex sect. *Nanae* (2% or less). Herb-dwarf shrub pollen makes up 56% to 89% of total spectra and consists of *Artemisia* (15% to 79%), Caryophyllaceae (8% to 35%), Gramineae (4% to 14%), and *Eritrichium* (up to 16%).

The Predporozhnaya weather station (*6*, Table 1 and Fig. 1) is situated in this region, a short distance outside the southern border of the forest-steppe. The data show that this is the warmest area in northeastern Yakutia and that it is the driest and least snowy part of the Indigirka drainage basin, but these climatic differences are insufficient to explain the unique nature and distribution of forest and steppe in this area. One possible interpretation is that there is a threshold of warmth and precipitation beyond which forest can no longer dominate the landscape, and that this threshold is exceeded only in the Inyali-Tebulyakh area.

The ice wedges associated with the polygons mentioned above are inactive and do not reach closer than 1 m to the soil surface (T. N. Kaplina, personal communication). A drier and still colder climate must have existed to allow the formation of these ice wedges just at the time when the wide dispersal of steppe plants occurred in northeastern Asia.

The soils of the steppes of the Yana and Indigirka drainages belong to the family of steppe soils (Naumov, 1973; Sokolov, et al., 1977). Soil ph is circumneutral in the A horizon and somewhat basic in the B horizon. The soils are completely base-saturated, mainly with Ca^{++}. Humus accumulation results mainly from the decomposition of roots rather than from peat accumulation above the mineral soil. A carbonate horizon usually occurs at depths of 15 to 25 cm. The silicate phase of the soil profile remains undifferentiated. Sokolov et al. (1979) state that the soil is a special type of cryoarid steppe soil, characterized by abbreviated profile, low base-saturation capacity, and a humus horizon somewhat enriched by slightly decomposed roots. Additional attributes are the humate-fulvate composition of the humus and the abundance of humines.

The standing crop of a montane *Koeleria* steppe with 60% vegetation cover in the vicinity of Tebulyakh Village is as low as 2 centners dry mass per hectare. That of a montane forb-sedge-fescue steppe with similar ground coverage near Ust-Moma settlement was 4 centners per hectare (Skryabin and Konorovsky, 1975).

The Kolyma Drainage, Taiga Portion

Extensive areas of steppe are extremely rare in the Kolyma drainage and are found only on south-facing slopes of large montane valleys. Steppe phytocoenoses have not been found on the coastal plain of northern Yakutia; the only known exception is a fragment of hemicryptophyte steppe dominated by *Carex supina* ssp. *spaniocarpa* on top of a pingo in the Kolyma Delta (Andreev and Perfilyeva, 1975) (*D*, Fig 1).

The dramatic decrease in the role of steppe vegetation in the landscapes of the Kolyma drainage in comparison to the Yana and Indigirka area can be attributed to the fact that winters are milder and have a considerably heavier snowfall. Year-around precipitation is also 1.5 to 2.5 times that of the Yana-Indigirka region. The result is an expansion of larch woodlands and especially *Pinus pumila* thickets onto the south-facing slopes.

The steppe element of the Kolymean flora is a slightly depauperate derivative of that of the Yana-Indigirka. There is also some resemblance to Chukotkan steppe communities in that both areas share endemic species such as *Potentilla anachoretica* and *Oxytropis schmorgunoviae*. Many of the dominant plants of the Kolymean steppes are also important in the Yana-Indigirka and Chukotka regions. Examples are *Festuca lenensis, Poa botryoides, Calamagrostis purpurascens,* and *Carex pediformis. Helictotrichon krylovii, Carex duriuscula, C. obtusata,* and *C. supina* ssp. *spaniocarpa* are extremely rare in the taiga portion of the Kolyma drainage.

In the Anyui Mountains, which extend from Cherski eastward along the Arctic coast [see endpaper map—*Eds.*], the xerophytic vegetation is represented largely by the xeropetrophyte groups except in the westernmost spurs which reach the Kolyma Valley. Although the Anyui Mountains are near the border of the Chukotkan tundra, many xerophytes are found which are not known to occur in Chukotka. Examples are *Agropyron jacutorum, Veronica incana, Lappula echinata, Campanula langsdorffiana, Artemisia gmelini* s.l., and *A. dracunculus.*

The Anadyr Drainage

Steppe associations are even rarer and more floristically depauperate in the Anadyr drainage, especially in the area dominated by large *stlaniks* or by taiga and *stlaniks.* This probably due to the reduced continentality of the climate. The annual precipitation is as much as 300 to 400 mm, and winter snow cover is deep (up to 73 cm at Mukhomornoye) (*18,* Table 1 and Fig. 1). Wind may also be a relevant factor. In the steppe-rich portion of the Yana-Indigirka region, mean wind velocity in January may be less than 0.5 m per second, and it is only about 2 m per second in July. Comparable figures for the large-*stlanik* region are 4.0 to 7.6 m per second in January and 3.2 to 4.6 m per second during July (Klyukin, 1960). July temperatures in the *stlanik* zone of the Anadyr drainage are comparable to those of the northernmost taiga zone, averaging 12.5° to 13.5°C. *Pinus pumila* is the main competitor with steppe vegetation on the south-facing slopes, and mild winters and deep snow appear to favor the pine.

In this region, patches of steppe are largely restricted to outcrops of solid rock and to the crests of buttes, especially those composed of basic or ultrabasic rocks. A few patches were found in the isolated larch woodland area on a spur of the Algan Mountains (*E,* Fig. 1) (Kozhevnikov, 1977). They are also known on the Ust-Belskiye Mountains and on the western slope of the Pekulnei Range (not shown on Fig. 1) in a transition zone between the large-*stlanik* subzone and the southern hypoarctic tundra. The number of species is low, although *Artemisia frigida* is present at one locality (*H,* Fig. 1) and *Chamaerhodos erecta* and *Carex duriuscula* were found at another (*I,* Fig. 1).

The Chukotka Tundra

The more or less regular occurrence of steppe phytocoenoses in the continental sector of the Chukotkan tundra would at first seem puzzling, since the climate is much less continental than that in the Kolyma-Omolon and Anyui taiga, or even that in the large-*stlanik* subzone. However, the annual precipitation and particularly the summer precipitation is lower than in the taiga and large-stem *stlanik* areas. Snow accumulation is commonly lower also, and the upper parts of south-facing slopes and the edges of terraces and hilltops are usually snow free in wintertime, especially in western Chukotka, where south winds predominate in winter. These are the optimum habitats for xerophytes in the tundra (Fig. 7).

These habitats are poor in moisture, partly because of perfect internal drainage, partly because of evaporation and evapotranspiration associated with the strong, warm, somewhat foehn-like winds which blow from the warmer interior of the continent. It may also be significant that many microthermous steppe plants are precocious cool-season perennials. They carry out the basic stages of their seasonal development before the onset of summer drought, and they develop normally even on Wrangel Island. Competition is also apparently a factor in the broad distribution of steppe associations in the Chukotkan tundra; the absence there of such powerful competitors as *Pinus pumila, Larix cajanderi,* and *Betula middendorffii* would seem to promote the expansion of steppe vegetation.

In moister, snow-protected sites at the foot of south-facing slopes, along swales, and in similar situations, the steppe vegetation is in contact with associations of low shrubs (Fig. 8), including *Salix glauca, S. krylovii, S. pulchra,* and sometimes *Betula exilis,* species typical of the tundra. Also found are cold-resistant boreal and hypoarctic species such as *Betula extremiorientalis, Juniperus sibiricus, Rosa acicularis, Ribes triste,* and *Spiraea stevenii.* Associations of dwarf hypoarctic shrubs such as *Arctuous* [=*Arctostaphylos–Eds.*] *alpina*, *Empetrum nigrum*, and *Vaccinium vitis-idaea* seem to be confined to shallow depressions in the slopes.

On the upper parts of the slopes, at least four other communities alternate with the true steppe communities (Yurtsev, 1974a). A sparse seral xeropetrophyte vegetation occurs on screes, bluffs, and cliffs. On moderately dry sites, cryophyte-microthermous steppe phytocoenoses such as those dominated by *Carex obtusata* and *C. rupestris* or Kobresieta rich in steppe species occur. Herb-dwarf shrub tundra-steppe associations occur on wind-swept gravelly protrusions of the slopes; they are commonly dominated by *Dryas punctata* [usually considered a subspecies of *D. octopetala* in North America–*Eds.*] and meadow-steppe plants. There are also dry-mesic meadows dominated by species of *Bromus*, *Roegneria* (*Agropyron* s.l.), and low bushes of *Rosa acicularis* around ground-squirrel burrows and in similar situations.

The true steppe communities here include a limited set of vascular plants. They occupy the protruding, driest, sunniest parts of the upper slopes, where the soil is usually stable and rich in fine-grained material. The presence of steppe "islets" in these tundra landscapes depends both on the low precipitation and occurrence of optimal relief forms. Low buttes with steep, south-facing slopes seem to be ideal. Here the summer deficiency of moisture is greatest because of the nearly complete loss of snow cover from wind and the negligible run-off.

The basic dominants of the Chukotka steppes are *Festuca lenensis*, *Carex pediformis*, *C. obtusata*, *C. supina* ssp. *spaniocarpa*, *Poa* species of the section *Stenopoa*, *Calamagrostis purpurascens*, *Pulsatilla multifida* [=*Anemone patens* ssp. *multifida–Eds.*], *Potentilla arenosa*, and in the warmest and

Figure 7. A landscape on the left bank of the Amguema River valley near the 176-km post of the Egvekinot-Iultin road in the inner part of the Chukchi Peninsula. The rugged surface of the 50–60 m terrace is dominated by dry *Dryas punctata* tundra, replaced by *Arctous* (=*Arctostaphyllos*) *alpina* heaths in shallow depressions. The upper parts of the south- and southwest-facing terrace slopes and the tops of the marginal low hills are occupied by cryophyte steppes replaced downslope by dry *Empetrum subholarcticum* heaths (dark contours). Mossy dwarf-birch thickets occupy slope concavities. The lower parts of the steeper slopes bear snow-bed meadows and, locally, small thickets of *Salix krylovii*. The slanting lower terrace surface is covered with hummocky and tussocky dwarf shrub-moss hypoarctic tundra dominated by *Eriophorum vaginatum*, *Carex lugens*, *Betula exilis*, and *Ledum decumbens*

driest sites, *Helictotrichon krylovii* and *Carex duriuscula.* Commonly found on petrophyte steppes are *Thymus oxydontus, Dracocephalum palmatum, Selaginella sibirica, Phlojodicarpus villosus,* and *Arenaria tschuktschorum.*

A number of species have a rather sporadic and apparently relict distribution. They are common enough on some xeric steppe-bearing slopes, but totally lacking on others. Examples are *Helictotrichon krylovii, Carex duriuscula, Cerastium arvense,* and *Alyssum obovatum* (=*A. americanum*). These species were probably restricted to a very few favorable sites during a critical Holocene period when humidity was higher, soil leaching most intense, and the more mesic vegetation types at their maximum expansion. After a reversion to drier conditions, these plants would then have reoccupied slopes adjacent to those few areas in which they had persisted.

Variants of cryophyte-microthermous steppe extend northward to the islands in Chaun Bay, where they are confined to the steep slopes and dry ridges of the higher sandy terraces—on islands that are remnants of the western Beringian shelf plain (*N, O,* Fig. 1). However, the richest steppe colonies, with *Helictotrichon, Carex duriuscula,* and *Eritrichium sericeum,* occur on the mainland to the south and southeast (*L, M,* Fig. 1).

In central Chukotka steppe phytocoenoses are known in montane valleys in the southern part of the Amguema-Kuvet Mountain massif (*P, R,* Fig. 1), where they extend eastward to the terraces on the left bank at the Egvekinot-Iultin highway crossing of the Amguema River (Fig. 7 and *S,* Fig. 1). This region is the eastern limit of several Asian species, including *Festuca lenensis, Poa filiculmis, Dracocephalum palmatum, Thymus oxydontus,* and *Polygonum laxmanii.* It is also the eastern limit in Asia of several amphi-Beringian steppe plants, such as *Carex obtusata, C. supina* ssp.

Figure 8. Close-up view of one of the largest steppe patches shown in Figure 7. Vegetation in the foreground consists of dry tundra dominated by *Carex rupestris* and *Dryas punctata,* replaced near the foot of the slope by tussocky and then hummocky tundra of dwarf shrubs and mosses. Two thickets of *Salix krylovii* can be seen at the foot of the terrace slope. The lower portion of the slope is occupied by dry *Empetrum* heath and *Empetrum-Festuca altaica* heath meadow. The upper and most extensive portion of the terrace slope bears different variants of cryophyte steppe

spaniocarpa, Silene repens, and *Cerastium arvense.*

The mountainous Amguema River basin displays the most frequent occurrences of steppe within the Chukotka tundra, probably because of the combination of winter wind conditions and optimal rugged microrelief involving hummocky moraines and high outwash terraces with excellent internal drainage (Fig. 7). Steppe patches several tens of meters in diameter are present, and several "islets" of *Helictotrichon krylovii* steppe and Helictotrichon-Dryas tundra-steppe have been found. Though poorer in species than steppe patches in western Chukotka, some American cryoxerophytes not found farther west appear in the Amguema steppe islets. *Smelowskia porsildii* s.l., *Potentilla rubricaulis* s.l., and *Erigeron compositus,* for example, were found by E. Yu. Norkina, T. V. Plieva, and the author in 1979. Most of the Amguema steppe patches consist of herb stands made up of three components: (1) tufted grasses (*Festuca* agg. *lenensis,* including all grades of transition between *F. lenensis* s.s. and the cryoxerophilous *F. auriculata; Helictotrichon krylovii; Poa glauca* and forms transitional to *P. botryoides*; and *Calamagrostis purpurascens*); (2) long rhizomatous sedges (*Carex obtusata, C. supina* ssp. *spaniocarpa,* the cryoxerophilous *C. rupestris*); and (3) taprooted perennial dicots, including some cushion plants in the coolest, most windswept sites (e.g., *Eritrichium tschuktschorum, Androsace ochotensis,* and *Minuartia obtusiloba*).

In the driest, most protruding parts of slopes, the structure of the herb stands resembles that of zonal dry steppes. A carbonate horizon forms in the soil, despite the acid composition of the gravel. Though the soil is saturated with roots, cover by higher plants is discontinuous. Space between tufts of herbs is filled by *Selaginella sibirica* and encrusting xerophytic lichens, some of which overgrow dead *Selaginella.* On slight slope concavities and, rarely, on narrow, dry terrace surfaces, stands of steppe sedges are denser, and the ground cover generally consists of the mesoxerophilous mosses *Rhytidium rugosum, Thuidium abietinum,* and *Tortula ruralis.* Both of these variants of sedge-dominated steppe can grade into cryophyte-microthermous associations codominated by *Carex rupestris* and *Festuca auriculata* and then into cryoxerophyte herbaceous vegetation with an admixture of steppe plants.

The basic dominant species of neighboring dry tundras, *Dryas punctata,* does not populate these steppe habitats, evidently because of dry soils and root competition. On moister sites, one can see tundra-steppe phytocoenoses in which spaces between patches of *Dryas* are occupied by tufts of *Festuca* agg. *lenensis, Helictotrichon krylovii,* sedges, herbs, and *Selaginella* (Fig. 8).

Some mesoxerophilous and xeropetrophilous plants typical of steppe and tundra-steppe associations in continental Chukotka occur eastward sporadically all the way to Bering Strait. Examples are *Selaginella sibirica, Potentilla arenosa, P. anachoretica, Androsace septentrionalis, Calamagrostis purpurascens,* and *Draba cinerea.* In the easternmost parts of the peninsula they tend to concentrate on outcrops of carbonate or basic rocks. They are found in rocky canyons, on dry fell fields, and especially on zoogenic meadows in the vicinity of ground-squirrel burrows or the feeding places of birds of prey. Some isolated relict populations of a few individuals have differentiated to the point where they are considered distinct taxa. Examples are *Potentilla beringensis, Arabidopsis bursifolia* var. *beringensis,* and *A. tschuktschorum.*

Wrangel Island

On Wrangel Island continental xerophyte and cryoxerophyte complexes are exceptionally diverse and rich in species. This is particularly evident in the southwestern and central parts of the island. Steppe plants constitute an appreciable proportion of many xerophilous variants of the forb-*Dryas* and forb-sedge-*Dryas* tundras. They are important in the colorful dry-mesic forb meadows, where the herbaceous cryoxerophyte vegetation is dominated by *Carex obtusata, C. rupestris,* and *C. duriuscula,* and in tundra-steppe communities (Petrovsky, 1967, pp. 339–40; 1973). Most sensational was V. V. Petrovsky's discovery in 1979 of *Carex duriuscula* as a local dominant in cryophyte-microthermous steppe areas in western and southwestern parts of Wrangel Island.

Since Wrangel Island is centrally located on the Chukotkan polar shelf, the xerophytes and xeromesophytes native to the island are important to developing an understanding of the past migration of xerophytes across the exposed shelf. Some 34 species are relevant to this point. Among them are *Selaginella sibirica, Calamagrostis purpurascens, Festuca lenensis, Carex obtusata, C. duriuscula,*

Cerastium arvense, Silene repens, Pulsatilla multifida, Potentilla arenosa, and *Aster alpinus.* While the majority of these xerophytes are also found in Alaska and the Yukon, a few are endemic to eastern Siberia. Examples are *Festuca lenensis, Lychnis sibirica* ssp. *villulosa,* and *Potentilla anachoretica.* The American relationships of the Wrangel Island xerophytic flora are emphasized by the presence of such species as *Erigeron compositus, Saxifraga monticola, Gastrolychnis apetala* ssp. *attenuata* [=*Melandrium apetalum* ssp. *attenuata*—*Eds.*], *Lesquerella arctica,* and *Carex hepburnii* (Petrovsky, 1978). It is interesting to note that there is a closely analogous group of species, mainly amphi-Beringian or purely American, in the southwesternmost part of the Canadian Arctic Archipelago (Porsild, 1964).

The highest concentration of xerophytes and cryoxerophytes on Wrangel Island occurs in the *Carex obtusata*—*C. rupestris*—*C. duriuscula* microthermous-cryophyte meadow steppes; these are found mainly on south-facing slopes and well-drained terraces with poorly sorted gravelly fine-grained soils on the inland part of the island (Petrovsky, 1967). Among the associated species, grasses are particularly prominent: 15 species are included, of which 7 belong to the genus *Festuca.* There are also 5 species of *Artemisia,* 10 legumes many Rosaceae (8 species of *Potentilla* alone), and many Cruciferae and Caryophyllaceae. *Selaginella sibirica* is also important. Some 50 or 60 species grow in these habitats, including certain mesophytes with a broader ecological range—notably, certain microthermous steppe herbs adapted to complete their development before the summer drought.

Betula exilis, Empetrum, and *Arctous* are completely lacking on Wrangel Island, and *Vaccinium vitis-idaea* ssp. *minus, V. uliginosum* ssp. *microphyllum,* and *Ledum decumbens* are extremely rare. However, there are extensive meter-high willow thickets (*Salix lanata* ssp. *richardsonii* and *S. glauca*) in the river valleys.

These landscapes have no modern analog on the Chukotkan mainland. They may be the closest living approximations of the northern variants of late Pleistocene tundra-steppe, though I would suggest a much decreased role of xeric phytocoenoses and an enrichment by various cryomesophytes in the present-day vegetation of Wrangel Island.

Of great significance for attempts to reconstruct late Pleistocene Beringian vegetation is a surface sample collected by V. V. Petrovsky from a vegetation plot 6 km inland from Somitelnaya Harbor on Wrangel Island (*T,* Fig. 1) (Davidovich and Ivanov, 1976, p. 23). This sample provides the first-known surface pollen spectrum displaying a close resemblance to ancient tundra-steppe pollen spectra. Spores of *Selaginella sibirica* constitute 97% of the total spore-pollen sum. Another 1% consists of locally produced pollen of dwarf *Salix* shrubs and single grains of *Betula exilis, Alnus fruticosa,* and *Pinus pumila* introduced by long-distance wind transport. Herb-shrub pollen accounts for only 2% of the total. Of this fraction, Compositae pollen accounts for 58% (*Artemisia* 50%!), Gramineae 15%, Caryophyllaceae 12%, Cyperaceae 5%, and Cruciferae 5%. Ranunculaceae, *Valeriana, Gentiana,* and Rosaceae together account for 10% of the herb-shrub pollen. This spectrum differs from the spore-pollen spectrum from the Inyali-Tebulyakh steppe in the lower abundance of wind-transported arboreal pollen and the absence of Chenopodiaceae.

On the whole, the xerophyte floristic complex of Chukotka is merely a depauperate derivative of that of northeastern Yakutia. It contains only half as many species (compare Tables 1 and 3). The depauperization is particularly marked among microthermous species, xerophytes, xeropetrophytes, and mesoxerophytes. The situation is markedly different in the case of the large group of cryoxerophytes. Of 78 species in the Chukotka flora, 24 are amphi-Beringian taxa (but absent in Yakutia), and 21 are Chukotkan or Chukotkan-Okhotian endemics. Among mesoxerophytes the proportion of amphi-Beringian species is nearly two-thirds. The proportion is similar among microthermous xeromesophytes, but only one of seven microthermous xerophytes (*Carex duriuscula*) is amphi-Beringian. All or nearly all of the amphi-Beringian species of the northeastern Yakutia steppe complex occur in Chukotka, the only possible exception being *Artemisia frigida.* Altogether, the amphi-Beringian taxa and those replaced in North America by vicariant species comprise some 70% of the xerophyte complex of Chukotka. The purely Asian element amounts to only 20% and consists mainly of the xerophytes proper (one-fourth are Asian endemics) and the xeropetrophytes (half are Asian).

DISCUSSION

As the situation in the middle Indigirka forest-steppe shows, a certain level of climatic continentality and aridity allows steppe vegetation to become dominant within the northern taiga subzone. Forests are then restricted to sites with better water supply and to north-facing slopes. Even these forested areas are more akin to steppe then to typical forests in terms of associated plants, animals, and soils.

A comparable situation in the southern hypoarctic tundra subzone, involving a cooler summer temperature regime, is well demonstrated in the inner periglacial portions of southwest Greenland. There, the zonal dwarf shrub-moss tundra, dominated by Ericaceae and dwarf birch, is confined to north-facing slopes and the moister valley areas. Elsewhere shrub tundra is replaced by xeric vegetation dominated by *Carex supina* ssp. *spaniocarpa, Kobresia myosuroides,* and associated halophyte communities (Bochner, 1949). A similar situation also exists in the colder, more northern, inner periglacial areas of Ellesmere and Axel Heiberg Islands (Beschel, 1963), and in Peary Land (Holmen, 1957). There some of the hemicryoxerophytes typical of hypoarctic steppes reach the northern limit of land. Examples are *Calamagrostis purpurascens, Draba cinerea, Erysium pallasii,* and *Potentilla chamissonis.*

Sokolov et al. (1979) show that there is a basic similarity among all arid-region soils, regardless of thermic zone. This similarity is noted in soils from

Table 3
Distribution of Northeast Asian and Amphi-Beringian Taxa among Ecological Subdivisions of the Xerophyte Complex of the Chukotka Tundra

	Number of Species and Subspecies						
	Total Xerophyte Complex		Asian Endemic	Amphi-Beringian			
Ecological Subdivisions	Number	Percent	Endemic to NE Asia	Represented by American Variant	Present in North America	Total
Microthermous	40	47.0	3 (7.5)[1]	9(22.5)	20(50)	29(72.5)
Xerophytes	6	7.1	1(16.7)	4(66.7)	1(16.7)	5(83.3)
Mesoxerophytes and Xeromesophytes, Combined	29	34.1	2 (6.7)	4(13.8)	18(60)	22(75.9)
Mesoxerophytes	19	22.4	1 (5)	4(21.1)	11(55)	15(78.9)
Xeromesophytes	10	11.8	1(10)	0(0)	7(70)	7(70)
Xeropetrophytes	5	5.9	0(0)	1(20)	1(25)	2(40)
Hemicryophilous	45	52.9	15(33.3)	7(15.6)	23(51.1)	30(66.6)
Xerophytes	16	18.8	7(43.7)	4(25)	6(37.5)	10(62.5)
Mesoxerophytes and Xeromesophytes	17	20	3(17.6)	2(11.8)	11(64.7)	13(76.5)
Xeropetrophytes	12	14.1	5(41.7)	1 (8.3)	6(50)	7(58.3)
Total sum of microthermous and hemicryophilous spp.	85	100	18(21.2)	16(18.3)	43(50.6)	59(69.4)
Cryoxerophytes	78	100	27(34.6	10(12.8)	44(56.4)	54(69.2)

[1]Numbers in brackets are percentages

such different areas as the ultra-cold "oases" in Antarctica and the high arctic, temperate steppe, hot steppe, and desert areas. These arid soils are characteristically base saturated and carbonate neutral to basic; humus formation is accumulative; and the transformation of mineral components takes place in situ. There is a profound difference between the humid and arid types of soils in every zone.

An analogous conclusion can be drawn regarding the vegetation and, in fact, the entire ecosystems of the humid versus xeric sectors of the different thermic zones. These observations have been formalized as the "Periodic Law of Geographical Zonality" by Grigoryev and Budyko (1956), manifested in the nearly continuous distribution of steppe associations in the Asian ultracontinental sector from central Asia to northeastern Yakutia and continental Chukotka. Another manifestation is the distribution of cryoxerophyte herbaceous associations from the alpine areas in the North American Cordillera northward all the way to Peary Land. Ecosystems within these belts have many features in common, including the structure of the vegetation, the floristic composition at family, generic, and sometimes even species level; groupings of life forms; and soils.

This concept gives us the key to understanding the nature of the Pleistocene cryoxeric landscapes. It also provides the theoretical basis for reconstruction of the vegetation and soils of these landscapes.

Latitutinal Zonation of the Plant Cover in Ancient Beringia

As can be seen in Tables 1 and 3, and as I have discussed in earlier papers (Yurtsev, 1962, 1963) there are many striking disjunctions or replacements by vicariant species among the microthermous xerophytes, that is, the true steppe plants, of northeastern Asia and northwestern North America, whereas amphi-Beringian distribution patterns are rare. Among the dominant species of the trans-Baikalian, Mongolian, and Kazakhstanian steppes and the North American Great Plains are *Artemisia frigida* and *Carex duriuscula* (=*C. stenophylla* ssp. *eleocharis* of American workers). The Beringian portion of the distributions of these two amphi-Beringian species are shown in Figures 9 and 10.

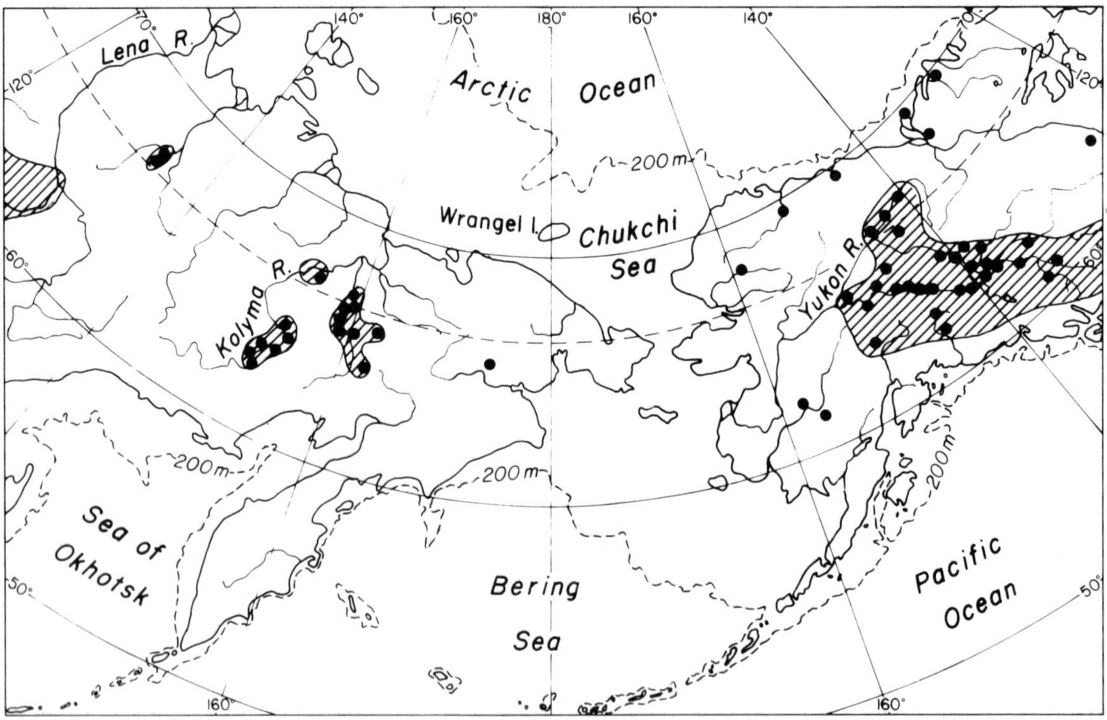

Figure 9. Distribution of *Artemisia frigida* Willd. in Beringia

At present, *Artemisia frigida* extends its range into the tundra zone only in the vicinity of the Mackenzie Delta; *Carex duriuscula* reaches the tundra zone only to the east of the Kolyma River delta and on Wrangel Island. The recent discovery of relict stations of both species in the middle reaches of the Anadyr drainage suggests the following speculation regarding the trans-Beringian migration of steppe xerophytes:

The migration could occur by either of two routes. One would be over the exposed polar shelf and adjacent hilly regions north of the montane glaciers and the other through the wide valleys and low mountains of the Omolon and Anadyr drainages, across the Anadyr Lowland, through the gap between the glaciers centered on the Chukotka and Koryak Mountains, and thence to the central Bering Shelf and finally to the Yukon Delta, Interior Alaska and Yukon Territory (Yurtsev and Korobkov, 1979). Migration could, of course, take place either eastward or westward.

Through the second route, sheltered from the north by high, glaciated mountains, the more thermophilous steppe plants could migrate. However, this could only happen if thickets of *Pinus pumila* and other *stlanik* and shrub vegetation, as well as *Eriophorum* bogs and tussock vegetation, had essentially vacated the area. With a subsequent rise in summer temperature and precipitation, these powerful competitors of steppe vegetation would rapidly have reoccupied the area, resulting in the enormous disjunctions in the ranges of many amphi-Beringian steppe plants. A particularly impressive example of this is the disjunction between populations of *Enneapogon borealis* of Selenga Dahuria and southernmost East Siberia and the probably conspecific *E. desvauxii* of Utah and Arizona.

The data suggest zonal differentiation of the xerophytic vegetation of Chukotka and the adjacent shelf plains at the time they were emergent. Landscapes were most diverse and complex south and southwest of the glaciated ranges of the Chukotka Mountains. These probably supported xerophytic and cryoxerophytic grasslands with an admixture of relatively thermophilous true steppe plants, as well as steppe-tundra and xeric-tundra associations and patches of hygrophilous and chionophilous arctic-alpine plants along the front of the glaciers. Finally, there would have been a

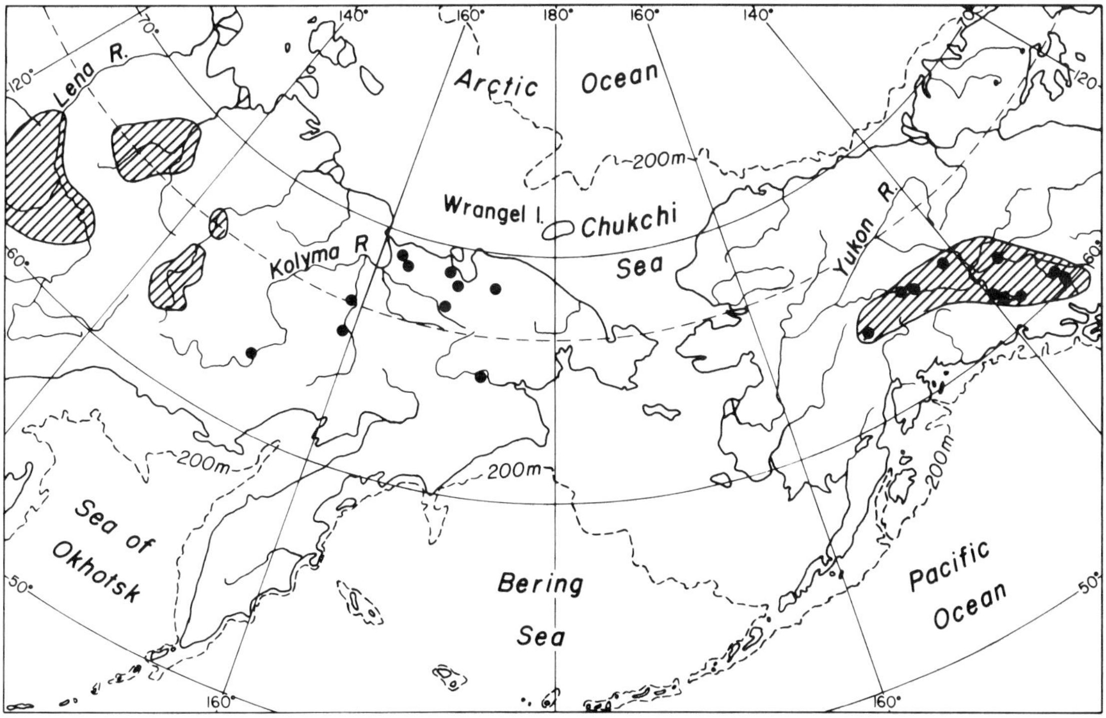

Figure 10. Distribution of *Carex duriuscula* C. A. Mey. in Beringia

few relict patches of *stlanik* and shrub thickets in sites protected by deep snow; it is also possible that scattered patches of larch woodland were present to the west of this area.

North of the glaciated mountain ranges, the steppe-plant group would have been progressively impoverished in thermophilous species. However, species such as *Festuca lenensis, Carex obtusata, Pulsatilla multifida, Potentilla arenosa, Cerastium arvense, Eritrichium sericeum* s.l., and *Arenaria capillaris* ranged at least as far north as Wrangel Island. Here they would have been mixed with the true arctic cryoxerophytes such as *Lychnis sibirica* spp. *villulosa, Erigeron compositus, Poa abbreviata,* and *Artemisia richardsonii.* The low northern spurs of the Chukotka Mountains would have supported such species as *Carex supina* ssp. *spaniocarpa, Poa filiculmis,* and *Polygonum laxmannii,* as well as xeropetrophytes such as *Dracocephalum palmatum, Thymus oxydontus, Alyssum obovatum,* and *Artemesia kruhsiana.*

The northernmost parts of the shelf plain extended for many hundreds of kilometers north of Wrangel Island. It was apparently a cold-arid desert similar to present-day Peary Land, with oases of vegetation along rivers and in areas of deep snow accumulation. Southward, toward Wrangel Island, the desert was apparently replaced on uplands by herb-*Dryas* dry tundra, tundra steppe, or sparse cryophyte steppe associations, dominated by cryophytes and xerophytes. The cryophytes would have included *Carex rupestris, C. hepburnii, Kobresia myosuroides, Koeleria asiatica, Festuca auriculata,* and *F. brevissima,* and the xerophytes would have included *Carex obtusata, C. duriuscula, Calamagrostis purpurascens, Selaginella sibirica,* and *Aster alpinus.* Lower parts of the dry slopes would have been partially occupied by low *Salix glauca* thickets and shrublands. Alluvial flats would have supported *Salix lanata* and, further south, *S. alaxensis.* Interspersed with these willow thickets would have been sparse floodplain vegetation and colorful forb-grass dry-mesic meadows. Herb-moss tundra with a layer of prostrate or, to the south, hemiprostrate dwarf shrubs such as *Salix polaris* and *Cassiope tetragona* apparently was restricted to moister, snow-protected sites; the admixture of hypoarctic Ericaceae would have increased southward. Seasonally wet, poorly drained lowlands could have been occupied by sedge-grass meadows, possibly with an admixture of halophyte associations. On dry, south facing slopes, various steppe associations predominated, and there would have been petrophyte steppe communities in mountainous areas.

Longitudinal Zonation of the Plant Cover in Ancient Beringia

A comparison of the floristic composition of xerophyte associations in northeastern Asia and northwestern North America indicates that there were floristic differences between these sectors in the past, as there are at present. The greatest differences are seen between the floras of northeastern Yakutia on one hand, and the upper reaches of the Yukon River drainage on the other. Here, the influences of the regional North Angaris and Cordilleran centra of xeric speciation would have been greatest. The differences must have been appreciable even on some of the polar-shelf plains to the east and west of Bering Strait. The presence on Wrangel Island and even Chukchi Peninsula, but not in Alaska, of Asian xerophyte taxa such as *Festuca lenensis,* one of the most common dominants of the eastern Siberian and northern Mongolian steppes, supports this conclusion. The flora of central Beringia was apparently characterized by the mixing of the most actively spreading xerophilous species from both continents; there would also have been periodic secondary contact of Chukotkan and Alaskan plant populations that had previously attained an amphi-Beringian distribution. The Bering Strait area must have functioned as a "bottleneck" for trans-Beringian migration of xerophytes, especially the more thermophilous or petrophilous ones, even at the height of shelf emergence. However, the more than 40 xerophyte species that presently occur on both sides of the Strait would have been enough to form diverse steppe, meadow-steppe, petrophyte-steppe, cryophyte-steppe, and tundra-steppe associations.

HOLOCENE CHANGES

The Holocene warming and humidization of climate brought about profound changes in northern xerophyte and cryoxerophyte complexes. This is especially true in the northernmost taiga and in the silty tundra lowlands and plains, where the late Pleistocene xerophyte vegetation was almost completely exterminated. The xeric vegetation has survived in a variety of refugial situations. In continental mountainous areas the relict vegetation is most commonly found in intermontane depressions, particularly on steep south-facing slopes and less commonly on dry floodplains. In the tundra zone, patches of relict xerophytic vegetation are found on the higher sandy terraces. In regions with a more maritime climate, relicts occur on outcrops of carbonate or basic rocks, and particularly on dry zoogenic meadows; they also occur on the drier saline coastal sites.

As I discussed above, different groups of xerophytes persisted in the taiga portion of the Kolyma drainage and in the continental sector of the Chukotkan tundra. The present day xerophyte complex in Chukotka is comparatively poor in thermophilous xerophytes and, particularly, in xeropetrophytes, but it is rich in some of the steppe or meadow-steppe species that prefer poorly sorted gravelly fine-grained soils. The survival of this latter group must be due largely to the absence of such competitors as *Pinus pumila* and *Larix cajanderi*. The removal by wind of snow from the upper parts of south-facing slopes during winter and the drying and warming of these sites by foehn winds during the summer are also of some importance.

As I mentioned, the extermination of xerophyte associations over huge areas has resulted in enormous disjunctions in the ranges of many species. However, a few of them have been able to survive as small relict populations in intermediate localities. Some of these small, isolated populations have had time to differentiate into new taxa of subspecific or even specific rank. Examples are *Plantago septata* ssp. *jurtzevii* Tzvel. on Big Rautan Island, and *Arabidopsis bursifolia* var. *beringensis* Jurtz. and *A. tschuktschorum* Jurtz. on easternmost Chukchi Peninsula. The ancestral continental populations of *Arabidopsis bursifolia* in Asia (which includes a population growing as far northeast as Big Rautan Island in Chaun Bay) and the American *A. mollis* are morphologically more similar to each other than they are to the Beringian populations, which should warn us there is no simple parallel between degree of morphological differentiation and time of mutual isolation of related populations.

A FINAL COMMENT

The author realizes that many of the above conclusions are, in fact, hypotheses. They need to be discussed and tested by workers on both sides of the Bering Strait. Further progress in the field is crucially dependent upon increased opportunities for cooperation between Beringologists working in the eastern and western sectors of Beringia.

10

THE VEGETATION OF LAND-BRIDGE BERINGIA

Steven B. Young

This paper offers a reconstruction of the vegetation of eastern Beringia during the last emergence of the Bering land bridge and traces the subsequent alterations in that vegetation. It is assumed that (1) the ancient vegetation was complex and variable over time and space, (2) it was not necessarily comparable to any modern boreal or arctic vegetation, and (3) data from a wide variety of sources are relevant. Evidence has been drawn from paleontological, phytogeographical, and ecological studies involving both plants and animals.

The reconstruction proposed here emphasizes a mosaic of vegetation types. Among the more significant are those adapted to mesic and arid situations, including a variety of woodland, brushland and steppe associations. Many of these probably do not have fully analogous modern counterparts; some of the plant species involved may no longer even occur in the area.

The reconstruction is compatible with the view that climate was colder than at present but also more continental, with comparatively warm summers over much of the area. It further suggests that the megafauna associated with Wisconsin Beringia could have inhabited the area continuously and in some numbers. In many areas the character of the vegetation was strongly affected by various types of natural disturbance. It is argued that the "birch rise" noted in many late Wisconsin pollen spectra is the result of catastrophic vegetation changes that were associated with the submergence of the land bridge and which destabilized the entire Beringian ecosystem.

Key word: Beringian vegetation

Now that the theory, which necessarily presented itself to me at first in outline, has been worked out in detail, I cannot but feel that in some places I may have pushed it too far. If this should prove to have been the case, I will readily acknowledge and retract my error as soon as it is brought home to me. Meanwhile my essay may serve its purpose as a first attempt to solve a difficult problem, and to bring a variety of scattered facts into some sort of order and system.

—Sir James G. Fraser,
Preface to *The Golden Bough*, 1890

INTRODUCTION

This essay is an attempt to develop a plausible picture of the nature of Beringian vegetation during the time of the most recent emergence of the land bridge; it also discusses the alterations in the vegetation that occurred with the final submergence of the land bridge. I am concerned mainly with the period from approximately 18,000 BP to 8000 BP and with the area of the Bering-Chukchi Platform, Alaska, north of the Alaska Range, and northern and central Yukon Territory. Questions

ISBN 0-12-355860-3

regarding the Asian sector of Beringia are not specifically discussed, but they are treated by Yurtsev elsewhere in this volume.

Our knowledge of the vegetation of Beringia during this time is still fragmentary and often equivocal enough to allow a variety of interpretations that agree only on the most basic points. The following synthesis is envisioned as a starting point, subject to refinement and, no doubt, alteration as we increase our understanding of the problem. Hence, I devote considerable attention here to describing the lines of reasoning from which my synthesis developed and the nature of the evidence used in supporting the assertions made.

Three basic comments are relevant. First, climate, relationships between land and sea areas, and the distribution of continental ice changed continuously during the time of our interest; hence, vegetation was probably in a constant state of flux. Second, the area is of enormous geographic extent and physiographic and climatic complexity; it cannot be presumed to have supported a uniform vegetation in the past, any more than it does at present. Third, the vegetation of an area at any given moment is the result of the interaction of a unique combination of many processes, both currently acting and historical. Any vegetation system is qualitatively as well as quantitatively distinct from any of its precursors or derivatives.

It should be obvious that no purely analytic, quantitative approach to a characterization of past (or present) vegetation patterns can have much hope of success in answering the basic question: what was the vegetation really like? Such answers as we can come up with must be synthetic and, to some extent, intuitively derived. It is, however, necessary to tie the answers to as many data points as are available. Any approach using data derived from a single discipline limits itself unnecessarily. Hence, in my reconstruction I have tried to utilize information from many fields. I devote some space in the next section to the relationships among these information sources, and generate some basic assertions that I believe are justified by information from the various sources.

SOURCES OF INFORMATION

Our sources of information can be usefully, if somewhat arbitrarily, separated into the following categories: paleontological, phytogeographic, autecological, synecological, and indirect evidence based on presumed requirements of zoological elements within the ecosystem.

Paleontological Information

Paleontological evidence can be subdivided into microfossil data and megafossil data. Most of the systematically acquired microfossil data is palynological and is treated extensively elsewhere in this volume (Ager, Matthews, Ritchie and Cwynar, Schweger). Only the basic points need be summarized here.

Pollen spectra dating from some 18,000 to 13,000 years ago in Beringia are remarkably similar. They are characterized by a low influx rate, high proportions of grass and sedge pollen, and exceptionally high, but also highly variable, proportions of *Artemisia* pollen. All samples from this interval have low percentages of spruce and other identifiable arboreal pollen. However, spruce pollen is present in sufficient quantity and frequency to suggest to me the possibility that the presence of sources of spruce pollen in land-bridge Beringia should not be altogether ruled out. Ager (this volume) reports a spruce percentage in modern samples taken a few tens of kilometers from spruce woodlands comparable to that of some full-glacial samples. It seems to me doubtful that these comparatively high values and similar ones from the Pribilofs (Colinvaux, 1967c) are entirely the result of a combination of low overall influx rates, long distance transport, and redeposition of pollen from older reworked sediments. On the other hand, inferred patterns of recolonization by spruce (Hopkins, Smith, and Matthews, 1981) are at variance with the concept of survival of spruce in central Beringia.

Full-glacial age samples characteristically include pollen from several taxa not normally found in treeless areas in Alaska today and not usually represented in modern pollen samples. The percentages are low, but their regular recurrence makes them, to my mind, highly significant. Among the most important taxa are Chenopodiaceae, *Plantago*, Labiatae, *Thalictrum*, and Umbelliferae. Whereas Umbelliferae and *Thalictrum* do presently occur on "tundra" areas, the species most likely to be implicated in any pollen rain are characteristic of more temperate areas. For example *Thalictrum alpinum* is a small, usually rare plant in alpine environments, while *T. sparsiflorum* is often large and abundant in lowland swamps and meadows.

Between 13,000 and 11,000 years ago, birch (*Betula*) pollen rises remarkably rapidly and consistently to high percentages in palynological sequences throughout Beringia. There is no question that this is indicative of a major, even catastrophic, vegetation change, and that birch, presumably dwarf birch, was the dominant source of pollen over much of our area for several thousand years. *Alnus* pollen is rare in full-glacial sequences, and it does not show any significant changes contemporary with the rise in birch pollen. Where *Alnus* percentages rise, the samples are generally at levels several thousand years younger than the birch rise, and their position is somewhat inconsistent with any other hypothesized changes in the vegetation. The rise in spruce begins about 11,000 years ago in easternmost Beringia. It seems to proceed steadily from east to west and does not reach some presently forested areas in western Alaska until about 5000 years ago.

The following assertions seem to be supported by the palynological data. First, land-bridge Beringia did not support any significant vegetation type in which spruce was an important component, although we cannot rule out the possible presence of relict gallery forests containing spruce. Second, the pollen influx was low and never comparable in land-bridge Beringia with that of modern temperate grasslands. Third, the pollen rain regularly included pollen from a number of taxa that are presently rare or absent in treeless vegetation beyond timberline. These taxa must have had some significance in the vegetation. Fourth, there was an immense and essentially simultaneous rise in birch pollen at the end of land-bridge times, and this can quite confidently be assumed to represent a major vegetation change. In fact, I believe that "catastrophic" would be the proper descriptive term and that this change was of much greater impact than any succeeding changes. Finally, there was a much slower and more sporadic colonization of the area by spruce and alder.

Evidence from megafossils provides some of the most intriguing data relevant to our problem. Megafossils are seldom distributed in such a way as to allow a systematic analysis, as is the case with pollen. While the megafossil evidence is fragmentary and must be treated anecdotally, it seems likely that it will ultimately provide the best source of data on the vegetation of land-bridge Beringia. Megafossils provide unequivocal evidence that certain plants occurred in land-bridge Beringia that have since become extinct in the area. Of particular interest are the remains of plants now confined to regions more temperate than present-day Beringia. Matthews (1974b) found fruits of *Alisma* in full-glacial sediments from the Isabella Basin, and Guthrie (this volume) reports the discovery of *Bouteloua* fragments in late-glacial ground-squirrel burrows. Considering the sparse research that has been carried out in this field, these discoveries are of great significance. A more systematic approach by Hopkins, Smith, and Matthews (1981) on a large number of dated wood samples provides invaluable evidence of the presence of large woody species. This work leaves little doubt that at least cottonwood, among arborescent species, was present in land-bridge Beringia, and that it was not confined only to the southern fringes of the area. It is difficult to evaluate the significance of *Larix* wood dated as 30,000 to 25,000 years ago from the Prudhoe Bay area (Hopkins, Smith, and Matthews, 1981), but it can hardly be ignored.

Studies of megafossils provide data to support the following assertions: first, they confirm that the land-bridge vegetation included species of plants that do not now occur in the area and that are characteristic of presently more temperate regions. Second, trees were present in land-bridge Beringia, including almost certainly cottonwood, quite possibly aspen, and, at an earlier time, larch.

Phytogeographic Information

Our phytogeographic information is from distributions of taxa that are disjunct, narrowly endemic, or rare and of local occurrence.

Disjunction in distribution theoretically can arise in one of two ways. Either the taxon surmounts a barrier to its continued steady migration by means of long distance dispersal, or a previously continuous range is disrupted by some event such as a change in climate. Normally there are clues that provide an answer as to which of the two is more probable; the clues are particularly useful in the case of a taxon that has an asymmetrical distribution, with a large main range and one or more small, local disjunct populations. If the plant is a recent invader, it would be expected to actively extend its range from its initial foothold. If it is a relict from a previously more continuous population, its present range can be expected to be stable or shrinking. Obviously, intermediate situations can also occur, as in the case of a plant that has great dispersal potential but little competitive ability; such a plant might colonize an area repeatedly but unsuccessfully. Also, the status of a species could oscillate between relict and active colonizer as climatic or other changes in the environment permit.

Colonization and extinction are long-term, multistage processes; direct information about them is difficult to gather. There are however some new concepts about colonization and extraction that are relevant to our understanding of the history of the Beringian vegetation. These ideas were largely initiated by the work of MacArthur and Wilson (1967), and are generally considered under the rubric of island biogeography, although their relevance is by no means limited to islands. Island biogeography addresses the question of the progression of the processes of colonization and extinction, and it provides both a theoretical framework and a body of data regarding what various stages in these processes might be.

Let us consider the process by which a species would change from being a widespread and significant member of an ecosystem to becoming extinct. (We are dealing here with comparatively short periods of geological time and can thus ignore the possibility of the evolution of a new, more competitive species.) To initiate the process, there must be some change in the environment that makes large portions of the original habitat unsuitable for the species. In extreme cases this situation could, of course, result in total extinction, as in the draining of a lake or the submergence of an island. The situation we are concerned with, however, is one that would result in the restriction of previously widespread species to a few small fragmented areas where conditions have remained more suitable for the continued existence of the species than in the surrounding environment. This would obviously be most likely in a complex environment, where "islands" of suitable conditions are likely to remain after, say, a major shift in climate. Once this restriction has occurred, it is probable that some local event will cause the ultimate extinction of the few remaining individuals on the "island." This is obviously a somewhat random factor. All other things being equal, there would be no way in which we could predict which "islands" would be deprived of the remaining individuals of a given species by an event such as a forest fire. In actual cases, however, there will be a number of nonrandom factors contributing to the likelihood of a particular species becoming extinct on a particular "island." These factors have to do mainly with size—of the "islands," of the area needed to support an individual, and of the population needed to assure breeding success.

These concepts become particularly important when the species of a flora (or biota) are considered collectively. In this connection, we can make the following logical assertions:

1. An area that has become an "island" will begin losing species from its original biota. Unless they are replaced by new species, the island's biota will become a diminished sample or subset of the original or potential biota.

2. The number of species in this sample will be roughly proportional to the size of the "island"; smaller islands will have smaller samples.

3. What species remain in the sample is predictable to some degree on the basis of individual species requirements for critical population size and on the needs of individuals for space.

4. Random extinctions will be more likely to take place among small populations, hence on small "islands."

5. It then follows that there will be an increasing dissimilarity among the biota of "islands" correlated with decreasing size of the island.

There is now a considerable body of data supporting these assertions. Much of it is derived from studies of organisms such as birds and insects (e.g., Diamond, 1973; Simberlof and Abele, 1976). Little work has been done on plants. Interestingly, some of the most important data have come from islands that are remnants of submerged land bridges. Beringia is a more or less untapped and potentially fertile field for this line of research.

Phytogeographically, Beringia is unique in that it contains an exceptionally large number of species. Many are highly disjunct and exist as small populations restricted to local areas. Furthermore, many of these local areas support populations of several species with highly disjunct ranges (e.g., Young, 1976; Murray, 1980b). These local areas exhibit precisely the set of characteristics one would expect from "islands" supporting the remnants of a biota whose earlier habitat had been disrupted by major environmental changes.

I assume that many of the disjunct distribution patterns associated with the present Beringian flora are the result of the processes I have discussed above. If my assumption is correct, we are dealing with true relict situations. It seems likely that, in most cases, the event which caused the initial disruption of previously continuous ranges was climatic change associated with the submergence of the Bering land bridge. Hence, the relicts, or at least a significant proportion of them, were probably widespread members of the flora of land-bridge Beringia.

Many of the species with relict distribution patterns in Alaska are typically inhabitants of well-drained, semiarid, or arid environments. Among the many examples that might be cited, *Eriogonum flavum, Potentilla multifida, Chamaerhodos erecta, Rosa woodsii, Podistera yukonensis, Phacelia franklinii,* and *Artemisia lacinata* are typical (range maps for these and other species listed below are given by Hultén, 1968). Other species that are more or less widespread in dry portions of interior Alaska and the Yukon have small disjunct populations in dry locations in western coastal Alaska. Examples are *Arctostaphylos uva-ursi, Plantago canescens,* and *Artemisia frigida.* Most of these relict populations have been discovered since Hultén's (1968) flora of Alaska and the Yukon.

Another group whose distribution patterns are relevant to this discussion includes those plants that are characteristic of temperate and boreal woodlands and meadows, including marshy habitats. Again, many examples could be cited, among which the following are representative: *Geranium bicknellii, Impatiens noli-tangere, Lysimachia thrysiflora,* and *Solidago canadensis.* There are also woodland elements that have disjunct populations in tundra areas of central Beringia, including in some cases marginal "High Arctic" areas such as St. Lawrence Island. Examples are *Lycopodium clavatum, Cornus suecica, Trientalis europaea,* and *Linnaea borealis.* Species comprising this woodland element tend to occur more sporadically than species typical of arid situations; it is less common to find several woodland species in an "association." One might argue that ancient woodland habitats of central Beringia were more totally disrupted than comparable arid environments of interior Alaska, and that "islands" supporting this element of the flora became almost totally depopulated. The recent discovery of woodland species such as *Populus balsamifera* and *Viburnum edule* in Chukotka by Soviet workers (Yurtsev, this volume) would also support the hypothesis that there has been a catastrophic disruption of the ranges of woodland species that may previously have had a trans-Beringian distribution.

Patterns of disjunction also show up in groups characteristic of disturbed areas, particularly those with saline or alkaline soil (e.g., *Suaeda occidentalis, Corispermum hyssopifolium,* and *Euratia lanata*). These tend, of course, to be weedy species with a strong potential for colonization, and it is difficult to evaluate their significance.

Finally, disjunct distribution patterns are quite common in species that typically inhabit moist, stony upland fell fields. Among the many examples from throughout our area are *Oxygraphis glacialis, Potentilla elegans,* and *Synthyris borealis.* It is generally difficult to discern any pattern in the distributions of this large group. If they are in any sense to be considered relicts, their preferred habitat must have been formerly more widespread and continuous. This would fit well with the contention of Ritchie and Cwynar (this volume) that fell-field vegetation was dominant in large areas of land-bridge Beringia.

The phytogeographic evidence discussed so far can be interpreted as supporting the following assertions: first, many if not most of the disjunct patterns of plant distribution in our area can be considered the result of extinctions of previously existing populations in intermediate areas. The disjunct populations are commonly relict populations. Second, areas that support relict populations, particularly if there are several relict species represented in an "association," can be considered as samples of a preexisting vegetation, although the sample from any single location can be assumed to be depauperate. Third, a pooling of data from many of these samples can be useful in reconstructing a picture of the vegetation type of which they are a remnant. Incidentally, species considered to be relevant are by no means confined to those thought to be relicts. Most and perhaps all of the associated species may also have been present in the former vegetation. Fourth, the most significant relict areas appear to be those characterized by species adapted to arid or semiarid situations. Only slightly less significant are those that include species characteristic of woodland and meadow situations.

From the phytogenic point of view one additional comment is in order on two of the most common vegetation types in Alaska today—tussock tundra and dwarf-birch scrub. Both these types are characterized by nearly complete dominance of a single species over immense areas. This is precisely the situation one would expect if this type of vegetation had been confined to small, relict areas, subjected to the loss of most of its original constituents by extinction, and had then been able to rapidly recolonize broad areas following a climatic change.

Autecological Information

Autecology refers to the relationship between an organism and its surrounding physical environment. To a large degree, this relationship can be thought of in terms of energy and moisture requirements and tolerances for negative factors such as cold, drought, and disturbance.

In an evolutionary sense, these environmental characteristics can be equated with selection pressures. Species requirements and tolerances can be expected to change over periods of geological time. Considering the relatively short time that has elapsed since the existence of land-bridge Beringia, I assume here that these evolutionary changes have been of little significance.

Much of the autecological information relevant to this discussion comes from observed correlations between the distribution of plants and climatic parameters. Although a cause-and-effect relationship is implied, the chain of causality is obviously exceedingly complex and can usually only be guessed. This may make the significance of the correlations difficult to interpret, particularly in the case of an individual species. However, when species whose distributions show similar correlations with physical parameters are treated as a group, some important assertions regarding tolerances and requirements can be made. A classic example of this approach is the study of the relationship between timberline and the 10°C isotherm for the warmest month of the year (e.g., Hustich, 1953). More recently, a similar approach has correlated the distributions of many wide-ranging arctic plant species with somewhat more precise indicators of available energy during the growing season (Young, 1971). Climatic data (mainly summer air temperatures) recorded at standard meteorological stations proved to be adequate for documenting the correlations. For purposes of the present discussion, the implications of my earlier (1971) study are as follows: first, most plant species of the Arctic have specific and predictable requirements for summer warmth, and they will not occur in an area in which these minimum conditions are not met. Second, the summer temperature regime of an area can be predicted on the basis of the presence of the least tolerant species that is known to occur there. Third, the presence of a single species with a known requirement for a given summer temperature regime implies the likely presence of a group of other species with similar requirements. The potential of this concept for providing proxy data for paleoclimatic reconstructions is obviously great. At present, its utility is limited by the fact that most of the paleontological information on the vegetation of our area is derived from palynology, which allows identification of taxa only to the genus or family level; this is not precise enough to allow meaningful correlations to be made. The necessity for more megafossil data in this connection is obvious.

It should be emphasized that the absence of a given species from an area does not necessarily mean that its specific temperature requirements are not met there. Hence, the absence of spruce in the land-bridge environment cannot automatically be assumed to be evidence for a mean summer temperature of less than 10°C during land-bridge times. Evidence of species with summer temperature requirements similar to those of spruce would indicate that the absence of spruce was due to some other ecological or historical factor. In other words, the absence of spruce as a major component of the vegetation of land-bridge Beringia by no means justifies the assumption that we are dealing entirely with a cold "tundra" environment, and that vegetation types more characteristic of temperate environments could not have occurred there.

Although summer temperature regimes may be the dominant physical parameter controlling the limits of plant distribution on a circumpolar scale, there are many other factors that may be of predominant importance on a regional scale. These would include available moisture, and disturbance due to freezing and thawing, including depth of thaw in permafrost, wind, and deflation or deposition of loess and sand. The interactions among these factors are complex and involve a number of feedback loops; space does not allow a detailed discussion of them. We may summarize these considerations by saying that data from a variety of sources support the view that in general land-bridge Beringia was drier and more continental than the area is at present (Barry, Gal-Chen, this volume) and that there is extensive evidence of much more active deflation and deposition of sand and loess (Hopkins, this volume). This information has obvious implications in any effort to reconstruct the past vegetation of the area.

Synecological Information

Synecology is the branch of ecology that deals with the relationships among individuals, populations, and species in a natural or artificial environment. Its relevance to this discussion involves the grouping of species into various aggregates that have been termed communities, associations, vegetation types, and formations, among others, and the stability of these aggregations. I should say at the outset that I do not view the spectrum of plant species that characteristically occur together in a habitat as being associated with each other in any obligate sense. Nonetheless, certain species do tend to occur together over broad areas and, presumably, over long periods of time; I shall refer to these groupings as "types." I am concerned mainly with predictions regarding their probable stability over time, given our other knowledge of the environment of land-bridge Beringia and the events that led to the modern environment of the area.

First, we need to examine what has become essentially an article of faith among many ecologists. This is the concept that the two major "biomes" of the northern cold regions are the tundra biome and the boreal-forest or taiga biome, and that these are separated by the "timberline," which is in turn correlated with the 10°C isotherm for the warmest month of the year (Hustich, 1953). This concept was developed largely by workers whose field experience was in Canada and Scandinavia, areas that were subject to continental glaciation in the late Wisconsin, and where the flora and vegetation are much less complex than in Beringia. Much of the present vegetation of Beringia is neither tundra nor boreal forest, and all indications are that, in the past, even less vegetation could be thus categorized. To give only one of many examples, cottonwood stands extend over comparatively large areas in Alaska many kilometers beyond timberline, if timberline is defined as the poleward or seaward limit of arborescent coniferous growth. To consider these stands, which are certainly woodlands if not, in fact, forests, as an expression of the tundra biome is absurd.

It should also be noted that the present location of timberline is strongly correlated with the relationship between land and sea, and with the present shoreline. Lowland "tundra" is normally distributed as a relatively narrow strip bordering the coast of the Arctic Ocean and smaller bodies of water such as Hudson Bay and the Bering Sea. Even at the highest latitudes, tundra is replaced by forested regions within at most 200 to 300 km from the coast. Since the relationship between land and sea in land-bridge Beringia was so different from that of today, it seems unlikely that any analog to modern lowland tundra would have been of much significance within the area, except perhaps along the southern coast of the land bridge.

The point I wish to emphasize here is that inclusive, extensive vegetation types, such as those that traditionally characterize a biome, are neither uniform over space, nor are they likely to remain stable over extended periods of time, particularly during major changes involving climate and land-sea relationships. In other words, the boreal-forest and tundra biomes are phenomena of the present, and they do not necessarily have any close late Wisconsin equivalent. With regard to a reconstruction of the vegetation of land-bridge Beringia, we should essentially ignore the traditional concepts of tundra and boreal forest and any associated climatic or other implications.

The question then arises as to whether any vegetation type could have remained more or less stable over the time period of interest, and, if any have remained stable, at what level in a hierarchy ranging in inclusiveness from a local "association" to formation, or biome should we look for these types? This is an important question since, if it were necessary to reconstruct the past vegetation solely on the basis of a pool of potential species, we would have to admit that it would be a hopeless task.

I suggest that the vegetation of land-bridge Beringia consisted largely, but by no means entirely, of a mosaic of more or less mesic to xeric vegetation types that presently have modern analogs in the area. These could be characterized at the level of, for example, cottonwood woodlands, riparian willow thickets, and alpine fell field. This is not to say that the composition of these types was necessarily identical to the modern analogs in all respects. I also suggest that modern wet to mesic vegetation types, such as tussock tundra and dwarf-birch scrub, in particular, were of little significance. In fact, they may well have been confined to relict areas, as some phytogeographic evidence mentioned earlier might indicate. I suggest that there were widespread xeric vegetation types that are presently represented only by more or less depauperate "samples" in relict xeric habitats. These must be reconstructed on the basis of scanty evidence. A subset of these vegetation types would be those that developed in highly disturbed areas subject to glacial outwash, wind erosion, and the deposition of blowing sand and loess. Fairly good modern analogs may exist for some of these subsets, and reconstruction of them is probably less problematical than that of types in stable xeric areas.

As a final comment, I assume that the species pool from which past vegetation types can be reconstructed was essentially similar to that in the area at present. However, certain species such as spruce may have been extinct locally during land-bridge times. There is also clear evidence that species of genera such as *Alisma* and *Bouteloua* were available as potential components of the ancient vegetation. It should be emphasized again that these genera are of temperate, rather than arctic, affinities. Their presence might be considered to be of equivalent importance to the modern presence of *Picea* in indicating the non-"tundra-like" nature of the ancient vegetation.

Information from Zoological Elements of the Ecosystem

Information on the insect fauna of land-bridge Beringia has been developed mainly by Matthews (e.g., 1974a, 1974b, this volume). For our purpose, the major significance of this work lies in the fact that late Wisconsin insect assemblages in our area are markedly different from those of the present; there is a mixture of genera that do not now characteristically occur together. This underscores the concept of the overall dissimilarity between the ecosystem of land-bridge Beringia and that of the present time.

To many people, the most intriguing and characteristic feature of the land-bridge Beringian ecosystem is the Pleistocene megafauna. The presence of a number of species now extinct either locally or on a worldwide basis is amply documented by fossil evidence (Guthrie, this volume). Evidence concerning population sizes and the exact times that these animals were present is much less conclusive. The presence of large herbivores is a highly relevant consideration for attempts to reconstruct the vegetation in that it places lower limits on the possible productivity of the ecosystem (Bliss and Richards, this volume). However, many of these considerations will remain highly speculative until we are better able to make estimates of population size and, particularly, the chronology of the megafaunal elements.

I assume that most and probably all of the species in the late Wisconsin megafauna persisted throughout the time of the last emergence of the land bridge. If this was so, then populations must have been large enough to resist random extinction. This implies that there were adequate supplies of appropriate fodder for all species during all seasons of the year. Finally, whether or not man is implicated in the final extinction of the megafauna (P. S. Martin, this volume), it seems likely that the end of land-bridge times was a period of dimunition of the habitat on which the large herbivores depended.

A SYNTHESIS: VEGETATION OF LAND-BRIDGE BERINGIA ON A REGIONAL BASIS

For purposes of this synthesis, I have taken the sector of Beringia between 140° and 180°W and divided the unglaciated portion into five regions (Fig. 1). These regions do not correspond to any probable physiographic or phytogeographic provinces. They are simply areas in which the degree of environmental uniformity might have been such that the mosaic of vegetation types represented would not vary excessively. The boundaries shown in Figure 1 are arbitrary; they do not presume any clear discontinuity between the vegetation of adjacent regions. Furthermore, they would presumably have changed with climate. The vegetation typical of Region 2 (south-central Beringia) would probably have retreated toward the southern coast of the land bridge during full-glacial time and expanded northward again during the late glacial.

Region 1: South Coastal Beringia

South coastal Beringia must have been similar in many respects to western coastal Alaska of today, but it was probably subject to more extensive disturbance due to constantly shifting shorelines and the formation of very extensive estuarine and delta systems. I see no reason to suggest that the vegetation of the area was much different than that of coastal regions today. Dune fields must have been extensive; their vegetation would have been dominated by coarse grasses, mainly *Elymus arenarius,* as they are at present. Deltas and more stable flats probably supported a closed sedge-meadow vegetation, similar to that of the coastal portions of the Yukon-Kuskokwim Delta of today. Backshores and depressions between dunes and beach ridges would have supported a more varied community, including such species as several coastal Carices, *Salix fuscescens* and its relatives, *Atriplex gmelini, Stellaria humifusa, Cochlearea officinalis, Potentilla egedii, Angelica lucida,* and *Artemisia tilesii.* There must also have been extensive systems of fresh-water marshes, ponds, and slow-moving streams. These would have supported aquatic species such as various *Potamogeton* and *Sparganium* species, as well as emergent species such as *Hippurus vulgaris, Calla palustris,* and *Iris setosa.*

Ager (this volume) points out that there is no

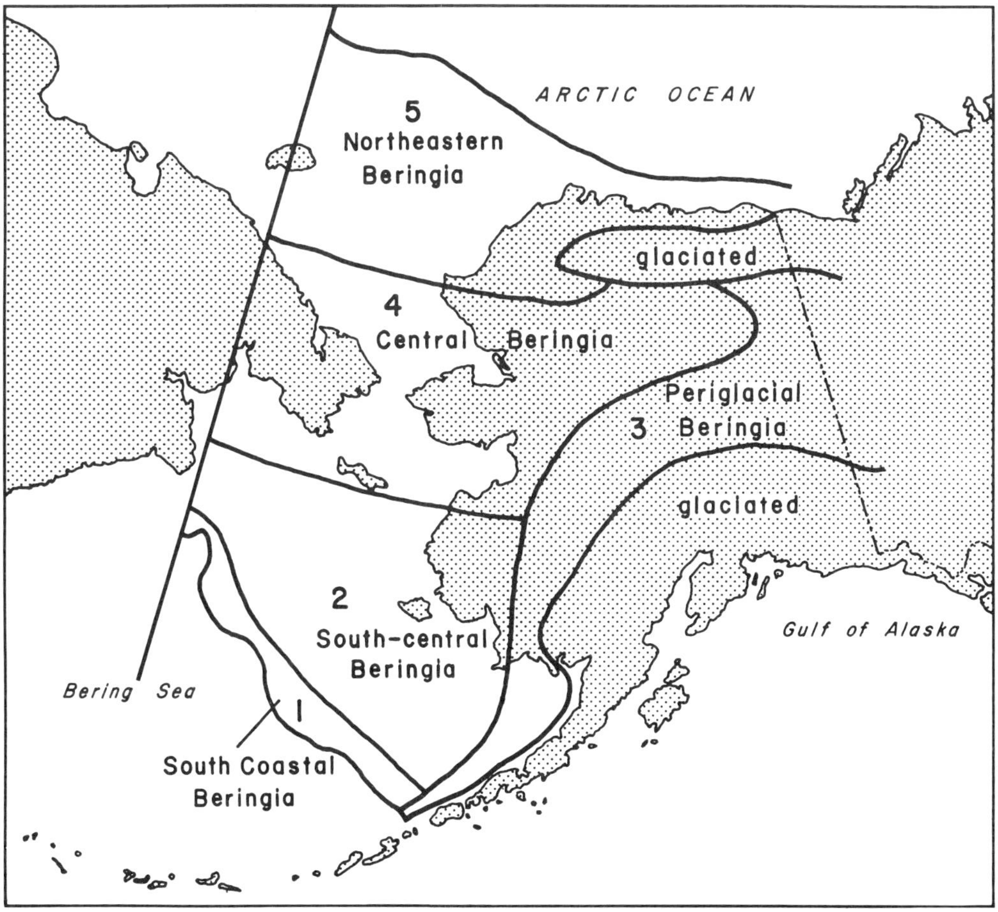

Figure 1. Beringia divided into regions of possible vegetation types, 18,000 years BP

palynological evidence to support the conclusion that there were extensive areas of coastal vegetation on the land bridge. However, I cannot visualize a terrain of the southern border of the land bridge that does not include extensive coastal formations. It seems unnecessarily complicated to postulate a series of totally different vegetation types inhabiting the area when conditions cannot have been too much different from the present Bering Sea coast.

Region 2: South-central Beringia

With respect to megafauna and possible human migrations, this region may have been the heartland of Beringia. I visualize the area as having supported a mosaic of vegetation types, including deciduous woodland, tall brush, meadow and marsh vegetation, and some sort of grassland or steppe.

Deciduous woodland probably would have been uniformly dominated by cottonwood, but would have been otherwise variable, depending on whether it was developed on river banks and bars or on somewhat rolling and better-drained terrain away from rivers. Modern analogs for the first situation occur extensively in the lower Yukon River valley. Upland cottonwood stands in the Togiak Bay area, the Seward Peninsula, and in the Brooks Range, are probably similar to the second type.

It is worth noting that the presence of cottonwood stands would not necessarily imply either a warmer annual temperature regime, or even a warmer summer temperature regime than presently occurs in coastal Beringia a few kilometers inland today. Cottonwood presently occurs at least 1000 km north of the region I am calling south-central Beringia.

Tall brush in land-bridge Beringia, as in Alaska today, was probably mainly associated with riparian situations. These must have been extensive in south-central Beringia, because major rivers such as the Yukon, Kuskokwim, and Anadyr must have crossed the coastal plain. At least at some times the rivers must have carried more water than they do at present. They were probably also subject to tremendous seasonal variations in flow, due to the summer melting of glacial ice. Hence, floodplains must have been of comparatively great extent. Small, local drainages and tributaries would also have supported stands of tall willow brush.

Plausible modern analogs for this vegetation type are well developed as far north as the upper Colville River and are particularly well developed in the upper Noatak drainage, where cottonwood stands also occur. We can assume with some confidence that there would have been extensive stands of *Salix alaxensis* on relatively stable river bars and shores, while species such as *S. niphoclada* probably would have dominated the sites exposed to flooding and siltation. *S. glauca* would have dominated stabilized silty river flats and formed extensive stands of open, parklike dwarf woodland, as it does in the upper Noatak today (Young, 1974).

These extensive willow stands may have been of critical importance to the survival of large herbivores that could subsist largely on browse during at least some parts of the year (see Bliss and Richards, this volume).

The vegetation in interfluve areas can be reconstructed with less confidence, and modern analogs, if they exist, are fragmentary and depauperate. The interfluve vegetation would have been developed on marine sediments, perhaps analogous to those of the Gubik Formation of the present Arctic Coastal Plain of Alaska. The very limited topographic relief would have consisted of areas that are presently the Bering Sea islands and the much lower minor undulations in the old sea floor. Microrelief associated with the establishment of massive ground-ice deposits similar to those presently found on some portions of the Alaskan Arctic Coastal Plain may have lent variety to the landscape. We can envision extensive areas of gently rolling terrain that would have supported the dominant vegetation of south-central Beringia. This vegetation type was presumably the major contributor to the pollen spectra typical of land-bridge Beringia, which are characterized by high levels of grass, sedge, and *Artemisia* pollen. What might this vegetation have been like? Under present climatic conditions, we would expect terrain for the type described to support either wet tundra and sedge meadow, tussock tundra dominated by *Eriophorum vaginatum,* dwarf-birch shrub, or a combination of these. However, as we have previously discussed, there are solid reasons for eliminating each of these types from consideration. I believe that we are left without an extensive modern analog of this vegetation type; rather, we have partial depauperate analogs that do not adequately reflect the complexity of the ancient vegetation.

There seems to be little question that the vegetation type we are trying to re-create was adapted to semiarid, highly continental conditions. Summers were apparently warm enough to allow growth of shrubs and trees where moisture was adequate. The summer temperature regime must have been at least as warm as in the present middle Colville Valley. There must have been deep thawing of the permafrost, and the deep active layer during the summer must have resulted in a well-drained, dry upper surface of the soil. These conditions are presently mainly confined to more or less steep slopes and to the few areas of active dunes in Alaska. Again, in terms of vegetation, these can be considered as only partial analogs.

It seems most likely that the vegetation had the general aspect of an open, rather thinly vegetated steppe. The most conspicuous plants were probably bunch grasses (e.g., *Festuca altaica, F. rubra, Calamagrostis purpurascens, Agropyron* sp., *Bromus* sp., and *Elymus innovatus*), various sedges (e.g., *Kobresia* sp. and upland *Carex* sp.) and a variety of forbs, among which one or more species of *Artemisia* would probably have been the most important. I believe that *Artemisia alaskana* (closely related or conspecific with the Asian *A. kruhsiana*), *A. arctica, A. borealis,* and *A. furcata* are the most likely candidates, with *A. frigida* and *A. tilesii* probably locally important on more disturbed areas. In addition, a large number of other taxa were probably prominent in the overall vegetation, providing a floristic richness and diversity presently unmatched in any xeric vegetation type in the area today. Many of these species are still present but are confined to isolated xeric habitats in Siberia, Alaska, and the Yukon, and some probably no longer exist in the region.

There were also probably more or less extensive areas of marsh and shallow pond in south-central Beringia. In view of the surrounding xeric habitats and the initial salinity of the soil, these marshes probably were not comparable to the common aquatic habitats of northern and western Alaska, where there is generally a build-up of peat and other organic matter, and where conditions are generally acid. Conditions probably favored the existence of grass-dominated wetlands, where species such as *Calamagrostis canadensis, Glyceria maxima,* and *Puccinellia* species probably flourished. These wetlands may have been quite productive in comparison to the various types of wet-sedge meadows commonly found in the area at present. They may have played a significant role in providing fodder for some of the large herbivores. There was also probably a wide variety of aquatic species other than grasses, quite possibly including taxa such as *Alisma* sp. that no longer occur in the area.

Region 3: Periglacial Beringia

The term "periglacial" is used here to denote proximity to large glaciers and does not refer to the climatic regime. The environment of much of what is now interior and western Alaska must have been greatly affected by the presence of extensive glaciated areas in the Alaska Range and smaller glaciers in the Brooks Range. The most important factors would have included the presence of immense, unstable outwash plains, the occurrence of powerful katabatic winds from the glaciers, the deposition and redeposition of aeolian deposits, and the probable occurrence of major floods during periods of rapid melting of the glaciers. All of these factors would have combined to create a landscape that was continuously disturbed. This, of course, would have had a profound effect on the vegetation.

The vegetation of outwash plains was probably similar to that of comparable present-day sites, although the area involved would have been much greater. Species such as *Epilobium latifolium, Hedysarum mackenzii,* Oxytropis sp., *Castilleja* sp. and a variety of grasses and *Artemisia* species would have been dominant. Willow thickets probably occupied extensive areas. The apparent lack of *Alnus* in the pollen record is puzzling, since alder is a common pioneer on outwash plains. However, *Alnus* is much more conspicuous on outwash in humid coastal situations. The highly continental climate of interior Beringia may have militated against its spread through the area. It would not be surprising to find that aspen (*Populus tremuloides*) inhabited the more stable outwash areas, perhaps in the form of dwarfed and isolated stands of woodland.

There is less basis for confidence in reconstructing the vegetation of unstable aeolian deposits, since these are rare in our area at present. We can envision the more active dune systems as being largely unvegetated, or supporting sparse stands of such species as *Elymus innovatus.*

There presumably would also have been

extensive areas of more or less stabilized aeolian deposits. I suggest that a vegetation type that might best be termed "loess steppe" would have inhabited these areas. It would have been a complex vegetation whose species composition depended on a variety of factors, the most important of which would have been the degree of stability and the amount of available moisture.

There can be little question that one or more species of *Artemisia* was of major importance and probably locally dominant in the loess-steppe vegetation. Based on modern distribution patterns and habitat preference, the most important species is likely to have been *Artemisia frigida*. This species has broad distribution in interior eastern Asia and western North America. It is presently the dominant species on more or less unstable loess slopes throughout much of interior Alaska. Its more sporadic distribution to the west is suggestive of a disrupted range. At present, *Artemisia frigida* is an aggressive colonizer of dry disturbed areas such as roadcuts.

Present-day loess slopes and unstable bluffs and cliffs seem to provide a modern analog to the loess-steppe vegetation. Although these areas are generally small and localized, and support only a depauperate flora, they seem to provide a sufficiently consistent picture to allow some descriptive generalizations.

On exposed, more or less active loess deposits, the main codominants with *Artemisia* would have been grasses. *Calamagrostis purpurascens* and *Agropyron spicatum* would have been most important, with *Poa glauca* and possibly some *Bromus* and *Festuca* species also of some significance. A number of other species, rare on the most disturbed areas, would have become important on lower slopes and on surfaces where rates of loess accumulation were relatively low. These species are too numerous to list exhaustively here. Among the more important probably would have been *Selaginella sibirica*, some *Carex* and *Kobresia* species, *Pulsatilla patens*, weedy species of Chenopodiaceae and Cruciferae, *Epilobium angustifolium*, one or more species of Umbelliferae, *Galium boreale*, and a variety of compositae. Exposed rock ledges and outcrops would have supported a somewhat different flora. Important species might have included *Artemisia alaskana*, *Arctostaphylos uva-ursi*, and a variety of herbaceous species such as *Saxifraga reflexa*. In more stable areas, shrubby species such as *Juniperus communis*, *Rosa acicularis*, *Potentilla fruticosa*, and *Sheperdia canadensis* probably became important.

I suggest that vegetation of the type described above dominated vast areas of interior Alaska in land-bridge times, and that it covered much of the region at low and middle elevations that is now white-spruce forest and black-spruce muskeg, as well as area that is now sedge meadow and other low vegetation. This concept is consistent with the pollen and distributional evidence, and with the idea that a highly continental, semiarid climate prevailed. However, this vegetation is not consistent with any climatic model that would assume that temperatures in the area were significantly cooler during the summer than they are at present. Most of the species involved are not tundra plants and do not presently extend their ranges significantly beyond timberline. I would propose that the climate of the area was characterized by bitterly cold winters and short, warm, sunny summers with little precipitation. Low precipitation would have inhibited the development of moss or bog vegetation cover, allowed soils to thaw to considerable depth in summer, and provided perfect drainage in all except the flattest places.

Region 4: Central Beringia

The vast interior plains of central Beringia probably supported a rather sparse, open vegetation as the dominant type. The vegetation may have been intermediate between what we now think of as fell field and a more steppe-like character.

The climate of the plains must have been highly continental, but the substrate would have been mainly marine sediments. There is presently no analogous environment in Beringia, and hence there is probably no good modern analog for the vegetation. Some of the undulating, ice-rich terrain along the north coast of Alaska might provide some semblance of an analog. Uplands in this area support a thin vegetation consisting of upland sedges, *Luzula*, and a wide variety of herbaceous species such as *Papaver* sp. Valleys contain sedge meadows that include neither the tussock-forming *Eriophorum vaginatum* nor dwarf birch. It seems doubtful however that the summer climate of central Beringia could have been as cool as that of the present Alaskan Arctic coast.

We can also envision central Beringia as something of a tension zone between different types of vegetation that would have had their most typical expression elsewhere. During warmer intervals, the vegetation of southern Beringia probably would have tended to colonize northward, while full-glacial conditions would have tended to eliminate or restrict these pockets of more temperate vegetation. Elements of the loess steppe probably also would have waxed and waned as climate changed.

Region 5: Northeastern Beringia

Whatever conditions might have been like in the central and south-central areas of land-bridge Beringia, there can be little doubt that the northern fringes were truly arctic. In view of the assumed continentality of the climate, it seems likely that the vegetation was more or less comparable to that which we presently term polar desert, and which is best developed on some of the islands in the Arctic Ocean and the Canadian Arctic Archipelago. The species represented would have been mainly the extremely cold-tolerant circumpolar arctic species, including *Juncus biglumis, Luzula confusa,* a few *Carex* species, probably some grasses, and a number of herbs such as various species of *Saxifraga* and *Draba.* Sedge meadows probably would have had only the most local occurrence. Arid elements, which still persist on Wrangel Island, Ellesmere Island, and in northern Greenland probably would have been rather better represented than in most high-arctic situations today. It seems unlikely that this type of vegetation would have been of much significance to the more important mammalian (and human) inhabitants of the Bering land bridge.

AFTERWORD: THE BREAK-UP OF THE LAND-BRIDGE BERINGIAN ECOSYSTEM

The single most significant event in the late-glacial environment of Beringia seems to have been the well-documented nearly simultaneous appearance throughout the area of massive sources of birch pollen. In my judgment, the only plausible interpretation of this event is the rapid invasion of much of the unsubmerged terrain of Beringia by vast stands of dwarf-birch scrub. I interpret this as being the result of a rapid shift from a highly continental climate associated with the vast land areas of the Bering land bridge to a more maritime climate caused by the expansion of the seas.

Dwarf birch was probably a widespread but relatively uncommon element of the land-bridge Beringian vegetation. Most parts of Beringia simply would have been too dry to allow extensive stands to form. In somewhat moister areas such as south-central Beringia, the other native species were probably too competitive to allow dwarf birch to dominate large areas.

The dwarf-birch invasion was probably most marked in the more mesic, southern portions of our region. A present-day analog seems to be in the McKinley Park-Denali Highway area of Alaska, where thousands of square kilometers are dominated by what is nearly a monoculture of dwarf birch. To the north and west of south-central Beringia, the birch invasion probably coincided with invasions of tussock tundra dominated by *Eriophorum vaginatum* and sedge meadows characterized by *Eriophorum angustifolium* and *Carex aquatilis.* With the widespread appearance of these "new" vegetation types, there also would have been profound changes in soils and other features. The new soils would have been much richer in organic matter, less well drained, more acid, and less subject to deep thaw during the summers. The beginnings of the bogs, muskegs, and organic-rich sedge flats so common today probably date from the time of the birch scrub and tussock tundra invasion of the end of the Wisconsin. At a later date many of these areas were in turn invaded by spruce, tree birches, and alders. While it took as long as seven or eight thousand years for the remnants of land-bridge Beringia to undergo the more or less complete transformation to the present vegetational situation, the initial stages were nearly simultaneous throughout the area and took place over such a short period of time as to be catastrophic. This vegetation change, coupled with the disappearance of much of the terrestrial ecosystem under the sea, must have had immediate and profound effects on the other elements of the ecosystem, including particularly the large mammals, and, presumably, the human cultures.

PALEOCLIMATE

If paleogeography can be taken as a given, then climate, in a very real sense, becomes the ultimate mediating factor that controls the nature of an ecosystem. Unless certain minimum climatic requirements are met, soils cannot form, plants cannot grow, and animals cannot eat. Furthermore, climatic requirements are rather specific for different plant species, so that the nature of the vegetation varies in a way that can be directly attributed to climate even though the precise nature of the interaction between climatic factors and the physiology of the plant may not be known. Hence, it is obvious that accurate information concerning the climate or paleoclimate can provide a powerful tool for predicting what an ecosystem is or was like.

The computer age has made possible large-scale but coarse-grained models with which the atmospheric and oceanographic phenomena that underlie large-scale global climatic fluctuations and cycles of glaciation and deglaciation can be investigated. Unfortunately, because they must generalize critical aspects of local geography, these recently developed global or hemispheric climatic models are relatively insensitive and provide only very general or even inaccurate portrayals of mesoscale, regional climates. They are blunt tools, indeed, for prediction of the character and productivity of regional ecosystems.

The construction of mesoscale models which treat regions the size of Beringia has just begun to interest climatic modelers. It is more difficult and much more complicated to attempt a mesoscale climatic model in which short-term and local variations are included, but subtle factors such as seasonality, continentality, and year-to-year variance may be more relevant to the nature of the ecosystem than averaged parameters such as mean annual temperature and mean precipitation.

It is not presently possible to begin with a reconstruction of the climate and proceed to a reconstruction of the vegetation and its productivity. On the contrary, as Barry points out in the first paper in this section, biological proxy data are essential for the construction of meaningful mesoscale paleoclimatic models. Thus, we find ourselves in the position of being asked to provide paleogeographic parameters and biological proxy data as a first step toward construction of mesoscale climatic models which we then propose, skating on the edge of circularity, to use to provide ourselves with paleoclimatic constraints or limiting factors governing our reconstruction of the ecosystem. This requires that climatic modelers acquire a rather sophisticated understanding of the strengths and limitations of both the climatic and the paleontological approaches, integrating two disciplines that traditionally have not been closely related.

Gal-Chen's contribution shows how the biological component of the system may provide significant feedback modifying local or even regional climates—a concept that heretofore has not received much consideration. However, in polar ecosystems where temperatures and length of growing season are critical, such microclimatic factors can be decisive.

Barry and Gal-Chen address both the state of present knowledge of the climatic regime of land-bridge Beringia and the nature of recent efforts to reconstruct past climates. It is clear that on a global scale the late Pleistocene period in which we are interested was characterized by a climate cooler and drier than that of the present. It also seems clear that this provides a reasonable starting point from which to attempt to reconstruct the ancient climate of our region. However, as is the case with so many of the questions that we address in this volume, it is equally obvious that increased data and improved techniques have uncovered new problems and new complexities. Both Barry and Gal-Chen may help us most in elucidating the shortcomings of our present knowledge and in suggesting ways in which improved paleoclimatic reconstructions might evolve. It will be immediately obvious that continued close cooperation between workers in a variety of disciplines will be crucial for this effort.

—Steven B. Young

11

APPROACHES TO RECONSTRUCTING THE CLIMATE OF THE STEPPE-TUNDRA BIOME

Roger G. Barry

Three possible approaches to reconstructing the paleoclimatic controls of the steppe-tundra environment are discussed. Simulations using general circulation models are shown to be of limited applicability at the present stage of model development. Synoptic and topoclimatic analogs based on present day conditions are rather more useful, but empirical estimates based on various proxy sources, combined with local energy budget (climatonomic) models, may provide the best approach at the present time.

The weight of evidence suggests a regime of cold, dry winters, especially in the interior valleys and in the vicinity of the modern Bering Strait and Chukchi Sea. Summers appear to have been drier and cooler than now, allowing more of the annual precipitation to fall as snow. Local topoclimatic differences may have been sufficient to provide a variety of habitats allowing a biotic mosaic to exist. However, caution is expressed in the tendency to equate general treelessness and herbaceous tundra with a particular type of "steppe climate." The dramatic late-glacial–early-postglacial biotic changes appear to reflect a more rapid vegetational response to climatic amelioration than occurred in Europe, presumably due to species availability.

Key words: paleoclimate, atmospheric circulation models, paleoecological evidence, topoclimate, treelessness, (paleo) temperature.

INTRODUCTION

Reconstruction of paleoclimates is usually attempted either on a broad global scale, from general climatological/meteorological principles, or at specific sites, using detailed knowledge of certain proxy climatic indicators. For both scales there have been major recent advances in techniques. Global-scale questions can now be examined by applying numerical modeling methods to suggest possible differences in global atmospheric circulation and climate, but they have only limited capability to depict regional details (Barry, 1975). Site-specific data on pollen or faunal assemblages can also be converted to climatic estimates using transfer function methods (Imbrie and Kipp, 1971; Sachs, et al., 1977; and Webb and Clark, 1977). This approach also enables spatial paleoclimatic reconstructions to be made at a specified time horizon if a suitable network of core sites exists and the dating control is adequate (CLIMAP Project Members, 1976). However, the transfer function method is unreliable when the fossil assemblage is non-existent in the modern environment–the "no analog" situation.

Clearly, the problem of reconstructing the climate of a *regional*-scale, "extinct" biome is one that cannot readily be solved by application of the above methods. For these reasons, the procedure adopted in this review is to outline possible empirical and modeling approaches to paleoclimatic reconstruction for the particular case of the steppe-tundra environment of Beringia during the

ISBN 0-12-355860-3

late Pleistocene. The climatic factors that need to be considered involve three scales. First, there is the nature of the general circulation and the radiation inputs appropriate to the period involved. Second, there are differences in regional geography due to sea level changes to be considered. Third, there are topo- and microclimatic factors related to local terrain, vegetation cover, and permafrost occurrence.

PALEOECOLOGICAL EVIDENCE

As a preliminary to consideration of possible paleoclimates associated with the steppe-tundra, it is essential to begin by reviewing the paleoecological evidence. The recognition of an extinct arctic steppe, steppe-tundra, or periglacial steppe biome in Beringia (the Alaska-Yukon area and northeastern Siberia) has been based on studies of pollen and faunal records by American and Russian scientists (Colinvaux, 1964; Hopkins, 1967a; Guthrie, 1968a, b; Yurtsev, 1972a; and Matthews, 1976, among others). Guthrie (1968b) notes the concordant evidence of the small mammal and large mammal assemblages and suggests extensive grass-sedge meadow on the Yukon-Tanana Upland, with some heath tundra at lower elevations. However, caution has been expressed about the interpretation of the faunal collections (Schweger and Habgood, 1976) and of the pollen assemblages (Ritchie, 1977; Ritchie and Cwynar, this volume). Ritchie suggests from studies near Inuvik, N.W.T., that slow soil development following deglaciation may have encouraged vegetational elements such as lichens, ericads and *Dryas* that are underrepresented in the pollen records. He concludes that this early tundra type of assemblage either was "similar physiognomically to the modern high-arctic polar desert" or was "a continuous tundra vegetation with an extensive component made up of . . . elements such as lichens, mosses, Leguminosae, Caryophyllaceae, Saxifragaceae, Cruciferae, and others. There is no direct evidence that these spectra represented steppe-tundra." Ritchie notes that his conclusions are similar to those of West (1964) that the late Quaternary plant communities were "merely temporary aggregations under given conditions of climate, other environmental factors, and historical factors." Since pollen types such as *Artemisia, Salix*, Gramineae and Cyperaceae are represented by species occurring in environments ranging from arid grassland to high arctic and alpine tundra (Ritchie, 1976), their value as paleoclimatic indicators may be very limited and, worse, potentially misleading. Moreover, *Artemisia* pollinates in late summer and may be over-represented in lake sediments since snow melt can cause lake overflow with consequent flushing of the pollen derived from early flowering plants (Maher, 1972).

In considering the paleoecological record it is worth noting the suggestions of Schweger and Habgood (1976) and Young (1976) that the vegetational associations may have comprised a mosiac related to the local conditions of slope, drainage, permafrost and topo- and microclimates. Young, for example, associates *Artemisia* species with warm, dry, south-facing slopes. This whole question requires more detailed modern and paleoecological research.

In an evolutionary context, lowland tundra had its origins in the Pliocene time according to Löve and Löve (1974), but the Pleistocene history is sketchy. Giterman and Golubeva (1967) note that the arid Asian steppes were adjacent to the northern tundra steppe zone during the Samarov (Illinoian-Riss) Glaciation, although Matthews (1974b, Table 6) identifies *herbaceous* tundra for this period on the Seward Peninsula, Alaska. He distinguishes steppe-tundra during Late Wisconsin time, when the maximum extension of the Eurasian cold dry steppes toward Beringia took place (Grichuk, 1971). This steppe-tundra appears to have undergone rapid decline in late Pleistocene/early Holocene time in association with the spread of shrub tundra communities. There is a risk of interpreting this spread as directly synchronous with climatic amelioration (warmer and wetter), although warming may have begun earlier with the vegetational change lagging due to the time needed for plant migration and soil development. Coope and Brophy (1972), for example, have demonstrated the failure of the vegetational (pollen) record to document brief warm intervals during late glacial time in England. Coope, et al. (1961) also raise the possibility that heavy herbivore grazing pressure may have helped to maintain steppe-tundra vegetation by suppressing pioneer woody species.

PALEOCLIMATIC RECONSTRUCTION USING ATMOSPHERIC CIRCULATION MODELS

Attempts to reconstruct past states of the atmospheric circulation have been revolutionized in the last five years through the availability of general circulation models (GCMs). A brief review of their basic characteristics and limitations is in order. GCMs solve the thermodynamic and hydrodynamic equations for the global atmosphere over a three-dimensional grid with a spacing of about 2½°-5° latitude/longitude (see Table 1) and, according to the model, from 2 to 11 levels in the vertical. The time step in the model is typically 5 minutes of simulated time; thus, experiments are rarely integrated for more than about 90 days due to the large demand on computer time. Usually, the geography of land and sea, and mountain effects are represented, as well as sea ice extent, sea surface temperature, snow cover extent, and surface albedo. Solar radiation generally simulates a diurnal cycle, while cloudiness and other climatic parameters are computed. Typical experiments involve an analysis of simulations of present-day winter and summer conditions (*control* cases), for which several "runs" are preferable in order to infer a noise level in the results. These are followed by one or more experiments incorporating some change in internal boundary conditions or external forcing. For example, sensitivity tests have been carried out to examine model response to a change in solar constant (Wetherald and Manabe, 1975), atmospheric volcanic dust content (Hunt, 1977) and sea ice extent (Herman and Johnson, 1978). Model calculations using different extraterrestrial radiation values, related to the Milankovitch orbital effects, have so far been carried out with simple energy balance models. It is worth recalling that in the high latitudes of the northern hemisphere, solar radiation in mid-July was 10 W m^{-2} (2%) less than now around 15,000-20,000 years ago and approximately 45 W m^{-2} (95%) greater

Table 1
Basic Features of Three General Circulation Models

	GFDL[1]	NCAR[2]	Rand[3]
Horiz. Grid	265 km (2½° lat.)	5° lat./long.	4° lat. 5° long.
Levels in Vertical	11 levels (weighted towards lower troposphere	6 levels (3 km spacing)	2 levels (800 and 400 mb)
Upper Boundary	9 mb	18 km (75 mb)	200 mb
Geography	Yes	Yes	Yes
Orography	Yes	Yes	Yes
SST	Fixed	Fixed	Fixed
Sea Ice Limit	Fixed	Fixed	Fixed
Snow Cover	Calculated	Fixed	Fixed
Solar radiation	Seasonal cycle only	Diurnal cycle; constant 15 Jan. or 15 July	Diurnal and seasonal cycle
Cloud	Calculated	Calculated at 3 and 9 km	Calculated at 800 mb or between 800 and 400 mb
Ice Age Experiment	20 Apr. - 1 Sept.	Jan./July (80 days each)	1 May - 31 July
Analyzed Period	August	Last 30 days	July

[1]Geophysical Fluid Dynamics Laboratory (Princeton, New Jersey) (Manabe and Hahn, 1977).
[2]National Center for Atmospheric Research (Boulder, Colorado) (Williams, et al., 1974; Williams and Barry, 1975).
[3]Rand Corporation (Santa Monica, California). This model evolved from the original Mintz-Arakawa model at the University of California, Los Angeles, and is currently being further developed by W. L. Gates and colleagues at Oregon State University (Gates, 1976).

than now about 10,000 years ago (Berger, 1978). The anomalies do not exceed 4% of the present caloric summer[1] insolation at 60° N, but their timing and seasonal distribution strongly suggest their role in triggering major changes in global climate (Kukla, 1975). Specific paleoclimatic experiments with GCMs have concentrated entirely on the last glacial maximum about 18,000 BP.

In examining model results it is essential to keep in mind the following considerations: (1) the models are global-scale and ignore small-scale processes or take account of them statistically by parameterization; (2) the grid resolution implies that many regional-scale climatic features will be omitted; (3) the experiments are performed for only *one* realization—possible year-to-year differences have not yet been examined; (4) many surface characteristics are "fixed," thus ignoring important feedbacks between atmosphere-ocean-cryosphere; (5) the models do not necessarily depict the present-day climate correctly in the control cases; (6) the interpretation is best concentrated on (control cases—experiment) differences rather than absolute values. Nevertheless, used with caution, these model experiments for the first time provide a largely independent source of meteorological information that can be compared with the field evidence or other interpretations.

Two experiments have used the CLIMAP data base (CLIMAP Project Members, 1976) for summer conditions at 18,000 BP (Gates, 1976; Manabe and Hahn, 1977). An earlier study (Williams, Barry, and Washington, 1974; Williams and Barry, 1975) used a simplified input compiled from literature sources (Barry, 1973) for both winter and summer conditions. On the hemispheric scale, the 18,000 BP summer conditions are shown to be generally drier and cooler than at present (Williams, 1978), but on the regional level the three experiments show major discrepancies. Thus, the NCAR and Rand results suggest less summer precipitation over Beringia, while the GFDL experiment indicates slightly more, except towards the interior areas. The NCAR model indicates much lower summer temperatures (although it may be biased by the cloud-radiation parameterization) while the Rand model shows a cooling of about 4° in the interior areas but no change over the present location of the Bering Sea. All three experiments suggest a predominance of east to southeasterly (overland) airflow over the area in summer (Figure 1), although the NCAR and Rand models show a mean low center at 170°E, 50-60°N whereas the GFDL results lack any comparable feature. The only conclusion that can be drawn is that at present the models are inadequate to allow reliable assessments to be made for specific regions.

[1] Caloric summer and winter are defined so that their lengths are equal and therefore secularly constant and so that the solar radiation receipts show a maximal contrast.

EMPIRICAL STUDIES OF CIRCULATION CONDITIONS

Paleoclimatic reconstructions for the last glacial maximum and for intervals in the Holocene have also been made by Lamb and colleagues using a semi-empirical method (Lamb, et al., 1966; Lamb and Woodroffe, 1970). The procedure starts from the pattern of mean July temperatures inferred from paleoecological evidence and from these data the 1000-500 mb thickness field (which is proportional to the mean temperature in the lower troposphere) is estimated through regression equations appropriate for the climatic regime at each location. This field is then used to depict areas of cyclonic or anticyclonic development based on R. C. Sutcliffe's "development theory" (1947). Finally, this result combined with the pattern of thickness, which indicates the steering of cyclonic systems, is used to sketch in the mean sea level pressure pattern (see Lamb, 1977, pp. 40-44). The pattern suggested by Lamb and Woodroffe for July during the last glacial maximum is much more cyclonic over the North Pacific, Beringia and the western Arctic Ocean than any of the model simulations. Their pressure map indicates more southerly flow entering the Arctic from the Pacific across Alaska. The corresponding map for January shows high pressure and east to northeasterly airflow and resembles the simulation for January 18,000 BP of Williams, Barry, and Washington (1974, Figure 7). Lamb and Woodroffe (1970) provide similar reconstructions for 11,500 and 10,500 BP which show generally similar summer airflow over Beringia. The comparable maps in Lamb, et al. (1966) for the Holocene do not adequately cover Beringia although the authors note that the circulation probably became generally weaker, more zonal, and cyclonic in high latitudes over North America after about 8000 BP.

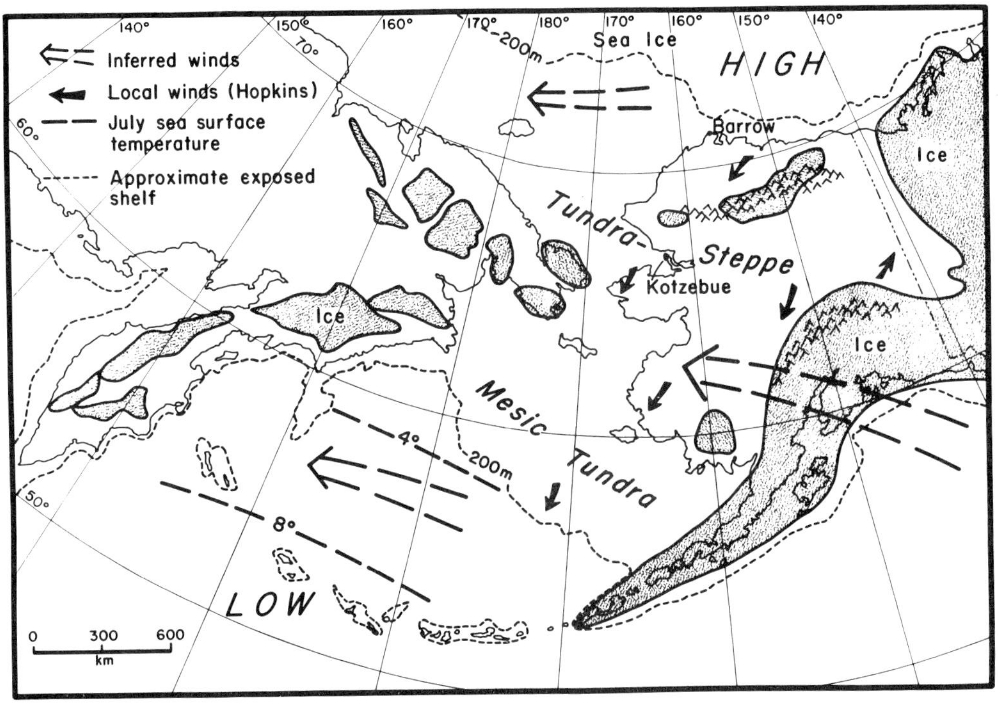

Figure 1. A sketch of the hypothetical Beringian environment at about 18,000 BP (partly after Hopkins, 1972; and CLIMAP Project Members, 1976). LOW and HIGH denote average summer pressure

REGIONAL STUDIES

The most detailed regional reconstructions of Alaskan paleoclimates are those of Péwé (1975) based on geological evidence of past glacial and periglacial activity. For the last glacial maximum (the Wisconsin), he derives snowline altitudes based on the height of the base of empty north-facing cirques. Figure 2 illustrates these past snowlines, compared with present, for two transects across Alaska. The analysis shows a snowline lowering below present of 300-400 m in western Alaska and 450-600 m in the interior and east. The lowering below sea level in southern Alaska is due to subsequent tectonic activity. The similarity of the overall patterns leads Péwé to conclude that the moisture sources must have been essentially the same as now, but with a reduced contribution from the Bering Sea. He estimates a reduction in July mean temperature of at least 2°C in the vicinity of Nome and Kotzebue, and of 4°-5°C inland (Péwé, 1975, p. 113-14). The presence of active Wisconsin ice wedges in the Fairbanks area suggests a mean annual air temperature of -7°C. Reconstructions of possible regional paleoclimates must take account of the major changes in glacial and postglacial sea level. World sea level fell about 120 ± 40 m during the last glacial maximum, exposing large areas of the continental shelves of the Beaufort-Chukchi-Bering seas. This produced the Bering land bridge and is the primary rationale for the regional name Beringia (Hopkins, 1967b, Figure 3). The Strait of Anadyr and Bering Strait were severed about 14,000 BP (Creager and McManus, 1967; Hopkins, 1973), although the shoreline of western Alaska still remained up to 100 km west of its present location about 10,000 BP (Hopkins, 1973).

Sergin and Shcheglova (1976) have attempted to estimate the climatic effects of the increased land area in Beringia during the last glacial

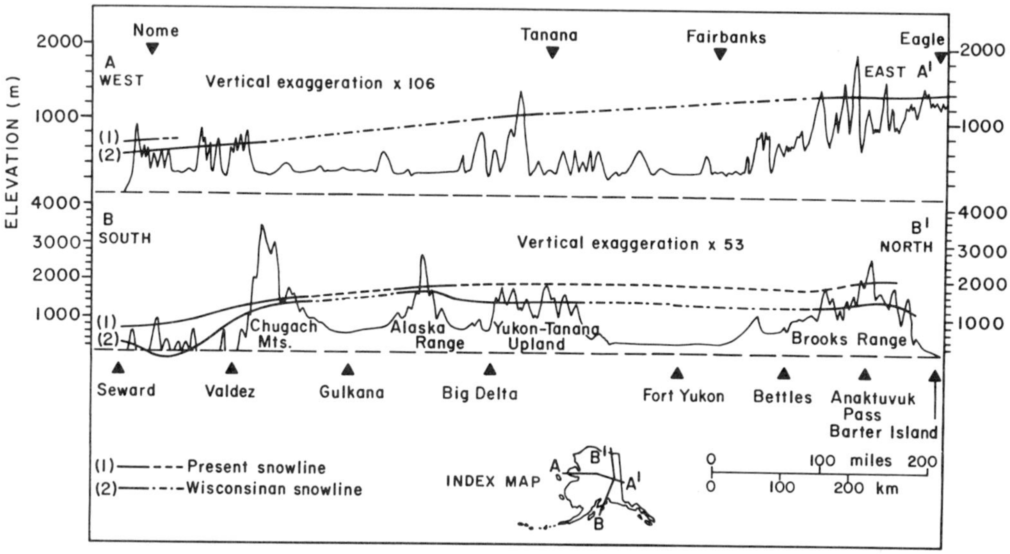

Figure 2. Profiles east-west (A-A′) and south-north (B-B′) of present and estimated-Wisconsin snowlines in Alaska (after Péwé, 1975)

maximum. They assume that the values of temperature and precipitation on a profile inland from the former coastline of Beringia (expressed as relative values in relation to the figure at the coast itself) are proportional to an average of 18 modern profiles in eastern Siberia, western and eastern North America. They adopt a 5°C reduction in mean annual temperature and a 20% decrease in annual precipitation at the coastline for this glacial period. Table 2 summarizes their results. These figures may provide a suggestive guide, but it should be noted that the gradients and annual regimes of these climatic elements are likely to have been quite different under a global glacial climate from those existing in high latitudes at the present day.

Studies in northern Alaska at present indicate that the coastal marine effect in summer is of very limited extent (Clebsch and Shanks, 1968). Recent investigations using supplementary stations in the vicinity of Prudhoe Bay confirm this finding and show that most of the inland increase in summer temperatures is concentrated in the first few kilometers away from the coast (Walker, et al., 1979). Interestingly, this gradient is closely paralleled by the thickness of the active layer as well as by changes in plant phenology and stature (Walker and Webber, 1979). This zonation would presum-

Table 2
Climatic Estimates for Beringia during the Last Glacial Maximum for a Transect North-northeastward from the Coastline at 63°N, 180°. (From Sergin and Shcheglova, 1976.)

Distance from the Coast (km.)	Mean Annual Temperature	Mean July Temperature (°C)	Annual Temperature Range (°C)	Annual Precipitation (mm)
0	-9	6	15	400
400	-11	9	21	290
800	-15	8	24	210

ably have existed in unglaciated areas of northern Beringia in late glacial time. In central Alaska, flow crossing the Alaska Range from the south would have undergone adiabatic warming, due to descent north of the mountains, and this would have accentuated the general effect of continentality.

The regional climatic effects of a large increase in land area during glacial times have been examined in detail previously only for the Torres Strait area of Australasia (Walker, 1972; Webster and Streten, 1978). In the case of Beringia, it is likely that the exclusion of Pacific Ocean water from the Chukchi Sea by the land bridge would have increased the severity of the sea ice regime in the Beaufort Sea sector of the Arctic Ocean, with less summer breakup than now, apart from the obvious increase in continentality in western Alaska. Surface temperatures of the North Pacific were lowered about 4°C compared with now off the coasts in summer according to CLIMAP Project Members (1976, Figure 2), except in the Sea of Okhotsk where the cooling was about twice as great, and northeast of Hokkaido where it was only 2°C lower than present. These estimated differences from present are much smaller than in the glacial North Atlantic. The Rand simulation suggests a difference of 5°C (Gates, 1976) between the two ocean areas. Consequently, the temperatures in maritime air masses affecting southern Beringia in summer would have been less extreme than those affecting land areas around the eastern North Atlantic. It is unlikely that ice persisted in the Bering Sea in summer.

An analog approach to glacial atmospheric circulation conditions, specifically in interior Alaska, has been used by Streten (1969, 1974). From the comparison with present-day situations in extreme years, he suggests that warmer than average winters require stronger mean flow in the lower troposphere, sufficient to disrupt the surface temperature inversion. More normal (cold) winters occur with frequent stable high pressure situations, cloudless skies and light winds. Dry, warm summers in the interior are associated with a strong upper ridge over Alaska with south to southeasterly airflow which is depleted of any moisture in crossing the coastal ranges and the Alaska Range. Wet summers occur with a deep upper trough which allows warm moist air from the western North Pacific to flow parallel to the major ranges.

From an analysis of two recent winters, Bowling (1979) infers that the high-amplitude wave pattern of the 1976-77 season, which caused southerly flow at 700 mb with heavy snowfall on the coastal ranges and the Alaska Range, and dry, warmer than average conditions in the lowlands, provides a synoptic model for the last glacial maximum. This is not out of line with the surface airflow in summer proposed in Figure 1. Another study by Fahl (1975), of conditions likely to be favorable or inimical to present glaciers demonstrates that summer hot spells at the Gulkana Glacier in the Alaska Range are associated with an enhanced anticyclone in the Beaufort Sea and anomalous easterly wind components; a rather similar pattern also favors warm conditions at the McCall Glacier in the Brooks Range. In both cases the important feature is the occurrence of overland flow of air. This feature, which also appears in the GCM simulations, should imply dry summers in the lowlands and glacier decay.

Parallel findings derive from a synoptic climatological analysis by Warmerdam (Barry, 1977, 1978). Data at Kotzebue show that a cold July on the Chukchi Sea coast of Alaska is associated at the present day with less frequent cyclone activity in Bering Sea-Bering Strait. The path of cyclone movement tends to be displaced eastward with cold air advection from the northwest. The usual summer pattern is for westerly airflow to set in following the collapse of the Aleutian low and the northward shift of the Pacific subtropical anticyclone, assisted by the higher temperatures and consequent tendency for thermal low development in interior Alaska; earlier onset of persistent westerly flow at Kotzebue (between mid-April and mid-June) gives rise to a *lower* summer total of accumulated thawing degree-days. For 1953-76, Warmerdam finds a +0.67 correlation between these two variables. Interestingly, the 500 mb trough and ridge locations on the actual dates of westerly wind onset at Kotzebue in five selected years closely match those shown by Streten (1974) for months with anomalously warm conditions in interior Alaska. This points to the role of land-sea temperature differences in helping to determine the airflow regime over the Chukchi Sea coast. Presumably, during the exposure of the continental shelf there would have been a southerly component due to this thermal gradient.

The local winds shown on Figure 1 are based on

dune orientations (Hopkins, 1972). Most likely these reflect predominant winds in early autumn when the ground would tend to be dry. They are not markedly out of line with inferred large-scale geostrophic flow in summer.

TOPO- AND MICROCLIMATIC EFFECTS

The general problem of micro-/macroclimate relationships is one that has received little quantitative investigation (Holmes and Dingle, 1965), although it is of vital importance for paleoecological/paleoclimatic reconstructions. Generally, in Arctic regions the low plant cover experiences a more favorable microclimate than conventional climatic observations, made at 1.5 m or higher, would suggest (Corbett, 1972). Wind speeds are considerably less at the level of the plant mat and within tussocks as demonstrated by Warren-Wilson (1959) and tundra canopy temperatures are 2-3°C higher in summer than in the air immediately above (Weller and Holmgren, 1974). Analysis of site variations at the International Biological Program Tundra Biome stations by French (1974) shows that, in general, environmental homogeneity among subsites in a given locality *decreases* in the more severe tundra and polar desert climates. Thus, there may be strong differences in energy budgets and microclimate over short distances.

More important in terms of paleoclimatic considerations relating to steppe-tundra is Young's (1976) suggestion regarding the importance of a mosaic of topoclimates. Observational evidence in central Alaska at the present day shows rather small differences in air temperature from north-facing to south-facing slopes, but major contrasts in potential radiation and temperature at the ground surface. Potential radiation estimates on north-facing slopes of about 12° at 65°N in the Yukon-Tanana uplands have an "equivalent latitude" of 80°N, whereas on corresponding south-facing slopes the equivalent latitude is 50°N (Slaughter and Long, 1975). Surface temperature differences of 4°-7°C were observed in summer 1970 and the mean daily range of surface temperature in June 1971 was 20.6°C on the south-facing slope compared with only 12.1°C on the north-facing slope. Climatonomic model (see also Gal-Chen, this volume) estimates by Lettau and Lettau (1975) suggest that local "warm spots" in the tundra environment could occur in areas with bluffs and deep channels. Steep slopes, especially those with a southerly aspect, will have a lower albedo in spring, while channels will accentuate runoff and lessen evapotranspiration. Their calculations indicate that a typical tundra environment with a July air temperature of 9°C could be warmed 1.8°C if evaporation were reduced by 10% due to greater runoff. Conversely, if snow cover were to persist a month later than usual, this temperature would be reduced by 8.8°C. A tenfold change in net radiation following snow melt has been demonstrated at Barrow by Weller and Holmgren (1974).

In interior Alaska, the winter climate of the valleys is dominated by the low-level temperature inversions (Bilello, 1974). Streten (1969) illustrates the occasional persistence of these at Fort Yukon when warming intervals occurred at Tanana and Fairbanks. In these situations, snow depths of about 50 cm would effectively insulate tundra plants and small mammals from the temperature extremes registered at the snow surface (Corbett, 1972). Of more interest from a paleoecological point of view is the occurrence of nocturnal inversions in these same valley locations in summer (Haugen, et al., 1971). The effect of cold air drainage in the upland valleys of the interior is evident in the growing season and frost occurrence data presented by Searby and Branton (1975). Nevertheless, around Fort Yukon there are as many growing-degree-days as in the Matanuska Valley near Anchorage, despite the area's shorter growing season, as a result of high daytime temperatures.

These general patterns of local, topo- and macroclimatic differences should have occurred in the past although the relative magnitude of the differences would be particularly dependent on the degree of cloudiness and regional wind velocities. With our present limited knowledge of the regional climate and atmospheric circulation in the late glacial, further speculation would be of limited value.

TIME AND SPACE PERSPECTIVES

Our view of paleoecological-paleoclimatic relationships should not be simply a static one, based upon the premise of near-equilibrium conditions. Recognition of the likelihood that even major climatic changes occurred rapidly during the Quaternary implies that the dispersal of plant species, especially arboreal ones, and the associated development of suitable substrates, may lag considerably behind a climatic amelioration. Reference was made earlier to Ritchie's (1977) suggestion that late glacial plant communities in the Mackenzie Valley may have been merely temporary aggregations of species. Insects, on the other hand, can respond much more rapidly to favorable micro-habitats (Coope and Brophy, 1972). Thus, Matthews (1974b) notes the occurrence of *prairie* insects at Deering, Alaska, about 12,000-13,000 BP. However, a crucial problem in our paleoclimatic interpretations is the tendency to equate tundra vegetation with a specific tundra climate, or steppe vegetation with a specific steppe climate, in the sense of the classical climate types of W. Köppen. It may be that the critical environmental feature of the steppe-tundra was a general absence of trees.

Péwé (1975) proposed that the xeric aspect of the vegetation may have indicated merely the widespread occurrence of treeless flood plains, alluvial gravels and areas of windblown sands rather than a drier climatic regime, although there is considerable argument to support the case for drier conditions. At the same time, the presence of some woody vegetation in the valleys seems to be essential habitat for mammoth (Olivier, this volume). Péwé suggests that the tendency for a higher permafrost table should also have given rise to local boggy sites during glacial and late glacial time, although alternatively a sparser vegetation cover may have caused deeper summer thaw.

Periglacial tundra was at its most extensive about 20,000 BP in Europe (V.P. Grichuk, 1973, Figure 20) and rapidly disappeared after 10,000 BP (Dolukhanov, 1978, Figures 57, 58). There, the trees had a much greater distance to advance during the late glacial amelioration than in Beringia, where local enclaves of aspen or cottonwood may have persisted on sunny slope sites and possibly larch and willow in sheltered valleys throughout the late Pleistocene (Hopkins, 1979). Consequently, the inferred rapid decline of the steppe-tundra biome in Beringia may reflect a closer temporal response of vegetational recolonization to the continued climatic amelioration than occurred in Europe (cf. Coope, 1977). Ager (1975) finds that steppe-tundra was replaced by birch-shrub tundra, suggesting increased summer warmth and moisture, as early as 14,700 BP at Birch Lake (64°19′N, 146°40′W). While later dates have been cited (Hopkins, 1972), if allowance is made for dating uncertainty, the rise of birch dominance seems to have been essentially isochronous south of the Brooks Range, suggesting the presence of small birch populations prior to this time. In eastern Fennoscandinavia, where there are east-west differences in the pollen sequences due to the oceanicity-continentality gradient, Vasari (1976) considers that the transition from a late glacial *Artemisia* maximum to the rise of *Betula* is time-transgressive. Again, therefore, the process involves vegetation dynamics with pioneer species and the specific role of climate is not easily isolated.

SUMMARY

Possible approaches to reconstructing the paleoclimate of the steppe-tundra environment of Beringia have been discussed. It is apparent that atmospheric circulation models are presently unable to shed light on the possible climatic determinants of this environment. Analog studies and topoclimatic analyses of modern conditions seem likely to be more helpful in the short-term in building up a picture of the likely range of local variability of paleoclimatic temperature conditions. However, there is a need to try and check such estimates with respect to specific locations where good paleoecological data are available. In a parallel context, Jacobs and Sabo (1978) have illustrated the significance of topoclimate for site selection by Thule Culture people on Baffin Island. Proxy data need to be interpreted with careful attention to their local topoclimatic, or macroclimatic, significance although, as implied in the preceding discussion, continual testing of modeling results against both proxy evidence and modern day analogs leading to improved models can be a fruitful means of refining our understanding of the steppe-tundra environment.

A critical problem that has not been definitively resolved concerns the apparent discrepancy between the calculated snowline lowering during the last glacial maximum (Figure 2) and a regime of cold dry winters and drier, somewhat cooler, summers than at the present day. The degree of summer cooling inferred by Péwé (1975) and

Sergin and Shcheglova (Table 2) would lower the mean July temperature at Fairbanks from its present value of 15°C to the threshold level for tree growth. Thus, interior lowlands may have resembled the modern upland tundras of Alaska; the mean July temperature (1970-73) is 9°C at Eagle Summit (1103 m) near Fairbanks, for example (R.K. Haugen, personal communication, 1974). With cooler summers, more of the total precipitation would fall as snow, at least maintaining the upland ice caps. Heavier snowfalls similar to present would probably have prevailed over the more maritime Alaska Range, especially in autumn and winter. In contrast, Hopkins (1979) postulates short, but warmer and drier, summers than now with warm air advection across Beringia, and long cold winters with thinner snow cover than at present. However, the evidence for summer warmth, other than locally, where föhn descent and/or favorable south-facing slope aspect were involved, does not yet seem compelling.

The possibility exists that the steppe-tundra was not a climatically determined biome in the sense of being in quasi-equilibrium with a specific climatic regime. Our recent recognition of the rapidity of climatic change makes it entirely plausible to envisage that the plants present in the area represented a fortuitous assemblage determined by the rates of plant dispersal and colonization which themselves were influenced by the pattern of glaciation, sea level changes, and the evolution of climate on a time scale of 10^2-10^3 years. However, while the *specific* arctic steppe biome was apparently unique to the Late Wisconsin, analogous herbaceous tundras are identified for earlier glaciations (Matthews, 1974). In this sense, the broad environmental conditions seem to have been controlled deterministically by climate and paleogeography. The Late Wisconsin was nevertheless distinctive in its particular faunal assemblages and in the presence of man. It would seem that close comparative study of faunal and vegetational change at the close of the penultimate Illinoian glaciation should clarify the extent to which the decline of the megafauna at the end of the last glaciation was due to climatic change and vegetational succession or to man.

ACKNOWLEDGMENTS

I wish to thank Dr. J.T. Andrews, INSTAAR, and Dr. R. Coope, University of Birmingham, England, and Dr. P. Calkin, SUNY, Buffalo, both visiting scientists at INSTAAR in spring, 1979 for their comments on a first draft of this paper. Questions posed during the conference materially assisted in sharpening its focus.

12

APPROACHES TO MATHEMATICAL MODELING OF THE STEPPE-TUNDRA PALEOCLIMATE

Tzvi Gal-Chen

Some aspects of climate dynamics and climate modeling are presented which are considered relevant for the understanding of the past and present climate in Beringia. The general concepts of transitivity, intransitivity and almost intransitivity are employed in order to explain a regional climate with no apparent modern analog. Feedback processes are intimately related to transitivity/intransivity. These feedbacks include, e.g.: vegetation-albedo-temperature; vegetation-albedo-vertical velocities-precipitation; vegetation-evaporation-temperature; surface cover and orography-convective fluxes of heat and moisture-planetary waves-precipitation. Some elements of Milankovitch (1930) ice-age theory are presented to illustrate the possible role of external processes—in this case changes in orbital parameters—in initiating the warming in Beringia some 14,000 years ago. A probable (but speculative) climate scenario for Beringia some 18,000 years ago includes: reduced precipitation year-round, colder winters, late springs and, depending on proximity to the continental ice sheets, either warmer or colder summers. General circulation models (GCMs) do not resolve, or even properly parameterize, many regional effects peculiar to Beringia. It is argued that mesoscale models that resolve explicitly smaller-scale processes, at the expense of specifying (rather than predicting) the larger-scale dynamics, may be more useful.

Key words: transivity; intransivity; climatic feedbacks; vegetation, effect of albedo on; vegetation, effect of evaporation on; ice age, Beringia; general circulation models

INTRODUCTION

One of the fascinating and difficult aspects of a natural system is that very often it is unique and cannot be duplicated. The complexity of natural systems makes it unlikely that an application of one scientific method (or even discipline) can lead to a satisfactory understanding, or even to a detailed observation, of a particular system.

The case of the climate of Beringia during the last glacial maximum is a good example. It appears to be qualitatively different from modern analogs. The climate undoubtedly depended on the global circulation system at that time, but, judging from the situation as we know it in modern tundra environments, it also depended upon a complex interaction with the surface—particularly the vegetation.

For a natural system of that type, mathematical modeling may be extremely useful. Unlike the natural system, one can deliberately change the parameters that control the model's climate (e.g., introduce a large lake, a shoreline change, large continental ice sheets, sea level changes) and by inspecting the model's response one can get an insight into the physical controls of the real climate. In the absence of any reliable laboratory,

ISBN 0-12-355860-3

or of present-day natural analogs, mathematical modeling has the potential of becoming a vital link between theory and observation.

The main purpose of this paper is to review the basic approaches to theoretical climate dynamics and climate modeling that might be relevant to studies of past and present climate in Alaska-Yukon and northeastern Siberia. This paper attempts to provide some plausible hypotheses regarding the climate of Beringia. It is recognized that great uncertainties exist in the theoretical understanding, as well as observation, of present and past regional climates. Nevertheless in this paper some speculations about the regional paleoclimates in Beringia are presented.

THE CLIMATE SYSTEM

General Overview

The major controls on climate are rooted in the general circulation, and are hence global in character. The effects of global circulation iterations and fluctuations are strongly modulated, however, by local physiographic, ecological and hydrological influences.

In describing the properties of the system, I have drawn extensively from two authoritative reports on the causes of climatic variation: *Understanding Climatic Change* (1975) by the U.S. National Academy of Sciences (NAS), and *The Physical Basis of Climate and Climatic Modelling* (1975) by the Joint Organizing Committee of the Global Atmospheric Research Program (GARP).

Figure 1 displays schematically the components of the climatic system. These are broadly grouped as (1) the thermal properties, which include the temperature of the earth's surface, oceans and atmosphere; (2) kinetic properties, which include wind, ocean currents, and motion of ice masses; (3) aqueous properties, which include humidity, cloudiness, ground water, lake levels, and the water content of snow and ice; (4) static properties, which include pressure density, salinity, topography and the composition of the air, in particular minor constituents like CO_2, O_3 and H_2O, which affect the thermal properties; and (5) the living cover (biota).

It is also useful to classify the components of the system, which are atmosphere, hydrosphere, cryosphere, land surface and biosphere. Although each of these components is treated by a different discipline, they are in fact highly interactive. From the point of view of climate modeling, the only justification for uncoupling the system's components is because time scales relevant to them differ greatly.

The atmosphere adjusts itself (i.e., achieves new equilibrium) to external changes in a month or so.

The hydrosphere includes the oceans, lakes, rivers, and subsurface water. Of these, the world's oceans are the most important for climatic variations. The upper layer of the oceans interacts with the overlaying atmosphere on time scales of seasons to years, while the deeper layers have an adjustment time on the order of centuries.

The cryosphere comprises the world's ice masses and snow. The changes of snow cover on land are mainly seasonal and respond primarily to solar radiation and the atmospheric circulation. Glaciers and ice sheets respond much more slowly, and have their own internal dynamics.

The land surface of the earth includes the mountains, surface rocks, and soil. These features change over the longest time scale. The changing of sea level and the building of ice masses 2-4 km thick during glacial times represent significant changes in the lithosphere and may also be viewed as earth-ice-ocean interactions. These interactions change the general circulation (GARP, 1975).

The biota include the plant cover on land and ocean and the animals of air, sea, and land—including man himself. The response time of the biosphere differs widely. Natural changes in surface vegetation, for instance, occur over periods ranging from decades to thousands of years, in response to changes in temperature, precipitation and incoming solar radiation.

In Figure 1 a distinction is made between external processes and internal processes. Except for solar radiation at the top of the atmosphere, which is truly external, the distinction is based merely on different time scales. For instance, in the July simulations of 18,000 BP by Gates (1976b) and Manabe and Hahn (1977), the sea-surface temperature and the distributions of topography (including ice sheets) and the albedo of the continental surface as determined by the CLIMAP Project Members (1976) are specified. Whatever the cause of these changes in the surface conditions, once established they persisted for more than one

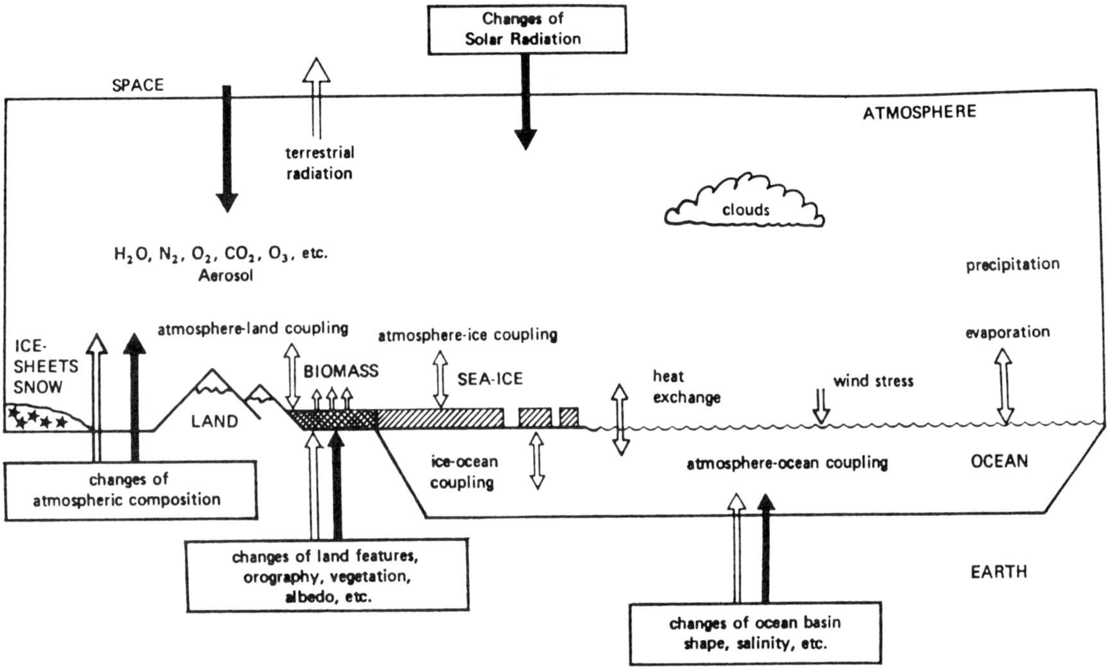

Figure 1. Schematic illustration of the components of the coupled atmosphere-ice-earth climatic system. The full arrows (⇒) are examples of external processes; open arrows (⇒) represent internal processes in climate change. Note some processes are either external or internal, depending on the time scale. Source: GARP (1975)

month and thus allowed the atmosphere enough time to reach an equilibrium.

Transitivity, Intransitivity and the Uniqueness of the Climate System

The climate system dynamics can, in principle, be formulated as a conventional initial boundary-value problem in fluid dynamics, namely:

$$\frac{\partial \vec{u}}{\partial t} + L\vec{u} = \vec{f} \quad \text{; in the interior,} \tag{1.a}$$

$$\frac{\partial \vec{u}}{\partial t} + H\vec{u} = \vec{g} \quad \text{; in the boundaries.} \tag{1.b}$$

where $\vec{u}$ is some vector (e.g., velocity, temperature, pressure, etc.), L is some nonlinear differential operator (in space) applied on $\vec{u}$, $\vec{f}$ is some specified forcing function (e.g., atmospheric composition), H is some nonlinear differential operator (in space) applied on the boundaries, g is the forcing function at the boundaries (e.g., solar radiation at the top of the atmosphere) and t is time. The solution is, in general, time dependent because of the initial conditions and because the forcing is, in general, also time dependent.

In the deterministic view, the behavior of the system depends on the forcing function ($\vec{f}$, $\vec{g}$) and on the initial conditions. The predicted values u are the internal system and the specified forcing (f, g) are the external system. As has been already mentioned, the solar radiation at the top of the atmosphere is the only genuine external forcing. Depending on the time scale of interest, however, one can specify the geography and the land-ice masses as part of the external forcings. Most atmospheric models remove the hydrosphere and the biosphere from the internal system, although it is not entirely justified to do so.

If the system (l-a), (l-b) were linear, then it is known from the theory of partial differential equations (e.g., Courant and Hilbert, Vol. 2, 1962) that a solution exists and is unique. Furthermore, as $t \to \infty$ the solution "forgets" its initial conditions; that is to say, its behavior is determined solely by the external forcing ($\vec{f}$, $\vec{g}$) and the spatial (linearized) operators (L, H). If the forcing is time

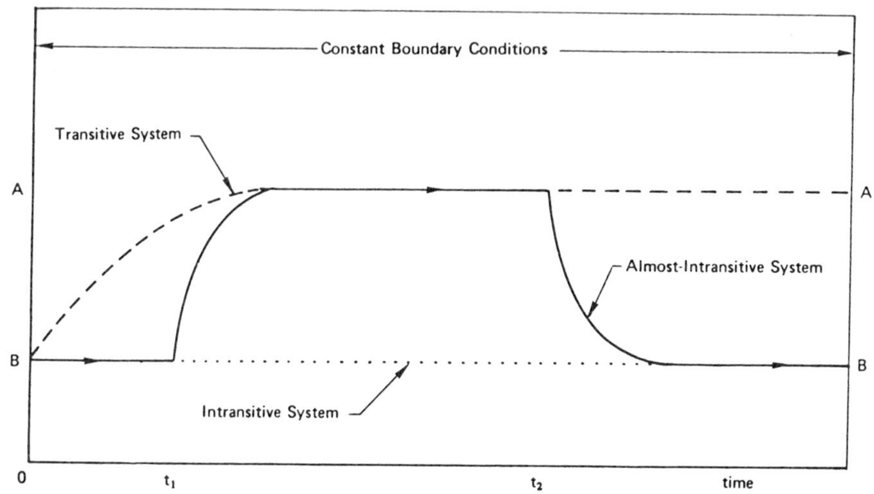

Figure 2. Illustration of transitivity, intransitivity and almost-transitivity in a climate system with time independent forcing. Source: NAS, 1975

independent, then the steady state solution (i.e., $\partial\vec{u}/\partial t = 0$) is attained as $t \to \infty$. The case of periodic external forcing is not complicated; the solution contains only those periods that are part of the forcing.

The situation is dramatically different for the nonlinear case. Even in the case of a dry fluid with constant boundary conditions and with constant forcing ($\vec{f}$, $\vec{g}$), a steady solution is often not possible. Nonsteadiness is the most characteristic property of turbulent fluids (Landau and Lifshitz, 1959), and the atmosphere is undoubtedly turbulent.

Deterministic initial value problems are termed by Lorenz (1975) as prediction of the first kind. For a turbulent fluid, it does not make much sense to talk about long-range predictions of the first kind. Two seemingly similar atmospheres, one with, say, a butterfly and one without a butterfly would "rapidly" (in a matter of weeks) diverge from each other and the nonperiodic components would become essentially uncorrelated (Lorenz, 1969). It is still possible, however, to talk about climate predictions in a statistical sense. We consider an ensemble of all the realizations (i.e., time series) associated with a given forcing. If the probability distribution is known (or can be computed from a model), then one can construct various statistics (average, variance, covariance, etc.). These statistics or probability distribution may be then called "climate." The statistical ensemble average, for instance, may still be time dependent (in response, for example, to the seasonal cycle), but the hope is that, in comparison to individual realizations, it is a much smoother function of time. The central problem of theoretical climate dynamics is to construct out of the deterministic initial and boundary value dynamic systems, a statistical dynamic system (Leith, 1975).

It is not always the case that the solution of such a statistical dynamic system is unique; for any given unique history of external forcing (e.g., complete specification of the diurnal, seasonal and longer term changes of orbital parameters) there might correspond more than one probability distribution. Such a system is termed by Lorenz (1968) as intransitive.[1] If, to each set of external forcings, there corresponds a unique probability distribution, then such a system is termed transitive.

Transitivity and intransitivity for time-independent forcing are illustrated in Figure 2. In a transitive system after a sufficiently long time only the statistics associated with state A are stable and realized. In the fully intransitive system

[1]Lorenz discusses the simpler case in which the external forcing is time independent.

at any given time either state B or A (or more) are acceptable. There is a third kind of behavior which is termed almost-intransitive. This is the case where, even for constant (time-independent) forcing, the solution is oscillatory. The system may spend some time at state B, then move rapidly to state A, and so forth. The climate exhibited by such a system would thus consist of several quasi-stable states, together with periods of transition between them. If we suppose that these transitions are brought about by some catastrophic processes, an almost-intransitive system—with the catastrophes modeled—may be approximated by a simpler intransitive system, with the catastrophes omitted.

Some speculations about the regional climate in Beringia and its transitivity or intransitivity are now called for. Since catastrophes are more likely to happen in regions where the variability is already very high, such as desert margins or arctic tundra, it is possible that the climate in Beringia around 14,000-12,000 BP (when a sudden rise of dwarf birch occurred—e.g., Ritchie and Cwynar, this volume) was simply a manifestation of another possible equilibrium. Alternatively, it could have been a transition climate from glacial to interglacial periods.

The treelessness in a high latitude may be related, as suggested by Hare and Ritchie (1972), to the mean annual net radiation at the surface, defined as

$$R = I(l\text{-}a) - \Delta I_r. \qquad (2)$$

Where I is the solar (short wave) radiation arriving at the ground (before reflection), a is the surface albedo and ΔI_r is the net infrared (long wave) radiation; I(l-a) is thus the absorbed solar radiation at the surface. If the net radiation is the main control of productivity in the Arctic (Fig. 3), then treelessness maintains itself via the albedo feedback; high albedo → low net radiation → low productivity → absence of forest → high albedo.

If, however, what controls productivity is mean summer radiation, then the climate in Beringia could perhaps better be explained by changes in external forcings; in this case, because of changes in orbital parameters (Milankovich, 1930). Table 1 illustrates the point. In the first line selected years before present are printed. The second line displays the amount of solar radiation received at the

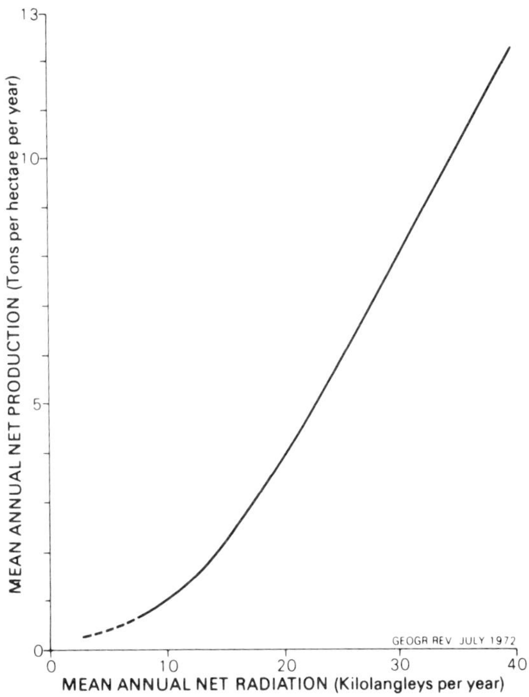

Figure 3. Normal relation between mean annual net radiation and net production of dry matter. Source: Hare and Ritchie (1972)

top of the atmosphere (in watts/m^2) at latitude 65°N during the "summer" (March 21-September 21) for various years BP. The values are adapted from Vernekar (1972) based on a presumed "solar constant"[2] of 1360 Wm^2. The third line shows the percentage change in summer radiation compared to present. The fourth line displays the equivalent latitude (i.e., the latitude which at *present* receives the same amount of summer solar radiation as was received by latitude 65°N in the appropriate year BP). Thus, for instance, 25,000 years ago latitude 65°N received the same amount of radiation which today is received at latitude 68.86°N. To the extent that summer radiation exerts a major influence on climate one can argue that the climate of latitude 65°N 25,000 years BP was similar to the climate of 68.86°N today. For calculating line 4, the summer solar radiation at latitudes 70°N and 60°N at present is taken to be 355.26 w/m^2 respectively. It is to be emphasized

[2]There is now a considerable debate among solar physicists as to whether or not the sun has been a "constant star" in the last billion years. See, e.g., Eddy (1976).

that changes in orbital parameters would not significantly affect the annual solar radiation; increase in "summer" radiation is compensated by decrease in "winter" (September 21-March 21) radiation. However, there is little productivity during the arctic winter; therefore, summer radiation may be a better indicator of vegetational changes. It is also worth nothing that the percentage change of solar radiation on the ground may be higher or lower depending mainly on the atmospheric and surface albedos.

From Table 1 it can be seen that 35,000 BP summer radiation was similar to today's. It then started to decline and reached a minimum (2.24% below present) 25,000 years ago. It then rose again, reaching a maximum (4.23% above present) 10,000 years ago. It can be deduced, for instance, that the birch rise at ~ 14,000 BP, presumably signifying a major change in vegetation (Ritchie and Cwynar, this volume), also corresponded to a significant change in summer solar radiation. The fact that at 18,000 BP the summer solar radiation was close to that of present, while glaciation was extensive (Hopkins, personal communication) does not pose a major paleoclimatological impasse, since ice sheets, once established, can persist for several thousand years under regimes that are unfavorable to their existence. From the last line of Table 1 one may conclude that, to the extent that incoming summer solar radiation exerts a major influence on productivity, fluctuations of the order ± 5° latitude in the potential position of the tree line are consistent with changes in orbital parameters. For instance, 10,000 years BP latitude 65°N was radiationally similar to 59.84°N today and, therefore, a northward shift of 5° in the position of the tree line is implied (provided that the relation between radiation and vegetation is real).

SOME ASPECTS OF LAND SURFACE PROCESSES AND THEIR EFFECTS ON REGIONAL CLIMATES

Energy Equation

The equation that determines the surface temperature and also describes the rate of exchange of energy across a unit area of the air-land surface can be written

$$\epsilon_g \sigma T_g^{4\uparrow} + H_{sg}^{\uparrow\downarrow} + LE_g^{\downarrow\uparrow} - I(1-a)^{\downarrow} - R_{\ell}^{\downarrow} = -G + S - M \tag{3}$$

Here the arrows stand for the direction of the flux, upward toward the air, downward toward the surface. Double arrows indicate that the flux can go in both directions, depending on the differences between the air and the underlying surface. $\epsilon_g \sigma T_g^4$ is the infrared flux from the surface (basically the Stefan-Boltzman law). ϵ_g is the ground emissivity (always ⩽1 and usually >0.9), σ is the Stefan-Boltzman constant (= 5.6696 x

Table 1
Summer Radiation Values in Relation to Present at Various Selected Years (BP) at Latitude 65°N

The term summer refers here to the period March 21 - September 21. The term equivalent latitude is used to denote the latitude which *at present* receives as much radiation as latitude 65°N in the appropriate year BP. Refer to text for further discussions. (Values of solar radiation are adopted from Vernekar, 1972).

1000 Years BP	35	30	25	20	19	14	10	07	02	01	Present
Latitude 65°N "Summer" Radiation (Watts/m²)	365.42	360.10	357.68	363.48	365.42	378.00	381.39	378.49	368.81	367.36	365.90
Percentage Change from Present	-0.13	-1.59	-2.24	-0.66	-0.13	3.30	4.23	3.44	0.79	0.40	0.00
Equivalent Latitude (Deg.)	65.23	67.73	68.86	66.14	65.23	60.06	59.84	60.80	64.03	64.52	65.00

10^{-8} $Wm^{-2}deg^{-4}$). H_{sg} is the heat flux due to convective motion.[3] It is termed sensible heat flux. E_g is the flux of water vapor, L is the latent heat of vaporization (+2.5 x 10^6 J kg^{-1} at 0°C). Thus LE_g is the heat lost by the ground due to evaporation or the heat gained due to dew formation. This is termed latent heat flux, because the water vapor is used to heat the atmosphere only upon condensation—a process which takes place only when the air parcel is sufficiently uplifted. As is indicated by the arrows, the sensible and latent heat fluxes could, in principle, be up or down depending on the temperature and moisture gradient. It is, however, mostly up; the most noticeable example for downward transfer is during temperature inversion (which is most common in the Arctic in winter— and during dew formation. I(l-a) is the short wave radiation, both direct and diffuse, that is absorbed by the ground; a is the albedo and 1 is the short wave (solar) downward-directed flux. $R\ell$ is the infrared (long wave) flux from the atmosphere to the ground (the atmospheric greenhouse effect). It depends on the CO_2 and water vapor content of the atmospere, the latter being highly variable. G is the soil heat flux and M is the heat used for snow melt. S stands for processes of presumed secondary importance (photosynthesis, heat transfer by warm rain, oxidation of biological material, fires). G is a seasonably-reversing term integrating to zero over the year. For land surfaces, it is also a common practice to neglect G for shorter terms.

In many applications the net radiative impalance, R, defined by (see also Equation 2),

$$R \equiv 1(1\text{-}a) - \epsilon_g \sigma T_g + R\ell \qquad (4)$$

is considered the *driving* mechanism for nonradiative processes. R is also called *net radiation.*

Surface Albedo

Of particular importance is surface albedo. Table 2 gives some typical values for various surfaces. Broadly speaking, albedo of land is higher than that of oceans; albedo of vegetated land is lower than that of bare soil; high zenith albedo is higher than low zenith albedo; and, *most important,* snow- and ice-covered surfaces have much higher albedo than their equivalent ice- and snow-free areas. Forest areas are effective, compared to open areas, in increasing net radiation due to their lower albedo. In high latitudes, they also determine whether the snow cover will be effective as a reflector.

Obviously, surface albedo is a very important climatic parameter, but before discussing its profound effect on the regional and global dynamics (and, hence, on the surface climate) a note of caution is appropriate. The relative importance of surface albedo depends on how effective the atmosphere is in scattering, reflecting, and absorbing the solar radiation. At sufficiently large solar zenith angles, the solar radiation must penetrate more scattering and absorbing atmosphere (aerosols, cloud droplets, water molecules), and therefore less radiation arrives to the surface. Clouds would have the same effect because of their higher reflectivity; the cloud effect, however, is somewhat compensated for by their increased infrared flux to the surface (Schneider, 1972).

The most obvious effect of increased albedo due to snow and ice is the reduction of surface

Table 2
Mean Albedos of Various Surfaces

Note the high albedo of snow and ice, and the low albedos of water and thickly vegetated land.

Surface	Albedo %
Water, plane surface	2.4
Water at equator (high sun)	6.0
Water at 60° latitude in winter (low sun)	21.0
Soil, dark	5 - 15
Soil, dry light sand	25 - 45
Forest and thickly vegetated land	15 - 18
Dry grassland and savannas	20 - 24
Snow-covered coniferous forest	30 - 39
Snow-covered open woodland	50 - 60
Snow-covered forest tundra	60 - 65
Snow (clean, dry)	75 - 95
Snow (wet and/or dirty)	25 - 75
Ice, sea	30 - 40

[3]Convection is the transport of various quantities (heat, moisture, etc.) by macroscopic air motion. The other modes of transfer are conduction (or diffusion), accomplished by microscopic molecular motion, and radiation, accomplished by photons.

temperature. To the extent that this reduction leads to a longer-lasting (because of less summer melting) and more extensive snow and ice cover, it would further increase the global albedo and reduce the surface temperature. The effectiveness of this snow and ice albedo-temperature feedback depends, globally, on evaporation rate, which is reduced during ice-age conditions due to lower sea surface temperature and more permanent sea ice. It is also dependent on the dynamics of the general circulation, which will favor less precipitation over the continents (Manabe and Hahn, 1977). It is now firmly established that glacial periods are generally drier than interglacial periods (CLIMAP, 1976).

To understand the profound effect that land-surface albedo has on the general circulation, it is instructive to look at the radiation balance of the earth-atmosphere system. For reasons to be explained further below, increasing the land-surface albedo, due to change in vegetation, *may or may not reduce the surface temperature.* However, an increase in land-surface albedo will undoubtedly cool the earth-atmosphere system, everything else being equal. Thus a vertical column above an affected land area will be colder with increasing surface albedo. In summer time, that means that tropospheric land-sea temperature differences are reduced; therefore there will be less inflow from the oceans, enhancing aridity and increasing albedo.[4]

Charney (1975) was apparently the first to demonstrate analytically (with a theoretical model) the link between surface albedo and large-scale, vertical velocity; the latter is highly correlated with precipitation. Manabe and Hahn (1977) demonstrated, using a general circulation model (GCM), that the aridity of an ice-age climate (at least in the tropics) is not only due to reduced evaporation from the ocean (as discussed before), but also is due to increased surface albedo over non-glaciated land surfaces (i.e., meager vegetation).

[4]The troposphere is the lowest 5–10 km of the atmosphere (depending on latitude), where the temperature usually decreases with height. Above the troposphere lies the tropopause, where the temperature is roughly isothermal, and above this the stratosphere, where the temperature increases with height.

Convective Fluxes of Heat and Moisture

It has been mentioned before that removal of vegetative cover, and hence increased surface albedo, may or may not decrease the surface temperature. Removal of vegetation reduces the capacity of the soil to hold moisture and also increases its resistance to evaporation. This decrease in evapotranspiration (Equation 3) is partially compensated by increased sensible heat flux, but this has its limits and, in order to achieve balance, the infrared flux from the ground, $\epsilon_g \sigma_g{}^4$, must also increase; thus *higher* surface temperature is obtained. Because GCMs apparently do not account for vegetative cover and its role in regulating evaporation and surface temperature, they usually show, in response to removal of vegetation, decrease in surface temperature. Experience shows (e.g., Ripley, 1976) that the surface temperature and the sensible heat flux usually rise.

Based on micrometeorological theory and observation, the convective exchange of heat and moisture is given by Businger (1973),

$$H_{sg} = \rho_a C_p C_H |\vec{U}_a| (T_g - T_a), \qquad (5.a)$$

$$LE_g = w_t L \rho_a C_E |\vec{U}_a| (q_g - q_a). \qquad (5.b)$$

Here ρ_a is the air density; C_p specific heat of air at constant pressure (- 1004 J $deg^{-1} kg^{-1}$); T_g is ground temperature; T_a is air temperature at anemometer level (10 m above the surface); q_g is moisture content of the ground; q_a is moisture content of the air (at anemometer level); $|\vec{U}_a|$ is the magnitude of the wind vector at anemometer level; L is the latent heat of vaporization (defined before); C_H and C_E are dimensionless coefficients termed by Deardorff (1972) "bulk transfer coefficients" for heat and moisture, respectively. These coefficients depend on the surface roughness (thus also on the vegetation) and the Richardson number, Ri, which is a nondimensional measure by the stability of the atmosphere, given by

$$Ri = \frac{g}{T_a} \frac{\frac{\partial T_a}{\partial z} - \frac{g}{C_p}}{\partial |U_a| / \partial z}. \qquad (6)$$

w_t is a wetness parameter ($0 \leqslant w_t \leqslant 1$) and is, in general, dependent on soil moisture content (and therefore on the ground hydrology), on soil properties and vegetative cover. Potential evaporation E_{gp} is realized when $w_t = 1$.

Planetary (Rossby) Waves and Their Relationships to Surface Processes

In the middle and high latitudes, the tropospheric land-sea differences (both thermal and orographic) produce strong standing disturbances (long westerly waves); these waves are superimposed on a vast circumpolar westerly flow. The waves show themselves most clearly in the 300 mb (monthly mean) charts (Figure 4) of geopotential heights (approximately real heights at pressure level 300 mb). Velocities are given by the so-called geostrophic relation, which expresses balance between height gradients and the Coriolis force:

$$u = -\frac{g}{f}\frac{\partial z}{\partial y}\,;\, v = \frac{g}{f}\frac{\partial z}{\partial x}\,;\, f \equiv 2\Omega \sin\phi\,. \qquad (7)$$

Here g is the acceleration of gravity (= 9.81 ms^{-2}), z = geopotential height, Ω = angular velocity of the earth's rotation ($T = 2\pi/\Omega$, T = 1 day), ϕ = latitude, x, y distances toward the east and north respectively, u, v components of the flow from west and south respectively. From the geostrophic relation (Equation 7) it can be deduced that the wind blows parallel to the contours, at a speed proportional to the gradient; in the northern hemisphere lower heights are to the left of the wind vector.

Although not obvious from the 300 mb mean charts, these long waves are related to the average location of the arctic front. The average location of the arctic front in July coincides rather remarkably with the boreal forest/tundra boundary in North America and Eurasia (Barry and Hare, 1974). In view of the crucial role played by the surface albedo and cloudiness (associated with the front) in determining the amount of solar radiation that is actually absorbed by the ground, it is not surprising that a correlation also exists between net radiation (both mean annual and summer) and the position of the tree line (Figure 5). It seems to me, however, that the latter correlation is physically more significant, particularly in view of the fact that the position of the arctic front is related to the structure of the long wave westerlies which are driven by tropospheric land-sea differences. On the other hand, thawing conditions today reach Fairbanks, Alaska, near 65°N about April 25—*before* they reach the 50th parallel in Ontario. The reason is that higher sunshine in the west is associated with the long wave ridge (in winter and spring) in the westerlies (Fig. 4a).[5] The planetary waves influence the surface processes (vegetative cover etc.) and also are influenced by them.

The trough over Alaska in July (Fig. 4b) is a major element in controlling summer precipitation there. In the surface map this trough manifests itself as moist southwesterly flow from the Pacific (Barry, this volume). If the winter and spring ridge persists through the summer, the surface flow will be dry southeasterly (or easterly) flow, presumably the conditions in Beringia 18,000 BP (Barry, this volume). Again one might speculate that the presumed dry arctic steppe at least partially maintained itself through aridity → treelessness → high albedo, reduced evaporation → reduced tropospheric land/sea contrast → flow more zonal → ridge not replaced by trough → reduced summer precipitation → aridity. Of course other reasons, more global in nature, might have contributed as well (e.g., lower sea surface temperature, orographic influences due to the presence of the ice sheet).

Implication for Beringia

We shall now employ the considerations of the previous sections to speculate about the paleoclimate in Beringia, say 18,000 BP. These speculations are subjective in the sense that I start with the assumption that an arctic steppe did exist and then try to see whether this is consistent with the feedback mechanisms discussed before. The possibility exists that the above mentioned mechanisms are not as efficient as I postulate them to be, or that other processes, not discussed in this paper, counterbalance them. It should also be emphasized that from a climatic point of view it is nearly impossible to equate arctic steppe with something similar to the African savanna. If anything, the climatic consideration discussed below implies

[5]Ridges are associated with surface highs and sinking motion; hence no precipitation.

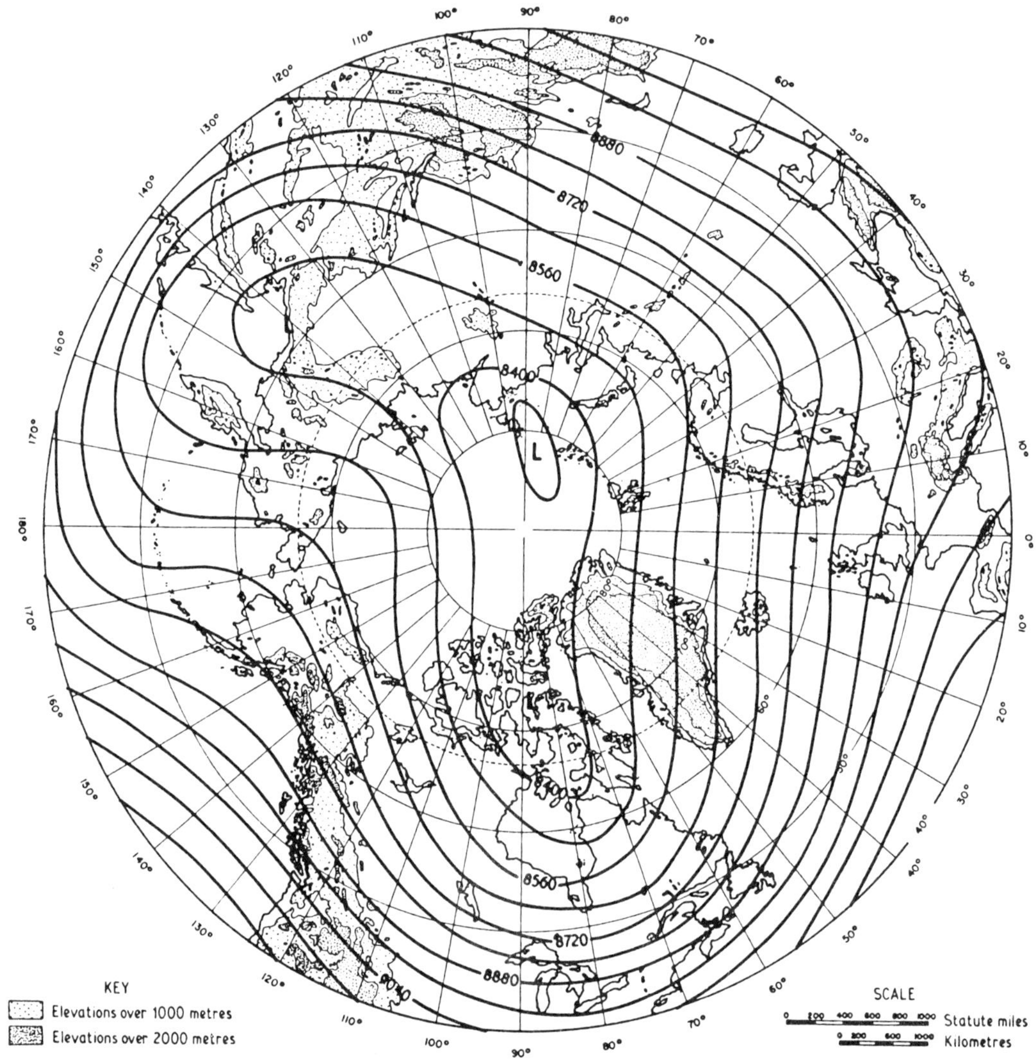

(a)

Figure 4a. Mean heights (in geopotential meters) of the 300 mb surface for January. The wind flows almost parallel to the contours at a speed proportional to the gradient, with lower heights to the left of the wind vector (in the Northern Hemisphere). Source: Barry and Hare (1974)

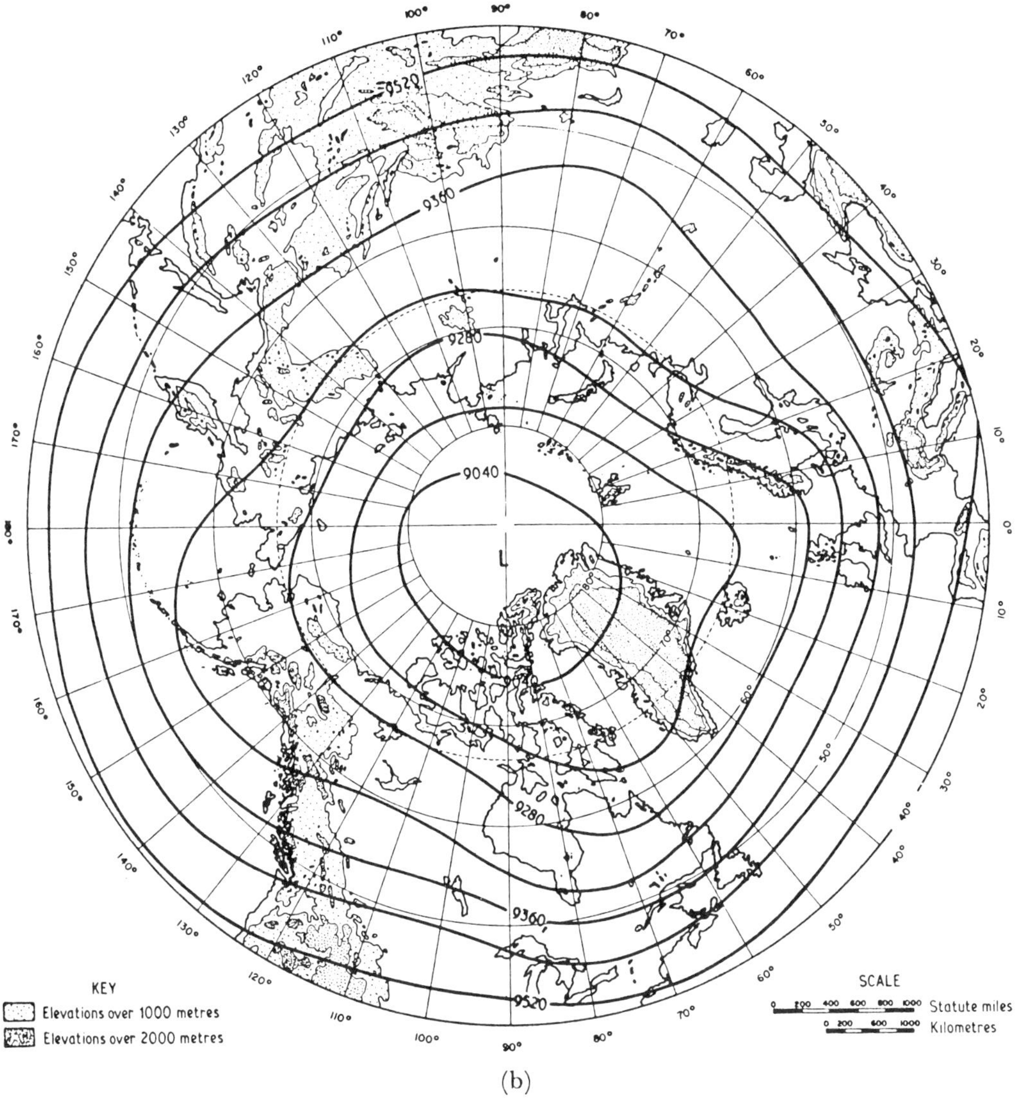

(b)

Figure 4b. Mean heights of the 300 mb surface for July

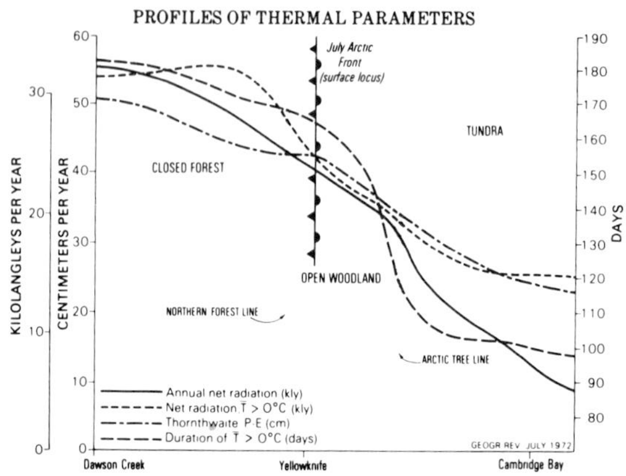

Figure 5. Profiles of various thermal parameters along a section passing through four vegetation zones from Dawson Creek, B.C. to Cambridge Bay, N.W.T. Parameters are mean annual net radiation, mean net radiation for season with air temperature above 0°C, Thornthwaite's mean annual P-E function and the duration of the season with air temperature above 0°C. Note the sharp decline in net radiation in passing from forest areas to the tundra. Units of net radiation are in kilolangleys/year. Langley is an energy unit = 4.81 x 10^{-4}J m^{-2}. P-E (evaporation) is given in centimeters/year. Source: Hare and Ritchie (1972)

patchy vegetation. I use the term steppe as opposed to tundra to emphasize that the treelessness, at least in some parts of Beringia, was due to aridity rather than cold summers.

With that cautionary preface let us turn to some of the central paleoclimatological questions of Beringia.

(1) *Is the concept of arctic steppe with no modern analog consistent with climate theory?* To the extent that regional climates, especially in marginal areas, can be intransitive (Flohn, 1975) the answer is positive (see more detailed discussion in the section on intransitivity).

(2) *What can be said about the severity of winters?* Paleoclimatic data as well as general circulation models (Williams, 1978) indicate colder and drier winters. The southward migration of the arctic front, the lower sea surface temperature (leading to cooling and reduced evaporation) the larger extent of sea ice (reducing evaporation) the chilling dry easterly wind from the continental ice sheets, are all (presumably) contributing factors.

(3) *Did spring (particularly the thaw) come earlier or later than today?* The general dryness of the Pleistocene suggests less cloud cover; this would increase the amount of short wave radiation reaching the ground (before reflection). The light snow cover and the possibility of spots with bare ground (Guthrie, this volume) would undoubtably accelerate the thawing. These considerations are relevant as long as the albedo does not change appreciably.

Inspection of Table 2 and Figure 5 reveal, nevertheless, that the ice-albedo feedback is far more effective in a snow-covered tundra (albedo 75-95) than in a snow-covered forest (albedo 30-39). It is therefore more likely that spring and thaw arrived later than present.

(4) *Was summer surface temperature higher or lower than today? Was summer precipitation higher or lower than today?* This probably depends on location. In the part of Beringia which was closer to the continental ice sheets, the prime factor in determining surface temperature was probably cold air advection (due to the presumed southeasterly wind).[6] Under that circumstance, of all the components of the surface energy balance (Equation 3) air properties such as temperature, humidity and speed (Equation 6) were the most important. Almost regardless of incoming solar radiation, the surface temperature was forced by the cold dry easterly wind. As the wind progressed westward, it was undoubtably modified by fluxes of sensible and latent heat from the ground. At some point solar radiation was able to resume its major role. Assuming less cloudiness and precipitation (because the prevailing wind was easterly) more solar radiation was absorbed at the ground, resulting in higher surface temperature.

As discussed before, reduction of vegetation can also contribute to decrease in precipitation, as well as to increase in the surface temperature.

Overall, those parts of Beringia sufficiently close to the continental ice sheets were characterized by warmer and drier summers.

MODELING APPROACHES

In order to test hypotheses and speculations such as those discussed in the previous sections,

[6]Advection is the meteorological jargon for transport by horizontal (usually large-scale) air motions.

one must have effective mathematical models. These models must be capable of predicting at least some of the consequences of changes in external forcing. If used cautiously, mathematical modeling could be a viable tool in reducing some of the ambiguities of the proxy record. For the case of Beringia, it is probably essential to specify the ice sheets that surrounded the area, and the exposure of the continental shelf and the land bridge. In lieu of modeling oceanic circulation, one should specify the sea surface temperatures (at, say, 18,000 BP). Since the treelessness and the relative dryness were presumably essential for the survival of the fauna, the vegetative structure must be part of the internal (predicted) system. However, models which specify the surface albedo (hence the vegetation) would still be useful. By inspecting the model's generated surface temperature, precipitation, and evaporation, one can check whether these are consistent with the presumed surface albedo.

For regional studies, it is imperative that the model reproduce the relevant ingredients of the present regional climate. For instance, one might require that, for the particular region studied (e.g., Beringia) the model's generated and observed Koppen climate classification will be the same. One can tolerate inconsistencies in other regions if they do not affect seriously the global climate. At present the GCMs fail to resolve present regional climate in Alaska and Eastern Siberia.

The failure of the global models to reproduce the present regional climate in Beringia and also their apparently conflicting results concerning the past climate in that area (Barry, this volume), may be due to the neglect of smaller-scale features. Such features may include details of the orography and microclimate, particularly the effect of vegetation. To resolve smaller-scale features, models are needed with a spatial resolution refined to perhaps 10-20 km in the horizontal (typical horizontal grid in current GCMs is 200-250 km) and 100 m in the virtical grid. Fortunately, this kind of resolution is not required everywhere. For large oceanic areas far removed from the region of interest, the current resolution is adequate. In the vertical, the 100 m resolution is required only in the boundary layer (the lowest 1 km or so of the atmosphere). In the free atmosphere a 0.5-1 km resolution is probably adequate. Unfortunately, the present techniques by which the GCMs are actually solved require a fixed grid. Variable grid models or finite element techniques (Strang and Fix, 1973) are currently under development at several research centers (e.g., Kurihara, et al., 1979; Staniforth and Daley, 1979), but this is technically a very difficult problem. In the meantime, other types of models might be quite useful. These are the so-called mesoscale or regional models.

Mesoscale models resolve explicitly only the region of interest; the larger scale motions are *forced* by specifying some or all of the lateral boundary conditions. This creates a great deal of "numerical noise" which must be either "filtered" or allowed to propagate away from the boundaries. How to do it properly is still an open question, but some progress has been achieved (Orlanski, 1976; Perkey and Kreitzberg, 1976).

Apart from the problem in the boundaries, mesoscale models can, in principle, easily accommodate all the sophistication of the current GCMs and more. In practice, because these models were originally designed to improve the very short range forecast (5-10 hours) or to study pollution dispersal, the major thrust of development has been in mesoscale dynamics and on much improved convective precipitation parameterizations (Kreitzberg and Perkey, 1976). Radiation and surface processes have been emphasized less although more recent models such as that of Anthes and Warner (1978) contain surface energy balance of the type discussed in the previous sections (Equation 3). The models still specify, rather than predict, soil moisture. It seems, however, that based on the experience gained from GCMs, it will be comparatively easy to introduce more sophisticated ground hydrology and radiation processes. Some mesoscale models have been tested, because of their ability to simulate mesoclimate. Thus, for instance, Pielke (1974), Pielke and Mahrer (1978) and Tapp and White (1976) simulated the sea breeze in Florida.

I see no reason why mesoscale models should not be used to simulate the paleoclimate in Beringia, particularly those features which are due to interaction between climatic and geographical features (orography, vegetation, etc.) on scales which cannot be resolved by global models. It does require, however, a dedicated effort of two to three individuals and substantial computer resources.

SUMMARY AND CONCLUSIONS

The climate system is vast and highly interactive with many feedback loops. The existence of feedbacks suggests that the climate system does not respond uniquely to a unique forcing and may be intransitive or almost-intransitive.

The albedo feedback, through ice, snow and vegetative cover, may be the key to the intransitive behavior and to the maintenance of the treelessness and relative aridity in Beringia. The surface fluxes of heat, moisture and momentum (frictional dissipation) are also modified by the vegetative structure. The above-mentioned surface processes and the orography create tropospheric land-sea differences which are driving the planetary (Rossby) waves. These waves, through their effects on temperature and precipitation, partially determine the surface cover.

A possible (speculative) climate scenario for Beringia at 18,000 BP as compared to the present, includes: cold dry winters, late springs and, depending on geographical locations (mainly proximity to the continental ice sheets), summers which are either colder and drier, or, warmer and drier.

GCMs are presently unable to reproduce the present regional climate in Alaska and northeastern Siberia. Mesoscale models may be capable of doing so and may also be used as an aid in reconstructing the regional paleoclimate of the arctic steppe.

ACKNOWLEDGMENTS

Throughout the preparation of this paper, discussion with F. Kenneth Hare of the University of Toronto and Roger G. Barry of the University of Colorado were extremely helpful. I am grateful to Dr. Steven B. Young of the Center for Northern Studies for helpful comments on the manuscript. The position papers and the discussions held at Burg Wartenstein Symposium No. 81 on the Paleoecology of the Arctic Steppe-Mammoth Biome were helpful in preparing the final version. I am thankful to all the participants of this symposium. I am grateful to the Wenner-Gren Foundation and especially to Lita Osmundsen for the opportunity to participate.

PRIMARY PRODUCTION AND THE PLEISTOCENE UNGULATES–THE PRODUCTIVITY PARADOX

When we consider the large-mammal fauna of Pleistocene Beringia, we find ourselves confronted immediately with the "productivity paradox." The abundance of large-mammal remains, including occasional finds of well-preserved carcasses, has drawn attention since the earliest days of exploration. The Beringian large-mammal fauna is famous for its diversity, gigantism, and predominance of grazing species—a startling contrast to the present-day depauperate, generally small-bodied, and predominantly browsing ungulate fauna. The contrast is all the more vivid when one realizes that Beringian taiga and tundra are areas of markedly low present-day productivity. The former presence of larger and more diverse ungulate populations in northern regions of periglacial climate at a time when primary productivity presumably was still lower constitutes the paradox. To oversimplify, how does one keep a mammoth alive and well under the seemingly impossible conditions of ice-age Beringia?

To resolve the productivity paradox, we must examine the climatic constraints and the nature of the ancient vegetation and its ecology, discussed in earlier sections, and then temper this with a consideration of the biology of surviving relatives of the Pleistocene ungulates. The rich fossil mammal record itself, however, provides clues to adaptations and thus insights into the nature of the paleoenvironments in which Pleistocene Beringian mammals lived. The seven papers that follow reflect this methodology.

Productivity in the grassland and tundra biomes of the present day is reviewed by Redmann and by Bliss and Richards, respectively. Both papers go on to demonstrate how modern ecological data and theory can be employed to solve the Beringian productivity paradox. Redmann points out that a "rich" environment is not necessarily a diverse one and that the presence of a diverse ungulate fauna does not imply any particular level of overall primary productivity. Bliss goes on to model a hypothetical Beringian ecosystem that involves most elements of the known Pleistocene ungulate fauna cropping vegetation components represented in different proportions in present-day Beringian tundra. Information on the interactions of species diversity, population levels of individual species, quality and diversity of herbage, and frequency of disturbances reviewed by Redmann will be useful in future attempts to elaborate and refine Bliss' pioneering model of the Beringian steppe-tundra ecosystem.

As Redmann notes, grazing successions play a role in maintaining ungulate diversity in temperate latitudes and may have been important in ancient Beringia. Jerome Martin uses his background in ungulate nutrition and range management to review the relationships between digestive physiology, grazing strategies, and niche separation.

We turn now to the Beringian mammal fauna, itself. Vereshchagin provides an inventory of Siberian soft-part remains, an evaluation of the adaptive significance of their morphology, and a review of the ecology of the nearest living relatives. Kubiak discusses the biology of a single species, woolly mammoth, the dominant member of the Beringian late Pleistocene mammal fauna. Olivier draws upon the ecological methods of inquiry reviewed by Redmann, upon Kubiak's morphological data, and upon his own first-hand knowledge of modern Asian elephants to frame questions about the ecology and behavior of the mammoth in Beringia. Finally, Guthrie explores the usefulness of fossil mammals as paleoenvironmental indicators. More than anyone else, Guthrie has thought about the implications of the dietary requirements, the dietary regime, and the social behavior of Beringian mammals and the ways in which these implications can be used to reconstruct the ancient environment.

During recent years, the problem of the productivity paradox has underlain the hottest disputes about the nature of Pleistocene Beringian landscapes, and the productivity paradox was at the heart of our Burg Wartenstein conference. Too many unanswered questions remain for us to claim that the paradox is solved, but the seven papers in this section point the way toward resolution.

—*Charles E. Schweger*

13

PRODUCTION AND DIVERSITY IN CONTEMPORARY GRASSLANDS

R. E. Redmann

Primary productivity in arid grasslands and savannas is controlled mainly by drought; nutrient limitations are most apparent in more humid regions. The quantity and quality of herbage available to large grazers can fluctuate widely over the year due to drought and winter conditions. Woody browse species, important in the diets of most year-round resident herbivores, are present throughout the grasslands and savannas.

Removal of green plant tissue by herbivores may reduce primary production. Detrimental effects of herbivores often are counterbalanced by beneficial effects. Large herbivore consumption is small in some grasslands; much of the primary production passes to saprophagous organisms, or is consumed by fire.

Species diversity arises from competition and niche divergence along resource gradients. Two species cannot occupy the same niche; under equilibrium conditions, one must be excluded. Competitive exclusion in natural communities is not easy to measure or prove. Disturbances prevent competitive exclusion and result in the maintenance of diversity.

A review of field data, and a theory-based model show that, in general, primary production determines the large herbivore biomass that can be supported. Animal size and dominance are also important. Herbivore diversity can be high in communities with low productivity and low biomass, but only where dominance is prevented. Dominance tends to be strongest in tundra communities, and weakest in tropical grasslands and savannas. Diversity of large herbivores in temperate regions is based mainly on variety of habitat, which is most pronounced in mountainous and periglacial areas. Large-herbivore diversity in Beringia could have resulted from patterning in the vegetation due to physiographic, edaphic and microclimatic variations, as well as fire and the intensity of utilization by the herbivores themselves.

Key words: productivity, grassland, grazing, browse, herbivore, habitat diversity

INTRODUCTION

Paleoecological evidence presented in other chapters of this volume suggests that the flora and fauna of Beringia during the height of the last Pleistocene glaciation was different from present-day tundra. Matthews (this volume) indicates that the tundra was steppe-like in uplands and lowlands over large areas of Beringia. This vegetation was best developed in Siberia, where a cold and extremely continental climate prevailed. Matthews proposes that a vast "steppe-tundra" biome extended from the Ukraine as far east as the Yukon.

The mammal fauna might have been extremely diverse (Guthrie, 1968a; Matthews, 1976a; Geist, 1978), and "almost certainly included more large

ISBN 0-12-355860-3

grazing ungulate species than occur in contemporary tundra environments" (Matthews, this volume). The fauna included mammoth, horse, bison and caribou. In addition, saiga antelope and other steppe mammals also may have been present, although the paleontological evidence is equivocal. A major paleoecological problem is to explain how a relatively diverse, steppe-like ecosystem functioned in the cold, dry environment of Beringia.

My objective in this paper is to describe aspects of the structure and function of contemporary grassland ecosystems that may be relevant to the paleoecology of the steppe-tundra. The description will include the comparative productivity of grassland ecosystems, particularly the quantity, quality, and seasonal patterns of primary production and its consumption by large herbivores. I also will attempt to summarize current ecological theory explaining the generation and maintenance of diversity in biotic communities, and the relationship between productivity and diversity. Although the emphasis is on grassland and savanna, I will include relevant information from forest and tundra biomes.

PRIMARY PRODUCTION AND ITS UTILIZATION

A General Trophic Model

The model of trophic interactions which serves as the framework for this paper (Fig. 1) is a modification of those given by Wiegert and Owen (1971) and Golley (1972). In this model, the ambiguous terms consumer and decomposer are replaced with biophage and saprophage. Both categories of organisms are heterotrophic, i.e., they degrade organic matter in order to obtain energy. Biophages utilize living organic material; saprophages consume non-living organic matter. The ultimate source of both types of food is the autotrophs, or primary producer trophic level.

Primary biophages (herbivores) differ from

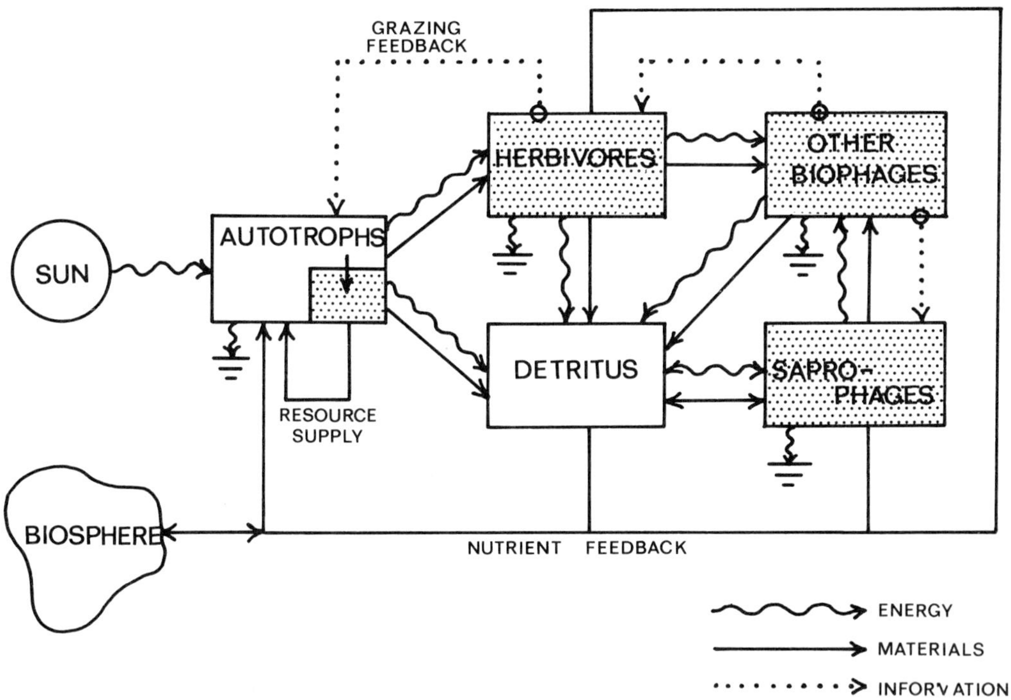

Figure 1. Patterns of flows of energy, materials and information in an ecosystem. Heterotrophic tissue is stippled. Modified from Wiegert and Owen (1971) and Golley (1972)

saprophages in a fundamental way: biophages, in utilizing the plant resource, directly affect the rate of resource supply. Saprophages by feeding do not directly affect the rate of food supply. According to Wiegert and Owen (1971), "a relatively low utilization of green plants by terrestrial grazers [is] the maximum permissible level that yet avoids severe long-term oscillations of plant populations caused by frequent escape of grazers from control. . . . A greater mean density of biophages and a higher utilization might result in periodic destruction of the plant community, local extinction of dependent populations and a total breakdown of the system."

There should be rapid evolutionary selection among autotrophs for resistance to herbivores. A variety of chemical toxins and morphological features have evolved among plants in response to the "grazing feedback" shown in Figure 1 (Whittaker and Feeny, 1971). Despite their defenses, plant populations are exploited and even stimulated by grazers in apparently stable systems (McNaughton, 1979). About 50% of the annual shoot production can be removed from managed rangeland before plant species composition shifts (Lodge, et al., 1971; Stoddart, et al., 1975).

The saprophages appear to be relatively inefficient at recycling the primary production in grassland. Without large grazing animals there tends to be accumulation of detritus, which adversely affects microclimate, nutrient cycling and primary productivity. Large herbivores in particular seem to facilitate "nutrient feedback" (Fig. 1) by speeding up turnover of plant material. This results from their trampling of standing dead material and breaking down of plant structural materials in the digestive system. Animals and the microorganisms in their rumens operate symbiotically to decompose what is often a low-quality food resource. The most successful herbivores in an ecosystem, over the long term, will be those whose *facilitation* of autotrophic productivity equals or exceeds their *inhibition* of the process.

Stoddart et al. (1975) are skeptical about the beneficial effects of animals in "manuring the range," especially in dry climates. On the other hand, Strugnell and Pigott (1978) judged that urine and feces increased productivity in East Africa: " . . . the grasslands are pale green, but patches affected by urine become dark green and grow more vigorously." Animals can redistribute nutrients in the landscape if their pattern of defecation differs from that of feeding; there is frequently concentration of feces around watering areas (Stoddart, et al., 1975).

The extent to which nutrient turnover increases productivity depends on other constraints on the process. The primary factor limiting productivity in most grasslands is water, not nutrients. Grasses support a large heterotrophic root system which is necessary to extract water from the soil. Many studies have demonstrated a close relationship between water supply and production in grasslands (Redmann, 1975; Coe, et al., 1976; Sims and Singh, 1978b).

Utilization of Primary Production by Large Herbivores

McNaughton (1976) observed that dense concentrations of wildebeest greatly reduced the green plant biomass before migration out of the Serengeti Plains of East Africa. The herbage removed totaled 85% of the initial standing crop. This heavy grazing prevented senescence and stimulated net primary productivity. Thomson's gazelle entered the areas of vigorous regrowth and high forage density that followed wildebeest grazing. The association of gazelle with wildebeest-grazed stands was maintained throughout the dry season of 1974. McNaughton concludes that the coexistence of wildebeest and gazelle "is a consequence of coevolution that has partitioned the grassland exploitation patterns during the critical dry season. Rather than competition, there is facilitation of energy flow into the gazelle population by the wildebeest population through the impact of the latter on the plant community."

Sinclair (1975) examined the utilization of herbage by ungulates, small mammals, and grasshoppers in two Serengeti grasslands. These groups made up most of the herbivore trophic level. Ungulates consumed 18.8% of the total annual grass production in the tall grassland; the other herbivores took 8.8% (Table 1). Of the balance, 19.2% went to the detritus-feeding organisms and 53.3% was removed by fire. In short grassland nearly 90% of the net shoot production went to the herbivores or saprophages.

Strugnell and Pigott (1978) used the differences in shoot production between grazed and ungrazed plots to estimate large herbivore consumption in

Table 1
Fate of Herbage Production in Two Serengeti Grasslands
(Sinclair, 1975)

	Tall Grassland		Short Grassland	
	$g\ m^{-2}\ y^{-1}$	% of Total	$g\ m^{-2}\ y^{-1}$	% of Total
Annual Grass Production	598	–	470	–
Ungulate off-take	112	18.8	160	34.0
Small mammal off-take	6.9	1.2	0.4	0.1
Grasshopper off-take	45.6	7.6	19.4	4.1
Removed by saprophages	115	19.2	232	49.5
Removed by fire	319	53.3	58.6	12.5

Rwenzori Park, Uganda. Results showed that one-third to one-half of the annual shoot production of this grassland went to the large herbivores. Another estimate of herbivore energy expenditure and biomass in the same area (Coe, et al., 1976) is much lower, indicating less herbivore pressure on the plant resource. Other factors, such as the extent to which herbivores consume browse plants (not included in Strugnell and Pigott's estimate), also affects estimates of plant utilization by herbivores.

Seasonal herbage production in 13 vegetation types in South Africa ranged from 0 g to 272 g m^{-2} (Hirst, 1975). Ungulate biomass was about 40 kg ha^{-1} in the wet season, rising to 65 kg ha^{-1} in the dry season. Ungulates consumed less than one-quarter of the net primary production during the wet season, but more than four-fifths over the full year. Woody plant production was not determined.

Coleman et al. (1976) partitioned the energy flow in a lightly grazed short grassland in Colorado (Table 2). In this system over 97% of the energy flow in heterotrophic organisms moved through the detritivores. The authors conclude that even in ecosystems exploited by man for grazing, the preponderance of energy flow follows detrital-saprophagic pathways. Only 15% of the net primary production remained above ground and available to large-mammal grazers. Many studies have shown that a large proportion of net primary production in grasslands is transferred below ground, as indicated by the root biomass and turnover in these ecosystems (Sims and Singh 1978a, 1978b).

These studies demonstrate that, in general, large amounts of primary production in grasslands go to the saprophagic organisms, not to the grazers. Any accumulation of above-ground dead plant material beyond that utilized by herbivores or saprovores is liable to be burned. Fire is an integral factor in grasslands and savannas (Daubenmire, 1968). Eltringham (1976) determined, for example, that over half of Rwenzori Park, Uganda, burns at least once in three years. Fire increases productivity in humid grasslands (Vogl, 1974), probably by removing excess amounts of dead plant material, but decreases plant productivity in semiarid grasslands (Redmann, 1978).

The data showing that most of the plant production in ecosystems is not consumed by herbivores have led to the hypothesis that herbivores

Table 2
Energy Flow in a Lightly Grazed Shortgrass Prairie in Colorado During a 154-Day Growing Season
Modified from Coleman et al. (1976)

Parameter	Energy Flow kcal m^{-2}
Net Primary Production	3452
Above-ground	517
Below-ground	2395
Heterotroph Energy Utilization (Above- and below-ground)	
Herbivores	
Arthropods and nematodes	53.7
Cattle	25.4
Carnivores	6.7
Saprophages	3021

are not limited by their food supply. This hypothesis precludes competition and niche separation among the dominant herbivores which, in turn, would affect species diversity, as discussed later in this paper. Sinclair (1975) points out a weakness in this hypothesis: most plant species are nutritious for only part of their life span. In the Serengeti, for example, there was a one- to four-month period during the dry season when green herbage was sufficiently low in abundance to limit herbivore populations. If only a short period of food-supply stress is sufficient to limit the herbivores, the total annual herbage production is irrelevant. "Trophic levels composed of small animals with high turnover rates and little time lag in their response to conditions will be limited by shorter periods of resource restriction than trophic levels composed of larger animals with greater population inertia . . . The temporary shortage of the food component of the primary production limiting the herbivores could be a feature common to all terrestrial vegetation-herbivore systems" (Sinclair, 1975). The limitation in food supply is not necessarily in terms of quantity, rather it is often a limitation in quality, as determined by protein content or digestibility.

Seasonal fluctuation in herbage biomass also is apparent in temperate grasslands. The tall-grass prairies of North America can produce large amounts of herbage at a rapid rate (Osage, Okla., Table 3). However, the seasonal duration of this live biomass production is short (Fig. 2). During the winter the quantity and quality of herbage is reduced in all the temperate grasslands. Plant biomass often is buried beneath a winter snow cover. In the bunchgrass ecosystems of the Northwest,

Table 3
Herbage Production in Grasslands and Savannas in North America and Africa

Location	Grassland Type Dominant Species	Annual Herbage Production and Year ($g\ m^{-2}\ y^{-1}$)	Maximum Productivity ($g\ m^{-2}\ d^{-1}$)	Reference
ALE, Wash., USA	Northwest bunchgrass *Agropyron spicatum* *Poa secunda* *Artemisia tridentata*	82 (1971) 114 (1972)	1.2	Sims and Singh (1978a)
BISON, Mont., USA	Mountain grassland *Festuca scabrella* *F. idahoensis* *Lupinus sericeus*	272 (1970)	4.5	Sims and Singh (1978a)
MATADOR, Sask., Canada	Mixed grassland *Agropyron dasystachyum* *A. smithii* *Artemisia frigida*	191 (1970) 135 (1968–72)	3.7	Coupland (1974)
OSAGE, Okla., USA	Tall grass prairie *Andropogon scoparius* *A. gerardi* *Panicum virgatum* *Sorghastrum nutans*	331 (1970) 416 (1971) 290 (1972)	6.2	Sims and Singh (1978a)
RWENZORI, Uganda	Savanna grassland *Themeda triandra* *Hyparrhenia filipendula*	553 (1969–70)	3.5	Strugnell and Pigott (1978)
SERENGETI, Tanzania	Savanna grassland *Themeda triandra* *Digitaria macroblephora*	598 (1966–72) 470 (1966–72)	3.9	Sinclair (1975)

snow cover can be less, and shrubs such as *Artemisia tridentata* provide a year-round food source for large-mammal herbivores. Low shrubs like *Artemisia frigida,* which occur in several temperate grassland types, are important in the winter diet of pronghorn (Mitchell and Smoliak, 1971).

Biomass of Large Herbivores

The greatest biomass of large herbivores has been recorded in the area of Rwenzori Park, Uganda, near the ecotone between the Congo forests and the plains of East Africa (Bourliere and Hadley, 1970). Estimates vary, but a biomass over 200 kg ha^{-1} (Table 4) appears to be conservative. Eltringham (1974) reported the biomass of large mammals in Rwenzori to be 336 kg ha^{-1}, up to three-quarters of which was made up of elephant and hippopotamus.

Biomass of large herbivores in temperate regions also can be high (Table 4). The optimum biomass in a boreal aspen forest in Alberta is

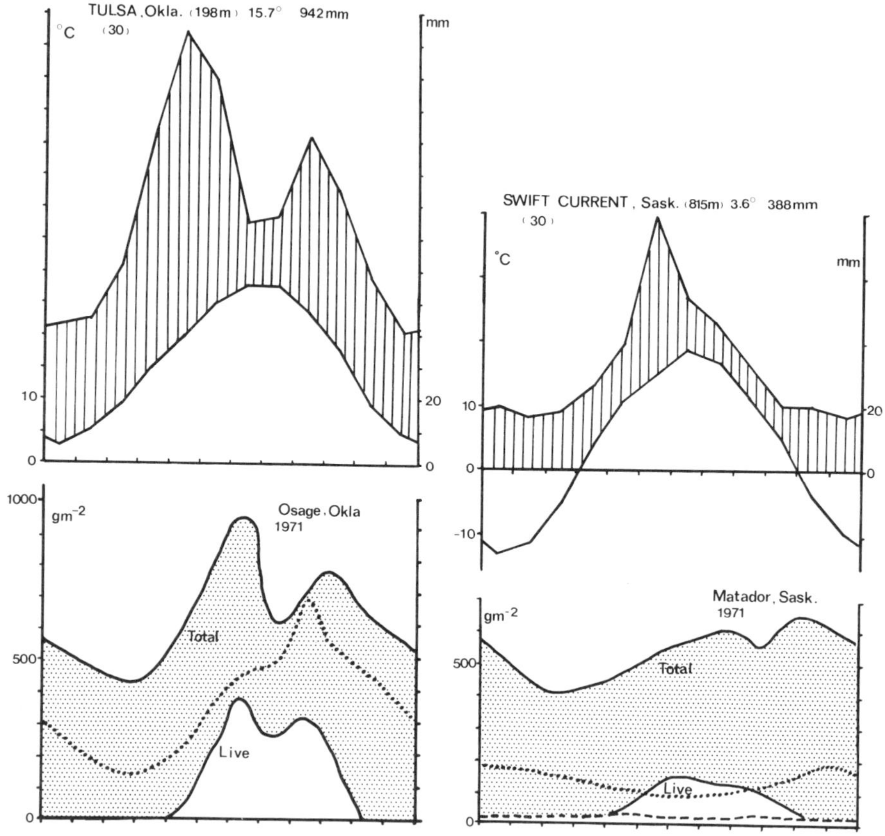

Figure 2. Seasonal pattern of standing crop biomass (grams per square meter) of primary producers in grasslands (lower portion of each graph). Live is separated from total biomass (live plus dead). Dashed lines for Ale, Wash., and Matador, Sask., are *Artemisia* species. Dotted lines are for total biomass on areas grazed up to the year prior to measurement (except Matador, Sask., which was estimated). Smooth curves were fitted by eye, and are meant to represent general trends and magnitudes. Osage, Ale, and Bison Range from Sims and Singh (1978a); Matador from Coupland (1974). Top section is a Walter and Lieth-type climate diagram for stations near the study sites *(Figure continues on next page)*

estimated at 57 kg ha^{-1} (Telfer and Scotter, 1975). The maximum density, with some winter feeding, was 98 kg ha^{-1}, which is as high as any African area except western Uganda. Biomass of saiga antelope in semidesert habitat of the USSR is 52 kg ha^{-1} (Table 4), "showing that highly adapted species in simple landscapes . . . are able to utilize food efficiently and produce considerable biomass" (Bannikov, 1967). Bannikov cautions that errors of census-taking in national parks are caused by protection of the animals, e.g., by reduction of predators, resulting in artificially high populations. This criticism probably applies equally to several other studies reporting high populations of large herbivores. That some of these ecosystems cannot sustain large herbivore biomasses over the long term is indicated by the reported degradation of the vegetation resulting from overgrazing.

Study of the biomass of large-herbivore communities in North American grassland is difficult because the fauna was altered drastically after colonization by European man. Historical accounts indicate that immense numbers of bison inhabited the pristine grasslands (Roe, 1970; Dary, 1974). Robert M. Wright, a buffalo hunter, recalled sitting in on a discussion about how many buffalo there were in the vicinity of Fort Dodge, Kansas:

> "Taking a strip of land fifty miles east and fifty miles west, they had first made it ten billion buffalo. But General Sheridan said, 'That won't do.' They figured it again and made it one billion. Finally they reached the conclusion that there must be one hundred million buffalo. But you know, they were afraid to give out these figures. They didn't think anybody would believe them. Nevertheless, *they* believed them" (Dary, 1974, p. 27).

Ernest Thompson Seton probably came up with a considerably more accurate estimate of bison populations: 70 million bison on an original range of 2 million square miles of prairie and plains (Seton 1910, Life Histories of Northern Animals). From these figures, the average year-round stocking rate would be 7.4 ha per animal, which is essentially identical to the carrying capacity for cattle on the most productive western Canadian rangeland in excellent condition (Lodge, et al., 1971). Assuming an average bison weighs 450 kg, the overall biomass was about 60 kg ha^{-1} for the

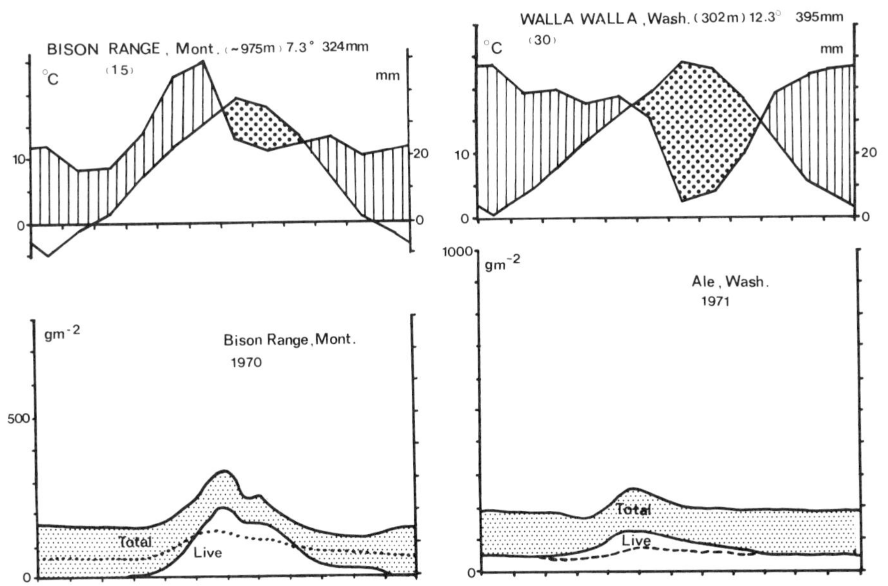

grassland as a whole. Petrides (1956) gives the biomass of cattle on overgrazed rangeland in Kansas as 131 kg ha^{-1}.

Geist (1978) calculates the standing crop biomass of large mammals in the St. Elias mountain region of the Yukon as 0.89 kg ha^{-1}, which he considers relatively high for a tundra region. "If the more productive lowland sites were included, and horses, elk, muskoxen and bison were reintroduced and allowed to stabilize their populations, the standing crop would probably be close to 1.75 kg ha^{-1} in a climatic regime similar to that found today . . . [this] periglacial ecosystem appears more productive by far than the tundra at

Table 4
Biomass Densities and Diversity in Large-Herbivore Communities

Location	Number of Species	Census Area (km^2)	Biomass Density (kg ha^{-1})	Dominance Index*	Reference
		Tropical Savannas, Grassland and Woodland			
(1) Uganda	9	600	235	0.34	Bourliere and Verschuren (1960) in Bourliers (1963)
(2) Tanzania	12	20000	201	0.45	Sinclair (1975)
(3) Uganda	7	49	189	0.25	Field and Laws (1970)
(4) Uganda	6	1670	187	0.38	Bere (1960) in Bourliere (1963)
(5) Kenya	17	116	131	0.29	Ellis in Bourliere (1963)
(6) Rhodesia	11	76	56.9	0.18	Taylor and Walker (1978)
(7) Uganda	6	26	51.4	0.51	Field and Laws (1970)
(8) Rhodesia	11	125	43.2	0.25	Dasmann and Mossmann (1961) in Bourliere (1963)
(9) Kenya	12	4150	42.7	0.51	Leuthold and Leuthold (1976)
(10) Sri Lanka	6	1865	7.98	0.41	McKay (1973)
(11) Sri Lanka	6	580	7.11	0.28	Eisenberg and Lockhart (1972)
		Temperate Grasslands and Fbrests			
(12) S. Dakota, USA	4	106	36.4	0.46	Drury (1950, Calahane (1952) in Petrides (1956)
(13) Wyoming, USA	6	75	33.6	0.49	Primary data in Bourliere (1963)
(14) Alberta, Canada	5	130	25.3	0.36	Flook (1970)
(15) Montana, USA	4	75	22.3	0.69	Ruth (1939) in Petrides (1956)
(16) Nebraska, USA	4	78	9.19	0.66	Ruth (1939) in Petrides (1956)
		Arctic and Alpine Tundra			
(17) Finland	2	–	4.30	0.87	Kallio (1975)
(18) Alaska, USA	3	31400	2.98	0.94	Guthrie (1968a)
(19) Norway	3	–	2.33	0.60	Ostbye (1975)
(20) Siberia, USSR	1	–	0.81	1.0	Vassiljevskaja, et al. (1975)
(21) Yukon, Canada	6	7160	0.80	0.44	Geist (1978)
(22) NWT, Canada	1	5600	0.36	1.0	Parker (1975)

*Simpson's Dominance Index (D) based on biomass of each large herbivore species in sample area (b) and total biomass (B): $D = \Sigma\ (\frac{b}{B})^2$. See Whittaker (1975)

comparable latitudes and altitudes with similar weather . . . not only quality, but quantity, of plant food available to ungulates [is] very great" (Geist, 1978, p. 204).

Large, slow-growing animals have lower food energy requirements per unit of biomass than smaller animals. Daily herbage consumption (C) is related to body weight (W) as follows (Sinclair, 1975):

$$C = 0.09\ W^{0.75}$$

For example, a pronghorn (46 kg) would consume 1.59 kg d^{-1} of herbage; a bison (450 kg) would consume 8.79 kg d^{-1}. With an annual herbage production of 2500 kg ha^{-1}, of which the recommended 50% is consumed, 2.15 pronghorns or 0.30 bison could be maintained per hectare. The total biomass of pronghorn would be 99 kg ha^{-1}; that of bison 191 kg ha^{-1}. Maintenance of a herd of pronghorns would require 1.3 times as much plant production as a herd of bison of equal biomass.

This type of analysis demonstrates that the larger the herbivore, the greater the biomass which can be supported with an equivalent impact on the vegetation. Actual data from Africa support this point. Taylor and Walker (1978) compared herbivore biomass and vegetation use by wild animals and cattle in different sections of a ranch in Rhodesia. The total biomass of the 10 wild large-herbivore species was approximately the same as the cattle. Field surveys showed that both herbaceous vegetation and browse were more heavily utilized in the game section: "Overall, 'range condition' of the game section is poorer than that of the cattle section." I calculated herbage consumption by the game and cattle using the equation above relating consumption to weight. The larger number of smaller animals in the game section was estimated to consume 21% more herbage than the cattle, even though biomass was the same on both sections. This could explain why the vegetation was more heavily utilized in the game section.

DIVERSITY IN BIOTIC COMMUNITIES

Diversity and Niche Separation

Contrasting diversities among biotic communities have interested ecologists for decades, but there is still uncertainty about the concept of diversity, how it is to be measured and what it is dependent upon. The number of species in a sample area is one measure of diversity termed species *richness* (Peet, 1974). Another aspect of diversity involves the concept of *evenness:* a community with 10 equally abundant species is more diverse than another community, also with 10 species, but with one species comprising 91% of all individuals. Diversity may apply to a single plant community or a whole landscape. In this paper I usually refer to landscape diversity, particularly in the case of large-herbivore communities which often range over large portions of the region and exploit several different plant communities.

The concept of the ecological niche refers to a species' position in a community in relation to other species, or the way a species population is specialized within a community (Whittaker, 1975). The idea that two species cannot coexist permanently in the same niche is known as the principle of competitive exclusion. Whittaker (1975) summarizes the idea in the form of three statements: "(1) If two species occupy the same niche in the same stable community, one will become extinct. (2) No two species observed in a stable community are direct competitors limited by the same resources; the species differ in niche in ways that reduce competition between them. (3) The community is a system of interacting, niche-differentiated species populations that tend to complement one another, rather than directly competing, in their uses of the community's space, time, resources, and possible kinds of interactions."

As species are added along a resource gradient, the niche "widths" of the species already there are reduced. Whittaker (1975) concludes: "(1) Species evolve toward specialization for different parts of a resource gradient, by which competition among species is reduced, and (2) given time, species evolve toward even spacing of their adaptive centers along the gradient. (3) Niche sequences can be lengthened by fitting in new species between old ones, while reducing the niche widths of the latter, but (4) there are probably limits on the number of species that can be 'packed' into the sequence along a given resource gradient." The more species that can be "packed" along each resource gradient in a community, the greater the species diversity in that community.

Resources such as food, water, or light usually are not divided evenly among the species in a community. The extent to which species monopolize resources is an expression of their dominance in the community. General mathematical hypotheses have been proposed to explain dominance-diversity relationships (see Whittaker, 1975, p. 89).

Evolution of diversity in plants makes possible evolution of diversity in animals, and vice versa. Grazing animals may increase plant diversity by preventing any one plant species from becoming too dominant. In this way, interacting groups of species can facilitate a general increase in diversity, or, according to Whittaker (1975), "diversity begets diversity". Species diversity is a "self-augmenting evolutionary phenomenon."

A large body of ecological literature is devoted to explaining the bases for the wide range of diversity in natural communities, particularly the decrease in diversity along the latitudinal gradient from tropical to polar latitudes. Simpson (1964) described the increase in species diversity of mammals from north to south in North America: There were 13 species per unit sampling area on Melville Peninsula (lat. 70°) and 163 species per quadrant in Costa Rica (lat. 10°). Simpson reviewed several possible explanations for the latitudinal differences and concluded that "The most important single element . . . is probably the frequent occurrence of killing winter frosts . . . that has many concomitants and effects, such as the fact that more temperate than tropical angiosperms are annual or deciduous and that there is *great reduction in diversity and quantity of winter food*" (italics added). In the tropics the great diversity of plants provides a greater variety of foods for the animals, available for longer periods, and permits more specialization in food habits.

Huston (1979) also reviewed explanations of differences in diversity, including the environmental stability, predictability and productivity hypotheses. All three of these parameters increase from temperate to tropical regions. Unpredictable and severe environments could have low diversity because unpredictability forces organisms to have broader niches which allow fewer species to be "packed in". Also, there is a greater chance of extinction for marginal populations (Pielou, 1975a, 1975b). Yet Huston (1979) cites examples of low-diversity communities in stable environments (e.g., coastal redwood forest and freshwater marshes) and communities with high diversity in environments which appear to be severe and unpredictable (e.g., Sonoran desert). Sclerophyllous shrub communities may have a higher diversity than some tropical rain forests (Richards, 1969, cited in Huston, 1979). Studies relating diversity to productivity also are equivocal.

"A General Hypothesis of Species Diversity"

The title of this section and the ideas which follow were taken from a paper by Huston (1979), who bases his approach to the problem of species diversity on nonequilibrium interactions of competing populations. In this case, differences in diversity are strongly influenced by variations among communities in the rates of competitive displacement. Factors such as relative competitive abilities and niche partitioning may not be particularly important. It should be emphasized that this approach deals with the maintenance rather than the generation of diversity.

Huston strongly criticizes the heavy dependence of current ecological theory on the concept of competitive exclusion and the assumption that communities exist at competitive equilibrium. Competitive exclusion can be observed in the laboratory when competition occurs in a uniform environment for the same limiting resource at the same time. In nature, fluctuations in the physical environment, predation, etc., are constantly changing population sizes and the nature of competitive interactions. Huston cites "growing" evidence that natural systems are rarely, if ever, at competitive equilibrium.

Huston's hypothesis is based on differences in the *rates* at which populations of competing species approach competitive equilibrium. The reduction or exclusion of some species at equilibrium reduces species diversity, but "when competitive equilibrium is prevented, a dynamic balance may be established between the rate of competitive displacement and the frequency of population reduction, which results in a stable level of diversity."

Huston uses the Lotka-Volterra competition equations in his model, but examines their dynamic, rather than equilibrium behavior. Under nonequilibrium conditions, variations in the carrying capacity (K) and competition coefficient (∝) have much less effect than under equilibrium conditions. Determination of ∝ for use in equilibrium solutions is difficult, and besides, "constant competitive coefficients may well be meaningless except in uniform stable environments which never exist in nature." Species may compete for many different resources, but do not compete at all times for all possible resources. There may be reversals of competitive dominance, when environmental conditions change. In short, equilibrium models of competition have not been very successful when applied to natural communities.

If competitive equilibrium is prevented by any disturbance which reduces population sizes, the outcome of competition may be completely different from that predicted by the equilibrium solution. The nonequilibrium model shows that maintenance of diversity is strongly related to interactions between: (1) frequency of population reduction (Fig. 3a) and (2) rate of competitive displacement, which is mainly controlled by population growth rates (Fig. 3b). At a low rate of competitive displacement, diversity will be low at very low frequencies of disturbance. This will allow a time period sufficient for equilibrium to be approached. Increases in the frequency sufficient to prevent competitive equilibrium will permit maximum diversity. A still higher frequency of disturbance will result in the inability of some competitors to recover, and result in lower diversity.

When the rate of competitive displacement is high, diversity will be low at low frequencies of population reduction due to the rapid approach to competitive equilibrium. Increasing frequency of disturbance increases diversity (the community is kept further from equilibrium) until the frequency becomes sufficiently high that some populations cannot recover. The hypothesized interactions between displacement and disturbance are summarized in Figure 4.

Huston's model is general, and is successful in explaining the patterns of diversity in the cases where previous hypotheses have failed. For example, desert and sclerophyll shrub communities have low productivities, suggesting low rates of competitive displacement, and are exposed to periodic disturbance, e.g., drought or fire. Huston's hypothesis predicts relatively high diversity in these communities, other hypotheses incorrectly predict low diversity. Low-diversity communities in stable environments, such as the coastal redwood forests, have high productivities that suggest rapid population growth rates. These forests exist for long periods of time with little or no disturbance. Huston's model predicts low diversity, whereas the productivity, stability, and predictability hypotheses incorrectly predict high diversity.

A nonlinear relationship between diversity and productivity is predicted by Huston's "dynamic equilibrium" hypothesis; high productivity generally reflects high population growth rates, and low diversity. Lack of diversity might also result from low productivity, e.g., nutrient deficiency; in this

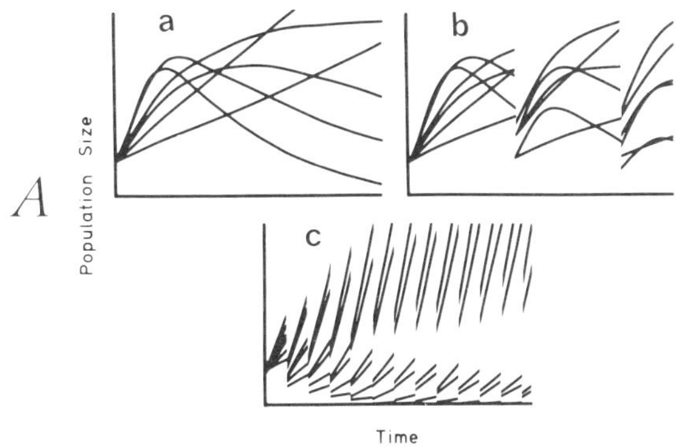

Figure 3a. Effect of frequency of population reduction on maintenance of diversity, based on computer simulations of Lotka-Volterra competition equations for six species. Values for r's, K's and ∝'s are identical in all three simulations; all population reductions are 0.5. (a) No reduction; diversity is reduced as the system approaches competitive equilibrium. (b) Periodic reductions; diversity is maintained for longer time than in a. (c) High frequency of reductions (5 times greater than b); diversity is reduced as low r populations are unable to recover

Figure 3b. Effect of growth rate on maintenance of diversity under nonequilibrium conditions. All parameters are identical in both simulations, except that all r values are doubled in d. Diversity is reduced more rapidly at higher r values. Reproduced from Huston (1979)

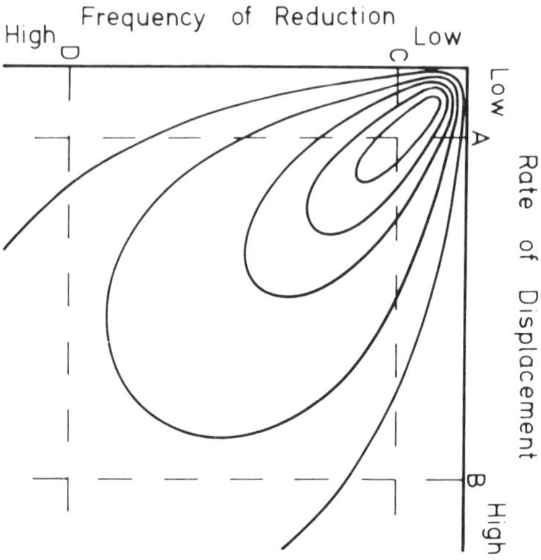

Figure 4. Generalized contour map of the dynamic equilibria between the rate of competitive displacement and the frequency (or magnitude) of population reduction; diversity is on the axis perpendicular to the page, and is represented by the contour lines, with the highest diversity within the inner ellipsoid (in the upper left corner). The dotted-line transects demonstrate the predicted changes in diversity when one parameter is held constant and the other varied. Reproduced from Huston (1979)

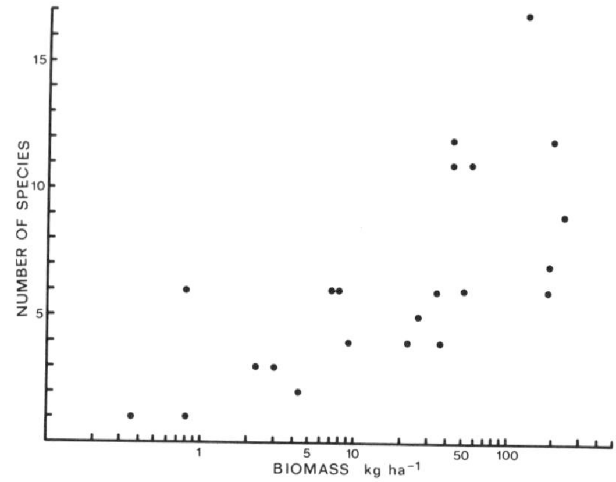

Figure 5. Relationship between biomass and species richness in large herbivore communities. The data points are described in Table 4

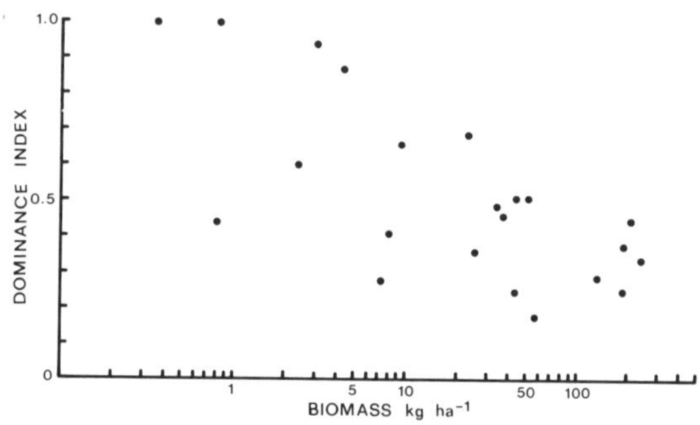

Figure 6. Relationship between Simpson's Dominance Concentration Index and biomass in large herbivore communities. The data points are described in Table 4

case additions of nutrients will increase productivity and result in higher diversity. The latter case may be especially pertinent for steppe-tundra communities. Slightly increased nutrient inputs and decreased frequencies of disturbance could have increased the diversity. Even slightly ameliorated environmental conditions, relative to present-day tundra, could promote increased community diversity. Also, a high diversity of ungulates in Beringian communities might reflect a lack of competition among species due to nonequilibrium conditions.

Diversity in Ungulate Communities

Huston's general hypothesis predicts that up to some optimum level diversity increases with increasing productivity. Above the optimum, diversity decreases. To test this, published data on species diversity and biomass of large herbivore communities were assembled with the assistance of J. V. Matthews (Table 4). Despite the differences in census methods, and the probable errors in biomass estimates, the results (Fig. 5) suggest that there is a general relationship between biomass (an indicator of productivity) and species richness. Dominance concentration is inversely related to biomass (Fig. 6).

There is a large variation in both components of diversity, richness (number of species) and evenness (as indicated by the dominance index) at any one biomass value. Relatively high diversity is possible in tundra (Table 4, Community 21), and low diversity in tropical savanna (Table 4, Community 7).

A trend toward reduced diversity in ungulate communities at higher latitudes can be shown using dominance-diversity curves (Fig. 7). The lognormal distribution for herbivores in Kenyan savanna is characteristic of rich communities in a diverse landscape, as discussed by Whittaker (1975). Finnish tundra, on the other extreme, is an example of a geometric series based on strong niche preemption by one dominant herbivore. The two other sites, in grassland and tundra, are intermediate.

High diversity can result when a mosaic of different habitats occurs in the sampling area. This habitat-based diversity ("gamma" diversity of Whittaker, 1975) should be distinguished from diversity arising from niche divergence within a single habitat type ("alpha" diversity). Competition between closely related species exploiting a particular vegetation type (e.g., several antelope species in a tropical savanna) is more likely than competition between more distantly related species in distinct habitats (e.g., moose and bison in the boreal forest fringe).

A peculiar aspect of Pleistocene ecology is the apparent mixing of species that now are in distinct ecological zones, for example horse and caribou. The special environmental conditions near present-day glaciers can result in peculiar faunal mixtures similar to those from the Pleistocene (Geist, 1978, p. 201). Geist describes periglacial ecosystems as "pulse-stabilized" systems maintained in early stages of plant succession by climatic and edaphic factors, implying frequent disturbance. According to Huston's nonequilibrium theory, such conditions should maintain diversity.

Niche separation in temperate ungulates based on food preferences may be less pronounced than for tropical species. "In complex, particularly mountainous landscapes of northern and moderate latitudes, as distinct from the tropics, various species make use not so much of different layers of vegetation and different groups of plants, as of various biotopes [communities] which make up

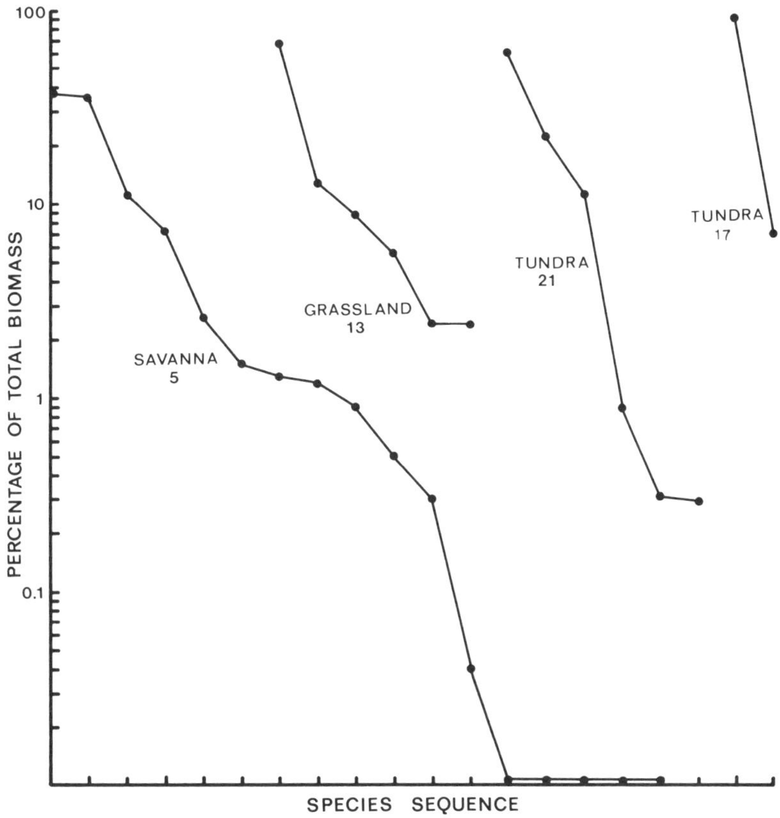

Figure 7. Dominance-diversity curves for a range of ungulate communities. Details are given in Table 4 (Communities Nos. 5, 13, 17, 21). Dominance is expressed as a species fraction of the total biomass in the community

these landscapes" (Bannikov, 1967). Different species living in the same habitat tend to compete: increase in red deer in Bialowieza Forest results in decreased moose populations. In complex topography moose occupy valleys and red deer occupy slopes. The biomass of one species of ungulate in simple landscapes (level taiga) is approximately equal to the biomass of several species of ungulates in complex landscapes such as mountain taiga (Bannikov, 1967).

The niches of bison and pronghorn within the grassland were separated, in part at least, by the preference of the latter for forbs and shrubs rather than grasses. My own examination of historical photographs shows the low shrub *Artemisia frigida* to be very abundant in a grassland of southeastern Saskatchewan in 1875 (Fig. 8), prior to intensive use of this region as rangeland. This species is recognized now as an indicator of heavily grazed, "deteriorated" rangeland in the northern Great Plains and in Canada (Forest Service, 1937). *Artemisia* species are important constituents of the diet of antelope (Mitchell and Smoliak, 1971). England and De Vos (1969) have proposed that antelope were dependent upon habitat resulting from heavy grazing by bison. This represents a niche separation similar to that proposed for wildebeest and Thomson's gazelle in Africa (McNaughton, 1976, and others). There is little reference to excessive slaughter of antelope, as in the case of bison . . . "Therefore, if the antelope were as numerous as estimated, and if they were not slaughtered to the same extent as bison, why did they decrease so greatly? Perhaps it was due to the reduction in suitable habitat brought about by the extermination of bison" (England and De Vos, 1969). Such interdependence of large herbivore species may also have significance in relation to the extinction of Pleistocene grazers.

The North American grasslands and savannas have the potential to support a relatively large biomass of large herbivores (Table 4), but diversity is lower than in tropical grazing lands. Only seven species—bison, pronghorn antelope, elk, mule deer, whitetail deer, moose and bighorn sheep—occupied the grassland and its margins in North America (England and De Vos, 1969). Mountainous areas tend to have the highest diversity of mammal species (Simpson, 1964). The ranges of nine ungulates overlap in the Rocky Mountains and foothills of Montana and Alberta: bison, pronghorn antelope, mule deer, whitetail deer, elk, moose, bighorn sheep, woodland caribou and mountain goat (Burt and Grossenheider, 1976).

Figure 8. North American Boundary Survey camp on North Antler River (Southeastern Saskatchewan) in 1873. *Artemisia frigida* is very abundant, probably indicating a history of heavy grazing. (Photograph C 73303 from the Public Archives of Canada)

Geist (1978) describes the mammal diversity of the St. Elias range in southwestern Yukon as "remarkable". Included among the 47 species of mammals are Dall's sheep, mountain goat, caribou, moose, mule deer, blacktailed deer and feral horse. The ratio of large herbivores to small herbivores in the St. Elias Range is similar to that of the Great Plains prior to European settlement and to that of the Eurasian Steppes (Gulday, 1967, cited in Geist, 1978). Geist indicates that at the end of the Pleistocene the regional fauna could have included musk oxen, modern bison, horses and elk, and perhaps also forest musk oxen, *Cervalces, Camelops,* longhorned bison, yak, mammoth, mastodon and saiga antelope. The late Pleistocene fauna in the region might have had almost as many species of large mammals as small mammals, a situation found today only in some African ecosystems. Several observations are given suggesting that ". . . the periglacial ecology of the St. Elias Range appears to be pertinent to an interpretation and understanding of the dry periglacial ecosystem of Würm and Wisconsin times" (Geist, 1978, p. 199).

A MODEL OF PRODUCTION AND DIVERSITY IN LARGE-HERBIVORE COMMUNITIES

The biomass and density of large herbivores supported in a hypothetical ecosystem can be modeled using the concepts discussed in previous sections of this chapter. The structure and assumptions of the model are as follows:

The annual above-ground primary production (herbage) is set at 150 g m^{-2}, which is similar to dry cold-temperate grasslands (Table 3). Herbivore biomass and density at other values of herbage production may be estimated assuming a linear relationship; i.e., halving plant production will also halve herbivore production. On average the grazers consume 50% of the herbage production; the system is stable over the long term.

Annual herbage consumption (C) by the grazers is related to body weight (W) as follows:

$$C = 32.85\ W^{0.75}$$

All other nutritional requirements (e.g., protein, minerals, vitamins) are satisfied. Biomass is determined only by food energy supply, which is available throughout the year. The five herbivore species range from very large (1500 kg) to small (50 kg). Paleoecologists may wish to think of this as the range from mammoth to saiga.

The herbage resource is appropriated by each species according to its dominance in the community, which may be strong, weak or nil (Fig. 9). Five cases are described based on degree of dominance and size of the dominant species. The biomass of each herbivore is determined by dividing the herbage appropriated by the herbage required to support the species (C).

The absolute values of biomass emerging from the model are less important to us than the comparisons among the five cases (Table 5). The highest total herbivore biomass is found in the community strongly dominated by the large species (Case 1). The lowest biomass occurs where the community is dominated by small herbivore species (Case 2). In both cases the biomass is concentrated in the dominant species (biomass fraction in Table 5), which reduces the evenness component of diversity in these hypothetical communities. The dominance concentration indices are 0.72 and 0.57 for Cases 1 and 2 respectively. These fall in the range of the tundra and temperate grassland communities discussed previously (Table 4).

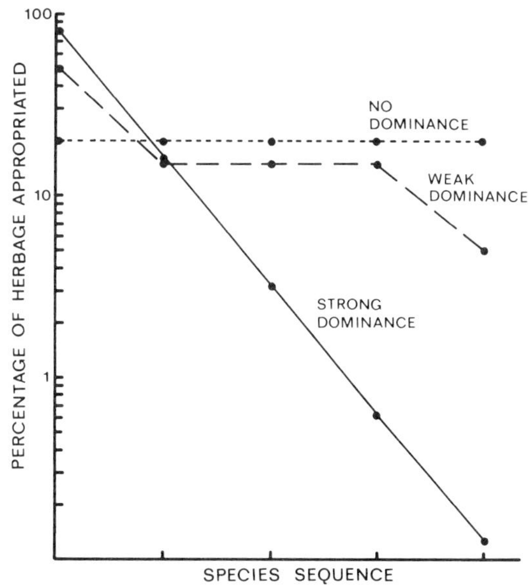

Figure 9. Hypothetical dominance-diversity relationships used in production-diversity model. Dominance is expressed as the proportion of the total herbage resource which each grazer appropriates

Table 5
Output from Model of Biomass-Diversity Relationships in Hypothetical Ungulate Communities

Species	Herbage Appropriated: Fraction	Herbage Appropriated: Weight (kg)	Animal Weight (kg)	Biomass Density (kg ha^{-1})	Biomass Fraction	Density (km^{-2})
Case No. 1 Strong Dominance by Large Species						
1	.80	600	1500	114	0.84	7.6
2	.16	120	550	17.7	0.13	3.2
3	.032	24	400	3.27	0.024	0.82
4	.006	4.5	250	.54	0.004	0.22
5	.0013	.975	50	.08	0.0006	0.16
			Total	135	0.72*	
Case No. 2 Strong Dominance by Small Species						
5	.80	600	50	48.6	0.72	97
4	.16	120	250	14.5	0.22	5.8
3	.032	24	400	3.3	0.049	0.82
2	.006	4.5	550	.66	0.009	0.12
1	.0013	0.975	1500	.18	0.003	0.01
			Total	67.2	0.57*	
Case No. 3 Weak Dominance by Large Species						
1	.50	375	1500	71	0.59	4.7
2	.15	112.5	550	16.6	0.14	3.0
3	.15	112.5	400	15.3	0.13	3.8
4	.15	112.5	250	13.6	0.11	5.5
5	.05	37.5	50	3.04	0.025	6.1
			Total	120	0.40*	
Case No. 4 Weak Dominance by Small Species						
5	.50	375	50	30.4	0.37	61
4	.15	112.5	250	13.6	0.16	3.0
3	.15	112.5	400	15.3	0.18	3.8
2	.15	112.5	550	16.6	0.20	5.5
1	.05	37.5	1500	7.10	0.086	0.47
			Total	83.0	0.24*	
Case No. 5 No Dominance						
1	.20	150	1500	28.4	0.28	1.9
2	.20	150	550	22.5	0.22	4.1
3	.20	150	400	20.4	0.20	5.1
4	.20	150	250	18.2	0.18	7.3
5	.20	150	50	12.1	0.12	24
			Total	102	0.21*	

*Dominance Index, see Table 4 for method of calculation. The biomass fraction is $(\frac{b}{B})$

Hypothetical total biomasses for communities with weak dominance are 120 kg ha^{-1} (large-species dominant) and 83 kg ha^{-1} (small-species dominant). Dominance indices are 0.40 and 0.24, respectively. The evenness is greatest in Case 5, the community with no dominance (0.22). Although the herbage is appropriated evenly by all 5 species, the differences in metabolic energy requirements result in a greater concentration of biomass in the larger species.

Strong dominance and a geometric decrease in herbage available to the remaining species could cause decreased species richness as well as decreased evenness. The small population size predicted for species 5 in Case 1 could make such a species vulnerable to extinction in a region where the dominants appropriate most of the food resource. Competition would either force extinction, or result in niche divergence, as described earlier. The absence of competition, due to either niche separation or frequent population reductions would produce more even appropriation of resources and greater richness and evenness in the community. Diversity in the vegetation could militate against dominance and result in increased herbivore diversity.

The results of this analysis agree reasonably well with observations on natural communities of ungulates. In relatively simple vegetation with severe environments and small numbers of herbivore species (e.g., modern tundra), dominance is strongly concentrated, as in Cases 1 and 2. In less homogeneous conditions, where a wider range of environmental conditions and communities are encountered (e.g. mountainous and periglacial regions, savanna and woodlands) dominance is weaker. The modeling exercise helps set limits on the production and diversity of ecosystems, but the challenge for the paleoecologist remains to determine where Berginian steppe-tundra fits in the range of production-diversity relationships discussed here.

14

PRESENT-DAY ARCTIC VEGETATION AND ECOSYSTEMS AS A PREDICTIVE TOOL FOR THE ARCTIC-STEPPE MAMMOTH BIOME

L. C. Bliss and James H. Richards

Present-day low- and high-arctic vegetation and plant production are described in addition to herbivore density and vegetation utilization. For the more important arctic rangelands, above-ground net annual production ranges from 70 g to 200 g m^{-2} (cryptogams comprise 30-50%). Large and small herbivores utilize only 1-2% of the vegetation in modern arctic ecosystems. Based upon food consumption rates of Alaskan and Canadian Inuit, the land and sea can support on a sustained-yield basis only 20% of the present coastal human population in Alaska and 7% in the Canadian western Arctic.

An ecosystem model is presented for Beringia 25,000 to 11,000 BP based upon the dominance of four major plant communities and six large herbivores. Assuming daily forage requirements of the herbivores, net annual plant production and forage harvest rates for a sustained yield, an estimated herbivore biomass and annual production is calculated as well as a human population density (0.02 - .03 km^{-2}). It is believed that only a diverse vegetation assemblage could have supported a diversity of herbivores and thus a human population. It is postulated that the earlier concept of an upland sedge-grass steppe-mammoth-man ecosystem could not have maintained a diversity of herbivores and carnivores without a diversity of plant communities.

Key words: vegetation, plant production, Beringian ecosystem, human populations, animal production

INTRODUCTION

The concept of an arctic steppe-mammoth biome with a diversity of small and large herbivores and carnivores is difficult to imagine in light of the modern arctic and northern boreal forest biomes and their rather limited faunas. A diverse fauna implies a diversity of plant communities and a diversity of feeding strategies, e.g., browsers vs. grazers, utilization of different species and probably the utilization of different plant parts of the same species by different herbivores (utilization of mature vs. young leaves or leaves vs. leaf-shoot systems).

It is not enough to ask what kinds of plant communities might have been present; rather we must ask what was the net primary production of a diversity of plant communities and what was their potential carrying capacity for a diversity of herbivores and a human population. Only partial answers can be given and those based only upon the modern analogs of plant communities, herbivore feeding habits and range utilization, and human food requirements.

The objectives of this paper are to: (1) set forth

ISBN 0-12-355860-3

the pattern of modern low- and high-arctic vegetation and its production; (2) describe the energy flow to the herbivores and human residents of present-day ecosystems; and (3) to speculate on energy flow within inferred ecosystems, including ones that involve man, 25,000 to 11,000 years BP, when it is assumed that the climate was warmer and drier in summer than at present.

MODERN VEGETATION PATTERNS

Open Woodland and Forest-Tundra

The open woodlands or lichen/shrub woodlands and forest-tundra of the taiga are important habitat for large herbivores in Beringia. These lands with their open-grown *Picea mariana* on shallow-thaw soils and *Picea glauca* on soils with a deeper active layer predominate in Alaska. *Picea glauca* extends farther north and west than *P. mariana* in contrast to most of northern Canada, where *P. mariana* dominates at and near treeline (Viereck, 1975; Hustich, 1966). In eastern Siberia *Larix dahurica* (*L. cajanderi*) dominates to the north with *Pinus pumila* on mountain slopes and *Chosenia macrolepis* and *Populus suaveolens* in valleys (Hustich, 1966; Tikhomirov, 1970). In the mountains of Kamchatka, *Betula ermani* dominates at 700-850 m. In places, *Larix dahurica* and *Picea jezoensis* grow to a height of 8-16 m (Gorchakovsky and Shiyatov, 1978).

As in the closed boreal forest, fire plays a major role within these lichen woodlands and forest-tundras (Rowe and Scotter, 1973; Viereck, 1973; Black and Bliss, 1978). From 1940 to 1969 the average annual burned area in Alaska was 377,000 ha. Most areas have burned at intervals no longer than 200-250 years (Barney, 1971). Conifer forest (mostly black spruce) and treeless vegetation (grasslands, mires, and tundra) accounted for 79% of the burned lands. In the Yukon Territory from 1962-73 an average 102,000 ha burned per year, while an average 185,000 ha burned per year from 1950-73 in the Mackenzie Valley region, N.W.T. (see Hernandez, 1974). The frequency of fire in black-spruce forests and open woodland results from the abundance of ericaceous shrubs, low branches of spruce frequently covered with inflammable lichens, and a thick carpet of lichens and moss on the forest floor (Viereck, 1975). Fire in the open woodland has no long-term effect on the species of vascular plants, but there are significant successional changes in the cryptogamic species (Black and Bliss, 1978). From 15 or 20 years to 120 years after burning, *Cladonia* spp. and *Peltigera apthosa* dominate with increasing amounts of *Aulacomnium palustre, A. turgidum, Hylocomium splendens* and *Peltigera canina.* From 120 to 200 years *Cladonia mitis* and the mosses *Dicranum angustum* and *D. undulatum* dominate. Stands 200- to 300-years-old are dominated by a ground cover of *Cladonia alpestris, C. mitis,* and *C. rangiferina.*

Fire also has a major effect on *Picea mariana* by altering the thermal regime of the soils and increasing the depth of the active layer (Rouse and Kershaw, 1971; Mackay, 1970); by providing a mineral-soil seed bed (Ahlgren, 1959); and by opening the semiserotinous cones (Vincent, 1965). Black and Bliss (1980) calculated that only a 6-7 year "window" occurs for seedling establishment due to seed availability after fire in the open woodland. Seed germination is dependent upon relatively high soil-surface temperatures, and seedling establishment can therefore be prevented during a cooler climate period. Massive fires could force treeline 100 to 200 km south during the cooler climate periods that have followed the Hypsithermal. Black and Bliss (1978) reported five sites in the Inuvik, N.W.T., area where black-spruce seedlings were absent following recent forest fires (1965-68). In the same area, Ritchie (1978) has described vegetation changes following deglaciation—a sequence of species invasions believed to represent plant succession rather than regional climatic change. The relative role of fire vs. climatic oscillation in the failure of tree establishment will require further study, but such researches are essential in order to understand the past boundaries and oscillations of treeline.

Low-Arctic Vegetation

Arctic vegetation of Alaska, Yukon Territory, eastern Siberia, and Chukotka can all be classified as tundra, with the exception of the cushion plant-lichen communities of the mountain slopes and benches (polar semidesert) and the nearly barren upper slopes (polar desert).

Tall-Shrub Tundra

Tall-shrub tundra occurs along most arctic rivers and streams as well as along lake margins and steep slopes at low elevations. In Alaska and the Yukon Territory *Salix alaxensis* predominates on sand and gravel bars and terraces of major rivers. These shrubs may be 3 to 5 m high near the mountains and in the foothills, but their height decreases northward toward the coast and they disappear entirely within 5 to 20 km of the river deltas near Prudhoe Bay and Barrow (Bliss and Cantlon, 1957; Spetzman, 1959). An understory shrub layer often includes *Salix lanata* ssp. *richardsonii, S. glauca, Betula glandulosa* and *Alnus crispa.* Various combinations of the latter group of species occur on fine-textured soils of small streams and cutbanks (Bliss and Cantlon, 1957; Drew and Shanks, 1965; Komarkova and Webber, 1981).

Moose (*Alces alces*) extend their range into the Arctic within these communities where they browse summer and winter on *Salix* and feed on emergent and floating aquatic plants in lakes and ponds in summer and on sedges in winter and summer.

In the province of Chukotka *Salix alaxensis, Betula nana* ssp. *exilis* and *Salix pulchra* are common along drainages (Andreev and Aleksandrova, 1981).

Low-Shrub Tundra

Low-shrub tundra often predominates in medium- to well-drained sites contiguous with forest-tundra. These same shrub species comprise much of the understory within the forest-tundra and for the first 50 to 100+ years following fire in the open woodland (Black and Bliss, 1978). In western Alaska, valleys within the Brooks Range, and the mountains of the Yukon, various combinations of *Betula nana* ssp. *exilis, Salix glauca, S. pulchra,* and *S. lanata* ssp. *richardsonii* provide an open canopy 40 to 60 cm in height. Clumps of *Carex bigelowii,* and *Eriophorum vaginatum* occur interspersed. On the continuous moss and lichen mat are varying amounts of dwarf shrubs (subshrubs), mostly species of heath. These include *Vaccinium vitis-idaea, V. uliginosum, Empetrum hermaphroditum, Ledum palustre* ssp. *decumbens, Arctostaphylos rubra, A. alpina, Cassiope tetragona,* and species of dwarf *Salix* (Hanson, 1953; Hettinger, et al., 1973). Winter snow cover is generally 25 to 50 cm. Caribou (*Rangifer tarandus*) feed in these communities, especially on sedges and lichens, and ptarmigan (*Lagopus lagopus*) are common feeders on willows.

In eastern Siberia and Chukotka similar shrub communities are common. These species include *Betula nana, Salix lanata, S. phylicifolia* and many of the heath subshrubs mentioned above (Andreev and Aleksandrova, 1981). Because of the floristic and structural similarity to the sub-Arctic, Soviet scientists commonly refer to this community or zone as subarctic tundra (Aleksandrova, 1970).

Cottongrass Tussock-Subshrub Tundra

Cottongrass tussock-subshrub tundra is an important community across Siberia, Chukotka province, western and northern Alaska, and northern Yukon (Andreev and Aleksandrova, 1981; Britton, 1966; Churchill, 1955; Corns, 1974; Hanson, 1953; Komarkova and Webber, 1981). In the U.S.S.R. these communities are termed hummocky tundra. They typically occur on rolling topography and small valley floors where soil drainage is intermediate. *Eriophorum vaginatum* dominates throughout, but the total cover of low shrubs of *Betula nana* in North America and *B. nana* ssp. *exilis* in Chukotka and the heath species *Vaccinium uliginosum, V. vitis-idaea, Empetrum hermaphroditum, Arctostaphylos rubra, A. alpina, Ledum palustre* ssp. *decumbens,* and *Cassiope tetragona* is usually greater. Dense carpets of mosses and lichens occur between the tussocks. In the Kolyma subprovince of Siberia and eastward in Chukotka, *Carex lugens* hummocks occur with *Eriophorum.* In Alaska *Carex lugens* and/or *C. bigelowii* tussocks are common in this plant community.

Cottongrass hummocky or tussock tundra is a common feeding ground for caribou, especially the nutrient- and protein-rich new shoots that appear in May and June (Hemming, 1971). *Eriophorum, Carex,* and various forbs are important plants in the summer diet of these animals. Voles (*Clethrionomys, Microtus*) and lemming (*Lemmus, Dicrostonyx*) are commonly found in these communities as are ptarmigan and numerous ground-nesting birds.

Sedge-Moss and Sedge-Grass-Moss Tundra

Sedge-moss and sedge-grass-moss tundra are

common throughout these lands. Mires are extensive in the Yana-Indigirka and Kolyma subprovinces of Siberia. These include wetlands dominated by species of *Carex* and *Eriophorum* with *Arctophila fulva* meadows in shallow lakes (Andreev and Aleksandrova, 1981). In Alaska and the Yukon Territory only limited areas of wetland sedges, grasses and mosses occur in the mountains and upper foothills, but they dominate the Coastal Plain in Alaska (Britton, 1966; Webber and Walker, 1975; Komarkova and Webber, 1981), the coastal areas of the Yukon Territory, and the Mackenzie delta of the N.W.T. (Hettinger, et al., 1973; Corns, 1974).

Sedges of shallow waters to moist soil include *Carex aquatilis, C. chordorrhiza, C. rariflora, C. rotundata, C. membranacea, C. scheuchzeri,* and *Eriophorum angustifolium.* Grasses of moist soil include *Dupontia fischeri* and *Alopecurus alpinus;* of standing water, *Arctophila fulva;* and of better-drained sites, *Deschampsia caespitosa, Poa arctica,* and *Arctagrostis latifolia* (Britton, 1966; Corns, 1974; Webber and Walker, 1975). On the rims of raised polygons, old shorelines, and other elevated sites, dwarf heath shrubs and low shrubs of *Betula* and *Salix* may be found. These wetlands are important summer feeding grounds for caribou and they support large numbers of shore birds and waterfowl each summer.

Grasslands

Grasslands, while not a part of the Arctic, are important in western Alaska on coastal sand dunes and on uplands in numerous inland areas. Hanson (1951) described coastal-grass and grass-forb communities from the Gulf of Alaska and Kodiak Island to Kotzebue Sound. Various combinations of *Elymus mollis, Festuca rubra, F. altaica, Calamagrostis canadensis, Achillea borealis, Lathyrus maritimus,* and *Lupinus nootkatensis* predominate.[1]

The upland grassland communities range from *Deschampsia beringensis, Poa glauca, Carex macrochaeta,* and *Calamagrostis canadensis* on sea cliffs of Kodiak Island to tall grass-forb and mid-grass-forb communities inland. *Calamagrostis canadensis* is present with the associated forbs *Epilobium angustifolium, Heracleum lanatum, Angelica* spp., and *Geranium.* Mid-grasses dominate many areas including the Matanuska Valley. These include *Festuca altaica, Agropyron trachycaulon,* and *Calamagrostis purpurescens* along with numerous forbs (Hanson, 1951). These grasslands, often adjacent to forests, are important habitat for deer and other large herbivores (Klein, 1965).

Cushion Plant-Herb-Lichen

At higher elevations in the mountains of the Yukon Territory and Alaska, cushion plant-herb-lichen communities predominate on windswept ridges and slopes. These communities are dominated by *Dryas octopetala, D. integrifolia,* numerous grasses and sedges, and small clumps of species of *Draba, Saxifraga, Arenaria,* and *Cerastium* (Britton, 1966; Drew and Shanks, 1965; Johnson, et al., 1966; Hettinger, et al., 1973). Species of the lichen genera *Cladonia, Cetraria, Alectoria,* and *Cornicularia* are also important.

Although standing crop and plant cover are greatly reduced in these communities compared to the previously described types, they are nevertheless utilized by caribou and arctic ground squirrels *Spermophilus parryi kennicottii).*

Upland sites in the Kolyma subprovince and to the east in the Chukotka province are also dominated by cushion plants and dwarf shrubs including *Dryas punctata, Diapensia obovata,* and *Salix polaris.* Dry-site sedges, grasses, lichens and mosses are also common (Andreev and Aleksandrova, 1981).

Fires

Fires are known to be an integral part of the taiga, including the open woodlands and forest-tundra, but they are also important in northern grasslands and the low-arctic tundra (Rowe and Scotter, 1973; Viereck, 1973; Black and Bliss, 1978; Wein and Bliss, 1973; Wein, 1976). Of the lands that burn annually in Alaska, 43% are non-forest (tundra, bog, grassland), but few of these lands are tundra. Wein (1976) reviewed the literature on reported fires, supplemented by personal observations of researchers. In the Yukon Territory, 10 fires were reported in 1968–72 and 7 of these were 1 to 5 ha in size. The largest was 1 km^2 in shrub tundra. Comparable data for

[1] Yurtsev (this volume) describes relict steppe communities existing within the subarctic and low-arctic zone of eastern Siberia. Young (1976) and others discuss some similar sites supporting steppe plant species in central Alaska and Yukon Territory.—*Eds.*

Alaska (1948-74) document 9 fires; the largest was 1600 km^2, an aggregate of three simultaneous fires on the Seward Peninsula.

Four sites were examined within three recently burned cottongrass tussock communities by Wein and Bliss (1973). They found little or no indication of plant succession after fire. Numerous seedlings of *Eriophorum* were found at each site, but few seedlings survived more than one year. *Eriophorum vaginatum* and *Carex bigelowii* are the first species to send up new shoots from old clumps. *Calamagrostis canadensis* and *Epilobium augustifolium* grow abundantly following fire in some sites. The shrubs *Ledrum palustre* ssp. *decumbens, Vaccinium uliginosum,* and *Rubus chamaemorus* recover within two to five years, but other shrub species appear to take longer. Above-ground vascular plant annual production approached production levels of control plants in only two years, and the levels of N, P, K and Mg were higher in the tissue of plants on the burned than on the unburned sites. From these limited studies, it appears that fires in the tundra are small and infrequent and that recovery of vascular plants from old shoot bases is quite rapid. Recovery rates for lichens would be slower as they are in the taiga. The general lack of plant succession following fire is not surprising considering the basic growth form of the dominant species, which favors sprouting from old shoot systems of both woody and graminoid species.

High-Arctic Vegetation

The Low Arctic, with its essentially complete cover of plants and diverse wildlife, contrasts sharply with the High Arctic. The narrow strip of coastal tundra from Barrow to Prudhoe Bay, much of the Keewatin District west and north of Hudson Bay, and the Arctic Archipelago of Canada and central and northern Greenland comprise the High Arctic (Bliss, 1979). Broad units of subdivision include small areas of tundra (low shrub, subshrub, cottongrass tussock, sedge-moss) and vast landscapes of polar semidesert (cushion plant, moss-lichen-herb) and polar desert (scattered herbs).

Tundra

Areas of low-arctic tundra form a minor part of the landscape (generally about 5% or 6%, but less than 2% within the Queen Elizabeth Islands)(Bliss, 1977). The most common are the sedge-moss meadows of lowland areas. The most abundant vascular plants are the sedges *Carex stans, C. membranacea, Eriophorum scheuchzeri,* and *E. triste,* and the grasses *Dupontia fischeri* and *Alopecurus alpinus* (Brassard and Longton, 1970; Muc, 1977; Bliss and Svoboda, unpublished). Mosses of the genera *Distichium, Cinclidium, Tomenthypnum, Drepanocladus, Campylium,* and *Orthothecium* (Vitt and Pakarinen, 1977) are abundant. In slightly raised moss hummocks small mats of *Salix arctica* and *Dryas integrifolia* may be found. Biologically, these meadows are the most productive sites in the High Arctic. They support musk ox (*Ovibos moschatus*), Peary's caribou (*Rangifer tarandus pearyi*), brown lemming in the southern arctic islands (*Lemmus sibiricus*), and numerous species of waterfowl and shore birds. The areas of low shrub, subshrub, and cottongrass tundra are seldom more than one to a few hectares in size and are of little or no importance in terms of wildlife distribution. Plant cover in all of these areas ranges from 80 to 100+%.

Polar Semidesert

The polar semideserts occupy upland sites where there is some winter snow cover but where soils are usually medium- to well-drained in summer. In the Keewatin District and the southern arctic islands of Canada, cushion plant-lichen communities predominate in soils that range from sands and fine gravels to silts and silt loams. Cover of flowering plants generally averages 10% to 20% with lichens and mosses providing 20% to 50% additional cover. The dominant species are *Dryas integrifolia, Saxifraga oppositifolia, Salix arctica, Carex nardina* and numerous small rosette species of *Draba, Saxifraga,* and *Minuartia.* Characteristic lichens include *Cetraria cucullata, C. nivalis, Thamnolia subliformis,* and the mosses *Rhacomitrium seudeticum* and *Ditrichum flexicaule* (Svoboda, 1977; Bliss and Svoboda, unpublished).

In the northwestern Queen Elizabeth Islands where sandy to silt and clay loam soils predominate, and where summers are cooler and cloudier, cushion plants are replaced with tufted small clumps of *Luzula confusa* and *L. nivalis,* open mats of *Alopecurus alpinus,* and scattered rosettes of *Papaver radicatum, Ranunculus sabinei, Cerastium arcticum,* and species of *Draba* and *Saxifraga.* Most plants are 1-2 cm high and flower rather abundantly, but few species set much viable seed (Bell and Bliss, 1980). Mosses of the genera *Ditrichum, Rhacomitrium, Tomenthypnum, Orthothecium* and the lichen genera *Thamnolia, Cetraria, Lecida,* and *Pertusaria* are common (Bliss and Svoboda, unpublished).

These semideserts support Peary's caribou, a few musk ox in summer, and very small populations of collared lemming (*Dicrostonyx groenlandicus*) along with a very few pairs of snow bunting (*Plectrophenax nivalis*) and parasitic and long-tailed jaegers (*Stercorarius parasiticus, S. longicaudus*).

Polar Desert

The polar deserts are very barren lands where a combination of fine- to coarse-textured substrates, dry summer soil conditions, limited soil development, and limited winter snow cover result in vascular plant cover of <2-5%, plus a very few lichens and mosses. The most common species are *Papaver radicatum, Cerastium arcticum, Draba corymbosa, Saxifraga cernua, S. oppositifolia, Alopecurus alpinus,* and *Puccinellia vaginata* (Bliss, et al., unpublished). The very limited plant cover provides no food base for mammals and birds, other than the rare colony of collared lemming.

Within the High Arctic of North America, an estimated 41% is polar desert, 45% polar semidesert, 8% ice, and 6% tundra. In the Queen Elizabeth Islands these percentages are 49%, 25%, 25% and 1% respectively (Bliss, 1977).

This detailed summary of plant communities and associated major species of wildlife will serve as the basis for discussing present and past ecosystems in terms of their carrying capacity for wildlife and people.

PLANT BIOMASS AND NET ANNUAL PRODUCTION

Subarctic and arctic environments with their short growing season, cool summers, and cold soils with a shallow thaw layer are not conducive to high rates of plant production. On the other hand, their rates of production are not as low as one might predict when comparing them to more temperate regions.

The available data for plant biomass or phytomass (alive standing crop) and net annual production are quite limited, yet they give a reasonable picture of range conditions and carrying capacity for large and small herbivores.

Forest-Tundra

There are few data available on the above-ground phytomass of these subarctic plant communities. Norin and Ignatenko (1975) reported that an open *Larix dahurica* woodland at Arys-Mas along the Novaya River (72° 30′N) on the Taimyr Peninsula had an above-ground phytomass of 3230 g m^{-2}. The understory shrubs *Betula exilis, Salix glauca, S. reptens, Ledum decumbens, Vaccinium vitis-idaea, V. uliginosum, Empetrum, Cassiope, Dryas punctata,* and herbs including *Carex casifolia* contributed 650 g m^{-2} and the moss and lichen layer 1490 g m^{-2}. In more open sites with stunted trees 15–30 m apart, phytomass was much less (Table 1). No estimate of net annual production was given, although a conservative estimate based upon low-shrub tundra data from elsewhere would be 625 g m^{-2} above ground.

Forest-tundra sites at Hardangervidda (785 m) in southern Norway and at Kevo in northern Finland (130 m) are dominated by *Betula tortuosa* with an understory of *Vaccinium vitis-idaea, Empetrum hermaphroditum,* and various forbs, lichens and mosses. Above-ground phytomass averaged 800 g m^{-2} at Kevo (Kallio, 1975) and 2440 g m^{-2} at Hardangervidda (Østbye, et al., 1975). Above-ground plant production averaged 590 g m^{-2} at the latter site (Table 1).

Arctic Tundra

Data on a tall-shrub community, dominated by *Salix alaxensis, S. lanata* ssp. *richardsonii,* and *S. glauca,* are available from the Meade River, Alaska (Komarkova and Webber, 1981). These willow communities, near the northern limit of this community type, are probably less productive, 185 g m^{-2}, than are the taller and higher-density stands to the south.

Low-shrub tundra has been sampled in Greenland, the Harp Station, U.S.S.R., the Mackenzie River delta region, N.W.T., and in northern Alaska. The data (Table 1) show that above-ground phytomass ranges from 300 g to 2200+ g m^{-2} and net annual plant production from 35 g to 800 g m^{-2}.

Table 1
Phytomass and Net Annual Primary Production of Subarctic and Arctic Vegetation
All data are for above-ground material (g m^{-2})

Plant Community	Phytomass (g m^{-2})	Net Annual Production (g m^{-2})	References
Forest-Tundra Low Arctic-Alpine			
Larch woodland	3230	—	Norin & Ignatenko (1975)
Larch open woodland	1390	—	,,
Tundra Communities Low Arctic-Alpine			
Birch forest-tundra	2440	590	Østbye, et al. (1975)
Birch forest-tundra	800	—	Kallio (1975)
Tall-shrub tundra	1150	185	Komarkova & Webber (1981)
Low-shrub tundra	2240	800	Lewis (pers. comm.)
Low-shrub tundra	1380	35	Gorchakovsky & Andreyaskina (1972)
Low-shrub tundra	285	63	Haag (1974)
Low-shrub tundra	270	100	Komarkova & Webber (1981)
Cottongrass subshrub tundra	366	170	Wein & Bliss (1974)
Cottongrass subshrub tundra	260	105	,,
Cottongrass subshrub tundra	490	98	Komarkova & Webber (1981)
Wet sedge-moss tundra	195	73	Haag (1974)
Wet sedge-moss tundra	195	70	Webber (1978)
Wet sedge-moss tundra	166	79	Komarkova & Webber (1981)
Subshrub tundra	415	235	Chepurko (1972)
Subshrub tundra	440	176	Østbye, et al. (1975)
Subshrub tundra	368	92	Komarkova & Webber (1980)
Tundra High Arctic			
Wet sedge-moss tundra	1000	81	Muc (1977)
Wet sedge-moss tundra	1040	70	Bliss & Svoboda (unpublished data)
Polar Semidesert			
Cushion plant-lichen	750	50	Svoboda (1977)
Cushion plant-moss	80	15	Bliss & Svoboda (unpublished data)
Moss-herb	920	20	,,
Polar Desert			
Herb barren	7	0.3	Bliss, et al. (unpublished data)
Herb barren	3	0.9	,,
Herb barren	10	0.3	,,

Cottongrass tussock-subshrub heath tundra communities have been sampled in the Yukon Territory and in northern Alaska. Above-ground phytomass of vascular plants, lichens, and mosses range from 250 g to 500 g m^2 with an above-ground net annual production of 100 g to 170 g m^{-2} (Wein and Bliss, 1974; Komarkova and Webber, 1981).

Phytomass and plant production have been more intensively studied in wet sedge-moss communities than in any other major type. These studies reveal that above-ground phytomass averages 170 g to 1,000 g m^{-2}, the great range resulting from variation in bryophyte cover and phytomass. Annual plant production ranges from 70 g to 80 g m^{-2} (Table 1).

Subshrub heath communities are generally of minor extent in most tundras; the species are most common as understory in low-shrub tundra and in cottongrass tussock tundra. The few measurements that have been made indicate that above-ground phytomass ranges from 350 g to 450 g m^{-2} and net annual production from 100 g to 250 g m^{-2} (Table 1).

No estimates are available on phytomass or net production for the Alaskan grasslands. Based upon grasslands in more temperate regions, above-ground production might average 200-400 g m^{-2}.

These production data indicate that for the more important arctic tundra rangelands of today, above-ground net annual production generally ranges from 70 g to 200 g m^{-2}, eliminating the three lowest and three highest values. Of this total, bryophytes and lichens may make up 30% to 50%, but these plants are little used by herbivores except for consumption of mosses in winter by *Lemmus.*

Phytomass and above-ground primary production data are included for three major landscape types in the High Arctic. Similar communities may have predominated during the glacial maximum in Beringia. It can be seen that only in the sedge-moss tundra meadows of lowlands is phytomass and above-ground plant production (1000–1100 g m^{-2} and 70–80 g m^{-2} respectively) comparable to communities in the Low Arctic. Elsewhere, plant cover, plant production and therefore herbivore density and production are greatly reduced. The barren polar deserts have essentially no plant cover and plant production, and consequently no food base for wildlife.

The caloric, lipid, and nutrient contents of arctic plants are relatively high. Evergreen subshrubs average 4900 calories to 5500 cal g^{-1} ash-free dry weight and deciduous shrubs 4600 to 5100 (Bliss, 1981). Herbs are generally lower (4500) and bryophytes in the Arctic relatively high (4600 cal g^{-1} ash-free d wt) (Pakarinen and Vitt, 1974). Data on lipid content ranges from 2.7% for one high-arctic species to 3.9% for nine arctic-alpine species (Svoboda, 1977; Bliss, 1962). Crude protein values for graminoids include 7.0% in *Arctagrostis latifolia* near Inuvik, N.W.T. (Scotter, 1972), 13.6% in *Carex aquatilis* at Barrow (Chapin, et al., 1975), and 19% in live and 9% in winter-cured *Carex stans* on the Truelove Lowland, Devon Island (Muc, 1977).

HERBIVORE DENSITY AND VEGETATION UTILIZATION

Accurate estimates of herbivore density in the Arctic are difficult to obtain because of the problems of estimating numbers and determining the areal extent of the populations. Animal migration and the uneven use of large landscapes also makes these estimates of questionable value. Calef (1974) estimated density of the Porcupine caribou herd to be 0.5 km^{-2}, caribou in all of Alaska 1.0 km^{-2}, and in all of Arctic Canada to be 0.3 km^{-2}. Calef goes on to state that previous estimates of arctic caribou densities are too high (2 to 3.7 animals km^{-2}) and should be nearer 0.4 km^{-2}. In the High Arctic, Hubert (1977) estimated the density of the musk-ox population to be 0.8 animals km^{-2} on northeastern Devon Island and 0.6 animals km^{-2} in east-central Greenland (Ferns, 1977). Based upon these limited data, large-herbivore densities appear to range from 0.4 to 1.0 animals km^{-2} in those regions with relatively lush vegetation.

Elk Island National Park in the aspen parkland of Alberta has a mosaic of vegetation (wet sedge meadows, fescue grassland, groves of aspen bordered with shrubs, and some mixed woods [aspen-white spruce] and black-spruce bogs). The area supports a mixture of large herbivores whose density and biomass have been studied over the years (Blood, 1973; Telfer and Scotter, 1975). Moose densities ranged from 1.1 to 4.6 km^{-2} and wapiti from 1.3 to 2.7 km^{-2} for a number of years

(Blood, 1973). Bison reached densities of more than 7.5 km^{-2}. While the feeding habits of these three herbivores and Virginia deer overlap, competition for forage seems minimal (Telfer and Scotter, 1975). The combined biomass of these four herbivores ranges from 2250 kg to 9800 kg km^{-2}, with an optimum 5710 kg km^{-2} quite comparable to African savannas. These values are much higher than the estimates for present-day tundras. Klein (1970) estimated 17 kg km^{-2} of herbivore biomass for the approximately 2,850,000 km^2 of tundra in North America. Guthrie (1968a) estimates 300 kg km^{-2} of ungulates in the Nelchina Basin, Alaska, and Hubert (1977) estimates 300 kg to 400 kg km^{-2} for wet sedge-moss tundra and cushion plant-lichen polar semidesert during summer and fall on Devon Island.

Small herbivores have much higher densities and often a much greater impact on the vegetation. *Dicrostonyx torquatus groenlandicus* reaches maximum populations of 1 to 7 ha^{-1} on Devon Island, 1 to 6 ha^{-1} at Prudhoe Bay, and 1 to 27 ha^{-1} at Barrow. The closely related *Dicrostonyx torquatus* at Tareya, U.S.S.R., ranges from 1 to 123 ha^{-1} as an annual maximum. *Lemmus sibiricus trimucronatus* is a more intensive breeder, reaching annual peaks of $\cong$ 1 to 155 ha^{-1}. *Lemmus obensis* at Tareya also maintains high densities (24 to 175 ha^{-1}). The arctic vole *Microtus oeconomus* has much lower peak densities, 5 to 8 ha^{-1}, at Eagle Summit, Alaska. These data are all derived from the IBP synthesis (Batzli, 1981). Arctic ground squirrels *(Spermophilus undulatus)* have a much lower density (0.2 to 1.4 ha^{-1}) in the limited areas where they are found, sandy and gravelly bluffs and dunes (Batzli and Sobaski, 1980).

Not all plant species are equally used by herbivores. *Dicrostonyx* and *Microtus* prefer the dicotyledonous plants including *Salix alaxensis, S. pulchra, Vaccinium uliginosum, V. vitis-idaea,* and *Dryas* (Batzli and Jung, 1980). *Dicrostonyx* in the High Arctic feeds on *Dryas, Saxifraga* and *Salix arctica,* avoiding sedges and mosses (Fuller, et al., 1977). *Lemmus* however prefers monocotyledonous plants, especially *Carex* and *Eriophorum.* Mosses become more important in their winter diet (Batzli, 1975; Batzli and Jung, 1980).

The evergreen shrubs *Empetrum hermaphroditum, Ledum palustre,* and *Cassiope tetragona* were unpalatable to all three rodents tested at Meade River, Alaska, probably because they contained tannins, alkaloids, anthraquinones, and terpenes (Batzli and Jung, 1980). *Dicrostonyx* maintained body weight when fed *Salix pulchra* but *Microtus* and *Lemmus* lost weight and died. This willow contains considerable tannins. All three species maintained body weight on *Carex aquatilis.* Caribou and reindeer also prefer highly digestible forage, i.e., *Salix* shrubs, *Carex aquatilis, Eriophorum vaginatum,* forbs, and lichens. The ericaceous evergreen dwarf-shrub species are avoided (White and Trudell, 1980). These are the first reports of the effects of secondary chemicals on the feeding habits of arctic herbivores. It can be expected that mammals of the late Pleistocene were also selective in their grazing and browsing habits because of secondary chemicals.

Forage consumption by arctic herbivores was studied during the IBP, though data are available from only Prudhoe Bay, Alaska, and the Truelove Lowland, Devon Island (White, et al., 1975, 1981). These data, coupled with those for moose (Gasaway and Coady, 1974) show that forage consumption is much higher in summer than in winter (Table 2). The data on musk ox indicate that these animals concentrate in those areas with the greatest amount of forage in late winter and early spring and that in summer the animals often graze in more marginal lands where forage is limited. Based upon the daily intake requirements of musk ox and the known phytomass of standing vegetation, it was estimated that only 1-2% of the herbage was removed per year. However, in the meadows grazed, 15% to 20% of the forage was removed in winter and in any given crater pawed in the snow, 80% to 85% of the potential forage was harvested (Hubert, 1977).

From numerous studies (Pruitt, 1959; Kelsall, 1968; Kelsall and Prescott, 1971) it is evident that depth, hardness, and density of snow are highly limiting to large herbivores. Spring and fall icings of the surface can be even more devastating to animal survival (Miller, et al., 1975). Thus, to support a diversity of herbivores, late Pleistocene climatic conditions must have resulted in winters with relatively shallow snow depths and infrequent hard crusts.

Table 2
Body Weight and Daily Forage Consumption for Three Northern Herbivores

	Summer			Winter			
Species	Body Wt. (kg)	$g\ kg^{-1}\ da^{-1}$	$kg\ da^{-1}$	Body Wt. (kg)	$g\ kg^{-1}\ da^{-1}$	$kg\ da^{-1}$	Reference
Caribou	80	31–56	2.5–4.5	70	16–42	1.1–3.0	White, et al. (1975) White, et al. (1981)
Musk ox	225–280	17–22	3.8–6.2	200–250	13–15	2.6–3.8	Hubert (1977)
Moose	380–430	23–31	9.9–11.8	300–400	11.5–18.3	4.6–5.5	Casaway & Coady (1974)

ENERGY FLOW IN MODERN TUNDRA ECOSYSTEMS

Digestive efficiency of arctic animals is not especially high. Batzli (1981) reported 30% to 40% for *Lemmus trimucronatus* feeding on grass and sedges and 25% when fed mosses. Fuller et al. (1977) reported 55% for *Dicrostonyx groenlandicus* feeding on dicotyledonous plants. Both *Dicrostonyx* and *Lemmus* feed primarily on green rather than brown shoots in winter. Efficiencies of other animals include 63% to 65% for *Ovibos moschatus* on winter hay and 75% for sedges in summer (Hubert, 1977) and 51% to 56% for summer feeding of *Lepus arcticus* (Smith and Wong, 1977). These efficiencies for arctic herbivores are quite comparable to those of temperate region cattle and bison (Richmond, et al., 1977).

Within the high-arctic ecosystem of the Truelove Lowland, 98% of the total-system living biomass is contained in the vegetation, less than 0.2% in the above-ground and below-ground herbivores and carnivores, and 1.0% to 1.3% in the saprovores (protozoa, nematodes, enchytraeids, Collembola, and Diptera) and microflora (bacteria and fungi). The data on net production are similar, 90% to 91% for the primary producers, 0.3% to 1.6% for the herbivores and carnivores, and 7.5% to 9.4% for the saprovores and microflora (Bliss, 1977). Data on the carbon budget and energy flow are presented in Table 3 for the environmentally and biologically contrasting sedge-moss-musk ox meadows (tundra system) and the cushion plant-lemming system related to the polar semidesert landscape. These data indicate that both systems are nearly in equilibrium in terms of system net production, yet the systems are very different in terms of the size of the carbon budget, the magnitude of net production in the various trophic levels, and the herbivores each system supports.

Table 3
Carbon Budget and Net Primary Production within a Cushion Plant-Lemming and a Sedge-Moss-Musk Ox Subsystem in the High Arctic (Bliss, 1977)

Parameters	Cushion Plant-Lemming	Sedge-Moss-Musk Ox
Carbon Budget (g carbon $m^{-2}\ yr^{-1}$)		
Gross primary production	373	3325
Autotrophic respiration	339	3060
Net primary production	34	265
Heterotrophic respiration	32	251
Ecosystem respiration	371	3311
Ecosystem net production	2	14
Net Production (cal $m^{-2}\ yr^{-1}$)		
Autotrophes		
Above-ground	4.68×10^4	2.22×10^5
Below-ground	1.96×10^4	4.73×10^5
Heterotrophes		
Herbivores-vertebrates	70	2950
Herbivores-invertebrates	380	130
Carnivores-vertebrates	+	+
Carnivores-invertebrates	145	497
Saprovores-invertebrates	6000	1.3×10^4
Bacteria and fungi	8100	7.6×10^4

The limited diversity of plant communities and the single-species near dominance in each of the major plant communities has not significantly reduced the kinds of herbivores and carnivores supported by these systems. The only high-arctic land mammals missing from these systems are Peary's caribou (*Rangifer tarandus pearyi*) and brown lemming (*Lemmus s. trimucronatus*). Animal density is low, especially of arctic hare, lemming and the three carnivores: short-tailed weasel (*Mustela erminea*), arctic fox (*Alopex lagopus*) and arctic wolf (*Canis lupus arctos*).

The actual percentage utilization of these communities by large and small herbivores is low (1-2%) and food seldom seems to be the limiting factor. The density-independent factors of spring and fall surface icings and winter snow depth and hardness appear much more critical at these latitudes for both musk ox and lemming.

Present-Day Arctic Ecosystems and Human Populations

Inuit settlements of modern Alaska and the general area of the Mackenzie delta, Canada, are mostly along the coast, because the seas are generally more productive than the lands. In the early 1970s, there were about 11,400 people in 33 settlements along the coast in Alaska north of 64°N. The two inland settlements of Kobuk and Anaktuvuk Pass had only 160 people (Patterson, 1974). In the Mackenzie delta region including Sachs Harbour, Banks Island, and Holeman, Victoria Island, there were 3600 people in 7 settlements (Koring, 1977).

The subsistence harvest data of Patterson (1974) for the period 1969-73 for Alaskan settlements and the comparable data for the Mackenzie delta region settlements for 1969–75 (General Hunting License Returns, 1969-70 to 1974-75; Berger, 1977) presented in Table 4 indicate the maximum food available for human consumption. These data assume that no meat is fed to dogs and that no meat is wasted, assumptions that are unjustified. The data indicate that more meat is harvested per person in the Alaskan Arctic than in the Canadian Arctic. The lands and seas of Alaska are more productive in numbers of species, size of populations, and, in some cases, animal size. Some communities also harvest berries which add to the diversity and nutrition of their diet.

Table 4
Consumable Country Meat (Mammal, Fowl, Fish) and Berries Harvested by Arctic Inuit Settlements (1969–73 Alaska, 1969–75 Canada)

Settlement	People (#)	Meat (kgx10^3)	Berries (kgx10^3)	Potential Human Consumption (g person^{-1} da^{-1})
Alaskan Settlements—Coastal				
Kotzebue	1696	484.2	7.6	780
Barrow	1904	615.4	–	890
Noatak	293	143.6	2.0	1340
Wainwright	307	215.2	–	1920
Alaskan Settlements—Inland				
Kobuk	60	39.7	3.5	1810
Anaktuvuk Pass	97	73.2	–	2070
Canadian Settlements—Coastal				
Inuvik	1025	102.3	–	273
Tuktoyaktuk	627	75.9	–	330
Aklavik	630	111.9	–	487
Sachs Harbour	152	41.3	–	744
Holman	260	85.7	–	903

It is estimated that daily meat consumption by adult Alaskan Inuit was 3100 g to 3600 g da^{-1} during early European contact (1880s). An equal amount of meat was fed to their dogs (Spencer, 1959). Using present-day estimates for meat harvest, and assuming that in pre-contact times people lived entirely on meat and berries, that the average person consumed 3000 g da^{-1} of meat, and that dogs, other carnivores, and scavengers consumed an equal amount of meat, the northern 8 coastal settlements in the Alaskan Arctic would support only 20% of the present population (1123 people) and the 2 inland settlements only 34% (51 people) of their present population. The comparable figures for settlements in the western Canadian Arctic are 7% (190 people) of their present population of 2700 people. It is evident from these rather conservative calculations, as well as from those by Riewe (1977) for Grise Fjord, Ellesmere Island, and by Kemp (1971) for Lake Harbour, Baffin Island, that presently relatively little of the daily food consumed by northern peoples is country meat. Present-day Inuit population levels are much larger than could be supported by native wildlife. This results from carbohydrates having become a very important part of their diet—in sharp contrast to pre-European times, when protein accounted for nearly all of their caloric intake. A further restriction would occur if all peoples lived inland and did not have access to marine mammals.

Ecosystems and Human Populations in Beringia During Late Pleistocene Time

The concept of an arctic steppe-mammoth biome is based upon: (1) a drier and colder climate, but with warmer summers (Streten, 1974; Gates, 1976; Hopkins, 1979a); (2) a tundra vegetation dominated by species of grass, sedge, and *Artemisia* (Colinvaux, 1964, 1981; Matthews, 1974a); and (3) a diverse herbivore fauna (Guthrie, 1968a; Harington, 1978). Although we will never know what this inferred biome was really like, we can speculate on its plant communities, herbivore biomass, and human carrying capacity. For this modeling effort, the time period from 25,000 to 11,000 BP was chosen on the grounds that Beringia was then near its maximum extent.

In calculating animal production and human carrying capacity, several modeling approaches were considered. The use of ecological efficiencies for each trophic level or the use of animal densities as the driving function in the modeling effort would yield a very low degree of resolution. The efficiencies of energy transfer at the herbivore and carnivore levels within arctic ecosystems are too poorly known, as are modern animal densities.

The larger and more accurate data base on plant production and the daily food requirements of modern large herbivores is believed to provide a higher degree of resolution in estimating herbivore carrying capacity than could be achieved by the other approaches. This model utilizes estimates of above-ground phytomass and net plant production, assuming the presence of steady-state plant communities. Forage utilization rates for an "average" monogastric and ruminant herbivore and mean reproductive rates for modern analogous herbivores are used. The annual harvest rate of herbivores is assumed to maintain a steady state population. The apportionment of meat yield to humans is based upon a percentage of animals harvested per year, percentage of meat consumed vs. lost to spoilage and carnivores, and the daily meat requirement of people in an arctic hunting ecosystem.

Late Pleistocene Ecosystem 25,000 to 11,000 Years BP

Reconstruction of vegetation in Beringia from lake-sediment cores (Colinvaux, 1964, 1967a, 1967b, 1981; Ager, 1975), colluvial and alluvial cores (Colinvaux, 1967b; Matthews, 1974a), exposed sediment sections (Matthews, 1974b, this volume), and patterns of modern low-arctic plant communities, and floristics of modern plant communities (Yurtsev, this volume; Young, this volume) leads to the assumption that four major habitat types occupied regional landscapes in Beringia. Tall-shrub communities of *Salix alaxensis, S. pulchra,* and *S. arbusculoides* occupied floodplains and low river terraces. A rich understory of herbs and grasses in open areas no doubt occurred as with present shrub communities (Bliss and Cantlon, 1957).

In poorly drained lowlands, sedge-moss meadows prevailed with a species composition and plant structure similar to those found in similar habitats today (Britton, 1966; Komarkova and Webber, 1981).

The rolling uplands that were much better drained contained a mosaic of upland sedge (not *Eriophorum* tussock tundra), grass, and *Artemisia* (Colinvaux, 1964, 1967a). Based upon the evidence that the Pleistocene ungulates had relatively small hoof sizes compared to modern species (Guthrie, this volume), wetlands or soft substrates are presumed minor and winter snows probably not deep nor compacted. Modern moose, caribou and musk ox are adapted to pawing through snow $\cong$ 50 cm deep, but horses are not. These assumed low rolling uplands with their tundra-graminoid vegetation must have been very important herbivore habitat during summer and winter.

At higher elevations with well-drained soils, little snow cover, and considerable influence of wind, cushion plants such as *Dryas, Saxifraga, Draba,* and dry-site sedges, rushes, and lichens predominated. These communities would have been similar to wind-exposed mountain slopes in the

Low Arctic and the polar semideserts of the High Arctic today.

Based upon modern Alaskan landscapes and their dominant plant communities, it is assumed that in a landscape unit of 1000 km^2, 5% would have been tall shrublands, 10% wet sedgelands, 65% upland sedge, grass and *Artemisia*, and 20% upland cushion plant-lichen vegetation (Fig. 1). This landscape is assumed to occupy a valley system, extending 10 km either side of a river to a ridge crest along each side of the valley and 50 km downstream.

Using the information of Guthrie (1968a, this volume) and Harington (1978), it is assumed that this late Pleistocene ecosystem had six dominant large herbivores. Mammoth (*Mammuthus primigenius*) and horse (*Equus*) were monogastrics; ruminants included caribou (*Rangifer*), musk ox (*Ovibos*), bison (*Bison*) and moose (*Alces*). The assumed percentage of time each species spent in the different plant communities is shown in Figure 1. Different foraging strategies of the browser (moose), mixed browser-grazer (mammoth, caribou, musk ox) and grazer (horse, bison) would have reduced interspecific competition as in modern systems with a diverse fauna (see Redmann, this volume). The assumed percentage of forage obtained in each community by each herbivore species is based upon feeding strategies of modern analogous species.

Based upon modern arctic ecosystems, the above ground phytomass and net primary production (NPP) rates of net annual plant production were assumed for the hypothetical Beringian communities (Table 5). Production rates for the tall-shrub community at Meade River on the North Slope were doubled, since this location is near the northern climatic limit of the type. Production rates for the other communities or habitat types were similar to the modern analogous types in Table 1.

The percentage utilization of native vegetation by large herbivores is difficult to estimate, let alone extrapolate from these data to Pleistocene time. Telfer and Scotter (1975) estimated that moose, elk, and bison utilized 20% of the herbaceous and aspen vegetation available in Elk Island National Park, Alberta. Hubert (1977) reported that musk ox utilized 5% of the available sedge phytomass or 7% of net production in the High

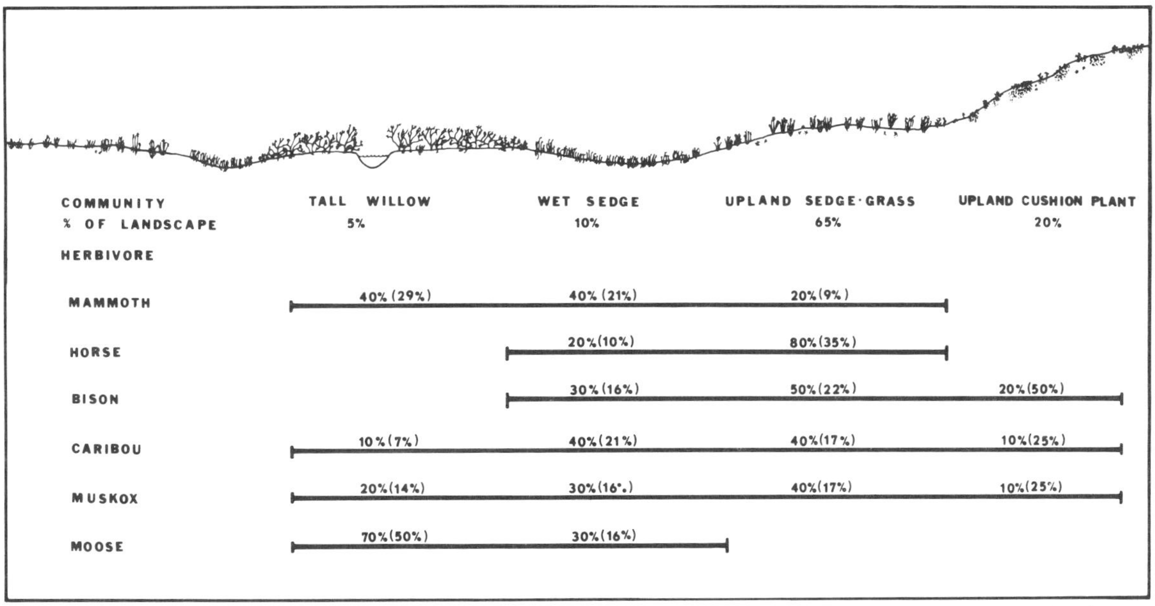

Figure 1. Theoretical ecosystem in Beringia 25,000 BP showing the four dominant plant communities and six dominant herbivore species. The first value is the assumed percentage of the total forage that a given herbivore obtains from a given plant community (add across). The second figure ( ) is the percentage contribution that a given community contributes to the total diet of an herbivore species (add down). The percentage of forage taken in each plant community by each herbivore is based upon the feeding strategies of its modern analogous species

Arctic. Phillips (1975) estimated that elephants in the Tsavo National Park in East Kenya utilized 4% to 12% of the net annual production. For these calculations estimates of 5% of net annual production in cushion plant, 5-15% in the upland sedge-grassland, and 15% for the other communities were used (Table 5). They result in a total annual harvest of 7625 to 14,125 t by large herbivores for the 1000 km^2 landscape unit.

Daily forage consumption as a percentage of live body weight has been calculated for modern large herbivores. Based upon the above-mentioned literature and a personal communication (1979) from R.J. Hudson, daily food consumption was assumed to be 4% for monogastrics and 2.2% for ruminants. Mean live body weights (kg) are based upon a population including young and immature individuals (Table 6).

Given these assumptions about forage available for harvest per habitat, the percentage of forage consumed by each species per habitat (Fig. 1), and the daily food consumption per species as a percentage of live body weight (2.2, 4.0%), times the days per year, it is possible to calculate the maximum herbivore biomass per species (Table 7). For example, moose consume 50% of the tall willow forage (1500 t) plus 16% of the wet sedge forage (180 t) or 1680 t. If moose consume 803% of their live weight per year, a moose biomass of 209 t (1680 ÷ 803% = 209) results. These biomass

Table 5
Estimated Above Ground Phytomass, Net Annual Plant Production (NPP), and Percent of NPP Harvested by Large Herbivores in Beringia in the Late Pleistocene

Habitat Type (Community)	Area (km^2)	Aboveground Phytomass (t km^{-2})	Net Primary Production (t km^{-2} yr^{-1})	Total NPP on 1000 km^2 (t yr^{-1})	NPP Harvested by Herbivores (%)	Forage Consumed 1000 km^2 (t yr^{-1})
Tall Shrub	50	1,000	400	20,000	15	3,000
Wet sedge meadow	100	200	75	7,500	15	1125
Upland grass, sedge, low shrub	650	400	100	65,000	5–15	3,250–9,750
Upland cushion plant	200	100	25	5,000	5	250
Total	1,000	—	—	—	—	7,625–14,125

Table 6
Mean Daily Forage Consumption and Mean Body Weight for Beringian Herbivores, Using Modern Analogues

Species	Daily Food Consumption (% Live Wt) Summer	Winter	Mean Value Used (%)	Mean Body Weight (kg)	Reference
Mammoth	6	4	4	2230	Laws, et al. (1975) Phillipson (1975)
Horse	4	—	4	150	Haenlein, et al. (1966)
Bison	2.6	1.4	2.2	450	Telfer & Scotter (1975)
Caribou	3–5.6	1.4–3.8	2.2	100	White, et al. (1975)
Musk ox	1.7–2.2	1.3–1.7	2.2	180	Hubert (1977)
Moose	2.7–2.9	1.4–1.5	2.2	300	Telfer & Scotter (1975)

estimates range from 95-135 t 1000 km^{-2} for mammoth to 130-300 t 1000 km^{-2} for bison. Total herbivore biomass would be 800-1450 t 1000 km^{-2} of Beringian landscape. The range of values results from the 5-15% range in the estimate of forage utilization in the upland steppe sedge-grass-sage community with scattered low shrubs.

The final calculations involve the percentage of herbivore biomass that can be harvested per year and maintain a sustainable yield (wildlife population). Rates of net animal production vary from species to species. Conservative harvest figures have been used that assume possibly high mortality rates and a need to maintain a sustained harvest (a steady-state herbivore biomass). The assumed harvestable biomass per year (percentage of biomass) ranged from 4% in mammoth (because of its low reproductive rates) to 9-10% in bison, horse, and moose (Table 8). Of the 60 to 110 t wildlife harvestable per year from this landscape, 50% would be usable meat (Kelsall, 1968). Modern Inuit require 1.1 t meat yr^{-1} when living entirely off the land. Therefore, the maximum human carrying capacity would be 30–50 people per 1000 km^2, assuming no meat wastage, which is unreasonable. More likely 10–20% of the meat would have been fed to dogs and lost to wild carnivores and 20-40% lost to spoilage. Assuming a 50% loss of meat to dogs, other carnivores, and spoilage, a more realistic estimate of 15-25 people might have been sustainable with a steady-state plant and herbivore production and a harvest rate that would maintain wildlife populations.

All of these calculations are based upon assumed sustainable harvest rates, assumed herbivore consumption rates, and herbivore biomass. This model (Fig. 2) does not utilize animal numbers and density, but these values can be calculated as a check on validity of parameter inputs, provided that mean body weights (Table 6) are utilized.

The calculated densities for caribou, musk ox, moose, and bison (Table 9) are similar to the modern densities of these species in the Arctic for caribou and musk ox (Calef, 1974; Hubert, 1977) and in aspen parkland for moose and bison (Blood, 1973; Telfer and Scotter, 1975). Herbivore biomass for this Beringian ecosystem (800-1450 kg km^{-2}) is significantly higher than estimates for modern arctic ecosystems (300–400 kg km^{-2})

Table 7
Estimated Annual Food Consumption and Herbivore Biomass for Six Beringian Species Based Upon Their Percentage Utilization of the Landscape Habitats (see Fig. 1)

Species	Food Consumption (% Live Wt) Daily	Food Consumption (% Live Wt) Annual	Forage Harvested All Habitat Types (t yr^{-1})	Herbivore Biomass (t 1000 km^{-2})
Mammoth	4.0	1460	1,399–1,984	96–136
Horse	4.0	1460	1,229–3,439	84–236
Bison	2.2	803	1,020–2,450	127–305
Caribou	2.2	803	1,062–2,167	132–270
Musk ox	2.2	803	1,216–2,321	151–289
Moose	2.2	803	1,680	209
Total			7,606–14,041	799–1,445

Table 8
Estimated Annual Harvestable Biomass, Based on Maintenance of a Steady-State Herbivore Biomass

Species	Herbivore Biomass (t 1000 km^{-2})	Annual Harvest (%)	Harvestable Biomass (t 1000 km^{-2})
Mammoth	96–136	4	3.8–5.4
Horse	84–236	10	8.4–23.6
Bison	127–305	9	11.4–27.5
Caribou	132–270	7	9.2–18.9
Musk ox	151–289	6	9.1–17.3
Moose	209	10	20.9
Total	799–1,445	—	62–114

Table 9
Herbivore Numbers and Density Per 1000 km^2 Landscape, Based Upon Herbivore Biomass and Mean Body Weight Per Species

Species	Herbivore Biomass (t 1000 km^2)	Mean Body Weight (t)	#'s 1000 km^{-2}	#'s km^{-2}
Mammoth	96–136	2.23	43–61	0.04–0.06
Horse	84–236	0.15	560–1573	0.6–1.6
Bison	127–305	0.45	282–678	0.3–0.7
Caribou	132–270	0.10	1320–2700	1.3–2.7
Musk ox	151–289	0.18	840–1605	0.8–1.6
Moose	209	0.30	697	0.7

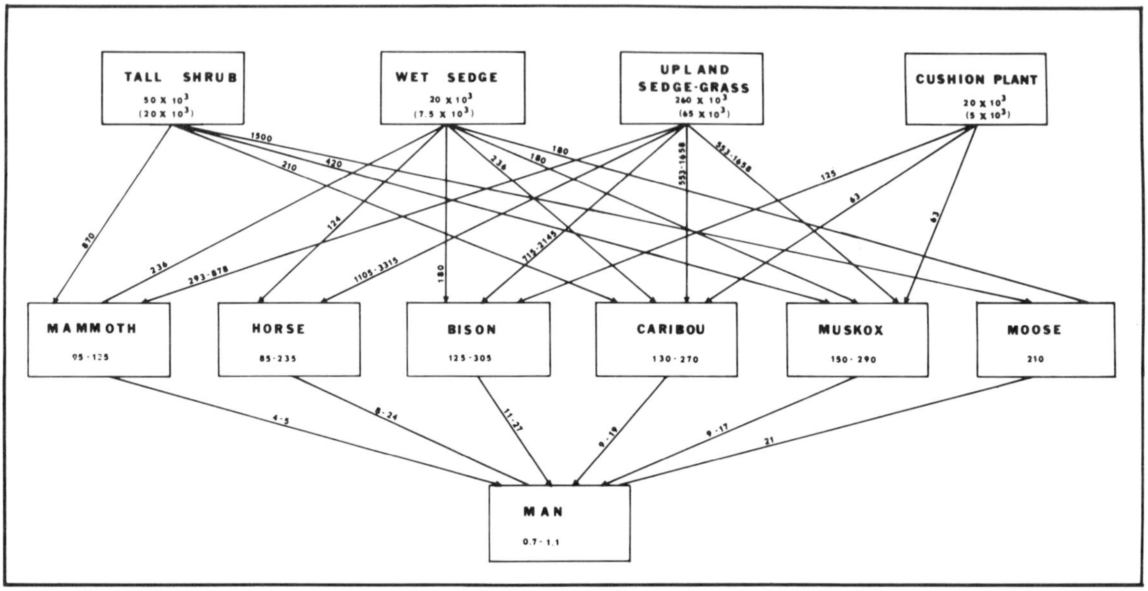

Figure 2. Energy flow within a theoretical ecosystem in Beringia, 25,000 BP. Living above-ground biomass and net annual production ( ) for the major plant communities and the living biomass of herbivores and people are contained in the boxes. Net production consumed is presented as transfers. All data are metric tons per 1000 km^2 with transfers expressed on an annual basis

(Guthrie, 1968a; Hubert, 1977), systems that are low in herbivore species diversity. Estimates of herbivore biomass from 12 studies in East Africa (Coe, et al., 1976), range from 3000–5500 kg km^{-2}. The herbivore biomass estimates for East Africa are only 3.5–4 times the estimates for Beringia.

Although the pollen and macrofossil record indicates little if any *Salix* shrubland, willow pollen is often poorly represented in pollen diagrams (Colinvaux, 1964). This model shows how important only a 5% habitat type would be to total herbivore production and human carrying capacity. The estimated 60-110 t of harvestable herbivore production would be capable of supporting 15-25 people. Without this plant community the system would have yielded only 35-85 t of herbivore production and supported a maximum 8-20 people. Moose and mammoth would have been much less important because willow forage was estimated to have provided 70% and 40% of their respective diets. Without *Salix* shrubland and wet sedgeland habitats, the herbivore diversity and human population would have been significantly reduced.

DISCUSSION AND SUMMARY

The ecosystem model presented here indicates that the arctic landscapes of Beringia could have supported a diversity of herbivores, assuming that the landscape had a diversity of plant communities. In this ecosystem, low rolling uplands with nutrient-rich, well-drained soils, and a deep active layer (≅65-100 cm) predominated. Plant communities of dry upland sedges, grasses, herbaceous sages and low shrubs were most prominent. However, because of their relatively high productivities, the limited land areas assumed to have been covered with tall shrubs (5%) and wet sedge-moss (10%) communities would have been essential for the maintenance of a diverse herbivore fauna of grazers and browsers and a greater density of people.

The efficiency of biomass transfer (standing crop ÷ net production consumed) from plants to herbivores and herbivores to carnivores (humans) is 0.4% and 0.1% respectively. The natural ecosystem on the Truelove Lowland in the High Arctic had herbivore and carnivore harvest efficiencies of 0.3% and 0.2% respectively. It is assumed that people harvested more herbivore biomass than would have other carnivores. Wolves, fox, and saber-toothed tigers would probably have harvested <1-2% of the herbivore biomass, or no more than 8-15 t. This represents enough usable meat to support an additional 2-3 people.

Densities of individual herbivore species are estimated to have been similar to densities of their modern boreal and arctic analogs, but the herbivore biomass of Beringia, due to greater species diversity, must have been larger (2.5–3.5x) than its modern counterpart. Human density is estimated to have been 0.02-0.03 km^{-2}. Modern population density in northern Alaska would be 0.01 km^{-2} for coastal people and 0.0003 km^{-2} for inland people if they consumed only 50% of available country meat (based upon Table 4 and 1.1 t $person^{-1}$ yr^{-1}), and 0.002 km^{-2} for native people in the Mackenzie River delta, and 0.0003 km^{-2} for Banks Island. These estimates of Beringian population density are comparable to what estimates for modern Alaskan coastal settlements would be if the people lived only from the sea and land. The estimates are two orders of magnitude greater than for modern settlements in inland Alaska or on Banks Island, where the people live off the land rather than from the sea. Hominid population densities for parts of Africa 2.8-1.7 million years ago have recently been estimated from results of excavations in Ethiopia (Boaz, 1979). The densities, ranging from 0.001-2.58 persons km^{-2}, are very broad estimates, providing little perspective for the estimates of human populations in this much younger Beringian landscape. Clearly, Beringia was capable of supporting reasonable numbers of people.

Human populations could have reached sufficient size to greatly influence herbivore density and species diversity as developed in the overkill hypothesis (P.S. Martin, this volume). However, this is questioned because the percentage of avian genera that became extinct at the end of the Pleistocene was equal to the percentage of mammalian genera that became extinct at that time (Grayson, 1977).

Based upon the implied importance of at least limited areas of tall shrubland and wet sedgelands, the landscape of Beringia must have had much more than a dry upland sedge-grass vegetation if it were to support a diversity of large herbivores. The very limited numbers of mammoth that could be supported with or without a 5% shrub vegetation indicate to these writers that the term arctic steppe-mammoth biome is not the most appropriate descriptive term for Beringia in late Pleistocene time. The original term steppe-tundra, with its implication of arctic rather than steppe climate and vegetation, appears in keeping with the pollen record and other data available (Colinvaux, 1981).

As a final point, a Beringian landscape emergent from the sea and covering 1 x 10^6 km^2 might have supported a maximum of 10,500,000 people assuming uniform density of people (25 people 1000 km^{-2}), 33 year generation time, a 14,000 year period of time, and no population growth. Realistically, human population density must have been much lower in Beringia during this time span and slow population growth would have occurred. The real challenge to archaeologists and paleontologists is to find in situ evidence of people plus herbivore remains in Beringia during this time interval, and through carbon dating and other techniques, estimate density of people and herbivores per site within a given span of years. It may then be possible to estimate the density of sites within an area. Such data would enable a refinement of this model to better estimate species diversity of herbivores and human carrying capacity from Beringia ecosystems—ecosystems which must have been very different from their modern analogs.

ACKNOWLEDGMENTS

Robert J. Hudson provided much of the background information used in the modeling effort. The manuscript has benefited from discussions with J. C. Kraft on past human populations levels and from reviews by P. A. Colinvaux, R. J. Hudson, and E. S. Telfer. The many discussions at Burg Wartenstein initially led to developing an ecosystem model. The writers assume responsibility for the values used in developing the model.

15

DIGESTIVE AND GRAZING STRATEGIES OF ANIMALS IN THE ARCTIC STEPPE

P. Jerome Martin

Speculation with respect to the physiological and digestive strategies of large herbivores on the arctic steppe may be made more productive by examining the strategies used by species which appear to be modern analogs of these animals. Horses and elephants are monogastrics with large cecums; they are able to exist on low-protein, high-fiber forages by consuming relatively large quantities per unit of body weight and by using microbial protein from cecal fermentation. Ruminants such as bison can consume less feed per unit of body weight than horses or elephants and, hence, require forage which is somewhat higher in protein and lower in fiber. Herbivores use physiological and behavioral adaptations to survive in very cold weather. Selection would likely favor large animals that could grow rapidly and attain a large mature body size. The demise of the arctic steppe was likely due to change in the vegetation and, hence, in the supply of food.

Key words: ruminants, monogastrics, forage, digestibility, adaptation, forage quality, urea recycling, mammoth, horse, elephant, bison

INTRODUCTION

Hypotheses with respect to the past are perhaps as difficult to formulate as those pertaining to the future! The assessment of digestive and grazing strategies of animals which no longer exist in an environment that was perhaps unlike any with which we are familiar is a difficult task.

My background in nutrition and management of domestic species dictates my approach to this question: I shall examine the digestive systems, diets, and some aspects of behavior of what appear to be modern analogs of animals that existed on the arctic steppe. Guthrie (1968a) stated that the most common large animals in interior Alaska during the Pleistocene were horses (*Equus caballus*), bison (*Bison priscus*) and woolly mammoths (*Mammuthus primigenius*); he concluded that the habitat was largely grassland. Hopefully, knowledge of present day horses, bison and elephants holds some clues about the arctic steppe and its demise.

MODERN ANALOGS OF LARGE PLEISTOCENE MAMMALS

Horses

The horse is a monogastric herbivore (Fig. 1). Since fermentation of food occurs only in the enlarged cecum and colon, which are located posterior to the stomach, microbial cell bodies which accrue from fermentation are not exposed to enzymic degradation and subsequent absorption in the stomach and small intestine.

Robinson and Slade (1974) concluded that microbial protein arising from cecal fermentation can be degraded in the cecum and can yield amino

ISBN 0-12-355860-3

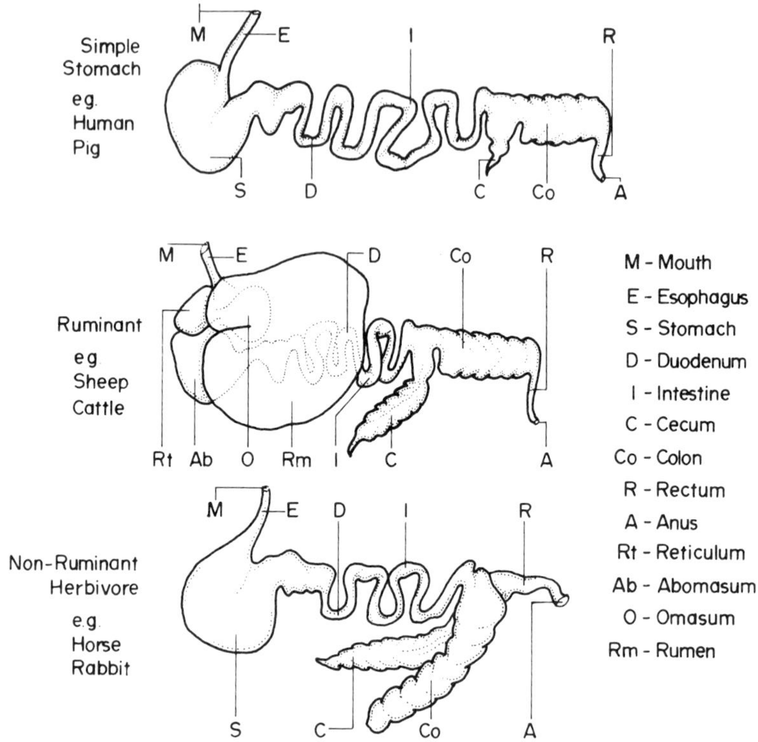

Figure 1. Comparative anatomy of gastro-intestinal tracts. (Adapted from Robinson and Slade, 1974, with permission)

acids which are absorbed and utilized by the animal. They cite work which indicates that mature horses, like mature ruminants, are "indifferent to protein quality" and that feral equines can survive on low-quality roughage that is low in nitrogen and essential amino acids.

The monograstric digestive tract of the horse is not as efficient as that of the ruminant in utilizing nutrients in high-fiber feeds, but is at least as efficient in the utilization of nutrients in low-fiber feeds (Table 1).

Hintz (1969) concluded that when high-fiber feeds ($>$ 15% crude fiber) were fed, the average digestibility values for organic matter and crude fiber for horses were 85% and 75%, respectively, of those obtained with sheep and cattle. However, a horse tends to consume more feed than a ruminant of comparable size. The ruminant must reduce the particle size of the ingested food sufficiently to allow the particles to pass through the reticulo-omasal orifice: the more fibrous the feed, the slower the rate of passage, since the more fibrous, more highly lignified feeds take longer to degrade. The rate of passage and, hence, the feed intake of horses is not limited by any similar constraint (Janis, 1976). Heinlein, et al. (1966), found that horses consumed more loose hay per unit of metabolic body weight ($W_{kg}{}^{.73}$) than sheep (79g vs. 43 g).

Table 1
Average Digestibility (i.e., % retained by animal) of Dietary Components in Horses and Cattle

Item	Organic Matter (%)	Crude Protein (%)	Crude Fiber (%)
High-Fiber Feeds			
Horses	51	56	38
Cattle	60	56	52
Low-Fiber Feeds			
Horses	79	79	56
Cattle	79	75	24

Adapted from Hintz, 1969, by Robinson and Slade, 1974

The maintenance requirement of horses, in kcal of digestible energy (DE) per day, is estimated to be 155 ($W_{kg}^{.75}$) (National Research Council, 1973) as compared to 137.5 ($W_{kg}^{.75}$) for cattle (Hironaka, 1976), a difference of 12.7%.

Domestic horses in North America are fed a variety of diets: some are simply provided with pasture, summer and winter, while others are fed very complex and expensive diets consisting of hay, grain and supplements. Mature horses can consume enough nutrients from good quality pasture to meet maintenance requirements during severe winters even if there is a deep snow cover. Horses will paw through the snow and consume large quantities of mature forage (3-5% protein). Salter and Hudson (1979) found that feral horses made craters in snow as deep as 60 cm. Pregnant mares will usually maintain their body weight throughout the winter on this feeding regime and will produce normal foals, even though they have not received supplemental feed or water during the winter.

Salter and Hudson (1979) reported that feral horses in the outer foothills of the Rocky Mountains, west-southwest of Sundre, Alberta, spent approximately 75% of daylight hours foraging during winter and spring. Grasses comprised 48-65% of the diet during the winter months and 50-70% during the remainder of the year. Sedges and rushes comprised 25-45% of the yearly diet and forbs less than 3%. Browse consumption varied from 0% of the diet in June to 9% in March. The horses grazed preferentially around trees and shrubs, and on hummocks, where snow depth was reduced. In spring, they selected new growth from areas which had been grazed previously during winter.

Bison

Bison live in a variety of environments in North America, ranging from short-grass prairie to the woodlands of the Slave River Lowlands in northern Canada. Flerow (1967) suggested that the Great Slave Lake area is somewhat similar to the cold forest-steppes that existed in Siberia, Alaska and Canada at the end of the Pleistocene period and that the Canadian wood bison (*Bison bison athabascae*) has "very close morphological affinity" with the extinct Pleistocene species *Bison priscus.*

Wood Buffalo National Park, in northern Alberta and the Northwest Territories, contains a large herd of bison (*Bison bison* x *Bison bison athabascae*). Fuller (1962) described the vegetative types in the park and their use by bison as follows:

(1) *Forest* (White spruce, jack pine, aspen and black poplar). This area is used primarily for shelter, although some summer forage is obtained from the floor of the aspen forest.

(2) *Upland Prairie.* Grasses such as *Calamagrostis canadensis, Poa pratensis, Agropyron* spp. and *Koeleria cristata* predominate in openings of the forest, constitution an important summer range.

(3) *Lowland Meadows.* The delta plain of the Peace River carries the greatest concentration of bison. *Carex trichocarpa* var. *aristata* (*Carex atherodes* Spreng.), the predominant species in the meadows, is a staple food for the bison in both summer and winter, but is especially important in the winter because it "cures into hay on the stem and produces a green shoot, rich in stored nutrient." Fuller stated that muskeg areas are generally avoided.

Bison in Wood Buffalo National Park are not affected by extreme cold per se and can be seen grazing on prairies at temperatures of -44°C on calm days. However, the combination of extreme cold and even light winds (over 300-370 cm/sec) will send them to the shelter of the forest, according to Fuller (1962). Fuller said that heavily crusted snow severely hampered the grazing efforts of the animals and that under such conditions the bulls would break a trail for the rest of the herd. He cites accounts of a severe "thaw and freeze" in about 1865 which apparently resulted in a marked decrease in bison numbers. Bison are said to be quite effective in pushing snow aside with their heads in order to gain access to forage. If necessary, they will also paw through the snow.

The bison is a ruminant. Fermentation in the ruminant digestive tract occurs primarily in the rumen, which is anterior to the abomasum and small intestine. The microbial protein which is produced by rumen fermentation is the main source of protein for the ruminant. Since microbial protein is high in quality (i.e., high in essential amino acids), the ruminant is able to utilize low-quality protein or even synthetic sources of non-protein nitrogen to provide high-quality protein for absorption in the small intestine.

Many workers have attempted to compare the digestive efficiency of bison with domestic ruminants. Schaeffer, et al. (1978), compared digestive functions, digesta retention times and plasma thyroid hormone concentrations in three breeds of cattle, American bison and Tibetan yak. Digestibilities of dry matter, energy and nitrogen were not significantly different among species. Apparent nitrogen absorption (fecal N minus feed N) per 100 g of dry matter digested was higher in the bison and yak groups than in the cattle groups. Mean retention times of digesta were 65-69 hours (h) for cattle versus 79 and 78 h for the bison and yak. There was no evident relationship between retention time of digesta and digestibility of dry matter. The authors concluded that: "Differences in the digestive efficiency of bison and yak relative to the domestic cattle when consuming poor quality feeds may be the result of an improved nitrogen economy in the gastro-intestinal tract resulting from longer retention times of digesta in the sites of microbial fermentation".

Bison digested nutrients more efficiently than cattle when high-fiber, low-protein ($< 7\%$) forages were consumed in the fall and winter (Peden, et al., 1974). There were no differences in digestibilities between species when higher quality forage was consumed in spring and summer. In this trial, bison were less selective than cattle with respect to the quality of forage consumed.

Hawley (1978) reported that slough sedge (*Carex atherodes* Spreng.) was the most important forage for bison in the Slave River lowlands. The mean apparent digestibility coefficients (%) of slough sedge when fed to bison and cattle, respectively, were: dry matter, 51 and 46; energy, 51 and 45; protein, 38 and 28; and acid detergent fiber (ADF), 47 and 42. The mean daily dry matter intakes for bison and cattle in a summer trial were each 1.6% of body weight. However, in a winter trial, the dry matter intake of cattle was 2.0% of body weight, which was significantly higher than that of bison (1.6% of body weight).

Elephants

Elephants in East Africa use grass as 80-90% of their diet (Laws, et al., 1970; Petersen and Casebeer, 1971). The daily food intake of elephants, other than lactating females, is 4% of body weight and the average passage time of digesta is 12 h (Laws, et al., 1970).

The elephant, a monogastric, has a very large cecum; thus, as in the horse, fermentation products may be available for absorption posterior to the stomach, thus minimizing the elephant's requirement for high-quality dietary protein. Like the horse, the elephant can effectively utilize large quantities of high-fiber, low-quality forage.

Petrides and Swank (1965) concluded that elephants do not digest foods very efficiently and produce only four-tenths as much biomass per acre as beef cattle.

ADAPTATIONS TO VARIATIONS IN FEED QUALITY, FEED SUPPLY AND CLIMATE

Low Protein Levels in Forages

Protein concentrations in plants decrease with increasing maturity of the plants. Animals require specific amounts of protein to meet requirements for maintenance, reproduction and lactation. An inadequate protein intake in the ruminant may cause decreased feed intake, weight loss and failure to conceive. Animals can store some protein in the blood, liver, and muscles for subsequent use, but rely heavily on daily consumption of this very important nutrient.

Protein requirements, as a percentage of the diet, vary directly with rate of growth and level of production. Young, rapidly growing animals need more protein as a percentage of their diet than do adult animals. Protein in excess of growth requirements is used by the animal as a source of energy.

Urea Recycling

The majority of protein that ruminants consume is degraded to ammonia, carbon dioxide and volatile fatty acids. The rumen micro-organisms use ammonia as a nitrogen source for their own cell synthesis. The resulting high-quality protein is eventually used by the ruminant: microbial cells are flushed into the abomasum by periodic ruminal contractions and are digested there and in the small intestine, in the same way that dietary protein is digested in the monogastric.

Some ammonia is absorbed from the rumen and other portions of the gastro-intestinal tract by the portal blood system, converted to urea by the liver and, after entering peripheral circulation, excreted via the urine or recycled to the rumen through the saliva or by diffusion. Kennedy and Milligan (1978) found an inverse relationship between the rate of urea recycling and rumen ammonia concentration in sheep. Since rumen ammonia concentration is directly related to the amount of protein consumed, the rate of recycling apparently increases as protein intake decreases. Kennedy and Milligan reported that the transfer of plasma-urea-N to the rumen in sheep was 6.2-9.8 g when (low protein) bromegrass was fed as compared to 1.2-1.3 g when (high protein) lucerne was fed. This adaptive response may be more pronounced in wild species than in domestic animals; but until more work is done in that area one must consider that the main factor affecting the magnitude of nitrogen recycling in ruminants is nitrogen intake.

Adaptation to Cold

According to Young (1978), the primary effects of pronounced cold on domestic animals are: (1) increased resting metabolic rate and hence increased energy requirements for maintenance and (2) increased rate of passage of digesta which results in lower digestibility of dry matter. Domestic livestock tend to eat more feed during cold weather, while bison tend to eat less feed during the winter than during the summer (Hawley, 1978).

Figure 2 shows how dietary energy is utilized by an animal. Note that all food energy consumed will be transformed into material wastes, body products or heat.

The energy requirement of an animal varies with metabolic body weight (body weight $^{.75}$). As indicated earlier, the maintenance requirement of cattle, in kcal/day, is approximately 137.5 ($W_{kg}^{.75}$). Thus, the maintenance requirement of an 800 kg animal is 20,700 kcal while that of a

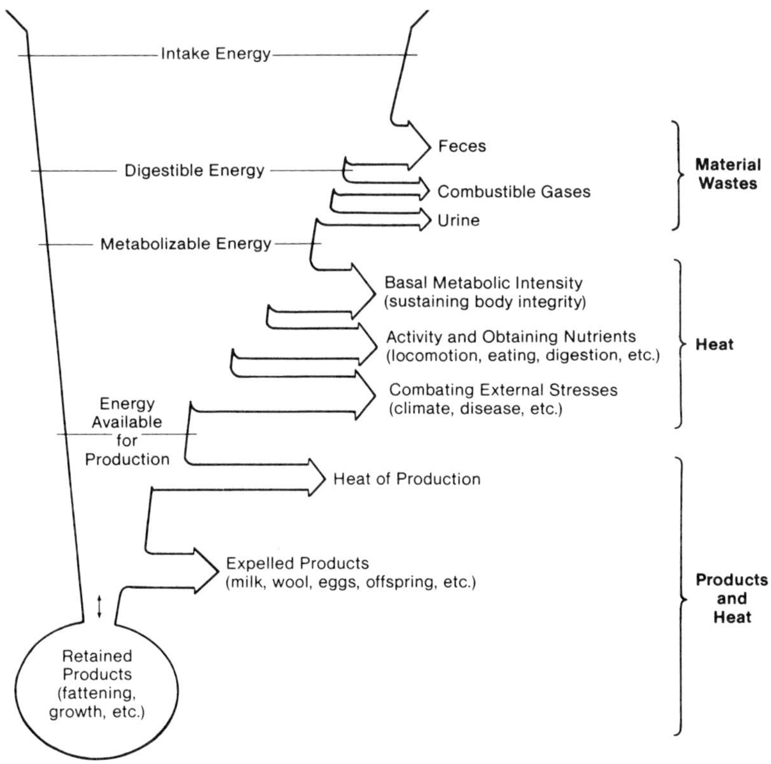

Figure 2. Utilization of energy by animals (from Young, 1977, with permission)

400 kg animal is 12,300 kcal. Hence, as body mass doubles, maintenance requirements increase by only 68%; the larger animal has a smaller surface area:mass ratio and, therefore, will not lose heat as rapidly as the smaller animal.

Heat is lost to the environment by conduction, convection and radiation. The net rate of heat loss depends upon: (1) the thermal demand of the immediate environment, as determined by temperature, wind, and precipitation and (2) the resistance to heat flow from the body, which is influenced by fat cover and hair coat.

Animals can modify heat losses by seeking shelter and by vasomotor control of blood flow to superficial tissue (Young, 1978).

Animals can adapt to cold by increasing their resting heat production (and, hence, increasing maintenance requirements). However, due to the amount of heat produced during digestion ("heat increment"), animals consuming high-energy diets (i.e., diets high in grain) seldom have to increase resting heat production, even during severe cold.

The effects of cold on digestibility of diets were not known until recently. Christopherson (1976) found that apparent dry matter digestibility was depressed by approximately 0.2 units for every drop of one degree C. The changes are apparently related to increased gut motility, increased rate of passage through the gastro-intestinal tract and elevated concentrations of thyroid hormone in plasma (Kennedy, et al., 1977). Westra and Christopherson (1976) reported that the mean retention time of feed in the rumen of sheep decreased significantly from 38.5 h in a warm environment (21.2°C) to 32.5 h in a cold environment (1.3°C). Increased appetite may compensate for the effects of cold weather on maintenance requirements and digestibility, providing that sufficient food is available. Young (1977) summarized the effects of cold on maintenance requirements and digestibility: "From a base of comfortable effective temperatures of 15° to 20°C the increase in maintenance requirement of feed energy is about 1 percent for each 1°C reduction in the effective temperature." This statement applies to cattle: some wild ungulates may be less affected by cold than Young's "one degree-one percent" maxim suggests.

Hawley (1978) found that bison ate less food during the winter than cattle but utilized it more efficiently. He suggested that reduced feed intake may cause a slower rate of passage and, hence, increased digestibility and that this "adaptive strategy" of bison allowed them to survive the winter on a limited amount of low-quality feed.

FORAGE QUALITY AND ITS IMPORTANCE TO RUMINANTS AND MONOGASTRICS

Factors Affecting Forage Quality

Forage quality is determined primarily by two factors: species and stage of growth. As a forage (i.e., a grass or herb) matures, the following changes occur: (1) yield of dry matter per hectare tends to increase (Fig. 3); (2) the proportion of leaf to stem declines (Fig. 4); (3) the percentage of cell walls (hemicellulose) in the dry matter increases and the percentage of cell contents (soluble carbohydrates, etc.) decreases; (4) the protein content of leaves and stems decrease (Fig. 5); and (5) the digestible energy content declines due to the increasing ratio of cell walls to cell contents and the progressive lignification of cell walls (Fig. 6).

The digestible energy content of a forage when fed to ruminants is inversely proportional to its acid-detergent fiber (ADF) (lignocellulose) content. Similarly, the amount of forage which a ruminant will consume voluntarily is inversely

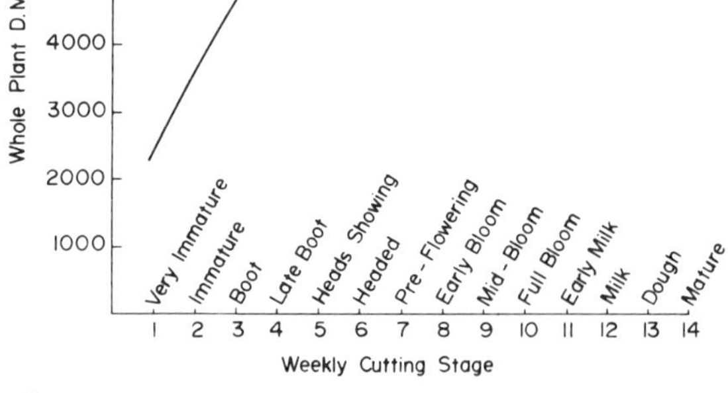

Figure 3. Dry matter yield of bromegrass at different stages of development (adapted from Kilcher and Troelsen, 1973, with permission)

proportional to the cell-wall constituents (lignocellulose plus hemicellulose, which is neutral-detergent fiber). Thus, as forages mature their digestible energy and protein contents decline; the more mature the forage the less a ruminant will eat of it, due to a decreasing rate of passage through the digestive tract.

Bell (1971) reported that in the Serengeti, zebras (monogastrics) consume the plants highest in cell-wall constituents from the top of the herb layer; topi and wildebeests consume leafy plants (higher in protein) further down the herb layer; and gazelles, the energy and protein requirements of which are high per unit of body weight, consume the high-protein fruits found near the ground. Bell stated that each member of the grazing succession prepared the vegetation structure for the next member. For example, zebras selected the tallest, most fibrous portions of plants, thus allowing other animals access to the higher quality portions of these plants.

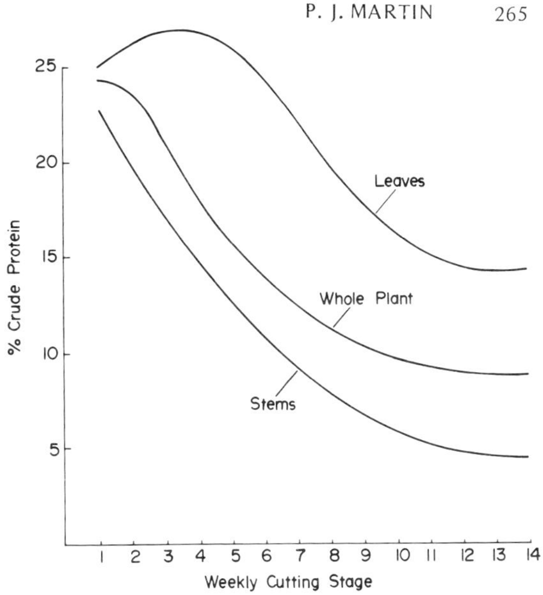

Figure 5. Protein content in dry matter of bromegrass at different stages of development (adapted from Kilcher and Troelsen, 1973, with permission)

"Secondary Compounds"

Recently, several workers have suggested that the concentrations of "secondary compounds" in plants are used as defense mechanisms, and that animals may minimize or avoid consumption of plants containing such compounds. "Secondary compounds" are described as toxic compounds and compounds which decrease the digestibility of specific nutrients. Tannins, alkaloids, non-protein amino acids and isothiocyanates are listed as secondary compounds by Cates and Rhoades (1977).

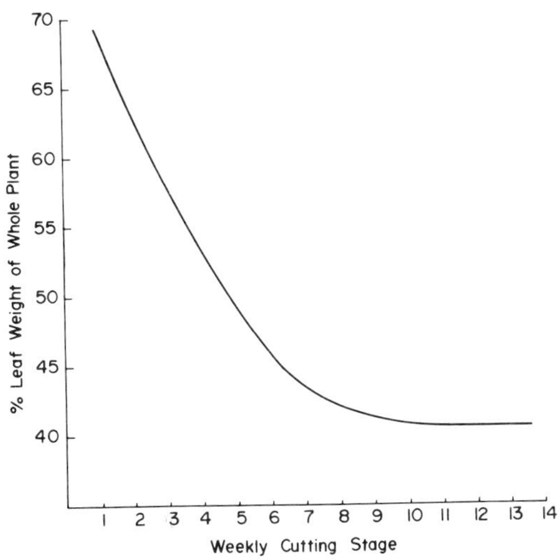

Figure 4. Percentage of leaf in the dry matter yield of bromegrass at different stages of development (adapted from Kilcher and Troelsen, 1973, with permission)

Domestic livestock often show little, if any, reluctance to consume plants high in alkaloids, cyanides or nitrates. Cattle will not consume significant amounts of *Astragalus pectinatus,* a milk vetch which often contains toxic levels of selenium, unless there is little else to eat, at which time they will eat this species and, hence, consume amounts of selenium which are often toxic.

Certain nutrients can be antimetabolites if they are present in inadequate or excess quantities: selenium deficiency decreases the availability of vitamin E and causes a number of clinical disorders, whereas selenium toxicity can cause death; and molybdenum, if present in excess, can cause copper deficiency. Thus, when assessing compounds in plants which may be toxic or which may interfere with the availability of nutrients, one cannot consider only "secondary compounds."

Herbivores choose foods primarily on the basis of digestible energy content, protein content and palatability. While secondary compounds and

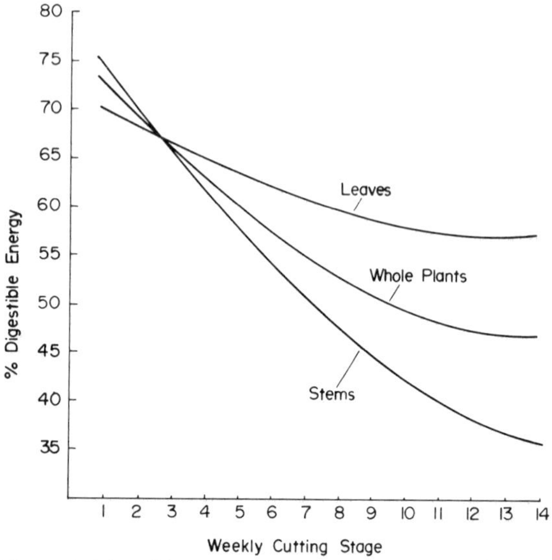

Figure 6. Digestible energy content of bromegrass plants, stems, and leaves at different stages of development (adapted from Kilcher and Torelsen, 1973, with permission)

toxic levels of specific nutrients have importance as evolutionary plant defense mechanisms, they likely have little effect on an herbivore's daily choice of food. There is very little evidence that herbivores discriminate against "secondary compounds." Freeland and Janzen (1974) stated that: "Herbivores should prefer to feed on plants or plant parts containing small amounts of 'secondary compounds'". Similarly, humans should discriminate against potatoes, spinach, carrots, radishes, bananas and tea which contain solanine, nitrates, carotatoxin, goitrogens, pressoramines and tannins, respectively (Gibson, 1980): but there is no evidence of such discrimination.

SPECULATIONS AND OBSERVATIONS

1) Bell's model of successive grazers suggests that during the growing season, horses and mammoths utilized the tall portions of the grass which were highest in fiber while bison and smaller mammals may have followed them and used the higher-quality portions of the plants.

2) During the winter, little opportunity existed for selective grazing: the mature grass was low in protein, high in cell-wall constituents and, hence, low in digestible energy. Mammoths and horses would have required large quantities of this low-quality forage daily, whereas bison most likely would have lowered their feed intake to a maintenance level and used nitrogen recycling in order to cope with the low-protein level.

3) Selection favored animals which could grow very rapidly and reach maturity quickly. While mature ruminants and monogastrics can apparently utilize low-quality forage to meet maintenance requirements, growing animals require higher quantities and qualities of food. However, animals can, to some extent, cease growing for a period of time without affecting ultimate body size. Under such conditions, they exhibit "compensatory" growth when feed supplies increase.

4) Large animals had some advantages in terms of survival in cold weather, due to a low surface area to mass ratio. The large animal has more resistance to cold and can survive periods of intense cold more easily than a smaller animal.

5) Animals which gained rapidly deposited a great deal of fat during the growing season and had an advantage. Fat is not just a source of energy for the winter; it is also a very efficient insulator. Also, a heavy hair coat would have provided a great deal of insulation.

6) Mammoths and horses, which ate large amounts of feed and digested it less efficiently than ruminants, converted a lot of mature plant material to forms which could be utilized by the vegetation for future growth. For example, plant protein was converted to urea which was used by the vegetation as a source of nitrogen. Perhaps heavy use by mammoths and horses provided for new growth of high-quality material for ruminants which might have followed in the grazing succession.

7) The demise of the arctic steppe was likely caused by a change in vegetation and, hence, a dècrease in the supply of food. Cold per se would not affect these animals although a very heavy snow cover could have hampered their movements and feeding ability.

ACKNOWLEDGMENTS

Drs. C.E. Schweger and B.A. Young deserve special thanks for their encouragement and assistance with this paper.

16

PALEOECOLOGY OF THE MAMMOTH FAUNA IN THE EURASIAN ARCTIC

N. K. Vereshchagin and G. F. Baryshnikov

An analysis of the morphology and ecology of the late Pleistocene mammoth fauna of arctic Eurasia indicates that they lived in a cold, dry climate in steppe and steppe-tundra biotopes and landscapes characterized by hard, frozen ground. The decimation of the mammoth fauna came as a result of temperature increases during interstades within the Valdai (i.e., Würmian of Europe, Wisconsin of North America) cold interval and the establishment of taiga and tundra vegetation at the end of this interval. The animals surviving the ecological catastrophe at the end of the Pleistocene (e.g., reindeer, arctic fox, marmot, souslik ground squirrel, and lemming) were able to persist in the severe conditions of present-day tundra as a result, in some cases, of their capacity for long migrations and, in others, of physiological adaptations that enabled them to cope with deep snow and occasional winter thaws. *(Abstract by V.C.S.)*

Key words: Pleistocene mammoth fauna, ecology and morphology of; reindeer; arctic fox; marmot, souslik; ground squirrel; lemming

INTRODUCTION

Siberian deposits of late Pleistocene (Zyrian, Kargin, and Sartan[1]) age contain a remarkable abundance of large mammal fossil remains. Since large animals are extremely scarce in the recent mammal faunas of the tundra and forest-tundra, conditions during the late Pleistocene must have been quite different. A consideration of the paleoecology of the late Pleistocene theriofauna is of particular interest because of the current rapid industrialization of arctic areas, the ongoing attempts to regulate utilization of the gene pool of such remaining animals as wild reindeer, polar fox, aquatic birds, and fish, and the recent attempts to enrich Siberian arctic faunas by introducing exotic species such as muskrat and reintroducing regionally extinct species such as musk ox.

The present paper is concerned only with zoological materials, including former geographic ranges, morphological characteristics, biology, and the probable ecological requirements of the extinct species. We also compare these data with similar information about species that have survived the Pleistocene. Special attention is devoted to the results of research on frozen fossil carcasses of Quaternary mammals, which have enabled us to revise earlier concepts of the adaptations and modes of life of the animals comprising the mammoth fauna.[2]

[1]The Zyrian, Kargin, and Sartan intervals of Siberia correspond to Würm I, II, and III of the Alpine climatic sequence widely used in Europe and to the Happy, Boutellier, and Duvanny Yar intervals of Hopkins (this volume).—*Eds.*

[2]We intend to devote a future paper to the similarities and differences between the landscapes and faunas of the Pleistocene steppe-tundra of Eurasia and those of the present African savanna.

ISBN 0-12-355860-3

ORIGIN AND COMPOSITION OF THE MAMMOTH FAUNA

Many aspects of the history of arctic theriofaunas remain obscure because of the lack of Tertiary fossils. Nevertheless, if we take into consideration the morphological conservatism of fossil mammals such as zebrine horses, the broad-fronted elk or moose, musk oxen, and trogontherian beavers found in early Pleistocene deposits in northeastern Siberia (Sher, 1971), we can say that a significant part of the arctic theriofauna must have developed locally within the boundaries of Beringia and the subarctic zone. The origin of arctic and subarctic theriofaunas has also been discussed recently by Pidoplichko (1954), Shvarts (1963), Sher (1971), and Kuzmina (1977). Like earlier zoogeographers, these authors have concluded that the recent fauna of the Arctic is a heterogenous aggregation consisting partly of endemics and partly of relatively late steppe colonists. This "steppe appearance" is especially well developed, as we shall see, in the late Pleistocene mammoth fauna.

The combined Pleistocene and Holocene mammal faunas of the Eurasian Arctic contain approximately 40 species, excluding the Pinnipedia and Cetacea (Table 1). The indicator species for the late Pleistocene mammoth fauna in arctic Siberia include the mammoth (*Mammuthus primigenius*), Chersky horse (*Equus lenensis*), woolly rhinoceros (*Coelodonta antiquitatis*), reindeer (*Rangifer tarandus*), steppe bison (*Bison priscus*), Baikal yak (*Poephagus mutus*), musk ox (*Ovibos moschatus*), saiga anteleope (*Saiga tatarica*), bighorn sheep (*Ovis nivicola*), Pleistocene hare (*Lepus tanaiticus*), Kamchatka marmot (*Marmota camtschatica*), long-tailed souslik or ground squirrel (*Citellus parryi*), true lemmings (*Lemmus* spp.), hoofed lemmings (*Dicrostonyx* spp.), narrow-skulled vole (*Microtus gregalis*), polar bear (*Thalarctos maritimus*), wolf (*Canis lupus*), arctic fox (*Alopex lagopus*), cave lion (*Panthera spelaea*), and wolverine (*Gulo gulo*). Most of these animals prefer open, treeless landscapes. Details of the morphological and ecological adaptations of most of these species are discussed below.

EXTINCT SPECIES IN THE SIBERIAN ARCTIC

Mammoth (*Mammuthus primigenius* Blumenbach, 1799)

Geologic age and history of range.—Tusks, teeth, bones, and skulls of mammoth are known from many hundreds of localities in all of the land-

Table 1
Composition of the Late Pleistocene and Recent Theriofaunas of the Eurasian Arctic

Orders	Number of Species Known from their Fossil Remains: Total	Became Regionally extinct	Survived the Pleistocene	Known in Modern Fauna
Insectivores	?	?	?	4
Duplicidentata (Lagomorpha)	2	1	1	2
Rodentia	7	2	5	12
Carnivora	7	1	6	7
Proboscidea	1	1	—	—
Perissodactyla	3	3	—	—
Arctiodactyla	7	5	2	3
TOTAL	27	13	14	28

scape zones of northern Eurasia. Occurrences range in age from middle Pleistocene (Riss) to Würm III and the beginning of the Holocene.

Morphology.—In body proportions and tooth structure, mammoths were closer to the Indian than to the African elephant. They were characterized by a cervical sinus, high withers, and a strongly sloping posterior. The ears were small, one-tenth to one-twelfth the size of those of the African elephant and one-fifth to one-sixth the size of those of the Indian elephant.

The woolly coat was thick, consisting of a fine underfur 15 or 16 cm long and guard hairs up to a meter in length. Hair hung in a long skirt-like mane from the dewlap and the abdomen. The trunk, ears, tail, and feet above the soles were completely covered with hair.

The tusks of old males were overdeveloped and weighed as much as 100 to 120 kg. They were 3.5 to 4.5 m long and curved upward and inward, ultimately crossing. The tusks of females were weaker, thinner, shorter, and straighter. The ends of the tusks began to wear down at the outer side when the animal was two or three years old. We have come across tusks that had been broken during the life of the animal and that eventually were abraded smooth again at the broken end. We have also seen tusks with what must have been very abnormal growths.

Ecology.—Only the summer feeding habits are known thus far (Ukraintseva, 1979). The Beresovka mammoth from the Kolyma River basin (Hertz, 1902; Pfitzenmaier, 1939) (Fig. 1) fed on sedges, grasses, and mosses. The Shandrin mammoth (Vereshchagin, 1975) fed on cottongrass, sedges, and mosses, as well as willow, alder, and dwarf-birch twigs. Winter food probably consisted of dry grasses, supplemented with willow, birch, alder, and larch twigs. The report by the engineer Beckendorf (1900) that a mammoth stomach from the Indigirka River contained spruce needles is quite a mystery, because there has never been any spruce along the Indigirka. Very large individuals of mammoth would have required 200 to 300 kg of succulent feed daily, and they must have caused considerable damage to floodplain vegetation.

The pervasive grinding down of the ends of even the smallest tusks suggests that they may have been used for stripping bark from trees and for scraping ice off vertical surfaces. The broken and subsequently smoothed ends of some tusks may be evidence that tusks were also used to break ice so that the animals could quench their thirst during dry and snowless winters.

No detailed fertility data are available, but the reproduction rate was probably lower than that of present-day elephants. Among the remains of some 140 mammoths excavated from the Berelekh "mammoth cemetery", 30% are juveniles 10 years old or less, and 2% are foetuses (Vereshchagin, 1977; Baryshnikov, et al., 1977).

Mammoths may have undertaken long seasonal migrations similar to those of African elephants. They probably made their way northward along the river valleys in spring, perhaps reaching the coast of the Arctic Ocean only in summer, and then probably moved southward for the winter.

Landscapes and biotopes.—Morphoecological data indicate that mammoths inhabited open landscapes of meadows and steppes with shrub thickets in the river valleys. Winters were probably extremely severe and cold with little snow and no watering places.

Taphonomy.—The largest concentrations of bone remains are observed in the ancient alluvium of broad river valleys. The carcasses of dead animals were carried by spring floods into the quiet water of larger rivers, where they became macerated. Some of the carcasses were deposited directly on the floodplains. Some mammoths died in landslides on hills and river banks, others became mired in mud, and still others fell into pits thawed into ground ice in the permafrost. Secondary anthropogenic burials of bones and skulls are characteristic of some Late Paleolithic encampments in European Russia and southern Siberia.

Whole carcasses and skeletons and partial skeletons with soft tissue attached have been preserved in permafrost in the tundra and forest zones of Siberia, especially in northern Yakutia. Most finds are between 11,000 and 45,000 years old. The latest find comes from the sharply cut valley of Kirgilyakh Creek near Susuman in the upper reaches of the Kolyma River basin (Shilo, et al., 1977). There, in 1977, a yearling mammoth that had starved to death some 40,000 years earlier was found buried in mud and rocky slope debris at the intersection of three ice wedges. An important discovery outside of our area was a skeleton with attached oil-soaked tissues and ligaments from the Staruni tar seeps in the eastern Carpathian Mountains of Poland (Niezabitowski, 1914a).

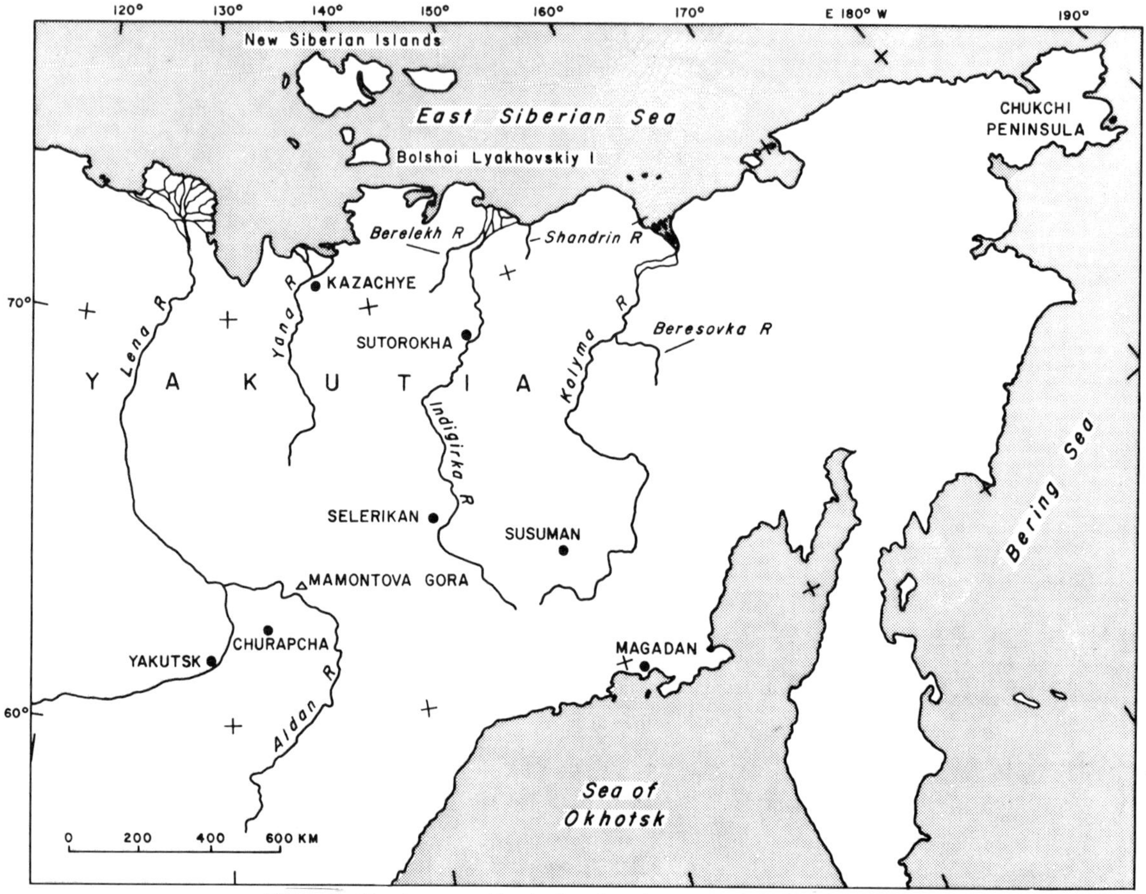

Figure 1. Northeastern Siberia showing places mentioned in the text

Chersky Horse (*Equus lenensis* Rusanov, 1968)

Geologic age and history of range.—The Anthropogene (late Cenozoic) deposits of Arctic Eurasia have yielded three or four species of horses. The Chersky horse was present during the late Pleistocene and ranged over all of northeastern Siberia from the Taimyr to the Chukchi Peninsula. Extinction probably occurred at the beginning of the Holocene, although it has been suggested that a small population of wild horses may have survived in northern Yakutia until the historical epoch (Vereshchagin and Lazarev, 1977).

Morphology.—Judging from the 37,000-year-old frozen carcass of a stallion found at Selerikan in the Indigirka River basin, the Chersky horses were low (134 to 136 cm at the withers) but heavy (Lazarev, 1977a, b). The hooves were massive, especially the front ones. The edges of the horny shoe-like ridge rimming the front hooves of the Selerikan stallion were noticeably worn down. Functionally, the Chersky horse occupied an intermediate position between draught animals and race horses.

The Selerikan stallion was a bay—i.e., coffee-brown with a dark streak running down the spine and a black mane and tail. The coat was thick and as much as 50 or 60 mm long on the sides.

The mummified remains of a pregnant mare have been found in frozen sediments in the Indigirka River basin (Garutt and Yurijev, 1966).

Ecology.—The stomach and intestines of the Selerikan horse contained herbaceous plants (90%), woody plants (5% to 7%), and mosses (1% or 2%). *Festuca,* a grass, and *Kobresia,* a sedge, predominated among the herbaceous plants; *Salix* (willow) and *Betula nana* (dwarf birch) twigs were the principal woody plant materials; and remains of green *Polytrichum* and *Sphagnum* predominated among the mosses (Ukraintseva, 1979).

Landscapes and biotopes.—Judging from the habitat preferences of domestic Yakutian horses and the wild Przewalski horse, late Pleistocene Chersky horses must have preferred open country but could have entered sparse, light coniferous forests of larch and pine if the snow in winter were loose and not more than 50 or 60 cm deep.

Taphonomy.—Remains of Chersky horses are abundant in the alluvium of Siberian rivers and in redeposited lacustrine loess-like loams, as well as in loess of Sartan age. Frozen carcasses occur in silty rock detritus in small intermontane valleys, where the animals evidently perished in mudflows.

Woolly Rhinoceros *(Coelodonta antiquitatis* Blumenbach, 1807)

Geologic age and history of range.—Bone remains, primarily of skull fragments of woolly rhinoceros are known from all of the landscape zones of northern Eurasia excepting the desert (Vereshchagin and Baryshnikov, 1980). Woolly rhinos first appear at the end of the middle Pleistocene in deposits of Riss age. No bone remains are known younger than the end of the Paleolithic.

Morphology.—The appearance and body proportions of the woolly rhinoceros were similar to those of the African white rhinoceros (*Diceros simus* Burchell, 1817), but differed, of course, in having a hairy coat. Short legs and hypertrophy of the horns were characteristic features, especially in the female. The front horn reached a length of 135 cm, and the posterior one, 50 cm. The woolly coat was soft and thick, dingy yellow and reddish-brown, and 10 or 15 cm long on the trunk. The ears were about three-quarters the size of those of the living species. The hooves had a small bearing surface, evidently an adaptation to firm, solid substrates.

Ecology.—Compressed woody remains of broadleaf species such as willow and alder and also mineral particles have been found in the deep alveoles of the upper teeth of the woolly rhinoceros. Lumps of excrement found in the area of the large intestine within the skeleton of a rhinoceros at Churapcha settlement in central Yakutia contained shoots of grasses, cottongrass, and sedges (Lazarev, 1977b). Pollen found in the foeces consisted of Gramineae (89%), Compositae (4.5%), and *Artemisia* (2.5%). No data are available on their winter feeding habits, but it is quite clear that the woolly rhinoceros was predominantly a grazer (Garutt, et al., 1970).

Nothing is known of reproductive patterns.

Landscapes and biotopes.—The woolly rhinoceros probably inhabited open areas in the cold and snowless steppe-tundra. In valleys and river floodplains it probably kept to shrub thickets.

Taphonomy.—Floods killed many rhinos and buried some complete carcasses in river valleys. Carcasses and other remains were also buried in the petroleum seeps of Staruni in the Carpathian region (Niezabitowski, 1914b). Skin fragments have been found in the Nizhneudinsk Cave in the Sayan Mountains (Chersky, 1879). Rhinos are only a minor component of the faunal material found in Paleolithic camp sites and never exceed 3% of either the total number of bones or the minimum number of individual animals in the well-studied Paleolithic encampments of the Russian lowland.

Steppe Bison *(Bison priscus* Bojanus, 1807)

Geologic age and history of range.—*Bison priscus* ranged over all of northern Eurasia and North America during middle and late Pleistocene time. Bison disappeared from the arctic zone of Eurasia at the end of the Pleistocene but persisted in the steppe zone until the Middle Ages.

Morphology.—*Bison priscus* was heavily built. The head was large and protected by massive horns. The animal stood high at the withers, but its posterior was relatively gracile. The hooves were small and adapted to a firm substrate.

The hairy coat was heavy and as long as 75 cm at the front of the trunk, furnishing protection against steppe winds (distribution of hair on the trunk of the recent European forest bison, *Bison bonasus*, is more uniform). The tail was short and covered with short hair. The coat on the carcass of a female *B. priscus* preserved in frozen ground at Sutorokha settlement on the Indigirka River was similar in color to that of a modern female Canadian forest bison (*Bison bison athabascae* Rhoads, 1897) (Flerow, 1977, 1979).

Ecology.—Recent bison are predominantly grazers. The hypsodont molars and the wide cochleariform incisors of *B. priscus* are similar to those of modern bison and suggest that, like modern plains bison, *B. priscus* fed on coarse, dusty, steppe grasses. Modern bison migrate in herds numbering several hundreds or even thousands under pressure of summer droughts or snowy winters. This may also have been true of *B. priscus.*

Reproductive rates are low in the living American and European bison; the first calf is generally born only when the female reaches an age of two or three years. Interrupted growth bands on horn sheaths recovered from permafrost in northeastern Siberia suggest that *B. priscus* females typically brought forth 11 or 12 calves during a lifetime.

Landscapes and biotopes.—*B. priscus* inhabited open steppes and arctic steppe-tundra landscapes characterized by thin winter snow cover. Present-day American bison inhabit prairies and also mixed prairie and taiga in southern Canada.

Taphonomy.—Recent European bison have been known to die because of excessive snow cover in the Beloyezhe Sanctuary and in the Caucasus, and bison in North America have perished during winter blizzards and periods of icy, crusted snow. Pleistocene *B. priscus* in the Arctic also died during snowstorms, during floods in river valleys, and by falling through thin ice. Carcasses and skeletons were then buried in oxbow lakes under layers of silt.

In the lowlands of Eurasia, bison remains are commonly found in loess and loess-like loam. The female carcass from Sutorokha settlement was buried in frozen loess-like loam. Articulated legs with skin and hair have also been found in water-laid loess and rock detritus at Kazachye village on the Yana River and Struyka Creek on the Indigirka (Flerow, 1977).

Yak *(Poephagus mutus* Przewalski, 1883)

Geologic age and history of range.—The present-day range of yak is confined to Tibet. However, a related and probably ancestral species (*P. bunnelli* Frick, 1937; *P. baicalensis* N. Vereshchagin, 1955) has been encountered in the Pleistocene of the Kuznetsk Basin, the Altai, Transbaikalia, and central Yakutia in Siberia, and in the Fairbanks district of Alaska.

Morphology.—The wild Tibetan yak is the largest of the present-day wild cattle. Tibetan yaks have massive bodies, high withers, and relatively short legs. They are covered with shaggy black hair and an abundance of underfur. Fluffy hair hangs almost to the ground from the legs, sides, and abdomen, enabling the animals to sleep on cold stones or snow. The tail, like that of a horse, is completely covered with long, coarse hair.

Ecology.—Recent yaks feed on herbaceous vegetation. In winter, they are capable of obtaining dry grass from beneath thin snow. Yaks forage alone or in groups of two or three. They visit watering places regularly in winter but can get along on snow if necessary.

Mating takes place in October and calving in June and July. The female gives birth to a single calf.

Landscapes and biotopes.—Yaks presently inhabit the rocky, treeless, semiarid, high-altitude landscapes of Tibet. Yaks reached Alaska during the Pleistocene by way of the treeless plateaus of northeastern Siberia.

Taphonomy.—Fossil yak remains have been discovered in cave deposits in the Altai Mountains. Yak bones are also sometimes encountered in lowland river alluvium. Yak remains in the Kuznetsk basin, and in Transbaikalia, come from colluvial slope debris.

Musk Ox *(Ovibos moschatus* Zimmermann, 1780)

Geologic age and history of range.—During middle and late Pleistocene time, musk oxen ranged throughout the Arctic, reaching England in the west, France and Moldavia in the south, and Alaska, Canada, and the northern conterminous United States in the east. Musk oxen probably survived until the historical period in northern Siberia and were present until the 19th century in Alaska. They are presently native to northern Greenland, part of northern mainland Canada, and the Canadian Arctic Archipelago. They were reintroduced to Nunivak Island, Alaska, during the 1930s, to several other places on the Alaskan mainland during the late 1960s, and to Taimyr Peninsula and Wrangel Island in Siberia during the 1970s.

Morphology.—Musk oxen have a thick-set body. Ears, neck, legs, and tail are very short. Hooves on the lateral second and fifth digits are small and high above the ground. Blood circulation in the legs is relatively sluggish, an adaptation to extremely cold winter temperatures.

Musk-ox fur is very thick; downy underfur 10 to 15 cm long comprises 60 to 80% of the coat. Guard hair on the chest and sides hangs almost to the ground, reaching a length of 60 to 90 cm. The fur is dark grayish-brown in winter and dark brown in summer.

Ecology.—Summer diet consists of meager shoots of grasses, sedges, cottongrass, willow, and dwarf birch. In winter, musk oxen feed on lichens, dry grass, and the twigs of shrubs. They are capable of securing food from beneath snow 15 or 20 cm deep, and remain quite fat at the end of winter. Because of their exceptional ability to utilize pasture economically, musk oxen do not migrate long distances.

Musk oxen can easily endure temperatures of -50° to -60°C. Their greatest threat is from deep snow, especially with an icy crust. A heavy snowfall can result in the death of all the young animals and of many adults.

Musk oxen live in small herds. Females calve once every two years. Yearling calves typically constitute 8% to 12% of Arctic herds.

Landscapes and biotopes.—Living musk oxen spend their winters in tundra areas on plateaus, among coastal dunes, and on rocky coastal slopes—places where snow is blown away by the winds. During the late Pleistocene, musk oxen inhabited lowland steppe-tundra landscapes.

Taphonomy.—In northeastern Siberia musk-ox remains are encountered in alluvial deposits of large rivers and in loess-like blanket loams. *Ovibos* is common in Late Paleolithic archaeological sites along the tributaries of the Dnieper and upper Don Rivers in the European part of the Soviet Union.

Saiga Antelope *(Saiga tatarica* Linnaeus, 1766)

Geologic age and history of range.—Fossil *Saiga tatarica* first appears in middle Pleistocene beds. The Pleistocene range of saiga extended westward to England and France, northward to the upper reaches of the Pechora River and the New Siberian Islands, northeastward to Alaska and coastal northwestern Canada, and southward to northern China and the Crimea. Northeastern Siberia and Alaska were inhabited by a northern subspecies, *Saiga tatarica borealis* Chersky, 1876 (= *S. ricei* Frick, 1937). *S. tatarica* is presently confined to the steppes and semiarid regions of the Kalmyk SSR, Kazakhstan, and Mongolia.

Morphology.—The saiga is a small, thin-legged antelope with a large head. It has a proboscidiform and gibbous snout with a large nasal cavity which is thought to function to remove dust from inhaled air and perhaps to warm this air during winter. The hooves are narrow, with small bearing surfaces, making them suitable only for very firm ground. The front hooves are considerably larger than the hind ones; this is evidently an adaptation for winter foraging beneath thin snow.

The color of its short summer fur is pale sandy yellow. Its winter fur is almost white and very thick, the staple up to 70 mm in length.

Ecology.—Saiga feed on steppe vegetation, predominantly on grasses and chenopods. They live in large herds, migrating seasonally in search of fresh forage or snow to quench thirst in winter. They are capable of running 60 or 70 km/hour.

Mating takes place in December. The female gives birth in May to two or, very rarely, three young.

Deep snow poses the greatest danger to saiga, and they die in large numbers during exceptionally snowy winters. Mass deaths also occur occasionally while saiga are crossing thin ice on rivers and lakes.

Landscapes and biotopes.—Recent saiga inhabit dry steppe lowlands and semiarid regions. They keep to flat country with firm, solid substrate, resolutely avoiding even slightly broken terrain. During the late Pleistocene, saiga inhabited the open expanses of steppe-tundra and steppe landscapes.

Taphonomy.—Fossil bones of *Saiga tatarica* are commonly found in the alluvium of lowland rivers. In Europe and the Crimea, they have been recovered from tar seeps and are commonly encountered in Pleistocene loam and in Paleolithic archaeological sites in river valleys and caves.

Cave Lion *(Panthera spelaea* Goldfuss, 1810)

Geologic age and history of range.—Bone remains of these lions have been found in much of northern Eurasia and North America. In Eurasia, they appear in Eopleistocene beds and disappear at the end of the late Pleistocene during Würm III, but small populations may have persisted into Holocene time in the Caucasus and in Tibet. Cave lion remains are known in North America only from deposits of Illinoian and Wisconsin age.

Morphology.—Cave lions were large, relatively long-legged cats similar in form to hybrids of tiger and lion. The opening of the internal nares is wider than in tigers, indicating a potential for powerful respiration and an ability to run short distances at high speeds (Vereshchagin, 1971).

Its winter fur was probably quite fluffy. It is unlikely that the cave lion had either a lion-like mane or a distinct tuft on its tail.

Ecology.—The cave lion was a eurytopic, cold-tolerant species which, like other members of the mammoth fauna, was poorly adapted to movement in deep snow. Prey, judging from the associated fauna, probably consisted of horses, donkeys, giant deer, red deer, reindeer, bison, aurochs, yak, musk oxen, sheep, goats, and saiga. Young mammoths and rhinoceroses were probably also attacked.

Reproductive patterns are unknown but probably were similar to those of other large cats, which produce two to four young per litter.

Landscapes and biotopes.—The cold steppe-tundra, forest-steppe, and gallery forests of river valleys would have been the usual habitat of cave lions. They ascended to altitudes of more than 2000 m and used caves as shelters.

Taphonomy.—Bones of *Panthera spelaea* are commonly found in caves and less commonly in alluvium, colluvium, and loess. They are also common in Paleolithic cave and valley occupation sites. Remains of cave lions are especially numerous in the tar seeps of southern California, where they are third in abundance after dire wolf and machairodont.

SPECIES IN THE SIBERIAN ARCTIC THAT SURVIVED THE PLEISTOCENE

Reindeer *(Rangifer tarandus* Linnaeus, 1758)

Geologic age and history of range.—The reindeer's fossil record begins in the early Pleistocene of Europe. During late Pleistocene time, its range extended westward to the British Isles and the Pyrenees, southward to the Alps and the Crimea, and eastward across Beringia to northeastern North America.

Morphology.—The reindeer is a medium-sized deer of moderately heavy build. Its eyes are large and vision is excellent. Reduction of the lower incisors is a distinctive dental feature.

Its feet are adapted for movement across compact snow and boggy surfaces. The phalanges are set wide apart, the lateral digits move apart freely, and the hooves are saucer-shaped. These adaptations enable reindeer to crater snow up to 80 cm deep to gain access to food (Formozov, 1946; Flerow, 1950).

Both males and females bear antlers through the winter. It is possible that antlers are used to keep competing individuals away from winter food.

The winter coat is very thick and impermeable to wind and snow during blizzards. Guard hairs are thick and brittle, with hollow medullary cells. The animals are light-colored in winter.

The antlers of taiga and forest reindeer are more weakly developed than those of tundra populations. Like moose, forest reindeer have relatively long legs, possibly as an adaptation to the loose, deep snow of the forest environment. Pleistocene antlers from Denmark and the Angara region of Siberia are similar to those of modern tundra reindeer (Yermolova, 1978). The late Pleistocene reindeer of Eurasia have, compared with modern reindeer, relatively short legs with digits set wider apart, and more flattened and widened hooves (Kuzmina, 1971).

Ecology.—A number of adaptations (the large eyes and excellent vision, the structure of the metatarsus, the ability to feed on lichens, and the strong herd instinct) have made reindeer well adapted to life in open, treeless spaces where the snow is not too deep. The great mobility and endurance of these animals enables them to undertake long seasonal cross-country migrations.

Reindeer feed on grass and semishrubs in summer and on reindeer lichen which they dig from under the snow in winter. Occasionally they feed on lemmings, voles, and birds' eggs. Forest reindeer also consume tree lichens. C. C. Flerow (1935) postulated that Pleistocene reindeer also fed on steppe grasses, and that the partial reduction of the lower incisors occurred during Holocene time in response to a softer diet.

The reproductive cycle is similar to that of other subarctic deer. Mating takes place in September, and single or twin calves are born in June.

Landscapes and biotopes.—Reindeer now prefer sparse coniferous forests and moss bogs in the taiga zone and moss-lichen vegetation in the tundra zone. In June, they migrate to the shores of the Arctic Ocean for relief from mosquitoes and flies. At the end of August, they begin the migration to the forest-tundra for winter feeding on reindeer-lichen pasture. During the Pleistocene, reindeer evidently inhabited open, tundra-like landscapes.

Taphonomy.—The largest concentrations of reindeer remains are found in loess and loess-like loam associated with Late Paleolithic encampments. Bones and antlers of *Rangifer tarandus* are constantly encountered in northern Eurasia (though never abundantly), in alluvium, hillside colluvium, peat, and other perennially frozen sediments.

Long-tailed Souslik or Ground Squirrel (*Citellus parryi* Richardson, 1827)

Geologic age and history of range.—The present-day range of the souslik covers the tundra, forest-tundra, and northern forest zones of northeastern Siberia, Alaska, and Canada. Three mummified carcasses of *Citellus glacialis,* a related species which inhabited Beringia probably during late Pleistocene time, have been found at a depth of 12.5 m in permafrost in the Dirin-Yuryakh River valley (not shown on Fig. 1) in the Indigirka basin (Vinogradov, 1948). Skeletal remains of late Pleistocene sousliks are also known from the northern part of the maritime lowland of Yakutia and from Bolshoi Lyakhovsky Island.

Morphology.—The long-tailed is the largest of the Paleoarctic sousliks. Summer pelage is a distinctly spotted brownish-ocher on the body; the head is darker, and the winter coat is lighter. Pleistocene sousliks were larger, and their tails were shorter.

Ecology.—Present-day sousliks feed on grass, berries, mushrooms, green mosses, insects, and the flesh of vertebrates. Mating takes place in spring, and 10 or 12 young are born in each litter.

Landscapes and biotopes.—In the lowland tundra and forest-tundra of Yakutia, sousliks are found in areas where the soil thaws deeply in the dry meadow and steppe segments of south-facing hill slopes and river valleys. Here they excavate their holes and go into winter hibernation.

Some behavioral traits of long-tailed sousliks are unique among tundra mammals. That they burrow deeply, hibernate in winter, and restrict themselves to a few local biotopes are indications of the relict nature of this species in the Arctic and Subarctic.

Taphonomy.—Bones of sousliks are encountered in loess and washed silts. The frozen carcasses found by Vinogradov (1948) bore winter fur, and their pose was typical of hibernating animals, indicating that they must have died in their nesting chambers.

True Lemmings (Genus *Lemmus* Link, 1795)

Geologic age and history of range.—Two recent species, *Lemmus* Linnaeus, 1758 and *L. sibiricus* Kerr, 1792 (= *L. obensis* Grants, 1827) inhabit the lowland and mountain tundra and forest-tundra of the Holarctic. A third, *L. amurensis* Vinogradov, 1924, lives in open spaces in the forest zone of eastern Siberia. Lemming remains are found in Pleistocene deposits in central France, Czechoslovakia, the Ukraine, and the Altai Mountains, far south of their present-day range.

Morphology.—Lemmings are medium-sized voles with very short tails and small ears concealed by fur. The winter coat is fluffy and up to 22 mm in length. The soles of the feet are covered with hair. Siberian lemmings turn a lighter color as winter approaches, but only the lemmings of the New Siberian Islands turn pure white (Ognev, 1948).

Ecology.—Siberian lemmings live in colonies and excavate shallow tunnels. They feed on the common species of sedges and cottongrasses. In summer, they also consume berries, mushrooms, mosses, and the bark of shrubs. They are active during winter, burrowing under the snow in search of food. They may even reproduce during winter.

Abrupt population fluctuations characterize Siberian lemmings. Spontaneous mass migrations of the young are a characteristic phenomenon in stress situations (e.g., flooding or freezing of their tunnels).

Landscapes and biotopes.—Lemmings are typical rodents of the present-day tundra. They prefer swampy areas with herbaceous vegetation and a moss cover or hummocky moss and sedge marshes along lakeshores. Arctic lemmings have adapted to life in the moss and turf stratum of the tundra where air and soil temperatures are low, permafrost is shallow, and snow is present during winter. Their ancestors probably inhabited forest swamps, shrubby sphagnum bogs, and sparse taiga, as does the present-day Amur lemming (*L. amurensis*). With the disappearance of the forests and the formation of open tundra landscapes, lemmings began to inhabit marshy tundra biotopes in periglacial regions.

Taphonomy.—Lemming remains are preserved in caves and in river alluvium.

Hoofed or Collared Lemming (Genus *Dicrostonyx* Gloger, 1841)

Geologic age and history of range.—The hoofed lemming lineage has been characterized by rapid evolution during late Cenozoic time. *Praedicrostonyx* (Guthrie and Matthews, 1971), the early Pleistocene ancestral genus, is known from western Europe, northeastern Siberia, and Alaska, and evidently dispersed throughout circumpolar regions during periods of cooling (Gromov and Polyakov, 1977). *Dicrostonyx* remains are found in late Pleistocene deposits in the foothill and lowland areas of Eurasia as far south as France, the Ukraine, the Altai foothills, and central Yakutia. Recent *Dicrostonyx* species have a Holarctic distribution in tundra and forest-tundra, extending even to islands in the Arctic Ocean.

Morphology.—Hoofed lemmings are medium-sized voles characterized by a short, hairy tail and small external ears in the form of skin folds. The summer fur of the Siberian hoofed lemming (*D. torquatus* Pallas, 1779) is brownish-gray. The winter fur is snow-white and clearly differentiated into underfur and guard hair that may be as long as 20 mm. The soles of the feet are thickly covered with hair. As winter approaches, the claws on the front feet enlarge rapidly, acquiring the form of small, sharp hooves which enable the animal to dig runways at the soil surface in dense snow.

Ecology.—Siberian hoofed lemmings feed predominantly on the shoots, leaves, and bark of dwarf birch and willow shrubs, as well as cloudberry and crowberry. They bring forth several litters of four to six young each year. Population fluctuations and the consequent migrations are not as pronounced in *Dicrostonyx* as in *Lemmus.*

Landscapes and biotopes.—Hoofed lemmings prefer the dry, elevated, well-drained parts of the tundra. They excavate their tunnels under stones, in crevices, and in hummocks.

Their morphophysiological adaptations indicate that they evolved in cold, dry, open, tundra-like landscapes in which permafrost was present and winter snow was dense and long-lasting.

Taphonomy.—Remains of *Dicrostonyx* are commonly encountered in owl scats in Pleistocene caves and in ancient river alluvium.

Narrow-Skulled Vole (*Microtus gregalis* Pallas, 1778)

Geologic age and history of range.—Fossil remains of this species are known from deposits as old as middle Pleistocene. During the late Pleistocene, its southern range extended to France and to the upper reaches of the Don River. Five frozen mummified carcasses were found in late Pleistocene deposits approximately 37,000 years old in the Dirin-Yuryakh River valley in the Indigirka basin (not shown on Fig. 1).

Morphology.—Narrow-skulled voles show clearly defined adaptations for burrowing, i.e., a narrow skull and shortened limbs. Their winter fur is thick and up to 22 mm in length. The coats of recent narrow-skulled voles vary from light ocher to dark grayish-brown in summer, turning white as winter approaches. The coats of the frozen fossil *M. g. egorovi* Fejgin from the Dirin-Yuryakh River valley (Gromov and Polyakov, 1977) were brownish-rust colored as a result of melanin pigment breakdown after death.

Ecology.—*Microtus gregalis* prefers legumes and grasses but in the Arctic feeds mainly on sedges, cottongrass, and prostrate willows. Grass fodder is stored for winter. These animals excavate complex runways under the snow at the ground surface. In the northern part of their range, they begin to reproduce at the end of June, giving birth to as many as three litters of nine offspring each year.

Landscapes and biotopes.—Narrow-skulled voles inhabit open landscapes ranging in character from semiarid steppes to lowland and mountain tundra and alpine meadows. They also inhabit meadow glades in the forest zone and in the alpine forest belt and are found in river valleys and along the shores of lakes in the tundra.

Taphonomy.—The bones of these animals are commonly encountered in the Pleistocene alluvium of Siberian rivers. They have also been recovered from bituminous loam near Karmalka in the Tatar SSR.

Arctic Fox *(Alopex lagopus* Linnaeus, 1758)

Geologic age and history of range.—Fossil remains of this animal are common in Quaternary deposits as old as late Riss. During late Pleistocene time, the Eurasian range of the arctic fox extended southward to France, Rumania, and the Crimea, Altai, and Baikal regions. They presently inhabit the tundra and forest-tundra of the Holarctic.

External form.—These small foxes have short legs and small ears. Pelage is grayish in summer and turns white in winter. The winter fur has long, soft guard hairs and a very dense underfur. The soles of the feet are covered with long, thick hair.

Ecology.—The arctic fox is ecologically similar to the steppe fox (*Vulpes corsac* Linnaeus). It is omnivorous, but its basic diet generally consists of lemmings and voles. Shelter is a hole in a gully slope. At the end of May or in June, the females bring forth 8 to 12 and sometimes as many as 20 pups; these develop more rapidly than do red-fox pups. Populations vary greatly from one year to another.

Landscapes and biotopes.—Arctic foxes presently inhabit open tundra areas. In winter, some follow the polar bears far northward over the ice of the Polar Basin. Others move southward to the forest zone, especially when lemmings are scarce. Arctic fox are poorly adapted for movement over loose forest snow, however. Those reaching the southern part of the forest zone are unable to return and die of hunger.

Taphonomy.—Fossil remains of Pleistocene arctic fox are common in alluvium. They are especially common in Paleolithic caves and occupation sites.

HABITAT OF THE MAMMOTH FAUNA

Having examined the species composition, functional morphology, and ecological requirements of the species comprising the mammoth faunal complex, we can now reconstruct with considerable confidence the ecological situation in the Arctic during the Wurmian-Sartanian-Wisconsin cold period. A comparison of the Pleistocene with the living arctic fauna leads to the following conclusions:

A climate colder and drier than that of the present is indicated by the density and form of the thick woolly coats of the mammoth, rhinoceros, horse, bison, musk ox, and yak. The pelts of all these animals had skirt-like fringes that would have been susceptible to a lethal wetting and freezing in cold, damp, maritime climates with frequent cycles of freezing and thawing. The fur of those mammals that have persisted in the Arctic to the present time—the lemmings, the arctic foxes, the reindeer and the bighorn sheep—is quite different in structure. Further proof of the severity of the late Pleistocene arctic climate is to be seen in the greater stature but smaller ears and tails of many late Pleistocene species compared with those of their living descendants.

That soils were generally firm and dry is indicated by the prevalence of fast-moving cursorial species such as horse, bison, and saiga antelope, by the relatively small hooves of many of the ungulates, and by the wide distribution of deep burrowers among the smaller animals. At present the soil thaws only to depths of 28 to 30 cm beneath the moss cover of lowland tundra at 70°N latitude in Yakutia. It could hardly have thawed more deeply under a grass cover during late Pleistocene time, but the ground must have remained firm and dry and swampy areas must have been at a minimum[3]. Areas of deeper thaw must have been present, however: sousliks, marmots, arctic foxes, and wolves now den either on south-facing slopes where the ground thaws and dries to depths of 60 or 70 cm or in the sandy alluvium of floodplains where the soil thaws to depths of 100 to 120 cm.

The snow cover was generally thin, and flat areas were probably blown bare by winter winds, making dried vegetation available to mammoths and ungulates during winter. A complete absence of snow during some severe winters seems to have

[3]If soils were drier, as they seem to have been, depths of seasonal thaw may have been much greater during late Wisconsin time than during the Holocene. Because of the greater heat capacity of wet soils and because large amounts of heat are required to thaw segregated ice, the depth of thaw in a given summer temperature regime is much greater in dry than in wet frozen soils.—*Eds.*

forced the large mammals to obtain ice from rock crevices and from vertical ice walls or else to migrate to other regions in search of snow and unfrozen springs.

The vegetation inhabited by the mammoth fauna can be shown to have consisted of open steppe on the watersheds and gallery forest and shrub thickets in the valleys. Most of the species comprising the mammoth fauna, including the mammoth itself, horse, rhinoceros, northern saiga, Baikal yak, bison, bighorn sheep, musk ox, long-tailed souslik, marmot, and cave lion, are or were inhabitants of open landscapes. Grasses predominated in the diet of the mammoths and of most of the ungulates. The patchy modern distribution of some of the small mammals, notably the long-tailed souslik and the Kamchatka marmot, and their occurrence in relict areas of steppe vegetation is further evidence that Arctic landscapes were once extensive steppe lands.

There seems to be a strong case for high biological productivity in the unglaciated Arctic during late Pleistocene time. The abundance of bone remains of mammoths, ungulates, rodents, and carnivores in the Quaternary deposits of northern Siberia seems in itself to indicate a high level of productivity for the plant associations of the late Pleistocene. Admittedly, an alternate explanation might be that bones as well as soft tissues are simply preserved better and occur more commonly in permafrost than in places where the ground has long remained unfrozen. However, high productivity for a possible modern analog of the Pleistocene steppe-tundra has been established by Tomirdiaro (1972), who reports hay yields of 4000 to 5000 kg per hectare in the grass and mixed-herb vegetation of the *alas* meadows which occupy drained Siberian tundra lakes for 15 or 20 years until superseded by mosses and sedges.

FACTORS IN THE EXTINCTION OR SURVIVAL OF ELEMENTS OF THE MAMMOTH FAUNA

Deaths due to old age or accidents associated with unfavorable weather conditions such as storms, glazed frost, snowless winters, floods, and other stress situations should not be confused with extinction. Extinction, as a rule, is a lengthy process associated with a low reproductive rate and a death rate that exceeds the natural increase of the population. Lacking statistics, we must limit ourselves to an analysis of environmental factors that were detrimental to some species but had little effect on others.

Bone remains of the large mammals comprising the mammoth fauna are scattered diffusely in loess, loess-like clayey loam, marine and lacustrine mud, and fluvial sand in the vast tundra regions of Siberia. Mass "burial grounds" containing thousands of bones representing hundreds of individual animals are rare and generally represent fluvial reconcentrations in oxbow lakes in river valleys and deltas (Vereshchagin, 1974, 1977, 1979).

Radiocarbon dates on bones and soft tissue point to the occurrence of two intervals of intensive animal destruction, 45,000 to 38,000 and 12,000 to 10,000 years ago (Vereshchagin, 1972; Orlova, 1979); these intervals coincide with the warmest phase of the Kargin Interstade, 42,000 to 37,000 years ago, and the end of the Sartan Glaciation, 11,000 to 10,000 years ago (Kind, 1973). The warm intervals of the Pleistocene saw an increase in winter precipitation, depth and looseness of snow cover, frequency of winter thaw, depth to permafrost, and abundance of lakes and swampy areas in the Arctic. Forests spread to the shores of the Arctic Ocean, and the extent of areas of herbaceous vegetation diminished drastically (Ukraintseva, 1979). This habitat deterioration led to an increase in mortality among the large mammals. A change in the quality of the diet is indicated by study of stomach contents of frozen carcasses: the food consumed by mammoths and ungulates began to consist predominantly of wet-ground plants such as sedges, cottongrass, sphagnum, and green mosses—plants that are nutritionally inferior to the plants of dry habitats and the mixed herb cover of meadows (Ukraintseva, 1979).

Climatic warming and humidification were detrimental to the mammoth and to several other species, and their populations evidently diminished drastically in northern Siberia during the Kargin Interstade. The increase in the role of herbaceous communities during the Sartanian cooling created better conditions for the mammoth and its associated species, but their numbers were already so diminished that mammoth and rhinoceros became extinct; horse, bison, saiga, yak, and lion disappeared from Arctic regions, and musk ox disappeared from Eurasia.

Reindeer, bighorn sheep, lemmings, narrow-skulled voles, arctic fox, polar bear, and a number of other species successfully survived the shocks of late Pleistocene environmental changes and have survived into the Holocene Arctic, either because of their capacity to undertake long seasonal migrations or because high reproductive rates and morphofunctional adaptations enabled them to cope successfully with deep snow. Finally, the sousliks and marmots have been preserved as Pleistocene relicts on small, persisting patches of steppe-tundra.

It is natural to ask whether the former productivity of the Eurasian Arctic could be restored and an abundance of ungulates once more maintained through the aid of artificial cultivation of the primary biomass and reintroduction of horse, bison, and musk ox. The answer probably must be no as long as the present climate persists and until a new glacial epoch sets in.

17

MORPHOLOGICAL CHARACTERS OF THE MAMMOTH: AN ADAPTATION TO THE ARCTIC-STEPPE ENVIRONMENT

Henryk Kubiak

This article deals exclusively with the morphological characteristics of the mammoth. A detailed summary is given of the present knowledge of mammoth features which could result from adaptation to the arctic steppe environment. Some morphological characteristics, among them, for example, the fat hump, the flap at the tail base, etc., are critically discussed, although these features are not assumed to be adaptive. However, most of the mammoth features analyzed here are adaptive and indicate: 1) Protection against the cold air and wind from the double-coated fur (long outer hair and thick underwool) covering the whole body, including the trunk, ears, tail, and limbs, and from the thick layer of subcutaneous fat. 2) Prevention of the loss of body heat by small ears entirely covered with hair. 3) The ability of the mammoth to deal with tundra vegetation due to its relatively long body and short limbs; the special structure of the tip of its trunk, high molars and great number of dental plates; and enlarged twisted tusks. 4) Its advantage in marshy summer pastures because of its broad feet and relatively small body size (compared with that of the mammoth's ancestors).

Another important subject mentioned is the role of physiological mechanisms, which undoubtedly were also significant in the adaptation of the woolly mammoth to the cold climate.

Key word: mammoth, morphological features of

INTRODUCTION

The woolly mammoth, *Mammuthus primigenius* (Blumenbach) belongs to those fossil animals of which more than skeletal remains are known. More or less preserved carcasses from the permafrost of Siberia, Alaska and even from ozocerite veins in Starunia (Eastern Carpathians)(Nowak, et al., 1930; Kubiak, 1969, 1971), as well as drawings and sculptures made by paleolithic man, allow us to recognize and study the anatomy of this extinct elephant in some detail.

However, one of the biggest obstacles to a comprehensive understanding of the mammoth, and especially of its adaptation to a cold climate, is the lack of a detailed summary of this subject. In contrast to scientific efforts, a number of popular and quasi-scientific articles and books have appeared in recent years, "in which fragmentary knowledge, folk tales, and science fiction are combined under the guise of verity—much to the chagrin of scientists and the confusion of the public" (Farrand, 1961). A recent example is that of Krause (1978), who comes to the conclusion that the mammoth was unable to live in a cold climate because its morphological features were not adaptive to the Arctic. Krause proceeds to fashion a thesis that the

ISBN 0-12-355860-3

presence of mammoth carcasses in Siberia is explained by the Biblical flood, and not by gradual extinction in glacial conditions.

The present article centers only on the morphological features of the woolly mammoth, well adapted to conditions of extreme cold and tundra vegetation, contrary to those of modern elephants living in rather warm environments.

The morphology of the mammoth is relatively well known because of the common finds of mammoth fossils: single molars, tusks, skulls, skeletons, carcasses, as well as drawings and sculptures by paleolithic man. However, not all of the details known are useful in this case, e.g., the interesting hyoid apparatus of the mammoth (Kubiak, 1980b).

The opinion that the woolly mammoth was a very large creature is very common. Even Sikes (1971, p. 7) recorded: "These creatures apparently grew slightly larger than our two species of modern elephant . . . . " In reality, the shoulder height of measured mammoth skeletons does not exceed 350 cm, or 320 cm according to Garutt (1964, p. 98-99). In contrast, the maximum shoulder height reliably recorded for the recent African elephant is 401.3 cm (Sikes, 1971, p. 15, 25). The modern Asiatic or Indian elephant is definitely smaller. Sexual dimorphism is evident in all of the species, including the mammoth, females being generally smaller. Thus, the woolly mammoth was not a giant within the elephant family. The opposite belief probably originated because early paleontologists did not distinguish between different species of fossil elephants. Some of the bones of older fossil elephants, such as the so-called "steppe" elephant, *Mammuthus trogontherii*, have been erroneously described as belonging to *Mammuthus primigenius*.

The morphology of the woolly mammoth differs from that of modern elephants in the following ways: (1) the mammoth had slightly different proportions, e.g., relatively shorter limbs and a longer body; (2) the head had a conspicuous topknot formed by large sinuses, and possibly lumps of excess fat, a trait which is also seen in prehistoric cave drawings; (3) the slope of the back, from shoulder to hip, was greater in the mammoth than in modern elephants; (4) its body was covered with long coarse hair and thick underfur; (5) under the epidermis of the mammoth was a layer of fat up to 9 cm thick; (6) mammoth tusks were larger and more curved than those of modern elephants; (7) dental plates and enamel layers in molars were thinner than those of modern elephants, and the mammoth had more molars and the highest number of ridge-plates; (8) the tip of the mammoth trunk ended in an upper fleshy "finger" and a lower, flattened, broad process; (9) the mammoth had the smallest ears, and (10) the shortest tail.

BODY PROPORTIONS

In comparison with modern elephants the mammoth had shorter limbs but a longer body. This was probably an adaptation to the acquisition of food. The mammoth was forced to adopt a grazing, or grass-eating, habit. The African elephants in their preferred habitat are primarily browsers, who feed mainly on arboreal plants and shrubs. Their limbs are longer and their body is shorter than the mammoth's. The proportions of the Asiatic elephant are considered as lying between those of the African elephant and the woolly mammoth. The Asiatic elephant is adapted to both grazing and browsing.

Fore- and hind limbs of elephants generally resemble cylindrical pillars, and, when the animal is at rest, there is little marked external evidence of the positions of shoulder, elbow and knee joints other than creases in the skin. Distally, they terminate in a circular manus or forefoot and a pes or hindfoot, both of which support isolated hooves. Each hoof corresponds in location to the appropriate toe. The elephant foot differs, however, from that of man (plantigrade) and of the dog or horse (digitigrade), in that the bones of the phalanges assume an intermediate position within the foot. These are thus said to be semi-plantigrade/semi-digitigrade in position, embedded in a cushion consisting of elastic fibers enmeshed with fatty tissue. The cushions are compressible and shock absorbing in character so that, as the foot receives the weight of the body, the cushion is compressed and the phalanges are extended somewhat laterally within it. This causes simultaneous expansion of the foot, enabling an adjustment to the surface of the terrain beneath it in a manner similar to a lightly inflated motor tire. "It is the possession of this shock-absorbing, internal cushion, and the self-adjusting sole, that enables this enormous animal to walk over rough terrain and be inaudible to the human ear" (Sikes, 1971, p. 32). The broad feet of the mammoth were also adapted to support its heavy body in marshy, poorly drained terrain. The best preserved feet are those of the mammoth carcasses from the Great Lyakhov Island, the Beresovka River, the Sanga-Yurakh River, the Berelekh "cemetery", and the Kirgilyakh River. A left hind leg, found in Berelekh with four preserved hooves (30-35 mm in diameter), has been described by Vereshchagin (1977, p. 27). The sole, or cornified skin (24 x 25 cm in size), is slightly convex and shows irregular fissures. The lower part of the leg is covered with reddish-golden dense hair, 30-40 mm long. In its upper part the hair has a length of 20-25 mm or more. Under the outer hairs is a brown-reddish underwool. The skin (15-18 mm thick) of the whole leg is preserved. Very remarkable are the small hooves. According to Guthrie (this volume) the small hoof size of many of the large mammal grazers suggests an adaptation to firm dry substratum, which indicates less insulative vegetation cover and a relatively deeper summer thaw in the Pleistocene Arctic.

The sole of the elephant foot, as well as that of the mammoth, consists entirely of cornified skin with distinctive fissures across it.

Modern elephants are pentadactyles, but according to Garutt (1964, p. 108), Farrand (1961, p. 729) and others, the mammoth forefoot was tetradactyl. In my previous study (Chmielewski and Kubiak, 1962, p. 14), however, it was concluded that the Siberian mammoths were tetradactyles, and the European pentadactyles. The small mammoth from Sanga-Yurakh River was even tridactyl because of the reduction of the first and fifth toes. The hindfoot in the Asiatic elephant has five toes, that of the African elephant and the mammoth, four toes.

The Head

The head of the mammoth was larger and more massive than that of any other elephant, and had heavy, long curved tusks and a relatively short neck. The conspicuous topknot in the mammoth head was formed of large sinuses and possibly lumps of fat and external hair. The head of the mammoth was higher and shorter than that of the recent elephants. The intermaxillary bones forming the alveolae of tusks were very long in the mammoth, and constricted in their middle part. This constriction is connected with the mammoth's strongly twisted tusks. In modern elephants with straight tusks the alveolae are nearly parallel.

The Shape of the Back

The general appearance of the line of the back is different in the two species of modern elephants. The highest point in the African elephant is in the region of the first lumbar vertebra. The middle portion of the back is concave, while the highest point in the Asiatic elephant appears in the middle of the back. These differences in the shape of the elephant back are connected with the length of the neural spines and their differentiated inclination on the vertebrae.

The known fossil material of the mammoth vertebral column is not adequate for an exact reconstruction of the back in this species. However, from the drawings and sculptures of paleolithic man, it is possible to recognize its general appearance. It was quite different from that of modern elephants. According to Vereshchagin and Baryshnikov (this volume), the mammoth was characterized by a cervical sinus, high withers and a steeply sloping posterior. The anterior part of the mammoth back formed the highest point and looked something like a hump. Posteriorly, the back line sloped steeply to the lumbar region. Between the head and the "hump" a concavity was visible. The neck was very short due to the cervical vertebrae, which were remarkably anteroposteriorly compressed. They were the shortest and most flattened cervicals in any elephant species. This external profile of the mammoth was recorded in nearly all of the paleolithic drawings. This curious profile of the back of the mammoth is in my opinion a result solely of the characteristic structure of the vertebral column. In spite of the nearly equal length of the neural spines of the thoracic vertebrae in the mammoth and in modern elephants, as mentioned above, the length of the spines in the mammoth decreased more from head to tail than in modern elephants. Accordingly, the anterior portion of the vertebral column appears higher than the posterior. Moreover, the density of longer hairs at the mammoth's neck caused the hump to appear even more enlarged.

Hair and Coat of Fur

According to the results of investigations by Vereshchagin and Baryshnikov (this volume): "The woolly coat was thick, consisting of a fine underfur 15-16 cm thick and guard hairs up to a meter in length. A long mane of hair resembling a skirt hung from the dewlap and abdomen. The feet (except for the soles), the trunk, ears, and tail were completely covered with hair."

A detailed study of the hair of the mammoth was made by Ryder (1974). The coat of the mammoth has been described as having an underwool, a longer, coarser yet fluffy outer hair, and even longer, more sparse, bristly but flexible guard hair. According to various authors, the longest hair was on the shoulders (about 50 to 100 cm). It hung along the flanks in a way similar to that of the yak or musk ox. The length of the hair on the forelegs was 25-30 cm, that on the hindlimb reached 4-12 cm. The hair on the sides of the head was 20 cm long. The terminal portion of the tail was furnished with a bunch of long bristly, coarse hairs.

According to Vereshchagin (1977, p. 28-29) three kinds of hair have been found at the Berelekh "cemetery". Individual hairs were 94, 102, and 115 cm long. These brownish, reddish, and even yellowish hairs probably hung along the flanks of adult mammoths. Their average diameter was about 250 μ. Single hairs could support 400-700 grams. There were also black, bristly hard hairs about 40-50 cm in length, probably from the neck and tail. The third kind of hair was that of the underwool, 10-20 cm long and yellowish in color.

Regarding the histological investigations of mammoth hair from Berelekh (Vereshchagin, 1977, p. 29-31) neither a medulla in cross-sections nor scales on the hair surface were visible. In the central part of the guard hairs brownish-yellow pigment granules occurred, although they were not detected in underwool hair. In guard hairs oval cross-sections occurred, while underwool hairs show circular cross-sections. In comparison, the guard hair of the woolly rhinoceros, *Coelodonta antiquitatis* living about 23,000 and 33,000 years ago, show a mosaic scale pattern, a circular cross-section, but no medullae. Some macroscopically black rhinoceros hairs appear dark-reddish and heavily pigmented under magnification (Kubiak, 1969, 1971; Kubiak and Dziurdzik, 1973). Hoyer (1914) also described pigment granules in mammoth and rhinoceros hair. The lack of medullae in mammoth hair, as well as in all rhinoceros hair (extant and fossil), is characteristic of and largely peculiar to these animals, although hairs without visible medullae do occur rather rarely in other mammals: in bats, in some breeds of sheep, and in some kinds of human hair.

The colors of mammoth hair have been described as yellowish, reddish, brownish and black. It has been concluded (Neuville, 1917) that no fading has occurred in the color of mammoth carcass hair. With electron spin resonance (Gillespie, 1970) a mammoth hair sample has been shown to have a concentration of free radicals equal to that in white wool, and so it was concluded that it had been white, and that the brownish color was a result of discoloration. It seems, however, unlikely that the mammoth was in fact white since it was a vegetarian like the musk ox, and only predators or prey in the Arctic turn white in winter. Although Ryder (1974), unlike Hoyer (1914), was unable to detect pigment granules, the coarser mammoth hairs were darker. This feature, common in double-coated mammals, suggests that the hairs were originally pigmented, the pigment having become diffuse with time (Ryder, 1974). Chemical degradation has been demonstrated in mammoth hair (Gillespie, 1970).

Sections cut parallel to the skin surface revealed sparse, round, non-medullated hairs with no glands or erector muslces. The hair density of the mammoth flank examined was 2 to 3 cm^{-2}, and hairs were of two sizes, suggesting origination from primary central (larger) and primary lateral (smaller) follicles. The presumed centrals were deeper than the presumed laterals, extending 4 to 5 mm into the skin. There were a few follicle groups with trios of larger "primary" follicles in which the laterals were 0.2 to 0.3 mm from the central. Between the central and the lateral were smaller "secondaries". Ryder (1974, p. 191) continues:

"In attempting to draw conclusions from this evidence, the tendency is to think of the modern elephant as having lost the hair seen in the mammoth, but the modern elephant must be regarded as the ancestral form. Thus the ancestral elephant probably lacked skin glands and hair erector muscles, and it seems that these were not developed by the mammoth. Erector muscles, but not the skin glands, would have been advantageous in a cold climate. Why the elephant lost these glands when they are retained in the similarly "hairless" rhinoceros is not clear.

"From the earlier descriptions it seemed that the longest and coarsest hairs of the mammoth could have grown in the central follicles of primary trios, and so be "bristles", while the shorter and less coarse hairs might have grown in the two laterals, creating "awns". Although the modern elephant has central and lateral primaries, the hairs differ little in size, and both compare with the longest and coarsest "bristles" of the mammoth. Evidence of even larger follicles was found in the mammoth, but these did not seem to belong to trio groups. Since at least some of the centrals in the elephant may be tylotrich follicles, the largest hairs of the mammoth could also have been tylotrichs. The development of these sensory hairs in the elephant could be associated with the thick and apparently insensitive epidermis, and they could have been retained in the mammoth. The mammoth acquired a greater "primary" follicle density not only by reducing the distance between the central and lateral follicles within trios, but by reducing the distance between the groups. The resulting greater hair density was associated with a reduction in the hair diameter of the "primaries", which is a common phenomenon in sheep.

"The modern elephant lacks an undercoat and so has none of the smaller secondary follicles that are developed later than the primaries. Such an adaptation to a hot environment is seen in the domestic "hair" sheep (for example, the southern Desert type of India) which has less underwool as a result of a reduction in the number of secondary follicles. The mammoth redeveloped an undercoat by the acquisition of smaller follicles lacking in the elephant.

"It is tempting to regard these smaller follicles as true "secondaries", having developed later in the foetus, but unfortunately the lack of glands removes one distinguishing feature. The undercoat usually forms a distinct group of hairs in the diameter distribution which are separated by a wide gap from the much coarser outer coat hairs, as, for example, in the wild sheep. In the mammoth it is difficult to detect any gap, so that the diameters seem to have a continuous distribution.

"It may be therefore that once secondaries have disappeared in evolution, they cannot be redeveloped, and so any new undercoat must be formed by increasing the number of follicles, and extending the diameter range, as seems to have happened in the mammoth. An objection to this interpretation is that the size difference between the primaries and secondaries is thought to be due to their development at different times and to follicle competition. If the smaller follicles of the mammoth were developed later, they could therefore be regarded as secondaries."

In spite of some differences in the particular descriptions of the mammoth hair it may be assumed that: 1) The mammoth was a double-coated mammal with underwool and outer hair. 2) The whole body of the mammoth was completely covered with hair, even the feet (except for the soles), the trunk, ears and tail. 3) The mammoth hairs were different in color: yellowish, reddish, brownish, and even black. The general appearance of the fur was most probably dark reddish-brown with the dark colors of the longest hairs covering the lighter underwool. 4) The length of the outer hair was different on particular parts of the body: the longest outer hair (about one meter long) occurred on the shoulders and hung down along the flanks. The terminal portion of the tail was furnished with a bunch of black, long bristly, coarse, flexible hair. The yellowish hair of the underwool was 10-20 cm long, and 35-130 μ thick. The brown-reddish outer hair ranged from 40 to 100 cm in length, and 170-370 μ in diameter. 5) Brown-yellowish pigment granules were present in the guard hairs, although absent in the underwool. 6) No medulla is visible in cross-sections of the mammoth hair. 7) No glands or erector muscles in mammoth hair have been detected so far.

Thus, despite the lack of hair erector muscles, it can be inferred from the above-cited features that the development of a long, dense, double-coated fur was a notable adaptation of the mammoth to a cold climate.

Skin and Fat Layer

The best studied skins are from the mammoth carcasses of the Great Lyakhov Island and the Beresovka River. Due to the thickness and roughness of their skin elephants are frequently described as "pachyderms" (thick-skin). The thickness of the epidermis, identical with that of modern elephants, varied in the mammoth in different regions of the body. The average thickness was about 3 cm. The skin was thinnest on the head and legs—about 2 cm thick, and thickest on the soles—5-6 cm.

The fat layer under the epidermis, however, was 8-9 cm thick. Detailed studies of the histology of mammoth skin are expected to be carried out on the Kirgilyakh mammoth calf.

The presence of a hump in the mammoth and questions about its function have given rise to considerable discussion and dispute. Because the neural spines of the thoracic vertebrae were not longer in the mammoth than those in modern elephants, some have suggested that the mammoth hump was a "reservoir" of fat. Indeed, this theory was common for years. However, there is no evidence for a fat hump in mammals living in the Arctic. Such a hump is well known in mammals of dry regions, e.g., in the camel and zebu. The fat layers in large arctic mammals like the musk ox, reindeer, and polar bear is more or less uniform in all parts of the body. This uniform insulation protects the animal against the loss of body temperature, especially during cold periods.

Tusks

Another feature characterizing the mammoth was a second pair of enlarged upper incisors or tusks. They were preceded by a pair of deciduous premaxillary incisors, or tushes, which probably represent also the second pair of incisors. The tushes (in the mammoth calf about 4 cm long) were replaced after one year by a pair of continually growing permanent tusks which developed in the mammoth to a very large size. The tusks of old male mammoths were overdeveloped, weighed up to 100-120 kg, were 3.5-4.0 m in length and curved upward and inward. The female tusks were thinner, weaker, shorter and straighter. The tusks of the woolly mammoth were considerably larger than those of existing elephants. Apart from their great size, the mammoth tusks were very different in shape. Their twist is the strongest in all elephant species, fossil and recent. They emerged from the upper jaw closely side by side, but then began to spread apart: downward and outward, next gradually spiraling on their axes they turned forward and inward to form a huge circle, sometimes with their tips actually crossing. In the opinion of some paleontologists (A. S. Woodward, O. Abel, H. F. Osborn, and others), such tusks were useless in the acquisition of food and as weapons. Moreover, one theory held that the overdevelopment of tusks was one cause of the mammoth's extinction. Carrington (1962, p. 131) spoke about tusks with crossed tips: "When this stage was reached they must have been singularly ineffective as weapons, and their only apparent use was to protect the trunk with an ivory girdle rather as a hard steel bumper protects the softer metal of a motorcar." It is known that the tusks assist the elephants in feeding, or at least for such purely utilitarian purposes as stripping the bark from trees. Garutt (1964, p. 105) points out that, according to the abrasion zones visible at the tips and lower surfaces of the mammoth tusks, those animals could use their elongated incisors as tools for removing the snow covering the plants in the arctic steppe. This author argues further that extraordinarily spiraled tusks did not twist in the alveolae and were more stable in them than straight tusks would have been.

Vereshchagin (1977, p. 22-25) speaks about tusks of 50-60 mm in length which had already shown use. These were tusks of mammoths 3-4 years in age. The ends of tusks about 50 cm long in mammoths aged 8-10 years became smooth at the tips. This was a result, in Vereshchagin's opinion, of the use of tusks as a plow share when the mammoth moved his head up and down at the vertical surface. Such an action might be repeated until the mammoth reached 25-30 years, when the tusks began to grow forward and inward and upward. From that age the function of the tusks changed and abrasion zones appeared at the lower surfaces of the tusks. But abrasion zones were observed also at the upper surface of the tusks. In old specimens the extraordinary spiraled tusks became useless, but due to their permanent growth and gradual spiraling on their axes the former abrasion zones rotated to the upper surface of the tusks and were preserved in this position.

Mammoth tusks have been found that have been broken off. The accidental loss of part of a tusk in extant elephants has been observed by many zoologists, and may have taken place in the mammoth, too. According to Vereshchagin and

Baryshnikov (this volume): "The broken and newly smoothed ends of some tusks may be proof that they were used for breaking off ice from crevasses so that the mammoths could quench their thirst during a dry and snowless winter". Such tusks are known from Berelekh (Vereshchagin, 1977). Another example has been described (Kubiak, 1980a, p. 40-42) in which the right tusk in a mammoth skull is broken off and the left one is strongly twisted with a downward pointed tip. At this tip a smooth abrasion zone is visible.

There are also mammoth tusks known with what must have been painful abnormal growth.

It should be mentioned that cross-tipped tusks represent a growth stage only in very old individuals of the mammoth. Thus, in my opinion, the tusks in the mammoth were as useful for them as the tusks in the African elephant are for digging up large tubers, roots, and soil. It may be that enlarged twisted tusks of the mammoth were more useful in the Arctic than short and straight ones. In fact, the characteristic constant tapering of even the smallest tusks indicates that they were used as tools and most probably assisted the mammoth to feed in the severe arctic conditions.

Molar Teeth

Six large molars or grinding teeth developed in each half-jaw and were composed of parallel ridge-plates of dentine covered with enamel and joined by cement. The six molars developed in sequence, the earliest forming towards the front end of each half-jaw, and the others developing progressively behind them. The whole progression of six molars along each half-jaw took place steadily in a forward direction throughout the life of the mammoth. Only one molar, or parts of two or three, were in use simultaneously in any half-jaw. As the front molars moved forward, they were worn down at the grinding surface and resorbed at the roots, leaving a thin, useless piece of dentine and enamel which fragmented irregularly, the pieces being dropped from the mouth or swallowed. Completion of the molar progression sets an absolute limit of the life-span of the mammoth at about 60 to 70 years as in modern elephants.

Each successive tooth was larger than the preceding one, and had a greater number of transverse plates. The number of plates and their thickness in each molar tooth varies considerably in the different elephant species. The African elephant has a maximum of ten thick plates, the Asiatic species nearly 20 thinner plates. The mammoth, however, had the greatest number of ridge-plates, i.e., about 25 flattened plates in the last molars. This is also the greatest number of transverse plates found in any elephant species, fossil and recent. Moreover, the molars of the mammoth had the highest crowns among the elephant representatives.

There were also abnormally developed teeth in the mammoth (Kubiak, 1965), but the significance of abnormalities is not yet clear. It seems that teeth with painfully abnormal growth appeared in particular regions and periods. The healthy condition of the dentition in the most recent mammoth fossils indicates that dental abnormalities were not the reason for the mammoth's extinction.

The specialized tooth structure reflects the animal's environment, as the additional hardness and strength imparted by multiple ridges would have enabled it to deal with the coarse tundra vegetation that formed its main diet. A study of fossil plants from Starunia (Eastern Carpathians)—a site where a woolly rhinoceros and a mammoth were found—shows the domination of such plants as *Betula nana, Salix reticulata, Dryas octopetala, Polygonum viviparum* and others which are known to the recent tundra of North Europe. According to Vereshchagin and Baryshnikov (this volume) only the summer feeding habits of the mammoth are known so far. The mammoth fed in summer on sedges, grasses and mosses as well as willow, alder and dwarf birch twigs. Its food in winter probably consisted of dry grasses, willow, birch, alder, larch, and pine twigs. Results obtained from pollen analysis of the stomach contents of the Beresovka mammoth (Tikhomirov and Kupriyanova, 1954) indicated the presence of 97.09 percent Gramineae. Zaklinskaya (1954) stated that on the basis of pollen analysis the Taimyr mammoth lived in an open area—a grass steppe, without any trees in the environment. The flora of the stomach contents is similar to that of the deposits enclosing frozen mammoth carcasses. According to Farrand (1961, p. 730), "the healthy and robust condition of the frozen cadavers indicates that the mammoths fared well on such a diet". Guthrie (this volume) concludes: "The larger the animal the more fiber it can tolerate. Perissodactyls and proboscidians can process food faster than ruminants of comparable size and can thus exist on a lower-quality/

higher-fiber diet. The gradient of increasing fiber tolerance is thus: small ruminants, large ruminants, equids, and proboscidians. All these grazers may live sympatrically where there is an adequate diversity along this protein/fiber spectrum". P.S. Martin (this volume) says that the mammoth appears to have favored open regions of either tundra, steppe, prairie, parkland or riparian habitat in grasslands.

On the adaptive features of the mammoth molars it may be said: 1) the teeth were highly specialized and had the highest crown and the greatest number of ridge-plates in any elephant species; 2) such molar teeth have enabled the mammoth to deal with the coarse tundra vegetation that formed its main diet; 3) the molar structure indicates a typical grazing animal; 4) the tooth structure reflects the mammoth's environment: tundra, steppe, parkland and grassland.

The Trunk

The trunk in the mammoth was probably thicker and more massive than that of modern elephants. It was also covered with hair which can be observed on more or less preserved mammoth trunks from the permafrost of Siberia. In Flerov's opinion (1931), the structure of the end of the trunk of the finger-like upper part and the lower flattened broad process, shows an excellent adaptation to the grazing habit.

The Ears

The ears of the mammoth were small, averaging 38 cm in length and 29 cm in width. Like the rest of the mammoths, the ears were covered with hair. There are only some mammoth ears known from the permafrost of Siberia (the Lena, Great Lyakhov Island) and the Starunia mammoths. The mammoth ear was 5-6 times smaller than that of the Asiatic elephant, which in turn is much smaller than that of the African elephant. In its general shape it was somewhat similar to that of the Asiatic species. According to Sikes (1971, p. 119-120), "in normal circumstances . . . elephants rest, gently flapping their ears back and forth with their characteristic *punkah* action. The fan-like veins on the medial surface become enlarged and lose heat to the atmosphere as the movement of ears creates its own local "breeze" on each side of the elephant's head . . . . it is noticeable how much lower is the ear temperature compared with the other parts of the skin. This is presumably due to the higher evaporation rate here. Any injury or disease which interferes with the ability of ears to flap . . . puts the individual elephant at a severe disadvantage as to body heat regulation". There is no way to contradict the view based on field evidence that, aside from their additional fuctions in behavior patterns, the ears of the African elephant are primarily important as heat regulators. Contrary to the African elephant, the mammoth was adapted to a cold climate where its small ears covered with hair protected it against the loss of heat.

The Tail

The tail of the elephant is long and cylindrical, supported by a variable number of caudal vertebrae. Its relative length varies greatly from one animal to another.

Only two well-preserved mammoth tails have been studied: from the Beresovka and Kirgilyakh mammoths. The tail from the Beresovka mammoth has 21 vertebrae. In modern elephants there are 26-33 caudals. The mammoth tail was also completely covered with hair. Its terminal portion was furnished with a coat of long, bristly, coarse hairs. The base of the mammoth tail is recorded to be very broad, forming a flap. Some scientists and students believe that this flap covering the anus was an adaptation to the cold. However, a similar structure of the tail base is present in modern elephants, so there is no reason for such a conclusion.

In summary, the woolly mammoth was apparently well-adapted to the arctic-steppe environment—its long hair, warm underwool, and thick layer of subcutaneous fat protected it against the cold air and wind. Its small ears covered by hair prevented the loss of body heat. The relatively long body and short limbs, the specially developed tip of the trunk, the high molars with a great number of dental plates, and the enlarged twisted tusks would have enabled the mammoth to deal with the tundra vegetation. Its broad feet and relatively small body size (as compared to that of its fossil ancestors) were advantageous in marshy summer pastures, although, as noted, its small hoof size might suggest a terrain of firm dry substratum.

It should be noted that the subect of physiological mechanisms by which acclimatization to the cold is achieved in mammals is not discussed here because we have no information on these questions concerning the mammoth. But changes in the basal metabolic rate, the acclimatization of specific tissues and inner organs, the thermogenic role of brown adipose tissue, and other causes certainly were also significant in the adaptation of the woolly mammoth to a cold climate.

ACKNOWLEDGMENTS

I wish to express my deeply felt gratitude to the organizers of the Wenner-Gren Burg Wartenstein Symposium: David M. Hopkins, John V. Matthews, Charles E. Schweger, and Steven B. Young for their kind invitation to take part in this meeting. I am also grateful to the Wenner-Gren Foundation and especially Lita Osmundsen for the opportunity to participate in this symposium for which this article was prepared.

18

ECOLOGY AND BEHAVIOR OF LIVING ELEPHANTS: BASES FOR ASSUMPTIONS CONCERNING THE EXTINCT WOOLLY MAMMOTHS

Robert C. D. Olivier

Through comparisons with the extant elephants, this paper seeks to generate assumptions about the ecology and behavior of the woolly mammoth and thus contribute to our overall paleoecological study of the Arctic steppe-tundra.

In an environment as markedly seasonal as the steppe-tundra would have been, food abundance must have fluctuated considerably over the year. On the other hand, the woolly mammoth may have had a much simpler and more economical food strategy than do living elephants, because the steppe-tundra appears to have satisfied both staple (grass) and supplementary (browse) dietary requirements simultaneously, allowing the mammoth to take in much greater weight of balanced diet at a time. This would be an important advantage if the summer months were a time of intensive feeding during which the mammoths had to accumulate large food reserves in order to survive the winter.

Riverine environments would have provided the optimal habitat for mammoths, as it does for living elephants. Seasonal flooding would favor habitat diversity, and help to maintain a mosaic of vegetation patches at different successional stages.

A model of mammoth extinction suggests two major factors: increases in the level of human predation, and changes in the availability of suitable substrata, habitat, and food.

Key words: mammoth, Asian elephant, African elephant, Quaternary

INTRODUCTION

Diversity within the Proboscidea has resulted in a wide range of forms and adaptations to a variety of ecological situations. The history of the order is characterized by a series of adaptive shifts that produced such varied groups as the mastodonts, gomphotheres, stegodonts, and elephants (Maglio, 1973).

In the following pages the term elephant will be restricted to the family Elephantidae, which under Maglio's (1973) systematic revision includes all the species relevant to this paper, namely the only two surviving proboscidians, *Loxodonta africana* (African elephant) and *Elephas maximus* (Asian elephant), together with the extinct *Mammuthus primigenius* (woolly mammoth).

The genus *Mammuthus* originated as part of the early Pliocene radiation of the Elephantidae in sub-Saharan Africa. The genus expanded rapidly northward, becoming extinct in its former range. The mammoth lineage in Europe and northern Asia is assumed to have become progressively cold-adapted, evolving through successive grades (*M. armeniacus* and *M. primigenius*). The woolly mammoth survived into the latest Pleistocene of both Europe and Asia (Maglio, 1973).

ISBN 0-12-355860-3

The *Mammuthus* group crossed the Bering Strait during the latest Villafranchian equivalent; the first appears in Irvingtonian faunas of Idaho, Kansas, and Nebraska. The successive forms *M. imperator* and *M. columbi* remained exclusively North American in distribution. During the late Pleistocene Rancholabrean stage, a second trans-Bering migration introduced the European woolly mammoth (*M. primigenius*) into North America, where it was a common element in northern deposits of Wisconsin age (Maglio, 1973).

At about this time the woolly mammoth is thought to have dominated a now extinct steppe-like habitat that stretched from northern Eurasia, through Beringia, into Alaska. Appropriately, therefore, the entire ecosystem has been variously termed "steppe-tundra" or "mammoth-steppe biome" (see other papers, especially Guthrie, this volume).

Through comparisons with the extant elephants, this paper seeks to generate assumptions about the ecology and behavior of the woolly mammoth and thus contribute to our overall paleoecological study of the steppe-tundra biome.

The living Asian elephant is probably more closely related, phylogenetically and anatomically, to the extinct mammoth than to the living African elephant. The very close similarities in the ecology and behavior of the two surviving species can therefore, I feel, provide an acceptably sound basis for assumptions about these features in the extinct mammoth.

EVOLUTIONARY STRATEGY OF THE ELEPHANTS

The early "proto-ungulate" stock (the order Condylarthra) from which the two main groups of ungulates (perissodactyls and artiodactyls) and the near-ungulates (or paenungulates, including the elephants) arose independently in the early Tertiary, are widely assumed to have been small, forest dwelling mammals, possibly omnivorous, and probably relying heavily on nonstructural plant parts low in cellulose, such as fruits (Eisenberg and Lockhart, 1972, p. 81; Janis, 1976).

This assumption is based firstly on the fossil record, and secondly on the fact that no vertebrate animal is known to be able itself to manufacture enzymes that can break down cellulose. One can envisage, therefore, that at some initial stage there were vast, untapped food resources, as embodied in high-cellulose structural plant parts, and a multitude of unoccupied ecological niches available to mammals that could successfully break down cellulose itself, as well as the digestible cell contents bounded by cellulose cell walls. Mammals have evolved such a capability by entering into a symbiotic association with cellulase-producing bacteria, and have provided fermentation chambers within the digestive tract where those bacteria can break down the cellulose.

The two major solutions to this problem have arisen independently several times in different mamammalian groups that subsist on high-cellulose diets, among which the ungulates and near-ungulates are included. The fermentation chamber has either evolved in the fore-stomach, as in the ruminant ungulates (artiodactyls), or in the cecum as among perissodactyl ungulates (rhinos, tapirs and horses), and the elephants (see P. J. Martin, this volume).

The massive and exponential species diversification noted in the fossil record of all three groups during the Eocene and Oligocene probably reflects the evolution of the ability to exploit high-cellulose diets, and the adaptive radiation that such a capability would hypothetically permit (Eisenberg and Lockhart, 1972).

This is not the place to discuss the possible relative advantages and disadvantages associated with these two different solutions to the problem of a cellulose diet. Instead, I wish to draw attention to general principles that apply equally to all types of herbivore, and particularly to the consequences of a cecal fermentation site as possessed by elephants. These principles can be used to construct a predictive model for the diet of elephants, which can then be related to their observed diet and ecological amplitude.

Firstly, one must consider the physiological consequences of a cellulose diet for animals of various sizes. According to Kleiber's Law, the maintenance requirements of an animal per unit body weight decrease with increasing body weight (Klieber, 1961). This means that among herbivores absolute body size is very important in determining the fiber/protein ratio that an animal can tolerate in its diet. Larger animals require proportionately less protein and so are able to tolerate a larger proportion of cellulose, and a more fibrous and/or less nutritious diet. Not only do larger herbivores have smaller metabolic requirements than do smaller ones, but their gut capacity relative to these requirements is greater. Thus larger herbivores have a relatively larger storage capacity and can tolerate a longer passage time.

Accordingly, a number of selective advantages for increasing body size can be identified. During the Tertiary profound climatic changes are believed to have taken place, in which the widespread tropical forest environment of the Eocene gave way in many areas to a cooler, drier Oligocene. Available herbage probably became increasingly seasonal in growth, and more fibrous during some parts of the year. While there may have been other selective pressures at work here, a larger body size is seen as the easiest way for animals to better exploit such a situation (Janis, 1976).

Whatever the true evolutionary reason, all groups leading to modern ruminants, perissodactyls, and elephants show an increase both in species diversity and body size during the Oligocene (Romer, 1966; Sikes, 1971), theoretically placing the largest, the elephants, at the top of the fiber tolerance range. That this indeed reflects a continued radiation into niches represented by increasingly fibrous diets is supported by a positive correlation in all groups between change in dental characteristics and size.

According to Janis (1976), an indirect estimation of the fiber content of the diet can be made from changes in the surface area of the post-canine teeth, and from changes that are likely to increase tooth longevity in the face of an increasingly abrasive diet. Janis cites several methods by which this surface area can be enlarged. One is to increase the overall size of the teeth, a second to increase the complexity of the enamel pattern on the occlusal surfaces, and a third to molarize the premolars. Trends of increased tooth size and complexity of the occlusal surface are perhaps more marked and have gone to greater extremes in the elephants than in any other group (Aguirre, 1969; Maglio, 1973), and similarly the pre-molars of elephants have become indistinguishable from their molars (Sikes, 1971).

Janis (1976) emphasizes the trend toward an increase in tooth life by means of increased hypsodonty (i.e., increased height of the tooth crown relative to its length or width). Hypsodonty is associated with an abrasive diet and is widely assumed to have evolved in the Miocene to deal with the abrasive silicates in grasses (e.g., Simpson, 1950). Perhaps it is significant that in the course of their evolution elephants have not only shown increased hypsodonty, but also an additional adaptation involving changes in the "angle of attack" of the occlusal surface of the molars, which is also interpreted as having increased longevity both of the tooth and the animal (Aguirre, 1969).

These trends in body size and dental features suggest that elephants are highly adapted to cope with abrasive, nutritionally poor diets of a high fiber to protein ratio, and that this capability is responsible for their success.

At this point the theoretical consequences of the elephant's possession of a cecal fermentation chamber must be introduced in order to qualify overall predictions regarding their diet and optimal ecological strategy. Firstly, the food intake of animals with cecal fermentation chambers is not dependent on the limitations of a sieve-like reticulo-omasal orifice (Janis, 1976). Thus, when faced with an increase in fiber/protein ratio in the diet, they are able to increase food intake and throughput in order to maintain a constant nutrient absorption per unit time and the same efficiency of cellulose digestion. When exploiting food resources of variable seasonal quality, such animals have a definite advantage over those unable to increase intake beyond a fixed limit.

Confirming this hypothesis are the numerous studies of the African elephant which suggest that food intake is relatively greater in areas of poor food quality (Williamson, 1975; Malpas, 1977). The relatively larger stomach fills in lactating females further supports this sort of reasoning (Laws, et al., 1975; Malpas, 1977). Laws, et al. (1975) have put forward a suggestion by Bell (unpublished) that there may be an inverse relationship between the amount of fiber in the diet and the rate of throughput (e.g., high-fiber content → low throughput), but since high fiber content is generally associated with a low crude protein content, this is unlikely according to the predictive model for nonruminants given above.

Cecal fermenters, such as elephants, are therefore primarily limited by the quantity, not the quality of food. One can thus expand the predictive description of their diet given above, and identify elephants with the high-fiber protein, nutritionally poor foods, that are usually abundant in terrestrial range lands.

Freeland and Janzen (1974) have described how the diverse bacterial and protozoan flora of mammalian gut fermentation chambers and associated alkali conditions can play an important and highly efficient role in degrading toxic secondary compounds in plants, and the ability of the animal to tolerate diets containing them. This cannot be true of cecal digesters, however, because the small intestine, where most toxins are assimilated by the body, lies anterior to the fermentation chamber. Thus, elephants can be expected to prefer abundant, fibrous foods that contain only minor amounts of toxic secondary compounds.

While the above is an adequate description of the types of diet to which elephants are best adapted, there are other selective pressures, suggested by their large body size and cecal fermentation, that favor a less restricted diet than the one described.

Firstly, one might expect that animals such as elephants cannot continue to increase intake indefinitely in response to decreased forage quality. At some combination of reduced quantity and quality, the elephant's ability to switch to an alternate food source should be a distinct advantage. Because an animal the size of the elephant requires such a large daily intake of food, albeit small in relation to body size, such a threshold may be reached comparatively quickly, and there is therefore probably a strong selective pressure for elephants to be able to exploit a wide variety of food resources when the normally abundant, low-quality ones become too scarce or too poor. Malpas (1977) and Barnes (1979) indicated that, where elephants are unable to so optimize their diet in the face of such a situation, their body condition is adversely affected.

Secondly, there is the pressure for dietary diversity that results from the fact that the products of full digestion are formed posterior to the small intestine. A restricted diet is possible when complex bacterial flora and an associated nitrogen cycle capable of synthesizing all the animal's requirements exist before reaching the small intestine; but, as indicated, this is not the case for the elephant. There may be a selective pressure for cecal fermenters like the elephant to include a wide variety of plants in their diet in order to fulfill their amino acid requirements. Janis (1976) believes that this is the explanation for the generally greater variety of plants in the diet of perissodactyls than of artiodactyls.

The intake of a variety of foods other than those to which the animal appears best adapted might itself have a consequence (the intake of foods containing toxic secondary compounds) that would reinforce this trend. In this event the prime detoxification system available to an animal with the fermentation site distal to the small intestine is through microsomal degradation (Freeland and Janzen, 1974). Microsomal degradation involves conjugation of the toxin with material provided by normal metabolic wastes or by the animal's tissues. The amount of any given food that can be safely ingested is set by the rate of microsomal degradation of any secondary compounds it contains, and by the availability of specific materials for conjugation. By consuming several different plant foods, and so using several different rate-limited detoxification pathways involving different conjugation materials, a mammal can ingest larger amounts of toxic foods per unit time.

If other selective pressures enforce the intake of toxic foods, this in itself imposes a third and supplementary pressure favoring a greater dietary diversity than the staple diet to which an animal may seem to be best suited.

To summarize predictions for the dietary strategy of elephants based on the above review of the relevant characteristics, one should assume that elephants will always select the most nutritious, least fibrous, least toxic plants or plant parts available. They appear, however, to be highly adapted to cope with abundant, non-toxic, fibrous foods of low quality, and forage of this nature, when available, should form their staple diet. However, in order to correct for any dietary imbalances associated with such a diet, and to compensate for temporal variations (either daily or seasonal) in forage supply, one should expect the staple food to be supplemented by a wide variety of different foodplants, especially if many of these are necessarily toxic. On the condition that food quantity is never limiting, one would expect elephants to show a very wide ecological amplitude (habitat adaptability).

In the following sections, observed elephant diets and ecological strategies are discussed in relation to this predictive model.

OBSERVED DIETS

Results of studies suggest that throughout East Africa grass is the predominant food of the African elephant, comprising a high proportion of its diet from a variety of vegetation types, including open grassland and forest. It is noteworthy that a number of these studies have shown that browse becomes important only in the dry season (Field, 1971; Guy, 1976; Barnes, 1979). Several reports, in particular those of Field (1971) and Field and Ross (1976), which itemized the diet according to plant species, show that African elephants take a wide variety of mostly woody species as food, even though these rarely dominate the diet.

In contrast, very few studies have been made of diet composition and feeding behavior of the Asian elephant. Eisenberg and Lockhart (1972, p. 97) concluded that this species was, like the African elephant, largely grass dependent. McKay (1973) recorded 88 foodplants, the majority of them trees and shrubs. Nettashinghe (1973) noted 79 species of foodplant, again dominated by plants other than grasses.

Vancuylenberg (1974), using the methods of Wing and Buss (1970), analyzed 117 droppings from different localities in Sri Lanka from October through November (wet season), and estimated grass to comprise more than 80% of the elephant's diet during those months. He also expanded McKay's (1973) food list to 225 species, yet again dominated by woody plants. Vancuylenberg (1974) also presents indirect evidence that indicates increased browsing and decreased grazing during the dry season (the reverse during the wet season) as has been shown for the African elephant in equivalent habitats. Finally, Kurt (1974a, p. 626) also concludes that grass is the Asian elephant's preferred food.

Olivier's (1978) study in Malaysia, the first to be carried out in habitats where grass availability is severely limited, is also in accord with the Sri Lanka data. Grass was the most preferred food, comprising up to 33% of the diet in habitats where it was available, and was the second most important constituent. Over 290 different plant species were identified in the diet of the elephant in Malayan tropical rain forest, but only 8% were grasses, sedges, and bamboos, and 18% were palms and pandans. This is the largest number of food species so far recorded in the diet of Asian elepants.

Thus, for both species of living elephants, the recorded diet agrees with the predicted diet. The diet of both species is generally dominated by grass, a food source which fulfills the theoretical staple diet qualities described above. For example, (1) grass has a relatively low nutritive value, generally providing at any time less crude protein per unit weight than browse (Dougall and Sheldrick, 1964; Field, 1971; McKay, 1973). Furthermore, the fiber/protein ratio of grass rises as it gets older, causing a progressive decline in its nutritional value (e.g., Bell, 1970; Field, 1971). In seasonal environments, particularly, this variation in the quality of the staple diet has profound effects on elephant behavior, obliging them to use a wide variety of foods in addition to the staple. (2) The high productivity of grass (in the sense of its ability to replace cropped biomass rapidly) causes it to be constantly available, even where its spatial coverage is limited. (3) Grasses are abrasive, containing large amounts of silicates. And finally, (4) grasses are generally relatively low in toxic secondary compounds.

An increase in size is one of the evolutionary trends seen in herbivores adapted to feeding on foods with the characteristics of grass. Thus, grass-eating large mammals face the problem of ensuring that the food to which they were adapted can still be brought to the mouth, since grasses are generally low-level plant forms. In many groups (e.g., horses) the problem has been solved by increased neck length. In elephants, however, the development of a trunk seems to be the solution (Spinage, 1972; H. Croze, personal communication 1977; Ukraintseva, 1979). The mechanical versatility of the trunk makes it suitable for taking a wide variety of foods, thus allowing elephants to supplement their staple grass diet with a diverse assemblage of plants.

Woody species are the most common of the supplementary food plants for both types of living elephants. It has been suggested that one reason these are taken is in order to optimize amino acid balances. (Browse, as noted above, is a relatively rich protein source.) Balances of such essential cations as sodium and calcium also appear to be critical for living elephants. Browse species in general are relatively rich in such ions, and the species selected by elephants are especially rich in essential cations in comparison to their staple food plants.

Field (1971), speculated on the importance of essential fatty acids in the diet, noting that fat levels are generally higher in browse leaves than in grasses. He also noted that milk production may be a drain on fatty acids. This observation as well as unpublished stomach content data (R.M. Laws, personal communication 1978), which show that females take relatively more browse than males, provide strong evidence that the function of browsing in living elephants is to optimize the nutritional balance of a basically poor staple diet. The incidence of vascular disease and lowered reproductive rates in areas where browse availability is so low that normal optimization is precluded (e.g., Kabalega Falls National Park, Uganda) supports this hypothesis (Field, 1971; R.M. Laws, personal communication 1978; Barnes, 1979).

Dietary optimization by elephants through intake of a wide variety of supplementary, particularly woody, food species might explain why in Malaya the plants in families that constitute most of the herbage in the primary forest (e.g., Dipterocarpaceae) are not utilized. Almost invariably such plants are found to be exceptionally rich in secondary compounds, particularly tannins (Janzen, 1975; Rhoades and Cates, 1976). Tannins precipitate both proteins and cations, the very components the supplementary food plants are supposed to provide. This may explain why, as several authors have noted, the dominant plants of primary rain forest provide so little food for the animals living in it (e.g., Medway, 1972; Whitmore, 1972).

It is not surprising, therefore, that Olivier (1978) found the woody species taken by elephants in Malaya to be mostly quick-growing pioneer species expected to contain "cheap" toxins rather than the energy-expensive, digestibility-reducing compounds like tannins (Rhoades and Cates, 1976). As theorized earlier, intake of such toxic plants would in itself reinforce requirements for dietary diversity, possibly explaining why elephants in Malayan rain forest select a much wider spectrum of plant species than has been recorded elsewhere.

Elephants may be expected to select the most abundant of the available woody plants that will satisfy their supplementary requirements. While pioneer species are necessarily scarce in a primary forest, as quick growers they are very productive, and are therefore relatively more abundant in a temporal sense than the majority of species.

To summarize, the foods which comprise the diet of the Asian elephant, the composition of which has evolved through natural selection, increase not only in abundance, but also in diversity, in mixed secondary habitats. The same is probably true of the foods of the African elephant, although its diet has not generally been discussed from this point of view. Grass, usually the most important food of the African elephant, is itself a highly successional plant form. Furthermore, there is evidence that the tree species utilized by *Loxodonta* are also fast growing and short-lived, i.e. secondary species, because in one forest study area 80% of them *never* exceed 46 cm d.b.h. (Laws, et al., 1975). Additional support for a preference for secondary over primary forest in this species also comes from Laws, et al. (1975), and Merz (1977).

Because both the staple and supplementary foods preferentially selected by living elephants are generally successional species, elephants appear therefore to be strongly associated with mixed secondary habitats.

The food of the woolly mammoth, as determined from remains found between the teeth and in the alimentary canals of frozen specimens, also fits well with the general predictions given above for the staple diet of the elephant. In addition, the woolly mammoth shows dental features more indicative of grazing specialization than those of either *Loxodonta* or *Elephas* (Maglio, 1973). It has been established that each of the frozen mammoths whose gut contents have been analyzed apparently died in summer; hence our knowledge of the diet of the woolly mammoth is restricted to summer forage (Vereshchagin and Baryshnikov, this volume). During summer the mammoth, like the living elephants, almost certainly subsisted on a staple diet of grass.

The most renowned and thoroughly analyzed stomach contents are those collected from the Beresovka mammoth found early in this century in northeast Siberia. The vegetal remains consisted mostly of grass and a few other higher vascular plants. According to Salensky (1905), detailed examination of large plant parts by Borodin showed that the food species gathered by the Beresovka mammoth just before its death consisted of various kinds of grasses (*Alopecurus*, *Hordeum*, *Agrostis*, *Puccinellia*, and *Beckmannia*). Sedges were represented by two forms of *Carex*. A mint (*Thymus*), pods and beans of a leguminous plant (*Oxytropis*), poppies (*Papaver*), and seeds of a buttercup (*Ranunculus*) completed the list. Sukachev's (1901) analysis of the same material lists more species identifications, but fails to record some of the above genera reported by Borodin while including others not mentioned by him. Excluded are the grasses *Agrostis* and *Puccinellia*, the mint and the poppy. Included is another grass *Agropyron*, and in small quantities, two mosses, *Hypnum* and *Aulacomnium*.

Pollen analysis of a 3 g sample taken from the Beresovka stomach contents revealed a number of plant genera (n=17) and species (n=23). Of the 8198 identifiable pollen grains so obtained, 97.07% were of grasses, 2.67% herbs, and 0.17% mosses and ferns (Tikhomirov and Kupriyanova, 1954). Unfortunately, I doubt that these figures can be correlated in any way with the proportional representation of plant groups in the actual diet of the animal. However, as Tikhomirov (1958) points out, the predominance of grass pollen evidently reflects the large proportion of the mammoth's range that was covered by grassy meadows and is in accordance with Sukachev's data indicating the large part played by grasses in the diet of the mammoth. For a recent summary of all the Beresovka analyses see Ukraintseva (1979), who also records that macro-remnants of *Larix*, *Betula*, and *Alnus* were found under the corpse, and may therefore have been part of the deceased mammoth's menu. Ukraintseva also reviews analyses of the stomach and the colon contents of the Shandrin mammoth in Russia. These also reveal a grass-dominated diet. Over 90% consisted of grasses and sedges, with the tips of branches of larch (*Larix*), willow (*Salix*), birch (*Betula*), and alder (*Alnus*) evident in the balance (see also Vereshchagin, 1975). These species are the only definite browse records for the mammoth known to me. Hibben (1946) records the latter three species in the stomach contents of an Alaskan mammoth.

These data confirm the expected dependence of mammoths on grass (for part of the year at least). The predicted species diversity of the diet is not apparent, but neither would the true diversity of the diet of living elephants be apparent from examination of only one or two samples of stomach contents.

Surprise has been expressed that the mammoth may have eaten larch, a conifer presumably rich in antiherbivory chemicals. However, as pointed out earlier, when there are few alternative resources for dietary optimization, elephants can (and do, e.g., in tropical rain forest) eat a variety of toxic plants (some of them extremely so) by taking relatively small quantities of each. It may also be significant that mammoths were apparently taking twig *tips*. As rapidly growing parts, twig tips are probably relatively lower in chemicals that directly and adversely affect dietary value (such as tannins, which precipitate amino acids and cations) than the more mature parts of any plant (Rhoades and Cates, 1976).

For those interested in steppe-tundra, the instructive aspect of the dietary strategy of the surviving elephants is that mammoths in steppe-tundra must have had a comparably diverse diet in order to optimize the grass staple. It is not clear whether the necessary supplements would have been provided by herbs or trees and shrubs, generally assumed to have been scarce, but the mere presence of an elephant in the ecosystem is evidence of considerable floral diversity.

Other workers have reached this conclusion independently. In a critique of concepts then current regarding the steppe-tundra or arctic steppe, Schweger and Habgood (1976) suggested that it must have had a mosaic nature with various regional combinations of steppe, grassland or tundra vegetation being determined by local drainage, permafrost and edaphic as well as microclimatic factors. Subsequently, Guthrie (this volume) shows that the diverse character of the large mammal community in steppe-tundra (he uses the term mammoth steppe) indicates that it was characterized by a patchwork of herbaceous vegetation types rather than homogeneous shortgrass communities. Similarly the palynological studies of Matthews, Sher and Giterman (this volume) and others point to a mosaic of habitats, each having different microclimatic conditions. They make note of several former plant communities possessing diverse ecological elements, and conclude that the floristic composition of steppe-tundra was richer and more variable than that of recent tundra.*

*See also papers by Yurtsev, Schweger, and Matthews (this volume). —*Eds.*

Already mentioned is the superiority of pioneer browse species over herbs as supplements to living elephant diets. In my opinion, the presence of mammoths in the ecosystem suggests that woody plants were probably much more abundant, even if localized or of shrubby stature, than has been generally supposed. *Salix* and *Alnus* (which are established supplementary browse food plants), and *Populus* (whose pollen is both easily destroyed or misidentified—Ager, 1975) would be species that meet these criteria, and it is interesting to note that deciduous gallery forests of poplar trees and willow-alder shrub occur today along most lowland rivers and in sheltered sites in uplands throughout much of the Yukon-Kuskokwim region of Alaska (Ager, this volume). Young (1976) argued that *Artemisia*-dominated slopes in Alaska are relicts of the steppe-tundra, but since the latter was generally thought to have been treeless, he found the presence of groves of *Populus tremuloides* in the supposed relict sites a confounding feature. Given current conclusions regarding the mosaic nature of the steppe-tundra, this is no longer a paradox.

Tikhomirov (1958) concluded that the presence of pollen from various trees and shrubs, including *Salix* and *Alnus,* allows little doubt that at the time of its death the Beresovka mammoth lived in a wooded zone, and that further consideration of species that grew in the mammoth habitat show a wide variety in ecological specialization, suggesting that natural conditions during the mammoth epoch were extremely varied.

In conclusion, I propose that lowland riverine areas and sheltered sites in uplands (particularly on south-facing slopes) were resource "hot spots" for mammoths in the mosaic of steppe-tundra habitats, providing them with the suitable browse that has been shown to be an essential supplement to the staple grass diets of extant elephants.

OPTIMAL HABITATS

Under natural conditions, elephants, like other animals, can be expected to express their greatest densities in areas with the greatest carrying capacity. Those will be areas in which high-quality staple and essential supplementary foods are both abundant and available in some optimal ratio. Abundance must be considered as the net product of spatial and temporal (productivity) components.

In which habitats, then, is the staple diet—grass—found and in which is it most abundant? In heavily forested areas of the tropics, grasses and grasslands are found naturally in riverine situations, particularly as components of the mosaic habitats on the alluvial floodplains of large seasonal rivers. Grass is most abundant, however, in the more extensive fire-induced savannas, particularly in areas formerly covered by deciduous forest. In these, the long dry season favors interference by man, as his use of fire is most effective there.

In which habitats are browse species (suitable as supplementary diet components) found, and in which are they most abundant? In view of the probable evolutionary demands for the essential incorporation of supplementary species into the diet, perhaps their most important requirement is a content low in protein- or cation-precipitating secondary compounds. Possibly a relatively high value of protein or certain inorganic ions is also important.

Janzen (1975, p. 35) gives the rationale for recognizing three gradients along which he predicts a plant's investment in "expensive" chemical defense (e.g., tannins), to drop, and along which I would predict the abundance of suitable supplementary foods for elephants to increase. The gradients are given below.

(1) *From habitats where plant production is aseasonal, to those where it is markedly seasonal.* One would thus expect species characteristic of deciduous forests to be generally more palatable than those of evergreen forest. Janzen shows that this is true for "livestock." Rhoades and Cates (1976, p. 199), arguing on the strength of their model of a gradient in "predictability," have also made this prediction.

(2) *From areas in which the plant has very little of what it needs for growth (e.g., water and/or inorganic ions) to those where such resources are abundant.* Soils in tropical evergreen rain forest are generally poorer in base cations than those in deciduous forest. Chemical defense apart, one could also predict that, with richer soils, the plants of deciduous forest may be expected to be generally richer in inorganic ions than in evergreen forest. In this context, a consideration of riverine habitats is particularly interesting and instructive. Exceptions to the basically infertile soils of the tropics are rare, and are usually geologically young. Allu-

vial soils often fall into such a category, and their fertility is greatly increased when they form from sediments carried downstream from geologically young, nontropical headwaters, as is the case in the alluvial plains of the Ganges and Brahmaputra.

The higher fertility of alluvial habitats suggests that not only will there be relatively less pressure on plants adapted to such situations to adopt expensive chemical defense, but that their tissues will also be relatively base rich.

(3) *From established forest or forest understory habitats to habitats characterized by plants of early successional stages.* This gradient may be envisaged as acting across both of those mentioned above. The significance of food plants from early successional stages as supplements to the diet of living elephants in terms of secondary compounds was detailed earlier and is fundamental to the predictive model being developed here. These plants occur in the same communities as grasses, the staple diet, which is not surprising, as grass itself is characteristic of the earliest successional stages.

Riverine situations also support growth of important elephant foods that represent later successional stages than grass. For example, both Fox (1972) and Whitmore (1972) note that many secondary tree species found in cut-over tropical rain forests are found only near streams in the virgin forest.

These different variables, whose gradients affect both the abundance and quality of staple and supplementary elephant food plants in different habitats, comprise two sets of counteracting, yet interdependent, variables. They may be considered as vectors in a "parallelogram of forces", with the resultant providing the most appropriate conceptual model for predicting the relative elephant carrying capacity of a habitat (Figure 1). In other words, elephant densities will be low in climax evergreen rain forests, since both staple and supplementary foods are apparently scarce and of low quality. Conversely, the model suggests that secondary deciduous forests will have the greatest potential carrying capacity, at least seasonally, since both staple and supplementary foods can be superabundant and of high quality.

Although the model presented in Figure 1 is undoubtedly an oversimplification, it has the advantage that each vector can be subjected to independent critical examination. The acceptability of the model can be tested by comparing separately the predicted action of each vector with available empirical evidence regarding elephant density, group size, habitat preference, and human cultural involvement. Olivier (1978) carried out this exercise and obtained extensive support for the model as a summary of the broad conclusions that can be drawn regarding optimal habitat for the living elephants.

The model, as a generalization, obscures the importance of adequate habitat diversity in predicting an optimal elephant range. For example, while elephant densities may indeed reach a seasonal peak in pure grasslands, such habitats cannot maintain elephants in isolation, and are temporary. The two extant species of elephants do appear best adapted to habitats with very similar diversities (grassland/forest ratios). This is indicated by similarities in observed diets, and in the effect of grass availability on seasonal movements. Futhermore, there is strong evidence that they would show the same feeding and habitat preferences under conditions of comparable availabilities.

In both Africa and Asia riverine situations provide the optimal elephant habitat, because such areas provide maximum interdispersion of cover types where gallery forest and alluvial plains interdigitate. As Sikes (1971) has noted, the arid semideserts and savannas of Africa are generally interspersed with areas of gallery forests, particularly in low-lying areas along water courses. She calls these the "primaeval habitat of loxodont elephants". Indeed in arid and grass-dominated environments gallery forests are the only source of essential water and supplementary browse.

In Asia, where elephant habitat tends to be dominated by trees rather than grass, riverine situations, particularly those associated with large seasonal rivers rising in geologically young catchments, provide not only habitat mosaics of optimal proportions through the action of flooding; they also apparently provide niches favorable for the growth of the preferred elephant supplementary foods (browse), and a high productivity of staple foods (grass). Furthermore, foods of both types are of relatively high quality, because plants of such riverine habitats are apparently under minimal pressure to evolve expensive chemical defenses, and because of their generally good soils.

My generalized model for optimal habitat was derived for tropical elephant ecosystems. It remains

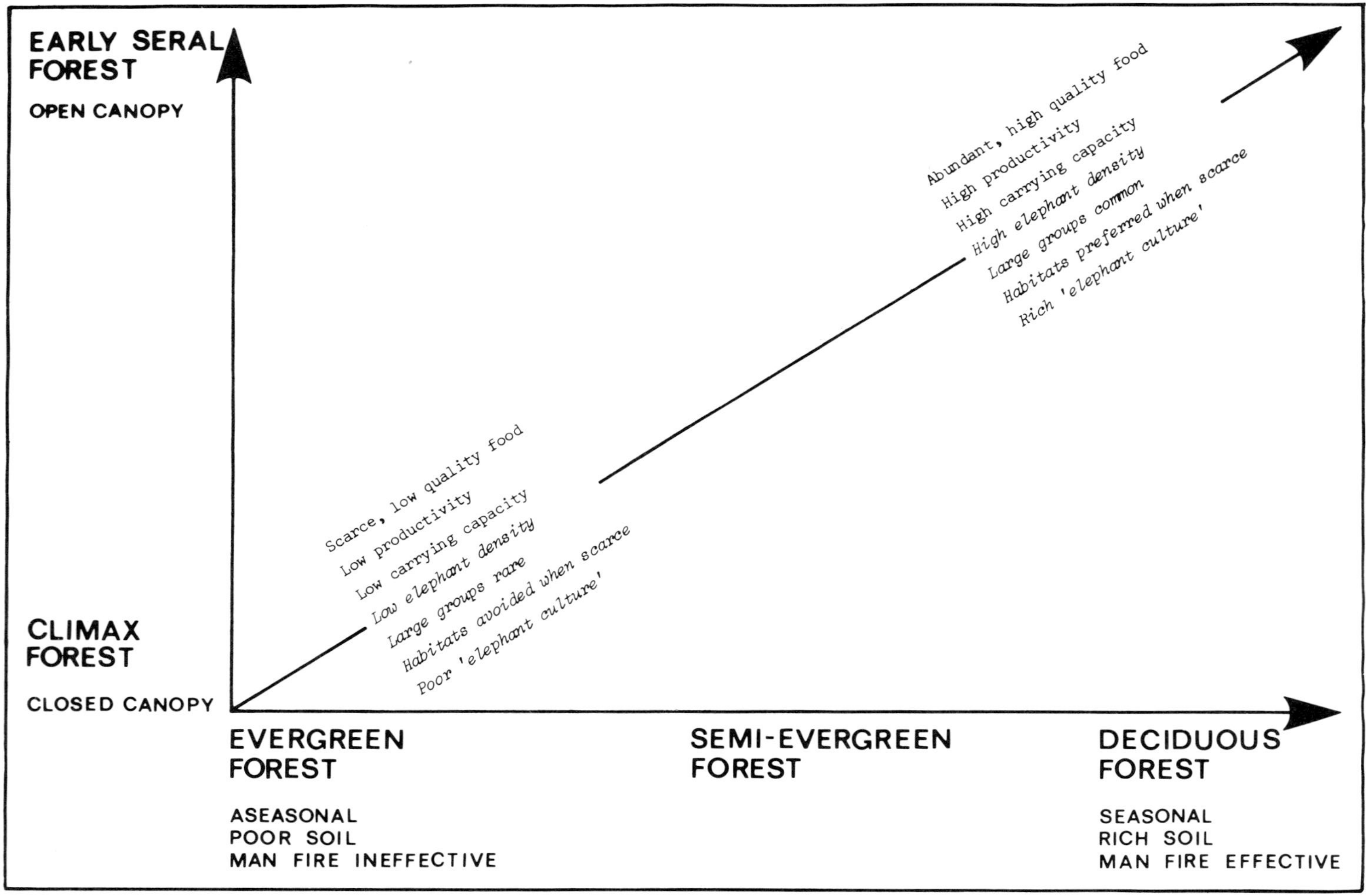

Figure 1. Model depicting the action of two sets of gradients in environmental variables, one set acting within tropical forest formations (vertical vector) and the other across them (horizontal vector), and their predicted effects on the quantity and quality of elephant food and environmental carrying capacity. These are shown along their resultant. Olivier (1978) examined the action of each vector independently, in each case collating evidence for those effects given in italics, thereby testing the model

to explore its relevance, if any, to the steppe-tundra biome. The ecological features of steppe-tundra responsible for its ability to support mammoths may become apparent by examining which, if any, of the recognized features of an optimal elephant habitat are likely to have occurred there.

The first prerequisite is an abundance of food. In an environment as markedly seasonal as the steppe-tundra must have been, food abundance must have fluctuated considerably over the year. This characteristic is considered below; it is worth stressing at this point that all the paleontological evidence points to an abundant supply of dietary staple, grass, during spring and summer. I assume that grasses occurred in an optimal ratio with supplementary foods. This was doubtless the case, but as discussed in the last section, the nature of the supplementary forage and its relative abundance is not obvious.

All three of Janzen's (1975) gradients along which supplementary foods suitable for elephants are predicted to increase in abundance are implied by the steppe-tundra evidence. Firstly, it was a markedly seasonal ecosystem. Secondly, the resources which plants need for growth were abundant, i.e., unlike the acidic soils of modern boreal and tundra areas, steppe-tundra soils were probably more alkaline and thus more conducive to herbaceous growth and nutrient release. Loess soils were also rich in available ions that would have facilitated rapid nitrification (Guthrie, this volume). Steppe-tundra thus seems to have been able to provide forage of high nutritive quality. Klein (1970) has found that plants adapted to northern ranges generally have greater quantities of nutrients per unit weight. Matthews (1976) has suggested that because of this, steppe-tundra regions may have supported their ungulate fauna at a much lower level of primary productivity than can tropical regions with an analagous fauna. Thirdly, it seems that much of the steppe-tundra landscape was characterized by "pioneer plants" maintained by loess and sand deposition, frost heaving and frost-boil formation, and strong winds (Ager, this volume). As with living elephants, mammoths probably played a role in maintaining (but not creating) these early successional communities through trampling and foraging (Olivier, 1978).

While speculation on the nature and extent of the mammoth's response to the seasonal cycle of food productivity is heavily dependent on the educated guesses of paleoclimatologists and paleoecologists regarding temperature, precipitation, length of growing season, etc., there can be little doubt that when food was available, it was of a very high quality. A picture emerges of a grass-dominated summer meadow, including a diverse assemblage of more scattered herbs. Because of the very high quality of the environment, the mammoth may have been able, by grazing nonselectively in such a pasture, to satisfy both staple and supplementary dietary requirements simultaneously, thus obviating the dependence on the type of browse species required by the living elephants. If this were so, then the woolly mammoth may have had a much simpler and therefore more economical food search strategy than do living elephants, allowing the former to take in much greater weights of balanced diet per unit time This would be an important advantage if the summer months were a time of intensive feeding during which mammoths had to accumulate large food reserves in order to survive the winter (see next section).

Given the conclusion that riverine situations provide the optimal habitat for both living elephants, was this also the case for mammoths? The answer appears to be yes. Seasonal flooding would favor habitat diversity, presumably helping to maintain a mosaic of vegetation patches at different successional stages. Loess would have formed mineral-rich soils, enhancing the nutritive status of the vegetation. My mind's eye is drawn once again to Ager's (this volume) description of deciduous scrub of *Populus-Salix-Alnus* occurring as gallery forests along the Yukon. If such "forests" fringed many of the rivers traversing the arid steppe-tundra zone of the Northern Hemisphere, they could have served as resource "hot spots" for supplementary foods. The analogy is Sikes' gallery forests in the midst of the extensive African savanna. Indeed Vereshchagin and Kuz'mina (1979) have concluded that the landscapes inhabited by the mammoth were represented by cold-tundra-steppe on interfluves and by river-valley forests and brush on the fluvial floodplains. In a separate paper, Vereshchagin (1979) pictures man hunting mammoths with spears in the "junglelike thickets of riparian willow".

RESPONSES TO FLUCTUATIONS IN FOOD SUPPLY

The increased use of grassland over forests during the wet season, when the quality of grass as a food is at its highest, is indicated by a number of independent observations of seasonal movements in living elephants. McKay and Eisenberg (1974) summarize data from Sri Lanka which show a general pattern of seasonal habitat utilization dependent on rainfall. Elephants use separate parts of their total range during wet and dry seasons. The prime factor determining movements from one part to the other appears to be availability of food, particularly grass. After the burning of extensive areas of grasslands in parts of their ranges, elephants either move to separate areas where fires have not been so extensive, or to primarily forested areas. Elephants move into burned and grassy areas as the latter become wet, evidently attracted by the lush regrowth of grass of relatively high nutritive value.

Similar evidence is available from the rest of seasonally dry tropical Asia. Singh (1969) found that in North India, some elephant populations move seasonally, while others do not, depending on whether there is a seasonal fluctuation in the availability of food and water. In India Lydekker (1924), and in Vietnam Bazé (1955) describe a general pattern of forest utilization during the dry season, followed by a movement to grasslands in response to an increase in quantity and quality of the grasses during the wet season. Sharatchandra and Gadgil (1975) state that the distribution of the elephant in South India appears to be governed by the supply of grass, bamboo, and water. During the dry season the elephants move from the grassy lowlands into the wetter hill forests.

A few studies record similar seasonal movements of the African elephant into forests during the dry season, back to grassland during the wet. Afolayan (1975) found elephant damage to forest plantation much more severe during the dry season than the wet. Laws, et al. (1975), showed that elephants use of a forest followed a definite pattern. There was a marked peak in the dry season, reflecting a real influx of elephants due to extensive burning of grasslands.

Thus we see that the living elephants are not restricted to a rigid annual cycle and, as a result of high mobility, can respond when necessary to proximal environmental factors (fluctuations in food availability due to rainfall, fire, and flooding) with appropriate movements. In this way they can maintain relatively high densities in seasonal areas. There is no reason to suppose that mammoths did not have the same potential capabilities, especially as they probably had a matriarchal social organization similar to that of the living elephants (see next section), which would allow the development of traditions that promote resource-oriented movements. Traditional use of trails, water holes, salt licks, patches of secondary growth in rain forest, and so on, is particularly evident among living elephants.

In such a markedly seasonal environment as steppe-tundra is assumed to have been, an abundance of food, particularly high-quality grasses, must have been considerably greater during the summer. Like living elephants, mammoths must have exhibited resource-oriented movements. The question arises as to the scale and timing of these movements, and it is interesting to speculate whether mammoths might have had separate summer and winter ranges analogous to the wet and dry season ranges of today's elephants. It seems to me that, unlike the living species, the mammoth would not have had the option of moving to areas with abundant leafy browse at times of the least grass availability, because such times (in a steppe-tundra situation) would have coincided with the leafless period of the deciduous woody food plants which I assume provided its preferred browse. I know of no evidence that mammoths browsed extensively in evergreen, coniferous forest during winter, and I know of no contemporary ungulates that could subsist on a diet dominated by coniferous foliage.

An alternative possibility is that mammoths were able to obtain significant amounts of "browse" in the form of bark. Certainly mammoth tusks commonly show wear patterns compatible with debarking activity (Vereshchagin and Baryshnikov, this volume). However, unless a tree is of substantial dimensions, elephants will not use tusks as a primary means of debarking. Bark can be stripped from branches or trunks thicker than a man's arm by using trunk, tongue and teeth alone. We must ask, therefore, whether trees substantial enough to warrant the use of tusks in debarking were available in the steppe-tundra, or alternatively, whether the observed tusk-wear is evidence that mammoths migrated seasonally to areas where such trees were available.

Apart from seeking an explanation for tusk wear created by debarking, other authors have suggested that mammoths must have migrated considerable distances in order to overcome presumed severe shortages of winter food. Churcher (1978) has suggested that this problem could have been completely overcome by Alaskan mammoths if they migrated up to 2500 km over a two-month period, twice a year. African elephants, while grazing, can cover an average of 0.5 kilometers per hour (Wyatt and Eltringham, 1974), and when walking purposefully can maintain 16 km/h for at least three hours (Sikes, 1971, p. 245). So even allowing a normal feeding regime of up to 18 hours per day, African elephants could certainly cover 2500 km in two months, using only one or two hours a day for concerted travel out of a possible six.

It seems likely that mammoths were capable of such a feat, but whether they actually covered such vast distances remains conjectural. There are fairly reliable records of former migrations by African elephants of up to 650 km biannually (Sikes, 1971, p. 245). Seasonal movements of this order of magnitude by mammoths seem much more plausible than ones exceeding 2000 km.

I suspect, however, that such migrations would do little to resolve the acute winter food shortage. Winter must have been a period of severe hardship for mammoths and, like most contemporary northern ungulates in winter, they probably gradually declined in weight and body condition until spring (Guthrie, this volume). For their winter survival mammoths may have depended on accumulating sufficient body reserves during summer, which implies nearly continuous summer grazing. The reported layer of subcutaneous fat in the woolly mammoths may therefore have had a double function—as protection against winter cold and starvation. The large fat deposits may even have provided water (as in camels), for as Schweger and P. J. Martin (1976) have pointed out, when snow is the water source, up to half the total caloric intake, in terms of body heat, may be expelled in order to convert snow to water.

It seems possible that mammoths eked out a precarious existence during winter, managing to survive until the following spring by a combination of the following strategies:

(1) *Burning body-fat reserves.* However, it is very unlikely that this could have played more than a minor role in winter survival.

(2) *Undertaking "elephant-size" migrations* to areas either warmer, or more wooded, or where food was more available (in the sense that snow cover was windblown or lighter and more easily cleared away with the tusks or feet) or a combination of these. Modern elephants are very adept at kicking up short-grass turf with their toenails, and at digging holes in loose sand. I see no reason why mammoths should not have been expert snow "craterers". If so, in winter they may have been able to obtain and exploit mature, poor-quality grasses that were above the fiber tolerance level of most other herbivores.

Twig tips, especially those of deciduous trees, may have provided another source of food, as they do for moose (*Alces*) during winter. Modern elephants eat bark and leafless branches in large quantities at certain times, employing their tusks to full effect. Bark and branch eating by mammoths during winter cannot be discounted, but it requires that edible trees were growing in the steppe-tundra regions (see above). Many aspects of the autecology of the modern elephants suggest that mammoths must also have used bark and branches, but as yet there is little supporting evidence. This is the single biggest paradox to emerge from my foray from tropical elephant ecosystems into that of the woolly mammoth.

SOCIAL ORGANIZATION

While the mother-offspring unit averages 2.75 animals in African elephant populations, the basic social unit observed and calculated in group-size frequency distributions consists of five to six elephants (Laws, et al., 1975). Such groups consist of two or three related mother-offspring units; that is, a matriarch and one or more mature daughters (or siblings) and their immature offspring. These are called *family units,* and larger groups are aggregations of family units. Bulls leave their parental family unit two or three years after reaching sexual maturity. Mature bulls are usually solitary, but may form temporary all-male groups which include on average two to three individuals (range

1-14) (Laws and Parker, 1968; Laws, 1969; Douglas-Hamilton, 1972).

The Asian elephant appears to have an almost identical social structure, based on a matriarchal family unit of comparable size (Lydekker, 1924; Anonymous, 1934; Eisenberg and Lockhart, 1972; McKay, 1973; Nettasinghe, 1973; Olivier, 1978). While sample sizes are not large enough to show a group-size frequency distribution conclusively correlated with multiples of five or six as in Africa, it is generally agreed that larger groups are aggregations of such family units (e.g., Sharatchandra and Gadgil, 1975).

As the latter authors note, in any given ecological situation, group size for most ungulates must be interpreted as a compromise between associated antipredator advantages (larger groups have better detective, deterrent, and repulsive qualities) and optimal resource exploitation (generally not favored by large groups). This is not the place to enter fully into a discussion of the matter but it is significant that groups smaller than the matriarchal family unit are rare in both species of living elephants, suggesting that the size of the group has strong selective advantages for elephants. The following few points are noteworthy:

Since cooperative defense of the young (Eisenberg and Lockhart, 1972), cooperative care and nursing of the young (Shebbeare, 1958; Eisenberg and Lockhart, 1972; Sharatchandra and Gadgil, 1975), and cooperative "rescue" (Olivier, 1978), have been observed in Asian elephants, there are obvious advantages in an adult female associating with at least one other adult. Owing to the slow development of its young, each adult may commonly have at any one time up to three young dependants, and this may explain why the size of basic family unit is in the middle of the 2-12 range.

Finally, the dependence in both living species of a family unit on its oldest member, the matriarch, illustrates the selective advantages of the cumulative experience and learning made possible by the evolution of longevity. This feature is probably important in terms of optimizing survival in adverse ecological conditions, or other dangerous situations, especially in animals who have very large range requirements and who live in seasonal environments.

Geist (1974), in his review of the relationship between social evolution and ecology in ungulates, has noted that closely related allopatric species may have very similar social behavior, even though their autecology is quite different. It is therefore not surprising to find such a similar social structure in both living elephant species, especially since in this case their diet and ecological strategy are also very similar. Since the diet and ecological strategy of the woolly mammoth seems to have closely resembled those of both extant elephant species, I believe that the mammoth's social organization was also similar to that described for African and Asian elephants.

This conclusion also lends further credence to Saunders' (1978) suggestion, after comparing the age structure of 14 family units of *Loxodonta* with that of *Mammuthus columbi* from the Lehner Mammoth Kill Site, that the latter represents a single family unit, catastrophically sampled. Family units of both surviving elephant species group together when threatened in open country, and both are extremely reluctant to abandon fallen individuals. The rest of the group under such circumstances repeatedly tries, and sometimes succeeds, in raising the victim to its feet (Eisenberg and Lockhart, 1972; Douglas-Hamilton, 1975; Olivier, 1978). Taken together with the undoubted ability of Clovis hunters to kill mammoths, probably by piercing the soft abdomen as do aboriginal elephant hunters of today, I find credible the suggestion that mammoths were procured as family units.

Subsequently, the age structure of two other *M. columbi* sample populations were considered corroborative of this procurement model, while another two were not. In the latter two cases it was postulated that a different procurement process was operating, biased towards mature, probably isolated individuals. From an understanding of elephant social organization and habits it seems reasonable to suppose that these were all adult males. Therefore, rather than imply local variation in the "procurement culture" of the hunters (i.e., specialization in lone bulls), it seems simpler to assume that certain areas were marginal for mammoths, and that hunters living there were therefore more likely to encounter lone adult males than family units. Several authors have suggested that it is adaptively advantageous for females to assume different food habits from males since they must maximize energy and nutrients for foetal growth and lactation. If so, this explains sex-related differences in habitat preference (see Geist, 1974, for review). Among elephants in Asia and Africa there is evidence for more consistent utilization of suboptimal habitats by adult males (McKay, 1973; Kurt, 1974a, 1974b; and Olivier, 1978; Laws, et al., 1975).

EXTINCTION OF THE MAMMOTH

Opinions expressed in some of the other papers in this volume suggest that interaction of many different factors were responsible for the dissolution of the steppe-tundra biome in general, and the extinction of the mammoth in particular. Figure 2 gives a much simplified summary of the interaction of the major factors thought to have been involved. The relative importance of the different arrows depicted remains, of course, the subject of debate. The model does however, clearly show that in the final analysis only two major factors can be directly implicated in the actual demise of the species (Arrows 1 and 2).

These are, firstly, increases in the level of human predation, until mortality from this cause exceeded natural recruitment. Predation by species other than man is assumed to have been negligible, as it is with extant elephants.

Secondly, disadvantageous changes in the temporal and spatial availability of suitable substrata, habitat, and food (i.e., niches), until the biomass of mammoths consistently exceeded the carrying capacity of the environment.

It is instructive that these two factors have independently brought the two extant elephant species into danger of extinction. With currently escalating ivory prices, it is the impact of man the hunter that is primarily responsible for alarming and rapid declines in African elephant populations (Arrow 1 in Figure 2).

On the other hand, it is the relentless reduction in the availability of suitable habitat, due primarily to the impact of man the agriculturalist, that has brought the world population of Asian elephants down to around 35,000 (Arrows 2 via 3).

If elephant species can clearly be driven towards extinction through the independent action of either one of these factors, then I consider the operation of *both* together, as I believe was likely the case during the close of the Pleistocene, to be quite adequate to explain the apparently "catastrophic" extinction of large numbers of mammoths over a relatively short span of time (Arrows 1 *plus* 2 via 4).

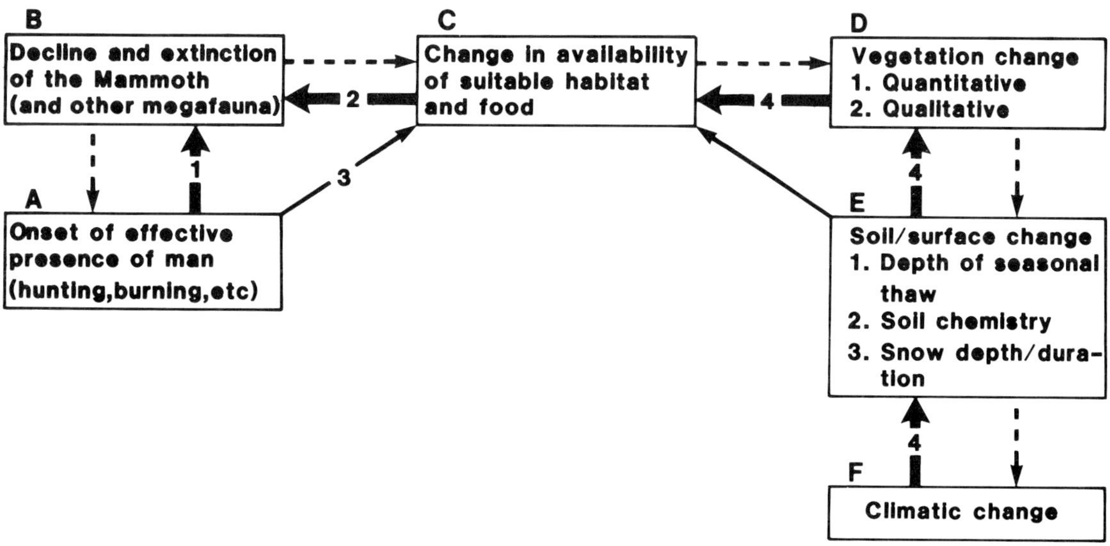

Figure 2. Model for the dissolution of the steppe-tundra and extinction of the woolly mammoth

19

MAMMALS OF THE MAMMOTH STEPPE AS PALEOENVIRONMENTAL INDICATORS

R. Dale Guthrie

The most puzzling feature of the mammoth steppe is its great diversity of large mammals. The reasons for their diversity, their dietary specializations and their gigantism are examined here, in conjunction with other information, as clues to an investigation of the general paleoenvironment.

The mammalian evidence suggests a specific vegetational mosaic with certain seasonal characteristics. Mammalian growth patterns indicate a long growing season, requiring diets high in available energy and nutrients. These features would require a seasonally rich Pleistocene vegetation in comparatively fertile soil that utilizes antiherbivory defenses quite different from those used by the dominant plant forms now present in the North.

The annual seasonal cycle reconstructed here is characterized by cold, but perhaps variable, winters with very little snowfall. Most of the annual moisture would have fallen during spring and little to none during summer. Summer soils generally were warm, dry, and had a deep thaw. This intraannual variability in the growth season produced hydric, mesic, and xeric plants within the same local communities. Windy conditions were most common.

A new model is presented to explain the high species diversity of Pleistocene mammals, their large social organs and gigantism, as well as the converse: Holocene dwarfing, shrinking social organs, range contractions and extinction with the demise of the mammoth steppe.

Key words: mammoth steppe; mammals, Holocene dwarfing of; Pleistocene

INTRODUCTION

Our Pleistocene heritage is rooted in a large mammal fauna which, for the most part, no longer exists. Nowhere is this more true than in Europe, northern Asia and Beringia. Within this broad zone lived an array of creatures more bizarre than the imagination could invent—giant woolly forms with grotesque antlers, horns or tusks, long manes, humps, and trailing tresses.

If we had found only caribou, sheep, elk and musk-oxen bones in the cutbanks on the Yukon and the Kolyma, this symposium about the Arctic Pleistocene environment would probably not have taken place. It is the mysterious fauna on the cave walls, the frozen carcasses of strange species no longer living and their scattered bones in human middens which excite our imagination and wonder. When learning of this exotic mixture of hyenas, mammoths, sabertooth cats, camels, horses, rhinos, asses, deer with gigantic antlers, lions, ferrets, saiga and other Pleistocene species, one cannot help wondering about the world in which

ISBN 0-12-355860-3

they lived. This great diversity of species, so different from that encountered today, raises the most obvious question: is it not likely that the rest of the environment was also different?

Given the immense size of the area involved there was a surprising continuity of large mammal species. Most of the species portrayed on the cave walls of southern France are also present as fossils in the muck deposits of Alaska and northeastern Canada. This continuity suggests a multicontinental mural composed of locally unique elements, each contributing to the character of the large mosaic (Fig. 1). An equally widespread distributional pattern seems to occur among the small mammals, birds, invertebrates and plants, but their fossil record is not as sharply resolved as is that of the large mammals.

As I see it, our present status in depicting this mosaic of related habitats depends heavily on the large mammals. Thus, in many past papers and discussions I have referred to these communities collectively as the *mammoth steppe*. The woolly mammoth seems to be a ubiquitous marker, both chronologically and spatially. Because of the steppic nature of the vegetation, this term seems to better capture the entirety than arctic steppe, tundra-steppe, steppe-tundra, loess-steppe, cold steppe and others that are more descriptively limited.

In this paper I use the fossil mammals as a focus to study not only how these species lived but also their environment in general. Historically, it was large mammal remains that first suggested the clearest outlines of the mammoth steppe. I

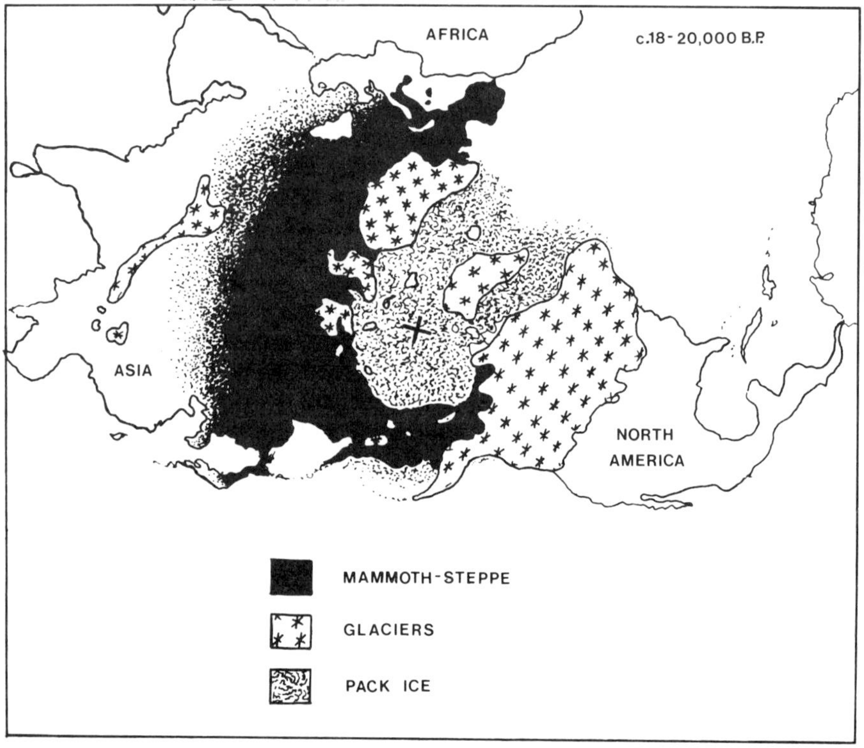

Figure 1. The mammoth steppe seems to have existed throughout the major part of the Pleistocene. Although there were spatial and temporal facies of diversity throughout this vast area, there was an identifiable integrity of xeric graminoid-forb dominance combined with a dominance of large grazing mammals, the 'mammoth fauna,' dominated by *Mammuthus*, *Bison*, and *Equus*. During the relatively short interglacials and interstadials, the mammoth steppe was disrupted by the northward invasion of more southern plant species and by the expansion of what had previously been minor community elements; both of these tended to be more mesic forms with better developed large mammal anti-herbivory defense systems

think the vegetation pattern of Europe and northern Asia is still misunderstood by many palynologists as a sort of low-latitude version of tundra or polar desert. In Alaska and northwestern Canada palynological interpretations of polar deserts (Cwynar and Ritchie, 1980; Colinvaux, 1980) are also at great divergence with the herbivore evidence for grasslands. The fauna, both taxonomically and in community composition, was quite different in the north during the Pleistocene—there was no homologous, nor even analogous, counterpart to extant zoogeographic biomes, provinces or life zones. This concept seems to be particularly heretical to many paleobotanists who are reluctant to abandon the "geoflora" concept and rather prefer to think in terms of cliseral shifts of biotic provinces altitudinally and geographically through time. However, there is accumulating evidence (mainly from macrofossils) that, like the fauna, plants during the Pleistocene occurred in distributions and associations that would be unthinkable today.

It is peculiar that the paleoclimatic indicators developed in more southern latitudes have not worked very well in the north. Northern herbaceous pollen lacks the arboreal diversity of southern woodlands, and graminoids are not distinctive at lower taxonomic levels (Johnson and Packer, 1967; Young, 1976). And because many northern plants reproduce vegetatively or are insect pollinated, the amount of pollen entering the fossil record may be an insensitive guide to diversity and biomass. Plant macrofossils and phytoliths, though common in some northern areas, have not been well exploited by paleoecologists. Because of their migratory abilities and tolerances of wide climatic regimes, large mammals are not necessarily the optimum paleoecological medium. In this case, however, it is the large mammals which create the puzzle, and as I shall argue, they are the best place to begin a search for the solution.

Our increasing understanding of the ecology of living large mammals can contribute much to our understanding of paleoenvironments. Using insights gained from extant mammal communities, I wish to present several new theories pertaining to the megafauna of the mammoth steppe and its environment. I will concentrate on a time during the last glacial maximum, around 17,000 BP, and then generalize this for earlier times, recognizing that our search at this point is for broad brush images.

We have thus far surmised several basic things about the "mammoth fauna." Foremost is that they were predominantly steppe-adapted species (e.g., Kowalski, 1967; Guthrie, 1968a). The pollen evidence also indicates predominantly non-arboreal plants (e.g., Colinvaux, 1967a).

There are several key pieces missing to the puzzle. One is why so much *species diversity* in large mammals? Only in central Africa do we find this megafaunal complexity today. How could such diversity have existed in the face of stringent interspecific competition? Did they adapt to different aspects of the topography or vegetation and, if so, how? A second piece is the question of *abundance*. Was this diverse fauna thinly spread or distributed at high densities? Proposals range from a rich, abundant fauna to a depauperate polar desert (see Ritchie and Cwynar, this volume). Most of the large mammals were well-dressed giants. So, our third missing piece to the puzzle is what central environmental feature could have been responsible for this *gigantism*? Many of the species would have been giants in any era, immense creatures, much larger than their modern counterparts. Fourthly, why where they *ornamented* with the physiologically costly social paraphernalia of shaggy ruffs, heavy horns, long tusks, or enormous antlers. These are traits wildlife managers recognize as indicators of high-quality habitat (e.g., Geist, 1971). How could this austere Arctic landscape, bounded by glaciers and frozen seas, be considered high quality?

Did these species have special adaptations which allowed them to tap diffuse northern nutrients; were the nutrients somehow greater in quantity than now, or was there possibly a greater turnover rate—a higher primary producivity? Were the soils more fertile? Did the vegetation use different antiherbivory defenses? Given a more herbaceous vegetation, are we dealing with a redistribution of extant northern species or intrusions of other species, and in what kind of communities? What kind of situation could maintain this herbaceous steppe—fire? Some type of climatic change? Was it windier, wetter, colder, dustier, cloudier—more arid? Was it continental—with the same seasonality of today?

In the fairly short interval of less than a millenium or two the megafauna changed dramatically (Guthrie, 1968a). There is evidence that the landscape, the vegetation, the climate, and the people also changed (Guthrie, in press). How all of this took place, and why, are also parts of the puzzle.

The answers to these questions must fit. They must intermesh with the other pieces of our puzzle. In fact, I often feel much like a jig-saw puzzler holding several pieces in my palm, sensing that they are keys to the fragmentary pattern, yet not quite knowing where each goes. Then once in a while, one fits and an entire segment jumps into focus, giving meaning to other segments.

MEGAFAUNAL ADAPTATIONS

Although mammals may not be good "indicators" of precise environmental situations, they are affected by their surroundings. Not only can certain mammals do better than others in a given environment, each situation results in subtle evolutionary changes in their morphology—a sort of "fine-tuning" to a specific habitat, which can be very informative to a mammalian paleoecologist. Mammals are also informative about some aspects of seasonality. In this section I will review some of the major environmental implications derived from my studies of the mammoth fauna, some of which are published (Guthrie, e.g., 1966, 1968a, 1976, 1978), others as yet unpublished.

1) One of the more informative features is body size. There is a general tendency for northern species to be larger than their southern counterparts—this was particularly true during the Pleistocene. Because size is such a major adaptive theme, I discuss it in a separate section.

2) The faunal evidence strongly suggests a dominance of herbaceous vegetation. Though its character is still debatable, most of the evidence points toward a xeric steppe. The type of herbivore present indicates a major bent toward dependence on grass and grass-like plants. In a competitive situation, the living counterparts of those herbivorous species that dominated the mammoth steppe are grazers. Also mature grasses and grass-like plants have abrasive phytoliths (which cause rapid tooth wear), as their major herbivore defenses. Woolly mammoths, northern equids, and bison all show indications in their dental anatomy of quite complex antiwear defenses, indicating a diet of grass (stomachs of frozen mammoths have yielded high percentages and a diversity of grass fossils—Tikhomirov, 1958; Ukraintseva, 1979).

3) The hooves of the Pleistocene Beringian fauna (horses, hemionids, saiga, bison, etc.) are not broadly splayed like those of caribou, moose, and musk oxen of today which use increased hoof surface to negotiate the swampy muskegs, tussocky terrain, and sphagnum meadows of the Arctic and Subarctic. The small hooves of Pleistocene species indicate a firm ground surface. As the mammoth fauna is ubiquitous in its distribution throughout the northern landscape, one can surmise that this firm substratum was a dominant feature.

4) Several of the ungulates in the mammoth fauna are intolerant of deep snow. Bannikov (1963) contends that saiga cannot tolerate snow depths greater than 20 cm. Horses and hemionids also find it difficult to get adequate food by digging through deep snow.[1] Thus I have concluded that winter precipitation through the mammoth steppe was probably quite light, and that the little snow that did fall was redistributed into drifts.

5) The singing vole (*Microtis miurus*) and pika (*Ochotona collaris*), instead of hibernating, cache summer hay and use it as supplemental food during winter. They now most commonly winter under snow drifts in windy alpine areas. However, from fossil evidence we know that they did live on rolling hills and in lowlands during the Pleistocene. From this one can conclude that the Far North was a windswept area with intermittent snow banks, rather than the powder-snow-covered taiga and hard-crust-covered tundra that characterize the Arctic today.

6) At any one time there was considerable megafaunal diversity in comparison to living Holarctic communities. We know that the present vegetation in the North could not support so diverse a grazing community. It is quite possible, then, that several important climatic-environmental aspects were different from those of today.

7) Extant mammals (e.g., Dall sheep, *Ovis dalli*, singing voles, *Microtus miurus*), now confined to disjunct alpine habitats, were then more widespread and present at lower altitudes. This suggests that alpine habitats retain remnant characteristics of the mammoth steppe (Guthrie, 1968a).

8) Much of the mammoth fauna that did not

[1] I once raised miniature asses (the same size as Pleistocene hemionids) in Fairbanks, Alaska, and they had considerable difficulty with the snow.

become extinct is now limited to more southern latitudes (e.g., hyena, lion, saiga, equids, bison, badger and ferret), which indicates that the ecological barriers of the present northern distributional limits were absent or located much further north than they are today.

9) At the beginning of the postglacial period the long-standing equid, proboscidian, and bovid dominance of the mammoth steppe was replaced by that of cervids. Cervids as a group are more mesic-adapted species, usually avoiding mature grasses in their diet and preferring leaves of dicotyledons, twig tips, and lichens. This shift is undoubtably indicative of a corresponding change in the vegetation patterns.

10) The survivorship curves of the large mammals are instructive about winter severity (Guthrie, unpublished). The curves are of the usual semi-rectangular form (type 1), with the most mortality occurring very early or late in life. This indicates that, once mature, animals were well adapted to winter conditions, but that conditions were sufficiently severe to keep juvenile mortality high.

There are additional phenomena on which I have not yet published. One involves the much wider distribution of Arctic ground squirrels (*Spermophilus parryi* and *S. undulatus*). Their fossils occur in areas where they do not live today. Ground squirrels prefer open country and require energy-rich seeds to cache for spring sustenance because they must emerge from hibernation while the ground is still snow covered in order to breed and bear their young at the time new plant growth emerges. They also require protein-rich forbs in order to rear their young to maturity before winter. Prime habitat seems usually to involve well-drained sites composed of irregular surfaces and cut banks. These provide the favored early succession plants and easier excavating.

Bear teeth may also furnish new insights into the ecology of the mammoth steppe. Grizzly bears (*Ursus arctos*) often graze in spring and midsummer (Herero, 1978), when grasses are abundant. Bears which fill their protein requirements by excavating for ground squirrels must wait until later in the season, when the annual frost has receded to the lower reaches of the burrow systems (Carl, 1971). Pleistocene bear (both *Arctodus* and *Ursus*) teeth from Alaska, Asia, and Europe show much wear from grass consumption and none of the dental caries characteristic of extant bears, which get most of their late summer carbohydrates from berry sugars.

The above mammalian features, taken as a whole, suggest a diverse fauna in an open landscape, firm in substratum, dominated by a complex assortment of herbs with many grass-like plants, and a windy winter. With these pieces of the puzzle in hand I shall now turn to the phenomenon and question of faunal diversity.

MEGAFAUNAL DIVERSITY

How was the diverse mammal community on the mammoth steppe adapted to a sympatric life? The heart of the issue is the way in which the secondary producers (herbivores) partition their use of the primary producers (plants). There are several ways for herbivores to avoid direct competition. Let us take a look at these and see which might be relevant to the mammoth steppe:

(1) *Specialization for plant form or species.* Herbivore species living together may have entirely different diets. This is the case with the browsing black rhino (*Diceros bicornis*) and the grazing white rhino (*D. simus*).

(2) *Specialization for plant parts.* Different herbivores may specialize by consuming different portions of the same plant. Bell (1970) showed that zebra in East Africa select for mature grass stems, wildebeest for sheathes and leaves.

(3) *Specialization for different growth stages of plant succession.* The grazing of plants by one species may make the same plants more palatable to another. McNaughton (1976) has shown that the rich grass regrowth occurring in the wake of a wildebeest migration is the cause of the ensuing gazelle migration.

(4) *Separation by habitat.* Different herbivores may overlap in distribution and diet but normally remain separated by habitat (Fig. 2). This is the case with mountain goat (*Oreamnos*) and sheep (*Ovis*) in North America (Geist, 1971). It seems also to be the case with reindeer (*Rangifer*) and musk oxen (*Ovibos*) (Bos, 1967), although with musk oxen there are also major season dietary differences involved. Some species may appear to be quite competitive but occur at different altitudes or in different types of terrain. The various

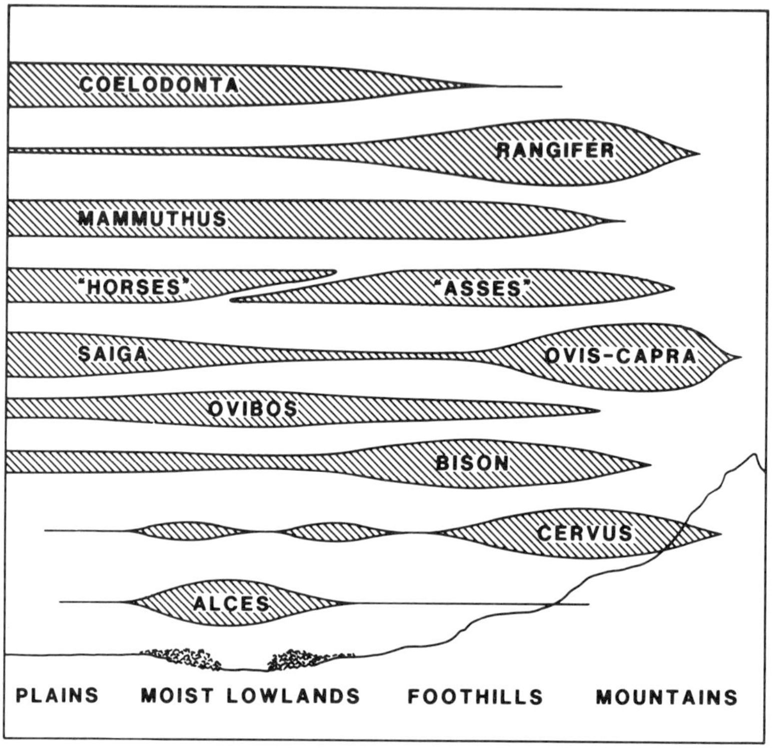

Figure 2. Mammalian herbivores are frequently separated by habitat preferences. We can assume that the same was true of the mammoth steppe and can reconstruct the rough outlines of large mammal habitat specializations by using distribution patterns of the modern analogs

Caprini of central Asia represent this kind of separation (Schaller, 1977).

It is a biological axiom that if phylogenetically (or ecologically) related species occurring together do not partition the available resources to avoid direct competition (as outlined above), one of the species is competitively eliminated. Schaller (1977) has proposed that the northern Caprini—goats, musk oxen, sheep, etc. are adapted to living in altitudinally partitioned areas, within which each species is a very adaptable feeder, occupying a wide dietary niche. Some component of the diversity of mammoth fauna seems to be explicable through spatial separation. The percentage of bison in Alaskan fossil localities increases with altitude (the same can be said for sheep), but the relative proportions of horses and mammoths decline (Guthrie, 1968a). Harington (personal communication) has observed that the frequency of horses among the equids is much greater in the lowlands (Old Crow Flats) and the percentage of asses (hemionids) higher in the hills (Dawson, Yukon Territory). Saiga (*Saiga tatarica*) seem to have a diet very similar to that of Dall sheep (*Ovis dalli*), but are specialized lowland runners, while sheep depend on steep upland escape terrain. Wapiti (*Cervus*) prefer high alpine meadows for spring and summer range (Murie, 1944), but normally graze below preferred sheep range.

There are other subtle spatial separations among living moose (*Alces*), wapiti (*Cervus*), musk oxen (*Ovibos*), caribou (*Rangifer*), and sheep (*Ovis*), where their ranges overlap. The same was probably true during the Pleistocene, and for many of the extant species as well—particularly in regard to seasonal use.

Nevertheless, topographic or altitudinal separations seem inadequate to explain all of the megafaunal diversity on the mammoth steppe. First of all, many, if not most, Pleistocene archaeological and paleontological localities on the mammoth steppe have a diverse fauna no matter what their location in terrain or altitude. Secondly, large mammal communities of analagous complexity which exist today seem to rely more on other methods of partitioning resources (McNaughton, 1976).

Unfortunately there are few living analogs to the ungulate diversity of the mammoth steppe. East Africa stands alone as a community of roughly comparable ungulate complexity. Several authors have emphasized various ecological gradients of specialization in East Africa. Estes (1974) contends that the central theme of the ecological spectrum is sociality, which increases from the solitary woodland browsers (duiker or *Cephalophus*) to the open plains grazers (wildebeest or *Connochaetes*). He relates this shift to increasing grass in the diet, the increase in body size, development of social organs, polygamy, decreased territoriality, and changes in body shape. Jarman (1974) and Bell (1970) came to similar conclusions but emphasized *feeding styles*. Geist in a series of papers (e.g., 1971, 1974, 1977) has discussed this same spectrum but has chosen to emphasize a different series of features, particularly the degree of community *successional maturity* and the stage of colonization—a rendition of the *r* and *K* evolutionary strategies. Leuthold (1977) has amalgamated the Jarman-Estes approaches and Janis (1976) has expanded the Jarman-Bell emphasis on diet to include proboscidians and equids in a general review of the various advantages and disadvantages of rumen-cecal digesters.

To summarize their conclusions about the East African ungulates, there is considerable specialization among the grazers in the community for plants of different protein/fiber content as well as for different plant forms. The smaller the animal the higher the metabolic rate and thus the higher the protein and soluble carbohydrate requirements. The larger the animal the more fiber it can tolerate. Perissodactyls and proboscidians can process food faster (gut transit time) than ruminants of comparable size and can thus exist on a lower-quality/higher-fiber diet. The gradient of increasing fiber tolerance is thus: small ruminants, large ruminants, equids, and proboscidians. All these grazers may live sympatrically where there is an adequate diversity along this protein/fiber spectrum. In addition, there are ungulate specializations in plant species of form which are somewhat independent of this protein/fiber spectrum (for example the browser-gazer spread).

With this brief review in mind, we can now exmine a few of the ungulates from the mammoth-steppe fauna. I should strongly emphasize that because of the abundance of new plant growth, there was probably little competition during spring. This is analogous to the rainy season in Africa (Bell, 1970) when the dietary specializations of the extant plains grazers also break down. Thus, we are looking for resource partitioning during the rest of the year. Utilizing the biology of the living counterparts, we can very briefly look at the ecological niches of the principal ungulates in the mammoth fauna:

Woolly mammoth. This species would probably have avoided competition with ruminants by eating a broad spectrum of herbaceous vegetation, principally coarse grass with supplements (Olivier, this volume). The extremely high-crowned, complex molars indicate a high percentage of mature coarse-stemmed grasses in the diet. The limiting parameter is sheer volume. Studies on African elephants have shown that they eat virtually around the clock and have a short gut-transit-time of around 12 hours (Laws, et al., 1975). Because sufficient volume is a potential problem the most common element in the mammoth's diet, aside from the early young spring growth, was probably the lower quality tall and mid-height bunch-grasses. Ceylonese elephants today are mainly grazers (Kurt, 1974); grass clumps are ripped from the soil—the roots shaken free of dirt and eaten whole. Mammoths, with their "fingered" trunks, would have been well suited to grazing. They were probably more migratory than living elephants are, but may have had traditional ranges on a large scale. Bulls were solitary or lived in small bands, and the females in reproductive-class herds (unlike the matriarchies of the African elephant, *Loxodonta*, and like the more closely related Asiatic elephant, *Elephus*) (Sykes, 1971). Mammoths had no obvious means of penetrating winter snow (the striations on tusks do not indicate wear from snow shoveling) for the quantity of necessary food and so probably used windblown winter ranges.

Horses and hemionids. (There may be several species, but their systematics are not yet clear.) In a complex grazing community equids specialize on grass stems of medium- to low-protein content, usually in the mid-level height range. Remember that we are not talking now about spring plant growth. The living related hemionids are found on the high arid plateaus of central Asia. Horses and those species with more ass-like adaptations may have been spatially separated throughout the mammoth steppe, but their habitats may have been interspersed in many areas as both species did occur in the same localities and, in one case, are pictured in the same frieze (Sanctuary Cave). Horses are normally found in large mare herds with a single stallion, while hemionids generally have sexes separate except during the breeding season (Ewer, 1968). Both groups crater through snow for winter food, but are then in a negative energetic balance, so cannot tolerate deep snow for lengthy periods. Thus, they also would seek out snow-free areas swept clean by winter winds. Living Asian hemionids have small, cupped hooves for irregular hard-substrate terrain (as did those on the mammoth steppe) and cannot exist on soft ground or deep snows. The late Pleistocene Beringian equids were petite and short legged—with not much over 60 cm clearance between brisket and ground. They were probably seasonally migratory, but had large traditional ranges.

Steppe Bison. In comparison to other ruminants, bison can tolerate considerable fiber in the diet (Peden, et al., 1977), but unlike horses, bison cannot live on the low-protein high-fiber stems alone; they require a high fraction of grass sheaths and leaves. In addition, they seem unable to exist on the taller grasses, which also have a lower-protein higher-fiber content than short grasses (see Guthrie, 1978, for a review). In a diverse community bison seem to be short-grass specialists using, in addition, tall and mid-level grass new growth, or regrowth after grazing by other ungulates. Thus, bison could benefit from prior moderate grazing by horses which would have removed stems and increased the relative leaf fraction by stimulating vegetative proliferation. In contrast to horses or proboscidians, bison are fairly well adapted to snow: they sweep the snow aside with their nose and cheek, then eat in a repeated cycle. However, snow does reduce range access, so they also could be expected to have concentrated in snow-free areas during the Pleistocene. The manner in which bison use the landscape would have meant a migratory habit, but they also could have had traditional ranges with only a loosely predictable migratory route. Judging from their large horns, bison males were separated from the females throughout all seasons except the rut; they probably grazed alone, or in small herds (Guthrie, 1978).

Sheep were more widespread than at present, and occurred at lower altitudes. They consume a wide range of plants (Hoefs, 1974), but prefer new growth of high-quality forage when it is available. They are lip-feeders and can feed selectively on low-growing herbs in small tidbits—taking regularly up to 14,000 bites a day (Hoefs, 1974). They cannot feed through deep snow, and in winter are confined to the windblown high-mountain ridges. Sheep are social, but herd size varies with season. They have great fidelity to home range.

Moose or Eurasian elk are high-brush animals feeding predominantly on deciduous leaves and pond weeds in summer and twig tips of willow in winter. Because deciduous trees are most common in the North in well-watered, disturbed sites, moose are usually found along streams or in the early fire-succession communities of the boreal forest. Moose are well adapted to deep snow. Their winter food, twig tips, is above the snow, and their long legs allow them to travel in snows of considerable depth. They are one of the least social of the ungulates thus far discussed, and least traditional in their home range use.

Musk oxen, in contrast to moose, can probably be best thought of as mesic hydric low-brush feeders. Low deciduous shrubs and sedges form the bulk of their winter diet. They seem to be animals of the open country. The females are quite social. They are nonterritorial but tend to stay in large areas of familiarity (Jingfors, 1980). They seem to be well adapted to the extreme windy cold weather of the tundra and the shrubby northern steppe.

Caribou or reindeer are well adapted to cratering for lichens through soft snow. Their summer diet is diverse. In winter, they are among the more social of cervids. They are not territorial, but migrate seasonally within familiar home ranges, the size of which—a few to hundreds of square kilometers—depends mainly on the size of the herd.

Wapiti or red deer or American elk are extremely adaptable. More than any other cervid they have moved toward an open-country grazer pattern. As a consequence, their teeth are complex and moderately high-crowned. They prefer grass, when it is available, to other plants (Murie, 1944).

Wapiti are not territorial and highly social, but they often undertake extensive seasonal migrations on familiar ranges. They are tolerant of moderate snow depths, shifting to twig browsing in stress conditions of deep snow cover. (Murie gives a figure of 75 cm as the depth to which they can crater for grasses.) They do not compete seriously with moose for winter range. (See Boyce and Hayden-Wing, 1979, for a review.)

Saiga is a selective herbivore, requiring mainly high-quality forage. It inhabits hard flat lands because it can run fast, and for long distances, but it cannot jump well nor is it a nimble negotiator of rough terrain (Sher, 1967). Saiga are quite social and extremely migratory when faced with inclement weather or seasonal range changes, although they generally occur in the same large areas year after year (Bannikov, 1963). Saiga are markedly tolerant of halophytes and could have done well in the sandy-arid late Pleistocene conditions in Beringia (Bannikov, 1963).

Several things emerge in this quick review. Most of the ungulates of the mammoth steppe were open-plains forms and quite social, all were migratory to a great degree, yet complementary in their dietary preferences. This dietary diversity is especially revealing, for combined with the fact that there must have been a diversity of herbaceous plant forms on the mammoth steppe spanning the spectrum from the small, high-quality forbs and leafy grasses required to support sheep and saiga, to the short grasses for bison and the taller medium-short grasses for bison and the taller medium-quality grasses for horses, and finally the large-stem, low-quality grasses apparently required by mammoths. These plant communities were spread across a diverse topography, allowing some local spatial separation of ungulates, and causing further diversity among the grazers. In addition, there were a few riparian tall willows for moose, although *Alces* was uncommon on the mammoth steppe (Vereshchagin, 1967; Guthrie, 1968a). There must have been some upland short-shrub growth in order to support musk oxen, and possibly, some lichen habitat for caribou (although there is evidence that cryptogams were rare). Each of these ungulates had specialized anatomical means of exploiting its food resources (Fig. 3).

The mammoth steppe had a high diversity of microtine rodents. Grant and Birney (1979) show that the percentage of microtine biomass and diversity is highest in grasslands with high vegetation cover (tall and mid-level grasses) and least in bunchgrass and shortgrass habitats.

All of this information can be taken to suggest that the mammoth steppe was not characterized by homogeneous communities but a diverse, rather fine- to medium-grained vegetational mosaic. This diverse herbaceous vegetation base required by the megafauna of the mammoth steppe will be examined in the following section.

PLAIDS VS. STRIPES: OR VEGETATIONAL DIVERSITY

At this point I should note that the greater diversity of large mammals during the Pleistocene, in comparison to extant faunas, is not unique to the mammoth steppe. Virtually all large mammal faunas throughout the Holarctic had a more diverse assemblage during the Pleistocene than they do today. Moreover, this principle holds true for small mammals as well (e.g., Guilday, et al., 1977; Graham and Semken, 1976). In fact, it is also true of plant communities (e.g., Leopold, 1967). In trying to solve the puzzle of high megafaunal diversity on the mammoth steppe we are dealing with a global phenomenon and solutions to our specific concerns may have a more general application.

In the preceding section I concluded that the striking diversity of the ungulate community on the mammoth steppe was primarily the result of a more fine-grained, "mosaic" distribution of flora communities than exists today. Again, this seems to have been a worldwide phenomenon. The markedly regional biome zonation we take for granted seems to have been a Holocene derivation. The integrity of most extant biomes quickly disintegrates as one looks backward into the Pleistocene (Guthrie, in press).

The upshot of this lack of biome integrity among local primary producers is, of course, a greater potential local species diversity among secondary producers when compared to modern communities. For example, caribou (*Rangifer*) and musk oxen (*Ovibos, Symbos*) moved far southward and mixed with tapirs (*Tapirus*) and peccaries (*Platygonus*) (Guilday, et al., 1975) and wild cattle (*Bos*) (Kurtén, 1968). Thus, the greater species diversity during the Pleistocene at any one locality comprised not just a greater number of species

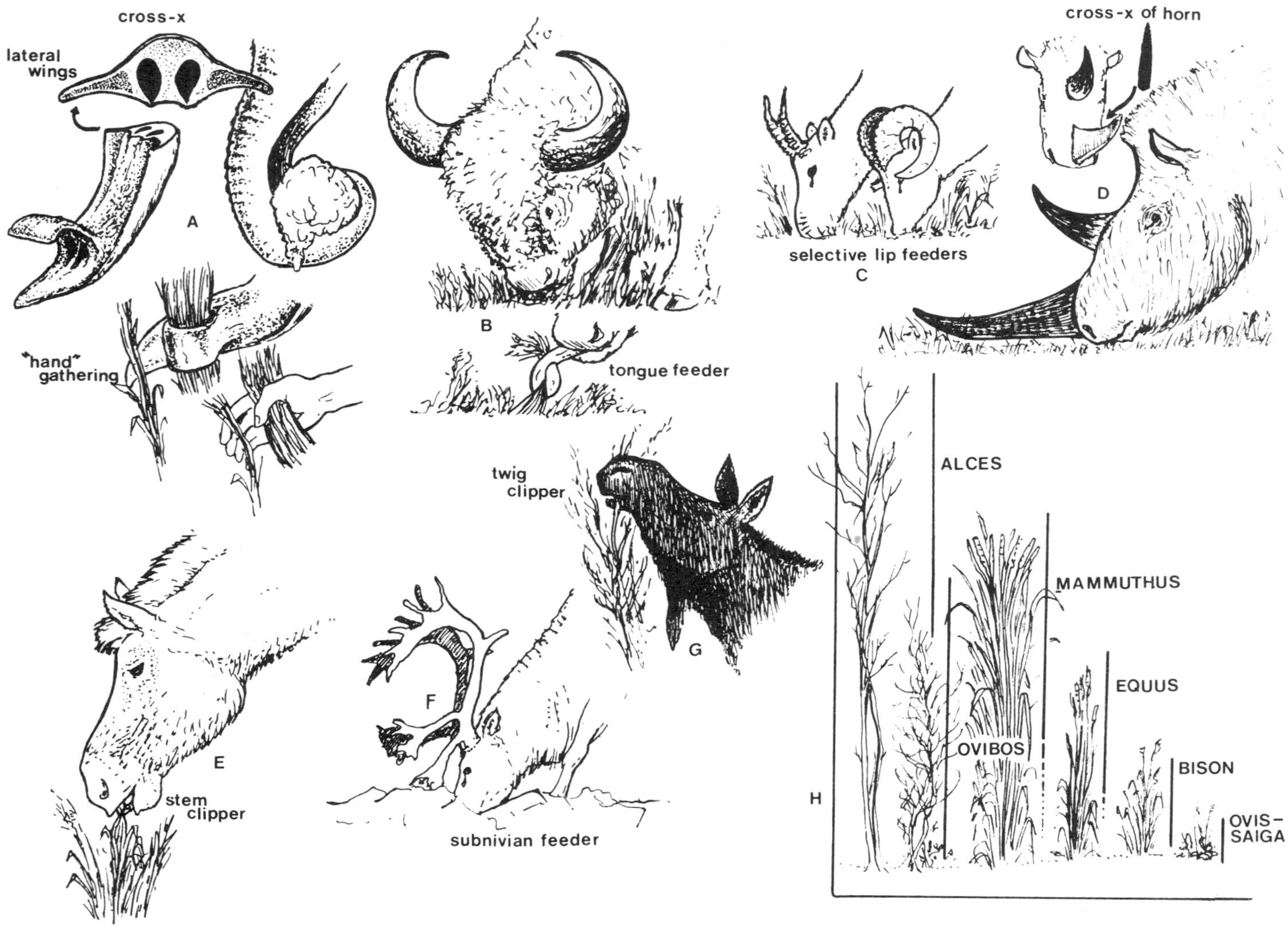

Figure 3. Large mammal species of herbivores are normally separated by the resources they use, at least during part of the year. This illustration shows some of these resources and anatomical specializations for their use by northern large herbivores. Most of the dietary specializations illustrated are for winter, the season when most resource separations occur

when the extinct ones are summed with extant ones, but it also was a distributional phenomenon resulting from the mixing of the extant biotic provinces. I suspect, then, that the solution to the causes of the greater biome blending during the Pleistocene will account for much of the greater species diversity and maybe most of it. So what were the causes?

Most animal physiologists would argue that neither temperature nor moisture deficit directly limits large mammal distributions. It is more a question of how the vegetation patterns affect interspecific competition. This then leads us back to ask why Pleistocene plant species may have been distributed in more diverse communities, in a more fine-grain mosaic, and how this would have affected competition among large mammals.

The reasons behind plant diversity are hotly contested, but there is a rough consensus that diversity increases with (1) increasing substratum fertility, (2) increased temperature equability (reduced seasonality), (3) diverse moisture conditions during different portions of the growth season, (4) a long stable evolutionary history, and (5) increased substratum and topographic diversity. These relations are not necessarily linear and their rank of importance may be contextual. The first three of these are particularly important in relation to the mammoth steppe.

(1) *Nutrient availability and recycling.* The soil mantle throughout most of the mammoth steppe consisted of loess (see Tomirdiaro, this volume). Its minerals and nutrients were thinly spread across the landscape by runoff water and wind. Unlike modern acidic boreal soils, mammoth steppe soils were probably alkaline and thus more conducive to herbaceous growth and nutrient release; newly formed loess soils were also comparatively rich in available potassium, phosphate, and calcium, which facilitated rapid nitrification. The pH in the Pleistocene loess averages around 8.2. By way of comparison, blueberry, *Vaccinium uliginosum,* so common throughout the North today, grows optimally at pH's of about 4.5. But more important than soil pH are soil temperature and plant growth strategies which determine nutrient turnover time (discussed later in this section). Increased nutrient availability would have decreased cryptogams and increased graminoids (McKendrick, 1980).

(2) *Length of growing season.* Equability, a concept used by several people (e.g., Axelrod, 1967), is central in importance, but needs some modification and elaboration for the North. I think that the most important component is *length of growth season,* i.e., number of days a plant can grow. Though many southern plant species would be killed by northern winter temperatures, the major *climatic* limitation for their growth and for that of most species is the abbreviated growing season on cool soils at high latitudes. The *biotic* component is even more important than climatic factors in creating the present zonation of plants and animals. Many southern forms are kept from moving northward by competition from species more precisely adapted to northern environments. A longer growing season on warm summer soils, especially one with changing moisture conditions, decreases competition because it allows for more sympatric sequential adaptations to growing, flowering, seeding, etc. A short growing season on cold summer soils thus intensifies competition by selecting for the taxa most precisely adapted to local conditions. I propose that the marked biome zonation during the Holocene was produced by shorter, and less diverse growing seasons throughout the Holarctic. Conversely, a longer growing season, on a fertile substratum, in the Pleistocene would have allowed forms with widely different adaptive strategies to reside together, thus decreasing zonation and fostering diversity.

Most of the Pleistocene physical evidence for the climate of the mammoth steppe relates to mean annual temperatures and tells us little about the growth season. It is the fossil biota which are most informative about seasonality. It appears that mean annual temperatures may be a nearly worthless ecological measurement; it is temperature patterns within the year, particularly spring to fall that are most revealing. There is evidence from the growth patterns of mammoth steppe ungulates that the growth season was longer than it is today. I will return later to the concept of the length of growth season during the Pleistocene because it is of critical importance in developing several other concepts.

(3) *Marked wet and dry conditions during different portions of the growth season.* As mentioned previously, the telling factor is not mean annual precipitation, but how and when precipitation

occurs throughout the year. Today, winter snowmelt quickly runs off of steep loess-covered terrain and remains in marshy flatland swamps all summer because of the poor drainage of the permanently frozen ground and the lack of summer aridity.

Though precipitation is very low in the North today it occurs in scattered rains throughout the growing season, generally peaking late in the summer. Thus it comes in the wrong form and at the wrong time to create and optimally maintain a productive steppe habitat with a solid substratum. For more herbaceous communities moisture limits quality and permits productivity to vary by several orders of magnitude during dry and wet growing seasons. There are several new lines of evidence which suggest a marked wet, and then dry, moisture regime throughout the North during the Pleistocene growing season. That late- and perhaps mid-summers were characterized by clear skies and arid conditions can be inferred from the presence, at least as far north as Fairbanks, Alaska, of numerous elements of the shortgrass plains fauna and flora (Guthrie, 1978, and in press): e.g., badgers (*Taxidea*), ferrets (*Mustela),* and gramma grass (*Bouteloua*). There is yet other evidence of heavy, warm spring rains manifest in the prominent silt erosional features throughout much of the mammoth steppe. Also, the high quality of individual ungulates during the Pleistocene would require significant spring moisture to maximize spring growth of herbaceous vegetation (Bunnell, 1978).

The nagging puzzle of large-mammal biomass invariably emerges from any discussion of the high diversity on the mammoth steppe. We may never know whether the actual primary productivity was quite different from now, but we can suppose that ungulate productivity was greater. A change in quality, not necessarily quantity, of the primary production must be the key. An herbaceous community has a much higher productivity than a woodland having the same biomass (in woodland, nutrients and energy are tied up in the form of wood). There is thus a higher turnover rate among herb biomass. Very few mammoth-steppe ungulates were adapted to eating plant parts from trees for their growth resources; rather, most are adapted to eating herbs or leaves of shrubs. The much greater sympatric ungulate diversity, combined with the more palatable and productive herbage, argues for a higher ungulate biomass on the mammoth steppe than is found today in the same areas. Generally, ungulate numbers are limited by winter range. The virtual lack of snow cover would have greatly increased the potential carrying capacity of the mammoth steppe.

I wish to propose that these changes we see at the end of the Pleistocene were a cycle or syndrome of events triggered by a seasonal change of moisture in the northern Holarctic which affected soil temperatures, floristic and vegetational composition, antiherbivory compounds, available nutrients in the substrate, litter decomposition, soil insulation, primary productivity, and finally, the kind, quality, and numbers of large herbivores.

Increasing snow cover and late summer moisture favored more mesic plants, which would tend to generate an insulative cover on the soil surface. Frozen winter soils insulated by this cover during the summer would show declines in summer temperatures and soil fertility, thus favoring nutrient-stress-tolerant plant species (Chapin, 1980). Among woody plant species so favored would be evergreens which produce much more toxic allelochemic antiherbivory defenses (Grime, 1979; Mattson, 1980). Among the deciduous species there are large quantities of tannins and resins. Ericaceous plants produce both digestibility-reducing and toxic antiherbivory compounds (Jung, et al, 1979). Because such toxic species are not eaten intensively the litter accumulates on the soil surface where decomposers are equally affected by its toxicity (Chapin, 1980). Summer moisture also favors the growth of cushion plants, which are at the same time shunned by large mammalian herbivores and are effective soil insulators.

The dominant northern graminoids today are also intolerant of grazing, but their antiherbivory defenses seem to involve a nutrient-deficiency strategy. They actively exclude sodium and calcium, which results in phenomenally low levels of these nutrients in their leaves (Chapin, et al., 1980). Also, Jung et al. (1979) found a wide distribution of various alkaloids among extant tundra grasses. Unbrowsed, and generally unburned because of the damp substrate, these graminoids, cushion plants, and evergreens accumulate as leaf litter, which creates more soil insulation. This further contributes to the cycle of colder soil temperatures and more nutrient-stress-tolerant plants. The accumulated litter and invasion by bryophytes clog drainage, magnifying the effect of the moisture, and further perpetuating the cycle.

Plants that are tolerant of nutrient-impoverished soils probably are not carbon limited and thus have extra resources to produce digestibility reducing compounds, particularly toxic resins (Mattson, 1980; Bryant and Kuropat, 1980). Another adaptation of these plant species to nutrient stress is slow leaf turnover time (evergreenness) and great root longevity (Chapin, 1980). However, the photosynthetic rate declines with leaf age, and nutrient absorption declines with root age, both of which greatly reduce productivity by slowing nutrient turnover rates. The slow vegetative turnover in the living plants due to their growth strategies and toxicity makes plant matter unavailable to herbivores and decomposers. The resultant considerable standing and litter biomass, and the decreased nutrient turnover in the soil due to coolness, must have created reduced primary productivity from the levels that existed during most of the Pleistocene. Even more critical to the large herbivores than the simple reduction in primary productivity was the increasing toxicity of the antiherbivory allelochemics and the decreasing nutrient quality. Thus for the Holocene large herbivore community the quantity of available utilizable resources must have declined several-fold from that of the late Pleistocene.

In contrast, plants adapted to more fertile soils, both the ruderals and those with more conservative strategies (Grime, 1979), capitalize on the spring flush of soil nutrients with a burst of growth and are generally not as well defended with toxic allelochemic compounds (Janzen, 1975). The large herbivores of the mammoth steppe were well adapted to these plants (Guthrie, in press).

In drier environments, like that of the northern Holarctic during the Pleistocene, grasses undergo considerable evapotranspiration, which moves both sodium and calcium into the grass leaves (Chapin, et al., 1980). Likewise, with warmer summer soil temperatures, deeper roots, and more rapid absorption rates, nutrients would have been more available. This would have shifted plant selection floristically and in vegetation patterns away from the more nutrient-stress-tolerant forms and toward a community that was more nutritious and more digestible (less defended with toxic allelochemics) for large herbivores. Such habitat would have been much more productive for the large mammals of the mammoth steppe.

As opposed to the view that glacial climate and soils are too hostile or rigorous for today's vegetation,[2] it is more likely that other stresses not now characteristic of the North were involved. That is to say the stresses to which the present vegetational dominants of the North are adapted were not intensified, but rather were actually less extreme in the glacial climates. It was not nutrient deficiency (the main limiting factor at present) but other stresses—most likely moisture stress—which made the mammoth steppe too rigorous for boreal forest and tussock muskegs. Most likely moisture stress was the limiting factor. On more fertile soils in climates with drier summers, other plants that are adapted to rapid uptake of soil nutrient pulses (Chapin, et al., 1980) outcompete the boreal nutrient-stress-tolerant species. Some of the same

[2]The "polar desert" interpretation of the glacial vegetation in Beringia by Ritchie and Cwynar (see this volume) is based on two main points: a lower than present total pollen influx during the Pleistocene and, secondly, no major influx differences of grass and wormwood, *Artemisia*, between Pleistocene and Holocene sediments. But what does this mean in terms of Pleistocene plant biomass or productivity? The plant biomass throughout most of the tundra and taiga today is relatively high—in fact, the vegetation is so thick as to be virtually impenetrable, making cross-country summer travel very difficult. In contrast, the productivity is quite low (see Brown, et al., 1980, for a discussion).

This plant biomass is so high principally because it consists largely of nutrient-stress-tolerant forms, which are mostly unusable by herbivores. Given the same biomass, but in utilizable plant forms, the environment would be an immensely rich rangeland.

This would hold true if northern vegetation during the Pleistocene had even 10% of its present above-ground biomass, in a more utilizable form to herbivores, and with increased productivity and greater winter access. These productive grazophilic communities would have produced much less pollen than is produced today in the tundra and taiga.

Grazing itself would also tend to reduce pollen production on the mammoth steppe. Most herbs respond to heavy grazing with vegetative growth and cease or limit flowering. For example, McNaughton (1979) found that ungulate grazing virtually prevented flowering in the Serengeti. If ungulates grazed one half of the mammoth steppe plants sometime during the growing season (theoretically removing a relatively small biomass—say in the range of 5-20%) the pollen production would have been considerably reduced. Pollen production, without a knowledge of the other forces at work, does not seem to be a good indicator of either plant biomass or productivity.

forces which now prevent spruce (*Picea*) in North America and larch (*Larix*) in the Old World from extending their ranges southward probably acted during most of the Pleistocene to prevent them from extending their ranges northward. These are the forces which determine their northern perimeter today.

In conclusion, the seasonally diverse growing season, the longer growing season than today, and the relatively fertile soils resulted in a complex vegetational mosaic of seasonally rich vegetation which was required to maintain the diversity of mammalian species and their inordinately high quality.

SUPERLATIVES ON THE MAMMOTH STEPPE

When one compares the Pleistocene large-mammal faunas to those of today there are several striking differences, foremost among them body size. Megafaunal gigantism was pronounced in almost all species, excepting some of the very late Pleistocene equids, mammoths and rhinos, which were smaller than their more southern contemporaries and closest modern analogs. I suspect that this was due to the generally more conservative growth strategies of equids, rhinocerids, and proboscidians. They can exist on low-quality forage (in sufficient volume), but grow slowly. The manner in which Pleistocene habitat was partitioned allowed species to coexist as long as some were able to use the less nutritous end of the vegetation spectrum during the summer. A second characteristic of the Pleistocene megafauna was the extreme development of social organs–extravagant horns, antlers, tusks, and tresses. Associated with these traits was a tendency for most forms to possess a large shoulder hump. I contend that these differences are central to an explanation of the character of the mammoth steppe and that a close examination of them can reveal much about their environmental features.

Wildlife managers have long recognized that "high quality" populations (large bodied, fast growing, with large trophy antlers and horns) are found where competition is light and food is both high in quality and abundant during the growth season (Geist, 1974). I have proposed elsewhere (Guthrie, 1976, and in press) that the seasonal distribution of competition and food quality and abundance is of major evolutionary importance.

It must first be understood that seasonality has both liabilities and assets. In the Far North, seasonality sets a narrow time limit on growth. Much of the year northern ungulates are in a state of obligate growth dormancy, because nutrients are insufficient not only for growth, but for maintenance as well. Northern ungulates gradually decline in weight and body condition during the winter. They lose the ability to grow during the winter even when they are artificially provided an abundant and nutritious diet (Wood, et al., 1962).

Yet, with the onset of spring and the subsequent explosion of new plant growth, northern ungulates have adapted by evolving an extremely rapid growth rate. In fact, so have some northern rodents. Not only is the arctic ground squirrel (*Spermophilus undulatus* and *S. parryii*) the giant of the genus, it has also, by far, the most rapid growth rate (Levenson, 1979). Klein (1970) has reviewed the advantages of northern ranges in detail. To summarize, he found that many northern plants have a greater quantity of nutrients per unit weight due to adaptations for rapid growth during the shorter season. Also, the absence of darkness results in less plant catabolic activity. Low sun angle and topographic relief result in a considerable variation in the chronology of new growth, allowing a long phenological sequence of high-quality forage for a mobile grazer–unlike in more southern latitudes, where new growth is more synchronized.

Additionally, Geist (1977) observes that the canalization of parturition (and hence rutting time) means that males have a long period to forage undisturbed by agonistic rutting behavior. Concentration of growth is heightened by the fact that ungulates feed less actively in hot sunny weather than on cool days (Leuthold, 1977; Fox, 1978). It should be remembered that most ungulates of the mammoth steppe had heavy pelts which probably were not shed until well into summer, after high-protein green growth was available to repay winter debts in order to afford moulting and new pelt growth. Females normally moult even later, due to the greater nutrient expense of nurturing young in spring and early summer.

Many ungulates decrease food consumption nocturnally (except on moonlit nights); however, in the Arctic during most of the growth season there is no real darkness, allowing a longer daily activity period. Proboscidians (Laws, et al., 1975) and equids (personal observation) feed around the clock when food is limiting, and ruminants have to take time out to ruminate, which often means that they too benefit from nocturnal feeding to refill the rapidly emptying rumen in spring (Leuthold, 1977). The mammoth-steppe ranges thus seem to have been able to provide high quality forage. The matter in question is the abundance and duration of the quality peak.

The diversity of ungulates living sympatrically tells us that Pleistocene vegetation was not as sparse as the Middle Eastern arid lands, High Arctic barrens, or simple high-alpine systems of today. The concurrence of a diverse, sympatric ungulate community and high ungulate quality argues for heterogenous and adequate rangeland that includes medium or tall herbaceous growth forms. We can conclude from their large body size that, for a wide array of ungulate species, neither range quality nor abundance was growth-limiting during the peak of plant-growth season.

I propose that the critical variable was the duration of the peak in range quality. As I mentioned earlier, northern ungulates have a limited time in which to grow—they may not (and cannot) do so outside the short growing season, but within the growing season there are physical limits on the rate of growth. The physical aspects of consumption, rumen turnover, assimilation, and transit time create inherent limitations on maximum potential growth rate. Thus, an extremely high-quality diet when limited to a very brief season is unable to produce large-bodied animals. The dynamics of natural selection preclude ungulate growth programs that try to reach maximum genetic potential with insufficient resources. However, a long season of growth on quality resources without high interspecific competition would select for large-bodied individuals with elaborate social organs.

There are several lines of evidence which point to increased length of growth season, particularly an earlier spring. Pleistocene mountain sheep were larger than those living today (Harris and Mundel, 1974). When raised in captivity, mountain sheep are larger bodied and larger horned than in the wild. This larger size is *not* due to increased growth rate nor is it due to growth being extended over more years. Rather, the larger body and horn size in captive animals results from growth beginning earlier in spring and lasting later in fall due to supplemental feeding. These captive sheep almost equal their Pleistocene counterparts in body and horn size (Guthrie, unpublished). In captivity Dall sheep (*Ovis dalli*) begin growing horns in late January, but in the wild, horn growth does not commence until late April (Bunnell, 1978).

What could have produced an onset of spring plant growth much earlier than its present scheduling? I suspect that a partial cause was the lack of uniform snow cover. There is presently abundant solar energy during the long spring days, but most of this is reflected by the snow. The onset of spring green-up currently varies more than a month, depending on snow cover. If, during the late Pleistocene, spring began with bare ground and mild air temperatures, there would have been enough sunshine to accelerate the onset of the season. New growth would have been particularly encouraged by early spring rains on a dehydrated landscape. Spring soils would not have been insulated with litter as they are today. More digestible forage (less toxic antiherbivory defenses) would have resulted in more intense winter grazing. Drier autumns may have produced more grass fires—an important consequence of which would have been a rapid warming of soils in the spring.

Adequate spring precipitation also seems to be a necessary element in the model. Bunnell (1978) was able to show a positive correlation between early spring precipitation and horn size in Dall sheep. My work with the captive Dall sheep has shown that most horn growth occurs before the end of June, when protein is highest in plants. Resources during the rest of summer, when plant minerals and fiber increase, are devoted to more somatic growth and fat deposits.

New spring plant growth on a snow-bare landscape would have been profuse and widespread. As mentioned above, spring plant growth is exceptional in quality (high in protein, phosphorus, potassium, etc., and low in fiber, phytoliths and antiherbivory compounds). This abundant new growth is eaten by all ungulates and even some carnivores (grazing grizzlies) without much interspecific competition or compartmentalization. The same was probably true on the mammoth steppe. In East Africa, Bell (1970) found that after the rains a number of ungulate species could be found eating similar short new growth on the high grazing catena. The frozen stomach contents of mammoths that died in spring bear this out. They reveal a diverse, relatively unselective dietary assortment of herbaceous spring growth (Tikhomirov, 1958). As these early spring plants began to mature and became shorter in supply, and as the later-blooming plants began to emerge, ungulate partitioning would have begun.

Although spring was likely of extreme importance in lengthening the growth season, a prolonged autumn may also have been involved. The decreased snow cover and accessible frost-hardy northern herbs could have provided growth nutrients well into early winter. Today many herbs are still green and growing (particularly those once grazed and on fertile soil) when they are first covered by snow in early October (personal observation).

During the end of the last glacial interval these conditions which promoted the biological superlatives on the mammoth steppe began to decline rapidly.

SOME RUMINATIONS ON CLIMATE

Scanning the various existing habitats in the Arctic and Subarctic, one sees a wide variety of seasonal patterns in temperature, moisture, wind, and weather. It is surprising that so few of these modern habitats are even approximate analogs to the mammoth steppe. One is thus reluctant to reject the conclusion that seasonal patterns in the Pleistocene must have been quite different, not just subtly different, to produce a change of so great a magnitude over such a large area.

In this section I shall briefly review the environmental implications of the large-mammal fossils and then speculate on the mammoth-steppe climate and its causes. The vertebrate faunal implications are:

(1) *A preponderance of herbaceous plants consisting for the most part of grass and grass-like species* is indicated by the grazing fauna (as well as by all the palynological evidence).

(2) *A long growing season of abundant high quality forage* is indicated by the well-developed social organs and gigantism of the large mammals.

(3) *A fine- to medium-grained mosaic of herbaceous plants* is indicated by the megafaunal diversity. Involved are plants which are tolerant of aridity as well as those adapted to more mesic conditions.

(4) *A light winter-snow cover* is indicated by the presence of some large mammals which are intolerant of deep snow.

(5) *Winds in both winter and summer* are indicated by the snowdrift adaptations of the small mammals occurring in combination with the large mammals which do poorly in snow, and by the relict aeolian features of the landscape.

(6) *Medium- to low-quality winter range and cold winters* are indicated by the ubiquity of cryogenic geologic features and by the deep annulae in the dental cementum of members of the mammoth fauna. Winter range was poor compared to that further south, however. Compared to that of the North today, Pleistocene rangeland was relatively good. Today it is virtually nonexistent for grazers.

(7) There was very little litter on the soil surface and *relatively high soil fertility,* in contrast to the modern syndrome of cold summer soil and nutrient-stress-tolerant plants.

(8) *Early spring snow melt,* critical in producing large body and horn size, are an expected result of the light snow cover (items 2 and 4) as are *postponed autumns.*

(9) The *firm, dry substratum* indicated by the small hooves of many of the ungulates means less insulative vegetation cover, which would promote *relatively deeper summer thaw.*

(10) A long summer, beginning with *spring rains* (very unlike the rain patterns which are usual in the North today), shifting to *clear skies with wind in late summer,* is indicated, resulting in *moderately warm and arid conditions long into autumn* (which would then be accompanied by very disparate day-night temperature cycles).

Assuming that this reasoning is sound, what could have produced this kind of climate? Undoubtedly, major cyclonic patterns were different than those of today.

It was probably windy throughout the year. Winter wind has several major environmental implications: (1) foremost, it would have made available to ungulates large areas of winter range. (2) Also, it would have speeded the first green growth of the spring by exposing the ground surface, thus eliminating the lag between first snowmelt and soil exposure, and exposing land to winter deflation with dust accumulation on the existing snow mantle, thus speeding melting. (3) Direct wind disturbance of the vegetation and soil, and its secondary effects on snow redistribution (in combination with varying macro- and micro-topographic relief), would have further promoted vegetational diversity on a more fine-grained scale than exists today. Not only would there have been varying stages of succession present at all times and continued wind "tilling" of nutrients, but the microhabitats would have provided a greater potential range to accommodate varying plant adaptations and hence greater species diversity in a complex fine- to medium-grained mosaic. (4) The present severely cold temperatures in the North are normally the result of temperature inversions which would have been occasionally upset by winter winds, thus producing slightly warmer weather at lower altitudes. (5) Wind would have tumbled ice crystals and abraded woody plants. There are many alpine areas where shrubs can escape winter mortality from wind scouring by growing only in snow-drifted swales. New growth is pruned annually by windblown ice crystals at the snow surface. (6) Summer wind would also have heightened aridity.

Like wind, some degree of aridity is a relative certainty for the ecological requirements of the fossil mammals and plants (Guthrie, unpublished). All aspects of the evidence suggest that snowfall was light and what little occurred was greatly redistributed by wind, as in mountain-sheep ranges today. Low winter moisture also suggests clear winter skies. Because of the well-developed cryogenic features one can safely say there had to be rather severe cold episodes (even without the insulation blanket of snow), though these could have been interspersed with relatively warm episodes.

The amount of summer precipitation is quite another question. Presently, northern habitats have a wet substratum—the result of snowmelt water and decreased percolation (due to the high frost table) and not simply high precipitation. We assume that the Pleistocene summer frost table was probably lower because plant cover and litter were less insulative. Precipitation may not have been less than the 15–30 cm presently experienced in northern continental interiors, but it seems to have been more concentrated in the spring. The precipitation curve must have shifted from the autumn peak of today to a spring peak—which would select against woody plants (S. A. Bowling, personal communication).

The seasonal moisture diversity throughout the growing season on the mammoth steppe makes such dichotomies as "continental" and "oceanic" almost inapplicable. Those who argue simply for increased continentality miss the fact that it is difficult to get greater continentality than, in say, central Yakutia today. The weather which would best produce the mammoth steppe would be characterized by little snow, spring rains, and semi-drought conditions for the remainder of the summer. Aridity during the summer and autumn would have inhibited the high frost table that now generates the muskeg bogs and, instead, would have created a firm substratum and encouraged grasses and grass-like plants. Spring rains would have provided adequate moisture for optimum plant development (as well as allowing for medium or tall herbaceous growth habits). This model would allow a more northern distribution of the specialized xeric grasses (*Bouteloua*[3], etc.) from the arid temperature grasslands. Yet the spring rains would allow the coexistence of more mesic grass-like plants which are around today, e.g., *Carex, Hierochloë, Poa, Festuca, Calamagrostis, Arctagrostis,* and *Dupontia.*

At least two interrelated features of grasslands seem to combine to produce the characteristic plains life-history strategy of ungulates. One is the comparative synchrony of plant growth in any one spot, and the other is seasonal disparity (De Vos, 1969). The former makes territorial defense unwarranted, as neither food volume nor quality are concentrated for any length of time. The resulting

[3] In an unfinished study of fossil ground squirrels (*Sphermophilus parryii*) conducted in Alaska, I have had nest materials identified at the Composition Analysis Laboratory, Colorado State University. Among the grass genera identified was *Bouteloua*, one of the dominants of the short grass plains. These finds from Wisconsin-age sediments are several thousand miles north of the present range of *Bouteloua*. This has important implications for northern paleoecologists because *Bouteloua* is a C_4 grass species. There are no native C_4 plants living in the Arctic today. C_4 herbs are adapted to summers having a long, warm, sunny, and arid segment in which water is quite limited. These C_4 herbs photosynthesize on a four-carbon molecule and open stomata at night, when the air moisture is highest.

animal mobility and the diffuseness of the forage reduces competition and gives rise to herding (for more opportunities to spot predators and better potential for group defense). The seasonal disparity characteristic of grasslands increases polygamy and promiscuity and reduces competition during the growth season, causing a selection for larger body size and more elaborate social organs. The dependence on a rapid canter, to efficiently cover long distances, results in a high shoulder hump (Guthrie, 1978). Combined with the phenomena of heavy antlers, tusks, and horns, shoulder humps become particularly exaggerated. This line of discussion demands a section in itself—but, basically, the direction of my argument is that the *mammoth-steppe ungulates exhibit these steppe adaptations more than any other faunas now living in open country.* This in turn suggests that the environmental factors, which contributed to the "grassland character" (e.g., Geist, 1974; Estes, 1974; Jarman, 1974; Bell, 1970) were also present to a greater extent on the mammoth steppe than in other areas at other times.

THE DEMISE OF THE MAMMOTH STEPPE

Some time around 14,000-10,000 BP—the exact chronology and sequence is still not totally clear—major changes occurred throughout the mammoth steppe. Large numbers of megafaunal species became extinct. Others greatly retracted their distributional ranges in a variety of directions, some to the south (e.g., *Saiga*) and some to the far north (e.g., *Ovibos*). Most species that remained in the Arctic and Subarctic experienced what has become known as "Holocene dwarfing"; they became smaller and their horns and antlers were greatly reduced in size and complexity. In the North there was a general increase in evergreen woody plant species to create the boreal coniferous forest and its various deciduous successional elements. Herbaceous communities remained dominant only in mountainous areas and lowland muskeg and even there, nutrient-stress-tolerant evergreens predominated.

I would like to outline and then discuss a series of propositions to form a general theory that accounts for the demise of the mammoth steppe:

1) The change was climatically induced.

2) It involved a change in the seasonal distribution of moisture particularly, but also of wind and temperature, and not necessarily great changes in their annual means.

3) The result was increased vegetation zonation and an increased abundance of plant species having strong ungulate antiherbivory defenses.

4) Concurrently, the wet swampy ground and deep winter snows made the remaining herbage inaccessible to all but the snow and muskeg specialists.

5) As a consequence, there was a radical decrease in food for ungulates—and increased competition for those that remained.

6) The ungulates most precisely adapted to the emerging vegetation zones and Holocene habitats outcompeted ungulate species with broader adaptations as well as those species closely adapted to the declining plant communities.

7) All potential ungulate niches in the area once occupied by the mammoth steppe are now filled to carrying capacity, except for a few special situations where reduced access to colonization or unstable environments of a long-term scale have limited colonization.

Evidence from a number of sources shows that a major climatic change took place during the recession of the Wisconsin-Sartan ice (Leopold, 1967). From what we can reconstruct, there was a less seasonal glacial environment in mid-continent North America and North Africa (Axelrod, 1967). I have proposed in earlier sections that something similar seems to be true of the mammoth steppe. Summers were long, having a wet season and a dry season, fostering a diverse, herbaceous landscape that contained only few woody plants. Plants, invertebrates and mammals were present that now are limited to more southern areas; these intermingled with more northern forms to produce heterogeneously complex floral and faunal communities. The greater nutrient-availability of the basic warm-summer soils resulted in high-quality herbs to which the grazing ungulates were adapted.

With the increased moisture of the Holocene—featuring wetter summers and more winter snows—cushion and evergreen woody plants were favored. Cool soils limited nutrient availability, which favored plants that produced more antiherbivory chemicals. For example, there are few mammals which can tolerate a diet high in alder (*Alnus*), dwarf birch (*Betula*), mature larch (*Larix*) or spruce (*Picea*) needles, (porcupines, *Erethizon*, climb above the defenses concentrated on the lower portion of spruce trees and eat the upper bark).

The shorter growing season and increased competition among woody plants has resulted in the present degree of vegetational zonation which existing herbivores exploit thoroughly. If musk oxen (*Ovibos*), which have been reduced in historical times by hunting, were to be restocked in their original habitats, tundra vegetation would be used by the mammals to its current winter potential, because winter range is the critical bottleneck for survival. Moose (*Alces*) and hares (*Lepus*) now exploit to its maximum the deciduous successional growth available to mammalian herbivores in the northern boreal forests. Additionally, the alpine areas are occupied in some form of vertical stratification by the Caprini. There are some local exceptions but, in general, the pattern holds. If other ungulates were introduced on a wide scale they would be outcompeted by the existing fauna.

Caribou (*Rangifer*) is the snow craterer without parallel. It can smell lichen and other plants through deep snow, and quickly paw down to them with its broad rounded hooves. A wide syndrome of adaptations including locomotor modes, respiratory structures, and high growth rate during a short season, etc., make caribou a tundra and mixed woodland-muskeg specialist without peer.

The same is true for moose (*Alces*) in the boreal forests where deep snows preclude ungulate subnivian feeding. Moose are especially adapted to walking in the deep snow and to browsing on the deciduous twigs above the snow. They also are especially adapted to the wet muskegs and summer ponds.

Musk oxen (*Ovibos*) are likewise well adapted to their windswept tundra ranges where low-growing woody plants and forbs are found. The same is true for sheep and goats on the highest alpine slopes. From everything we can gather about the extinct species, it is unlikely that they could outcompete extant species where the latter are confined to their zone of specialty. Thus when one removes these present habitats for which living ungulates are well adapted, there is virtually nothing left for any other ungulate species. There are no empty widespread ungulate habitats in the Far North today which could conceivably be filled by either living or extinct species.

The combination of short growing season, wet substratum, deep snows, and reduced distribution and complexity of nutritious herbaceous plants drove the distributional limits of many ungulates (e.g., *Cervus*, *Equus*, *Bison*, and *Saiga*) southward during the Holocene. The shorter Holocene growing season has also reduced the body size of almost all species. Horn and antler size have also diminished. The surfeit of nutrients ceased to be available, placing at a selective disadvantage those genomes which provided for the elaborate excrescences of social stature (Geist, 1974). Rather, because of limited growth resources, the more conservative strategies of survival and maintenance were followed and horns and antlers were reduced in size.

This climatic shift in seasonal patterns also greatly affected the distribution of other elements of the mammalian fauna. The collared lemming (*Dicrostonyx*) and Arctic fox (*Alopex*) are now confined to the High Arctic. Where the arctic ground squirrel (*Spermophilus*) abounded, now tree squirrels (*Sciurus*, *Tamiasciurus*, *Glacomys*) live. Very large badgers (*Taxidae*) and ferrets (*Mustella*) which lived in central Alaska and the Yukon Territory (Anderson, 1973) are now found thousands of kilometers to the south.

The extinctions of long-standing members of the mammoth steppe such as mammoth (*Mammuthus primigenius*) and woolly rhino (*Coelodonta antiquitatis*) indicate that climatic shifts in the Holocene were of a unique character. This was corroborated by the evolutionary dwarfing during latest Pleistocene time of the northern steppe bison (*Bison priscus*), red deer (*Cervus elaphus*), some sheep (*Ovis dalli*), and other species, including carnivores (Kurtén, 1968). Similarly, Leopold (1967) proposes unique climatic conditions for the more southern latitudes during the early Holocene.

MODERN ANALOGS

Could such a widespread biome have so recently existed and not left a sizable remnant? The answer seems to be yes. There are two major herbaceous biomes existing today: the alpine, or high-latitude tundra, maintained ultimately by summer coolness, and the temperate grasslands. Neither seem to approximate the mammoth steppe.

The grassland biomes today are at mid-latitudes and are maintained by a combination of wind and aridity through several controversial proximate forces (fire, root competition for moisture, etc.). They are extremely continental in climate (De Vos, 1969), which produces a number of features very similar to those of the mammoth steppe (e.g., grazing faunas) and a number of other features quite dissimilar (e.g., highly zoned vegetation, and an absence of the cryophilic elements). Lowland arctic tundra is even less similar to the mammoth steppe, as the melted winter snows remain puddled on the frozen ground through most of the summer, creating an almost impenetrable muskeg or bog.

High alpine tundra has some similarities to the mammoth steppe. The topographic heterogeneity creates a patchwork of herbaceous vegetation and, because of the slope, it is well drained with sharp nutrient pulses. Winter winds drift the snow, creating exposed rangeland interspersed with subnivian habitats. It is probably not coincidence that a large segment of the survivors of the mammoth steppe now reside in alpine areas.

But alpine habitat is of limited value in our search for analogs. The relief and limited extent make this habitat unsuitable for animals such as saiga or mammoth. The short alpine summer creates the prostrate growth habits characteristic of alpine tundra. More importantly, the season is too short for production of the ungulate giants which marked the mammoth steppe.

The high plateaus of central Asia, sometimes called the "alpine steppe" (Schweinfurth, 1957) come close to what the mammoth steppe must have been like, particularly parts of the Tibetan Plateau and areas to the north. Though the vegetation and large mammals are altitudinally zoned (Schaller, 1977), they are fairly diverse. The large yak (*Bos grunniens*) survives in this area as do the wild hemionids (*Equus hemionus*). Historically, saiga (*Saiga tatarica*) occurred in the northern portion of the plateau and there are still small populations in Mongolia, just southeast of Lake Baikal. In the recent past, wild horses (*Equus przewalski*) were common throughout this area, as were bactrian camels (*Camelus bactrianus*). Both species may still exist in remnant populations. Additionally, several species of sheep (*Ovis*), antelope (*Procapra, Pantholops*), and goat (*Capra*) are found there. Because of its altitude, alpine-steppe summers are windy and relatively dry, cool, and often cloudy, and due to the latitude, moderately long. Winters are windy, with little snowfall. Most of the woody plants are willows or aspen, but the vegetation is mainly herbaceous.

Though somewhat analogous to the mammoth steppe, this area certainly is not a homologous relict. The central Asian plateaus seem to have had a relatively stable late Quaternary history without the dramatic ecological change that occurred in the North. They can best be thought of as a late Pleistocene relict pattern which existed to the south of the mammoth steppe in central Asia during the last glaciation (Schaller, 1977).

Throughout the North there are a number of small, well-drained grasslands (Guthrie, 1968a; Young, 1976) in special edaphic conditions such as steep bank, southern exposures, and major mountain-pass outwash deltas, but these are probably not microcosms of the general mammoth-steppe conditions, although they may support some of the same flora and fauna. The mammoth steppe may be gone forever.

MAN IN ANCIENT BERINGIA

Beringia was almost certainly the gateway by which man first entered the Americas, and therein lies one of the bases for the abiding interest in ancient Beringian landscapes and climates. Yet, in 1982, we seem farther than ever from reaching a consensus on the time of man's first entry into the New World. The view that the bearers of Clovis industries passed from Beringia southward into an unoccupied continent about 12,000 years ago still finds strong advocates. Another group, at least equally numerous, favors hotly debated evidence for occupation thousands to tens of thousands of years earlier in both Beringia and American regions to the south.

The timing of man's entry into the Americas is an issue that can be discussed but certainly cannot be settled here. Relevant is a consideration of the environmental constraints imposed by life in a dry region with a prolonged, very cold, dormant season and of the earliest times at which man is known to have evolved the capacity to deal with these constraints. It is evident that large ungulates must have been the most substantial and most dependable food resource available to human bands living in Pleistocene Beringia and nearly inconceivable that Paleolithic men could have survived even briefly there without sophisticated hunting skills. Müller-Beck examines the Old World archaeological record for evidence concerning the time of emergence and increasing competence of human big-game hunting in the Old World and then goes on to discuss the timing of man's penetration into Eurasian regions of progressively harsher climate. Morlan and Cinq-Mars review archaeological evidence pointing toward a human presence in eastern Beringia at various times during the past 60,000 years. Haynes presents the case for a late arrival of man in central North America and discusses the problem of Clovis' Beringian and Old World origins.

The 11,000- to 11,500-year old Clovis industries of central North America provide a record of activities of big-game hunters equipped with skills that certainly were essential for survival in Pleistocene Beringia; thus Haynes' reconstruction of Clovis lifeways has great relevance for attempts to reconstruct late Pleistocene human activities in Beringia. The collapse and partial extinction of Pleistocene large-mammal faunas interests many of the contributors to this volume. It seems inevitable that man played at least some role in this drastic faunal change. Paul Martin here reconsiders and elaborates his Pleistocene overkill model in an attempt to accommodate possible evidence of lengthy periods of coexistence of man and mammoth in the Old World.

Archaeological assemblages of stones and bones by themselves provide only the most skeletal insights into the thoughts, feelings, beliefs, social relationships, and hunting strategies of vanished peoples. Yet, this intellectual and social repertoire played a crucial role in allowing humans to colonize regions as forbidding as Beringia. Marks, in our final paper, provides an analysis of social, behavioral, and psychic factors involved in the present-day hunting practices of a tribal group in Africa as a reminder of the as-yet unseen complexities in human hunter/ungulate prey interaction in Pleistocene Beringia.

—David M. Hopkins

20

LATE PLEISTOCENE MAN IN NORTHERN ALASKA AND THE MAMMOTH-STEPPE BIOME

Hansjürgen Müller-Beck

The adaptation of man to the steppe-tundra was evidently a slow and time-consuming process. There is evidence of man actively hunting elephants during the middle Pleistocene, as early as 400,000 to 300,000 years ago, in the steppe and grassland areas of the Old World. Man gradually adapted to the cooler regions of the steppe-tundra that covered vast areas of Central Europe; we have evidence of man hunting *Mammuthus primigenius* in this environment as early as 80,000 to 60,000 years ago. We know that Neanderthal man lived and hunted in the warmer regions of far northern Eurasia. By 35,000 years ago man shared the more temperate areas of the mammoth's habitat from France to Siberia; and by 14,000 years ago man lived, at least seasonally, throughout the mammoth's Old World domain.

Mammoth was not the only prey of early man, who obtained perhaps one specimen per hunter per year. More often *Rangifer tarandus* and *Equus caballus* or *E. przewalski* were hunted, as well as *Bison, Bos, Saiga, Cervus,* and *Capreolus.* There is evidence at kill sites of many medium-sized animals, and some small, such as *Lepus* and *Alopex.* Birds, fish, and plants were also consumed.

Man was well adapted to the more meliorated areas of the steppe-tundra by 40,000 years ago and had the technical resources to cross the land bridge from the Old World to the New. It is also reasonable to assume that if man had crossed the land bridge during an earlier time he would have adapted to the steppe-tundra conditions in Beringia as he did in Northern Eurasia.

Key words: Northern Eurasia, adaptation, mammoth hunting, Pleistocene

This paper is intended to give an impression of the adaptation of man to the steppe-tundra-mammoth biome in northern Eurasia, as far as the available data permit. At present, it appears that this adaptation developed mainly during the time span 80,000-20,000 years ago. It would seem that this was a necessary preliminary step to the dispersal of man into the New World by way of Beringia. I am not trying to reconstruct the roots of the earliest occupation of the New World based on typological reasoning. I prefer, instead, to concentrate on the adaptive mechanisms, independent of local artifact traditions, that enabled man to hunt and live in an environment evidently not used at all before late Pleistocene time.

We must also keep in mind that, in the northern latitudes, mammoths were probably not encountered too often and hunted rather rarely compared to smaller game. Perhaps, in fact, a hunter was lucky enough to kill such an enormous supply of meat only once a year. But we may also be certain that prehistoric hunters were impressed by this

ISBN 0-12-355860-3

huge creature, as shown by one of the earliest known animal figurines—a mammoth carving from Vogelherd Cave (Fig. 1).

We must also keep in mind that all living elephants are inhabitants of grassland areas, so it is likely that the mammoth (*Mammuthus primigenius*) also preferred a steppe-like environment, even though it could move at times into areas better designated as wet tundra or forest-steppe.

Modern city dwellers find it difficult to imagine the possibility of hunting large adult elephants with simple weapons of wood, stone, or bone rather than the high-powered guns used by modern elephant hunters. But according to explorers' reports from both Asia and Africa, elephants were hunted on foot and on horseback, using spears, swords, traps, and poisoned arrows, before the introduction of firearms (e.g., Lackowitz, 1899). Early observations as well as most modern studies of elephant behavior have shown that single young or adult elephants are not difficult to approach if well known to the hunter. Elephants have poor eyesight but excellent hearing and sense of smell and prefer to use fixed trails along which they can be ambushed from the side or from behind by an experienced hunter pushing a lance into the abdomen of the animal from below. This situation is especially likely in bushy areas close to streams, lakes, and waterholes. The only thing a hunter has to do is await his luck. Another possible technique in proper wind conditions involves one hunter attracting the attention of a single elephant from the front while another attacks the beast from the back (Lackowitz, 1899) (Fig. 2). This, naturally, is not possible with the well-defended herds of female elephants and their offspring on normal alert.

The earliest real kill site known thus far is Lehringen in northern Germany (Fig. 3). There, a

Figure 1. Carving of an adult mammoth from Vogelherd Cave (Aurignacian), 50 mm long

relatively primitive individual of *Paleoloxodon antiquus* was found in a marl pond (Adam, 1951). The pollen data show very clearly that this site belongs to the forest period of an interglacial or thermal interval, but that large grassy areas were present in the flat river valleys. More than likely, the Lehringen site dates from an interglacial within the *Mittelpleistozän* of German authors about 300,000 to 130,000 years ago. There is a less likely possibility that the site could be assigned to the Eemian, the last major thermal of Europe, but *Paleoloxodon antiquus* definitely disappeared from Central Europe more than 80,000 years ago. In any case, below the ribs and in the abdomen of this elephant a well-preserved lance 2.15 m long was found, made from the wood of *Taxus* (yew) with a well-cut tip hardened by fire. *Taxus* is still considered today to be the best European wood for making weapon shafts. Some 30 simple flakes produced by the Levallois technique were also found near the skull. Adam (1951) was able to reconstruct the events convincingly. Evidently the elephant, about 45 years old, was wounded by the lance, which was pushed into the chest or abdomen, and tried to escape (wounded elephants in Africa are known to reach a maximum speed of 25 km per hour). Possibly the animal was ambushed very close to the pond in which its remains were found, or perhaps it was wounded farther away and sought relief in the marly pond, was trapped there, and died. The distribution of the flakes suggests that not much more than the trunk and parts of the head could be reached by the hunters.

To try to understand our adaptation problem fully, however, we must look further back into the beginning of elephant hunting in the steppe regions of Eurasia. Here we encounter the difficulty of defining the different types of steppe and also that of defining the different species of elephants. The oldest form, evidently confined to warmer climates, was *Archidiskodon meridionalis*. Younger and adapted to cooler conditions was *Mammuthus trogontherii*, very similar to the New World *Mammuthus columbi* in both size and climatic range. The youngest form is *Mammuthus primigenius*, the woolly mammoth, the only elephant persisting in Central and Northern Europe after 80,000 years ago. *M. primigenius* was well adapted to a subarctic climate, occurring together with *Rangifer* (reindeer), *Alopex* (arctic fox), and even *Ovibos* (musk ox). If our stratigraphic interpretation in Central Europe is correct (and this is still open to question), then the earliest representatives of *M. primigenius* appeared during the second half of the middle Pleistocene during the so-called

Figure 2. Elephant hunting in Africa, middle of 19th century, by distracting the animal and attacking from behind (after Lackowitz, 1899)

Riss or Saale Glaciation (Toepfer, 1970) and may be dated around 150,000 years BP. There are earlier finds that have teeth transitional in size between *M. trogontherii* and *M. primigenius* (Adam, 1961), but even these are probably younger than 200,000 years.

We must also keep in mind that elephants are creatures able to move over great distances, which makes interpretation still more complex in Eurasia where a history of shifting vegetation zones in changing climates must be considered. It may very well have happened that elephants such as *Mammuthus primigenius*, considered to be well adapted to subarctic conditions, moved far into nonarctic areas during winter and were encountered and hunted there by man. Or, earlier, individuals of *M. trogentherii* may have penetrated during summer into subarctic vegetation. Thus, we need to know the season in which a kill was made.

The archaeological record indicates that man in the *Homo erectus* stage of evolution was a hunter not only in the tropics and savannas but also in the temperate steppes of the Old World (Collins, 1969). His presence is documented in the more open stream valleys and also at the grassy shores of the European seacoast. *Homo erectus* evidently

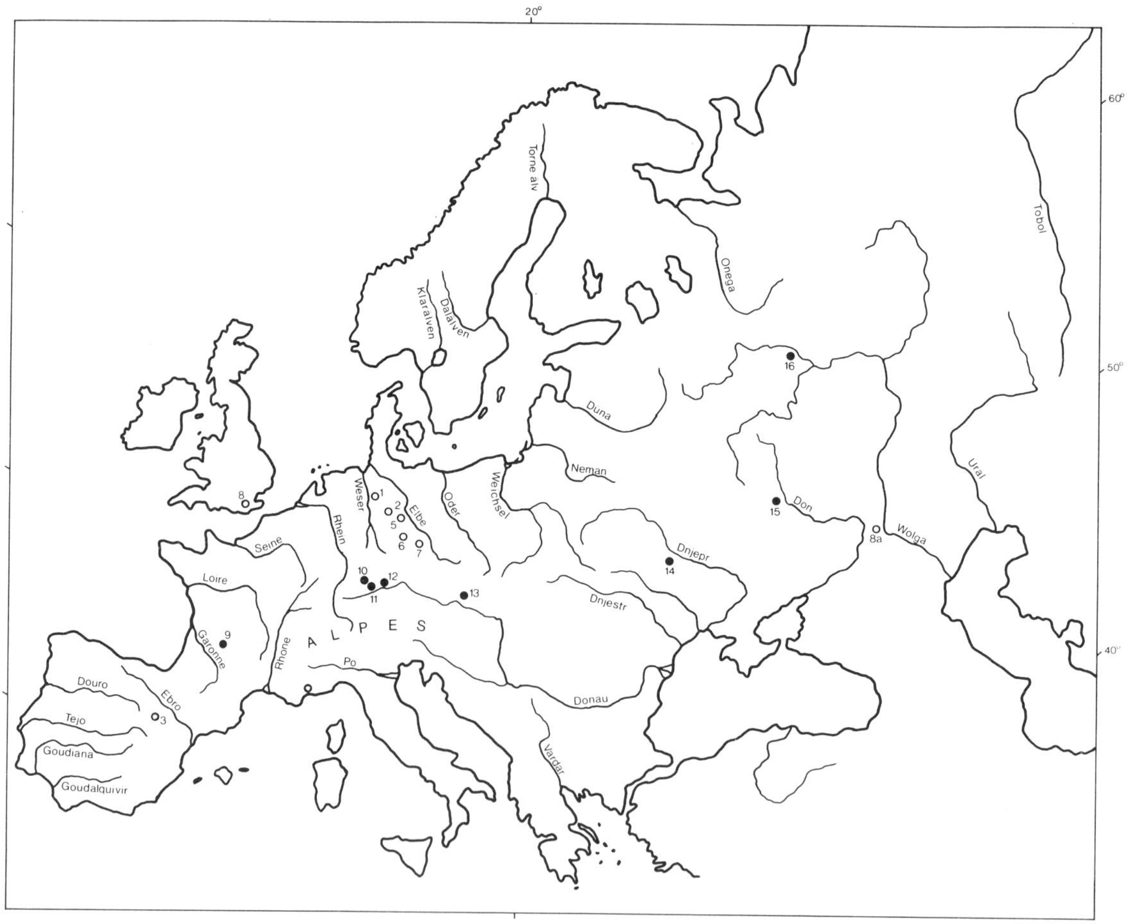

1 Lehringen
2 Salzgitter-Lebenstedt
3 Torralba
4 Terra Amata
5 Hundisburg
6 Königsaue
7 Markkleeberg
8 Crayford
8a Volgograd
9 Abri Pataud
10 Haldenstein
11 Vogelherd
12 Mauern
13 Pavlov
Dolni Vestonice
14 Mezhirich
15 Kostienki
16 Sungir

Figure 3a. European reference map. Open circles = sites older than 40,000 years. Points = sites younger than 40,000 years

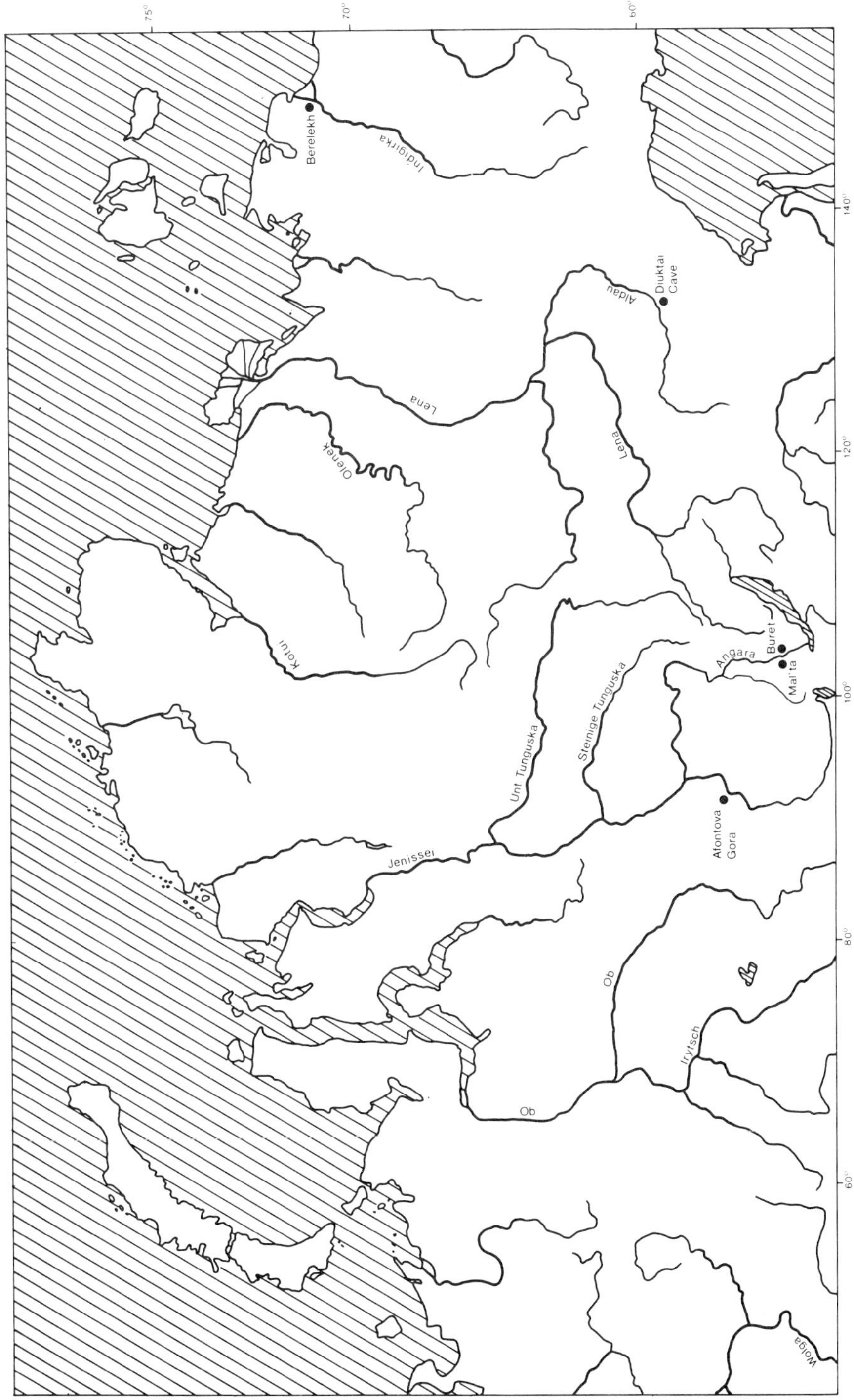

Figure 3b. Siberian reference map

avoided the denser forests as well as regions too cool for his level of technical adaptation. But, wherever his traces are reported and preservation conditions permit, there are food bones proving that large mammals were hunted. These sites commonly include the local forms of elephants.

Large-mammal hunting can be shown, for example, at Terra Amata, a site at Nice in southern France situated behind a raised gravel beach on what was then a narrow grassy slope below a dense forest extending up into coastal hills (de Lumley, 1966, 1976). Terra Amata is a well-observed camp, the oldest known with clear activity details anywhere in Europe. The artifacts document an early

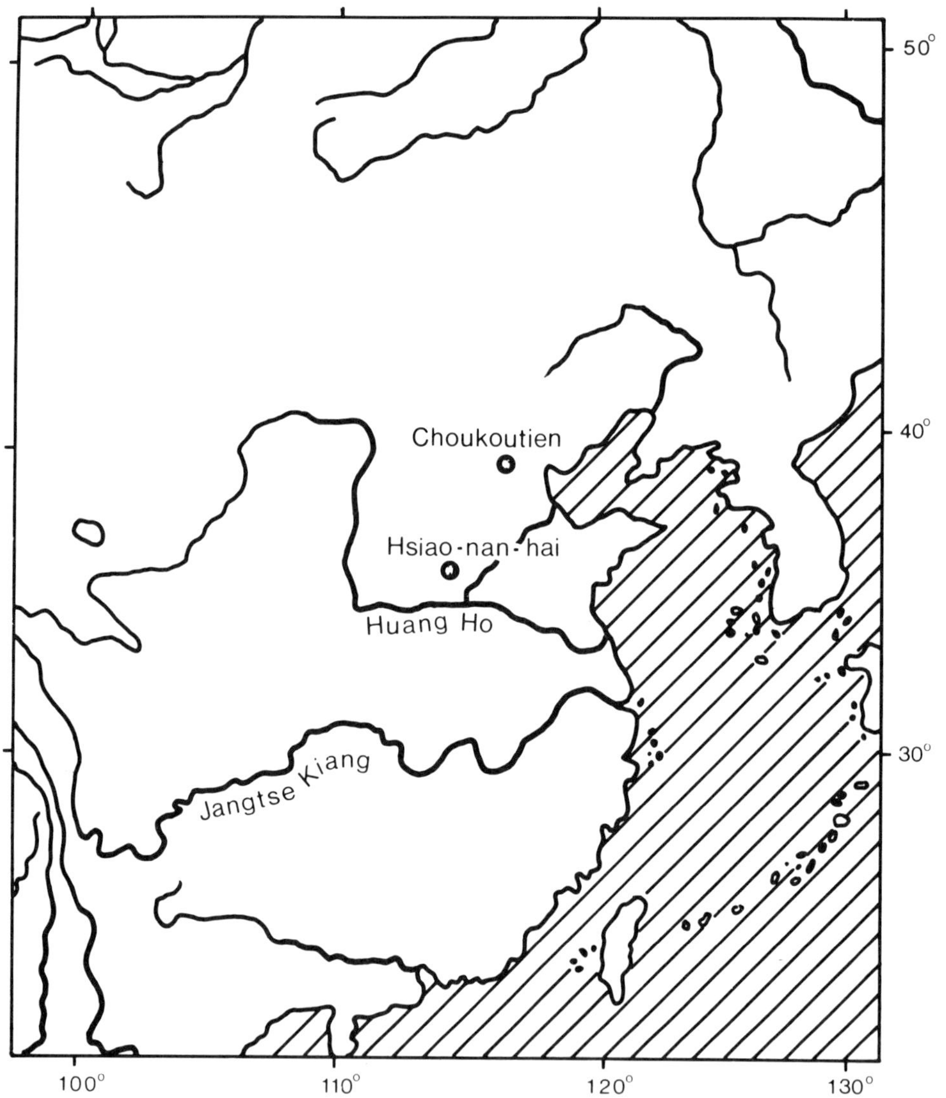

Figure 3c. Chinese reference map

hand-axe technology combined with relatively advanced flake tools, well-constructed fireplaces, and remains of hut-like shelters built of tree branches. Among the animals hunted were forest elements such as *Cervus* (red deer), *Bos* (ox), *Sus* (pig), and *Dicerorhinus kirchbergensis* (Merck's rhinoceros) as well as mountain species such as *Capra* (goat). The largest animal hunted was *Archidiskodon meridionalis,* an elephant that occupied areas transitional between forest and steppe.

Elephant remains are also found at Torralba and at nearby Ambrona, upland sites in Spain (altitude 1100 m) considered to belong to the same general time range (Freeman, 1978). There, the predominant faunal elements (including an early form of *Bos primigenius; Paleoloxodon antiquus,* the so-called forest elephant; and deer of genus *Cervus*) are indicative of forest or forest-steppe; all needed some woodlands for feeding and cover. This may also have been true of *Equus* (horse), present in the Torralba fauna.[1] In any case, among the elephants found at Torralba and Ambrona was at least one *Mammuthus trogontherii.* Also present were some individuals of *Dicerorhinus hemitoechus,* an inhabitant of areas transitional between forest and steppe (Adam, 1954). Thus, the Torralba and Ambrona fauna record the presence of steppe elements in uplands. This represents the first clear record of man's encounter with the forest-steppe ecotone outside the large stream valleys of Europe. Casts observed in marly sediments indicate that the Torralba hunters used weapons the size of thrusting lances, and throwing spears and hand axes of soft limestone which, though relatively crude, were evidently sufficiently useful. Also present were relatively refined flint flake tools used for more controlled and regular cutting.

Pollen studies at Torralba show that pine prevailed in the arboreal component over hazel, oak, and alder. Nonarboreal pollen values are high, as one would expect in a forest-steppe environment close to the wet areas in which the sites are situated. Grasses reach more than 40% and chenopods up to 20%, but sedges are very low. The pollen indicates vegetation typical of an oceanic steppe with dry soil conditions but with groups of trees on the more sheltered and better-watered slopes. Although it is impossible to reconstruct the vegetation in the immediate vicinity from pollen spectra, we must imagine a mosaic of different communities determined by topography as well as substrate differences.

That the Torralba and Ambrona sites were close to a pond and formed a promising hunting area for *Homo erectus* is beyond question. Here was a place where the prey could be met most easily and where experienced hunters could find sufficient cover to ambush single elephants, rhinos, bovids, horses, and deer. Butchering was done directly at the kill site. Most likely the bones found represent repeated hunts over an unknown span of time and probably different seasons.

The Terra Amata, Torralba, and Ambrona sites document hunting of elephants along with other large animals in Europe more than 300,000 years ago. Included is a site close to the seashore and others at the edge of the steppe at an altitude of 1100 m.

During warmer intervals toward the end of the Saale glacial sequence (if this dating is reliable, and there are still some doubts about that), hunters crossed the Elbe River as far as Hundisburg and Markkleeberg near Leipzig (Toepfer, 1970), and there, for the first time on record, encountered elephants clearly belonging to the species *Mammuthus primigenius* together with *Rangifer tarandus* (reindeer). But the remains recording these encounters were found dispersed in gravel beds and not in sealed living floors well controlled by archaeological excavation and documentation. There is little doubt that the artifacts and woolly mammoth are contemporary, but whether reindeer were also present during the same season as man is uncertain. For the moment, therefore, we cannot be certain that man hunted in the area of *M. primigenius* during the middle Pleistocene at any season other than summer.

In the Far East, locality 15 of Choukoutien belongs in about the same time range at the end of the middle Pleistocene. There, man-made tools have been found together with *Coelodonta antiquitatis,* the woolly rhino, which farther north and

[1] All modern horses are inhabitants of the open steppe, but Musil (1969) has shown that *E. taubachensis,* at least, had moved into relatively wet and thinly forested areas. The closely related *E. steinheimensis* (Adam, 1954) and, later, *E. germanicus* (one of the main prey species of Neanderthal man in Central Europe—Musil, 1969) also ranged at least partly into woodland areas.

west often occurs with *Mammuthus primigenius*, *Cervus elaphus* (red deer), *Canis lupus* (wolf), and two species of pig. This fauna suggests denser and warmer forest areas interrupted by wide, grass-covered river valleys. *Coelodonta* may have been only a seasonal visitor, and the same may have been true of man. Thus far we have no indication that man went farther out into the cold steppes of eastern Asia than he did in Europe during the middle Pleistocene. On the contrary, it appears that, in the East, man ventured less into colder climates (Aigner, 1969).

Another important early site is the sequence of travertine layers at Weimar-Ehringsdorf, generally considered to be of Eeemian age[2] (Steiner, 1979). There, we have *Mammuthus primigenius* at the base of the *auelehm*. In the overlying lower travertine, the presence of *Paleoloxodon antiquus* indicates a forested landscape and a thermal interval. Higher still, the upper travertine contains *M. primigenius* again.[3] During the periods in which the travertine was formed, the slopes of the Ilm Valley attracted animals to ponds in which they had some difficulty moving and could be ambushed rather easily by experienced hunters. The hunters there were early variants of Neanderthal man using well-made tools, often bifacially worked, including some very functional knives and points that represent early weapon tips (Müller-Beck, 1966). There is evidence that carnivores were present as well as man (*Crocuta*, spotted hyena, in the lower travertine and *Felis*, lion, in the upper). However, the carnivore remains are even rarer than those of men.

The faunal remains at Weimar-Ehringsdorf have been collected mostly during quarrying operations and only very rarely by planned excavation, but we can at least assume that the remains belong to elephants hunted by carnivores and man. Some figures on age classes from Günther (1975) are interesting (Table 1). He shows a clear difference in the age distributions of the *Paleoloxodon* and *Mammuthus* populations, but it remains open to question whether this reflects a change in human hunting techniques or, as seems more likely, differences in the behavior of the two elephant species. It should be mentioned that *Capreolus capreolus* (roe deer) is present in both the lower and the upper travertine, indicating relatively thermal local conditions during the accumulation of both layers. However, *Sus scrofa, Dicerorhinus kirchbergensis*, and *Dicerorhinus hemitoechus* occur only in the lower bed with *Paleoloxodon*. On the other hand, *Coelodonta antiquitatis* is found only with *Mammuthus* in the upper bed, and this may also be the case for *Rangifer tarandus*.

Our knowledge of the distribution of vegetation at the beginning of the last ice age (Würm in the Alps, Weichsel in northern Europe) is fairly good (Frenzel, 1968). Seventy or eighty thousand years ago the distribution of steppe-vegetation complexes had broadened under colder, drier, and, during the winter at least, more continental conditions. How far men ventured out into the steppe area for hunting is not yet clear. A slightly warmer interval that brought forest cover at least to valleys sheltered from harsh continental winter winds has left two sites where remains of the activity of man close to a river and a lake have been excavated and reported. Neither has a real living floor, but the sites are preserved as a result of hunting close to water.

Table 1
Abundance by Age-Class of Proboscidian Remains at Ehringsdorf-Weimar (Günther, 1975)

Age Class In Years	*Paleoloxodon antiquus*	*Mammuthus primigenius*
<5	10%	5%
5–20	28	15
20–40	28	60
>40	28	20
	100%	100%

[2] The age of the archaeological materials at Weimar-Ehringsdorf is still open to question (Steiner, 1979); for the moment, one cannot rule out the possibility that this interesting complex may be older than Eemian. However, this is of minor importance for our discussion, since Steiner's conclusions would just make our evidence 10,000 years older.

[3] Based on numbers of enamel plates per cm, Günther (1975) places some of the mammoth teeth at Weimar-Ehringsdorf in the subspecies *Mammuthus primigenius trogentherii*. Whether this represents a long-term genetic difference or just a local and temporary larger range of variation is another question. The latest mammoth tooth, at any rate, has a very low length:plate number index, putting it clearly into the true *M. primigenius* category.

The first is Salzgitter-Lebenstedt near Braunschweig, west of the Harz Mountains in northern Germany (Tode, et al., 1953). Here again we have only a preliminary report that leaves the precise age in doubt (conceivably Saltzgitter-Lebenstadt is older than 80,000 years and contemporary with Hundisburg and Markkleeberg). For us, however, the precise age is relatively unimportant; the site does illuminate the activity of man—*Homo sapiens neanderthalensis*, based on a skull fragment found in the site. The site consists of a midden formed by scattered bones of butchered animals that are washed down a slope near a small creek. Macrofossils in the archaeological horizon include several arctic plants: *Salix polaris* and *S. herbacea* (arctic willows), and *Armeria* sp. cf. *A. armeria* (thrift). *Pinus* (pine) dominates the arboreal pollen spectra; *Picea* (spruce), *Alnus* (alder), *Salix* (willow), and *Betula* (birch) are also present. Sedges dominate the nonarboreal pollen, but grasses are well represented and *Artemisia* (sage) is present only in very small quantities. We encounter a cold and relatively wet oceanic steppe environment which without doubt belongs within the range of steppe-tundra (Giterman, Sher, and Matthews, this volume). Although the abundance of tree pollen suggests subarctic taiga, pollen transport from the west and southwest may falsify the local picture; the site may have been just at the edge of the climatic "low arctic." In addition to well-made flake tools produced by the Levallois technique, the site yields somewhat asymmetrical hand axes of a late type. Also present are well-made bone tools, including lance points made from ribs of large mammals and a small point which was to be attached to a larger shaft, probably of wood. The fauna, according to the preliminary account of Kleinschmidt (1953), fits very well the vegetation described above: *Mammuthus primigenius* (minimum of 16 individuals), *Rangifer tarandus* (at least 80 animals, among them 15 juveniles), *Bison priscus* (6 or 7), *Equus* sp. (4 to 6), *Coelodonta antiquitatis* (2), and a few other species including some birds and fish that are represented by single specimens.

Although the Salzgitter-Lebenstedt site is still only partially excavated and the results only partly published, it is evident that *Mammuthus* was the most important prey, at least in terms of quantity of meat, and that it was hunted within at least the summer range of *Rangifer tarandus.* This is the earliest documented instance of mammoth hunting outside the range of *Cervus elaphus. Bison, Equus,* and *Coelodonta,* however, are present in the butchering area. The vegetation remains record a tundra influence, even if enough subarctic trees were present to attract *Coelodonta* to the site.

The other site dating from early Würm or early Weichsel time is the well-excavated Königsaue near the lake of Aschersleben (Mania and Toepfer, 1973). Königsaue consists of three archaeological horizons occupied during a warmer interval, very likely the Brörup Interstade of European authors, about 60,000 years ago. Mania ventures to give a rough reconstruction of the vegetation near the site itself (Fig. 4) and also of the general region of the Elbe/Saale valley in eastern Germany at that time (Fig. 5). This map, though highly speculative, gives an impression of the possible different levels of productivity in a vegetation patchwork (see also Redmann, this volume). Productivity was high enough to permit man to risk hunting and living in the region, at least seasonally, as early as 60,000 years ago.

The three levels at Königsaue are characterized by tools that are within the range of the Micoquoid-Mousterian technocomplex.[4] The food bones are rather badly preserved, and it is possible that hyenas were responsible for some of them. In any case, remains of only two species—*Equus* sp. and *Canis lupus*—are preserved in level C, the highest horizon. Level B has a richer fauna, including a minimum of 1 to 3 individuals each of *Mammuthus, Rangifer, Bison, Crocuta, Canis lupus, Felis leo spelaea* (cave lion), *Coelodonta, Equus (Asinus) hydruntinus, Equus* sp., and *Cervus elaphus.* Level A, the lowest, is richer and contains *Mammuthus primigenius* (minimum of 4 individuals), *Rangifer tarandus* (5), *Bison priscus* (3), *Equus caballus* ssp. cf. *E. caballus mosbachensis* (4), *E. (Asinus) hydruntinus* (1), *Coelodonta*

[4]Level C at Königsaue has bifacially retouched stone artifacts within the range of Micoquoid-Mousterian inventories; this technocomplex is a tradition of very long duration between the older Acheulean and the younger *Blattspitzen* industries. Level B is characterized by flake tools without bifacial trimming that are typologically Mousterian, but the predominance of flake tools is a function of the restricted activities at the site and does not put this inventory outside of the range of the late Micoquoid-Mousterian. Level A again is characterized by bifacial trimming of stone tools and thus is Micoquoid-Mousterian.

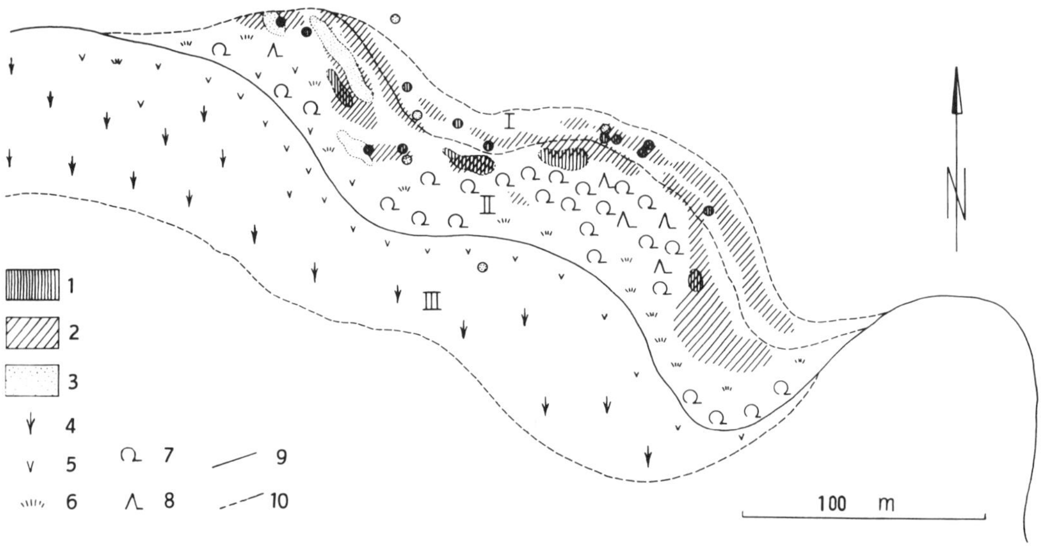

Figure 4. Vegetation at the Königsaue Campsite (Mania, 1973): 1 = artifacts from A. 2 = artifacts from B. 3 = artifacts from C. 4 = *Phragmites*. 5 = moss, mainly *Sphagnum*. 6 = sedges. 7 = deciduous trees ((*Alnus, Populus, Betula, Salix*). 8 = conifers (*Picea amorika* and some *Pinus*). 9 = shorelines. 10 = boundaries of vegetation zones (typical mosaic vegetation)

antiquitatis (2), *Cervus elaphus* (1), and *Crocuta spelaea* (1). The fauna is similar to Salzgitter-Lebenstedt, differing only in the presence of *Cervus elaphus* and *Equus (Asinus) hydruntinus*, which hints that the climate may have been slightly warmer. Vegetation documented by macrofossils in the archaeological horizons confirms this. Close to the former lakeshore, sedges are found. Farther away, there is *Sphagnum* and some tree remains, including *Betula, Salix, Populus* (poplar), *Alnus*, and *Picea amorika*, which are typical of wet areas. *Pinus* is scarce (Fig. 4). All of these taxa except *Salix* are present in a pollen spectrum from the site and pollen of *Tilia* (linden) is present as well. *Pinus* pollen is very abundant (62%), *Betula* comprises 26%, and *Picea* is much lower (3%), reflecting pollen influx from a broader area than that represented by the microfossils. In a sample taken from marl representing deposits farther out in the lake, *Pinus* comprises only 26% and *Picea* only 1%, but *Betula* comprises up to 61% of the arboreal pollen. The significance of the paleobotanical record is an indication that real forests extended east of the Harz Mountains, at least near lakes and rivers, during the Brörup Interstade (Fig. 5). Grasses, sedges, and *Artemisia* very likely were important in drier areas and areas at higher altitudes in which the cold, dry, continental winds from the east eliminated forests and opened up the vegetation to steppe communities.

Although the faunal samples from Salzgitter-Lebenstedt and Königsaue are still small, they provide documented proof that man—still in the form *Homo sapiens neanderthalensis*—adapted at least 60,000 years ago to the edge of the steppe-tundra in Central Europe. We have sufficient direct data to reconstruct the larger vegetation areas during this interval. Gallery forests existed close to rivers and lakes in the lowlands, and farther out in the drier lowlands there was a grass-rich continental steppe. Taiga was present in more hilly and higher terrain; in the still-more exposed mountains, alpine tundra prevailed (Fig. 5). That the traces of man are best preserved near lakeshores and in valleys is understandable in terms of higher sedimentation rates there. But if butchering activity was concentrated in the valleys, site visibility may be higher there, too.

At Volgograd, farther to the east, near the Don River in the Soviet Union, there is another important site, perhaps of a slightly later date.[5] The Vol-

[5]The Volgograd site, mentioned above, may even be contemporaneous with the Hengelo Interstade, because the high ratio of Chenopodiaceae there signals more southern and warmer influences. For the moment, at least, a late date for this site cannot be ruled out.

gograd stone tools are partly bifacially trimmed and are very similar to the inventories from Salzgitter-Lebenstedt and Königsaue, though generally thinner and more slender (Zamyatnin, 1961). On typological grounds we may estimate that the Volgograd site is probably between 60,000 and 40,000 years old. Among the nonarboreal pollen, chenopods are most frequent, and grasses, sedges, and *Artemisia* are less abundant. In the arboreal pollen spectrum, *Pinus* and *Picea* are present at 49% and 2% respectively, levels similar to those at Königsaue. So, again, the hunters were present in a steppe of continental character with gallery forests near large bodies of water sheltered from the cold and steady winds. The fauna gives the same picture: *Mammuthus primigenius* (minimum of 51 individuals), *Bison priscus* (366), *Equus caballus fossilis* (42), *Saiga tartarica,* the antelope living now on the drier continental steppes of northern Eurasia (36), and *Cervus elaphus* (15), but only 1 each of *Rangifer tarandus* and *Canis lupus.* Neither *Coelodonta, Crocuta,* nor *Felis* are present. We must keep in mind that at Volgograd, too, we do not know the season during which the fauna was hunted—if at any special season at all.

The sites that I have mentioned give only the outer limits for a very large area in Europe in which steppes of diverse character were used by human hunters during the older part of the classical Würmian or Weichselian between about 80,000 and 40,000 years ago. Three main steppe types occupied this area:

(1) An oceanic temperate steppe rich in grasses and populated by *Equus* and *Cervus* predominated in the west near the Atlantic coast.

(2) A subarctic steppe rich in sedges and *Artemisia* and populated by *Rangifer, Bison,* and *Coelodonta* lay more to the east; this steppe type became tundra-like to the north and at higher altitudes as well as during colder times.

(3) A continental cool temperate steppe rich in chenopods and populated by *Bison* and *Saiga* lay still farther east in regions practically without oceanic influence.

All of the above steppe types could be said to have been "mammoth-steppe" in which *Mammuthus primigenius* was present. At least during summer and early autumn, human mammoth hunters penetrated these steppe regions as far north as 52°N in Central Europe and 49°N in Eastern Europe.

In northeastern Asia during the same time range, a different steppe environment is documented by the fauna of the Hsiao-nan-hai Cave near Ayang on the border of the southern Hopei Plain, and a human presence there is indicated by the presence of flake tools (Aigner, 1972). The Hsiao-nan-hai Cave fauna includes *Equus hemionis, Coelodonta, Capreolus, Pseudoaxis* (a species of deer), *Gazella* (gazelle), *Capricornis* (goat), *Ursus arctos* (brown or grizzly bear), and even *Pongo* (chimpanzee). The presence of *Capreolus, Coelodonta,* and *Pseudoaxis* shows that some tree cover must have been present, but *Gazella* and *Equus hemionus* document the steppe aspect. Which kind of elephant—*Paleoloxodon* or *Mammuthus*—was present on this far-eastern steppe is unknown, but we shall see a little later that the range of *Mammuthus* came into contact with *Capreolus* in Western Europe. In the Far East, we have still another steppe type in which man was present: (4) A continental temperate type, probably rich in grasses and herbs with some tree cover, populated by *Gazella, Coelodonta,* and *Capreolus.*

Figure 3 shows that by 40,000 years ago, man had moved northward at least to 49° and 52°N in the west and just north of 40°N in the more continental east of the Old World. The northern limit of human occupation 40,000 years ago seems to have been slightly north of the 10°C mean annual isotherm of the present time and somewhere between the 0° and the 5°C isotherm of that time.

After 40,000 years ago, the situation was generally slightly different. During the so-called Hengelo Interstade, 39,000-37,000 years ago (Grootes, 1977), and the corresponding Kargin Interstade of Siberia, conditions were improving. Man and animals would have been able to move farther northward at the level of adaption reached by that time. There was limited reforestation similar to that during the Brörup Interstade (Frenzel, 1967). In the Central European highlands (*Mittelgebirge*), forest-steppe or forest-tundra rich in *Artemisia* must have prevailed, and at lower altitudes with good wind and shelter some taiga areas reappeared. Closer to the Danube Valley and further to the southeast, the former open loess-steppe-tundra became overgrown by denser grass-steppe beneath which a brown arctic soil formed. There were scattered trees in the moister areas near water and at the base of well-watered slopes. Farther west, a grass cover mixed with herbs increased to form a rich

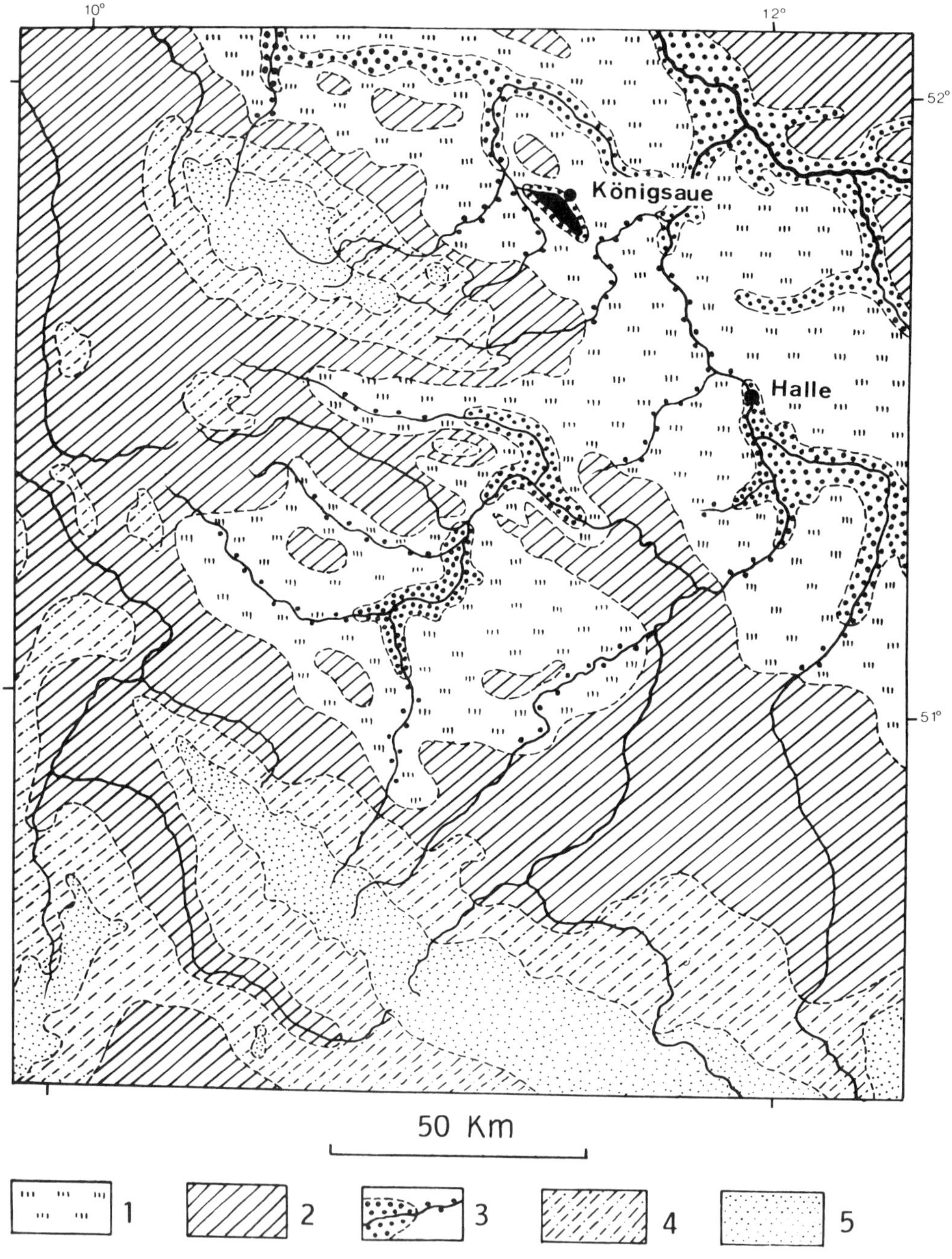

Figure 5. Reconstruction of the vegetation east and south of the Harz Mountains in the Brörup Interstade around the Elbe and Saale Rivers (Mania, 1973). 1 = Tschernosem grass-steppe. 2 = taiga. 3 = riverine forest. 4 = mountain forest-tundra. 5 = stony mountain tundra

hunting area with animals such as *Mammuthus, Equus, Bison,* and *Cervus.* The large *Ursus spelaeus* (cave bear) replaced *Ursus arctos* in many herb-rich areas. More refined bone points were used to hunt these large bears, and *Blattspitzen* (leaf-shaped stone points) were made for use as projectile points and knives. These projectile types mark the end of the Middle Paleolithic Micoquoid-Mousterian in Central and Eastern Europe; the Eurasian Upper Paleolithic started about 35,000 years ago. From the *Blattspitzen* level we have more than 40 different sites of which only two will be mentioned here.

The first of these is the Weinberghöhlen Cave site near Mauern at an altitude of 426 m in the midlands of the Fränkische Alb, close to the Danube Valley (von Koenigswald et al., 1974; Brande, 1975; von Koenigswald and Müller-Beck, 1975). There, in the *Blattspitzen* level (horizon IV), the following animals are present: *Mammuthus primigenius, Coelodonta antiquitatis, Rangifer tarandus, Cervus elaphus, Equus caballus, Bos* or *Bison, Capra ibex, Gulo gulo* (wolverine), *Crocuta spelaea, Felis leo spelaea, Ursus spelaeus, Castor fiber* (beaver) and *Lepus* sp. Among the small mammals there are *Lemmus lemmus* (Norway lemming), *Dicrostonyx torquatus* (varying lemming), and *Arvicola terrestris* (water vole). The minimum number of individuals in each species is difficult to estimate because part of the early excavation notes and materials were lost during World War II. Furthermore, animal predators, mainly *Crocuta,* were responsible for some of the bones carried into the cave.

At Mauern we are clearly dealing with a fauna that contained some arctic elements, but the overall impression is more subarctic or cool temperate, because of the presence of forest dwellers such as beaver and of species requiring some cover such as *Cervus* and probably *Coelodonta.* However, the woodland species may have been brought from the forested Danube Valley, which is visible from the entrance of the cave. On the adjoining steeper slopes and higher above the cave, arctic floral assemblages very likely prevailed, while a forested environment dominated the valley bottom and was probably responsible for the poorly developed *parabraunerde* (brown forest soil) still visible as a paleosol in nearby loess sections (Bleich, 1975).

The microfaunal elements signify a cool climate according to von Koenigswald. *Lemmus* and *Dicrostonyx* are not members of modern temperate habitats, but belong to the subarctic or lower arctic.[6] We should add that at the time of the Mauern occupation, Central Europe must have had warm continental summers: the high summer sun should generally have resulted in summers as warm as those of today. But summers must have been shorter and must have followed a cooler and often wetter spring and preceded a cold and brief autumn. The winter was also more continental and thus colder and longer than today.

According to Brande (1975), pollen records contemporary with the *Blattspitzen* horizon both inside the cave and outside in the valley sediments show considerable steppe components, while arboreal pollen declines (Fig 6). The difference between pollen records in the valley bottom and contemporaneous pollen spectra in the cave sediments reflects the different wind conditions inside the cave. A correlation is still possible, however, with the 20,000-year sequence documented by the valley core. *Pinus* always dominates the arboreal pollen, but *Picea, Salix, Betula,* and *Alnus* are also present. Sedges mostly dominate the nonarboreal pollen, and grasses follow, but *Artemisia* remains generally low. There is a period, however, in which grasses exceed sedges, and *Artemisia* increases to 15% of the total pollen, indicating that the normally cold and wet steppe situation changed temporarily to a drier one, though still of subarctic-oceanic character. It is very likely that this interval is identical with the Hengelo Interstade of Dutch authors, although the available radiocarbon dates do not make this correlation certain. Local chronostratigraphy and typological cross-dating of the artifacts suggest that man was present toward the end of or shortly after the Hengelo Interstade, about 37,000 years BP.

This also means that an increasingly oceanic climate can be assumed from Mauern westward, while the climate must have been even more continental to the east.

The second site to be mentioned here is the Haldenstein Cave near Urspring in the valley of the Lone River, a western Danube tributary in the

[6] Here I consider "subarctic" solely as a climatic definition, whose precision, however, leaves something to be desired as was evident in the discussion at our conference. Hopefully, we shall be able to be more precise with this terminology in the near future.

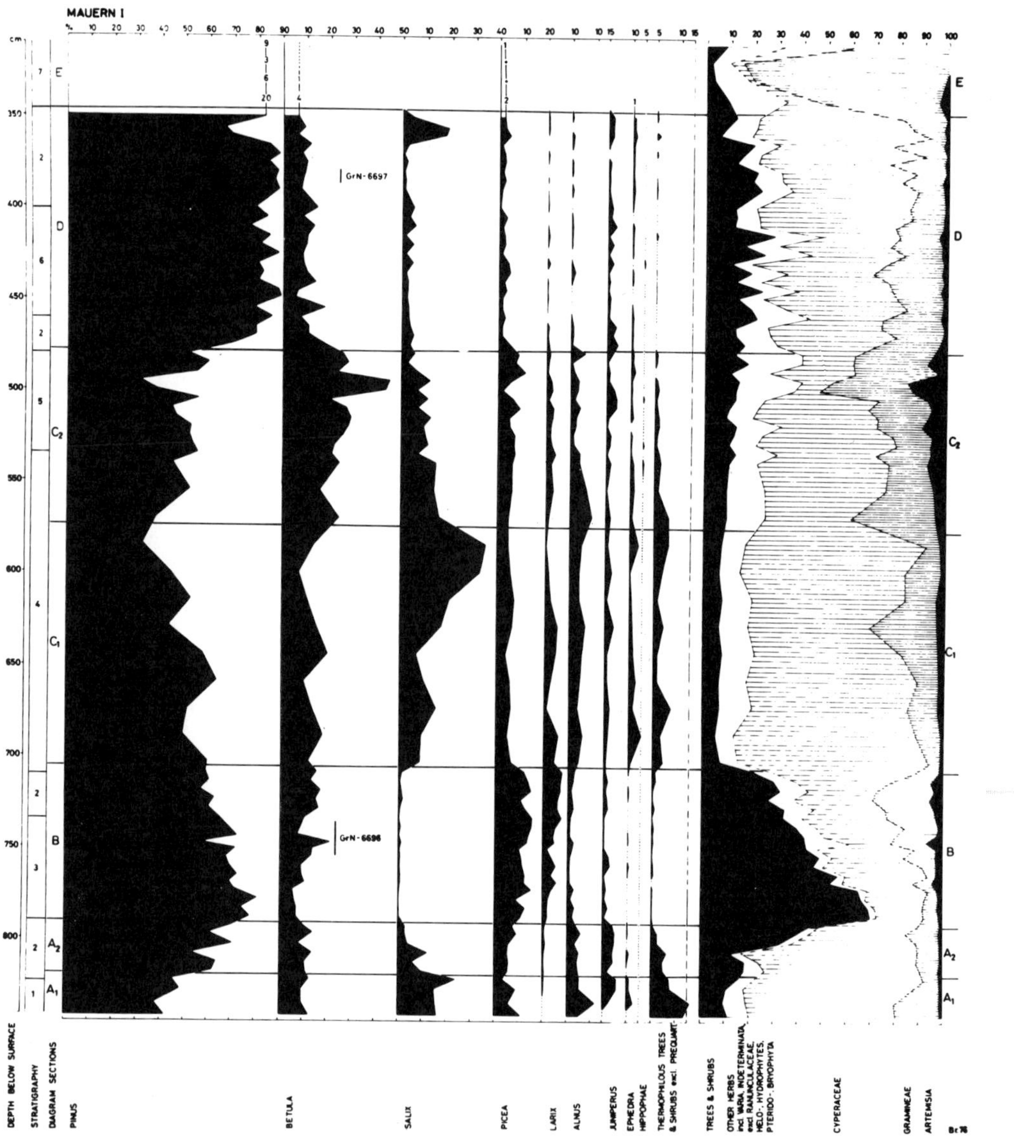

Figure 6. Pollen sequence from Mauern in the valley bottom in front of the Weinberg Cave (Brande, 1975). GrN 6697 = 29,290 ± 260 BP. GrN 6696 = 49,970, in the same position: GrN 7652 = 63,400 $^{+1300}_{-1100}$ BP. 800 - 700 cm = Brorup, 550–350 cm = Interpleniglacial (Hengelo, Deuekamp, etc.)

Schwabian Alb. Judging from the form of the few artifacts found (Fig. 7), Haldenstein is probably slightly younger than Weinberghöhlen (Mauern), between 37,000 and 35,000 years old. Here, at an altitude of 526 m, at the end of the valley close to the harsh, open plain of the Alb, the following animals are reported: *Coelodonta, Rangifer, Cervus, Equus (Asinus) hydruntinus,* and, for the first time, *Equus przewalski,* the horse of the cold, open, eastern Eurasian steppes (Riek, 1938). Unfortunately, no information regarding flora is available from this site, but the climate must have been

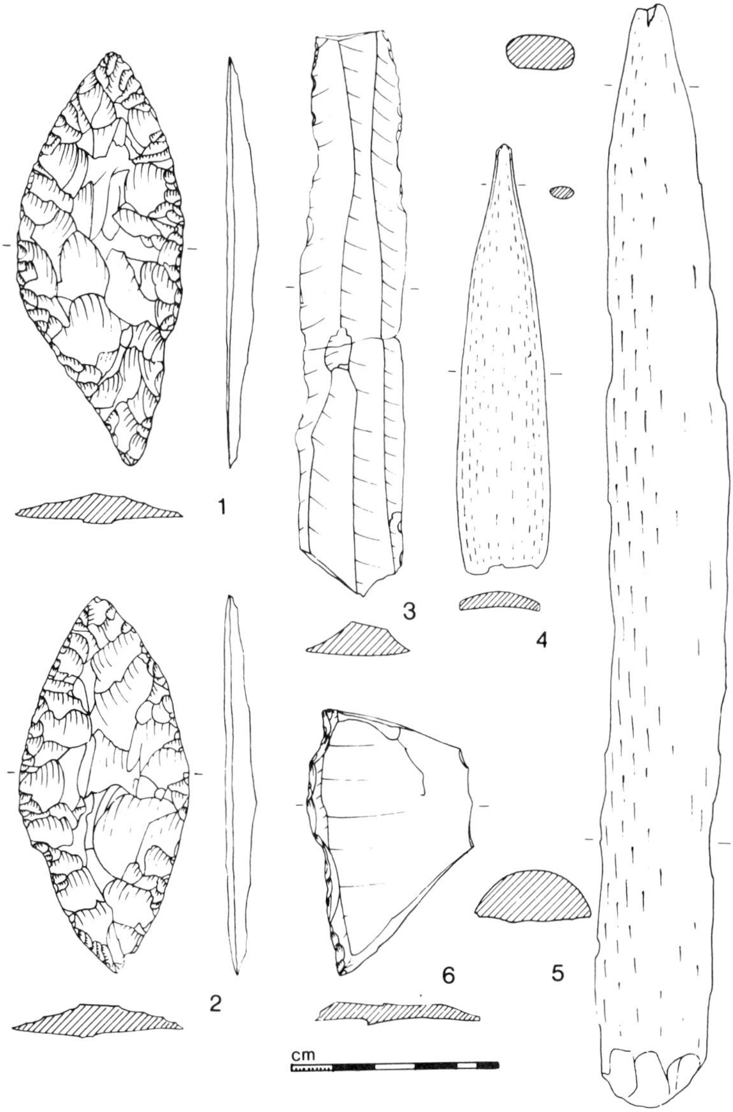

Figure 7. 1-3 = *Blattspitzen* and blade from Urspring-Haldenstein. 4-5 = bone points from Vogelherd VI. 6 = side scraper from Vogelherd VI

cooler and also drier than at Weinberghöhlen (Mauern). We may expect a more continental steppe environment at Haldenstein but not without some trees in the more-sheltered areas.

Layer VI in the multilayered Vogelherd site, closer to the Danube in the same valley, also contains traces of man's activities (Fig. 7) in about the same time range. More important, however, are the inventories from Vogelherd horizons V and IV (Riek, 1934). These are typical Aurignacian and thus Vogelherd is the first Upper Paleolithic site, sensu strictu, to be mentioned here. The two layers are approximately 31,000 years old (Hahn, 1977) and correlate roughly with the Denekamp

Interstade of the Central European floristic sequence. Among the hunted animals documented, *Mammuthus primigenius* with at least 24 individuals was predominant as a meat producer. *Equus przewalski* (minimum of 22 individuals) was also intensively hunted (as were horses during the European late Pleistocene). Also present are *Rangifer* (13), *Coelodonta* (8), *Cervus* (2), and *Sus scrofa* (1). The presence of pig signals a milder climate; summers, at least, lasted longer and were of continental character. There are no paleobotanical reports from the Vogelherd Aurignacian, but if we take the Mauern pollen profile to the east as a reference (Brande, 1975) (Fig. 6), then we can assume that the Lone Valley at this time supported a dry steppe with a little *Artemisia* and some trees in wind-sheltered spots.

Further to the east at Kostenki I, a site on the middle Don River in southern Russia (Rogatchev, 1957; Lazukov, 1957), level 5 may belong in the same time range as Vogelherd V/IV, although it could be slightly older. The fauna includes *Mammuthus, Equus caballus fossilis, Bos, Bison,* and *Saiga tartarica* (another indicator of the dry surroundings at that time). The presence of *Bos* most likely indicates relatively warm conditions, and this may also be true of *E. caballus.* There is no doubt that *Bos* requires some woodlands, so we must expect that gallery forests were present in the valleys of the Don region at the time of level 5 of Kostenki I. The higher sequence at Kostenki I (Lazukov, 1957; Klein, 1969) is also significant, as we may mention at least briefly. In level 4, *Cervus* is still present, but so is the arctic fox, *Alopex.* In the almost contemporary level 4 of Kostenki VIII Tel'manskaya (Klein, 1969), *Bos* and *Bison* are still present, associated with the highest black soil that can be considered to be older than the climatic minimum between 20,000 and 18,000 years ago. Level 3 of Kostenki I is in pure loess and contains *Cervus: Alopex: Rangifer* in the ratio 2:69: 205. This is the most extreme Paleolithic reindeer count known from southern Russia and, presumably, is also evidence for a colder, more severe climate during the time of level 3. The fact that level 3 was occupied during a period of loess deposition supports this climatic inference and indicates that the predominance of reindeer is not simply a reflection of winter hunting.[7]

Strangely enough, *Mammuthus* is totally missing from level 3 of Kostenki I. Possibly grass productivity was not sufficient to permit mammoth to compete with other herbivores such as the evidently large populations of reindeer—or perhaps hunting by man was so intense that the woolly mammoth disappeared from the Don Valley nearly 20,000 years ago. These are questions that must be left open for the moment.

That limiting factors existed for *Mammuthus* in earlier times can be shown by the data from Abri Pataud in southwestern France (Bouchud, 1975). There, in level 11 (early Aurignacian, about 33,000 years old), only one bone fragment of *Mammuthus primigenius* was found among 599 of *Rangifer* and 359 of *Equus caballus,* and only 4 of *Bos,* 2 of *Cervus elaphus,* and 1 of *Sus.* The pollen flora in Abri Pataud level 11 is dominated by *Pinus, Betula, Quercus* (oak), *Fraxinus* (ash), *Alnus, Juniperus* (juniper), *Artemisia,* and sedges and grasses (Donner, 1975), indicating some grassland and at least patches of denser forest that very likely were much larger than those indicated during the Brörup Interstade at Konigsaue.

In level 7 at Abri Pautaud (slightly later Aurignacian, about 32,800 years old), the following bone remains are reported: 2 fragments of *Mammuthus,* a predominating 395 of *Rangifer,* 80 of *Cervus elaphus,* 60 of *Equus,* 3 of *Sus,* and, surprisingly enough, 7 of *Capreolus capreolus* (in Central Europe roe deer are considered to be typical of interglacial forest conditions). The pollen flora is similar to that of level 11 except that *Artemisia* is not reported and *Abies* (fir) is present. *Abies* is confined to the warmer areas of Europe today, and was not present during the Aurignacian time range in Central Europe. The faunal remains indicate that Abri Pataud must have been near the western climatic limit of mammoth, so numerous at that same time on the open eastern steppes. The pollen sequence indicates that level 7 at Abri Pataud is contemporaneous with the warm Denekamp Interstade. Paleontologists noted as early as the beginning of this century that Aurignacian levels show an increase in *Cervus* and *Capreolus* between a previous *Rangifer* maximum in the "cold"

[7]It is likely that reindeer avoided the mosquito-infested Don River Valley during summer, moving instead to the cooler and drier north closer to the ice sheet. However, an analysis of the reindeer remains from level 3 of Kostenki I would be desirable in order to firmly settle the argument concerning the season of occupation.

Mousterian and a succeeding one in the Proto-Magdalenian.

Level 5 at Abri Pataud (Perigordian IV, 28,000 years old) accumulated during a cooler interval; the pollen flora lacks *Abies*, but *Artemisia* reappears. Intensively hunted *Rangifer* yields 21,119 bone fragments, and *Cervus elaphus* is represented by 172, *Bos* (48), *Equus* (44), *Mammuthus* (5), and *Ovibos* (1), but *Capreolus* is still present and represented by 10 bone fragments. The Abri Pataud record also indicates that both southwestern France and continental southeastern Europe in the Don region offered better living conditions for man than the north-central European corridor between the Scandinavian and Alpine ice sheets, forming a very well-used corridor between the West and the East (Fig. 8). But man adapted himself to all of the different European steppe conditions and he was present in large numbers at least as late as 25,000 years ago and again after 18,000 years ago in the north-central European corridor itself. Moreover, some contact existed between Eastern and Western Europe nearly all the time.

Level 3 at Abri Pataud (Perigordian VI, on the order of 23,000 years old) was formed in a time of a similar flora (Donner, 1975) without *Abies* but

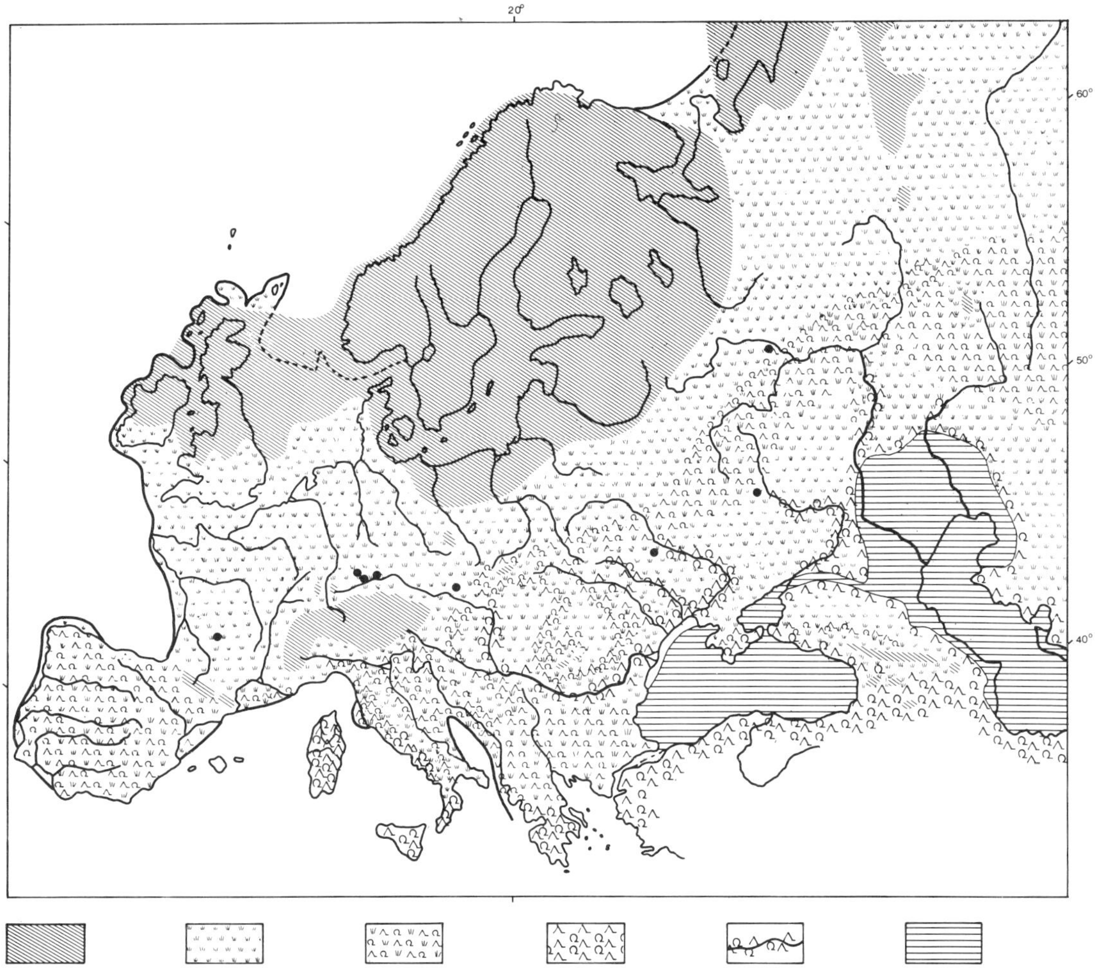

Figure 8. Last glacial maximum vegetation (based on Frenzel, 1968, with additions). Dots = sites mentioned in text (after 40,000 BP). 1 = ice sheet. 2 = predominant steppe and steppe-tundra. 3 = park-steppe and forest-steppe. 4 = taiga and forest. 5 = riverine and gallery forest. 6 = Black Sea and Caspian Sea

with *Artemisia.* The reported faunistic remains are *Rangifer* (2080), *Bos* (71), *Equus* (66), *Cervus* (19), *Capreolus* (1), and *Mammuthus* (1).

We can safely conclude that the ranges of *Mammuthus* and *Capreolus* overlapped in the Dordogne area between 33,000 and 23,000 years ago. Whether the impression is correct that both were rare is more difficult to decide. Because *Capreolus* was hunted in the same manner as other deer, its rarity can be assumed very reasonably. *Mammuthus,* however, may have avoided narrow valley areas such as that in which Abri Pataud is located because it had to compete there with a large range of species, including *Bos primigenius* and *Equus.* Mammoths may have kept to the open highland plains where hunting may have been more difficult. Still another factor may be reasonably considered: perhaps intense hunting by a well-fed and rather dense human population was responsible for the rarity of mammoth. One is tempted to conclude that competition with other species and intense human predation were the factors that placed the western limit of *Mammuthus primigenius* near Dordogne from 23,000 to 13,000 years ago.

Campbell's (1977) survey of the early part of the Upper Paleolithic of Britain—an interval that includes Aurignacian and Proto-Solutrean sites from about 33,000 to 20,000 years old—indicates that *Mammuthus* is present in 10 sites, *Equus przewalski* in 13, *Ursus arctos* and *Crocuta crocuta* in 12, *Coelodonta* in 11, *Rangifer* and *Bison priscus* in 10, *Megaloceros* (giant deer) in 9, *Cervus elaphus* in 7, and *Lepus timidus* (varying hare) in 5, but there is neither *Saiga* nor even *Bos,* evidently reflecting the influence of the cold sea ice to the west.

An interesting early Upper Paleolithic record in Central Europe is provided by the Pavlov and Dolni Vestonice sites in Czechoslovakia. Pavlov is about 23,000 years old, thus older than level 3 and more likely contemporaneous with level 4 of Kostenki I. At Pavlov we find a Gravettian variation defined as Pavlovian by Klima (1968). Among the hunted animals, *Mammuthus, Rangifer,* and *Equus przewalski* predominate; however, a complete description of this site is not yet available.

The stone tools at Pavlov include some few leaf-shaped bifacially retouched points, but similar points have not been found in the nearly contemporary site of Dolni Vestonice, a bit more to the west on the same broad slope of the Pavlov Mountains. Dolni Vestonice, however, yields a split-based bone point, a type common in earlier Aurignacian assemblages (Klima, 1963). At Dolni Vestonice, the main prey by far was *Mammuthus primigenius* (mostly young animals), while *Equus* and *Rangifer* are far less common. This very likely indicates a specialized mammoth-hunter's camp. My impression is that mammoth hunting at Dolni Vestonice took place mainly during the late autumn at the edge of the wide, swampy valley bottom of the river Thaya, where skillful ambush hunting may not have been too difficult. Meat was dried as a staple food on the windy and sunny slope, where large middens of elephant bones remain preserved in the loess in former ravines. Among the arboreal pollen, *Pinus* is once more dominant, but *Picea, Ulmus* (elm), and *Salix* are again present, providing another documentation of gallery forest in the sedge-grass steppe. At the end of a moist soil-forming interval, the loess sedimentation rate increased and buried the remains of the two large site areas.

The Pavlov and Dolni Vestonice sites probably were close to much-used mammoth trails. Some insights into the probable activities there are provided by excavations at the much younger North American site of Umingmak, a musk ox hunting site on Banks Island in the Canadian Arctic. At Umingmak the hunt was oriented, 3500 years ago, to the trails of *Ovibos moschatus.* Meat was obtained in the later summer as a staple provision for the still harsher winter in which man was able to survive as a consequence of 20,000 years or more of technical adaption (Müller-Beck, 1977, 1979). The hunted animals moved (and still move in an essentially unchanged landscape) on fixed and predictable trails through valley vegetation (Fig. 9), where they were easily intercepted by human hunters. A useful slope for dry meat processing is also near at hand.

Another important early Upper Paleolithic site is Sungir', east of Moscow and far north of the Kostenki sites. At Sungir', the blade cores were still made by an early technique, not too different from the Levallois principle and not quite as refined as typical Aurignacian cores, which are older in Western Europe. Bifacially retouched points and knives are present, as well as side scrapers and more conservative forms of end scrapers, but there are also very typical burins and backed blades

(Bahder, 1978). *Mammuthus* and *Equus* were hunted in a colder range of steppe environments, but *Bison, Canis lupus, Gulo, Alopex, Rangifer,* and *Lepus timidus* were most common (Ivanova, 1972). *Pinus, Alnus,* and *Betula* are represented by charcoal and also by pollen, along with *Artemisia,* sedges, and grasses. The vegetation evidently records a warmer interval and a steppe with at least some trees, in which mammoth but also reindeer were present at about 56°N and within a few hundred kilometers of the late Weichsel maximum position of the ice sheet—although the continental ice had not yet reached its late Pleistocene maximum at the time of the Sungir' occupation. Conditions were at least as severe as those at 65°N today, but there is no question that the region was

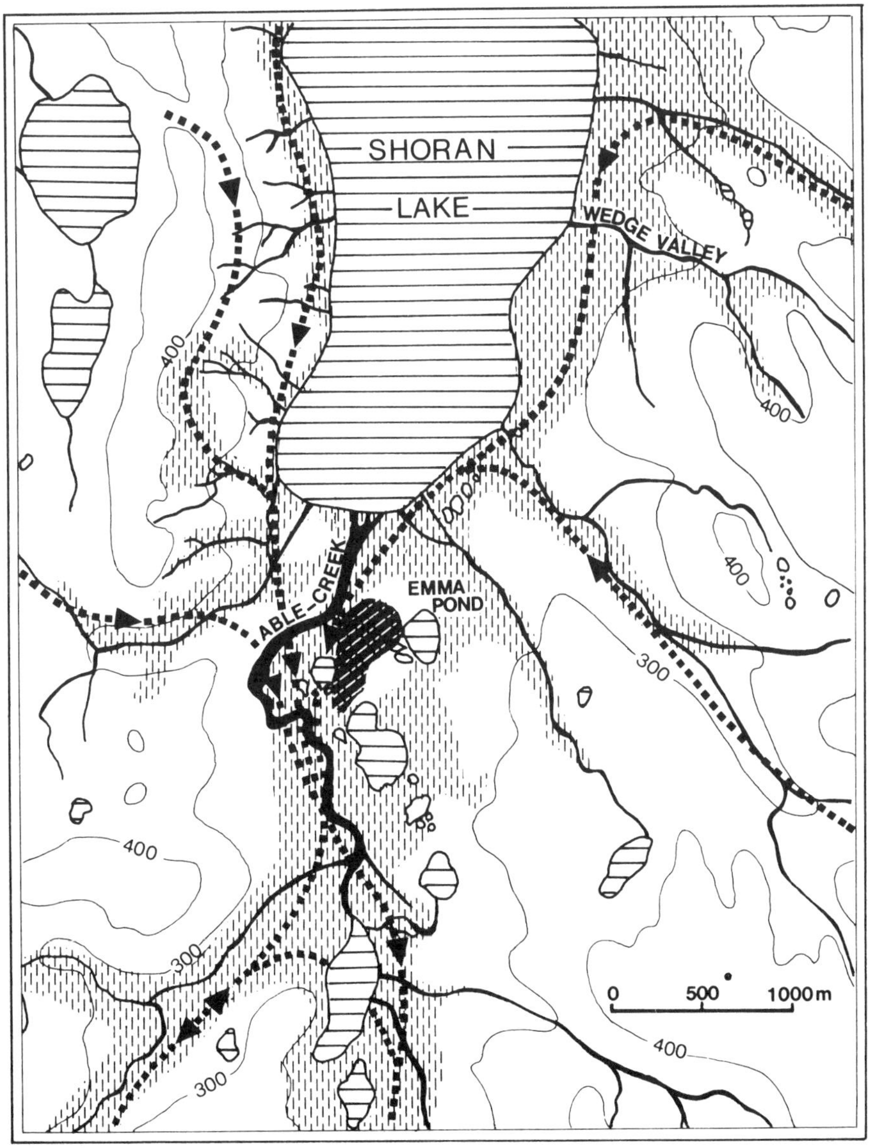

Figure 9. Musk-ox trails at Umingmak on Banks Island (Müller-Beck, 1979). Arrows = direction of trail use. Lightly shaded areas = sedge. Dark shaded area = Umingmak site (Pre-Dorset 3400 BP). Altitudes in feet

rich in game animals. It is even possible that Sungir' was a winter as well as a summer camp. Analysis of the reindeer remains will indicate whether this is true.

At Kostenki, the youngest levels (level 1 at sites I and VIII, both of which are in the uppermost levels of the loess and very likely less than 16,000 years old) again show a change in the faunal situation. *Mammuthus* is again more common among the food bones. *Rangifer* is still frequent and *Bison* and *Bos* are also present, but *Cervus elaphus* is absent. Apparently *Cervus* found it difficult to compete simultaneously with *Bison* and *Rangifer* until reforestation had progressed significantly. That reforestation had at least started is shown by the reappearance of *Bos.*

The mammoth-bone houses at Mezhirich in the Ukraine (Pidoplichko, 1976) also belong in this late time range, although precisely how old they may be is still open to question. They are younger than the last glacial maximum, but there is no reason to believe, as Pidoplichko asserts, that they are younger than 10,000 years and of Holocene age. The stone tools are rather late looking, and the bone needles as well as the high abstraction of such ornaments as female figurines indicate that the site was occupied after 15,000 years ago. The bone remains from the three house areas, including those from which the houses were constructed, comprise 1753 fragments representing at least 110 individuals of *Mammuthus primigenius.* Also present are *Rangifer tarandus* (38/10), *Bison priscus* (17/6), *Equus* sp. (2/2), *Canis lupus* (79/12), *Gulo gulo* (59/8), *Ursus arctos* (12/3), and *Lepus* sp. cf. *L. lepus* (2576/73). Mezhirich is evidently a very specialized mammoth-hunting camp in which the bones were also used for construction. Hare was very likely caught in snares but perhaps only during a part of the time that the camp was used. There seems to be no direct connection between the reported flora and the site level itself. It is not unlikely, however, that forest elements would have been prominent because the camp is near the floodplain of the Rosava and Ros' Rivers, where the presence of *Ursus arctos* and hare would document the presence of gallery forests. Away from the rivers we would still expect steppe vegetation, as the presence of *Equus* and also the far-ranging *Mammuthus* would suggest.

There is no doubt that a healthy mammoth population was still present in the Ukraine at the time of the Mezhirich occupation, but how large it was is not really clear. The bones used to build huts with areas of 23 m^2, 20 m^2, and 12 m^2 (Pidoplichko, 1976) could have come from animals hunted over a long span of time; some could even have been collected in the river sediments from a freshly eroded bone bed. Comparison with tent and hut sizes known in later contexts indicates that a shelter area of 12 m^2 is about normal for a nuclear family, so that the two larger Mezhirich houses could have been used by two families each, as were some early Paleo-Eskimo houses. Such heavy construction would make more sense for five families living together during winter time. We must also keep in mind that in nearly all arctic cultures, human activities are very much reduced during winter and that normal behavior involves the preparation and storage of meat late in fall. Snaring of hare would be typical for early spring, especially in years when winter snow lasted so long that the provisions prepared in autumn were used up before hunting on a larger scale could resume. If we assume that each Mezhirich family had one active hunter, then each would have had to kill about 20 elephants to build the huts found at the site. We might hypothesize that an average hunter was active for 20 years between the ages of 15 and 35. These assumptions would suggest that only one elephant was killed each year by each hunter. Dried or frozen, this would be a good supply of meat for the family dependent upon him.[8]

Interesting, also, are the data on the ages of the hunted elephants at Mezhirich: 3 were less than 2 years old, 34 were aged 3-9 years, 49 were in the 9-20 year class, 8 in the 2-35 year class, and 3 were older than 35 years. The age distribution of the mammoth remains at Mezhirich is strikingly different from the age distributions among *Mammuthus* and *Paleoloxodon* at Ehringsdorf (Table 1), which were much closer to a normal population distribution. It could very well have been that the younger elephants were overhunted at Mezhirich. This may have been a factor promoting the extinction of the species.

Turning for a moment to China, we note that the Upper Cave at Choukoutien contains *Paleoloxodon namadicus, Gazella, Equus hemionus,*

[8]For purposes of this discussion, I ignore the possibility of meat sharing among the families at Mezhirich.

Cervus, Capreolus, and *Sus* and that *Rangifer* is absent (Aigner, 1969). *P. namadicus* evidently adapted in China to relatively mild steppe conditions. The presence of *Sus* and *Capreolus* document the presence of some forest in the area. The traces of man in the Upper Cave include small bifacially retouched points, very likely used as projectile points. The southern limit of *Mammuthus primigenius* in the Far East is approximately defined by its presence in poorly dated but evidently younger sites in Manchuria in which blade tools were found together with pebble tools (Aigner, 1969). These sites also contain remains of *Alces* (moose), *Coelodonta,* and *Capreolus,* but *Rangifer* is still missing.

Very early occupation in Northeastern Siberia, presumably by mammoth hunters, is reported by Mochanov (1978). However, in the absence of more complete data, I am a little skeptical about the early dates attributed to some sites. A date as early as 35,000 years for the Dyukhtai Culture (Mochanov, 1978) seems inconsistent with the typology of the artifacts belonging to it. The small blades and wedge-shaped cores, especially, seem out of place. Although we cannot exclude the possibility, it will be very surprising if these stone tools should prove to be more than 10,000 years older than those of Europe. That mammoth was hunted in Siberia as early as the Sartan Glaciation is shown by Klein (1971, quoting Soviet authors), but even the earliest substage of this glaciation did not start earlier than 22,000 years ago. Sungir', Pavlov, and Abri Pataud are all older. In any case, keeping the Dyukhtai sequence in mind as well as the general situation to be expected in the more southern parts of Siberia, we may accept a date for the earliest possible hunting of mammoth there as about 35,000 years BP—at about the time of the transition from *Blattspitzen* complexes to the Aurignacian of Central Europe. Considering what we know so far, there is really no reason that man should not have been adapted to conditions on the Siberian steppes at this time.

Important too is the Mal'ta site, west of Irkutsk, somewhere between 25,000 and 13,000 years old (Gerasimov, 1935, 1941). Mal'ta yields a more typical Gravettian industry with some local peculiarities (Gravettian sites are 28,000 to 18,000 years old in Europe but some Eastern Gravettian sites are younger). Among the mammal remains, *Mammuthus, Coelodonta, Equus, Rangifer,* and *Alopex* are frequent, but *Bison* is rare. There is no question that Mal'ta lay within the reindeer-mammoth zone.

Better dates are available for the mammoth cemetery of Berelekh (Skarlato, 1977), which at least partly involves human hunting activities. Radiocarbon dates there range from 13,700 to 10,440 years; the older date is considered more reliable because it was run on mammoth flesh (comment by A.V. Lozhkin to D.M. Hopkins, August 1979). Excavations by Vereshchagin (1977) produced 8431 *Mammuthus* bones representing at least 140 individuals, *Rangifer* (31/4), *Gulo* (7/4, including one carcass), *Equus lenensis* (12/3), *Bison* (8/2), and *Coelodonta* (1/1). An occupation site a few hundred meters from the mammoth cemetery[9] contains faunal remains in quite different proportions: *Mammuthus* (79/2), *Lepus tanaiticus* (850/28), *Canis lupus* (18/4), *Rangifer tarandus* (2/1), and *Equus lenensis* (1/1).

It is clear that a healthy mammoth population was available in northeastern Siberia as late as 13,700 years ago.[10] Lozhkin (1977) shows that the Berelekh site, which now lies in larch forest tundra 150 or 200 km south of the coastal arctic tundra, was in a steppe-tundra environment when the mammoth cemetery accumulated. Pollen of herbaceous plants comprise up to 90% of the pollen spectra in the bone bed. The herbaceous component includes up to 60% cereals, up to 16% sedges, up to 21% Caryophyllaceae (pink family), and up to 20% *Artemisia.* Shrubs constitute less than 10% of the total pollen and consist mostly of willow (up to 73%). Altogether, Berelekh seems to have been a good place for elephants well adapted to colder conditions, and they were good game for men accustomed to such a climate, at least during summer.

At the time that Berelekh was occupied, mammoth was quickly disappearing in western Eurasia.

[9] It is quite possible that some archaeological remains have been overlooked at Berelekh and that at least some of the elephant bones are refuse from hunting in the valley, granting that they may have been rebedded by the stream.

[10] That mammoths also were abundant much earlier in Northeastern Siberia is indicated not only by the other, mostly very preliminarily published archaeological sites but also by dates for the mammoth carcasses found in the region, most of which are older than the Sartan Glaciation (Vereshchagin and Baryshnikov, this volume).

Table 2
Faunistic and Floristic Remains from Some Late Paleolithic Sites in Northern Eurasia

Site Name	Technocomplex	Date (BP)	Type of Site	Vegetation Pollen/Macrofossils	Type of Analysis of Foodbones
1. Salzgitter-Lebenstedt	Late Acheulean (Micoquian)	60,000 or older	Washed midden	*Pinus, Picea, Alnus, Betula Salix polaris, Salix herbacea*	Minimum-number count
2. Königsaue A	Micoquian	60,000	Lakeside camp	*Pinus, Picea, Alnus, Populus Betula, Salix, Tilia Artemisia, Sphagnum,* grasses, sedges	Minimum-number count
3. Königsaue B	Micoquian Flake-inventory	60,000	Lakeside camp	as Königsaue A	Minimum-number count
4. Volgograd	Late Eastern Micoquian	60,000 to 40,000	River area camp	*Pinus, Picea, Artemisia* chenopods, grasses, sedges	Minimum-number count
5. Mauern Weinberghöhlen	*Blattspitzen*	40,000	Cave —rare visits	*Pinus, Picea, Alnus Salix, Betula* grasses, sedges	Relative presence[a]
6. Urspring	*Blattspitzen*	35,000	Cave —rare visits	No records	Relative presence[a]
7. Vogelherd levels IV & V	Aurignacian	31,000	Cave —repeated visits	No records	Minimum-number count
8. Kostenki I,5	Eastern *Blattspitzen*	35,000? 30,000?	River area camp	"Loess-steppe"	Presence/absence
9. Kostenki I,3	Eastern Gravettian	Younger than 25,000?	River area camp	"Cold loess-steppe"	Minimum-number count
10. Abri Pataud level 11	Aurignacian	33,000	Cave —repeated visits	*Quercus, Alnus, Pinus, Fraxinus, Betula, Juniperus, Artemisia*	Number of bones
11. Abri Pataud level 7	Aurignacian	33,000	Cave —repeated visits	Similar to level 11, additional *Abies* without *Artemisia*	Number of bones
12. Abri Pataud level 5	Perigordian	28,000	Cave —repeated visits	Similar to level 11 including *Artemisia*	Number of bones
13. Abri Pataud level 3	Late Perigordian (Western Gravettian)	23,000	Cave —repeated visits	Similar to level 5	Number of bones
14. Early Upper Paleolithic sites in GB	Proto-Solutrean and Aurignacian	33,000 to 20,000	different	—	Number of sites with listed species
15. Sungir'	Sungir' *Blattspitzen*	28,000?	River area site	*Pinus, Alnus, Artemisia* sedges, grasses	Presence/absence
16. Kostenki I,1	Late Eastern Gravettian	After 16,000	River area site	"Loess-steppe"	Relative presence[a]
17. Mezhirich	Late Eastern Gravettian	After 16,000	River area site	No direct report	Number of bones/ Minimum-number count
18. Mal'ta	Siberian Gravettian	After 25,000	Open site	No records	Presence/absence
19. Berelekh bone bed	Unknown	14,000 to 10,500	River sediment?	*Salix* Caryophyllaceae grasses, sedges	Number of bones/ Minimum-number count
20. Berelekh site	Late Siberian Gravettian	Younger than 14,000	River area camp	As bone bed	Number of bones/ minimum number count

Mammuthus primigenius	*Rangifer tarandus*	*Cervus elaphus*	*Capreolus capreolus*	*Megaloceros*	*Coelodonta antiquitatis*	*Sus scrofa*	*Bison priscus*	*Bos primigenius*	*Saiga tartarica*	*Equus caballus*	*Equus przwalski*	*Equus* sp.	*Asinus hydruntinus*	*Lepus timidus*	*Lepus* sp.
16	80	–	–	–	2	–	6/7	–	–	4/6	–	–	–	–	–
4	5	1	–	–	1	–	3	–	–	4	–	–	1	–	–
1	2	1	–	–	1	–	2	–	–	3	–	–	1	–	–
51	1	15	–	–	–	–	366	–	36	42	–	–	–	–	–
X	X	X	–	–	X	–	?	?	–	X	–	–	–	–	–
–	X	X	–	–	X	–	–	–	–	–	X	–	–	–	–
24	13	2	–	–	8	1	–	–	–	–	22	–	–	–	–
X	–	–	–	–	–	–	X	X	X	X	–	–	–	–	–
–	205	2	–	–	–	–	–	–	–	–	–	–	–	–	–
1	599	2	–	–	–	1	–	4	–	359	–	–	–	–	–
2	395	80	7	–	–	3	–	–	–	60	–	–	–	–	–
5	21,119	172	10	–	–	–	–	48	–	44	–	–	–	–	–
1	2,080	19	1	–	–	–	–	71	–	66	–	–	–	–	–
10	10	7	–	9	11	–	10	–	–	–	13	–	–	5	–
X	XX	–	–	–	–	–	XX	–	–	–	X	–	–	XX	–
XX	XXX	–	–	–	–	–	X	X	–	–	–	–	–	–	–
1,753/110	38/10	–	–	–	–	–	17/6	–	–	–	2/2	–	–	–	2,576/73
X	X	–	–	–	X	–	R	–	–	–	–	X	–	–	–
8,431/140	31/4	–	–	–	1/1	–	8/2	–	–	–	–	12/3	–	–	–
79/2	2/1	–	–	–	–	–	–	–	–	–	–	1/1	–	–	850/28

[a] R = rare X = present XX = common XXX = abundant

At the well-documented Gönnersdorf site (Magdalenian and about 13,000 years old) (Bosinki and Fischer, 1974; Poplin, 1976), *Rangifer, Cervus,* and *Equus* as well as *Alopex* were regularly hunted, and *Coelodonta* was still fairly common, but mammoth remains are nearly absent. Mammoth was now as rare in Central Europe as it had been 20,000 years earlier in southwestern France, 1000 km to the west.

To summarize: after 40,000 years BP, we find man—now *Homo sapiens sapiens*—hunting in a wide range of steppe, forest-steppe, and tundra-steppe environments. Man was hunting over the full range of *Mammuthus* by 14,000 years ago, and he very likely traveled over most of it as early as 25,000 years ago. Within the mammoth-steppe, after 40,000 years BP, we can recognize the following provinces: (1) An oceanic subarctic type with *Rangifer, Bison, Coelodonta,* and *Equus caballus* or, locally, *E. przewalski.* More to the west, *Capreolus* was also present. Sedges, grasses, and *Artemisia* comprise up to 10% of the pollen spectra, but some trees were present in sheltered areas. (2) A continental subarctic type with *Saiga, Bison,* and *Equus.* The vegetation is poorly known but includes chenopods and *Artemisia.* (3) A cold subarctic type with *Rangifer* and *Alopex* and vegetation rich in *Artemisia* and sedges. (4) A cool temperate forest-steppe of more continental character with *Capreolus, Coelodonta,* and *Alces;* some denser forest areas must have been present.

The limited data available at this moment (and archaeologists as well as paleontologists must always stress this point) allow us to conclude that man adapted very well to the different provinces within the mammoth steppe more than 60,000 years ago. He may have started to enter the steppe in a small area in Northwestern and Central Europe as early as 150,000 years ago, well before the Eemian Interglacial, and there is no doubt that he hunted in subarctic steppe environments during summer, at least, earlier than 40,000 years ago. His weapons were improved stone projectile points and a full range of bone projectile points (Albrecht, et al., 1973), probably used with the *atlatl.* For winter housing, huts had to be constructed; elephant bones were used as building material. Even the religious world must have been more complex: the figurines found in late Paleolithic sites (Fig. 1) imply some forerunning development.

It is hard to believe that man did not conquer nearly all of the remaining mammoth-steppe territory soon after 40,000 years ago, with the exception, perhaps, of the most severe north during the coldest stages of the Würm/Weichsel/Sartan Glaciation between 22,000 and 16,000 years ago. Man may have reached a level of adaption that placed adventurous hunters and their brave wives in a position to move into Beringia as early as 40,000 or 30,000 years ago. There is no reason to doubt that they might have reached the eastern edge of the steppes in Beringia itself just at that time, if a dry crossing of the land bridge was possible. But if men really did penetrate Beringia at this early date, as the remains from Old Crow may indicate (Morlan and Cinq-Mars, this volume), it is difficult to understand why those hunters did not move beyond, toward a warmer south, even before the glaciers briefly closed the western Canadian ice-free corridor.

21

ANCIENT BERINGIANS: HUMAN OCCUPATION IN THE LATE PLEISTOCENE OF ALASKA AND THE YUKON TERRITORY

Richard E. Morlan and Jacques Cinq-Mars

Evidence for human occupation in eastern Beringia now extends back in time to the early Wisconsin (about 60,000 years ago), but for the earliest periods this record must be interpreted on the basis of altered bone, antler, and ivory specimens that have been recovered from redeposited contexts. Such interpretations depend on a broad analytical approach in which taphonomic concepts are employed to identify a more representative suite of bone-altering agencies. Pending the discovery and excavation of undisturbed archaeological sites of such ancient provenance, we have declined to define a technological complex on the basis of these materials. Their existence nonetheless raises a number of interesting questions concerning the time and means of human adaptation to northern latitudes. Not the least of these is the evolutionary status of the first Beringians whose identity may be shrouded in the little-known record of human evolution in East Asia.

A recently excavated deposit at the Bluefish Caves affords a new late Wisconsin record of human occupation during the most severe period of glaciation in North America. Although the artifact inventory from the Bluefish Caves is presently too meager to permit the cultural classification of the occupants, the assemblage, as well as other indicators of human activity, is adequate to show the direct association of man with a diverse ungulate fauna during full-glacial times. In fact the association of the ungulates in a primary burial assemblage is an important contribution to ongoing controversy concerning the nature of the vegetation and landscape during the full-glacial interval.

In view of the numerous recently published summaries of later Beringian archaeological manifestations, we provide only a brief discussion of final Wisconsin and early Holocene sites and complexes. There are still too few well-stratified, well-dated, and well-published sites in eastern Beringia to provide a sound basis for reconstructing the role and response of human societies to the rapid, perhaps even chaotic, changes that took place in the final millennia of the Wisconsin. Dry Creek component I and the Chindadn complex at the Village site may contain part of the early history of the fluted point complexes that became widespread in North America during those millennia, but additional recent finds of fluted points themselves have not appreciably aided the resolution of the many questions that pertain to their origins.

Finally, we comment on a number of themes concerning the peopling of the New World including the concepts of migration and colonization, the importance of the Bering land bridge and the ice-free corridor, and the likely role of mankind in late Wisconsin extinctions.

Key words: Eastern Beringia, Wisconsin, mammoth fauna, taphonomy, Old Crow, Bluefish, Dry Creek, Chindadn, fluted points, microblades

ISBN 0-12-355860-3

INTRODUCTION

In 1967, when *The Bering Land Bridge* was published, the region known as Beringia had produced only one site (Trail Creek, Larsen, 1968) possibly old enough to be relevant to the colonization of the New World. Nonetheless, every author touching on this question has adopted the quite reasonable working assumption that people entered the Western Hemisphere by traveling, probably on foot, through northeast Asia to Beringia and thence east and south to midcontinental North America. However, a number of more recent discoveries in Beringia now enable us to discuss a regional Pleistocene archaeological record of considerable antiquity. It is not yet clear just how these earliest finds can be synthesized into a view of cultural history or how Asian precursors and American descendants can be linked in an historically meaningful way, but it is encouraging to report that the Beringian archaeological record no longer begins in the Holocene.

Several themes in the literature on the peopling of the Americas need critical discussion but must be summarily treated here, because space does not permit a detailed critique. Many pertain to prevalent and sometimes contradictory views of Beringia espoused in reviews and summary statements of "New World origins." Included, for example, is the idea that Beringia was inhospitable during glacial periods, that the earliest evidence for human occupation in the New World will in some sense consist of crude tools, that Beringia can be construed as a pathway to North America without considering the requirements of human life in the region, and that the pathway was used primarily or only when dry land existed at Bering Strait and when the major North American ice sheets were sufficiently withdrawn to permit passage through Canada into the midcontinental area.

Many of these themes have persisted in the literature despite careful arguments by archaeologists who have worked in the north and have presented alternate views. Giddings (1960, p. 121), for example, opened a review of the archaeology of Bering Strait with the observation that Beringia has long been "a center, rather than a way-station, of circumpolar ideas." Elsewhere he championed the theory "that a very sparse Arctic population slowly spread over the belt of northern climate, neither pursuing nor evading pursuit, but simply existing and adjusting at random to the environment, the sons sometimes hunting beyond the range of their fathers but never really leaving home" (Giddings, 1954, p. 88). Campbell (1963, pp. 41ff) presented a lengthy discussion of the environmental characteristics that make northern latitudes quite easy to traverse and occupy for human societies adapted to conditions of winter cold and darkness, deep snow, and cold water bodies. Other papers in this volume show that some of the more difficult features of Holocene northern environments—boggy terrain in summer, for example—may not have been so widely developed during the Pleistocene and that Beringia may have then offered better conditions for northern-adapted hunting societies.

Nonetheless, a certain threshold in human technological development had to be crossed. Adequate clothing, shelter, means of movement over snow and cold water, and methods of food procurement and storage during the "winter bottleneck" were required before human groups could live in northern latitudes during all seasons. This threshold, albeit poorly defined, certainly did not permit continued reliance on crude tools or "flimsy water craft" (Haberland, 1969, p. 352). Yet MacNeish (1978, p. 491) refers to the earliest New World inhabitants as "unsophisticated technologically . . . perhaps unskilled as hunters," and Bryan (1978, p. 311) expects us "to find somewhere in Beringia evidence of a much more basic (i.e., technologically less specialized) stone and bone technology than is represented at Old Crow or in the Diuktai culture."

Perhaps these peculiar and misleading themes are partially responsible for the fact that four recent reviews of early man in the Americas (Bryan, 1978; Griffin, 1979; MacNeish, 1978; Rouse, 1976) bear little resemblance to one another although each purports to be an adequate summary of the existing evidence. We seem to be commited to a marathon debate in which an "archaeological oral tradition" takes the place of published reports as a source of "evidence" concerning the past, and each debater is avowedly prejudiced toward one or another view of paleo-American origins including a particular time for the arrival of people in the Western Hemisphere. Some writers (e.g., Jennings, 1974, p. 76; Griffin, 1979, p. 44) have espoused arbitrary standards for the guidance of judgment and the statement of a verdict, while others (e.g., Lorenzo 1978, p. 2; Alexander, 1978) have descried the "legalistic" approach to the defense of such views.

Much of the new evidence from eastern Beringia fails to meet the arbitrary standards that have been thrust upon us in recent years, but we cannot simply ignore the new finds even though they may be more difficult to interpret and even though their interpretation may leave much of the prehistoric picture incomplete. Some of the new evidence is of surprising antiquity; it involves bone alterations rather than lithic craftmanship; and it has been recovered from redeposited contexts rather than from "archaeological sites." Only two fundamental assumptions are required to interpret this material: that human behavior can be understood by studying the results of that behavior; and that at least some of the results of human behavior can be separated from the results of noncultural processes. An undisturbed archaeological site would make an invaluable contribution to our understanding of the earliest Beringians, but we must not forget that in the final analysis all of our evidence of past human behavior is derived from the study of its material manifestations and that the record is and always will be incomplete.

There are two aspects to the problem of separating natural and artificial alterations of objects that might be interpreted as the remains of human activity. The first, the misattribution of artificial causes to naturally modified pieces, has received the most attention and has created a world-wide problem that recurs with some regularity in the archaeological literature and the "oral tradition." Compare, for example, contrasting views on the authenticity of "artifacts" from the Chuqui complex of Chile (Lanning, 1970, pp. 91-94; Lynch, 1974, p. 360) or the Calico site in California (Simpson, 1978; Duvall and Venner, 1979). In Beringia, the best example of this sort of problem was published in a refereed journal and had such excellent illustrations that the natural origins of the stream-rolled gravels from Sedna Creek, in northern Alaska, is clearly apparent (Schlesier, 1967). Those who have worked in the area or seen the collections have unfortunately remained generally silent. Others (e.g., Davis and Panlaqui, 1978, p. 73) have cited Sedna Creek as a component of the poorly defined British Mountain complex and have thereby helped to reify this collection of natural gravel as an archaeological entity.

The second aspect of the attribution problem is the danger of incorrectly attributing to the human action all objects and alterations which occur in "archaeological sites." Clark (1972, p. 5) has rightly cautioned that the "rough nature" of the collection from Batza Tena is probably partly due to damage caused by geomorphic processes and trampling by man and other animals. Many other examples could be cited, and this latter aspect of the problem is probably more insidious than the first, since it is more difficult to detect without detailed field observations.

The separation of artificial and natural alterations on bone is a problem even more complex than that arising from stone. It appears that we cannot make a confident interpretation of many individual specimens or even of some kinds of assemblages, and the effort is made even more difficult by the unfortunate history of bone interpretations in the literature of paleoanthropology (Morlan, 1980a, pp. 3ff). As Frison (1974, pp. 51-52) has noted, some rather simplistic studies have seriously reduced the credibility of later researchers who attempted to interpret bone tools, and a recent review by Binford (1981) shows that such credibility has not been generally restored.

Many of the interpretive errors that continue to enter the literature on early man result from either too much reliance on common-sense attributions of prehistoric human behavior that are limited by the experience of the individual analyst, or from facile treatments of ethnographic observations that often fail to provide "*specific knowledge* of the particular effects that might be expected to result" from human behavior (Binford, 1978, p. 10, italics his). An ethnographic document that does not include an adequate description of the patterned material results of a human practice can be used only in a general way to account for apparently similar patterning in the archaeological record. We need not only more but better "baseline data" on which to base our inferences (Stanford, 1979a), and these data must be collected through both field observations and experimentation.

There are so many possible alterations and altering agencies that an integrative framework is needed to give structure and order to our knowledge. Such a framework has been devised by paleontologists on the basis of the concept of taphonomy, which was originally defined by Efremov (1940) as the "science of the laws of embedding." Taphonomy has been utilized more recently as a comprehensive approach to all factors affecting individual animal remains and assemblages from the time of death until the moment of discovery and collection (see Clark and Kietzke, 1967; Voorhies, 1969; Behrensmeyer, 1975; Behrensmeyer and Hill, 1980). Many aspects of this approach have been in frequent, if implicit, use in archaeology for many years, but the more systematic application of taphonomic analysis to archaeological problems has begun only recently. The concept of taphonomy encourages the analyst to consider a much broader range of factors pertaining to the specific condition of individual specimens and the composition of assemblages, and it has been used most effectively in the analysis of vertebrate fossils.

ORGANIZATION OF THIS REPORT

In this report we attempt an overview of archaeological materials dating from late Pleistocene and earliest Holocene time in eastern Beringia. We limit ourselves mostly to materials believed to be more than 10,000 years old, because the later remains have been summarized recently by several authors (Ackerman, et al., 1979; Anderson, 1978, 1980; Clark, in press; Dumond, 1977, 1980a, 1980b; Morlan and Workman, 1980; Powers, 1978; Workman, 1978, 1980a, 1980b).

The earliest materials, consisting primarily of bone and antler artifacts from the Old Crow region (Fig. 1), are not precisely dated but seem to range in age from early through final Wisconsin time (Fig. 2). None of the evidence for this earliest period has been obtained from undisturbed archaeological sites, and our presentation centers on the reasons for inferring human activity on the basis of alterations visible on redeposited, fossilized bone, antler, and tusk specimens. Several of these were found in situations that suggest close proximity to former archaeological sites.

The late Pleistocene and early Holocene collections consist primarily of stone artifacts with variable quantities and qualities of preserved bone. Only a few of these collections have been adequately published, and our discussion must therefore remain rather superficial. Enough is known, however, to arrange them in a tentative sequence and in groupings that can be discussed in terms of a larger frame of reference encompassing portions of western Beringia and midcontinental North America.

If the earliest evidence consists mostly of bones and the later material mostly of stones, the Bluefish Caves of the northern Yukon are a useful transition in which a few lithic artifacts were found in secure association with bones of Pleistocene animals. We have arranged the discussion accordingly, but this arrangement should not be construed as having historical significance.

The discussion closes with a thematic review of the likely role of the Bering land bridge and the so-called ice-free corridor in the peopling of the Americas. We also venture into the chaotic subject of late Pleistocene extinctions, but we would not pretend to have provided a solution to any of these complex problems.

THE EARLIEST EVIDENCE: MOSTLY BONES EX SITU

Old Crow Localities

Extensive geological, paleontological, and archaeological reconnaissance in the northern Yukon Territory during the past decade and a half has shown that three interior basins—the Bell, the Bluefish, and the Old Crow—have tremendous significance for late Pleistocene paleoecology and archaeology (Fig. 3). Although located in the unglaciated refugium of eastern Beringia, these basins were profoundly affected by advances of the Laurentide glacier system, which dammed and diverted the Peel and Porcupine drainages to create enormous lakes fed by glacial meltwater as well as by normal regional runoff. The sediments of at least two successive lakes have been recognized in the Old Crow basin where they are thought to record the Illinoian and the classical Wisconsin glacial advances, respectively (Hughes, 1969, 1972; Hughes, et al., 1981).

Very few vertebrate fossils have been found in the Bell basin, and only a modest number are known from the Bluefish basin. The Old Crow

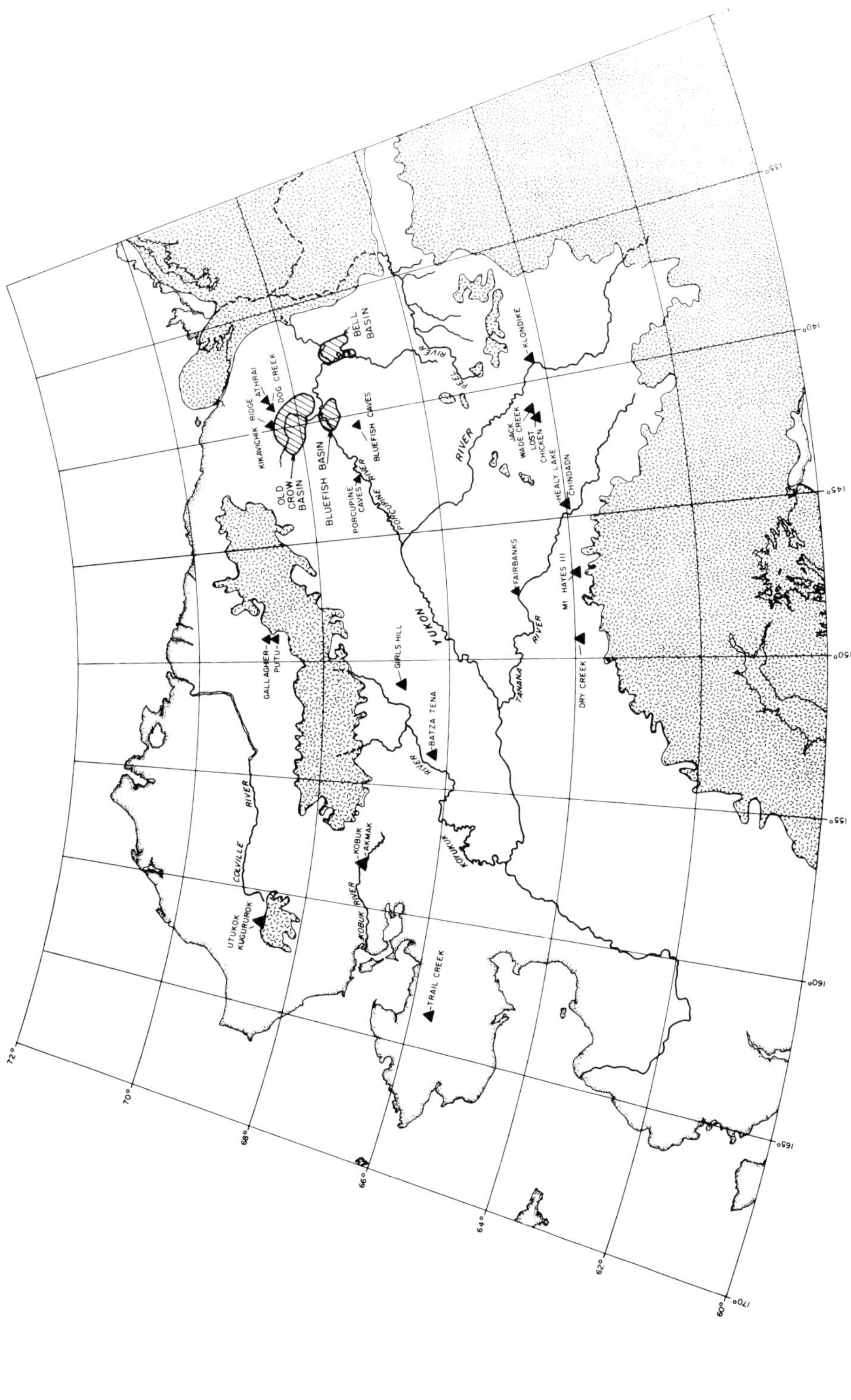

Figure 1. Eastern Beringia showing locations of selected Alaskan and Yukon sites discussed in this report. Stippled areas are areas of Wisconsin glaciation as mapped by Prest (1969) and Hughes (1972)

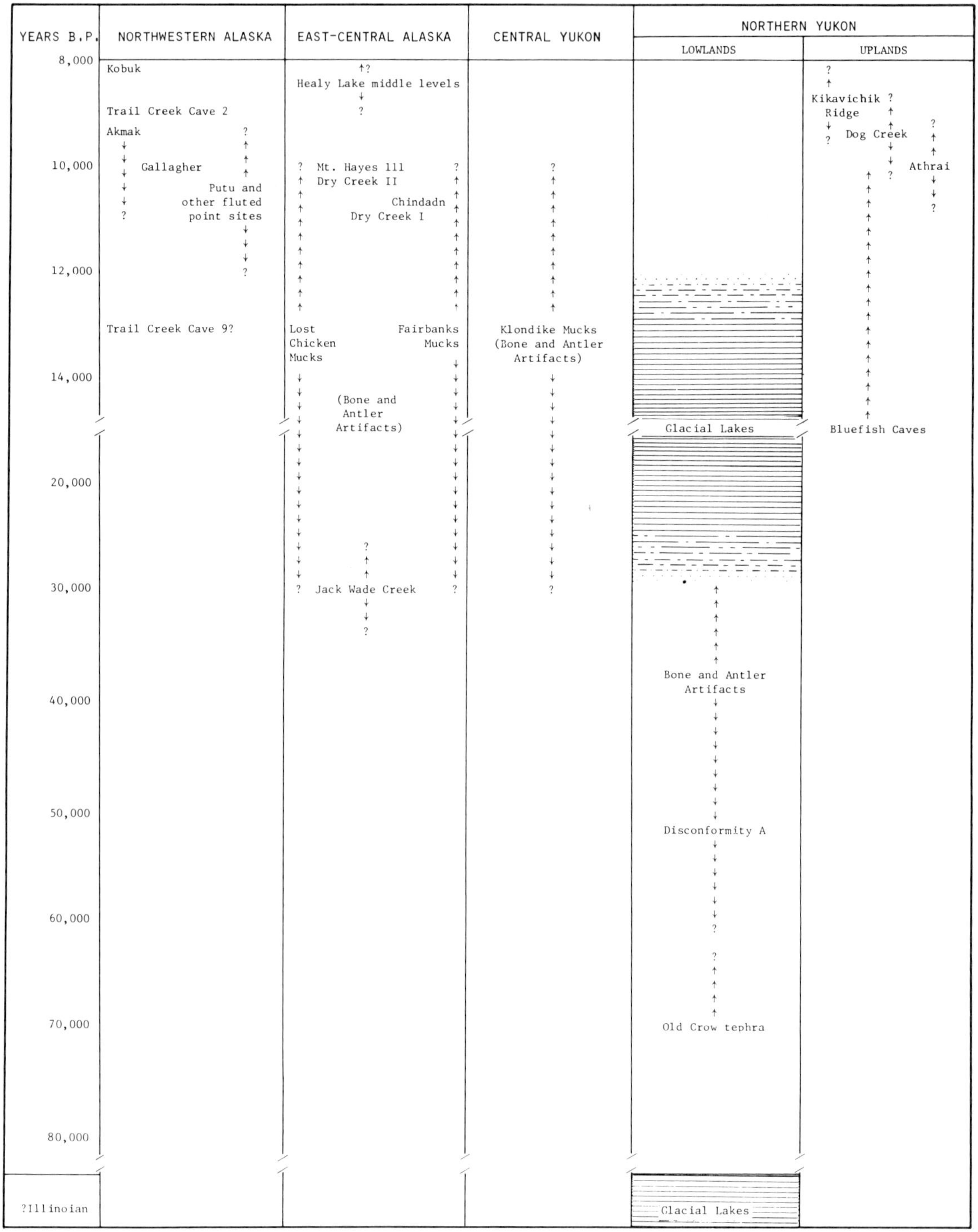

Figure 2. Chronological chart showing assumed chronological position of selected eastern Beringian archaeological manifestations

Figure 3. Gross topography of the Old Crow region of northern Yukon Territory is shown in the upper half of this illustration. The Old Crow Range appears in the foreground, the Porcupine River valley dissected into the Bluefish Basin is in the middle distance, and the Keele Range, a northern outlier of the Ogilvie Mountains forms the horizon.
The lower picture is a view of a major Pleistocene exposure on the Porcupine River (HH62-228) in which the dark colored sediments at the top consist of clays laid down in a glacial lake of classical Wisconsin age. The exposure is 55 m high

basin, however, has proven one of the world's richest repositories of late Pleistocene vertebrate fossils. Traveling along the Old Crow River, one cannot help but notice the thousands of fossilized bones, teeth, tusks, and antlers that literally pave many of the point bars. Tens of thousands of specimens have been collected since C.R. Harington (1977, 1978) began a long-term program of paleontological research in 1966. In that first year of field work, Harington found a permineralized caribou tibia that had been fashioned into a fleshing tool equipped with delicate denticulations on the working end. Such workmanship could only have been accomplished while the bone was fresh, and a new chapter in early-man research was opened when a radiocarbon date of $27{,}000\ ^{+3000}_{-2000}$ (GX-1640) was obtained on the apatite fraction of a large portion of the specimen (Irving and Harington, 1973). Two mammoth long-bone fragments that had been broken and flaked while fresh gave similar apatite dates.

These finds launched a long series of archaeological studies that continues to the present time (Irving, 1971, 1978; Bonnichsen, 1978, 1979; Morlan, 1978a, 1980a). New collections and reexaminations of older ones continue to enlarge the archaeological inventory from the Old Crow Pleistocene. These studies have also revealed the importance of establishing carefully documented analogs to guide the interpretation of alterations observed on bones and similar organic materials. The needed analogs have been developed thus far both through field observations and through a wide variety of experiments (Bonnichsen, 1973, 1978, 1979;

Harington, et al., 1975; G.A. Haynes, 1978, 1980; Morlan, 1980a; Stanford, et al., 1981). An important result is the conclusion that mammoth or mastodon bones can be fractured and flaked to make useful tools by techniques quite similar to those used in the flaking of stone. No other agency is known that could be responsible for the regular and patterned production of such specimens.

The majority of the artifacts from the Old Crow Pleistocene were made when the bones were fresh (Figs. 4 and 5). Collections now on hand in the National Museum of Man[1] include 116 mammoth or mastodon long-bone fragments that were artificially fractured while fresh, 54 bone cores that were flaked as well as fractured, and 27 bone flakes struck from such cores (Morlan, 1980a, Table 9.1). There are also seven specimens of mammoth or mastodon tusk ivory that were fractured and flaked by man while the ivory was fresh, and one ivory segment that was intensively polished on the working end of a possible fleshing tool (Fig. 6).

The collections also include many green-fractured bones representing mammals smaller than mammoths. Interpretation of these pieces is less clear because of the possibility that carnivores and other agencies may have been responsible for their fractures. Some of the smaller bones are also flaked in patterned ways that probably indicate human activity, but these markings are difficult to separate from the results of carnivore gnawing, which can chip back bone edges and produce effects very similar to core production.

A wide variety of scratches, scrapes, and ground and polished facets are also visible on some of the Old Crow fossils. In many cases it seems impossible to determine the manner in which these alterations were made. Some, however, are very complex and can be interpreted in terms of artificial use, but more experimental work is needed to improve our interpretive analogs.

A few of the Old Crow fossils exhibit cuts (Fig. 7) that precisely match those made experimentally on fresh bone and antler by stone tools. Most of these cuts are readily interpretable in terms of a butchering task or a procedure in artifact preparation, and they represent very firm evidence of human activity for which interpretive analogs are already well developed.

The Old Crow bone, antler, and ivory artifacts have been described in great detail in two recent publications (Bonnichsen, 1979; Morlan, 1980a). In the spring of 1980 a total of 251 specimens were in the National Museum of Man collections. An additional 18 specimens were probably modified artificially, and there are more than 200 pieces that might be interpreted as results of either natural or artificial alteration (Morlan, 1980a, Table 9.1). The alterations that permit the recognition of an archaeological record among these fossils—certain kinds of fracturing, flaking, grinding, polishing, scratching, scraping, and cutting—have been examined in the context of a holistic taphonomic analysis which has also included the recognition and recording of many naturally induced alterations, including exfoliation, split line formation, pitting, etching, scoring, furrowing, secondary mineral formation, permineralization, rounding, and certain other kinds of fracturing. Hence the archaeologically significant attributes can be viewed as indicators of a special class of "perthotaxic" factors (Clark and Kietzke, 1967, p. 117) that play a role in the history of the fossils following or accompanying death and prior to burial or destruction of the remains (Morlan, 1980a, Fig. 3.1; 1980b). Taphonomic analysis has proven a useful approach because of its potential for recognizing a sequence of both natural and artificial events in the "life history" of each specimen—a recognition especially important in the analysis of redeposited specimens.

Until 1977, we had very little knowledge of the probable stratigraphic levels from which the Old Crow vertebrate fossils and artifacts had been eroded for redeposition on or near the valley bottom. In 1977 and 1978, members of the Yukon Refugium Project (Morlan, 1977a) recognized and traced an erosional contact, provisionally called Disconformity A, which is now interpreted as the floodplain of a sizable stream that drained the Old Crow basin, incising a valley at least 12 m deep. Our earliest dated archaeological

[1] The inventory of artifacts cited here includes only those in the National Museum of Man collections in Ottawa. A much larger collection of comparable specimens is housed at the University of Toronto, where Drs. W. N. Irving and B.F. Beebe are responsible for their study and reporting. It is believed that the number of artificially modified specimens in that collection runs into the hundreds and perhaps into the thousands.

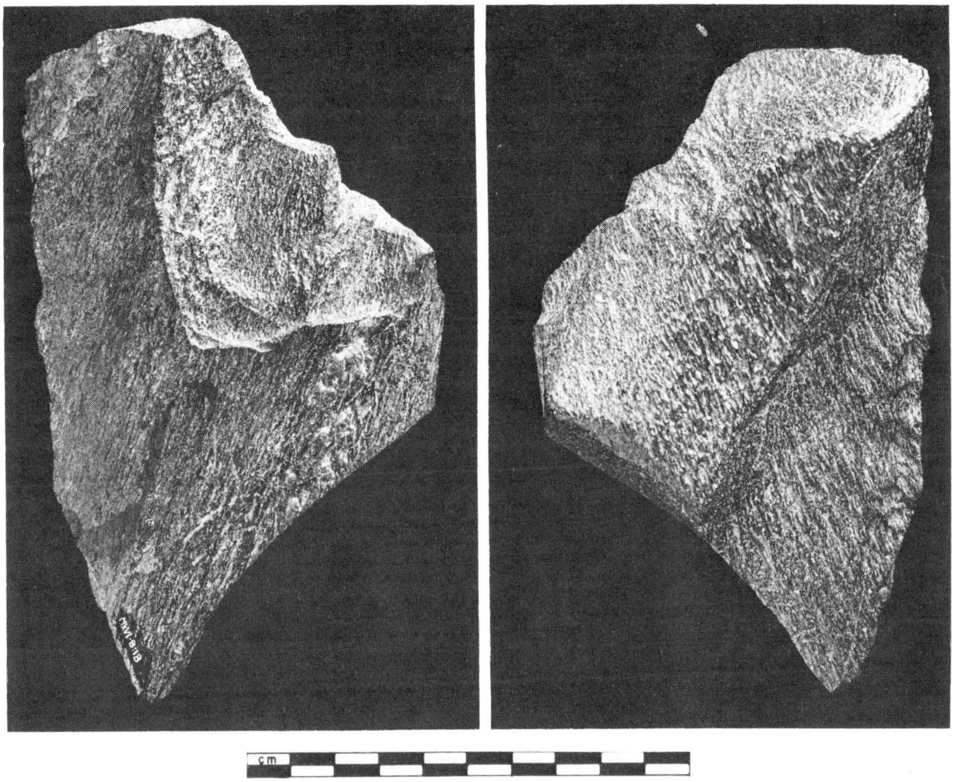

Figure 4. A bone core made on a mammoth (cf. *Mammuthus* sp.) limb-bone fragment and collected in reworked deposits on the floor of the Old Crow River valley. The outer face of the core, seen on the left, exhibits several large flake scars, and the inner face, seen on the right, reveals details of the core platform (at the top) and the fracture surfaces created when the bone was in a fresh condition

evidence comes from this floodplain and from the foreset beds of point bars that probably formed at the same time. Scattered probable artifacts have been found in several stratigraphic positions intermediate between Disconformity A and the overlying glacial-lake clays of classical Wisconsin age. It now appears likely that most artifacts found near modern river level were eroded from these intermediate strata during the extensive downcutting that took place when the glacial lake was drained near the end of Wisconsin time.

Although the Pleistocene archaeological record in the Old Crow basin now appears to be bracketed between Disconformity A and the classical Wisconsin glacial lake (Hughes, et al., 1981), the age of the disconformity has not been precisely determined. Limiting dates are provided by a radiocarbon assay on peat found in growth position above this ancient floodplain (>51,000 years BP) and both radiocarbon (>56,000) and fission track (<100,000) dates for the Old Crow tephra which occurs just below the disconformity in several sections (Morlan and Matthews, 1978; Morlan, 1979; Westgate, et al., 1981). These age estimates indicate that man entered the Old Crow basin late in early Wisconsin time, and we have every reason to believe that human occupation continued uninterrupted in the northern Yukon basins until they were flooded with glacial meltwater during the classical Wisconsin advance. While it is possible that human occupation began even earlier, the vertebrate fossils recovered from older deposits believed to represent the Sangamon Interglaciation are devoid of the kinds of alterations that seem attributable to human agency.

None of the artifacts attributable to the Old

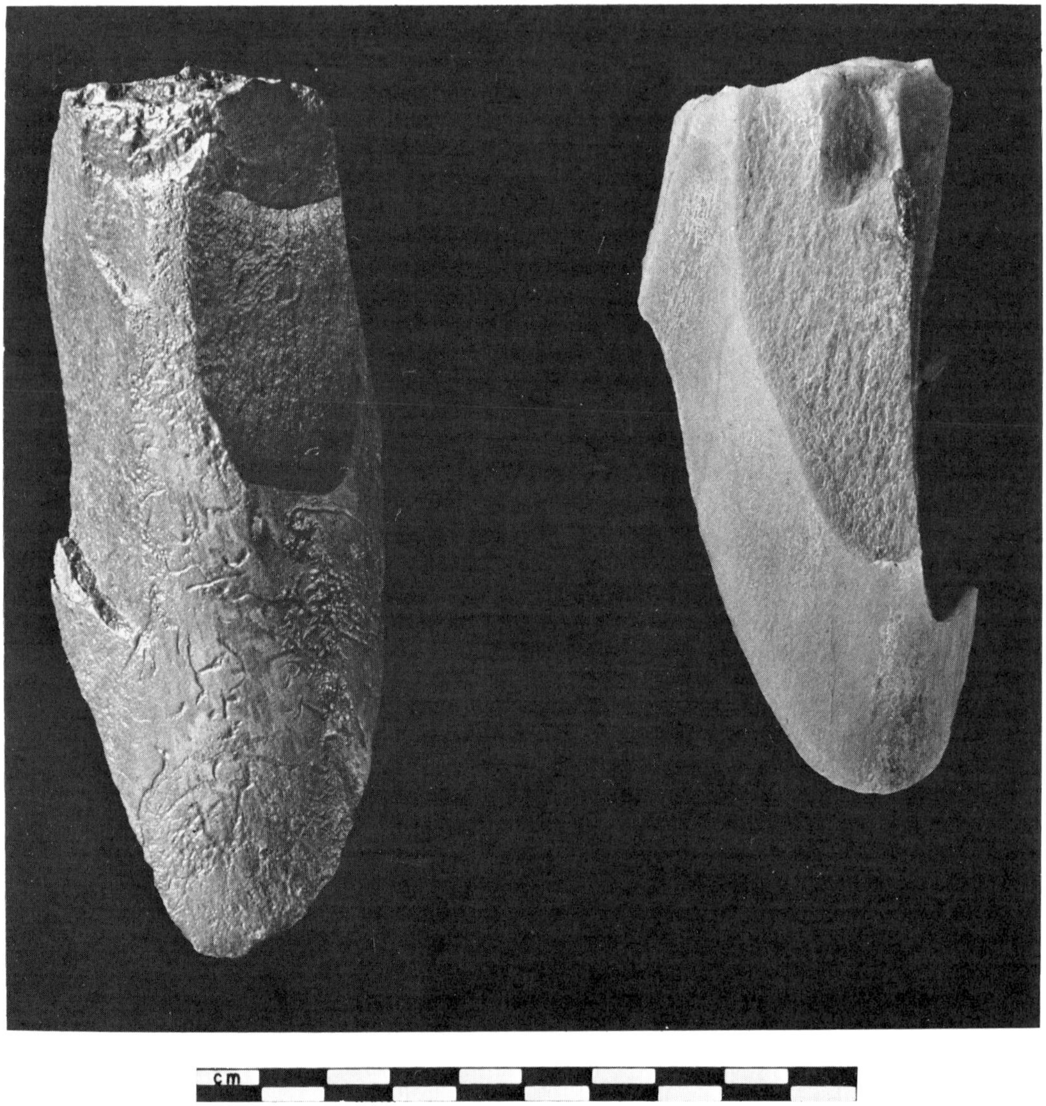

Figure 5. Bone flakes. Left: a flake made from a mammoth (cf. *Mammuthus* sp.) limb bone core, recovered from reworked deposits on the floor of the Old Crow River valley. Right: an experimental replica made from an African elephant (*Loxodonta africana*) limb bone core during the Ginsberg elephant butchering experiment (Stanford, et al., 1980)

Crow Pleistocene has been recovered from an undisturbed archaeological site, but consideration of the hydrodynamic characteristics of some of the fossils suggests that their recovery sites are not far removed from former primary archaeological deposits. The discovery, for example, of a nearly intact mammoth femur near several large fragments of a mammoth ulna and a huge tusk segment may indicate that the mammoth carcass was differentially sorted by fluvial processes and that these recovered elements represent part of a lag deposit. Three artifacts, including the ivory specimen illustrated here (Fig. 6), were found in association with these bones and could be part of the lag deposit, although they also could have been transported into a secondary association with it (MlVl-2, midsection, Morlan, 1980a, chapters 6, 9). At another locality (MlVl-13, Morlan, 1980a, chapters 7, 9), a tentatively identified soil profile that is equivalent stratigraphically to Disconformity A has yielded six bison bones of which two were still articulated and two exhibited clearly defined butchering marks made with stone tools (Fig. 7). These specimens were probably not moved by fluvial processes, but could have been disturbed by colluvial activity, assuming that local relief was greater than the remarkably flat erosional contact that elsewhere comprises the disconformity. Finally, a large portion of a bison radius, fractured when green, probably by man, occurred in fine organic silts which, judging from invertebrate and plant fossils, represent the bottom of a thaw lake. This bison bone must have been eroded from archaeological deposits on the shore of the lake, and a radiocarbon date on the enclosing matrix is >37,000 years BP (GSC-2792; see MlVl-3, Morlan, 1979; 1980a, chapter 7).

We believe that the search for primary archaeological sites should focus on a 3-5 meters-thick zone of early and mid-Wisconsin sediment just below the upper glacial lake of classical Wisconsin time. There are 60 major exposures along the Old Crow River valley which can be examined in a systematic survey for undisturbed human habitation and kill sites. Although we may not find such sites presently exposed in the valley walls, our search criteria—evidence of subaerial weathering beneath surfaces that once were stable and habitable, and presence of large pieces of bone and ivory that do not match the hydrodynamic potential of the enclosing sediments—should enhance our chances of discovering undisturbed sites if they are to be found. The samples now on hand, however, provide an insufficient basis for an attempt to define a technology that might represent a particular historical human group, because every known specimen could have come from a different archaeological site, while the specimens together could represent many different groups and span as much as 30,000 years of northern Yukon prehistory.

Figure 6. Polished fleshing (?) tool of mammoth (cf. *Mammuthus* sp.) ivory excavated from presumed early Wisconsin sediments in the Olc Crow valley (MlVl-2, midsection deposit)

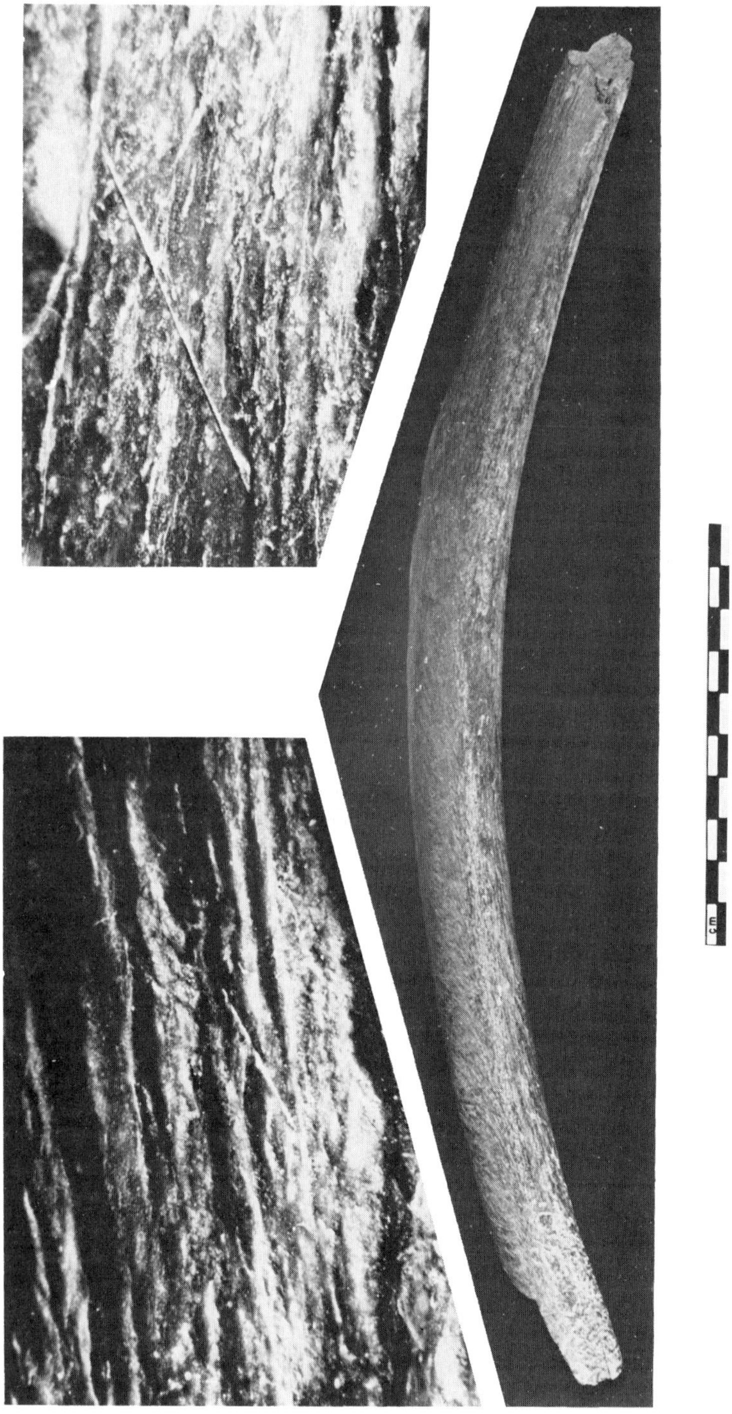

Figure 7. Bison (?) rib butchered with stone tools, excavated from a possible soil equivalent to Disconformity A in the Old Crow valley. The enlarged views of the cuts are magnified approximately 19 times in relation to the overview of the specimen

Other Eastern Beringian Localities[2]

Several other nonglaciated areas in eastern Beringia have produced vertebrate fossils ranging in age from pre-Illinoian through latest Wisconsin and Holocene time. Some of these fossils were modified while fresh in some of the same ways we have described from Old Crow localities. Most of these finds have been reviewed recently by several writers, and our discussion will be confined to brief characterizations of the artifacts and suggestions concerning probable age. We will offer few details on stratigraphy which, in most cases, is poorly understood.

In the Klondike region of west-central Yukon, deposits similar to the famous Fairbanks mucks continue to be exposed and destroyed by placer mining (Harington and Clulow, 1973; Péwé, 1975). An extensively cut caribou antler fragment, described as a "punch" (Harington, 1975; Bonnichsen, 1979, p. 171, Pl. IX-1), and several green-fractured mammoth and/or mastodon long bones and a cut bison skull (Morlan, 1977a, p. 4) represent secure evidence of human activity associated with Pleistocene fauna. A mammoth bone core discovered in 1979 represented the first such eastern Beringian specimen outside the Old Crow region, and additional examples were found in 1980 (J. Cinq-Mars and C.R. Harington, unpublished data). Harington (1978, pp. 58-60) suggests that the mammal faunas of the region date mostly from late Wisconsin time, and radiocarbon dates on the bones have ranged from relatively late in the Holocene to beyond the limit of conventional radiocarbon dating. As we discuss below, it is noteworthy that some of these dates fall within the period during which large glacial-meltwater lakes filled the interior basins of the northern Yukon.

Placer mining near Lost Chicken, in eastern Alaska, has produced a late Wisconsin fauna among which several artifacts have been reported. Among the most interesting are several cut caribou antlers; a *Bison crassicornis* tibia, recently radiocarbon dated as 10,370 ± 160 years old (I-8582), that was fractured while green; and a massive green-fractured mammoth limb-bone fragment (Bonnichsen, 1979, p. 172, Pl. IX-2, IX-3; Harington, 1980, p. 170, Fig. 2). The paleontological potential of this locality has been known since the early part of this century (Whitmore and Foster, 1967; Foster, 1969), and ongoing work there will undoubtedly enlarge the archaeological record (Lee Porter, unpublished).

Evidence at nearby Jack Wade Creek indicates that man was involved in the demise of a faunal assemblage dominated by sheep that also included bison, caribou, a small Yukon horse, and a moose (Porter, 1979; Porter and Hopkins, in press). A radiocarbon date of 29,700 ± 240 years BP (QL-1178) was obtained on collagen from a *Bison ? crassicornis* horn core (Porter, 1979, Table 1).

Artifacts of various kinds have been recovered from the famous Fairbanks muck deposits of east-central Alaska during the past half century (Hibben, 1943; Rainey, 1939, 1940; Bonnichsen, 1979, p. 174). Faunas ranging in age from Illinoian to late Holocene are thought to be represented, but most of the fossils date from late Wisconsin time (Péwé 1975, pp. 91-101; Harington, 1978, p. 75). Radiocarbon dates on bones and on well-preserved hair and soft tissues span all datable portions of the Wisconsin. Selective collecting, oriented primarily toward vertebrate paleontology, has probably significantly diminished the recoverable archaeological record from the Fairbanks localities; furthermore, many of the samples obtained by Otto Geist and others have never been analyzed from any standpoint. Our studies of Old Crow and Bluefish Caves materials have shown the importance of careful microscopic examination of fossil bone fragments (see below). A similar study of the enormous Fairbanks collections might reveal a rich record of human butchering activities. In fact, the size, diversity, and temporal span of the faunas from the Fairbanks area would render this kind of analysis important regardless of its outcome: it would be just as interesting to discover no evidence of artificial butchering as to delineate butchering practices.

Another context in which the evidence consists mostly of ancient bones is in the lowest level of Trail Creek Cave 9, where Larsen (1968, pp. 61-63) recovered a series of sheep and bison bones and a horse scapula. A bison calcaneus, believed to have been broken by man, was sacrificed for a

---

[2] Equivocal bone and ivory specimens have been found at Whitestone River in northern Yukon and on the Stewart River in central Yukon (Bonnichsen, 1979, p. 169; Morlan, 1980a, p. 274). A mammoth bone core was recently recovered from a placer mine near Carmacks in central Yukon and will be described elsewhere.

collagen-based date of 13,070 ± 280 years BP (K-1327), and the horse scapula gave a date of 15,750 ± 350 years BP (K-1210) on its organic fraction. These dates are quite important because they contribute direct evidence to the late Pleistocene history of three large herbivores (see Matthews, this volume), but the archaeological interpretation of this earliest Cave 9 material is open to doubt. Only the broken calcaneus has been advanced as evidence for human activity. The possibility that carnivores were responsible for its fracture was dismissed because of the absence of tooth marks on the bone (Larsen, 1968, p. 62). However, it has been shown in recent years that the limits of carnivore fracture capabilities are not yet adequately documented (G.A. Haynes, 1978, 1980; Morlan, 1980a), and the Trail Creek calcaneus cannot be regarded as proof of late Pleistocene human activity without supporting evidence that might narrow the list of alternate interpretations.

This brief review shows that several localities in eastern Beringia have produced evidence similar to some of the materials from the Old Crow region. Although more precise stratigraphic and chronological control is badly needed, most of the archaeological specimens appear to be of late Wisconsin age. It seems certain that the archaeological significance of existing collections from such places as the Fairbanks mucks has not been fully realized and should be further explored by means of field collecting and laboratory analysis of existing old and new collections. An interesting anomaly that appears to be emerging from the record thus far is the scarcity of mammoth bone cores and flakes in all areas except the Old Crow region. The few examples from the Klondike area suggest that other specimens will be found in future years, but we must entertain the alternate possibility that the distinctive procedure of fashioning large butchering tools by flaking bones has a relatively restricted distribution and is related to local scarcities of suitable lithic resources (see Morlan, 1980a, for a more detailed discussion). The shaping of bone and antler tools with stone implements, the creation of complex polished facets during tool use or production, and the butchering of animals with stone tools seem to have been widespread practices in eastern Beringia during late and middle Wisconsin time.

We have yet to find stone tools in direct association with the affected bone and antler pieces at the localities discussed above, but the distinctive tool marks on the osseous materials support the inference that stone tools were in use throughout this early portion of the record (Morlan, 1980c). One recently test-excavated site, Bluefish Caves, has produced the kind of direct association that is needed. The deposit, which corresponds in time to some of the Alaskan and Klondike localities just described, provides us with a clear primary contextual transition between the earliest evidence, based mostly on bone, and the later archaeological manifestations in which the vagaries of preservation have left little but stone tools.

THE BLUEFISH CAVES: STONES AND BONES IN SITU

The Bluefish Caves are located in the foothills of the Keele Range, in the vast and little-known uplands of northeastern Beringia. Just to the north lies the Bluefish basin, which was inundated by glacial meltwater during classical Wisconsin time. Three small caves have been recognized at an altitude of about 600 m along the western portion of a ridge of limestone and dolomite overlooking the middle course of the Bluefish River. The two caves test-excavated thus far (Caves I and II) are situated back-to-back on opposite sides of the end of the ridge and are oriented north and southwest, respectively (Fig. 8). Results of testing in 1978 have already been reported (Cinq-Mars, 1979). More extensive excavations in 1979 and 1981 greatly amplified the picture without requiring major alterations of the original chronostratigraphic interpretation.

The sediments within and just outside the caves consist primarily of aeolian and cryoclastic materials that show various degrees of organic enrichment. The profile consists of a basal frost-rived bedrock substrate (unit a) overlain by aeolian silt or loess that contains localized organic lenses (unit b), in turn capped successively by a humus-rich cryoclastic rubble (unit c), and modern humus and vegetation (unit d). This depositional sequence characterizes both caves, but a thickening of the loess in Cave I appears to have been accompanied by an influx of redeposited humic material and a thinning of the overlying humus-rich rubble.

The aeolian sediments of unit b have yielded a

Figure 8. The Bluefish Caves site. Above: aerial view taken from the north, showing the east-west trending dolomite ridge. Cave II is located just behind Cave I, facing in the opposite direction. Below: Cave I during the 1979 test-excavation. The large bone pit feature can be seen in the foreground

typical late Pleistocene fauna that includes the remains of fish, birds, and both small and large mammals. The large mammals include a canid, a mustelid, *Alopex, Vulpes, Equus, Rangifer, Cervus, Ovis, Alces, Bison,* and *Mammuthus.* In contrast, the overlying humus-rich rubble of unit c has yielded a rather impoverished vertebrate fauna in which all taxa are extant today in the northern Yukon Territory.

Paleobotanical evidence includes both plant macrofossils and, somewhat surprisingly, fairly well-preserved pollen. Pollen diagrams have been prepared for both caves; the two profiles are similar and generally reflect the sequence that has been reconstructed and dated elsewhere in eastern Beringia (Cinq-Mars, 1979, Fig. 12; Ritchie, et al., 1982). A treeless, herbaceous-tundra assemblage (zone I) is followed by a marked *Betula* rise (zone II) and then by the appearance of a regional boreal forest dominated in the pollen spectrum by *Picea.* Dates obtained elsewhere in the Yukon and Alaska suggest ages of 14,000 BP to 13,000 for the *Betula* rise in the middle of the aeolian unit and about 10,000 BP for the increase in *Picea* (Ager, this volume).

The chronology suggested by the pollen evidence gains support from two radiocarbon determinations on collagen extracted from bones found in the aeolian unit. An *Equus* femur excavated from the upper part of the Cave I loess has been dated as 12,900 ± 100 years old (GSC-2881), and a *Mammuthus* scapula from the lower half of the loess unit just beneath the Cave II dripline has yielded a date of 15,500 ± 130 BP (GSC-3053). Hence, the evidence furnished by stratigraphy, vertebrate paleontology, palynology, and radiometric dating together indicate that the Bluefish Caves preserves a record of the crucial final millennia of the late Pleistocene in eastern Beringia—millennia during which major environmental changes including the extinction of megafauna took place quite rapidly. Of even greater importance to this discussion, the caves have also provided definite evidence of human activity in association with the megafauna.

The archaeological evidence includes stone artifacts, bone artifacts and more general bone alterations, and spatial patterning in the bone distribution. All this evidence comes from within the aeolian unit; there is no sign of a prehistoric human presence in the overlying stratigraphic units that record the increased abundance of spruce pollen. The stone artifacts include an angle burin made on an edge-retouched chert blade (Fig. 9), a chert burin spall, and a variety of chert flakes, some of them bifacial trimming flakes (Cinq-Mars, 1979, Fig. 17). A microblade fragment was found but was out of context. All of the rock types represented among these specimens are exotic to the caves. Additional evidence of local flint-working activities has been derived from microscopic examination of sediment samples. Chert microflakes less than 3 mm across have been found in discrete stratigraphic zones that likely resulted from wind deflation of flint-knapping areas nearby, outside the caves. Finally, the lithic inventory also includes small pebbles exotic to the caves as well as one cobble-sized hammerstone (Fig. 9).

A second kind of archaeological evidence consists of two large-mammal tibiae, representing *Ovis* and *Rangifer,* respectively, that were split longitudinally and whittled and polished, primarily at one end. These specimens were found beneath the dripline in Cave I and outside the mouth of Cave II, respectively, and may have been used as fleshing tools (Fig. 9, Fig. 10:10).

The most important cultural indicators, numerically, are cutting, scraping, and piercing marks that have resulted from the use of stone tools for butchering, disarticulating, and defleshing the vertebrate skeletal elements found in the two caves. Traces of butchering are found on both cranial and post-cranial elements, and a quantitative analysis of their placement can be expected to aid interpretation of specific tasks. It is noteworthy that few if any of these alterations could be recognized in the field, because they are readily blurred or even completely masked by the thinnest trace of loess or other residue. Only careful laboratory cleaning and microscopic examination have made possible the recognition of the quantitative importance of those alterations that are so diagnostic of animal carcass processing. Even on the intensively root-etched surfaces of the 15,500-year-old mammoth scapula it was possible to identify six incisions and scrapes on the lateral and medial faces of the bone (Fig. 10:14, 15). Other kinds of surface alterations on the bones include striations and polished areas more difficult to interpret. Some may have resulted from

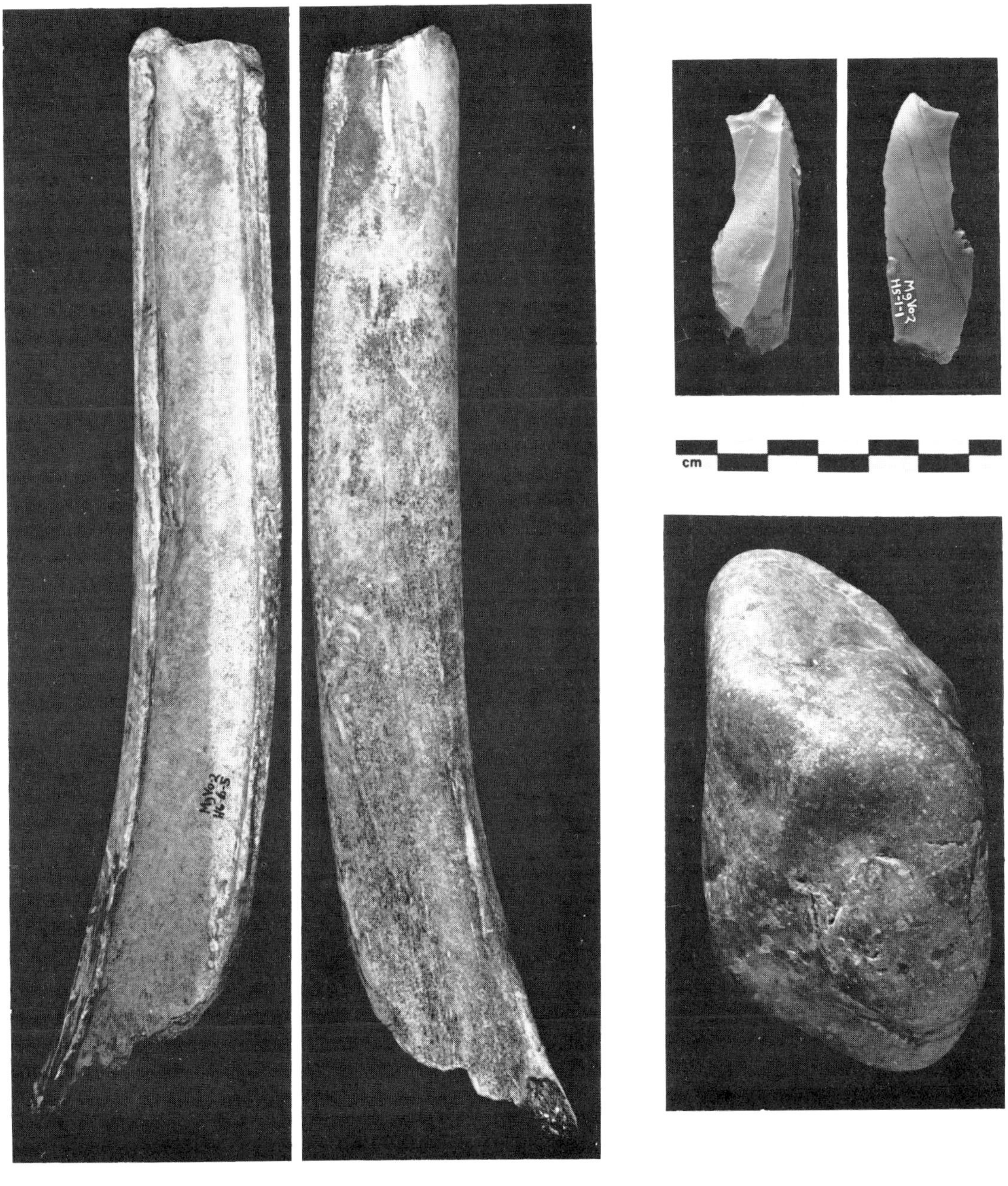

Figure 9. Artifacts from the Bluefish Caves site. Left: two faces of a fleshing (?) tool made on a split *Rangifer* tibia, and showing distal, polished areas as well as proximally located cut and shaving marks. Upper right: dorsal and ventral faces of an angle burin fashioned on a distally notched (truncated) blade made of bluish grey chert; the proximal end also exhibits notching, and edge-retouch is visible along the right-distal margin. Lower right: small cobble-size hammerstone

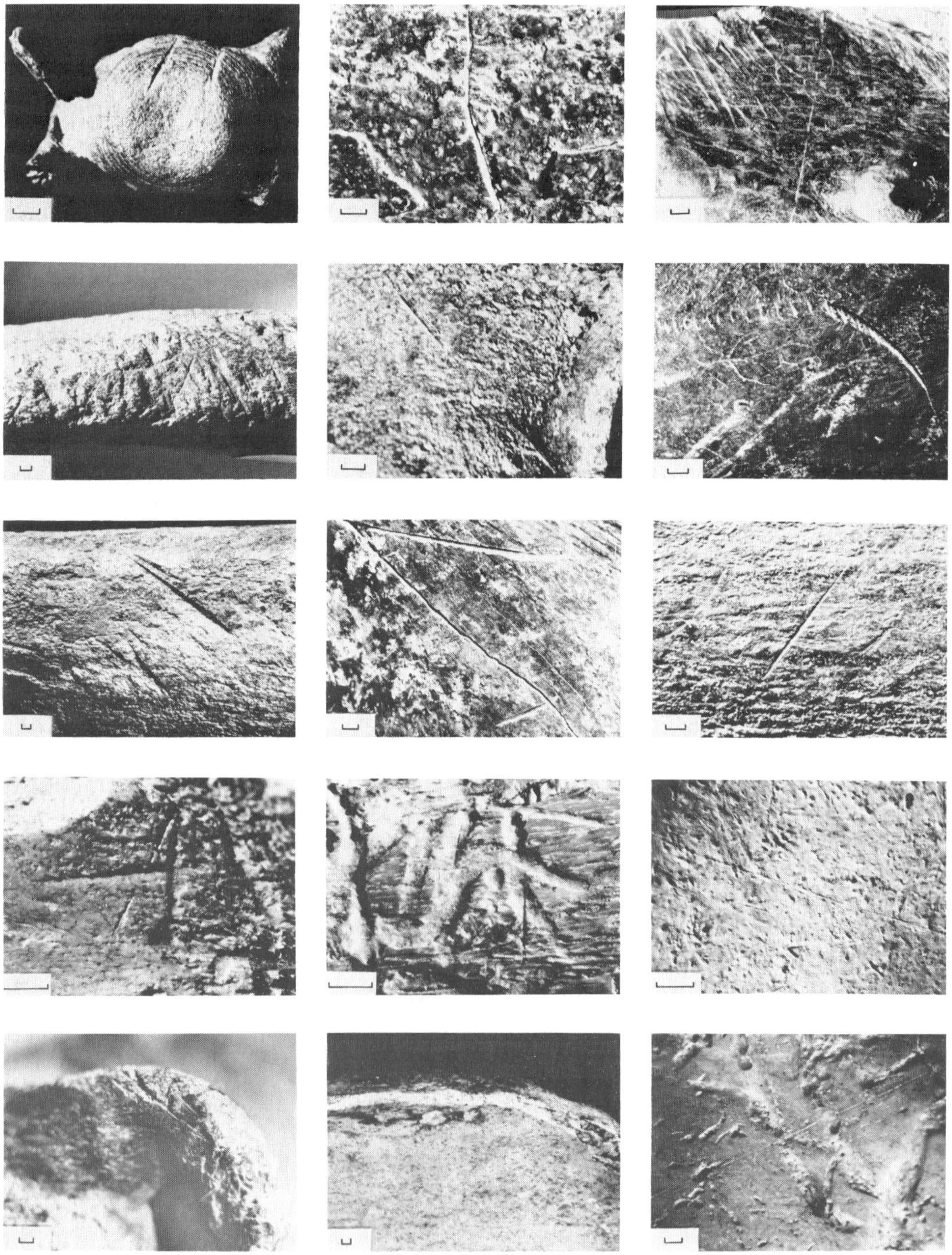

periosteum removal, but others could have been produced by carnivores acting either as predators or as scavengers.

Fracture patterns on bones from the two caves are also difficult to interpret. Ongoing examination of individual specimens, especially those of *Equus, Bison, Rangifer,* and *Ovis,* suggests a strong correlation between fresh bone fractures and traces of carnivore gnawing. Some of the spirally fractured bones may have been eaten by carnivores or may have been broken incidental to the ingestion of marrow and other soft tissues (G.A. Haynes, personal communication; Morlan, 1980a). On the other hand, some of the green-fractured bones may have been artificially broken and may represent the kinds of less formal butchering implements recently described by various researchers (Frison, 1974, 1977; E. Johnson, 1977; Bonnichsen, 1979; Morlan, 1980a). Obviously the expression of this sort of technological "expediency" behavior (E. Johnson, 1977) is likely to overlap with and may even be masked by the effects of primary and secondary carnivore activities.

Finally, as further evidence of cultural activity, some of the faunal remains at Cave I are arranged in a large pile that was uncovered in six adjoining 1 m squares in front of the cave (Fig. 8, lower). This pile probably resulted in part from the mode of dispatching game and in part from the manner of dismembering the carcasses. Many of the faunal elements in the pile exhibit butchering marks, and the excavation also yielded a few small stone flakes as well as the hammerstone seen in Figure 9. In a smaller concentration of bones in Cave II, the dated mammoth scapula and other elements likewise exhibit butchering makrs.

The Bluefish Caves can be scrutinized from several vantage points. From a local perspective, the data suggest that both cave areas have undergone a similar depositional history, and the contents of the aeolian unit appear to date entirely from late Pleistocene time, between 16,000 and 10,000 years ago. Cave I, at least, exhibits some degree of internal complexity. Some of its stone artifacts (the burin spall, the flakes, and possibly the microblade) occur stratigraphically within the humus-enriched and thickened aeolian units, as much as one meter above the basal loess member that itself has yielded a wide range of faunal elements in association with microflakes. It is the basal member that appears to be laterally equivalent to the bone-pile-bearing loess unit outside the cave mouth. Because the latter has also yielded culturally altered bones as well as stone flakes and the hammerstone, there may be a record, albeit discontinuous, of as much as 6000 years of human occupation condensed within this relatively thin profile. The Cave I bone pile itself may have formed over a significant period of time or within a very brief span, but in either case it indicates extensive cultural butchering activities that utilized the carcasses of many large mammal species. The size of the pile and its location at the base of a cliff terminating the bedrock ridge suggest that animals may have been dispatched below a "jump" and that the carcasses were processed where they fell. It is also possible that the natural amphitheater that delineates the front of the Cave I area was developed as a "surround" which could have been employed with or without the jump. We hope that detailed analysis (now underway) of the vertebrate remains will assist in interpretation.

A question of more general interest concerns the apparent paucity of artifactual materials and the absence of what one might call "diagnostic" implements. This is a question that seems often to perplex archaeologists, especially as we begin to look more closely at deposits that appear superficially to be merely paleontological in character. Examples range from Skaratki, Poland (Chmielewski and Kubiak, 1962), and Volch'ya Griva in western Siberia (Tseitlin, 1979, pp. 52-55), to

←

Figure 10. Selected series of butchering marks on bone and antler material from the Bluefish Caves site. From top to bottom and from left to right: 1) cuts (disarticulation ?) on caudal vertebra of medium-size mammal; 2) dense cutting and scraping scars on tine fragment of a *Rangifer* antler; 3) deeply incised cuts on beam fragment of *Rangifer* antler; 4) crosscut at the base of ascending ramus of cf. *Alces* mandible; 5) disarticulation marks on *Rangifer* metatarsal; 6) deeply incised cut on *Equus* sp. mandible; 7) discontinuous incision on *Rangifer* phalange; 8) two deeply incised cuts on diaphysis of an *Equus* sp. humerus; 9) two small cuts almost obliterated by root-etching, on diaphysis of a *Bison* sp. metatarsal; 10) elongated shaving facet along inside margin of split *Rangifer* tibia (see Fig. 9); 11) shallow incision on an *Equus* sp. cervical vertebra; 12) curved cut mark on an *Equus* sp. sacrum; 13) short, diagonal cut on diaphysis of a *Mammuthus* sp. fibula; 14) shallow incision on a *Mammuthus* sp. scapula; 15) linear scraping marks on a *Mammuthus* sp. scapula. (Scale bars represent 1 mm lengths)

Grundel, Missouri (Mehl, 1966), and the Dutton and Selby sites in Colorado (Stanford, 1979b). It is possible, of course, that Bluefish Cave I is only one of a number of activity-specific localities within a topographically complex site area. Cave II (as well as other unexplored shelters) could eventually produce evidence of complementary human activity that includes many more artifacts. It is obviously of great interest to determine as precisely as possible the presence or absence of human occupation in late Pleistocene levels of caves now being excavated in the lower Porcupine drainage in Alaska (Fig. 1; J.E. Dixon, D.C. Plaskett, and R.E. Thorson, unpublished manuscript). With or without evidence of man in the Pleistocene units, those caves can be expected to produce faunal assemblages that will be invaluable for comparisons with the record from the Bluefish Caves.

Modest though it is, the archaeological evidence from Bluefish demonstrates that people were living at what would then have been the eastern edge of Beringia during a period characterized by harsh periglacial conditions and by large windswept glacial lakes lying immediately to the north. The loess that enclosed this record may have been derived from the newly exposed borders of the lowland areas that were gradually draining and that left behind broad lake-studded plains. The rich fauna that is also enclosed in the loess can be regarded as a primary burial assemblage in which several ungulate taxa soon to disappear from the region are found in association with a variety of predators and the cultural remains of man. Hence, the Bluefish Caves indicate that the eastern Beringian archaeological record, when better known, will provide a basis for assessing man's role in late Pleistocene extinction and a foundation for understanding the emergence of later arctic and boreal cultural adaptations in northwestern North America.

LATER EVIDENCE: MOSTLY STONES IN SITU

Final Wisconsin environmental changes were as dramatic in eastern Beringia as elsewhere, and the shift from herbaceous or steppe-tundra to a moister shrub-tundra seems to have been accompanied by a more zonal, less heterogeneous composition of vegetation than had been present earlier. As Guthrie suggests (this volume), these changes probably are related to the demise of many species of large ungulates in the region, and they would therefore have had a profound effect on human subsistence strategies.

The earliest definable eastern Beringian cultural complexes found thus far belong to the 11,000 to 10,000 year range that R.D. Guthrie (unpublished manuscript) has dubbed "the critical millennium." Sites of this age are situated in upland areas that would have been within or near the margins of the receding and residual herbaceous or steppe-tundra habitats (Ager, 1975, pp. 87-88). These small refugia may have supported remnants of a rapidly declining late Pleistocene megafauna that could be exploited by human hunters (Morlan, 1977b, p. 98; D.L. Johnson, 1977, p. 151).

The locality that best exemplifies this combination of site location and habitat change is Dry Creek, a stratified multi-component site in the Nenana River valley along the northern foothills of the Alaska Range (Fig. 1; Thorson and Hamilton, 1977; Powers and Hamilton, 1978; C.E. Holmes; W.R. Powers; R.D. Guthrie; and J.F. Hoffecker, unpublished manuscripts). The earliest component (I) has been dated as 11,000 years old; it contains a range of cobble and flake tools, including burins, and thin bifacial knives and points, some characterized by triangular or subtriangular outlines and flat retouch (Fig. 11). Powers (unpublished manuscript) states that the recovered tool kit "is broadly comparable to Paleoindian tool-kits from the Plains of interior North America" and also exhibits similarities to the eastern Siberian Kukhtyi III assemblage (Mochanov, 1977, pp. 87ff, Pl. 28). It may be noteworthy that both Dry Creek I and Kukhtyi III lack the oft-abundant microblades that characterize so many Beringian sites in final Pleistocene time, but Kukhtyi III is nonetheless assigned by Mochanov to his Dyuktai tradition (Mochanov, 1977, Table D).

The site that has yielded material most closely comparable to that of Dry Creek I is found on the shore of Healy Lake in the middle Tanana River valley, Alaska (Cook, 1969, 1975; McKennan and Cook, 1968; J. P. Cook and R. A. McKennan, unpublished manuscript). There, the lowest levels of the Village site yielded a lithic assemblage called the Chindadn complex, dated as between 11,000 and 10,000 years old. At that time, the vegetation of the middle Tanana lowlands had already changed to shrub tundra (Ager, 1975), but herbaceous or steppe-tundra probably still occupied the foothills of the Alaska Range, not far to the south. Although Chindadn shares with Dry Creek I a number of biface forms, including subtriangular knives, small triangular points, and basally thinned concave-based points, the Chindadn complex also includes microblades.[3] Possibly the microblades and some of the other artifacts in Chindadn were introduced into apparent association with bifaces through post-depositional deflation of the rather thin aeolian sediments that constitute the Village site. In any case, the bifaces have a decidedly Paleoindian character (see C.V. Haynes, this volume) with at least two of the specimens exhibiting such extensive basal thinning that they can be considered to fall within the range of variation of eastern Beringian fluted points (Clark and Clark, 1975, p. 33).

Fluted points are now known from a number of localities, mostly in the northern half of eastern Beringia (Figs. 1, 12). In the northern Yukon they occur at the Kikavichik Ridge (Irving and Cinq-Mars, 1974) and Dog Creek sites, two localities situated along the northern rim of the Old Crow basin. Farther west, fluted points have been found in the Utukok and Kugururok River valleys of northwestern Alaska (Thompson, 1948; Solecki, 1951; Humphrey, 1966, 1970), at the Putu site (Alexander, 1974), at Girl's Hill (Cook, 1970; R. Gal, unpublished manuscript), at the Batza Tena obsidian source area (Clark, 1972; Clark and Clark, 1975, 1980), and at a number of surface localities examined during recent salvage archaeological surveys (Cook, 1970, 1977).

Although fluted points are widely used elsewhere as index fossils, in eastern Beringia they are generally of unknown age, having been found in surficial or, at best, shallow deposits.[4] Nor have the rare finds of fluted points in excavatable contexts provided a clear picture of their historical significance in Beringia. For example, the Putu complex excavated by Alexander (1974) produced charcoal dated as 11,470 ± 500 years old (SI-2382; C. V. Haynes, 1980, p. 119), but soil samples yielded later dates (GaK-4939: 6090 ± 430 BP; WSU-1318: 8450 ± 130 BP; Alexander, 1974, p. 25, personal communication, 1977). Other artifacts from the Putu complex include delicate gravers, scrapers, burins, one triangular basally thinned point which resembles those from Chindadn at Healy Lake, and approximately 140 microblades (Alexander, personal communication 1977). The relatively thin sedimentary sequence from which the Putu complex was excavated may not have provided adequate separation for materials of different ages, so the association of microblades and fluted points could be more apparent than real. It is interesting that the Batza Tena fluted point clusters, despite their surficial occurrences, have not produced microblades, although a rather bewildering array of other biface forms were found clustered with the fluted points (Clark and Clark, 1975, p. 33).

Most of the sites and components that appear near the top of our chronological chart (Fig. 2) have been assigned by one or more authors to the "American Paleo-Arctic Tradition" defined by Anderson (1978). These components are generally characterized by the widespread microblade and microcore technology, by certain forms of burins (e.g., Donnelly burins), and by variable numbers and forms of bifaces (Fig. 13). One of the most interesting of these manifestations is Dry Creek II (Powers, 1978; Powers and Hamilton, 1978; W.R. Powers; R.D. Guthrie; and J.F. Hoffecker, unpublished manuscripts). A series of artifact clusters both with and without microblades was carefully mapped during the excavation of an exhaustively studied sedimentary sequence (Thorson and Hamilton, 1977). Absence of the otherwise almost ubiquitous microblades in a few of the clusters raises questions as to the functional and historical significance of these distinctive artifacts. What is the meaning of their presence or absence in such other contexts as Dry Creek I, Chindadn,

---

[3] Cook's (1975, pp. 131-132) statement that Chindadn is similar to Dry Creek only with respect to microblade technology was based upon a comparison with the original Dry Creek sample (Holmes, unpublished manuscript) that later proved to represent Dry Creek component II. The Dry Creek I component, in which we see similarities to Chindadn, was discovered in later excavations.

[4] We agree with Clark and Clark (1975, p. 33) that the apparent associations between fluted points and artifacts of the Arctic small-tool tradition are probably misleading and not indicative of the age of the points (see Giddings, 1964, p. 233; Hall, 1969; Holmes, 1973).

and Putu? R. D. Guthrie (unpublished manuscript) suggests that the microblades were produced for the purpose of arming slotted antler or bone points (see also Anderson, 1968), although well-preserved examples of such points are relatively rare and do not occur at Dry Creek. Slotted antler points armed with microblades may nonetheless have been widespread during late Pleistocene and early Holocene times as suggested by their occurrence in such widely separated sites as layer III at Trail Creek Cave 2 in Alaska (Larsen, 1968) and numerous sites in the Yenisei basin of Siberia (Abramova, 1979a, 1979b). The absence of antler and bone specimens in the microblade complexes of Kobuk and Akmak at Onion Portage (Anderson, 1968, 1970), some of the middle levels at Healy Lake (J. P. Cook and R. A. McKennan, unpublished manuscript), the Denali complex as represented at Mt. Hayes 111 (West, 1967, 1980), and the Ugashik Narrows phase of the Alaska Peninsula (Dumond, et al., 1976; not shown on Figs. 1 and 2), is probably a result of poor preservation of organic materials rather than an indication that they were not used at these sites. All of these sites and complexes are believed to date to between 10,500 and 8000 years ago, and all contain objects closely similar to artifacts attributed to the latest

Figure 11. Selected artifacts from Dry Creek Component I. Upper: sub-triangular biface knife. Lower left: thin, lanceolate projectile point exhibiting extensive basal thinning on one face. Lower right: small, flat, triangular projectile point

Figure 12. Fluted points from eastern Beringia. Upper and lower left: specimens made of obsidian from Batza Tena, Alaska. Upper right: nearly complete basal fragment of a unifacially fluted point from the Kikavichik Ridge site, northern Yukon. Lower right: basal segment of a bifacially fluted point from the Dog Creek site, northern Yukon

Figure 13. Selected artifacts from Dry Creek Component II. Upper left: spatulate biface knife exhibiting both use and hafting wear. Lower left: lanceolate biface knife showing extensive edge grinding or wear. Upper right: nearly exhausted microblade core with a succession of five rearticulated core tablets. Lower right: small transverse burin

phases of the Dyuktai culture and the descendant Sumnagin culture of the Lena River basin in Siberia (Mochanov, 1977, 1978, 1980). These kinds of materials are very widespread in Alaska and the Yukon and often appear among the concentrations of artifacts found on the surface throughout the Brooks Range (e.g., Cook, 1970, 1977; Morlan, 1970; Anderson, 1972; Holmes, 1973; Hall and McKennan, 1973; Hall, 1975; Irving and Cinq-Mars, 1974; Cinq-Mars, 1978).

A number of final Wisconsin and early Holocene archaeological complexes cannot be readily accommodated in the groupings discussed thus far. Examples include the Gallagher Flint Station Loc. I, in the north-central Brooks Range, where quarrying activities may be responsible for the apparent divergence of the inventory (Dixon, 1975; Anderson, 1980); Anangula in the Aleutians, which may represent the coastwise spread of a distinctive culture not seen in the interior (Laughlin, 1967, 1975; Aigner, 1978a; Vasil'evsky 1975), or perhaps a regional descendant of the paleo-Arctic tradition (Dumond, et al., 1976); and the Groundhog Bay and Hidden Falls assemblages of southeastern Alaska, which may also be derived from the paleo-Arctic tradition (Ackerman, et al., 1979; Dumond, 1980a, p. 990, citing S.D. Davis, unpublished manuscript). A growing number of distinctive lithic assemblages is being recovered from the Dog Creek, Athrai, and other sites located on the northern periphery of the Old Crow basin as well as from other localities on the western flanks of the Richardson Mountains. These include large core and flake tools; a wide range of bifaces (both knives and projectile points); transverse, angle, and dihedral burins; and occasional large blades. Although these collections have not been dated, some may represent distinctive regional late Pleistocene/early Holocene manifestations; it is also reasonable to suppose that some of them may contain the lithic implements that might be expected to accompany the older bone inventories of the Old Crow basin (Cinq-Mars, 1978; Morlan, 1980c).

Clearly, by the end of the Wisconsin there was already considerable diversity among the cultures of eastern Beringia. Furthermore, the range of technological variation within each of the groupings discussed above is quite considerable and in each case implies a complex history merely glimpsed in the data now on hand. It is possible to subdivide these overlapping ranges of variation in many ways: while Dumond (1980a, p. 25) includes Dry Creek I and Chindadn in an enlarged "Siberian-American Paleo-Arctic tradition," Anderson (1980, pp. 235-238) prefers to place more emphasis on the presence or absence of bifaces in early Siberian and Alaskan sites and accordingly isolates the Dry Creek I and Chindadn complexes from the subsequent "American Paleo-Arctic." These differences suggest much more than an ongoing debate between, as Dumond suggested (1974), lumper and splitter. Surely there is some historical significance in the parallel development between, on the one hand, the bifacial Akmak and unifacial Kobuk complexes in Alaska and, on the other hand, the bifacial Dyuktai and unifacial Sumnagin in Siberia (Anderson, 1980, p. 236). Likewise there are large-scale parallels in the distributions of major microcore preparation procedures on opposite sides of Bering Strait (Morlan, 1978b).

There has been a dangerous tendency to treat all Beringian fluted points as a single indivisible grouping, whereas we would note that significant technological variations can already be seen in the small series known from Alaska and the Yukon. Furthermore, even if Dry Creek I and Chindadn can be construed as developmental facies of a fluted point tradition, there may be substantial differences between the two components. Our doubts concerning the reality of associations between such bifaces and microblade techniques show that we cannot yet confidently define two contemporary but contrasting traditions in Beringia and then simply derive fluted points from "the one without microblades" (C.V. Haynes, 1980, pp. 119-120). It is possible that the fluted points of eastern Beringia will be found to be truly associated with microblades at some localities but not at others. This kind of variation could arise if the fluted point complex were of sufficient duration to bridge the time of appearance of microblade technology, which may have intruded into Alaska and the Yukon from the west. It already appears that the temporal span of fluted points in the southwestern United States (C.V. Haynes, 1971, Fig. 6) bridges this period of eastern Beringian prehistory.

For some years now it has been apparent to most writers (e.g., Cook, 1975, pp. 132-133) that substantial time depth must be expected in the Beringian archaeological record in order to account for the divergence of so many regional variations. We regard the known distribution of late Wisconsin sites as adequate to show that most if not all eastern Beringian habitats were being exploited by human societies during the formation of late glacial and early post-glacial landscapes. We shall need many more well-stratified, well-preserved, and well-dated sites to resolve and enlarge this interesting prehistoric record and to provide the background for the various threads of subsequent development traced recently by many writers.

DISCUSSION

Archaeological and paleontological research during the past 15 years in eastern Beringia has begun to reveal a complex and interesting record of human occupation in a changing landscape in Alaska and the Yukon Territory. We still do not know when mankind first entered the region, but we begin to see the potential for a continuous record of human occupation beginning in early Wisconsin time and extending into the Holocene.

It is difficult to perceive a meaningful pattern in the scattered and poorly dated but apparently early occurrences of fossilized bone, antler, and ivory artifacts in eastern Beringia. The earliest evidence to come to light thus far is in the interior basins of the northern Yukon Territory, which have yielded radiocarbon dates that suggest that archaeological deposits represent most if not all intervals from 50,000 to perhaps 28,000 years ago (Morlan, 1980a, chapter 7). Because these northern Yukon basins were occupied by glacial lakes during classical Wisconsin time, one might expect that a significant interruption would be found in the record of human occupation in this lowland region. Other areas of eastern Beringia were continuously inhabited, however, by Pleistocene megafauna and probably by man throughout the period during which the northern Yukon lowlands were under water (Fig. 14). A long time gap in the series of collagen-based radiocarbon dates from the Old Crow basin coincides approximately with the period during which the glacial lakes were in existence, and one of us (Morlan, 1980a, pp. 268ff) has argued that the peak in abundance of dates around 28,000 years ago may represent increased large-mammal mortality during the rapid inundation of the basins in late Wisconsin time. Portions of this time gap are filled by dates on bones from the Klondike region of west-central Yukon and from the Bluefish Caves. A nearly continuous sequence of radiocarbon-dated large-mammal tissues from the Fairbanks mucks spans most of the classical Wisconsin interval (Fig. 14). Although most of the fossilized bone and antler artifacts were found in redeposited contexts, we can hope ultimately to derive a reasonably continuous archaeological record parallel to the nearly continuous record of fossil vertebrates.

The eastern Beringian bone artifacts recovered thus far hold little promise of revealing diagnostic historical connections with Siberia or other Old World areas. Serrated end-of-the-bone fleshing tools, for example, are widespread in time and space, as are antler wedges, antler billets, and bone awls. The antler billet from the Old Crow basin (Morlan, 1980a, Pl. 4.27) is similar in design to an example from cultural layer VIIa in Dyuktai Cave, which produced a rich late Pleistocene fauna (Mochanov, 1977, p. 14, Pl. 5:15). Similar billets have also been found in Gravettian sites of south-central Czechoslovakia (Morlan, 1980a, p. 278).

The technique of producing sharp-edged bone and ivory flakes seems to be a more widespread Eurasian trait than the published literature would indicate. Although bone preservation is rare in much of the East Asian mainland, there are interesting hints of bone-flaking techniques in some Paleolithic Chinese sites (e.g., Breuil, 1939; Chia, et al., 1979; Wei, 1979), and some of these are old enough to provide potential antecedents to the eastern Beringian materials. Likewise, in central Europe a long sequence of bone-flaking techniques can be seen in sites as early as the Micoquian levels of Kulna Cave (Valoch, 1979) and as late as the magnificent Gravettian occupations of Predmosti (Absolon and Klima, 1977), Pavlov, and Dolni Vestonice (Klima, 1963). Scattered examples of this kind of tool production have also begun to appear in Siberia. Mochanov (1977, p. 29, Pl. 5:9), for example, describes a leaf-shaped spear point made by retouching a massive ivory flake from cultural layer VIII of Dyuktai Cave. An artificially worked mammoth bone from Ust-Mil' II and a long bone flaked both longitudinally and transversely from Ikhine I have also been reported (Mochanov, 1977, pp. 36, 44, Pls. 9:8, 12:3).

A similar picture is developing in the New World. The Dutton site in eastern Colorado yielded a mammoth-bone core, a series of bone flakes, and differentially polished and cut bones (Stanford, 1979b). The Lamb Springs site, Colorado, produced a series of large bone flakes (D. Stanford, personal communication 1980), and a bone core may have been associated with a Clovis point at the Kimmswith site in Missouri (R. Graham, personal communication 1981). On the other hand, the abundant broken bone from Valsequillo, Mexico, does not seem to include cores and flakes (see Armenta, 1978), and the broken mammoth bones in Owl Cave were not treated in this way (see Miller and Dort, 1978). Although it would be premature to assign cultural or technological meaning to scattered bone-flaking practices, it is clear that bone cores and flakes should be considered along with lithic artifacts in our efforts to reconstruct prehistory.

The recognition of artifacts in deposits more than 50,000 years old in the northern Yukon has several implications that we do not now fully

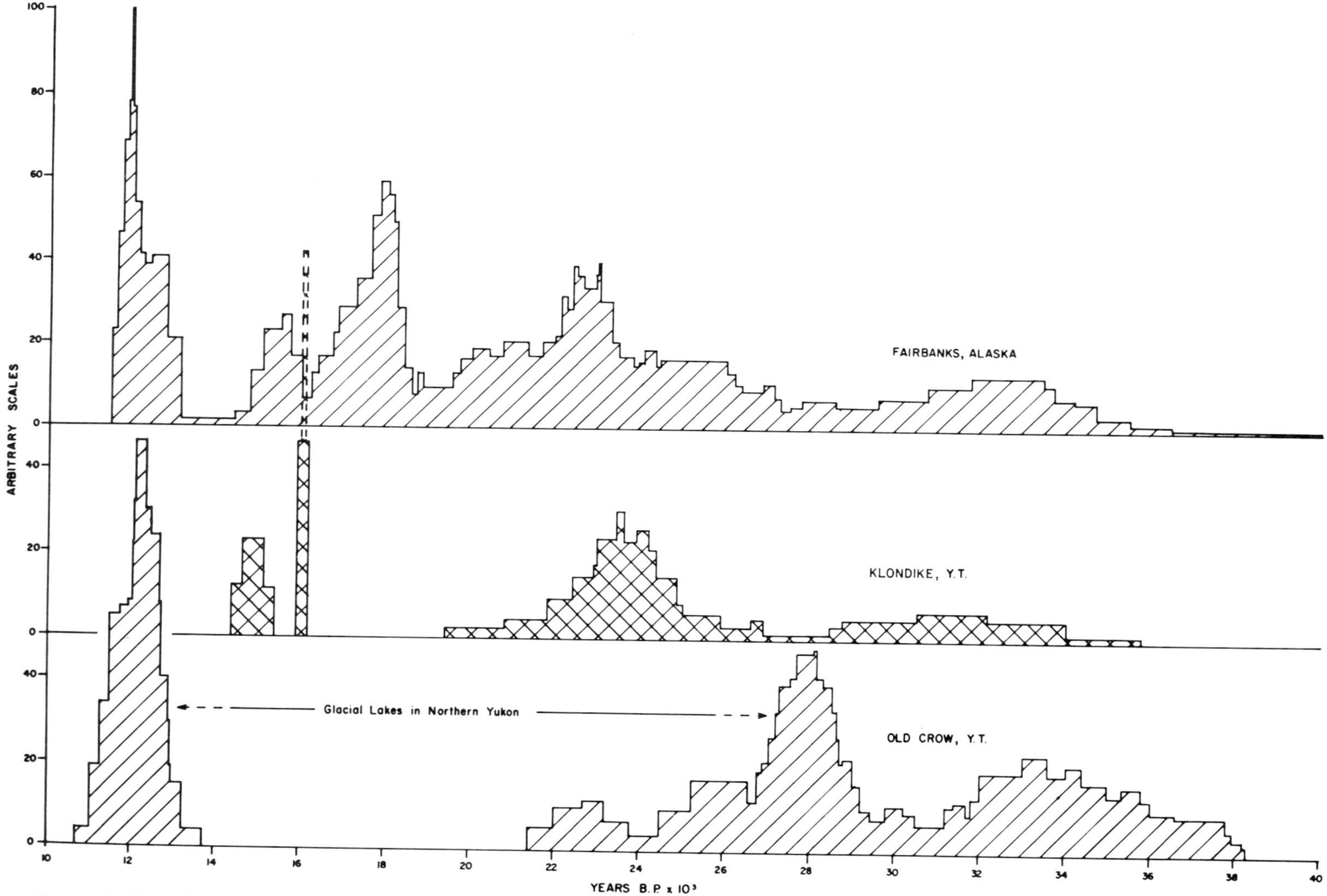

Figure 14. Three histograms showing the summed distributions of all finite dates older than 10,000 years BP on vertebrate tissues including bone collagen from the Fairbanks mucks (top), the Dawson area mucks (middle), and the reworked deposits in the Old Crow region (bottom). Each graph was constructed by converting individual radiocarbon dates to the form of small histograms of equal area (but differing shape depending upon the size of the one-sigma error) and then "stacking" the histograms on top of one another along the chronological scale as indicated by the center of each date range

understand but that we should not ignore. This eastern Beringian evidence shows that people had achieved the technological capacity to live in high latitudes by early Wisconsin times, an achievement perhaps based in part on earlier adaptations to glacial climates farther south, as Bordes and Thibault (1977) proposed. Klein (1976, p. 91) suggested that human occupation of the most continental climatic zones of Eurasia coincided with the emergence of fully modern humans and that we should not expect the kind of adaptations required for human residence in Beringia prior to that evolutionary development (see also Jelinek, 1980). Although fully modern man may have evolved in Africa as early as 60,000 years ago (Protsch, 1978, p. 5), modern humans appeared later in southwest Asia (around 40,000 BP, according to Jelinek, 1980) and Europe (around 35,000-30,000 BP: Howell, 1968, p. 155; Lasker, 1976, p. 302; Klein, 1976). Aigner's interesting review of Paleolithic China shows, however, that the paleontology, archaeology, and physical anthropology of Pleistocene East Asia can be modeled in ways that suggest the independent emergence, at an unknown but possibly earlier time, of fully modern man in that area. The apparent absence of a definable neanderthaloid grade in East Asia (Aigner, 1978b) indicates that we should not necessarily expect to encounter *Homo sapiens neanderthalensis* in the northern Yukon despite the antiquity of some of the artifacts.[5] In any case, we insist that evidence for early high-latitude adaptation must come from the archaeological record and not from evolutionary theory.

As we mentioned earlier, most discussions of the peopling of the New World assume that the Bering land bridge was exposed at the time of man's arrival (e.g., Müller-Beck, 1967; Bryan, 1969; MacNeish, 1976, p. 317; Lorenzo, 1978, p. 2; Griffin, 1979, p. 49), although a few authors have noted that the Australian record argues for a very early appearance of watercraft and suggest that a land bridge to the New World was not necessary (D.L. Johnson, 1970; Irving, 1971, p. 72; Hallam, 1977, p. 130). We find even more impressive the evidence for Acheulian and possibly earlier crossings of the Straits of Gibraltar (Bordes and Thibault, 1977, p. 124) which are more similar to Bering Strait, with respect to width and currents if not to water temperatures. More to the point, however, is the fact that humans could have found the Bering Strait in any of a wide variety of conditions—dry land, open water, shore-fast smooth ice, or nearly impassable jumbles of shifting sea ice—depending upon when they first reached the area. In any case, we doubt that the Bering Strait was ever a physical barrier to human movement, although it could have functioned as a "filter barrier" at many times in the past (Hopkins, 1979, p. 35). It is noteworthy, however, that a broad land bridge existed between Alaska and Siberia for at least a few thousand years beginning approximately 70,000 years ago (D.M. Hopkins, this volume). This bridge may have been crossed by the earliest human inhabitants of the northern Yukon Territory. Certainly the availability of such a dry platform is important if we are to envision the kind of gradual spread of human groups depicted by Giddings (1954, p. 88), for it is only in such circumstances that we can imagine the slow enlargement and locational shift of the customary territory of a hunting band preying upon spatially contiguous populations of game animals.

Most summaries of the peopling of North America have also concerned themselves with the possible presence and size of an ice-free corridor between the Cordilleran and Laurentide glacier systems and have variously depicted major physical barriers, inhospitable environmental conditions, or no barrier at all. It is now reasonably clear that portions of the corridor were closed by coalescence of the two major ice masses during periods of unknown length in both early (Jackson, 1980, Fig. 1) and classical (Matthews, 1980; Rutter, 1980) Wisconsin time, but we have too little archaeological data to understand how these conditions may have influenced human prehistory. Furthermore, an interval of some 40,000 years separated the main glacial pulses of the Wisconsin (D.M. Hopkins, unpublished manuscript). Conditions intermediate between full-glacial and interglacial environments would have characterized the corridor area during this interval, and human populations now known to have been resident in

[5] Although the investigators (Irving, et al., 1977) have suggested that a human mandible fragment from the Old Crow valley may date from the Pleistocene, the specimen was recovered from recent point-bar deposits, has not been dated, and in any case represents an 11½ year old child, too juvenile to exhibit significant evolutionary characters.

northern Yukon Territory since before 50,000 years ago could have moved southward along many different routes to the midcontinent of North America. Given the physiographic position of Beringia as a sometime extension of Siberia, however, our eastern Beringian archaeological evidence does not necessarily qualify as "proof" of the peopling of the "New World" as these terms are usually employed. Some of our eastern Beringians may never have "left home."

Much of this discussion has had a decidedly negative tone and has been extensively garnished with qualifiers and doubts, and we would like to end this review on a more positive note. Although we cannot offer a strictly data-based reconstruction of intersite and intercontinental relationships, we believe that the basic principles that characterize human capabilities and the processes of cultural change can be used to set limits on our thinking about the past. We believe, for example, that archaeologists have too often emphasized the concept of "migration"—implying that one place is a point of departure and another a place of arrival—in considering the peopling of the Western Hemisphere. Giddings' comments concerning the gradual shifts that never involve leaving home summarize a more likely process leading to the colonization of eastern Beringia more than 50,000 years ago. We can suppose that these "colonists" arrived by gradually enlarging their hunting territories across an enormous geographical area of northern Eurasia in which there were few, if any, major ecological boundaries requiring significant and rapid readjustments. The immigrants were well adapted to life in a single, widespread biome in which a mosaic of more local habitats afforded a variety of resources (see other papers in this volume). Once a part of the region became inhabited, there is no reason to doubt that human occupation continued there. The changes in habitat that we know to have occurred were so gradual that human societies should have been able to adjust without serious dislocations.

These people can be expected to have maintained active communication with their neighbors on all sides, and certain aspects of the northern environment—e.g., treelessness, seasonally frozen terrain—would have particularly favored communication over tremendous distances. Perhaps this is why C.V. Haynes (this volume) finds such interesting parallels between the Clovis culture of southwestern United States and those same Gravettian sites of Czechoslovakia (Dolni Vestonice, Pavlov) to which we referred in our discussion of ancient bone-flaking techniques. Perhaps, too, this accounts for the close similarities between Kostenki I, level 5, in eastern Europe, and the Llano culture of North America noted by Müller-Beck (1967). These similarities do not imply that any human group has traveled farther than its nearest neighbor, but they do imply that many human groups were in frequent contact across a large portion of the Northern Hemisphere. Doubtless there were also significant north-south contacts in East Asia, and we should expect to find in the New World long-established north-south ties reaching well into the midcontinent of North America.

The many dissimilarities among the sites and complexes of this broad region reflect the ongoing evolution of more local cultural traditions that inevitably developed distinctive styles and responded to local needs. These people must have been challenged by rigorous environments, and they must have been both inventive and receptive to new ideas. Communication must have been less regular at times, with human commerce (especially north-south) filtered by significant physiographic and environmental gradients. Furthermore, there were occasionally effective filter barriers such as the inundation of the Bering land bridge, the coalescence of Cordilleran and Laurentide ice, and the formation of glacial lake Mansi in western Siberia (Grosswald, 1980, Fig. 7; Volkov, et al., 1980).

Viewed in this light, we can take another look at problems mentioned earlier. Perhaps our confusion over the mixture of similarities and differences between the Dyuktai culture and the latest Wisconsin cultures of Alaska is due in part to the effects of the then-new Bering Strait filter barrier. Perhaps the earliest fluted points appear in eastern Beringia at a time when Bering Strait formed an increasingly effective filter barrier to the west while coalesced or closely adjoining ice sheets still reduced southeastward communication within North America. If this speculation is correct, then late Wisconsin dissipation of the glacial-corridor filter could have permitted a rather sudden diffusion of the fluted-point complex into the midcontinent. It seems quite likely to us that the widespread appearance of fluted points in central

North America was accomplished so rapidly because there were already resident populations in place to act as both recipients and carriers of these new ideas (cf. Dumond, 1980b, p. 991; Clark, 1978; Dixon, 1976). One of us has argued elsewhere (Morlan, 1977b) that fluted points may have spread from Beringia as part of a response to the deterioration of steppe-like conditions there at the close of the Wisconsin, but it seems unlikely that population movements were involved except for gradual resettlement of newly deglaciated terrain.

Obviously our view of ancient cultures in the New World is at variance with the view of P.S. Martin (1973, this volume; Mosimann and Martin, 1975) who attributes the extinction of the mammoth (and other Pleistocene megafauna) to the skillful predation of newly arrived human populations. We suppose, and sites such as Valsequillo, Dutton, Selby, Manis, Meadowcraft, and a few others seem to show, that people were present in much of North America before the 11,500 BP appearance of the Llano complex in the U.S. Southwest, and in at least some of these areas the earlier people were either preying upon mammoths or scavenging their carcasses. We would not, however, omit from consideration a role for man in the extinction process. In fact, we would note that the process was not restricted to the late Pleistocene but is ongoing and currently accelerating (see Hester, 1967; Myers, 1980). We would also acknowledge, however, the marked environmental changes that occurred in the late Wisconsin. Not the least and perhaps the most important of these is the highly zonal character of modern biomes (see Van Devender and Spaulding, 1979; Guthrie, this volume), which during their formation must have partitioned local populations in areas too small for their long-term support and which could have profoundly influenced the opportunities of enterprising hunting bands to overexploit fragmented game populations (D.L. Johnson, 1977, p. 151; Morlan, 1977b, pp. 100-101).

If the past decade has opened new archaeological vistas in eastern Beringia, the next can be expected to help us interpret them. Enormous areas of Alaska and the Yukon Territory are still essentially unexplored with respect to archaeology, and the nonglaciated character of much of this region holds promise for the gradual development of a long and interesting record of human occupation.

ACKNOWLEDGMENTS

We thank the Archaeological Survey of Canada, National Museum of Man, for encouragement and support during the preparation of this paper. Morlan's travel to the Wenner-Gren symposium at Burg Wartenstein, Austria, was provided by the Wenner-Gren Foundation for Anthropological Research, and subsequent travel in Europe was supported by the National Museum of Man. The paper prepared by Morlan for that meeting was published elsewhere (1980a) and contains many acknowledgments applicable to his portion of this paper. References to the Bluefish Caves and to sites on the Old Crow basin periphery resulted from Cinq-Mars' field work as a principal investigator (1975-79) in the Northern Yukon Research Programme, sponsored by the University of Toronto, the Social Sciences and Humanities Research Council, the Canadian Donner Foundation, the Department of Indian and Northern Affairs, and Canadian Arctic Gas Study, Ltd. This is Contribution No. 60 of the Yukon Refugium Project.

22

WERE CLOVIS PROGENITORS IN BERINGIA?

C. Vance Haynes

The Clovis culture represents the earliest clearly defined culture known in North America and is restricted to a period of only 500 years between 11,500 and 11,000 years ago. Clovis tool kits vary from site to site, and there are regional variations although samples are still woefully inadequate to determine regional differences in lifestyles. Clovis people appear to have been foragers but primarily were hunters of mammoth and bison for both food and material resources.

An Old World origin for Clovis seems probable on the basis of comparisons of their bone and stone technology with that of Europe and Siberia during the late Paleolithic. In eastern Siberia before 11,000 BP, there appear to have been two distinct lithic traditions, one dominated by microblades (Dyuktai) and one without microblades (Mal'ta-Afontova). Clovis may have been a descendant of the latter people, who crossed Beringia in pursuit of big game between 20,000 and 15,000 years ago when steppe-tundra united the two northern continents. Subsequent decline of steppe-tundra and Pleistocene megafauna after 15,000 BP may have led the hunting cultures farther south-eastward until they passed from the habitats of *M. primigenius* to those of *M. columbi* and *M. jeffersoni* about 13,000 to 12,000 years ago. The nearly explosive increase and spread of the Clovis culture thereafter may have been the direct result of this contact with game resources not previously exploited by man, or at least not with the intensity brought to bear by the Clovis big game specialists. The Clovis dispersal took place during a period of the greatest environmental change subsequent to the end of the Sangamon interglacial. The combination was more than many elements of the Pleistocene megafauna could withstand.

Key words: Paleoindian, Clovis, Quaternary extinction, radiocarbon dating, paleoclimates

INTRODUCTION

The Clovis culture, marked by distinctive fluted projectile points commonly found in situ with bones of large extinct mammals, represents the earliest clearly defined culture known thus far in North America (Hester, 1966). Stratigraphically controlled radiocarbon dating has shown that Clovis sites were occupied within a remarkably restricted period of only 500 years between 11,500 to 11,000 BP (Haynes, 1970, 1980). Statistical evaluation of 21 dates from the Lehner site and 10 dates from the Murray Springs site indicate Clovis in Arizona to be 11,000 ± 100 years old (Fig. 1). This evidence, combined with the lack of evidence of man south of Canada before Clovis, has led me to suggest that Clovis progenitors may have been a distinct infusion of late Pleistocene megafauna hunters from Siberia who, once south of the North American ice sheets, found abundant game and resources (Haynes, 1964). This abundance promoted rapid population growth and expansion of fluted-point-making

ISBN 0-12-355860-3

people throughout the Americas in the millenium 11,500-10,500 BP, coincident with the extinction of the Pleistocene megafauna (Haynes, 1966). P.S. Martin (1973) has carried this a step further by suggesting that Clovis man, being a superpredator, expanded as a concentrated wave-front leaving behind a land sparsely populated by man and devoid of mammoth, horse, camel, and some other elements of the megafauna.

The main alternate hypothesis for Clovis origin is that the culture developed within the Americas from a population present in the New World before late Wisconsin glaciation began 30,000 to 50,000 BP (Bryan, 1969; Bonnichsen, 1978; Stanford, 1978).

The difficulty with a hypothesized indigenous origin for Clovis is the lack of conclusive evidence of the presence of earlier cultures with a developmentally antecedent technology. The dating of some and the nature of other postulated pre-Clovis sites south of the ice sheets are still equivocal, and tool assemblages from these sites do not reveal a developmental sequence forming a continuum with Clovis assemblages. This could be the result of an insufficient number of radiocarbon-dated sites with adequate quantities of artifacts, but, on the other hand when Clovis traits are compared with those of certain Paleolithic cultures in the Old World, several unmistakable similarities become evident.

If we assume that Clovis progenitors came from eastern Siberia by way of Beringia, then what pathways did they use and how are Alaskan finds related to their migration into the New World? In order to develop pertinent hypotheses I will first examine the Clovis culture, then compare it to Late Paleolithic sites of the Old World, and finally examine the record in eastern Beringia for possible Clovis progenitors.

THE CLOVIS CULTURE

The single most diagnostic artifact that reveals the former presence of Clovis people is the lanceolate Clovis fluted projectile point. The Clovis

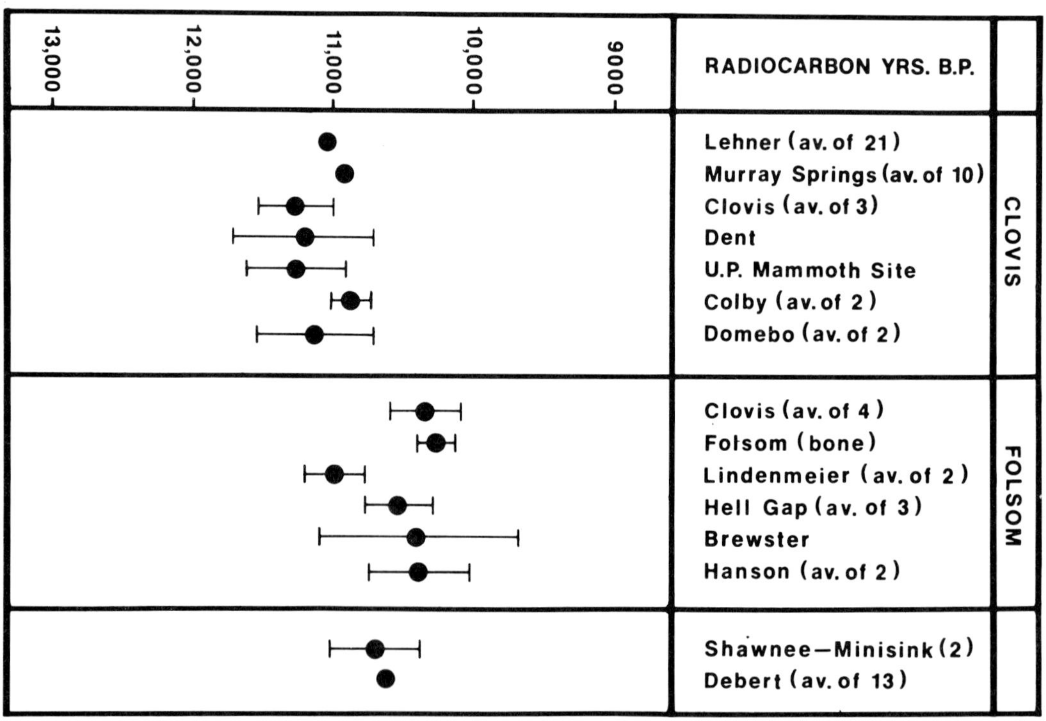

Figure 1. Comparison of radiocarbon dates at stratified sites of fluted point cultures in North America

point is characterized by one or more long, relatively narrow, longitudinal flake scars removed from both sides of the points by applying pressure or percussion to a specially prepared platform at the base. Distinctive Clovis points have been found on the surface from the Canadian plains to Central Mexico and from coast to coast. Where stratigraphic successions permit, correlations on the basis of culture, fauna, vegetation changes, and radiocarbon dating are consistent. Strata below Clovis levels are without archaeological evidence in spite of evidence of suitable game animals in the western successions. Levi Rock Shelter, Texas (Alexander, 1978), and the Dutton-Selby sites, Colorado (Stanford, 1979), are claimed to be exceptions, but the evidence is not convincing.

Studies of late Quaternary alluvium reveal that most streams in the United States were in a degradational mode during the final phase of deglaciation about 12,000 years ago. By 11,500 BP, Clovis people were occupying riparian sites on abandoned terraces along streams that had become considerably smaller in channel area than during the preceding 10,000 years. In the southwestern United States Clovis sites along these channels were buried about 11,000 years ago by clayey silts washed from adjacent slopes as aggradation occurred (Haynes, 1968).

Mammoths became extinct after 11,000 BP, and Clovis evolved into Folsom, Plano, and other cultures that developed great skill in hunting bison, one of the few target species of Clovis hunters to survive into the Holocene. After 7000 or 8000 BP many smaller tributaries, particularly in the Southwest, entered a cycle of gully cutting and filling every 2000 years or so; the cycles tended to be in phase (within the margins of error for radiocarbon dating) over large areas (Haynes, 1968). Because of this alluvial history many Clovis sites still lie buried beneath modern floodplains, such as that of the Shenandoah River at the Thunderbird site (Gardner, 1977). Most western sites such as Naco (Haury, et al., 1953), Lehner (Haury, et al., 1959), and Domebo (Leonhardy, 1966) were exposed by gullying during the middle Holocene as well as by the modern gullying that led to their discovery.

Some stratified Clovis sites in the West have been associated with springheads. Their proximity to water may be significant in that evidence from western sites indicates that water tables were lower during Clovis time than at any time during the previous 10,000 years (Haynes, 1968). This may have played a significant role in animal extinction at the end of the Pleistocene. Modern elephants follow traditional trails and have an affinity for certain springs (Olivier, this volume), and the concentration of animals around a reduced number of end-of-Pleistocene watering places may have enhanced man's role in extinction. Whether or not P.S. Martin's (1973) wave-front overkill theory is correct, it is apparent that Clovis hunters contributed to the extinction of Pleistocene megafauna.

Clovis tool kits vary from site to site, and there are suggestions of regional variations, but the samples are still woefully inadequate. The projectile points are invariably fluted on at least one side and vary greatly in size and shape. No better example of this variation is known than the eight Clovis points (Fig. 2) found with the single mammoth skeleton at Naco, Arizona (Haury, et al., 1953), and if the other Clovis sites within the San Pedro Valley are taken into consideration an even greater variation is seen (Haury, et al., 1959; Hemmings and Haynes, 1969; Hemmings, 1970). On the other hand, similar forms can be seen in widely separated regions.

Resharpening of Clovis points was a common practice, and impact damage to point tips clearly indicates that they were indeed propelled (Fig. 3). Broken points were reversed in reworking in at least one case, and one longitudinally split point from the Lehner site was refluted and the fracture edges ground before reuse (Haury, et al., 1959, Fig. 12f). Exotic materials were utilized for some points, and clear quartz crystals seem to have been a successfully met challenge to the knapper at the Lehner and Simon sites, and in Virginia.

The use of Clovis points, hafted in short foreshafts as knives, would seem logical but has not been conclusively demonstrated. Edge-wear studies in progress will be useful in this regard. Knowledge of hafting methods is still speculative in spite of replication experiments (Lahren and Bonnichsen, 1974), but several Clovis points display highly burnished spots on their faces, suggesting movement against a tight binding.

A few Clovis points appear to have been made from large flakes, but most were reduced from bifacial preforms. The collection from the Simon site in Idaho (Butler, 1963; Butler and Fitzwater,

Figure 2. Eight Clovis points found in direct association with a single mammoth skeleton at Naco, Arizona, display the variations in size and shape to be expected among the weapons of a single band. The longest point measures 116 mm. (Courtesy of the Arizona State Museum)

1965) is exceptional in that part of it represents a sequence of large, superbly reduced bifaces ending with the largest (18.1 cm) Clovis points known (Fig. 4). One of the Simon bifaces is remarkably large, especially when one considers that only half of it, measuring 17 x 15 x 1.7 cm, remains. The very large yet thin flake scars from this and other bifaces reveal great skill and force on the part of the knapper. Even large quartz crystals were reduced to well-formed bifaces. The other bifaces, 20 in all, as well as the five Clovis points are all made from five or six varieties of flint (conchoidally fracturable forms of SiO_2: e.g., chert, chalcedony, jasper, etc.) that are of the finest quality known and were obviously carefully selected for their exquisite coloring, translucency, smooth texture, and other aesthetic qualities. This selectivity is observable in many Clovis collections, but reaches a high point in the Simon collection.

Why this aesthetic selectivity and perfection in knapping in the Simon site collection? A clue is provided in the Anzick (Wilsall, Montana) site collection, where similar bifaces and Clovis points were found along with bevel-ended bone foreshafts and red ocher in what seems to have been a Clovis grave (Taylor, 1969; Lahren and Bonnichsen, 1974). The human bones are few, but represent two young children. They are not burned as stated previously (Lahren and Bonnichsen, 1974), but there may be some question about their association (Taylor, personal communication). No bones or bone tools were found at the Simon site (which was exposed a few centimeters below the surface by a road scraper) but it could also have been a grave, and a reddened area observed there (Butler, 1963) may have been ocher-stained instead of fire-reddened. In any case, the Anzick site suggests that large bifaces and Clovis points were considered to be important in the hereafter.

Another suggestion in regard to the selection of beautiful materials is that Clovis people may have believed that beautiful weapons were pleasing to the prey (R. E. Ackerman, personal communication).

From the Murray Springs, Arizona, site we have learned that bifaces were used as tools and reduced on the spot to produce other bifacial as well as flake tools as the need arose (Hemmings, 1970; Huckell, n.d.). This appears to have been done at the Clovis (Blackwater Draw), Williamson, Virginia (McCarey, 1975), and Wells Creek, Tennessee (Dragoo, 1973), sites, although the latter two are quarry sites as well.

In both the mammoth and bison kill areas at Murray Springs the incredible lack of disturbance of the Clovis occupational surface is revealed by the numerous concentrations of hundreds of bifacial thinning flakes. Detailed lithic analyses of these by Bruce Huckell (n.d.) reconfirms the emphasis on biface preparation and modification

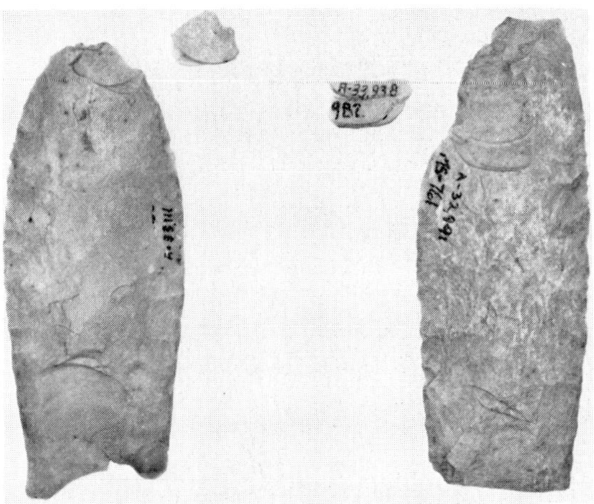

Figure 3. Two Clovis points from Murray Springs, Arizona, reveal impact damage. The point on the left was found in the bison kill area and the impact flake in the hunters' camp 138 meters away. The reverse situation holds for the point on the right where point, missing its base, was found in the camp and 47 meters from the impact flake in the bison kill area

Figure 4. Bifaces from the Simon site in Idaho are made from high-quality, beautifully colored flint and may represent a sequence of reduction by knapping ending in one of the largest Clovis points known (181 mm). (Photograph of Eichenberger casts)

suggested by Hemmings (1970). Numerous flakes from individual clusters of a particular material have been fitted together to reconstruct overshot flakes that extend from one side of a biface to the other and thus indicate biface width (Fig. 5). Similar bifaces replicated on the basis of these findings were used by Huckel to determine use wear from the skinning and butchering of a circus elephant that died in Tucson in 1975 (Huckell, 1979). Similar butchering experiments using modern-made Clovis points have been conducted at the Smithsonian Institution (Stanford, et al., in press).

Large Clovis bifaces were apparently made by direct percussion with a billet made presumably of antler, bone, ivory, or hardwood, but some bifaces, such as some of those from the Williamson site, appear to have been roughed out with hammer stones and thinned by fluting (Fig. 6) as in the Cattail Creek fluting tradition recorded by Painter (1970). Fluting of large bifaces and projectile points to make them thinner was usually accomplished by a combination of direct percussion from a platform that had been prepared by beveling the base and by pressure flaking, although some modern knappers can make narrow flutes by direct pressure applied by hand. Evidence of lateral- and basal-edge grinding is more common than not and presumably was done to prevent the cutting of sinew or whatever was used to bind the point to a foreshaft.

Most Clovis tools other than projectile points are made on large flakes and include side scrapers, end scrapers, concave scrapers, unifacial knives, flake knives, gravers, burins, and various combinations of these. Some are large primary flakes, others are bifacial thinning flakes and yet others are bladelike flakes. Similarities between some Clovis flake tools, particularly side scrapers, blades and tools on blades, and Old World late Paleolithic tools is unmistakable.

Clovis blades (Fig. 7) are large and similar to Aurignacian style blades in that they are triangular in the transverse section, thickset toward the distal

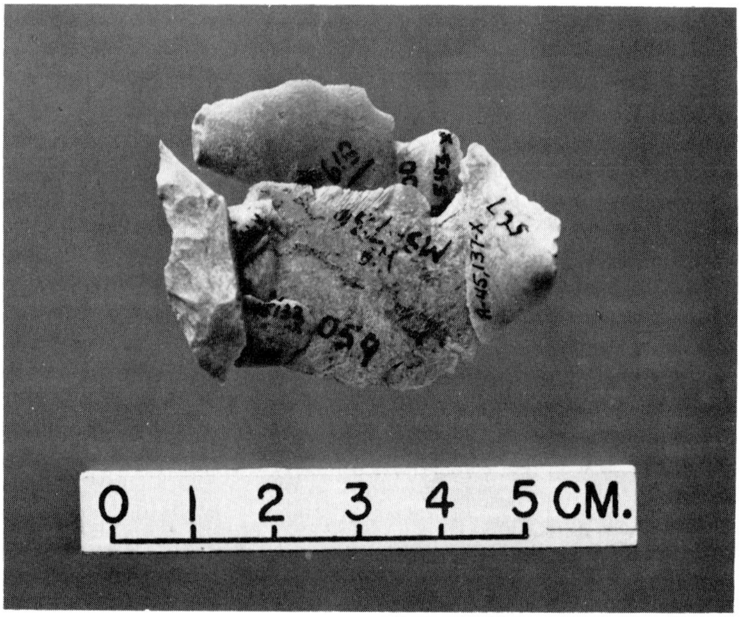

Figure 5. Reassembly of bifacial reduction flakes from the Murray Springs site, Arizona, has allowed determination of approximate widths for bifaces not found at the site. The overall length of this assembly is 47 mm

end, and have ground platforms with edge angles of 60° to 71°. Tools on such blades include end scrapers, side scrapers, concave scrapers, and combinations thereof. Blade cores are rare but are known from the Williamson site (Haynes, 1972). No microblades are known from Clovis assemblages, but objects from the Shoop site in Pennsylvania have been interpreted as microblade cores (Witthoft, 1952). Similar objects from the Debert site are considered to be wedging and grooving tools (*pieces esquilles*). MacDonald (1968) and Cox (n.d.) apply the same interpretation to Shoop specimens.

Burins are rare in Clovis assemblages and when present are commonly incorporated into other tools such as a flake knife from Murray Springs (Hemmings, 1970). This particular tool displays another Old World trait, removal of the bulb of percussion by pressure flaking (Fig. 8).

Bone tools were undoubtedly more important in the Clovis industries than the few that have been found would indicate. Examples include awls or punches, scrapers, fleshers, points, cylindrical points and foreshafts, a bead, and a shaft wrench. An inscribed bone was found at the Clovis site (Hester, 1972) and a cylindrical ivory point from Florida has a simple zigzag mark engraved on two sides (Fig. 9) (Haynes, 1976). This point and others like it were most likely made from fresh ivory because the axis crosses the cone-in-cone structure of the tusk from which it was made, and fossil ivory

Figure 6. The thinning of some bifaces at some Clovis sites was done by a combination of fluting and transverse flaking as displayed on these specimens: two from the Williamson site, Virginia, right; and one from Murray Springs site, Arizona, left. Middle specimen is 98 mm long. (Eichenberger casts)

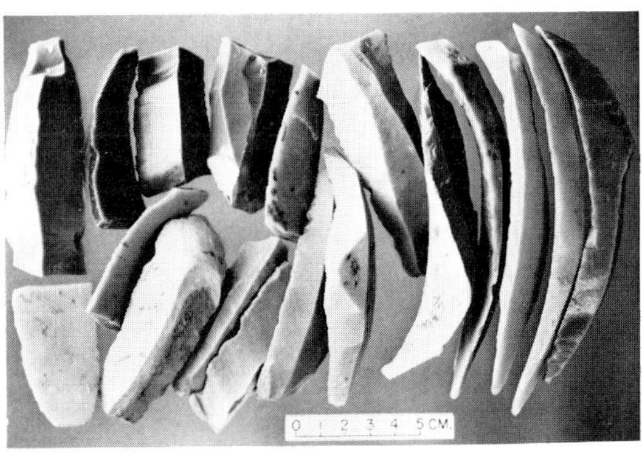

Figure 7. This cache of chert blades was uncovered in the Clovis horizon at Blackwater No. 1 locality by mechanical equipment in 1962 (Green, 1963). (Eichenberger casts)

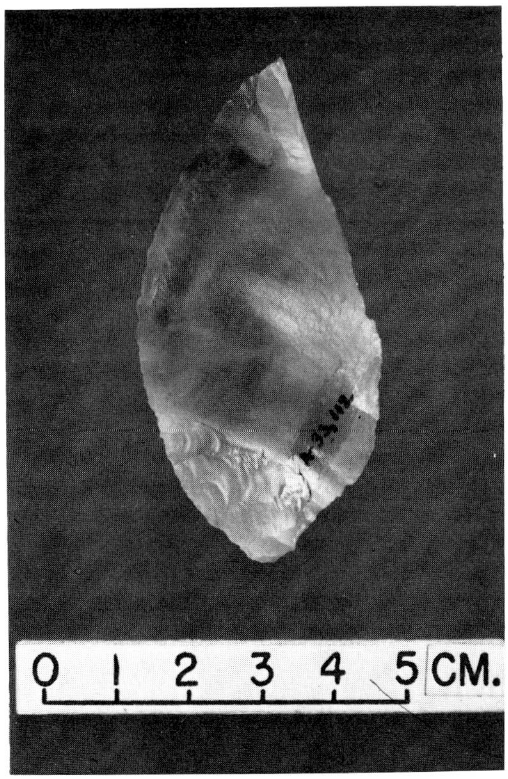

Figure 8. This Clovis combination tool from Murray Springs, Arizona, is made from a large flake with edges retouched for cutting and scraping, and an end made into a burin by special preparation and the removal of three burin spalls. These features as well as the pressure-flake removal of the bulb of percussion from the ventral side are common features on some artifacts of the Old World Late Paleolithic. Length of piece is 67 mm

tends to separate along these contacts. Most of the cylindrical bone or ivory artifacts have beveled bases that are crisscrossed with grooves to roughen the surface, presumably to increase friction at contact with an opposing bevel (Fig. 10). A mastic in conjunction with wrapping and sinew may have completed the scarf joint. Similar artifacts are known from Alaska and Oregon (Cotter, 1962) and are virtually identical to some from Siberia, even to the crisscross grooves (Haynes, 1976).

Clovis subsistence was clearly based on big game, with emphasis on mammoth and bison of forms now extinct. Proboscidian remains are present in every Clovis site where bones are preserved, and many Clovis bone tools are made from mammoth bone and ivory. At kill sites the mammoth carcasses were only partially disarticulated and only a few members removed, indicating only partial utilization. Some elements were placed in piles at the type Clovis site (Blackwater No. 1) (Hester, 1972), Murray Springs (Haynes, 1973), and Colby. Frison (1978) has suggested that the Colby piles may have been winter caches of meat. Most toe bones are conspicuously absent, which raises the question of a cultural cause. Sikes (1971, p. 17) has observed that the trunk and footpad meat of elephants are delicacies to some African elephant hunters. On the other hand toe bones may not preserve well and could be easily scavenged by animals.

Bison carcasses at Murray Springs and Clovis were more dismembered than those of mammoths, indicating more thorough butchering and more complete utilization of meat from bison than from mammoth.

At Murray Springs and Lehner there are indications that horse, camel, bear, tapir, and rabbit may also have been taken by Clovis hunters. The bear and rabbit bones were found in hearth areas at the Lehner site in 1974-75, and enough of them were calcined and charred to indicate roasting by Clovis people. The tapir jaw found in 1955 was also burned and presumably roasted (Haury, et al., 1959). On the other hand water holes and springs are attractive to animals and many die naturally at such places. Their bones get scattered and broken by carnivores and trampled by other large animals, and consequently spirally fractured bones in such locations should not necessarily be attributed to the efforts of man.

How Clovis people procured these animals, especially mammoths, is one of the more intriguing questions faced by archaeologists. There are only

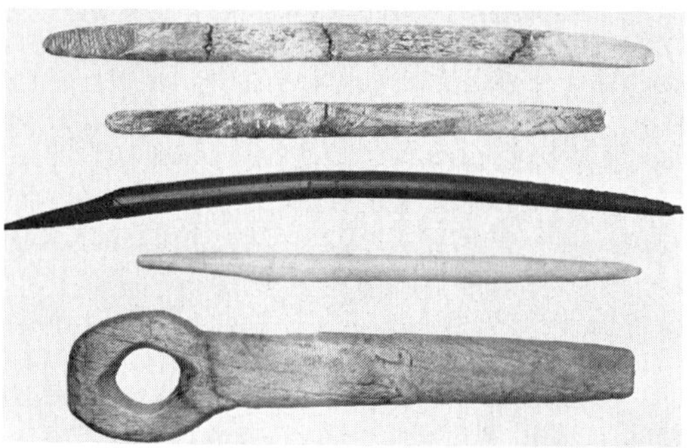

Figure 9. Typical Clovis bone and ivory tools include cylindrical bone foreshafts with beveled ends from the Anzick site (top two) and bevel-base cylindrical points. The black specimen is made from fresh ivory subsequently mineralized by manganese in the Aucilla River, Florida. The bone point below is from the 1936 excavations at the Clovis-type site at Blackwater Draw, New Mexico. The shaft wrench (lowermost) from Murray Springs, Arizona, is made from mammoth leg bone and is 257 mm long. Similar bone and antler tools are common in European collections of Late Paleolithic artifacts. (Eichenberger and Stanford casts)

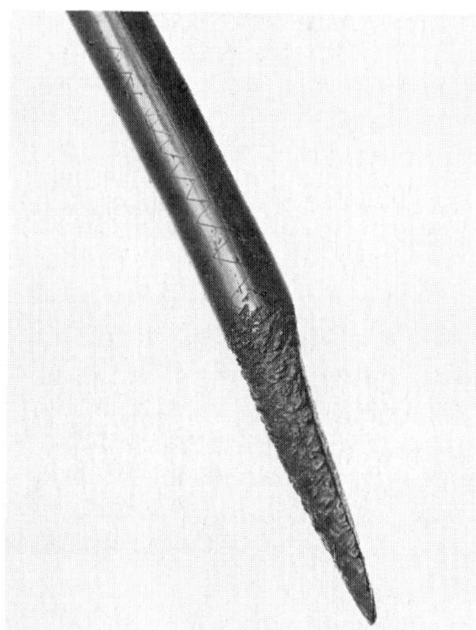

Figure 10. The cylindrical ivory point from the bed of Aucilla River, northern Florida (Fig. 9), bears a zigzag line design on two sides. Similar designs on similar tools are known from the Old World Late Paleolithic. Maximum diameter is 14 mm. (Eichenberger casts)

two known bison-kill sites, and both show evidence that the animals were in low areas with wet ground when they were apparently surrounded and attacked by Clovis hunters (Hemmings, 1970; Hester, 1972). Whether Clovis points were used on spears, lances, or darts propelled with the aid of a throwing stick (*atlatl*) is not known, but impact fractures on several Murray Springs examples from the bison kill (Fig. 3) indicate propulsion with a force upon impact at least as great as that applied by a knapper in fluting or striking off a moderate-sized thinning flake, and this after presumably penetrating several centimeters of hide and flesh. In the collection of Clovis points from the bison-kill area several were snapped in half and some had the basal corner or "ears" broken off. Proper understanding of Clovis-point damage will help us understand their function.

On the other hand, Clovis points from mammoth-kill sites show relatively little damage, although many appear to have been resharpened. Tips broken by lateral snapping represent the most common form of breakage. Do these observations imply different techniques for penetrating mammoth versus bison with Clovis points? Replication experiments are much needed in this regard, but judging from observations on the thickness and toughness of modern elephant hide (Huckell, 1979) it is questionable whether spears, even with the aid of an *atlatl*, could penetrate 2 cm or more of fresh hide and still reach deep enough to affect the heart. A neck shot severing the spinal nerve might kill, but the hunter would have to be standing on ground higher than the animal's neck.

Some of the bone and ivory points from Florida springs and rivers have almost needle-sharp tips and possibly could penetrate mammoth hide more readily. Two cylindrical bone points from the Blackwater No. 1 Clovis site have relatively blunt points by comparison, but this may be partly a result of post-depositional corrosion.

It has been assumed that mammoths were attacked one at a time by Clovis hunters, but recently Saunders (1977) questioned this. He found that the Lehner site assemblage is not one marked by specific age (i.e., young, tender, and unwary), but represents instead the normal age distribution of a family unit when compared to African elephants. On this basis he proposes that the Lehner assemblage was a single family of 13 mammoths from two to 30 years of age systematically killed at one time by Clovis hunters. If this were true one might further speculate that the single mammoths associated with Clovis points at Naco, Escapule, and Murray Springs sites are ones that got away from the Lehner slaughter. On the other hand, the simultaneous slaughter of 13 seemingly healthy mammoths by even a coalition of Clovis bands is difficult to imagine without more formidable weapons than shafts tipped with Clovis points of stone, bone, or ivory.

Mammoths of the Pleistocene were more like the living Indian elephant (*E. maximus*) than the African species (*Loxodonta africana*), but there is no way of knowing whether the behavior of mammoths hunted by Clovis people was similar to the defensive behavior of modern elephants. In the absence of a time machine, the observations on African elephants by the Douglas-Hamiltons (1975) and on Asian elephants by Olivier (this volume) offer the best information available. The Douglas-Hamiltons find that, aside from elephants being defensive and protective of one other, the only generalization that can be made about elephant behavior is that it cannot be generalized. Elephants are highly individualistic; and they are led by matriarchs who very much determine the temperament of their family unit. The attitudes of elephant groups range from timid and wary to threatening when approached by man. Individual variations in behavior are similar, and commonly reach extremes in which some individuals are belligerent and attack in earnest without warning. Some individual elephants when in trouble or wounded elicit great concern on the part of other elephants, who will rally to their aid even at the risk of death. Other individuals become practically outcasts and do not receive the concern that others might. Some elephants become trusting and approachable after repeated exposure to man in situations where no danger arises, and young bulls engaged in sham battles are commonly unaware of being approached by humans. However, the threat behavior of an individual—and just how far it will go in charging a human—is consistent and therefore predictable.

Marks (this volume) points out that the Bisa and other hunters in Africa "know" individual animals and can predict their behavior. The Bisa are sedentary whereas Clovis hunters most likely were far-ranging, but even so Clovis men probably could have determined which individual mammoths were the most vulnerable within a few hours or days of observation. These traits would have made the successful hunting of mammoths dependent upon keen observation of individuals in the family unit. Selection by Clovis hunters would appear to have been based on the behavior and not the age of the elephants. Because some behavior is independent of age, an accumulation of individual mammoth kills might appear to have the age distribution of a family unit.

Most of the mammoth kills took place on low ground at springs, along a creek, or in a pond, suggesting that the animals were ambushed at watering places. There is no evidence of mammoths actually being stuck, but the wet ground may have restricted their freedom of movement. It appears that the animals were observed at water holes until a suitable individual presented an opportune target. Primitive techniques, such as disembowelment or hamstringing, that work well with keen-edged steel probably would not work with stone, because a sawing action would be needed to cut through the hide. All indications are that the prey was dropped pretty much where attacked. Upon being attacked near a watering place it is likely that a mammoth would head away from deeper water. On the other hand, a wounded and dying animal might head for water. The use of poison in conjunction with Clovis points is a possibility suggested by the late Louis S.B. Leakey (personal communication), but it is impossible to evaluate at present.

I can imagine a hunter taking advantage of a mammoth's threat-charge. For instance, he could have used a long slender lodgepole set against a stop and tipped with a Clovis and raised point at just the right moment to receive the full force of the advancing pachyderm. In any case, I feel certain that in order to successfully hunt mammoths Clovis people studied and took advantage of the predictable behavior of individuals. The same probably held for bison, but for small groups instead of individuals.

There is no evidence of the use of traps, deadfalls or drives by Clovis hunters (Frison, 1978), and if *abatis* or barriers were used, as they may have been at the Murray Springs bison kill (Hemmings, 1970), they have long since decayed away.

Another Clovis characteristic alluded to earlier was their exploitation, for knapping, of flint of the highest quality and of considerable aesthetic beauty. They knew of and made use of many of the major aboriginal chert quarries known in the conterminous United States. Many collections reveal that they used multiple local sources within a few tens of kilometers of each other as well as material from as far as 300 km away. This figure of 300 km for the maximum distance from lithic sources turns up repeatedly in the literature. Considering the relatively brief time span of their culture and the lack of evidence of previous inhabitants, Clovis men must have been skilled prospectors. This should not be surprising considering that flint was probably the most important raw material to their economy and way of life. Procurement of other raw materials such as wood, bone, ivory, hide, and sinew was to no small degree dependent on an adequate supply of flint The discovery of a new and dependable source must have been cause for great joy, and all the more so if it had attractive colors or patterns. Such a find could open up new territories and extend their range of foraging. It could also provide new material for barter or trade. In this regard, an alternative explanation of the Simon site bifaces and Clovis points is that they may have been either a cache of materials to be traded or a tradesman's set of samples. Many of the ridges between flake scars are worn as if they had rubbed against each other during transport, presumably in a leather bag.

In a re-evaluation of the Shoop site, Cox (n.d.) points out the high degree of utilization (to the point of near exhaustion) of the flint tools and the fact that most of the raw material came from nearly 300 km away. This suggests that a local source of flint had not yet been located by the Shoop band, because closer sources were used by later Indians.

How chert or chalcedony was mined in Clovis time is unknown, but presumably techniques were similar to those of later Amerinds who used large stone mauls on outcrops and dug pits when surface exposures were depleted. It is doubtful that fire was ever used to deliberately fracture flint because heating flint enough to fracture renders it useless. On the other hand, there is evidence that some cherts were gently heat-treated to improve their knapping qualities (Purdy and Brooks, 1971; Fitting, et al., 1966).

Clovis sites are relatively small compared to later Folsom and Plano sites. Murray Springs has three principal activity areas: a mammoth-kill area, bison-kill area, and an associated hunting camp. Tools and points from all three areas total only 36 pieces. The mammoth-kill area of 300 m^2 contained 6 tools (including 3 points) and over 10,000 flakes in clusters indicating tool sharpening and biface thinning, resulting presumably from butchering. The bison kill area of 450 m^2 contained 10 points and 7 tools, plus about 2500 thinning and sharpening flakes. The camp area 40 to 100 m away covers about 200 m^2, and contains 6 broken projectile points, 1 resharpened point, 7 flake tools, 6 blade tools, 2 bifaces, and hundreds of thinning and sharpening flakes. The camp's relation to the bison kill is unequivocably displayed by an impact flake, found among bison bones, that matches a damaged Clovis point from the camp, and by the reverse situation as well (Fig. 3). Part of a mammoth long bone and a mammoth tooth from the camp area also indicate ties with the mammoth kill. On the basis of the evidence of 11 bison killed and estimates of the potential meat provided thereby, Hemmings (1970) proposes a band size of 50 to 100 people. This seems large, and judging from the small size of the camp area, I consider it a maximum. The Naco mammoth had eight Clovis points in its skeleton, four times more than have been found with any other mammoth. This suggests that it may have got away from its hunters only to die shortly thereafter. If it is assumed that each point belonged to a separate hunter, at least eight men had attacked. If there were two points per hunter, four would have been involved. The actual number probably falls between these figures and may represent one-fifth of a band population of 20 to 40 men and children.

The Folsom culture, which followed Clovis and obviously developed from it, specialized in hunting bison, the mammoth having become extinct or very nearly so (Owl Cave, Idaho, may be a Folsom-Mammoth association–Dort and Miller, 1977). Some Folsom sites are approximately the same size as Clovis sites, but others are much larger and apparently represent repeated rendezvous of several bands (Wilmsen, 1974). Bull Brook, Debert, Holcombe, Wapanucket, and Reagan are eastern sites of fluted point makers contemporary with Folsom. After Clovis time the increase in site size and number of sites indicates a significant increase in both population and cultural diversity. Extrapolating back in time we see few, if any, people south of the Mackenzie Valley in Canada before Clovis time.

To summarize: Clovis people appear in general to have been foragers, mainly hunting mammoths and bison as both food and material resrouces. A typical tool kit that might be found among a band of Clovis people would contain, in addition to the diagnostic projectile points, knives on bifacial thinning flakes (some with graver or borer tips); end scrapers (some with spurs) on both flakes and blades; side scrapers on large flakes and blades; notched blades; well-made bifacial preforms of various sizes that could serve as choppers; and a variety of bone and ivory tools of which only two forms, bevel-based cylindrical objects and a shaft straightener have been preserved in Clovis sites. Their big-game take was commonly a single mammoth which was only partially butchered and utilized, or a small group of bison which were dismembered and apparently more efficiently utilized than mammoth kills.

The small amount of available data indicates that the basic Clovis tool kit was the same, except for minor local components, whether in the eastern or western parts of the continent and regardless of environment. This and the wide-ranging flint sources confirm a high degree of mobility and lack of dependence on the resources of a restricted environment. Clovis people appear to have wandered over extensive areas looking for lithic sources and exploiting the megafauna, usually at watering places. The degree to which vegetation was utilized in their economy is unknown due to lack of preservation, but presumably it was great. Clovis campsites were small and located only a few tens of meters away from their game takes. Hearths were shallow basins less than 3 m in maximum dimension, no deeper than 20 cm, and without stones. Similar depressions without charcoal are common at Murray Springs. At kill sites, bifaces were used for butchering and were sharpened or reshaped on the spot, producing distinct piles or clusters of debitage. Hunting camps adjacent to kills contain end scrapers and point bases, indicating wood, bone, and hide-working and projectile repairs. Reverence for the dead and a belief in the hereafter is suggested by the magnificent grave goods and red ocher at the Anzick site.

MAMMOTH HUNTING SITES OF THE OLD WORLD

From Klein's (1969) review of the Late Paleolithic of the Kostienki-Borshevo region it is apparent that some mammoth-hunting sites of the Don Valley bear resemblances to Clovis sites and to Clovis artifacts as well, but there are considerable differences as might be expected, considering that the areas are 14,000 km apart. Sites in humus layers of the T-2 terrace of the Don River are relatively small compared to those in younger deposits. Many of the artifacts there reveal a tool technology comparable to Clovis, including bifacial projectile points, end scrapers, side scrapers, borers, and a few blades associated with many bones, including remains of *Mammuthus primigenius* (Müller-Beck, 1967). Hearths in the T-2 sites are commonly of similar size and configuration to the Murray Springs examples, and burned bone is present.

The age of these earlier occupations of the second terrace of the Don is uncertain, but 20,000 to 15,000 BP is a reasonable limit from the available geochronology and clearly confirms a Sartan age (Klein, 1969).

In eastern Europe, the mammoth-hunter camps of Pavlov and Dolni Vestonice, 26,000–24,000 radiocarbon years old, reveal a culture oriented toward the utilization of mammoths for everything from food to housing made out of mammoth bones (Klima, 1963).

The use of bone and ivory for tools and art work shows these materials to be equal in importance to the stone used to make blade, flake and bifacial tools. Red ocher is prevalent and many bone tools are similar to the Clovis beveled-base bone points, shaft straighteners, and a cylindrical ivory piece about 5 cm in diameter and 10 cm long found at the Blackwater Draw Clovis site in 1962.

Of even more interest in regard to Old World potential sources of New World cultures are the Late Paleolithic sites of Siberia (Klein, 1971). It is again apparent that there are no direct cultural ties with the Clovis sites. At Tomsk on the upper Ob River, end scrapers, flakes, blades, and prismatic cores with traces of fire were found in association with a single carcass of *Mammuthus primigenius* in terrace deposits of Sartan age. It may be significant that a microblade technology was not represented. Atchinsk is another man-mammoth site on the Ob River.

Farther east, on the middle Yenisei, are several mammoth-hunting sites, the most important of which appear to be Afontova Gora II and Kokorevo II in deposits of the second and third terraces. Horizon C-3 at Afontova Gora II produced bones of six mammoths as well as remains of reindeer, arctic fox, and arctic hare, which were even more abundant. These were associated with fire areas and numerous artifacts including bifaces, side scrapers, end scrapers, flake points, notches, backed blades, burins, borers, and large numbers of retouched flakes and blades. Bone artifacts included cylindrical projectile points (some with lateral grooves), polishers, awls, needles, ivory spheres, and antler shaft straighteners.

Associated with bones of *Mammuthus primigenius* at Kokorevo II were borers, end scrapers, flake points, wedge-shaped cores, plano-convex biface scrapers, bone projectile points, polishers, awls, needles, and ornaments. At both Afontova Gora II (C-3) and Kokorevo II flake tools are predominant. The grooved-bone points are characteristic and imply the use of microblade inserts.

The famous sites of Mal'ta and Buret I on the upper Angara River contained evidence of slab-lined structures, cache pits, a grave with red ocher and ivory ornaments, and a variety of flake tools and some blades. Long cylindrical bone projectile points with beveled bases are represented, and many art objects carved from bone or ivory were found with a mammoth at Mal'ta, level 2. Neither microblades and wedge-shaped cores nor slotted bone points were present at either Mal'ta or Buret I.

The recent discoveries at Dyuktai Cave and in related terraces of the Aldan River are of particular interest because of the occurrence in the lower levels of bifacial projectile points as well as oval and triangular knives; discoidal, Levallois, and wedge-shaped cores; multifaceted burins, large side scrapers, small end scrapers on blades, and numerous retouched flakes, all associated with many large animal bones including *Mammuthus primigenius*. However, the abundance of small blades and wedge-shaped cores in the collections on display in 1973 at the Bering land bridge conference in Khabarovsk is so unlike any known Paleoindian mammoth-hunting assemblage that I see no obvious relationship.

The Dyuktai Culture as defined by Mochanov (1976) lasted from 30,000 to 11,000 years ago and has left wedge-shaped cores and microblades as essential artifacts. On the other hand, the tool assemblages from Mal'ta, Buret I, and Tomsk are without microblades and wedge-shaped cores and represent a flake-using (as opposed to microblade-using) culture or cultures in part contemporaneous with Dyuktai.

Notably absent from stratified Clovis sites are any forms displaying art work, but red ocher is known from the Anzick site mentioned earlier, where Clovis points, large bifaces, and beveled-base bone foreshafts coated with red ocher were found concentrated in a small area and possibly in association with two child skeletons (Taylor, 1969, and personal communication; Lahren and Bonnichsen, 1971). The zigzag design on the Florida beveled-base ivory point also mentioned earlier may be the earliest example of such art known from the New World.

From this brief and cursory examination the most significant fact to emerge is that the Late Paleolithic assemblages at mammoth-bearing sites in Siberia represent at least two different traditions. One is a predominantly flake industry with large blades, although the presence of slotted bone points implies at least some use of microblades. The other is characterized by wedge-shaped cores and microblades. Bifacial foliate forms of various sizes occur in both. Cylindrical bone points found at some sites commonly differ from New World forms in not having beveled bases and in being slotted along two sides, presumably to accommodate microblades. Specimens from Mal'ta are an outstanding exception because those on display at the museum in Irkutsk show a remarkable resemblance to the beveled-based bone points from Clovis sites. Shaft straighteners from Afontova Gora II are made from antlers, in the central European fashion. The Murray Springs specimen, which more closely resembles those from Molodava and Pekarna Cave in Europe, is made from the wall of a mammoth long bone (Haynes and Hemmings, 1968).

Clovis artifact assemblages more closely resemble those from the Old World Late Paleolithic big-game hunting sites (Klein, 1969) than do possible pre-Clovis assemblages in the New World, such as Pikimachay Cave in Peru and the El Horno site in Mexico (MacNeish, 1978). Similarities include beveled-base bone points (Mal'ta), the use of grave goods with red ocher (Mal'ta), bifacial basally thinned projectile points (Kostienki), end scrapers (Kostienki), blades from prismatic cores (many sites), bone polishers (Afontova Gora), flake knives, and hearths in shallow depressions (Klein, 1971). To me it is unlikely that this assemblage of traits was derived in the New World independently of the Late Paleolithic of the Old World.

The sites of Afontova Gora II, Mal'ta, and Buret I suggest that some Late Paleolithic hunters had moved as far east as central Siberia as early as 20,000 years ago and overlapped there the microblade-dominated Dyuktai culture which may have reached eastern Siberia at about the same time from places farther southeast. Müller-Beck (this volume) concludes that mammoth-hunting cultures originally derived from the west could have reached eastern Siberia as early as 30,000 years ago, but he questions the existence there of the Dyuktai culture with a microblade technology at such an early date. In any case, the occurrence of an eastern-European-type mammoth-hunting culture (Clovis) in central North America 12,000 to 11,000 years ago is a reasonable basis for the hypothesis that Clovis progenitors from Siberia had reached eastern Beringia 14,000 to 12,000 years ago if not before.

THE BERINGIAN CONNECTION

If the Clovis culture represents a late Pleistocene infusion into America from Siberia, where is evidence of Clovis progenitors in Alaska and Yukon Territory, and how old should it be? In Siberia and far-eastern U.S.S.R., Mochanov (1978a) has defined the Dyuktai culture as a technology of wedge-shaped cores, microblades, and bifaces used by people hunting Pleistocene megafauna at least as early as 18,000 years ago. By 10,000 BP, a clearly related culture is in Alaska, where the Denali (West, 1976), Akmak (Anderson, 1970), and Gallagher (Dixon, 1975) assemblages are obviously derived from Dyuktai (if they are not, in fact, a part of it), but none bear much resemblance to Clovis assemblages. All that can be said with assurance about the maximum age of the arctic core-and-blade tradition in Alaska is that 10,690 ± 250 (SI-1561) seems to be a valid date for the core-and-blade component represented by Horizon

II at Dry Creek and that it is earlier than the date of 9857 ± 155 BP (K-1583) at the Onion Portage site (Hamilton, 1970). In the lowest cultural level of the Dry Creek site in central Alaska, Powers and Hamilton (1978) have found triangular bifacial points, Paleoindian-style scrapers, and flake tools in a level dated 11,120 ± 85 BP (SI-2880)(Thorson and Hamilton, 1977). Microblades are absent, whereas the next-youngest level contains a microblade and core assemblage. In Siberia, as we have seen, there are sites without microblades where flake tools, bifaces, and, in some cases, cylindrical bone points occur in association with bones of Pleistocene megafauna. Mochanov (1978b) recognizes the existence there of another cultural tradition, "Mal'ta-Afontova", which is without microblades and more like the early Late Paleolithic of eastern Europe. It is possible, therefore, that two distinct traditions also coexisted in Beringia before 11,000 BP. If Clovis was derived from one of these it most likely would be the one without microblades. Aside from the lack of compelling evidence, a problem with this hypothesis is the question it raises of how the "Mal'ta-Afontova" tradition could get to Beringia through eastern Siberia without being transformed by contact with Dyuktai. Some cultural mixing and transformation would seem likely and may be manifest in the occurrence of bone points slotted for microblades seen in the collections from Afontova Gora II and Kokorevo; but this does not explain the absence of microblades in Clovis assemblages.

The bone artifacts from the Old Crow Flats, if indeed they are all artifacts (Morlan, this volume), are pre-Clovis, but may be unrelated to Clovis. Until stratified occupation sites are found at Old Crow any relation to the problem of Clovis origin is equivocal at best.

Where the technology of fluting bifacial projectile points originated is another problematic question. If the combination of disappearance of steppe-tundra vegetation and human predation caused the extinction of the arctic-steppe megafauna by 12,000 BP, as is suggested by Morlan (1977a), the development of fluting could have taken place in Alaska or along an ice-free corridor sometime between 14,000 and 12,000 years ago when the Bering Strait had isolated North America from Siberia. Even though the strait may not have been a significant barrier (Hopkins, 1978), fluted points did not cross it in either direction. On the other hand, it has been argued that fluted points found in Alaska are post-Clovis and represent influx from the south (Bryan, 1969; Dixon, 1975). Typologically most of the Alaskan forms are small and have multiple fluting scars (Clark, 1978). Some closely resemble fluted points from the Holcombe, Michigan, site (Fitting, et al., 1966), and the Great Basin area (Davis, 1973; Touhy, 1968). There is no positive dating for this type, but the artifacts at the Holcombe site are believed to be the latest form of fluted points and of the same age as the second stage of Lake Algonquin, which existed between 11,000 and 10,500 BP (Karrow, et al., 1975). Indisputable Clovis points also occur in the Great Basin as surface finds, but are rarer and possibly older than the Holcombe type. On the other hand, two point bases from the Putu site in the Brooks Range of Alaska are similar and probably related to a charcoal date of 11,470 ± 500 BP (SI-2382) (Alexander, 1974, and personal communication). If this association is valid it makes at least these two Alaskan fluted points as old as the oldest fluted points (Clovis) from interior North America.

A different type of fluted point found in Alberta is sometimes called a triangular fluted point. There are four of these in the museum at the University of Alberta, of which three are from the Birch Hills area 320 km northwest of Edmonton (Bonnichsen, personal communication). Unfortunately, all are from the surface, so their age is uncertain, but they do have technological similarities to triangular points from Kostienki and the Dry Creek site.

CONCLUSIONS

At our present state of knowledge there are too few well-dated, stratified archaeological sites in Beringia to allow convincing conclusions to be drawn about (1) the earliest peoples to enter the New World from northeast Asia, (2) how many separate movements there may have been, and (3) how any of these movements were those of Clovis progenitors. Therefore, the hypothetical model of Clovis origin I will develop here is based on assumptions that I believe best fit the meager facts we have to go on, but there is so little evidence that my interpretations are little more than speculation.

Many of the Late Paleolithic mammoth-hunter

sites of Eurasia show similarities in artifacts and features such as dwellings, fireplaces, and ocher burials. While game takes were not exclusively mammoths, the socioeconomic orientation of the cultures toward the hunting and utilization of mammoth meat, hide, bones, and ivory is clearly manifest. The mammoth was to the Paleolithic hunter as the reindeer is to the Laplander or the inland Eskimo. On the basis of the similarities among Clovis bone and stone artifact assemblages and such Late Paleolithic hunting camps as Pavlov, Dolni Vestonice, level 5 of Kostienki I, and many others, I assume that the earliest root for Clovis stems from Europe during the early Late Paleolithic (Gravettian-Aurignacian), about 28,000 years ago. Mammoth hunters moving eastward gave rise to such sites as Afontova Gora II, Mal'ta and Buret I sometime between 25,000 and 15,000 BP. During this time cultural diversity had already developed in eastern Siberia, where Dyuktai people had acquired a core-and-blade technology. Bifacial points or knives at Dyuktai Cave and bone points slotted for microblades at Afontova Gora may be evidence of cultural mixing, but some people apparently did not adopt microblades. It may be these Mal'ta-Afontova descendants that entered Beringia with artifacts characterized by bifaces, blades, and cylindrical bone and ivory points between 20,000 and 15,000 years ago.

One of the major contributions that the Wenner-Gren symposium made to my thinking was the recognition that much of northern Eurasia during the last glaciation remained unglaciated and became covered with a loess (aeolian) substrate (Tomirdiaro, this volume) that supported a steppe-tundra mosaic (Giterman, Sher, and Matthews; Matthews; Young, this volume), stretching from Europe to Alaska under a cold dry, continental climate. Riparian habitats within this mosaic (Schweger, Young, Ager, this volume) may have provided the nutritional diversity favoring megafauna in general and *M. primigenius* in particular (Olivier, this volume). These conditions, in conjunction with the mobility of certain elements of the megafauna, would have led Late Paleolithic hunters with a technology adapted to the cold climate (Jelinek, 1965) farther eastward as their hunting success and mammoth-oriented economy led to more frequent band nucleation. Relatively rapid expansion eastward is indicated by radiocarbon dates of about 22,000 BP for Afontova Gora II, and Mal'ta—farther east—is believed to be approximately of the same age (Tseitlin, 1979).

Beringia reached its maximum extent as sea level approached its lowest level 20,000 to 15,000 years ago (Hopkins, personal communication). Steppe-tundra vegetation once again expanded eastward, and undoubtedly Late Paleolithic hunters followed it. Two unslotted bone points from the muck deposits of Goldstream near Fairbanks, Alaska (Rainey, 1939), may have belonged to Clovis progenitors.

By 15,000 BP inundation of the Bering Strait was underway as rapid melting of continental glaciers occurred. By 14,000 BP Alaska was separated from Siberia. Exactly when *M. primigenius* became extinct in Alaska is unknown, but soft parts have been dated as late as 15,380 ± 300 BP (SI-453). It is accompanied by a comment that an artifact may have been in association, but no further word has appeared (Stuckenrath and Mielke, 1970, p. 203). A skeletal part is dated as 13,500 ± 100 BP (QC1365) (Matthews, this volume, Fig. 2). This is the latest date for mammoth that I have been able to find in the literature.

After 15,000 BP deglaciation was rapid, and there appears to have been little in the way of glacial ice to obstruct the passage of man and megafauna through Canada between the western edge of Laurentide ice and the Cordillera (Rutter, 1978). So by the time megafauna were becoming extinct in Alaska perhaps 13,000 years ago, Clovis progenitors may have been able to continue finding woolly mammoths southward until about 1,000 years or so later when their descendants had reached the Canadian prairie plains and had encountered new species of megafauna, including two species of mammoths (*M. columbi* and *M. jeffersoni*). By 11,500 BP the distinctive Clovis point had been developed, and less than 1,000 years later, Clovis people had spread from Edmonton, Alberta, to Guadalajara, Mexico, and to the Atlantic and Pacific coasts. By 11,000 BP all elements of the megafauna except bison were gone, and the Clovis culture with it. In less time than the precision of radiocarbon dating can resolve, the Clovis culture had changed to Folsom, Midland, Plainview, and other cultures of regional distribution. This transition led southward through Central America to produce stemmed fluted points which may be as old as 11,000 BP at Tierra del Fuego, at the southern tip of South America (Bird, 1969).

The suggestion that man could walk through Canada 12,000 years ago and have descendants reach Tierra del Fuego a millennium later is unacceptable to some, but a thousand years can yield about 40 generations. I find no difficulty with such a rate of movement, especially considering the character of the people and the rapidly changing ecological conditions. As a band or two would move southward, successful hunting would promote population growth. Since there were few, if any, competing cultures, new bands would form and move into new areas. Furthermore, the exploration of new territory may have been just as much a cultural drive then as it has been in historical time, especially in the absence or near absence of other peoples. Each new valley would have been a potential source for new flint quarries, new family groups of mammoths, new plant resources, and new watering places. Within a year, a newly arrived band could be familiar both with the environment and the behavior of individual mammoths in several family units. In a few more years, they may have cropped most of the easily taken individuals and dropped that valley from their seasonal round in favor of another.

As the transition to Holocene climate progressed water tables fell, many streams became ephemeral, and reliable watering places were fewer. Megafauna under stress become more vulnerable to hunting. Although large mammoths would have become scarce, neither the scope nor the scale of the impending extinction would have been obvious. Longer forays could result in larger areas being covered. The southward movement would have accelerated through the narrows of Central America and along the west coast of South America until Tierra del Fuego was reached about 11,000 years ago. This scenario differs from Mosimann and P.S. Martin's (1975), in that theirs postulates a low human population behind their hypothetical wave front. On the contrary, in central North America I see a continuity in which the Clovis culture is transformed into Folsom, Midland, Plainview, and other Paleoindian cultures that maintain a bison-hunting economy once the mammoths are gone. The continued success and growing populations of those successor cultures are manifest in greater numbers of sites, larger sites, and in the greater cultural diversity that continued into the historic period (Haynes, 1967).

The spread and development of the Clovis culture throughout the United States, southern Canada, and northern Mexico took place during the period of greatest environmental change since the end of the Sangamon interglacial. Between 12,000 and 10,000 years ago there were major changes in global climate, local and regional vegetation, and erosional-depositional processes and hydrology, and many forms of Pleistocene animals became extinct. Whether climatic change or man was the prime cause of extinction is ardently debated.

Guthrie (this volume) recounts morphological evidence that the late Pleistocene megafaunas were under significant stress as the Pleistocene ended, but that megafauna had survived other periods of climatic stress during previous interglacials. Furthermore, man was present in the Old World during those earlier interglacials. Such conditions of stress have been, in fact, likely catalysts for natural selection and evolutionary change, but one significantly different factor at the end of the Pleistocene was that men were now highly skilled specialists in hunting large mammals. Theirs were hunting societies with skills and technologies adapted not just to surviving but to providing enough resources to allow time for religion, ceremony, and artistic endeavors that included aesthetic as well as functional forms. I agree with Jelinek's (1967) view that the presence of skilled Late Paleolithic hunters (superpredators as P.S. Martin called them in 1967) at a time of interglacial environmental stress provided the fatal combination that led to large-scale megafaunal extinction. The rock art of the Old World, depicting various elements of the megafauna, may have been Paleolithic man's appeal to the Gods to save his prey from extinction.

ACKNOWLEDGMENTS

My research on the geochronology of alluvial stratigraphy and Paleoindian sites has been generously supported by the National Science Foundation, the National Geographic Society, and the University of Arizona. This work has benefited greatly from discussion with colleagues of this symposium as well as others too numerous to mention individually, but I alone take full responsibility for its errors and defects.

23

THE PATTERN AND MEANING OF HOLARCTIC MAMMOTH EXTINCTION

Paul S. Martin

It is unlikely that climatic change at the time of the last glacial was the primary cause of mammoth and other megafaunal extinctions. The mammoth fauna had survived similar periods of climatic transformation. Moreover in western Europe mammoth extinction apparently preceded the disappearance of steppe-tundra, the mammoth's supposed habitat.

The global pattern of all extinctions of large land mammals appears to follow Paleolithic man's footsteps. The loss of many large herbivores in North and South America, apparently suddenly, between 12,000 and 10,000 years ago is consistent with this pattern. The presence of man, and of a hunting culture, in America during this period is well accepted by archaeologists. Computer modeling of the needs of a hunter population growing at the rate of only 1% per year and advancing southward from the Bering land bridge suggests that mammoth extinction from human predation alone could have taken place in less than 500 years.

Key words: Mammoth, extinction of; Pleistocene hunter populations; Bering land bridge

Climatic change and human impact are two theories commonly appealed to in the search for the cause of late Pleistocene mammoth and other megafaunal extinctions. The former broad range of the mammoths, extending throughout nearly all parts of the Holarctic region, the presence of several species within the genus, the quantity and quality of the fossils, the abundant association with Paleolithic archaeology in certain parts of Eurasia, and the sparse but unmistakable association with big game hunters in the New World make mammoths especially interesting to paleoecologists.

Humanity's efforts to account for bones of extinct mammoths must have given the gods many a laugh (Farrand, 1961, p. 729, citing Digby). Farrand aimed his remarks at catastrophists who would interpret the famous frozen mammoths of the Arctic as proof of a sudden widespread climatic crisis, preferably one generated by "worlds in collision" or some similar cosmic extravaganza. Data assembled by Farrand showed that the abundant Siberian mammoth remains, including the occasional record of a frozen mammoth, were of varied ages. Thus, the fossil record of frozen mammoths could not be combined and treated as evidence for a single instantaneous "quick freeze."

Pointing out that woolly mammoths were well adapted to extreme cold and that they had inhabited Arctic latitudes in pre-Wisconsin as well as in Wisconsin time, Farrand discounted climatic change altogether as a likely cause for their extinction. Beyond noting that the problem "... concerns other species of mammoth and many other large Pleistocene mammals as well," he left the matter unresolved.

It seems to me that Farrand's basis for discounting climatic change remains valid. Although climates were indeed changing during the late

ISBN 0-12-355860-3

glacial and early Holocene when mammoths were declining or already extinct, the Ice Age climate had changed many times before. Whether or not mammoths thrived especially well in some grassy, productive steppe-tundra during cold dry times of the last glaciation, they had also at least endured the last interstadial and various earlier interglacials. Advocates of climatic change as the cause of megafaunal extinction must demonstrate, among other things, that earlier interglacials were different from the Holocene.

Another difficulty in the climatic-cause theory presents itself. Mammoths occupied a broad range of habitats. Various species in the group ranged from arctic to tropical latitudes. Thus, the most plausible bioclimatic model for mammoth extinction in any one local region must still be provincial. A general theory that accommodates the global pattern of late Pleistocene extinction seems to require some approach other than climatic.

The global pattern of extinctions of all large land mammals appears to follow Paleolithic man's footsteps (Fig. 1). Africa and southern parts of Asia were affected first with losses among the suids, primates, equids, and giraffids at the end of the Acheulean, perhaps 200,000 years ago. Australia followed with the extinction between 40,000 and 20,000 years ago of twelve genera of giant marsupials and size reduction among the surviving groups of large marsupials. Between 20,000 and 10,000 years ago Europe and northern Asia were affected, to a relatively slight degree. North and South America were stripped of large herbivores, apparently suddenly, between 12,000 and 10,000 years ago, with only extinct species of *Bison* persisting into the Holocene. Intensity of extinction appears to have been heavier on the American and Australian continents than in Africa and Asia. Extinctions continued into the Holocene on oceanic islands, affecting birds and mammals of a smaller size than those lost on the continents.

Impressed with the sizable number of genera of large animals that vanished from the continents in late prehistoric time, I first explored implications of a model, called overkill, attributing extinctions to human activity (P.S. Martin, 1967). A second model, in which high rates of human population growth and high faunal kill rates simulate maximum impact, may be termed "blitzkrieg" (see P.S. Martin, 1973, 1974; Mosimann and P.S. Martin, 1975).

How could humans have exterminated the mammoth, specifically? Budyko (1967) approached the problem in a model based on differential equations and the assumption of a very slow growth rate for the hunting population. Mosimann and P. S. Martin (1975) modeled mammoth extinction in the New World using a higher population growth rate of 1% annually. Difference equations were generated to accommodate a computer simulation.

An average mammoth biomass of 5000 lbs. per sq. mi. capable of replacing numbers lost to human predation at a maximum annual rate of 5% is suddenly exposed to highly skilled hunters arriving in the vicinity of Edmonton, Alberta, from the north (Fig. 2). The hunters do not distribute themselves randomly throughout the 3 million sq. mi. of mammoth (and mastodon) range to the south; rather, they advance along a front that represents the expanding periphery of their range. The hunters double in numbers every 70 years. They destroy an average of 7000 lbs. of mammoth biomass per person per year or the equivalent of 6 mammoths per hunter. In 532 years a total population of 90,000 people will have destroyed a mammoth-mastodon herd of 3.4 million animals. The front of hunters invading the continent will have spread from southern Wyoming to southern Arizona in 120 years. Increasing the human growth rate to 3% would reduce time between arrival and extinction from 500 to less than 300 years.

This is only one of a variety of simulations that generate essentially the same result: the loss of large numbers of large mammals within a few years in any one region, making discovery of fossil evidence of the catastrophe improbable. A key point in the chronology falls around 11,000 years ago in southern Arizona, where mammoth-hunter overlap has been dated on the basis of detailed study by Haynes (1970, this volume). The blitzkrieg model of overkill can be applied to all extinct mammals of the New World that disappeared with the entry of Paleolithic hunters.

A major controversy exists regarding human origins in the New World, with many authors accepting ages for human invasion long before 12,000 years ago. The blitzkrieg model does not require absence of *any* people or even *any* hunters prior to the time of extinction, which appears to have occurred after 12,000 years ago. MacNeish (1976) and Bryan (1978) summarized what they regard as overwhelming evidence for pre-Clovis

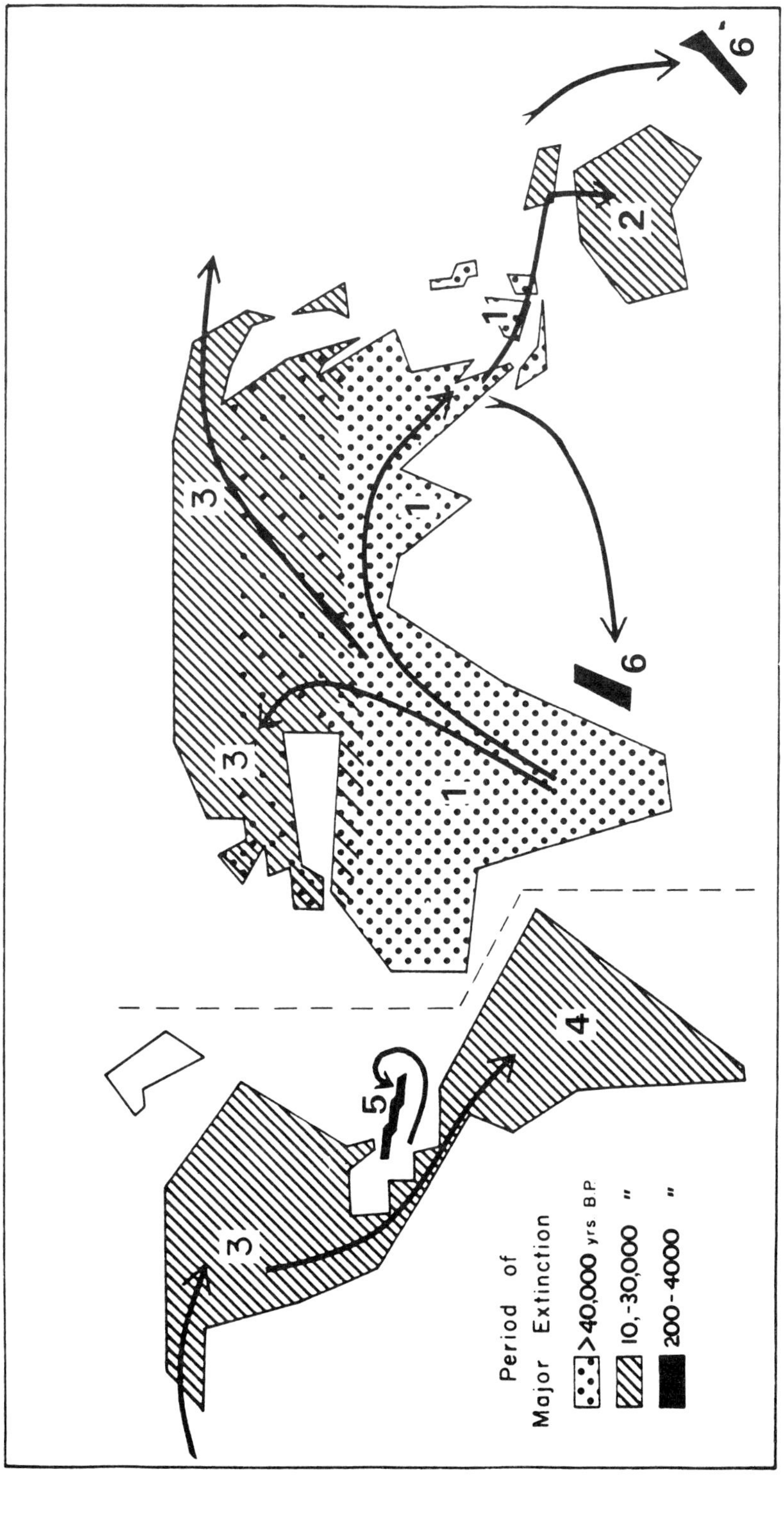

Figure 1. Global extinction of large mammals (revised from Martin and Wright, 1967)

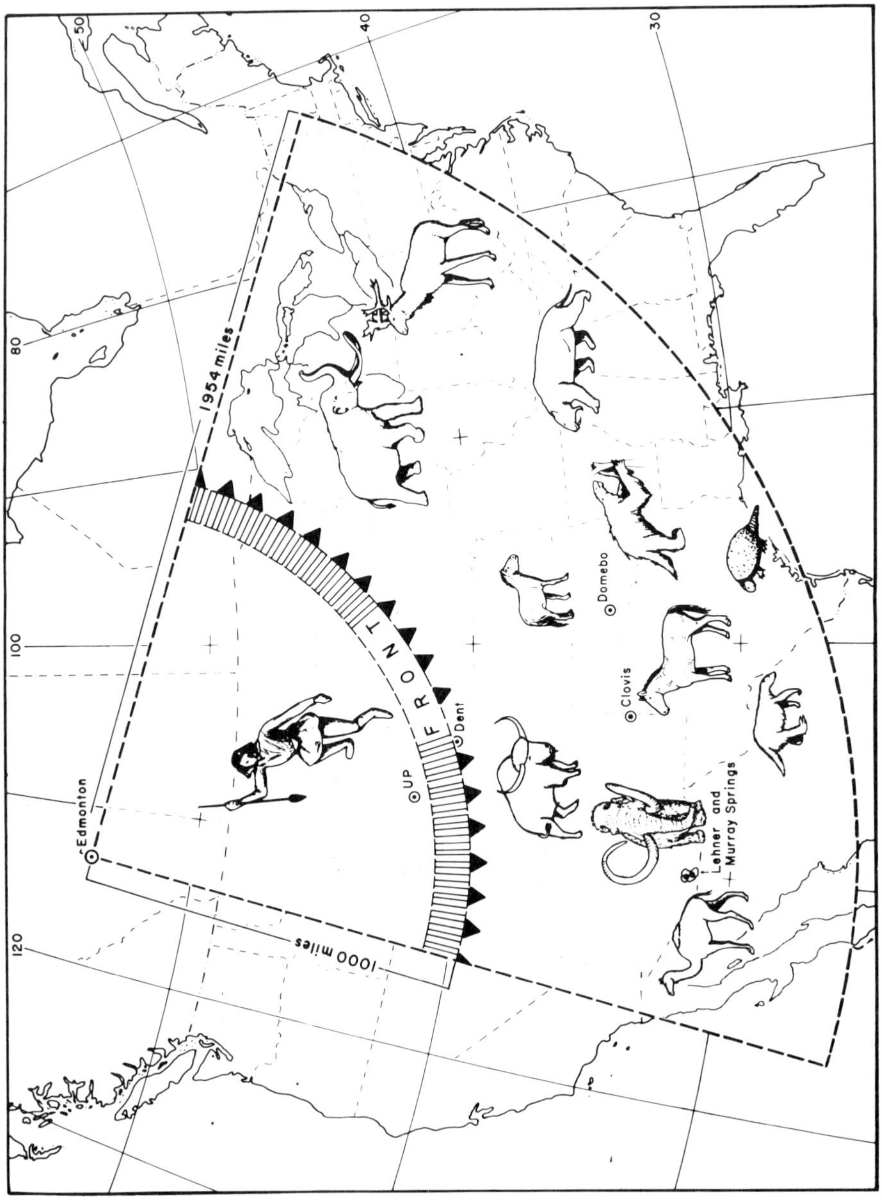

Figure 2. The wave-front feature of the blitzkrieg model of overkill

(pre-12,000 BP) peopling of America. Stanford (1979b) reports what he believes may be a pre-Clovis mammoth kill. An early occupation of the New World, if it can be established, would not be incompatible with extinction by overkill, provided that the earlier cultures were not dedicated to big game hunting for an appreciable time span.[1] That is, an economy based on fish, small game, and plants, or a brief, ephemeral occupation, lasting perhaps a century or less, by small numbers of Paleolithic hunters that briefly entered the New World perhaps to die out without issue, would not be discordant with a sudden, massive impact by mammoth hunters tens of thousands of years later.

It should be understood that no abundantly rich Paleolithic mammoth-hunting cultures have been found in the New World, including the Clovis culture, which might be comparable to those known in the Ukraine and Central Europe (Klein, 1973; Kozlowski, et al., 1974; Pidoplichko, 1978) or Siberia (Klein, 1971; Mochanov, 1978a). According to Saunders (1978), the mammoths from all American "Clovis" sites total only 38 individuals, less than might be found in one good archaeological site in the Ukraine. For example, a total of 1693 mammoth bones from at least 109 mammoths were used to construct three Late Paleolithic houses at Mezhirich (Pidoplichko, 1978). The fossil record of the New World has not yielded mammoth or other extinct animal bone carvings so frequently found in Europe (Jelinek, 1975, for example), and only one mammoth bone "shaft straightener" or "*baton de commandement*" (Haynes and Hemmings, 1968). While faith in a New World pre-Clovis American invasion seems to be growing, and interest in the subject is intense, there seem to be no sites to match, for example, the Russian site Molodova I, with its oval enclosure of mammoth bones around a great quantity of cultural debris, including roughly 29,000 pieces of flint, hundreds of fragments of animal bones ("kitchen debris"), 15 hearths, and a spot of red, ochreous pigment. Klein (1969) summarizes opinions on the possibility of the site representing ruins of a structure. Molodova I is one of 33 Mousterian sites in European Russia; it represents the time roughly 50,000 years ago when competent hunters camped both in the open air and in caves. They made stone tools, could control fire, and at least on one occasion, buried their dead (Klein, 1969). They and their more numerous Late Paleolithic followers had ample time to discover America. However, if they did, they apparently lost their capability for leaving highly visible sites in the fossil record. Even an extremely slow population growth rate for a group of 100 invaders, say 1.4% per generation of 20 years that Laughlin (1967) suggested for the peopling of America, would yield a human population of millions and, inevitably, an extremely rich artifact legacy in only 15,000 years. Few archaeologists claiming to have found early sites in America would also claim that their finds match the size or number of the contemporary sites of the Old World, including those of Siberia.

The abundance of cultural material associated with the mammoth in the Ukraine has encouraged some Russian investigators to attribute mammoth extinction to human agency. At the same time, the scarcity or absence of man-mammoth associations in other parts of Europe and Asia has turned paleontologists away from thoughts of overkill. For example, Vereshchagin (1967b) states: "Man's influence upon the mortality of mammoth, horse, bison and saiga in the Russian Plain was far greater than it was over the vast territory of Siberia. In the Pleistocene, in the north of Yakutia, an influence of primitive man upon animals has yet to be proved." Evidently, high archaeological visibility has been an expected outcome of massive human destruction of a fauna—mistakenly, I believe.

A paradox of the blitzkrieg version of overkill (Mosimann and P.S. Martin, 1975) is that maximum human impact will yield minimal visibility in the fossil record. There is abundant archaeological evidence for the hundreds of thousands of years in which the prehistoric population of Eurasia hunted large mammals—reindeer, giant deer, woolly mammoth, woolly rhinoceros, horse, bison, and musk ox. If this same population did exterminate Old World mammoth, rhinoceros, and musk ox, as theory suggests, the event would have happened too swiftly to be discovered easily by direct fossil evidence, We cannot expect to get a picture of extinction from the fossils found in hunters' camps. If Klein (1973, p. 59) is right that

[1]We have to note here that for the Arctic, and perhaps most boreal regions, there are no real alternatives to big-game hunting. For Wisconsin Beringia we must assume big-game hunting unless marine adaptations allowed the invaders to enter with a fishing economy.

aurochs, wild horse, and red deer were hunted more extensively than mammoth and rhino, it does not necessarily follow that mammoth and rhino did not become extinct through human impact.

During the time of the last glaciation a stasis occurred in which mammoths replaced their peripheral losses. At this time, the unglaciated Arctic regions of Eurasia and Alaska would have been a refuge from hunters for mammoth, reindeer, woolly rhino, musk oxen, wild horse and other species ranging north of 60°N latitude. Hunters would have been present only along the southern part of their range. Rough estimates of the boundaries for the northern limit of Late Paleolithic peoples during the Weichselian glaciation, around 24,000 to 14,000 years ago, and for the southern limit for woolly mammoth, are shown in Figure 3. Occasional summer camps or even permanent habitations may have been established locally by hunters ranging north of latitude 65°N. However, only south of this latitude has the Late Paleolithic yielded abundant archaeological sites. In his summary of Siberian prehistory, Klein (1971) lists roughly 50 sites, all but one south of the 60th parallel. Human occupation on the Aldan River at latitude 64°N has been reported by Mochanov (1978a) for Ikhine and Ust-Mil' as early as 30,000 years ago, and the area may have been inhabited during the height of the Sartan Glaciation. Possible migration of any group from the Lena basin to the Yukon (Old Crow Flats)(Morlan and Cinq-Mars, this volume) would theoretically have been easier during the last interstadial, when woodland was more widespread, than later, during glacial conditions, when woodland was more restricted.

On the Kamchatka Peninsula, a well-stratified site at Ushki offers a resemblance to ancient sites of North America (Dikov, 1978). At least until 14,000 years ago, the Koryak Coast to the northeast was covered by tundra and blocked to human

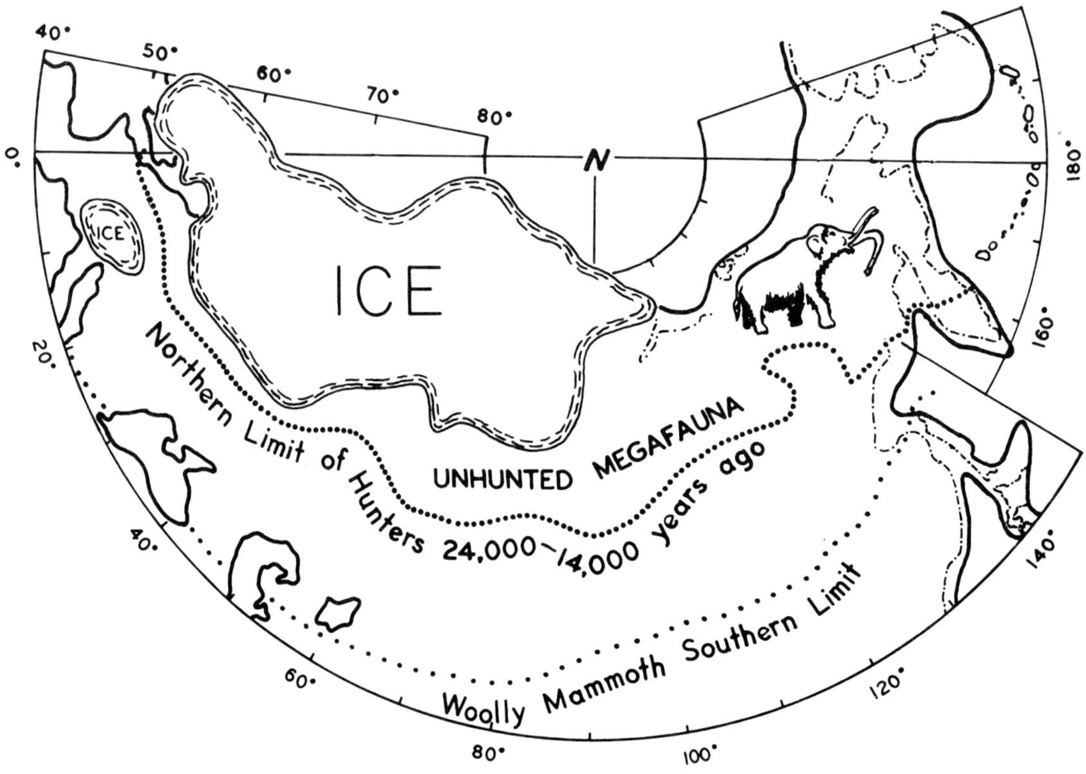

Figure 3. Approximate southern limit of the woolly mammoth (after Kurtén, 1968) and northern limit of hunters during the last glaciation

invaders by mountains harboring alpine tundra and valley glaciers. To reach the Bering land bridge, prehistoric hunters may have had to follow the Anadyr River north to 64°. While that latitude is below the Arctic Circle and is no greater than that of Ikhine, which was occupied over ten thousand years earlier, the environment on the Bering land bridge was a flat tundra, perhaps without firewood beyond the coast.

Somewhere along the 65th parallel, around or after 13,000 years ago, some inventive Paleolithic band managed to winter in the featureless and unprotected coastal plain of the land bridge. Whatever its nature, that adaptation allowed people of the Subarctic to extend their range eastward through the tundra of Alaska in search of major prey, including the woolly mammoth. A suggested chronology, based on rapid growth of a population expanding north and east at a rate of roughly ten miles per year, is shown in Figure 4. The entire mammoth population of the eastern Siberian Arctic (140°-180°E) could have been destroyed in roughly 100 years. Very little archaeological evidence could be expected under these conditions. The hunters, having occupied Arctic tundra, could then have been drawn easily to lower latitudes and to the American midcontinent by pursuing their familiar prey in regions in which their prey was not yet wary of man.

The model suggested here predicts mammoth extinction in Alaska at or around 11,600 years ago, in phase with the inferred arrival time of Paleolithic big-game hunters (see Figure 4). This is considerably later than the youngest known ^{14}C-dated mammoth site in Alaska (Matthews, this volume). It is concordant with the hunting of mammoth in northeast Asia which Mochanov (1978a) says continued until 11,500 to 11,000 years ago. While the blitzkrieg version of overkill could be stretched to accommodate a 500-year difference between mammoth extinction in northeast Asia and in the southwestern United States, it would founder if Mochanov's 11,000 date proves to be a better estimate.

Did the decline of tundra or steppe-tundra and its replacement by muskeg and taiga in the north

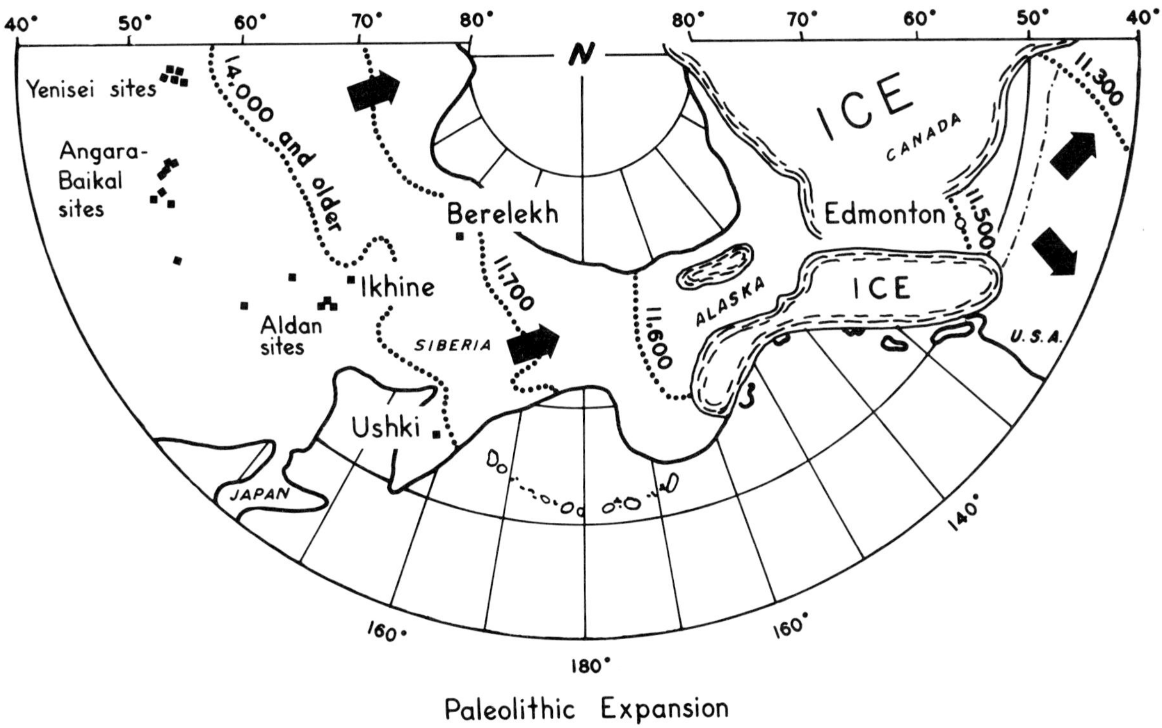

Figure 4. Mammoth extinction. Paleolithic-invasion isolines based on projected chronology for mammoth extinction in North America

doom the mammoth? This idea was advanced by Flerow (1967), is touched on by Klein (1971), and was much discussed at the Wenner-Gren Conference. Because the time of mammoth extinction in Alaska remains in doubt, critical questions about the environment at the time of extinction must go begging. However, some helpful insight about tundra and mammoths may be obtained from western Europe, where late-glacial paleoecology had its start.

In Scandinavia, a late-glacial steppe-tundra habitat for mammoth may have existed along the margin of the retreating Fenno-Scandinavian glacier. The total number of ^{14}C-dated Scandinavian mammoths is small. Fifteen dates from eight localities have been reported (Berglund, et al., 1976). Most apparently represent interstadial incursions of the mammoth during the Middle Weichselian. Discounting a Finnish date of 9000 (TX-127) that contradicts palynological evidence, the youngest Scandinavian mammoth is from southern Sweden represented by a tusk collagen-dated as 13,360, 13,260, and 13,090 years old (Berglund, et al., 1976).

Late Pleistocene elephant remains (including some straight-tusk elephants) were much more numerous in Denmark, where 85 finds have been reported (Møhl in Berglund, et al., 1976). Apparently all pre-date the late glacial. Many may be of mid-Weichselian age; Møhl reports one ^{14}C date of mammoth in Denmark of 40,000–35,000 BP (personal correspondence). Mammoths are missing from the late-glacial megafauna of Denmark, a fauna rich in reindeer that includes wild horse, steppe bison (*Bison bonasus*), and giant elk (*Megaloceros giganteus*). The environment featured a late-glacial tundra with a strongly continental phase during the earlier Dryas, starting 12,000 years ago (Van Geel and Kolstrup, 1978). The Danish environment had apparently been suitable for mammoth at an earlier time. The freshly deglaciated lime-rich soils of the Danish late glacial supported both tundra and steppe plants (Iversen, 1973) and should have provided suitable mammoth pasture. While a mammoth tusk has been found by Bosinski at a Magdalenian site, Gonnersdorf in West Germany (Bhattacharya, 1977), the abundant European Paleolithic sites of Lower Dryas, Alleröd and Upper Dryas age typically contain reindeer without mammoth. Stuart (1977) notes the scarcity of mammoth remains in Britain after the Middle Devensian, 14,000 years ago, at a time when grassy tundra persisted in Britain. Evidently tundra and park-tundra habitats stocked with abundant reindeer lingered in northern Europe several thousand years after the decline of mammoths. During the Holocene, the return of deciduous forest did not see the return of the straight-tusked elephants that had previously occupied the temperate deciduous forests during the interglacials. The point I hope to make is that the habitat of Europe in the late glacial appears to have been entirely suitable for mammoth, and that of the Holocene for the temperate-climate-favoring straight-tusked elephant.

Based on a sinew date of 11,450 BP (T-297) from the Taimyr River, the Siberian mammoths of the extreme north may have been the last on the continent. Perhaps the richest single mammoth-bone bed to be inventoried, the mammoth "cemetery" at Berelekh, northeast Siberia (north of 70°N.), contained at least ten tons of bones and about 140 individual mammoths, mainly females and half-grown males (Vereshchagin, 1974). Bones and skulls were intertwined with brownish strands of hair, and wool with pieces of skin. A tusk was radiocarbon dated as 12,240 ± 160 BP (Vereshchagin, 1974, p. 7). Vereshchagin and Kuz'mina (1977), in a more detailed account, reported younger dates from the Berelekh site on wood, and a 13,700 BP date on mammoth soft tissue. While an archaeological site lies close to the mammoth cemetery, its full stratigraphic and chronologic record remain to be disclosed in detail. It would appear, however, that in northern Siberia, north of the Arctic Circle, the mammoths apparently were thriving when they were scarce in or gone from Europe. I propose that they had never been hunted in this region until around 12,000 years ago. Heavy human predation on mammoth can certainly be postulated for the Ukraine (see Klein, 1973; Bhattacharaya, 1977) until roughly 13,000 years ago. During the last glaciation, losses from the Ukrainian herds could have been replaced by southward migration of herds of unhunted animals from the north. Had there been a similar unglaciated steppe-tundra refuge in Scandinavia, rather than the inhospitable ice sheet, the full-glacial mammoth herd of western Europe might have also flourished to a greater degree than it did.

What was the condition of the late-glacial elephants of America at the time that mammoths were waning in Europe? As was the case in northeast Siberia, it appears that they too were thriving. Over 160 mastodon localities and a much smaller number of mammoth localities are known from the state of Michigan (P.S. Martin, 1967). Aside from the occasional redeposited bone of interstadial or greater age, the Michigan fossil elephants represent mortality between the time of deglaciation, beginning before 14,000 years ago, to the time of mastodon extinction, perhaps 10,500 years ago. A similar abundance of mastodon remains is evident in post-Wisconsin deposits of New York State and in southern Ontario, where Dreimanis (1967) reported that mastodons postdated the Lake Warren shore of 12,400 years ago. North American mastodons and mammoths were apparently entering the fossil record in large numbers at a time when no mammoths are known to have entered the fossil record of Denmark and when mammoths were scarce or extinct in western Europe. Extinction of Holarctic proboscidea was time-transgressive.

Mastodons are believed to have been associated with forested habitats, including the spruce-pine and spruce-fir pollen zones widely encountered in the fossil record of eastern North America. Mammoth records in the East may represent invasion of the narrow periglacial parkland or treeless environment that eluded American palynologists until the last few decades.

Mammoths, both woolly and Columbian, were more numerous than mastodons in the western portion of North America. In addition, three other species have been found in western Canada: the Columbian (*M. columbi*), imperial (*M. imperator*), and southern mammoths (*M. meridionalis*) (Harington and Shackleton, 1978). Many of the radiocarbon-dated records are of interstadial and interglacial as well as late Wisconsin time. The majority of the records are from the prairie of southern Alberta and Saskatchewan, where Pleistocene deposits of various ages are well known. Harington and Shackleton attribute the small number of records in British Columbia and northern Manitoba to an environmental cause: that is, that the alpine valleys of British Columbia and the base-poor soils of the Canadian shield were unfavorable habitat.

Their distribution map suggests another possibility—that because of earlier deglaciation and thus a longer habitat exposure the late Wisconsin mammoths are more widely distributed in Alberta (37 records) and Saskatchewan (35 records), than in Manitoba (5 records). If mammoth extinction occurred before most of Manitoba and British Columbia emerged from under ice, there would have been little opportunity for fossilization in those provinces. The large numbers of late-glacial-age elephants in extreme southeastern Canada (Dreimanis, 1967) and the absence of any except interstadial-age proboscidian remains from farther north on the Canadian shield support the thesis that a crisis occurred for mammoths around 11,000 years ago. Had they survived to a later date, say 8000 years ago, their remains would be found in Holocene deposits considerably north of their present limit. That northern limit follows the 44th parallel in eastern North America and extends north into the Canadian prairies to the vicinity of Edmonton. Had they been slowly declining to the point of extinction, the northern limit of their range would yield fewer records than the southern, which does not appear to be the case (see maps in Dreimanis, 1967, and P.S. Martin, 1967).

Mammoths ranged widely through the Holarctic region during the Pleistocene. While their numbers cannot be estimated directly, their fossil bones are common over most parts of their range. Mammoths appear to have favored open regions of either tundra, steppe, prairie, parkland, and perhaps especially, riparian habitat in grasslands (Olivier, this volume). In North America, mastodons favored forests or forest-prairie borders. Old World mammoth remains are especially abundant in association with prehistoric artifacts of central Europe and the Ukraine. Cultural associations are less common in other parts of Europe and Asia and relatively rare in North America. After 14,000 years ago mammoths were rare or not present in the late-glacial tundras of western Europe; fossils of Siberian mammoths and American mastodons and mammoths from that time are very common. Western Europe provides the first evidence of a decline in numbers of elephants in the northern part of the Holarctic region. While data from Alaska are deficient, and potentially contradictory, elephant extinction appears to have been virtually synchronous in the region extending from eastern Siberia to the southern United States.

The vast difference in archaeological visibility

between the mammoth cultures of the Old World and the New invites explanation. A shift from a sustained yield to an exploitative economy starting 12,000 years ago could explain elephant extinction throughout the Holarctic region. It is not impossible that small numbers of people occupied America at intervals long before the time of mammoth extinction. Such an event would not preclude the possibility of a major cultural impact around 11,000 years ago, at the time human occupation was indisputably established. However, the abundant Late Paleolithic sites of the Ukraine, which are characterized by mammoth bones (Klein, 1973), stand in marked contrast to the abundant late Pleistocene bone deposits in western North America, which yield only elephant fossils, without obvious associated archaeological material. In America, well-documented associations of human artifacts with extinct mammoth remains come mainly from Clovis-age kill sites of 11,000 years age. These are far less archaeologically visible than the mammoth camps of Ukraine and eastern Europe. Apparently, elephant extinction in Ameica was rather sudden, later and more rapid than in Europe. If it were accomplished by highly effective hunters, as modeled by Mosimann and P.S. Martin (1975), extinction could have taken place in less than 300 years.

Finally, for those seeking test implications for Beringia itself, I would anticipate that (1) mammoth extinction occurred there around 11,600 years ago, (2) that most, if not all, of the other extinct Beringian large mammals disappeared at the same time, and (3) that despite the tremendous promise of the Old Crow region in the Yukon, no rich Paleolithic sites to match the age or content of those in the Aldan region of Siberia (Ikhine, Ust-Mil', Dyukhtai Cave) will ever be found in the Alaskan portion of Beringia. These predictions, which some may find overly bold, rest entirely on my concept of the meaning of Pleistocene extinctions in the Holarctic region.

24

ARGUING FROM THE PRESENT TO THE PAST: A CONTEMPORARY CASE STUDY OF HUMAN PREDATION ON AFRICAN BUFFALO

Stuart A. Marks

Our understanding of prehistoric events is based largely upon archaeological interpretations of artifacts and fossils found together in their ambiguous settings. Such interpretations may never be conclusive; rather, they reflect stages in our methodologies and in our processional conceptualizations of humans and their circumstances. This paper describes the cause and effect relationships of the main variables affecting human predatory behavior and their effects on a wild population of a major prey species, the African buffalo. Fluctuation of buffalo populations living in close proximity to developing human communities has been a function of disease, quality of range, and severity of human predation. The present effect of human predation on buffalo is a function of technology, transient cultural and social styles, and outside influences that increase demand for buffalo beyond local needs.

This case study provides an empirical model demonstrating the complexities which characterize human hunting activities now and in the recent past. It is suggested that if prehistoric hunters were associated with the Arctic steppe-tundra they were few, far between, and behaved as opportunistic and generalized hunters and foragers.

Key words: hunters, Valley Bisa, weapons, buffalo, prey populations, frequency of kills (hunter), hunting tactics, mammoth

INTRODUCTION

Although archaeological sites in North America (Frison, 1974) and in Africa (Klein, 1978, 1979) leave little doubt as to the prehistoric association between humans and the larger wild mammals, the nature of the association remains largely conjectural because the evidence is insufficient and inconclusive. As we move backward from the present towards the more distant past, we increasingly lose information essential for the reconstruction of ancient lifestyles. We may never be certain of the actual strategies used by hunters for containing and killing their prey, nor, for that matter, their degree of dependence upon wild mammals, since this evidence is not likely preserved at archaeological sites. In my view, it is instructive to remind ourselves of these uncertainties as we cautiously proceed in our attempt to interpret whatever evidence is uncovered at archaeological sites.

The abundance, gigantism, and diversity of the terrestrial fauna associated with the Arctic steppe-tundra biome (see Guthrie, Redmann, this volume) suggest the African savanna as a possible modern day analog. The interactions between subsistence hunters and a well-stocked faunal environment in Central Africa provide a perspective from

ISBN 0-12-355860-3

which to judge speculations on the role of prehistoric hunters on the Arctic steppe-tundra. Blessed by relatively little evidence, what shall we expect of prehistoric peoples associated with this extinct biome in terms of their capabilities as hunters and of their effects on wild mammalian stocks?

A main theme of this paper is that the form and context of the relationships between members of a subsistence society and the surrounding wild fauna are a function of that society's total socioeconomic strategy. Changes in the form or nature of a basic component of a society's relationships inevitably entail a reorganization to some degree of its other components. For example, changes in technology or political alignments manifest themselves in the ways resources are acquired and apportioned. This paper provides substance for this argument by providing a contemporary case study of human exploitation of a large mammal population in the Luangwa Valley of Zambia (Central Africa). It blends pattern with process in describing changes in the relationships between hunters in a marginal, subsistence society and their major prey species, the cape buffalo. The main variables and their cause and effect relationships are diagrammed systematically for the period 1966-73. Then the time frame is extended backward and forward to illustrate how these functions have interacted through time.

THE VALLEY BISA AND NABWALYA STUDY AREA

Since the Valley Bisa of Zambia and the Nabwalya Study Area have been described elsewhere (Marks, 1976), here only information relevant to the present article is reviewed. Today the Valley Bisa occupy an irregular rectangle of land some 3300 km^2 in the Luangwa Valley of Zambia, called the Munyamadzi Corridor. This corridor is surrounded on the north, south, and east by national parks and on the west by an abrupt escarpment. Until recently, these imposed and natural features bordering their homeland have led to the relative isolation of the Valley Bisa from mainstream political and economic events elsewhere in Zambia. Since it was not possible to study the entire Munyamadzi Corridor, a smaller site of some 60 km^2 near its center was chosen for intensive study. This site, referred to as the Nabwalya Study Area (Fig. 1), contains a mixture of the major habitat types and wildlife species found throughout the corridor.

The Valley Bisa are matrilineal in terms of descent and status, and matrilineality is the organizing principle around which their villages are formed. In terms of standard of living, access to labor and consumer markets, cash, effective political power, mortality, morbidity, and literacy rates the Valley Bisa are near the bottom of Zambian society. Their numbers increased from 3544 in 1910, when the first government census was taken, to 5783 in 1963. The most recent census shows a decline in population.

Today most Valley Bisa are shifting agriculturalists growing sorghum, maize, pumpkin, and millet on the fertile soils adjacent to the perennial rivers. One of these rivers, the Munyamadzi, transects the upper third of the Nabwalya Study Area; the Valley Bisa's habitations and current fields are located along the margins of this river. Little of what is grown locally is converted to cash; and the only customers are the few civil servants in residence in the valley. During the dry season a few men are employed locally by safari operators within the corridor. At this time, chickens, tobacco, and wild meat are hawked to government officials and others who frequent the valley. The 1973 census recorded 96 men, 165 women and 249 children in residence at Nabwalya. In the recent past, upwards of 70% of the men were away seeking or engaged in employment in the urban centers during the course of a given year. Since they enter the labor markets as unskilled laborers, the lower percentage of males away in 1973 and in subsequent years reflects the difficulties they were experiencing in obtaining jobs.

Because the tsetse fly is present in the Luangwa Valley, the Valley Bisa are unable to keep domestic livestock. Consequently, the pursuit by hunters of the abundant stocks of wild mammals is the major means by which the community's need for animal protein is met. During 1966-7, the community at Nabwalya killed 101 mammals, which together with the eight elephants (*Loxodonta africana* Blumenbach) killed by wildlife personnel, yielded the equivalent of 91.5 kg (carcass yield) of meat/adult/year. The main prey species, in numerical order, were cape buffalo (*Syncerus caffer* Sparrman), impala (*Aepyceros melampus*

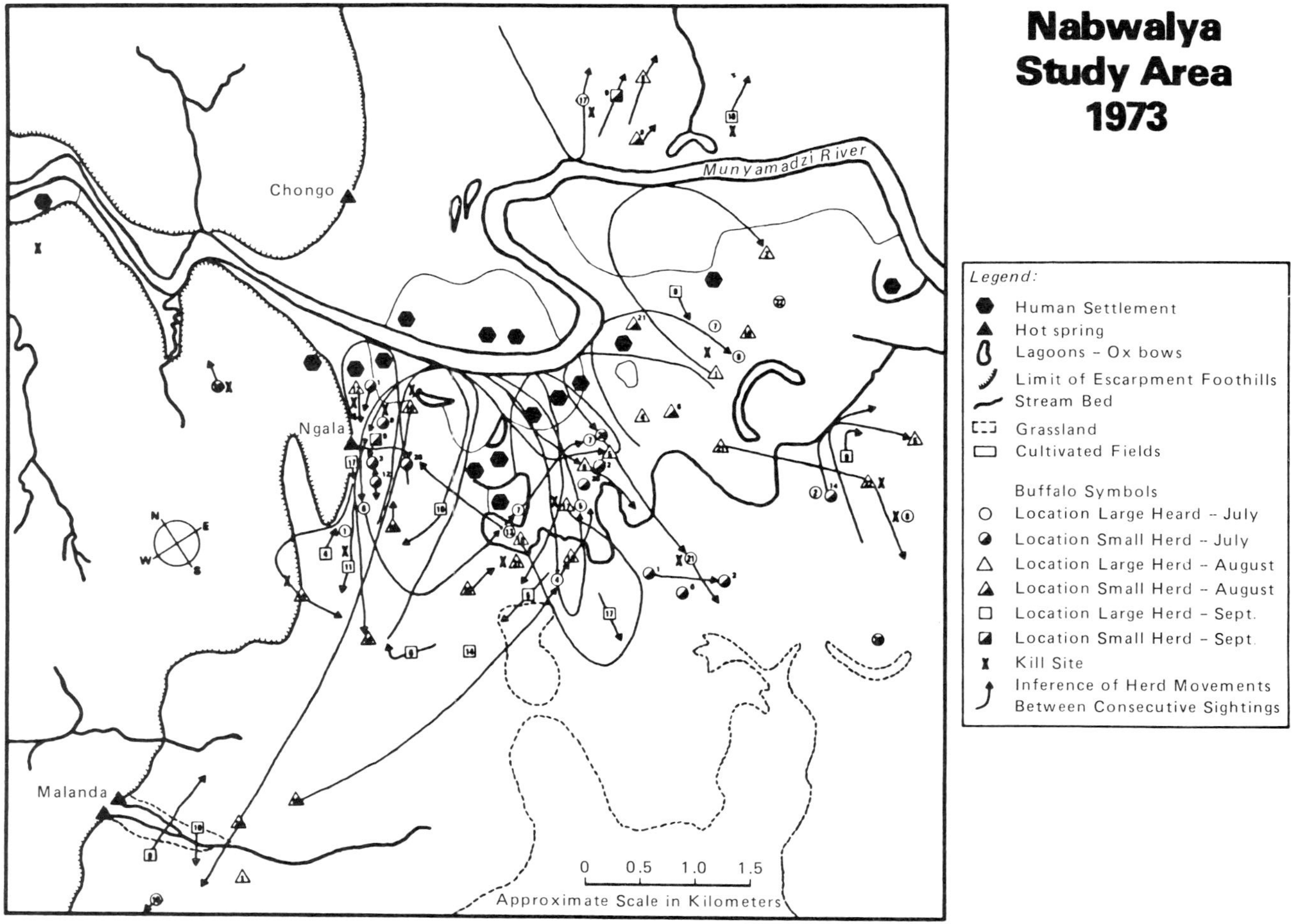

Figure 1. The Nabwalya Study Area showing location of villages, fields, and buffalo herd movements. The numerals within or adjacent to buffalo symbols refer to days of month

Lichtenstein), warthog (*Phacochoerus aethiopicus* Pallas), and waterbuck (*Kobus defassa* Rueppell). Although wildlife is abundant today around Valley Bisa villages, observations recorded in official documents show and the Bisa themselves testify that this was not always the case. Reasons for the current abundance of wildlife include the dependence of this society on subsistence agriculture and urban employment (which has reduced the number of hunters during this century); the outlawing by government and thus the gradual abandonment of bows and poison arrows and an indigenous arsenal of traps, pits, and downfalls; and the proximity of human-occupied lands to the surrounding national parks which serve as wildlife sanctuaries. Until recently, the main prey of Bisa hunters, particularly buffalo, impala and warthog, have maintained or increased their numbers commensurate with human activities that tend to favor the expansion of grassland, such as shifting agriculture, gathering and burning.

Only men hunt, but not every man is a hunter. In 1973, 25% of men in residence at Nabwalya spent some time hunting. Generally, hunters are individuals who have received a dream summons, have a guardian shade, have spent a period of training under an adept, possess one or more types of magic, and have accepted the social role of providing meat for their resident matrikin (Marks, 1979a). The relative isolation of the Valley Bisa during the current century resulted in the retention of many hunting traditions that diminished elsewhere in Zambia as "modernization" occurred.

A hunter usually pursues game alone; on those occasions when two hunters search for prey together, one plays a subservient role to the other. Although a few individuals own or have access to modern breech-loading weapons, the main hunting weapon employed by the Valley Bisa is the muzzle-loading gun. In recent years, the opening of the Munyamadzi Corridor to frequent commercial traffic during the dry season, the construction of development projects by the government at Nabwalya during 1977, increasing use of modern weapons, increasing contacts with outsiders, monetization of wild meat, and the increasing economic differentiation among valley residents have greatly enlarged the scope and spheres of influences which impinge upon local hunters.

A REVIEW OF BUFFALO BEHAVIOR AND ECOLOGY

This section reviews the observations on buffalo by Grimsdell (1969) in Uganda, Sinclair (1977) in Tanzania, and Pienaar (1969) in South Africa. The African buffalo is one of the biggest and most widely distributed ungulates on the African continent. Although it shows considerable geographic variation, it is considered a single species by taxonomists. Buffalo in east-central and southern Africa are large and black, the males weighing up to 500 kg. Adult males are identifiable in the field by their massive bosses and the shape of their horns.

Buffalo generally are found in herds whose size depends on environmental conditions, primarily rainfall, availability of drinking water, and vegetative structure. In the relatively open acacia woodlands and plains of the Serengeti National Park (mean rainfall 800 mm), herd size ranges from 50-2000 animals (mean size 350). In densely forested Arusha National Park, herds average 50 head. In drier regions, herds tend to be smaller.

Each herd inhabits a spatially defined, yet nonexclusive, home range whose dimensions vary locally and seasonally. The size of the home range during the dry season is limited by the availability of drinking water and adequate grazing land. Size of range is also related to herd size. Grimsdell (1969) described a herd of 130 buffalo in Uganda with a home range of 10 km^2; Leuthold (1972) gave a home range of 85 km^2 for a herd of 400 in Tsavo National Park; and Sinclair (1970) depicted a herd (size not stated) in Serengeti National Park which roamed over 450 km^2 during the course of three years. The home ranges of two Serengeti herds were considerably larger during the wet than during the dry season. Buffalo prefer riverine vegetation, which includes elements of both forest and grassland. Grass leaf is the preferred component of diet.

Grimsdell (1969) found that buffalo herds favor specific areas for different activities (particularly resting and ruminating) and had well-defined routes connecting these localities. The average daily movements for herds was about 9 km (range 5-13 km.). For the Serengeti area, Sinclair reported herd movements of up to 30 km a day. The spacing of herds in the Serengeti was significantly different than that to be expected from a random distribution; herds were uniformly spaced about 8 km apart during the dry season. Sinclair suggests that this spacing behavior by buffalo is a function of the fragmentation of their preferred habitat during this season.

Whereas all females and all juveniles less than two years of age live in herds, some adult males are found in small bachelor groups of 2-10 bulls. Very old males occur as solitaries. In the Serengeti, bachelor bulls comprise about 15% of the adult male population, the majority of males remaining with or near the breeding herds.

Within the breeding herds, family ties between mother and offspring last at least three years and perhaps longer for females. Males begin to break away from their mothers during the third year and, during the fourth and fifth years, form subgroups around the herd's periphery. Relationships between males within a herd are determined by rank. Grimsdell (1969) describes a linear hierarchy for 15 adult males in which the highest ranking bulls are closely associated with the females and provide most of the observed copulations with cows. Lower-ranked males are eventually expelled from the herds and form their own bachelor groups. Small bands of males also leave voluntarily during the dry season, returning to the herds at the onset of the rains.

Members of bachelor bands range in age from the occasional subadult to the very old. For his analysis, Sinclair (1974) divided the bachelor groups into two categories of bulls younger and older than ten years of age. The older males had left the herd permanently and were extremely sedentary in areas of suitable habitat. One group of males restricted its movements within a ½ km radius during a three year period. By comparison, he found the distance between end points in the range of a herd was 50 km.

Herds tend to break up into smaller units during the dry season. Pienaar (1969) suggests that the larger the herd, the further and more often it has to travel in order to find sufficient grazing at this time of year. In environmentally lean conditions, when only localized patches of suitable habitat are available, it is adaptive and advantageous for the herd to break up. By restricting their movements throughout the year, old bachelors appear to do well, relative to herding buffalo, in terms of obtaining energy and protein needs. The age-specific survival rate of old bulls is greater than that of cows; however, significantly more old males are killed by lions than would be expected from their numbers in the population (Sinclair, 1974, p. 683). Sinclair (1974) suggested that herding in buffalo evolved as an anti-predator behavior to compensate for their poor running and escape abilities, particularly those of young calves.

BUFFALO BEHAVIOR AND MOVEMENTS AT NABWALYA

Valley Bisa hunters recognize three classes of buffalo which in terms of size and composition are similar to the groups described by Sinclair (1977) for the Serengeti. According to scale and beginning with the smallest unit, these classes are *kakuli,* individuals or bands of bachelor bulls; *kasakaombe,* small groups of either sex containing up to 60 animals; and *ibumba,* large herds numbering in the hundreds. The bachelor bands are composed of old bulls which have retired or have been chased from the larger herds. These bulls tend to become sedentary and localized in their movements, often living along the river close to the villages where their locations are well-known to hunters. Small groups of juveniles and adults of either sex break from the larger herds during the dry season are said to follow movements similar to those of the larger herds. However, their smaller size makes their movements quieter and more difficult for hunters to follow, and consequently more difficult to predict, than is the case with the larger herds.

During 1973, hunters at Nabwalya identified four large herds whose ranges overlapped during the dry season. The range of the main and largest herd (475 buffalo) encompassed the center of the study area south of the Munyamadzi River and extended beyond its southern boundary to the Mupamadzi River some 10 km further south. These two rivers were the focal points in this herd's movements for water and for pasturage, and it spent several days or weeks, depending on disturbance, in each section of its range. Two other large herds, each containing fewer animals, were found on either side of this herd. A fourth herd roamed north of the Munyamadzi River. The movements on the study area of both large and small herds are predictable to within a few days, and hunters generally wait until buffalo are close to their villages before attempting to make a kill. Figure 1 shows the frequency of occurence for both large and small herds for a 79-day period.

The small home ranges of bachelor bulls near the river bring them into close proximity to humans and into almost daily contact with some form of human activity. These bulls become sensitive to human presence and difficult to approach; once wounded they are pugnacious and dangerous. Some bulls have been witnessed to withstand repeated assaults by hunters with muzzleloaders and shotguns, for in order to make a kill with these weapons, the hunter must be within 20 meters of his intended target and the bullet must be well-placed. On the other hand, buffalo herds cover a much wider range which may include several days or weeks without human contact. Such buffalo are typically more curious than belligerent with humans, for hunters rarely stalk a herd without being noticed and the last part of the stalk is typically under the scrutiny of many if not all herd members. However, once a herd is startled by a hunter, it becomes suspicious, restless, and generally unapproachable for a number of hours thereafter. Hunters realize that their chances of successfully concluding a stalk on a herd are high if they are one of the initial stalkers on a given day. However, if herds are disturbed too frequently, they may shift their movements to the southern portion of their range and out of human contact sooner than expected.

HUNTING TACTICS

Figure 2 compares the monthly frequencies of sightings for the main wildlife species around Nabwalya during three periods of observation. It indexes the numbers for each species I observed each month divided by the time I spend afield with hunters. The 1966-7 index reflects "expected" trends during an annual wet-dry season cycle. With wildlife dispersed and cover abundant during the rains (December through May), hunters must invest more time searching for and in securing prey. Records of hunters' activities during 1966-67 show that wet season hunts averaged 6-7 hours per foray while dry season hunts, when wildlife was close to

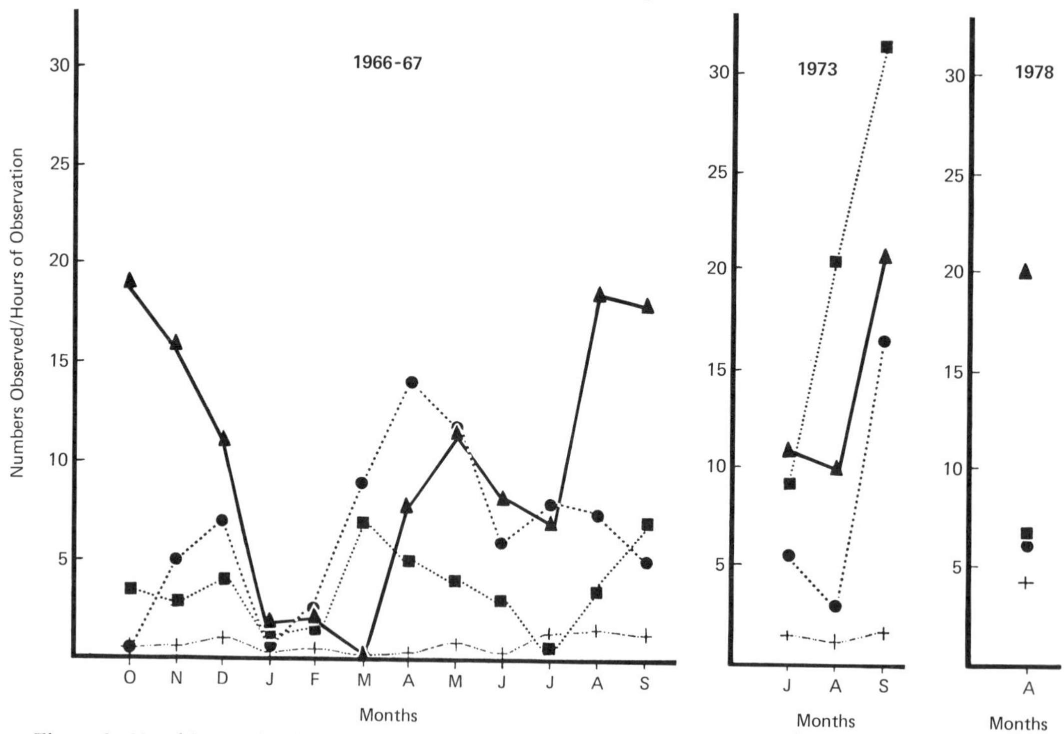

Figure 2. Monthly trends of major wildlife species on Nabwalya Study Area as a function of observation time: 1966–67, 1973, 1978

the river, averaged 2-3 hours. The majority of kills by hunters are made during the dry season.

The Valley Bisa have three hunting tactics for placing themselves within range of their prey: (1) Lying-in-wait behind blinds for game at water holes. This is not a common practice at Nabwalya. The main watering site, the river, is too open and continuous for it to serve as an appropriate site for ambush. Elsewhere in the corridor, however, this tactic is used late in the dry season at isolated water holes. (2) Ambushing is a commonly used tactic for buffalo herds. When the hunter ascertains the direction of a herd's movement, he circles in front and places himself close to a point where he anticipates the herd will pass (Fig. 3a, b). Accordingly, hunters are able sometimes to kill stragglers following a herd (Fig. 3c) or to approach single animals which are dispersed and feeding (Fig. 3d). Provided the herd has not recently been alerted, neither unfavorable wind direction nor inadequate cover preclude stalks. In 1973, on three separate occasions, I witnessed three different hunters, confounded by already alerted herds, run at right angles towards stampeding herds. Twice this tactic succeeded in stopping a segment of a herd long enough for the

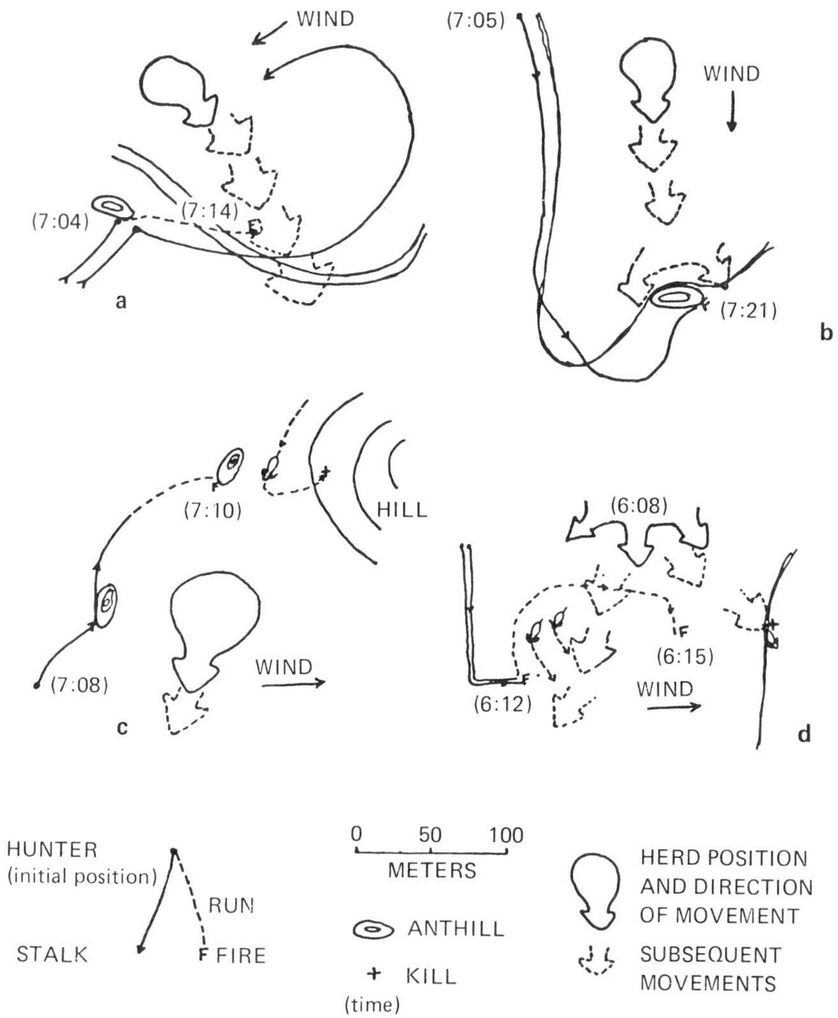

Figure 3. Examples of tactics employed by Bisa hunters against buffalo herds, 1973

hunter to fire his weapon. (3) The most common search tactic is a brisk walk averaging approximately 1.6 km for each 15 minutes of travel time. This pace is slowed slightly in thickets, in tall grass, or as the hunter approaches an area he suspects to harbor game. Search patterns may be described geometrically as a series of straight lines connecting points arranged in an irregular circle or ellipse. Hunters travel in straight lines to the places where they expect to find game.

During forays, hunters typically observe many more animals than they stalk. During 1973 I analyzed a sequence of 50 hunts. In this hunting sequence prey were observed on 562 separate occasions, but hunters attempted stalks on only 143 occasions (25%). The situational variables (number of prey, distance, cover, habitat) which hunters use to determine the vulnerability of prey was described as a species specific gestalt pattern (Marks, 1977a). Hunters observed buffalo on 55 occasions (9.8% of total observations) and made 40 stalks (40/55 = .73). This was the highest stalk response I recorded for any species. In contrast to other species, buffalo were infrequently observed, but were nearly always stalked when observed. Hunters were often able to complete their stalks of buffalo (45% of the time); of the 18 completed stalks, 13 ended in wounded buffalo escaping, two in weapon misfires, and three in kills. Although hunters seem to realize the odds against their making a kill every time, they are inclined to stalk buffalo no matter what the circumstances. The main variables in hunter-buffalo encounters recorded during 1973 are summarized in Table 1. Buffalo behavior and movements are usually predictable and if one's tactics end in a kill, the payoff is abundance of meat for the hunter to distribute for a rather minimal amount of time invested.

From August 1966 to August 1967, hunters killed 39 buffalo and were known to have wounded an additional 26 that escaped. From January 1973 through mid-September, hunters killed 21 buffalo and wounded an additional 19 that were not recovered. The sex ratio was two bulls killed to each cow. Ten buffalo were culled from the large herds (four bulls, six cows), two from the smaller herds (one of each sex), and nine bulls from the bachelor bands. However, a comparison between buffalo class frequency at Nabwalya and kill selection suggests the importance of factors other than buffalo presence in the selective process of hunters' decisions. Marks and Shea (manuscript) suggest that the hunters' cultural preferences for fat cows, their desire to compensate for their inadequate weaponry and minimize risk, together with their exploitation of sporadically occurring windfalls (chance encounters with small herds) all play a part in Valley Bisa selection of this species. For 1973, the daily probability of a buffalo kill for this community was nine chances in 100 attempts (Marks, 1977b).

The type of weapon a hunter uses is an important factor in determining the sex of the buffalo selected and method of disposal. In 1973, individuals with access to rifles killed exclusively cows; most of the meat was exported, and some of it sold. On the other hand, cows are generally unavailable to hunters with muzzle-loading guns. Although hunters with these weapons may try to obtain cows during the dry season (when cows remain fatter than bulls), they are unlikely to be able to slip beyond the periphery of herd bulls to get a shot at the cows. Hunters with muzzle-loading guns seem to recognize these odds and expend most of their efforts attempting to cull bulls.

A MODEL FOR HUMAN PREDATION

The main variables affecting the population levels of buffalo and the level of human predatory activities are presented as an oval diagram (Fig. 4). These variables are linked by arrows indicating cause and effect relationships.

Buffalo herd size is a function of the availability of water and quality of pasture during the dry season, the presence or absence of virulent epizootics (such as rinderpest), and the nature and extent of carnivore and human predation. Numbers of buffalo in the bull and small herd compartments of the diagram are related to herd size. Human subsistence activities such as clearing and abandoning cultivation sites, cutting of trees for construction of houses and other uses, and the setting of fires cause environmental changes which create grassland (pasture) rather than woodland. These more human-maintained environments are less favorable habitat for tsetse fly, the main vector of trypanosomias in humans. The presence of this fly precludes the raising of domesticated livestock by humans.

The structural and spatial distribution of the

Table 1
Main Variables in Hunter-Buffalo Interactions, 1973

Date	Hunter Code	Weapon[1]	Buffalo Group[2]	Wind Direction[3]	Cover Code[4]	Hunter Tactic[5]	Duration (minutes)	Final Distance (Paces)[6]	Comments and Outcome
7/10	H	s	H	var.	3	S	6	300	herd alerted
7/19	C	s	3B	var.	2	S	7	33	buffalo detect hunter, run 200m
7/19	C	s	3B	var.	2	S	2	55	buffalo run
7/19	C	s	2B	H→B	5	S	1	34	buffalo alerted by warthog spooked in tall grass
8/1	P	r	H	H→B	2	C&S	25	42	herd spooks
8/1	P	r	H	B→H	3	C&S	8	53	wound, escape
8/1	H	s	H	H→B	2	S	10	60	herd spooks
8/3	H	s	3B	var.	3	s	?	30	herd spooks
8/3	H	s	6B	var.	4	S	?	15	detect hunter
8/6	M	m	H	B→H	4	A	64	10	wound, escape, Figure 3a
8/6	H	m	H	?	2	C&R	6	?	
8/6	M	m	H	C	3	C	17	?	herd spooks
8/7	H	m	2B	?	4	—	—	10	hunter came upon buffalo suddenly
8/9	Lo	m	8B	H→B	3	C&S	9	40	group wary, contains wounded bull
8/10	H	s	H	H→B	2	S&W	2	25	kill subadult male, Figure 3c
8/17	P	r	H	H→B	2	C&S	6	38	wound, escape, Figure 3d
8/17	H	r	H	H→B	3	R	3	61	kill adult female, Figure 3d
8/20	G	r	H	?	1	A	?	45	kill adult female, moonlight night
8/23	H	m	1B	C	1	A	17	70	hunter declines already wounded buffalo
8/23	H	m	1B	var.	4	S	10	29	bull spooks
8/23	H	m	6B	H→B	2	S	10	38	wound, bull escapes with herd
9/9	Lo	m	H	none	3-4	S	?	16	weapon misfires
9/10	Lo	m	H	H→B	2-3	S&C	?	25	herd spooks
9/12	H	r	7B	none	1	S&R	12	61	wound, bull escapes
9/12	H	s	3B	?	3	R&S	13	34	wound, bull escapes
9/12	H	s	3B	var.	2	R&B	8	25	wounded bull escapes
9/12	H	s	1B	var.	2	S	9	18	wounded bull escapes
9/12	K	m	1B	B→H	4	S&B	10	3	wounded bull charges briefly, then flees
9/12	K	m	2B	var.	3	S	3	12	wound, bull escapes
9/13	K	r	1B	var.	3	S	7	15	kill wounded, adult male
9/13	Lo	m	4B	B→H	3-4	S	?	12	bulls spook
9/19	Ma	s	30	var.	2	R&S	10	80	herd detected hunter
9/21	D	s	H	var.	3	S&R	15	80	herd wary
9/21	D	s	4B	B→H	4	S	9	22	wound, escape
9/27	L	m	H	B→H	3	A	16	22	wound, escape, Figure 3B

[1] m = muzzleloader; s = shotgun; r = rifle
[2] H = herd; B = bachelor bulls
[3] H→B = wind from hunter to buffalo; var. = variable; c = cross
[4] Scale from 1 (open) to 5 (dense); 3 = adequate cover for stalk
[5] S = stalk; R = run; A = ambush; C = crouch; W = whistle; B = charged by buffalo
[6] Final distance between hunter and buffalo

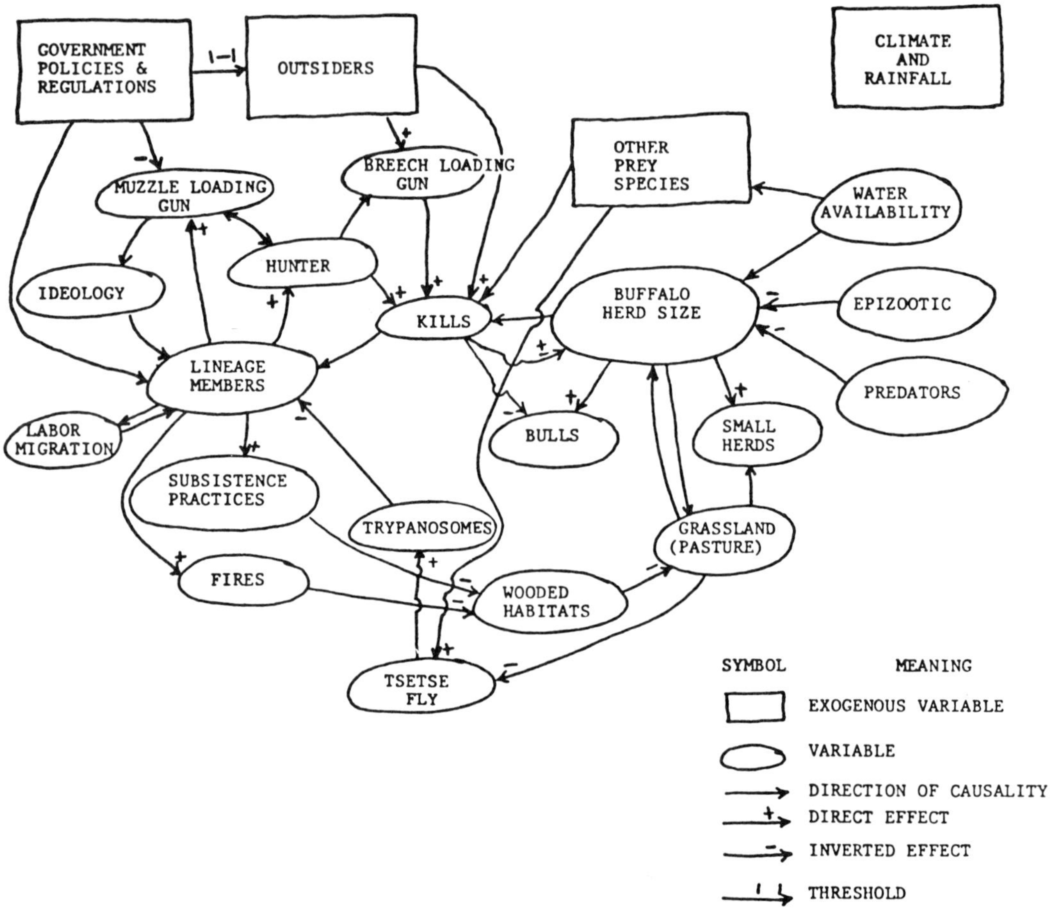

Figure 4. Oval diagram of human-buffalo system in the Luangwa Valley of Zambia. Minus signs suggest that higher values of the preceding factor result in lower values for the succeeding factor, while plus signs result in higher values

human population is determined largely by the matrilineage, which also controls the socio-economic activities of its members. Most lineage members live in the same village and cultivate land together. The lineage allocates roles among its members and has strong norms covering the distribution of assets from its economic activities. Prior to colonial impositions, these lineages were largely autonomous and subsisted on local resources, including wildlife. Its members engaged in local cultivation of domesticated plants, in gathering wild plants, in hunting and in trade. Intermittent warfare involving other ethnic and local groups was common.

The colonial government drastically reduced the numbers of men engaged in hunting, outlawed indigenous hunting implements and has become progressively stricter in the enforcement of its regulations. The lineages formerly adapted by designating a few of their junior members to become hunters, thereby providing meat for consumption by the members while providing some protection from wild animals to its crops. Lineage elders provided the young hunter with a muzzleloading gun and fostered an ideology of equitable distribution for the proceeds from the hunters' quarries. For a while, the relative isolation of valley villages permitted a closed cycle of subsistence hunting with muzzleloading guns, and the lack of people willing to pay cash for meat ensured an equitable distribution of the proceeds from hunters' quarries. The number of kills was not excessive; wildlife increased around villages, and buffalo multiplied as the danger from rinderpest diminished in the 1960s. Since 1974, new patterns have emerged due to the increasing use of modern weapons by both outsiders and local hunters. Some local hunters have responded to these new "frontier-type" conditions with an individualistic (capitalistic) ideology, improved techniques for harvesting buffalo, and by increasing sales of wild meat to outsiders. As a consequence, buffalo numbers at Nabwalya have diminished. The following case study provides the historical context for these changes.

THE MODEL IN RETROSPECT AND PROSPECT

The journals of European explorers provide a basis for analyzing human and animal life in the Luangwa Valley over 100 years ago. Explorers were followed by administrators concerned with affecting social, economic, and political changes that led to then-unanticipated environmental consequences for marginal societies like the Valley Bisa. Here I provide the broad outlines of events between 1830 and 1978 affecting buffalo numbers and human predatory behavior.

European explorers were impressed with the prosperity and self-assuredness of valley inhabitants. Although the valley floor was sparsely inhabited, villages were strategically located to control access by outsiders into the hinterland. Europeans also commented on the variety and abundance of wildlife. In 1831, Gamitto found the Valley Bisa obtaining much of their meat from scavenging buffalo carcasses killed by lions. The existence of hunting guilds and an ingenious assortment of techniques for harvesting wildlife, together with David Livingstone's observation in 1866 (Livingstone, 1874) of isolated granaries in the bush, indicates that valley people had mastered techniques for controlling wildlife numbers around their residences and gardens. Tools and techniques for the taking of large mammals included weighted spears thrown from trees over waterholes, downfalls, pitfalls, footsnares, poisons, bows and arrows and hamstringing. David and Charles Livingstone (1865) found buffalo near Lake Nyassa wary from being hunted with poisoned arrows and commented on the effectiveness of traps and snares in keeping wildlife wary.

In the past, the driving political ambition of most Bisa men was to succeed to a position of lineage leadership (today this role is formalized as village headman). Succession to this prestigious role occurred on the death of the incumbent and was the outcome of a formal selective process by elders acting under strong factional interests. Candidacy for the office depended upon political skill and the individual's achievements in trade, war, or hunting. With material possessions limited and ephemeral, all young people began their careers at the bottom of the social hierarchy. In pursuit of individual ascendance, a young person pursued a complex strategy in which, because competition was fairly open, the results were immediate and obvious.

The closing decades of the last century were turbulent in the Luangwa Valley, marked by increasing traffic in goods and slaves, and by increasing skirmishes with other ethnic groups. To protect themselves against these human incursions, the Valley Bisa left small scattered settlements for larger stockaded villages. There were no defenses against new diseases that at this time affected both human and wildlife populations. Humans were plagued by sand fleas, famines, and smallpox. Rinderpest decimated populations of many ungulates, particularly buffalo.

The establishment of a colonial regime at the turn of the century had far-reaching effects on Valley Bisa life and on wildlife. The consequent end to slavery, ivory trading, and local trading arrangements deeply transformed local political and economic power. The political system changed from one that had been fluid, competitive and autonomous to one fixed by a formal hierarchy and sanctioned by the state. The enforced payment of the hut tax, together with a desire for manufactured goods and prestige items (such as guns and bicycles), were factors that turned large numbers of men into migrants. Migrants found employment throughout central and southern Africa. Generally, they returned to their valley villages after middle-age. With the energies and talents of its young and middle-aged men siphoned elsewhere, village societies in the rural areas began to stagnate and to progressively lose their social dynamism. Yet the effects of colonial policies on the environment were even more detrimental in the long run (Vail, 1977).

The colonial administration acted quickly to control the guns and ammunition in African hands. At the same time, it declared illegal the taking of game with any of the traditional weapons or traps. The effects of these acts were to cut off the customary sources of animal protein without providing for its replacement, and to undercut a major means by which wildlife populations were kept in check. The prohibitions on traditional traps, snares, and defenses against damage by wild animal populations, together with the absences of large numbers of men in labor employment, set the stage for increasing wildlife populations and the spread of trypanosomiasis by its vector, the tsetse fly. Increasing wildlife populations during the 1920s and '30s forced valley residents to take initiatives to protect their crops and granaries. Their solution was to maintain two residences—one in the fields and one in the village, with the huts of lineage members ringing the cultivated site or the stored grain. In 1940, the government appropriated vast areas of the Luangwa Valley for game reserves and resettled people living within its boundaries.

When they assumed control, colonial administrators noted that the carcasses of buffalo that had succumbed to rinderpest were common and that the survivors were wary. In East Africa, rinderpest continued to cause heavy mortality among buffalo until the 1960s (Sinclair, 1977). Rinderpest disappeared as a result of successful government programs to innoculate against the disease in cattle, which had acted as carriers of the disease to buffalo. Subsequent to this successful campaign, buffalo populations throughout southern Africa have increased. This increase is apparent in Figure 2 for the Nabwalya Study Area. As buffalo numbers have increased, hunters at Nabwalya have shifted more of their efforts to the pursuit of this species.

Hunting by the Valley Bisa has undergone several transformations as their social and economic environments have changed. Until recently, most males were familiar with the use of bows and arrows, poisons, spears, and traps—weapons featured both in hunting and in warfare. The introduction of the state and capitalism outlawed the use of these indigenous methods and redirected the activities of most adult men away from their home villages and towards the cities. The imposition of these circumstances had far-reaching consequences and called for the reorganization of village life. With external demands on the productive energies of most of its men, lineage members designated one of their men as hunter to provide the lineage with wild meat and to protect their crops against raiding mammals. Those selected remained in their villages using their hunting skills and knowledge of local happenings to compete eventually for succession to the position of headman against the claims of their brothers and nephews, who through their employment elsewhere, returned in their middle years with money and a different set of skills. Thus, one of the major changes of the last century has been the differentiation of skills and restrictions in weapons and their ownership among lineage members.

Under the sanctions of the subsistence system, individuals who hunted for the matrilineage with muzzle-loading firearms could achieve ascendance within the social system through timely kills and through judicious distribution of meat among kinsmen. Strong norms mitigated against the sale or outside distribution of meat from animal carcasses killed locally. Furthermore, the hunter focused his pursuits on species whose consumption by humans carried few prohibited associations. A spectrum of species including hippopotamus (*Hippopotamus amphibus* Linn), zebra (*Equus burchelli* Gray), baboon (*Papio sp.*), ant bear (*Orycteropus afer Pallas*), and elephant shrews (*Petrodromus tetradactylus Peters*) were no longer sought as prey; the consumption of their flesh, for many people, became associated with the capricious occurrence of sores and leprosy. Hunters concentrated on pursuing buffalo, impala, and warthog. The theory that game animals possessed spirits dangerous to young hunters and their dependants insured the subordination of hunting and its proceeds to the control of elders who also monopolized the possession of weapons. Two circumstances contributed to the accumulation of guns in the hands of elderly men—succession to the headmanship of a village implied the inheritance of a gun, and the purchase of a gun became the standard investment for money earned through urban employment. Young and middle-aged men did most of the hunting, but the proceeds belonged to and were distributed among kinfolk by the elders. In time, the young hunter acquired prestige and followers. Given the right combination of circumstances, a hunter would eventually assume leadership of a faction within his village, and eventually of a village of his own.

Two points are worth mentioning about the hunting processes that I observed between 1966 and 1974. First, individuals varied in terms of the time they allocated to hunting, but the characteristic yields were quite high. During 1973, 50 hunts showed kills of 16 mammals. These yielded 1517 kg (butchered weight) of meat for a production figure of 9.4 kg of meat per hour of hunting time. Put another way, one mammal was killed for every 10 hours expended in hunting. Secondly, the sex of prey killed by hunters was such that there was little damage to the recruitment of wildlife populations. Both the species' behavior and hunters' use of muzzle-loading guns conspired to ensure that males of their prey were the predominant sex killed.

Capitalistic influences, together with government regulations and increasing accessibility of the Munyamadzi Corridor to outsiders, are today transforming the modes of hunting. Increased contact by the Valley Bisa with outsiders in the Corridor and with government civil servants who provide cash for meat and trophies (ivory, rhino horn), means that increasing numbers of animals killed are diverted from the distribution exchange system of the lineage into the cash market. These transactions give local hunters access to cash, a reward previously strongly denied them within the subsistence exchange context. This commercialization of wildlife meat and products affects village social life, differentially benefiting a few while many receive nothing, and influences the composition and abundance of wildlife.

During a ten-day visit in August 1978, the causes and consequences of these influences were evident (Marks, 1979b). The records of the National Parks and Wildlife Service, which administers the Munyamadzi Corridor as a game management area, show the following increases from 1974 to 1977: outside hunting permits issued, from 4 to 43; elephants killed under license, from 4 to 63; supplementary licenses purchased, from no information to 110. During 1977, the government maintained at Nabwalya a large work force whose mission was to drill a well and to construct an infirmary, a court, and a school building. Responding to cash incentives for meat from this contingent of outsiders, some local residents took to setting wire snares for buffalo, and others provided their hunting skills as guides for cash. Both the use of wire snares and breech-loading weapons increased predation on females. Sinclair (1977, p. 258) provides examples from the Serengeti of the destructive potential of wire snares on herds of buffalo. Increased offtakes from herds, including intensive predation of females together with the increased human traffic and activities around Nabwalya resulting from construction and from the amalgamation of villages around the chieftaincy, caused some of the remaining buffalo herds to shift elsewhere. In 1978, only two small herds remained (some hunters claimed it was one herd that had split). No bachelor bulls were observed during my stay. Local hunters blamed the safari hunters and outsiders, while admitting larger harvests than usual during previous years. Decreases in observed numbers for both buffalo and zebra are shown for 1978 in Figure 2. Today zebra are increasingly killed primarily for their skins, which brings up to K100 (US$150) in outside markets.

SUMMARY AND CONCLUSIONS

This paper describes a multiplicity of variables which over time have affected buffalo populations and human predatory behavior in the Luangwa Valley. I have not argued for a homeostatic system. Instead, I have sought to illustrate the dynamism of the recent history of both human and prey populations. Buffalo populations have fluctuated through time as a function of disease, quality of range and human predation. Human predation on buffalo was a function of technology, cultural style, and outside pressures which increased the demand for buffalo beyond the immediate society. This increased demand for buffalo is comparable to an overall increase in the density of the human population at Nabwalya.

To some extent in the past, the prey and human systems operated independently of each other. However, current human pressures have greatly reduced and will probably extirpate the buffalo from Nabwalya in the near future. The presence of buffalo elsewhere in the Munyamadzi Corridor and the Luangwa Valley retains the capability of this species to recolonize Nabwalya should recent developments there falter. This short-run view suggests that human predation on large mammals is largely opportunistic in nature.

What is the relevance of the current study to generalizations about prehistoric hunters of the now defunct Arctic steppe-tundra biome? First, I think it is important to remind ourselves of the speculative nature of our interpretations of past events and of the evidence upon which they are based. The case study reminds us that the linkages between humans and their resource environments are manifold and subtle even when their trends are documented from several different sources of information such as oral and written traditions and field research on both human and prey populations. Archaeological remains are not likely to enable us to establish conclusively whether or not man was the primary causal agent in exterminating a species. P.S. Martin (this volume) maintains that a shift in human use from a sustained yield to an exploitative economy was sufficient to explain the extinction of many large mammals in North America. He is also aware that his blitzkrieg theory of overkill is not likely to produce positive evidence in the fossil record. Haynes (this volume), based upon his own and others' excavations, describes the Clovis people as mainly foragers who hunted mammoths and bison for food and material resources. Currently, we have little evidence that humans were an important constituent of the Arctic steppe-tundra. My own proclivities incline me towards the following view. I suspect that prehistoric peoples associated with the defunct steppe-tundra were few, relatively insignificant, and thinly scattered at points on the landscape where they could advantageously tap into both floral and faunal production. These people were probably more dependent upon small and medium birds and mammals than upon plants, as current subsistence analogues in the Arctic suggest (North Slope Borough Contract Staff, 1979). Their small numbers inclined them more towards a generalized hunter and forager mode than towards being specialists on the larger forms. The opportunistic kill of a mammoth or the finding of a salvageable carcass was an occasional event certain to loom large in their collective memories as surely as it does in the memories of hunters today. As Olivier (this volume) suggests, these men must have been knowledgeable about mammoths and other large-mammal species and their respective vulnerabilities. But I do not see these prehistoric hunters as active causal agents in the extinction of their prey and armed with technologies and social structures capable of outstripping the regenerative abilities of their basic resources. I prefer to reserve such outcomes as a recent manifestation of the pervasive activities of industrial man. To attest to man's capacity for killing mammoth is not to conclude that he was the primal agent in its demise.

Secondly, human predation on wild mammals should be studied and reviewed within the framework of community ecology. The activities and prey selection of other predators provides information applicable to that of humans. The principle of competitive exclusion alerts us to expect differences in the selection by predators subsisting on the same or similar stocks of resources. In national park studies (Pienaar, 1969; Schaller, 1972), lions (*Panthera leo* Linn) are shown to select and kill more adult bull buffalo, which by virtue of circumstances are more vulnerable to predation than cows. However, lions at Nabwalya kill more cows, apparently accommodating the kills of bulls by the Valley Bisa. In this instance, Valley Bisa selection was more a reflection of their antiquated weaponry and vulnerabilities than of their cultural preferences for fat cows.

Thirdly, the ingenious assortment of traps and poisons in use by contemporary hunters, together with the rapidity with which they learn and relearn prey behavior, cautions us not to limit the repertories of prehistoric peoples to the more durable forms of their culture. It is difficult, if not impossible, to determine for a given site whether humans actively preyed upon the mammal, chanced upon it under vulnerable circumstances, poisoned or drugged it elsewhere, or scavenged its remains.

In conclusion, an ecosystem in the process of shifting states would increase the vulnerabilities of mammals and people alike. Mammals in marginal habitats were readily picked over or picked up by discriminating hunters, who in the process learned to transcend the collapse of their present.

ACKNOWLEDGMENTS

Materials for this article were collected while I was a research affiliate or associate with the Institute for African Studies at the University of Zambia and supported by the Foreign Area Fellowship Program in 1966-7, by the joint Committee on African Studies of the Social Service Research Council and the American Learned Societies in 1973, and under sponsorship of the Zambia Desk and the Development Studies Program, U.S. Agency for International Development in 1978.

The materials were organized and written for the Burg Wartenstein Symposium while I was a recipient of a Grant for Independent Research from the National Endowment for the Humanities in 1979 and a fellowship from the Harry Frank Guggenheim Foundation in 1980. The opinions expressed in the article are my own and do not reflect the policy of any organization which has supported my research. I also wish to thank the Wenner-Gren Foundation for Anthropological Research for a travel grant enabling me to attend the symposium and Martha S. Marks, whose love and loyal support for a subsistence wage earner continues to make his pursuits both meaningful and worthwhile.

PALEOECOLOGY OF BERINGIA–A SYNTHESIS

INTRODUCTION

To recreate a landscape, green with life or windswept and barren, and then to repopulate it with animals and men is a formidable task. That, however, is just the task that we have set ourselves in this book. Building on discussions at Burg Wartenstein and upon the papers collected in this volume, we now undertake to assess the paleoecology of Beringia during the last 40,000 years.

Pleistocene Beringia attracts our attention in part because of the paradoxical former abundance and diversity of large, gregarious ungulates in a region that now supports very few large mammals—a paradox heightened by the apparent presence of this larger, more diverse ungulate fauna at a time when colder world climates would seem to have made for ecosystems less productive than those of the present time. Hibbert reminds us that the substance of this paradox has awaited our attention for more than 50 years.

The effort to explain this productivity paradox has led to dramatically different reconstructions. At one extreme is Guthrie's postulate of a late Pleistocene arctic steppe whose richly productive grasslands supported herds of ungulates interacting with the vegetation and each other in a manner similar to game populations in the grassland vegetation of the present-day African Serengeti

Plain.[1] At another is the reductionist picture of Ritchie and Cwynar which postulates a Beringian Pleistocene landscape supporting only sparse, discontinuous vegetation of herbaceous tundra on upland sites and local sedge-grass meadows on lowlands—a landscape that seemingly could hardly have supported large ungulate populations.

Failure to resolve this debate previously has been at least partly due to a lack of precision in discussions of Beringian chronology and stratigraphic correlation. Was there really a productivity paradox, after all? The Pleistocene was hardly a period of constant climate and unvarying landscape. Ritchie and Cwynar question whether the rich and well-known ungulate fauna was really contemporaneous with the depauperate low-influx pollen spectra of late Pleistocene time. Pollen spectra generally have been more precisely dated than vertebrate faunas, most of which consist of bones collected out of context from beaches, river banks, and floors of placer pits. Also uncertain has been the time relationship and degree of contemporaneity among pollen spectra, macrofossil floras, vertebrate remains, glacial advances and retreats, times of activity of the widespread stabilized dunes, and times of activity of the equally widespread features recording formerly more intense frost-heaving and solifluction. Elsewhere in this volume, Hopkins undertakes to provide the requisite conceptual framework by defining a series of time-stratigraphic units for Pleistocene Beringia and Matthews and Giterman and colleagues address these questions of correlation and chronology in an effort to sharpen our appraisal of the extent to which there really was a productivity paradox.

We shall begin by reviewing our current knowledge of what was where and when. We focus our discussion upon the situation in Beringia between 40,000 and 14,000 years ago, because the reality of the productivity paradox is quite evident within this interval and because our picture of the flux of climatic, paleogeographic, and biotic changes at earlier times is still sparse and also severely handicapped by the imprecision of available geochronometric methods. One insight that immediately emerges from this survey is that although we have a rich vertebrate, insect, and plant macrofossil record from the latter part of the Boutellier interval and also from the time of the birch pollen-assemblage zone, we have very little macrofossil material clearly dating from the Duvanny Yar interval. Most of the lacustrine pollen records, on the other hand, begin within the Duvanny Yar interval or even later.

[1]Citations of authors' names without accompanying dates refer to papers in this volume.

THE SCENARIO

Beringia During the Late Boutellier Interval

During the last ten or twelve millenia of the Boutellier interval, Beringia saw a gradual and probably fluctuating deterioration from the conditions of a relatively mild, mesic interstadial that peaked about 40,000 years ago toward the conditions of a severe cold, dry glacial and periglacial stade, the Duvanny Yar interval. At the beginning of this transition, sea level lay not far below its present position. The land bridge was reduced to a narrow isthmus and may even have been briefly flooded. The coastal mountains were heavily glaciated, but mountain masses farther north may not have held glaciers much larger than the present ones. Thermokarst ponds and lakes were scattered in lowlands and major river valleys. Peat bogs, however, were rare, and the poorly drained, acidic soils that characterize so much of Beringia today must also have been rare. The great dune fields so active during the Duvanny Yar interval were generally quiescent, 40,000 years ago, but loess was accumulating slowly on vegetated lowland surfaces in several and perhaps in many places.

Spruce, larch, cottonwood, the shrubby stone pine, and possibly aspen may have reached their present-day geographic limits 10,000 years ago, but seem to have been distributed in rather dry, open woodlands quite distinct from the mixtures of forest and open bog that characterize present-day Beringian taiga. Willow thickets evidently were present, but alder shrubs either were lacking or restricted to a few, small, isolated refugia. Dwarf birch was present, possibly with about the same cover value as in present-day forest-tundra areas. Ericaceous dwarf shrubs and *Sphagnum* moss played only minor, local roles in the ground cover. Pollen spectra and assemblages of insect and plant macroremains dating from 40,000 to 30,000 years ago exhibit mixtures of forest and open-ground species. Species indicative of firm, dry substrates are prominent, and included are some taxa with steppe affinities. No pollen influx data are available for pollen sequences as old as 40,000 years, but by 30,000 years ago, pollen-influx rates amounted to only two or three hundred grains/cm^2/yr—or only a small fraction of Holocene pollen-influx rates at the same sites (Cwynar, 1982).

These observations depict a vegetation quite different from that of modern Beringia. Instead of today's mosaic of closed forests and muskegs within the taiga and vast expanses dominated by *Carex* marsh, sedge-tussock tundra, and dwarf-shrublands above or beyond the forest limit, the vegetation seems to have consisted of extensive meadows and steppe-tundra on upland sites; shrub thickets on floodplains, on lower slopes near persistent snowbanks, and on other sufficiently moist sites; open woodlands on favorably situated slopes; and lines of trees along riverbanks and beside smaller water courses. The lack of alders, Young suggests, may reflect the generally dry climatic conditions.

Insect faunas of Boutellier age consist of mixtures of mesic and dry-ground species but are dominated by species now found in open, treeless areas. Some of the East Beringian assemblages show high relative abundances of *Aphodius,* a genus which consists predominantly of dung feeders.

Many of the dated mammal faunas and many of the preserved large-mammal carcasses in Beringia date from 40,000 to 30,000 years ago (Table 1; also see Matthews, Fig. 2). The ungulate fauna during the latter part of the Boutellier interval was indeed rich and diverse and quite different from modern Beringian faunas in either tundra or taiga regions. In addition to mammoth, horse, and bison, which evidently continued to share dominance through the Duvanny Yar interval, the fauna of the Boutellier interval included abundant reindeer/caribou (*Rangifer*) as well as mountain sheep (*Ovis*), steppe antelope (*Saiga*), and moose (*Alces*). Camel (*Camelops*) was present in eastern Beringia, and woolly rhinoceros (*Coelodonta*) in the west. Perhaps it was during the Boutellier interval that badgers and black-footed ferrets, both denizens of present-day North American steppe regions, extended their ranges northwestward into Beringia.

During the final 10,000 years of the Boutellier interval, sea level gradually lowered, and a progressively larger expanse of the Bering land bridge became exposed. By 30,000 years ago, Beringia was essentially treeless, many sand-dune areas had been activated or reactivated, and loess was accumulating at an accelerating pace. Nevertheless, there seems to have been little, if any reduction in the diversity of the ungulate fauna up to that time.

The climate of the Boutellier interval was clearly more mesic than that of the ensuing Duvanny Yar interval, but just as clearly, Boutellier climates were drier than present-day Beringian climates. The distinctive steppe-tundra mammalian fauna was already in place. The numerous occurrences of well-preserved large-mammal carcasses in redeposited loess and stony, organic-rich colluvium suggest that individuals died in winter and became buried beneath deposits of spring snowmelt floods and gully washouts. Probably the climate of all except the earliest part of the Boutellier interval was characterized by stormy, snowy winters and dry, sunny summers.

Beringia During the Duvanny Yar Interval

Sea level fell to its Wisconsin minimum about 18,000 years BP and then recovered enough by 14,000 years BP to breach the Bering land bridge (Hopkins, Fig. 3). When sea level was at its minimum, the land bridge was a plain more than 1000 km wide north to south connecting North America and northeastern Asia, Alaska and the Yukon became the easternmost province of an unglaicated zone that extended westward across Siberia and ultimately into central and western Europe.[2] This wide corridor assured a high degree of biotic exchange, limited only by the size of the species pools available at each end and by the limited range of terrain types and climatic situations on the land connection itself.

Although the exposed marine shelves were relatively featureless, there was great topographic diversity in the Yukon, Alaska, and northeastern Siberia. Extensive glaciers and icecaps covered southern Alaska and Yukon Territory as well as the Koryak Mountains and Kamchatka. The northwestern margin of the Laurentide ice sheet never lay far from the easternmost boundary of Beringia. Glaciers of more limited extent advanced in the

[2]Most of the locations mentioned in the text can be found on the endpaper map or on Hopkins' Figure 1.

Table 1
Well-dated Late Pleistocene Beringian Faunas[1,2]

	BOUTELLIER INTERVAL										DUVANNY YAR INTERVAL				BIRCH ZONE								
	Old Crow Flats[4]	Canyon Creek[5]	Kirgiliakh Cr.[6]	Selerikan Cr.[7]	Ezhantsy[8]	Fairbanks Area[9]	Ust-Mil' II, B, C[10]	Jack Wade Cr.[11]	Tyonek Creek[12]	Ikhine II, B, V[13]	Duvanny Yar[14]	Porcupine Cave[15]	V. Troitskaya[16]	Alyoshkina Z.[17]	Trail Cr. Caves[18]	Berelyekh R.[19]	Teklanika R.[20]	Bluefish Caves[21]	Dyukhtai Cave[22]	Old Crow Flats[23]	Ust-Mil' II, A[24]	Dry Creek[25]	Ushki I, Lev. VI[26]
Mammuthus primigenius (mammoth)[3]	138	1	1		8		9		1	7			8	X		140	X	X	788		4		
Coelodonta antiquitatis (woolly rhino)					9		4			4			3		1	1							
Bison (bison)	21	2?			8	1	5	3		72		X	23	X		2		X	15	X	3	X	X
Ovibos (musk ox)																		X	1				
Ovis sp. (mountain sheep)		1						11				X			1			X	13			X	
Saiga (saiga)														X									
Cervus elephus (wapiti or red deer)															?			X				X	
Alces alces (moose)	1	1						1										X	41				
Rangifer tarandus (caribou or reindeer)	5	1(cf.)			10			2		8			2	X		4		X	42				
Camelops hesternus (yesterday's camel)		1																					
Equus sp. (horse)	1	2		1	20		7	1		64	1	X	9	X	1	3		X	16		9		
Spermophilus parryi (ground squirrel)	6	X									X							X					
Marmota monax (woodchuck)	2																						
Lepus arcticus (arctic hare)	22	1														827		X	196				
L. tenaiticus (steppe hare)														X				X					
L. americanus (snowshoe hare)	70																	X					
Dicrostonyx torquatus (collared lemming)	75																	X	18				
Lemmus sibiricus (brown lemming)	3																	X	10				
Ondatra zibethicus (muskrat)	25																						
M. miurus (singing vole)	1																						
Castor canadensis (beaver)	4																						
Castoroides ohioensis (giant beaver)	3																						
Felis canadensis (Canada lynx)	1(cf.)																		9				
Panthera leo (lion)	1																						
Alopex lagopus (arctic fox)	3									1								X	47				
Vulpes (red fox)										1			1					X	18				
Canis (gray wolf)		1														X			8				
Gulo gulo (wolverine)																4							
Mustela erminea (short-tailed weasel)	1																						

NOTES

[1]All of the faunas listed in this compilation were either found in situ or in situations where their original in situ position could be clearly established. Most are associated with radiocarbon-dated material or include soft parts or antlers that have been directly dated. A few rely upon bone-collagen dates that are confirmed by radiocarbon dates on organic material stratigraphically above or below.

[2]X = taxon present. Numbers in italics = number of elements (bones, teeth, antler or tusk fragments). Numbers in boldface = minimum number of individuals.

[3]All proboscidian fragments are assumed to be mammoth.

[4]Old Crow Flats, Yukon Territory, Unconformity A, more than 50,000 years BP (Morlan, 1980, Table 6.1).

[5]Canyon Creek, Alaska, 40,000 years BP or older (Weber, et al., 1981).

[6]Kirgiliakh Creek, Siberia, ca. 40,000 years BP (Shilo, et al., 1981; Arslanov, et al., 1981).

[7]Selerikan Creek, Siberia, 38,000–35,000 years BP (Arslanov and Chernov, 1977).

[8]Ezhantsy, Siberia, 40,000–35,000 years BP (Mochanov, 1977).

[9]Fairbanks area, Alaska, ca. 36,000 years BP (Guthrie, 1981).

[10]Ust-Mil' II, layers B and C, Siberia, 33,000–30,000 years BP (Mochanov, 1977; Tseitlin, 1979).

[11]Jack Wade Creek, Alaska, ca. 30,000 years BP (Porter, 1979).

[12]Tyonek Creek, Alaska, ca. 29,000 years BP (Thorson, et al., 1981).

[13]Ikhine II, layers B and V, Siberia, 32,000–30,000 or 26,000–24,000 years BP (Mochanov, 1977; Tseitlin, 1979).

[14]Duvannyy Yar, Siberia, less than 30,000, more than 14,000 years BP (Sher, et al., 1979; Hopkins, this vol.).

[15]Porcupine River cave LPC-1, Alaska, ca. 19,000 years BP (E. J. Dixon, Jr., written communications 1978, 1979).

[16]Verkhne Troitskaya, Siberia, ca. 18,000–14,500 years BP (part of the lithic collection from V. Troitskaya comes from undated strata underlying the dated fauna (Mochanov, 1977; Tseitlin, 1979).

[17]Alyoshkina Zaimka, Siberia, 15,000 years BP and slightly younger (Sher, et al., 1979; Hopkins, this vol.).

[18]Trail Creek caves, Alaska, 16,000–13,000 years BP (wapiti at Trail Creek was found in a different cave from the bison and horse and its age is uncertain) (Larsen, 1968).

[19]Berelyekh River, Siberia, ca. 13,500 years BP (Vereshchagin, 1977; Mochanov, 1977).

[20]Teklanika River, Alaska, ca. 13,500 years BP (Norman Ten Brink, written communication 1980).

[21]Bluefish Caves, Yukon Territory, 15,000–12,000 years BP (Cinq-Mars, 1979; Morlan and Cinq-Mars, this vol.; R. E. Morlan, personal communication 1982).

[22]Dyukhtai Cave, Siberia, 14,000–12,000 years BP (Mochanov, 1977; Tseitlin, 1979).

[23]Old Crow Flats, Yukon Territory, filled-gully locality, MkVI-9 and MkVI-12, 13,000–11,000 years BP (Morlan, 1979, Table 9.4 and p. 267–8).

[24]Ust-Mil' II, layer A, Siberia, ca. 12,000 years BP (Mochanov, 1977; Tseitlin, 1979).

[25]Dry Creek, Alaska, ca. 11,000 years BP (R. D. Guthrie and W. R. Powers, unpublished manuscript 1979; R. D. Guthrie, personal communication 1981).

[26]Ushki I, level VI, Kamchatka, ca. 10,500 years BP (Dikov, 1977).

Brooks Range and in many smaller mountain massifs throughout Beringia. Lakes blocked by glacial disruption of river systems and enlarged by glacial meltwater occupied lowland basins in Yukon Territory and Alaska. Lakes were also present in two tectonic basins on the exposed shelf of northern Bering Sea (Hopkins, 1979a, Fig. 1.4).

The glaciers fed meltwater streams which braided across broad floodplains, aggrading valleys with alluvium and outwash. Reworking of sediment led to extensive loess deposition. Sand exposed on the unstable floodplains and on the exposed and thinly vegetated continental shelf was whipped into active dune complexes that came to cover large areas of the land bridge; the present-day coastal lowlands of northwestern Canada, Alaska, and northeastern Siberia; and broad interior alluvial valleys and basins.

Beringian climates were most severe during the Duvanny Yar interval. The Polar Front would have been displaced southward at all seasons, and probably lay south of the Beringian shoreline during much of the year. Hopkins' reconstruction based on aeolian deposits confirms Barry's suggestion that summer circulation was predominately southerly and southeasterly rather than southwesterly as at present, and that it was perhaps controlled by a high-pressure cell over the Laurentide ice sheet and a low over the northwestern Pacific Ocean or the southwestern Bering Sea. Winter circulation was easterly and northeasterly and evidently controlled by a persistent low over the Gulf of Alaska and perhaps also by katabatic winds off the Laurentide ice sheet.

Lowered surface temperatures on the North Pacific Ocean, reduced land areas, and the presence of a seasonal ice cover on the Bering Sea would have reduced the value of these water bodies as evaporative moisture sources during the Duvanny Yar interval. The great breadth of the exposed Bering and Chukchi shelves and the interception of moisture in air masses crossing the glacierized southern mountains contributed further to diminish the precipitation reaching central and northern Beringia at all seasons. The predominately northward and northwestward trend of summer storm tracks should have made the diversified terrain of the Chukotsk Peninsula and the Anadyr Mountains the best-watered region of Beringia and may account for the ambiguous evidence that *Populus* and *Alnus* persisted somewhere west of central Alaska (Anderson, 1981). On the other hand, moist air masses would have penetrated least often to the Arctic coastal lowlands of northern Alaska and Yukon Territory. This region probably was, as Young suggests, polar desert during the Duvanny Yar interval. All of these factors contributed to a drastic reduction in precipitation. Away from southern coastal regions, present-day Beringia has a modest 100–400 mm of annual precipitation, of which more than half falls during the summer months; annual precipitation during the Duvanny Yar interval was probably reduced to half these amounts. Thus, summers were dry and sunny, and winters were dry and windy. Snow cover was thin and discontinuous throughout unglaciated Beringia and almost absent in what is now northernmost Alaska and Yukon Territory.

The very modest reductions in snowline in Beringia during the Duvanny Yar interval (100 m in the eastern Brooks Range and 200 to 300 m in the western Brooks Range and on Seward Peninsula according to Hamilton, 1981, and R. E. Nelson, unpublished data, 1974) dramatize the very dry climatic conditions of that time. Nevertheless, snowline *was* lowered and, as Barry emphasizes, this reduction in the face of widespread evidence of very low snowfall must mean that average annual temperatures also were substantially lowered.

Plant macrofossil accumulations of Duvanny Yar age are very rare, and so our picture of the vegetation must be reconstructed largely from pollen evidence. Invertebrate animal fossils are also rare, as are vertebrate remains clearly assignable to the Duvanny Yar interval. Nevertheless, the evidence is sufficient to allow us to state that the biota present during the Duvanny Yar interval represent a depletion of the elements comprising the Beringian ecosystem during the Boutellier interval rather than a transformation and enrichment of an impoverished ecosystem like the present one.

To the extent that negative evidence allows conclusions, Beringia seems to have been almost treeless (Hopkins, Smith, and Matthews, 1981). Cottonwood persisted, among other places, in riverine sites in northwestern Alaska, judging from the persistence of *Populus* in pollen spectra of Duvanny Yar age from the Noatak and Kobuk River drainages (Anderson, 1981). Giterman believes that *Pinus pumila* also persisted locally in northeastern Siberia. Alder was either absent or restricted to refugia near present-day Bering Strait.

Dwarf birch was present but infrequent, and ericaceous shrubs and subshrubs were also rare. Bryophytes, so important in the region today, had drastically lower cover values, and *Sphagnum*, one of the dominants of many Holocene lowland biotopes, was virtually absent (Janssens, 1981). *Selaginella sibirica* was a prominent ground cover in Siberia but not in Alaska and the Yukon. Plants with steppe affinities persistently appear as minor elements in pollen spectra and also in the very rare plant macrofossil assemblages. Examples include Chenopodiaceae pollen and seeds of *Chenopodium* and *Corispermum*. Halophytes seem to have been more common—note Matthews' report of *Ranunculus cymbalaria* and *Carex maritima* in the east and Giterman's report of *Eurotia* and *Salicornia* in the west.

The low pollen influx rates of Duvanny Yar time (100 ± 50 grains/cm^2/year—rates comparable, as Ritchie and Cwynar emphasize, to influx rates in modern fellfield and polar-desert communities) point toward an incomplete plant cover. This is confirmed by the fossil insect assemblages. The widespread evidence of redistribution of sand and silt by aeolian processes also confirms the lack, in many places, of a continuous vegetation cover. The general scarcity of organic detritus, roots, and buried turf layers in alluvium, pond sediments, and loess of Duvanny Yar age supports the inference that the vegetation was mostly sparse and of low stature and indicates a general lack of net organic accumulation.

Thus, we arrive at a reconstruction of Duvanny Yar vegetation generally similar to Yurtsev's and somewhat less optimistic than Young's (although we believe that Young does provide a "best-case scenario" for local areas within each former Beringian vegetation formation). Our reconstruction differs from Ritchie's and Cwynar's mainly in the greater emphasis placed upon moisture as a limiting factor and upon ruderals and other species tolerant of drought and disturbance as major floristic elements. It seems that graminoid-dominated vegetation intermixed with a variety of herbs covered large areas of Beringia including most of the land bridge. Judging from the percentages of *Artemisia* in pollen spectra of Duvanny Yar age, the genus was more important in the open, discontinuous plant communities of Duvanny Yar time than in most present-day Beringian plant communities. Mountainous uplands and high plateaus had little plant cover at all. Shrub thickets were certainly present, but consisted mostly of willows and were restricted to those few biotopes where moisture relations were especially favorable. Schweger's postulate that valley bottoms and snow-bed areas supported willow thickets and relatively productive sedge-dominated meadows emphasizes the importance of those areas as productivity "hot spots." Young is probably correct in suggesting that tall-shrub communities, including *Populus balsamifera* and *Salix alexensis* but not *Alnus* could be found locally around springs and seepages and along riverbanks.

The Duvannyy Yar type section has yielded a fairly rich series of insect faunas, summarized for us by Giterman and her colleagues. East Beringia has thus far produced only a few insect faunas of Duvanny Yar age, of which the best documented is Matthews' 24,000-year-old sample 21 (see his Fig. 1). All of these faunas show a dominance of taxa indicative of dry substrates and treeless conditions. True steppe species of Mongolian origin are prominent in the Siberian faunas but have not yet been recognized with certainty in insect faunas in eastern Beringia. The Cleoninae fossils require further study, however; some may represent the Mongolian species recognized in East Beringia assemblages. Aquatic beetles are almost lacking, in contrast to their abundance in most insect faunas of older and younger age. Beetles of the genus *Aphodius* (dung beetles) are consistently present.

Some of the dominant insect species of East Beringian fossil assemblages survive today in calcareous fellfield sites at higher altitudes in the Yukon. The factors that cause them to be presently restricted to such areas are not known. It is certain, however, that regosolic soils are not the prime factor because the same species were evidently abundant in the alluvial lowland areas during both the Boutellier and Duvanny Yar intervals.

Mammalian remains dating from the Duvanny Yar interval, although scarce, are nevertheless sufficient to clearly establish the presence in parts of Beringia of mammoth, bison, horse, caribou, mountain sheep, and saiga (Table 1). If radiocarbon dates on bones are to be believed, both barren-ground musk ox (*Ovibos*) and the so-called forest musk ox (*Symbos*) were also present (Matthews, Fig. 2). An ungulate fauna different from and more diverse than present-day Beringian ungulate faunas was, indeed, present in the region during the Duvanny Yar interval.

Beringia During the Time of the Birch Zone

An abrupt climatic change 13,500 ± 500 years ago resulted in hydrologic and vegetational changes that swiftly transformed the landscape of Beringia. Rapidly rising sea level drowned ever-widening areas of the continental shelves of the Bering and Chukchi Seas. Dwarf birch proliferated, evidently in response to the more mesic climate; climatic warming at that same time is indicated by rapid deglaciation. The new climatic and landscape regime persisted, broadly speaking, through the next 5000 years, although the record is complicated by evidence, still poorly understood, for a series of climatic oscillations reminiscent of the very late Pleistocene and early Holocene climatic oscillations recorded in Western Europe. The record is also complicated by the relatively slow dispersal across Beringia, during this interval, of forest species that evidently had been eliminated earlier by the severe climatic conditions of the Duvanny Yar interval.

Beringian vegetation during the time of the birch zone must have been lusher and may have been more productive than at present. Pollen influx rates, for example, increased rapidly from the low levels of the Duvanny Yar interval, reaching peak rates several times those of the Holocene.

A rich record of the vegetation and invertebrate faunas of the birch zone is preserved in Beringian thaw-lake deposits and terrace alluvium dated 13,500 to 8500 years old (e.g., McCulloch and Hopkins, 1966). Abundant macrofossil and pollen evidence indicates that the cottonwood (*Populus balsamifera*) became widely distributed, extending its range into present-day tundra areas where cottonwood trees no longer grow. Aquatic biota, including the cattail (*Typha latifolia*), the large freshwater mussel (*Anodonta beringiana*), and the beaver (*Castor canadensis*) extended their ranges well beyond later Holocene limits. In spite of the evidence, however, for a hydrologic regime altered in such a way that lakes and through-flowing streams became abundant in what had previously been a very arid region, landscapes during the time of the birch zone seem to have been much less waterlogged than most present-day Beringian landscapes. Peat accumulations dating from the time of the birch zone are rare. *Sphagnum* is only a minor element in spore-pollen spectra; *Artemisia* and Chenopodiaceae continue to be significant pollen taxa; and chenopod seeds are abundant in some of the macrofossil collections. Evidence of continuing sand movement, albeit at reduced intensity, during the time of the birch zone testifies to an incomplete plant cover on sandy soils.

East Beringia has produced many insect faunas of birch-zone age, but most have not yet been published. Several from deposits 13,500 to 12,000 years old are listed by Matthews (Fig. 1, *9–11*). These resemble, in most respects, insect faunas from deposits of late Boutellier age. They consist of mixtures of mesic and dry-ground species and are dominated by taxa now found in open, treeless areas. *Aphodius* (dung beetles) are numerous in one sample. Interestingly, insect faunas associated with macrofloras that contain an abundance of Chenopodiaceae also contain weevils of the family Cleoninae, a group that usually feed on chenopods.

The paleoclimatic proxy data of birch-zone age all seem to point toward a climate characterized by snowy winters, a rapid spring snowmelt with resulting floods, mudflows and gully washouts, and warm, dry, sunny summers.

Whatever the cause, the time of the birch zone was also a time of rapid and dramatic changes in Beringian large-mammal faunas. Mammoth disappeared soon after 14,000 and horse soon after 13,000 years BP. Bison persisted three or four thousand years longer but then either disappeared or persisted only in diminished and localized relict populations. Reindeer/caribou were present, probably in greater abundance than during the Duvanny Yar interval, and moose (*Alces*) appeared. There is no record of the presence of *Symbos* or *Saiga,* but mountain sheep were widely distributed until 10,000 or 11,000 years BP. Beaver, apparently absent since the mesic, peat-forming episode that ushered in the Boutellier interval, reappeared in Beringian streams and ponds about 10,000 years ago.

Especially interesting is the record of wapiti/red deer (*Cervus*). After an absence that began some time beyond the reach of radiocarbon dating, wapiti/red deer reappeared and became widely distributed in Beringia no earlier than 15,000 years BP, then disappeared again no later than 10,000 years BP. Perhaps the abundance of browse in the generally open, treeless landscape of early birch zone time was an environment more favorable for wapiti/red deer than either the dry steppe-tundra landscapes of the late Boutellier and Duvanny Yar intervals or the generally waterlogged tundra and taiga landscapes of the Holocene.

THE PRODUCTIVITY PARADOX

Our close and critical examination of the paleobiological, paleoclimatic, and geochronologic evidence shows that the productivity paradox was, indeed, real. Even during the Duvanny Yar interval, the worst of times, at least six and possibly eight large ungulate species—mammoth, horse, bison, reindeer/caribou, mountain sheep, saiga, and perhaps also tundra and forest musk ox—were present in Beringia. This ungulate fauna lived at a time when the climate was cold and arid and when the complex of biotopes that we term steppe-tundra—a mosaic of barren mountain slopes, discontinuous cryophytic vegetation on uplands, and treeless, graminoid-dominated meadows and willow thickets in sites where moisture was least limiting—extended right to the south coast of the land bridge.

Despite the problems posed by flexible dietary requirements and variable behavioral repertoires, Guthrie, Olivier, and Vereshchagin and Baryshnikov all show us that important paleoenvironmental inferences can be drawn from the presence of these six to eight ungulates. All are gregarious. Although we can say almost nothing about Beringia-wide population densities, the ungulates were present in numbers, where present at all. Olivier informs us that the mammoth population, for example, would have included matriarchal nursery herds numbering at least six to eight cows and calves as well as the more solitary bulls. The presence of mammoths, then, implies concentrations of fodder capable of supporting and (in summer) fattening groups of six to eight animals. These concentrations must have been spaced closely enough so that the nursery groups could move from one productive spot to the next without excessive weight loss. Analogous inferences can be drawn from the presence of the other ungulate species.

Mountain sheep and musk ox are more or less sedentary and tend to cluster within familiar areas a few tens of kilometers across, but the other ungulate species were wide-ranging, essentially nomadic, and capable of moving hundreds or even thousands of kilometers over the course of a single year. The presence of sheep and musk ox implies that there were areas a few tens of kilometers across that had sufficient productivity to sustain groups of at least ten to fifteen individuals from one year to the next. The presence of mammoth, horse, bison, saiga, and caribou, on the other hand, may simply imply the former presence of productive hot spots that might have been visited and intensively exploited only once in a decade, allowing time for extensive recovery of the plant cover between episodes of intense grazing and trampling.

We note that sheep evidently inhabited diversified upland landscapes, but their wide distribution as fossils indicates that during the late Pleistocene they were much less dependent than at present on availability of nearby rugged escape terrain. Modern Eurasian saiga, on the other hand, are dependent on their ability to flee over flat, unbroken terrain. The almost complete lack of saiga remains in upland areas indicates that ancient Beringian saiga populations were denizens of the broad, open valleys and coastal lowlands.[3] The distribution of fossil remains suggests that mammoth, horse, bison, and caribou/reindeer probably utilized both the hilly uplands and the lowland flats, although Guthrie (1968a) notes a bias in the Fairbanks area toward abundant horse remains near the Tanana valley and abundant bison remains deeper in the uplands. There may be paleoenvironmental significance in the fact that horse remains are disproportionately abundant in late Pleistocene deposits at the inner edge of the Arctic Coastal Plain of northern Alaska and that vertebrate remains of any kind are rare in the stabilized sand-dune and coastal-plain areas still farther north.

Feeding mechanisms and dietary strategies differed among the ungulate species (Guthrie). Two of the dominants—mammoth and horse—were cecal digesters, primarily grazers, and adapted to survive on low-quality forage in large quantities. Olivier, Redmann, Jerome Martin, and Bliss and Richards each note the requirement shared by all ungulates for a balanced variety of plant foods.

[3]The finds of saiga in placer excavations near Fairbanks lie within a few kilometers of the open Tanana Valley lowland and probably represent individuals or groups that took temporary shelter in upland valleys during winter blizzards.

They emphasize the need for browse even among species such as horse and mammoth that we commonly conceive as grazers. Neither horse nor mammoth was a selective feeder. The lumbering mammoth must have been the species least prepared to be discriminating in its choice of food items, and probably the species that had the most devastating effects upon the plant cover in the areas where it fed. Bison, the third dominant, is notable among ruminants for its ability to process relatively low-quality forage, although at a limited rate. Sheep, musk ox, reindeer/caribou, and saiga are small-mouthed, selective feeders which differ, however, in their favored food items. Sheep and musk ox capitalize on the most tender and nutritious plant parts available on the landscape. Sheep, especially, depend upon ruderals in mesic meadows. Reindeer/caribou are catholic in their food choices but unique in their ability to locate and utilize lichens in their winter range; they do well in the present-day mesic tundra and seem to have been less numerous during the Duvanny Yar interval than at present. Saiga utilize a spectrum of plant species but are notable for their ability to make effective use of harsh and bitter browse, including halophytes—food resources unpalatable to most other ungulate species.

The range of mobilities, terrain preferences, feeding choices, and digestive strategies among the ungulates of Duvanny Yar time suggests, as both Redmann and Bliss and Richards point out, a diversity rather than a great density of productive biotopes. It suggests the possibility of a partitioning of resources in areas that were at times jointly occupied by two or more ungulate species. And it suggests that Matthews (1976a) was probably correct when he postulated that Beringian ungulates interacted in ways analogous to the well-known grazing successions of central Africa. Redmann points out that these same interactions can be perceived among the less diverse modern ungulate faunas of the Great Plains of North America.

The morphology of the late Pleistocene ungulates is informative. Large stature and extensive investment in social ornamentation among most of the ungulate species suggest to Guthrie the impact of a winter bottleneck, that is, a winter stress period which limited recruitment, kept populations low, and reduced intraspecific competition for what must have been the abundant food resources of the summer growing season. (It is interesting though that late Pleistocene Beringian caribou were generally smaller than the living forms.) We may speculate that the winter bottleneck did not involve an inability to cope with the extreme cold but was rather a resource limitation—a long season during which food was scarce.

The productivity paradox, then, resolves itself into a question of how the sparsely vegetated Beringian landscape of Duvanny Yar time could have supported clusters of large-bodied herbivores that, no matter how scattered, still required local richly productive forage areas spaced closely enough to allow energy gained during feeding periods to be balanced, over the course of a year, by energy lost while moving from one feeding area to the next. Differences in dietary preferences, including differences in preferred forage species and preferred plant parts must have permitted constructive interactions among different species utilizing single forage patches. But no matter how significant these interactions may have been, it is clear that the Beringian steppe-tundra of the Duvanny Yar interval must have contained well-dispersed productivity hot spots. The crux of our reconstruction of the Beringian ecosystem must be an understanding of these patches of significant productivity.

Climatic and Microclimatic Factors

Gal-Chen reminds us that incoming radiation has fluctuated considerably during the last 40,000 years at the latitude of Beringia. The amount of incoming summer radiation received 35,000 years ago was about the same as present. A summer-radiation minimum occurred early in the Duvanny Yar interval, 25,000 years ago, and a peak occured at the beginning of the Holocene, 10,000 years ago.

Direction of slope exposure becomes important at the high latitude of Beringia. Barry points out that differences in the amounts of radiation received on moderately steep north- and south-facing slopes in the Yukon-Tanana uplands correspond to "equivalent latitude" differences of about 30°! This means that even during the summer-radiation minimum of 25,000 years ago the more hilly parts of Beringia would have included significantly warm areas on south- and southwest-facing slopes.

Important differences in winter microclimate result today from the persistent low-level temperature inversions that affect many areas of Beringia. To the extent that winter air is still, valley bottoms and basins become the sites of extremely low temperatures. On the other hand, wind chill is much more severe on hilltops and in open lowland areas. The winter-active sand dunes indicate that many areas of Beringia suffered windy winters during the Duvanny Yar interval and that these must have been severely cold. However, the duration of the dormant season, accessibility of forage, and variations in the thickness, density, and continuity of the snow cover probably were more important factors governing the nature of the Beringian ecosystem than were ambient temperatures.

Discontinuity of winter snow cover throughout most of Beringia during the Duvanny Yar interval is a certainty. Lack of trees and rarity of shrubs meant that the reduced snowfall would have been redistributed by winds, allowing upland sites to be blown clear at least in patches and causing large drifts to accumulate on lee slopes. Cratering by ungulates in search of buried vegetation must have contributed further to the patchiness of the winter snow cover. To be sure, the long winters were periods when there was no new growth, but the thin and patchy snow cover meant that the persisting above-ground stems and leaves remained accessible to the ungulates.

The reduced albedo of a patchy snow cover should have led to earlier thawing and, as Guthrie suggests, to early onset of spring growth. Windblown sediment deposited on snow also should have played an important role in reducing albedo. Gal-Chen doubts that spring growth was initiated much earlier than at present because, despite the evidence of reduced snow cover, "the ice-albedo feedback is far more effective in a snow-covered tundra . . . than in a snow-covered forest." Dirty snow, however, has a lower albedo and thaws much more rapidly than does clean snow. The net effect of the thinner, patchier, and dirtier snow cover of the Duvanny Yar interval was to bare and warm the soil earlier in the spring and thus move the peak in plant production closer to the annual radiation peak.

The northerly and northeasterly winds of winter gathered snow, as they still do in the treeless regions of Beringia, in large lee-slope drifts that during the spring and early summer radiation peak lay on or near the warmest spots on the landscape. Thus, persistent moisture sources were juxtaposed with areas having the most favorable radiation climate. Guthrie's speculation that spring precipitation was an important part of the late Pleistocene climatic pattern seems unlikely and may be unnecessary.

Instead of a single south-to-north regional wave of spring growth, there would have been many local "spring" protein flushes distributed throughout the region during a period of several months, as the patchwork of snowbanks melted. We think it reasonable to assume that patches of protein-rich hot spots may have been available in various places on the Beringian landscape throughout an annual growing season lasting perhaps four or five months. Groups of phenology-tracking ungulate species thus might have reenacted a grazing succession in one seasonal hot spot after another.

Edaphic Factors

Late Pleistocene soils and substrates differed in important ways from those of the present. A persistent observation that recurs in many contributions to this volume is that for late Pleistocene Beringia there is much direct evidence of dry, firm, easily excavated soils and an absence of surficial organic accumulations. A dry climate and lack of organic accumulation at the soil surface would interact to favor warmer, more deeply thawed soils of high pH compared to present-day Beringian soils.

Loess accumulation greatly enhances fertility, bringing fresh, finely comminuted mineral matter to the soil surface. A deeper annually thawed layer means that fewer nutrients are lost to recycling through retention in permafrost. A deeper active zone also enhances microbial activity, an important factor in maintaining rapid nutrient turnover. Finally, a deeper and more nutrient-rich soil favors the deep-rooted graminoid and herbaceous species that are so prominent in the palynological record.

Interactions and Feedback Effects of the Ungulate Fauna

The ungulates themselves played an important feedback role in maintaining rapid nutrient cycling and an optimal level of primary productivity. While consuming standing vegetation, the ungulates manured the landscape, and their feeding constantly pruned and removed surplus plant production, prevented organic accumulations, and contributed to the maintenance of warm soils during the growing season. Their trampling and feeding augmented the tendency for nutritious, pioneering ruderals to remain prominent in the vegetation and helped to suppress the more slow-growing late-successional species characterized by a high proportion of underground tissue and heavy investment in antiherbivory compounds in above-ground plant parts.

Interactions among the ungulates must have been important. Cratering of snow by bison and caribou and, to the extent they were able, by mammoth and horse may have cleared the ground for smaller, more selective feeders. In summer, the monogastric mammoth and horse, which must ingest large quantities of low-quality fodder, probably created or helped to maintain grassy openings in the vegetation. The result was a grazing succession in which each ungulate species created and maintained a level of environmental diversity that helped to serve the needs of the coexisting species. Indeed, the Beringian landscape would have been exceptional if constructive interactions among the ungulate species had not been present.

It is probably wrong to conclude that the diverse ungulate fauna of the Boutellier and Duvanny Yar intervals formed a single community, densely clustered on the landscape like some great zoo menagerie, as is shown in the idealized and oft-cited National Geographic Society mural of J. H. Matternes (in Guthrie, 1972). A more realistic assessment of the evidence suggests quite a different panorama—one which at a superficial glance might seem as empty of large mammals as is present-day tundra. But if we look more closely, we find the animals, at special sites and not all present simultaneously together. The late Pleistocene fauna was one that lived within a complex but largely treeless environment—an environment which, the evidence insists, contained more habitats suitable for large ungulates than does present-day Beringian taiga and tundra. We can speculate (as Bliss and Richards do) on range apportionment and carrying capacity of the complex mosaic that we call steppe-tundra, but unfortunately many details are missing, among them those most critical for understanding man's place in the system. Some of these may remain obscure for many years to come.

LIMITATIONS ON LATE PLEISTOCENE DISPERSALS

When sea level was at its lowest—during the Duvanny Yar interval, 18,000 years ago, for example—Alaska and Yukon Territory constituted the easternmost province of an unglaciated land mass that extended westward to Europe. This wide corridor assured an opportunity for extensive biotic exchange. When sea level was higher—during much of the Boutellier interval, for example—the Bering land bridge was narrower, a bottleneck limiting dispersals and reducing gene flow between populations already established in eastern and western Beringia. But even when the land bridge was at its broadest, climatic or edaphic factors that we do not yet completely understand inhibited free biotic interchange between East and West Beringia.

A long-term imbalance in exchanges of mammals across Beringia was noted more than a decade ago by Repenning (1967). This imbalance evidently results in part from the location of the evolutionary centers for steppe- and arctic-adapted species predominantly on the large Asian land mass and in part from the repeated merging of treeless steppe and arctic landscapes during cold, dry intervals on the Asian continent at a time when East Beringia was largely if not entirely closed off by continental glaciers from steppe areas further south in North America (Hopkins, 1976b). But some mammalian species that reached either East or West Beringia never crossed the land bridge. Among the ungulates, camel (*Camelops*) was present in East Beringia during the Boutellier interval (Weber and others, 1981) but never ranged much north or west of the Fairbanks area. Woolly rhinoceros (*Coelodonta*) was present in West Beringia during the Boutellier interval and also during the time of the birch zone, but never penetrated much east of the Kolyma River valley.

Several of our contributors (Yurtsev, Giterman and colleagues) emphasize differences in floristics and invertebrate faunas that also indicate limitations affecting free interchange of plants and animals across Beringia. Causes are obscure. Why, for example, was *Selaginella sibirica,* a club moss, such an important spore producer and presumably so important in the ground cover on the Siberian side and such a minor element in Alaska and the Yukon? The insect fossils, especially the beetles, also differ in ways that suggest there was little continuity of steppe habitats across the land bridge during the Duvanny Yar time, or if steppe was continuous, that the environment was too severe in other ways for many steppe species to survive. We voice this conclusion with caution, however, because studies of the modern Siberian and Alaskan/northwest Canadian insect faunas also suggest that there was little continuity of riparian faunas on the land bridge even though the river that once occupied the Hope Sea valley received tributaries from both eastern and western Beringia (Hopkins, 1979a).

MAN'S ROLE

The antiquity of man in North America south of the ice sheets has been hotly debated for more than a generation. The conservative view, well formulated for us by Haynes, has man appearing about 11,500 years ago and thereupon dispersing rapidly throughout the Americas. At another extreme are serious proposals that man was present in the Americas hundreds of thousands of years ago (e.g., Steen-McIntyre, et al., 1981). A similar but less completely articulated argument has recently been in progress in East Beringia. Morlan and Cinq-Mars' review informs us that the known primary archaeological record begins 15,000 years ago at the Bluefish Caves, but that damaged bone interpreted as modified by humans occurs in contexts more than 50,000 years old. Jopling and others (1981) have ascribed human workmanship to damaged fossil bone found in a context they consider of an age greater than 125,000 years. The issue of man's antiquity in Beringia and in the Americas will probably remain unresolved for many years to come; it is not an issue that we can hope to decide here. Nevertheless, some discussion is needed.

Two qualitatively different sets of questions, it seems to us, often become confused in this discussion. The simple but important question whether people were present at a given time and place needs to be separated from the more complex questions of who they might have been, from where they came, how they may have lived, and with what they might have been equipped. As Morlan and Cinq-Mars note, the simple discovery of an acknowledged and firmly dated artifact or even a well-dated collection of animal bones showing unambiguous traces of butchering is all that is required to inform us of man's former presence, but a coherent occupation site or activity area is required to answer more detailed questions about the origins, affiliations, and adaptations of an ancient human group. To many North American anthropologists and archaeologists, the presence of a scattering of bones showing butchering marks or a handful of flint debitage and a utilized flake is no more interesting than a shadow on a wall unless those fragments can be related to some fuller record of human activity. To the paleoecologist, however, the presence or absence of man is a critical parameter, for it seems inconceivable that modern man could be present on any landscape without having significant impact upon the ecosystem. Fire frequency must have increased drastically, for example, after human bands appeared on the American scene. Furthermore, the presence of large-ungulate remains in just about every American or Beringian archaeological site in which soil conditions permit bone preservation indicates that big-game hunting has been an important activity for tens of millenia (e.g., as early as 21,000 years BP at Valsequillo, according to Irwin-Williams, 1967, 1978).

In Pleistocene Beringia, lack of alternate resources must have obliged human bands to rely largely upon large ungulates not only for protein but also for vitamins and other essential nutrients and also for fuel (dung and fragmented bone), clothing, covering for shelters, and perhaps even structural elements (ribs, long bones, tusks, and antlers) to support shelters.[4]

---

[4]We base this argument on the evidence that sea-mammal hunting seems to be a Holocene development, that little is known about the history of fish utilization prior to 10,000 years BP, and that the berries that are such an important diet supplement in the ethnographic present are mainly produced by the ericads—plants that were rare in Beringia prior to the time of the birch zone.

It has long been recognized that arctic bands subsist almost exclusively on animal foods and that they manage to fill essential dietary needs by making complete use of animal carcasses. Ethnographic accounts and personal experience (C.E.S.) indicate that ruminant stomach contents consisting of partially digested or fermented plant remains are popular and eaten with great gusto after a kill. This activity should have great significance for human nutrition

Müller-Beck concludes his review of the evolution of man's ability to exploit cold regions with the statement that "it is hard to believe that man did not conquer all the remaining mammoth-steppe territory soon after 40,000 years ago." The evidence reviewed in the next section convinces us that man was widely distributed, though not necessarily numerous, in Beringia by 35,000 years BP.

Sites such as Sequim in western Washington (Gustafson, et al., 1979), Fort Rock Cave in eastern Oregon (Bedwell, 1973), Wilson Butte Cave in Idaho (Gruhn, 1961, 1965), Meadowcroft Rock Shelter in Pennsylvania (Adovasio and others, 1978), Valsequillo in Puebla, Mexico (Irwin-Williams, 1967, 1978), Tlapacoya in the Valley of Mexico (Mirambell, 1978) and Taima-taima in Venezuela (Bryan, 1978) seem to us to provide a convincing body of evidence indicating that man was also present south of the ice sheets as much as ten millenia earlier than the time of the Clovis hunters. Who these people were, where their Old World origins lay, and what technological and stylistic traditions might have been at their command, however, can hardly be reconstructed from the extremely fragmentary evidence available to us now.

Man In Beringia Prior to 20,000 Years BP

Man's presence in Yakutia at the western edge of Beringia late in the Boutellier interval seems to be firmly established in Aldan River valley sites described by Mochanov (1977, 1978).[5] Ust-Mil', which has yielded small numbers of indisputable artifacts and associated large-mammal remains, is radiocarbon-dated between 35,000 and 30,000 years old. Ezhantsy, a much richer site, lacks published radiocarbon dates but seems to be reliably assigned to the same time interval on the basis of its stratigraphic position in the floodplain alluvium and uppermost gravel of Aldan River Terrace III. Another early site of less certain age is Ikhine, which has yielded a small inventory of stone tools as well as large-mammal remains. Wood from two levels at this site has been dated by three laboratories: seven specimens dated by two of the laboratories yield ages clustered between 24,000 and 27,500 years; a third lab obtained dates of 30,200 ± 300 and 31,200 ± 500 (GIN-1019 and -1020), respectively, from the upper and lower levels. Our guess would be that the 27,000- to 24,000-year-old wood dates the occupation and that the older wood is redeposited, but Tseitlin (1979) prefers the older dates.

Faunal remains indicate that the early inhabitants of the Aldan Valley utilized the large ungulates, but little more can be said thus far about their former subsistence base. Mochanov groups the inventories from Ezhantsy, Ust-Mil' and Ikhine in his Proto-Dyukhtai culture (here termed the Proto-Dyukhtai industry). Lithic inventories from Proto-Dyukhtai sites include wedge-shaped cores and microblades presumably derived from them but no obvious unifacial or bifacial stone projectile points. It would seem that Proto-Dyukhtai hunters must have relied upon lances tipped with slotted bone, antler, or ivory points armed with microblades when hunting large mammals.[6]

In eastern Beringia, the early record is very sparse but not nonexistent. No lithic tools have yet been found that can be shown to be older than 12,000 years. Flakes and cores formed from mammoth bone and a cluster of bison bones exhibiting what seemed to be clearly defined cut marks, all found on disconformity A in the Old Crow Flats are interpreted by Morlan and Cinq-Mars as evidence for a human presence in the Old Crow Flats as early as 60,000 years BP. Others, however, are unwilling to accept these ancient, damaged bones as indisputable proof of human activities (e.g., Binford, 1981; Guthrie, 1980, 1981), and Morlan (in press) has recently proposed possible alternate interpretations of all of the possible artifacts in question. Whether or not the older material is accepted, however, Morlan states that "on the basis of redeposited and directly dated bones that we can defend as artifacts, people were present in the Old Crow Flats as early as 35,000 BP and as

in northern subsistence economies, because the stomach and especially the rumen is a rich source of vitamins that would be otherwise unavailable in an exclusively meat diet.

[5]Of the Aldan Valley archaeological sites, only Dyukhtai Cave is shown on the endpaper map. Other Aldan valley sites discussed here are shown in Figure 1 of Mochanov (1978).

[6]We accept R. D. Guthrie's suggestion (in a 1979 manuscript cited by Morlan and Cinq-Mars) that wedge-shaped cores were secured in a forked stick for the purpose of pressing off microblades and that a major use for these microblades was the arming of slotted antler, ivory, and bone projectile points.

late as 25,000 BP," (R. E. Morlan, oral communication 1982). In Alaska, a pair of obviously butchered caribou skulls from firmly dated context at Jack Wade Creek (Porter, 1979; Porter and Hopkins, in press) provides definitive evidence for the presence of human hunters about 30,000 years ago. None of these East Beringian occurrences, however, tell us anything whatsoever about the origins, affinities, or abundance of this shadowy former presence.

The sketchy archaeological record for late Boutellier times in the Aldan Valley cannot in itself establish whether a year-around occupation is represented or whether, as Müller-Beck suggests, the Proto-Dyukhtai visitors might have moved south each winter to a region where foraging possibilities were better. Farther east in the 4000-km-wide span of Beringia, however, there could have been no southern winter refuge. Steppe-tundra extended southward to a winter-frozen sea or to glacierized coastal mountains. Human bands reaching eastern Alaska and Yukon Territory near the end of the Boutellier interval must have been year-around residents in Beringia. Given the butchered bone and bone artifacts and given the inadequacy of alternative food resources, we must assume that these people were predators upon the large-mammal populations. Their dependence must have been greatest during winter—i.e., during the season when the large mammals were in the most difficulty in their own right. It is interesting that the large-mammal fauna persisted, almost undiminished, for another ten or fifteen thousand years.

A Hiatus During the Duvanny Yar Interval?

There is no record in any part of Beringia east of the Aldan River valley of a human presence between about 25,000 and about 15,000 years BP. Even in the Aldan Valley, evidence of a human presence during this time span is sketchy, but a lithic inventory assignable to the Dyukhtai industry seems to record a human presence at Verkhne Troitskaya about 20,000 years BP. Whether the hiatus from 25,000 to 15,000 years BP is apparent or real—whether it is simply a product of the generally poor fossil record in deposits of the Duvanny Yar interval or whether, as Müller-Beck postulated in 1966, it records the protracted abandonment of Beringia by human bands—cannot presently be determined. The hiatus ends, in any case, with the appearance of the highly visible Dyukhtai industry in northeastern Siberia and the equivalent Paleoarctic tradition (Anderson, 1970) in eastern Beringia.[7]

Late Pleistocene Human Occupation in Beringia

Paleoarctic sites featuring butchered bones and associated lithic materials are scattered from Dyukhtai Cave in the west to the Bluefish Caves in the east, and poorly preserved surface sites of this tradition are even more abundant in eastern Beringia. The oldest Paleoarctic site seems to be Mochanov's Verkhne Troitskaya site, perhaps 20,000 years old. Other notable northeastern Siberian sites include Dyukhtai Cave (ca. 13,500 to 12,000 years BP), Berelekh (ca. 13,500 BP), and Ushki Layer VI (Dikov, 1977, 1978) (ca. 10,300 BP). The deepest flint debitage and butchered bones in the Bluefish Caves of eastern Beringia are at least 15,000 years old, but, as Morlan and Cinq-Mars tell us, the other East Beringian sites known thus far are several thousand years younger.

Despite regional and age-dependent differences, the lithic inventories from Paleoarctic sites seem to us to form a coherent group. Consistently present are abundant blades and microblades, wedge-shaped microblade cores, and large ovate or asymmetrical bifaces. Also consistently present though not abundant are bifacial and symmetrical or slightly asymmetrical objects that resemble projectile points, though some may have been used as knives. These appear earliest in the Verkhne Troitskaya site. Most commonly, they are fairly thick and of bipointed willow-leaf or laurel-leaf

---

[7]For want of a better term, we adopt the usage of Dumond (1979), which broadened the term Paleoarctic Tradition to include the Dyukhtai culture of Mochanov (1977, 1978). The reader should be warned, however, that Morlan and Cinq-Mars and many other Arctic archaeologists consider this use of the term and concept to be excessively broad.

We exclude from the concept Level VII at Ushki (Kamchatka), which seems to have little in common with Paleoarctic sites, though Level VI (about 10,300 years old) seems to be an excellent example of one. Morlan and Cinq-Mars would also place the Kukhtui Site of Mochanov (1977) and Level I at Dry Creek (Powers and Hamilton, 1978) outside the Paleoarctic Tradition.

outline; others are contracting and square based. These objects would not seem out of place in the American stemmed-point tradition of Bryan (1980).[8] Dyukhtai Cave in Siberia and Healy Lake and Dry Creek Level I in Alaska have yielded projectile points or point-like knives of another type: thin, bifacial, symmetrical, and triangular. They resemble triangular points or knives occasionally found in late Paleolithic sites in west Siberia and European Russia (Bader, 1978, pp. 201–2 and Figs. 86, 122; Powers and Hamilton, 1978; J. P. Cook and R. A. McKennan, unpublished manuscripts cited by Morlan and Cinq-Mars.[9]

A variety of activities are represented in the various Paleoarctic sites. At Dyukhtai Cave, the nature and abundance of faunal and cultural remains and, especially, the consistent presence of hare remains indicate repeated, lengthy occupations, probably during winter. At Berelekh, the enormous quantity of mammoth bones, including pairs of pieces in odd and evidently artificial arrangements, suggest the remains of bone-and-hide houses; the presence of a wolverine mummy amongst broken mammoth bones and of a tusk engraved with the image of a mammoth suggests ritual activity; and the very abundant hare remains, again, suggest winter subsistence activities (Mochanov, 1977; Vereṣhchagin, 1977).[10] Ushki Layer VI, with its remains of semisubterranean houses with entrance tunnels and its abundant salmon bones (Dikov, 1977), was also a base camp that was certainly occupied during winter and perhaps also at other seasons.

It is not evident that any of the East Beringian Paleoarctic sites were occupied for lengthy periods. The Bluefish Caves were used to butcher game that perhaps had been driven into a surround or over the limestone cliff into which the caves are indented, according to Morlan and Cinq-Mars. Dry Creek was a camping and butchering site; clusters of well-rounded cropstones suggest that ptarmigan may have been utilized and that occupation took place during late summer or early autumn (W. R. Powers and R. D. Guthrie, unpublished manuscript 1979). The Trail Creek Caves seem to have been briefly occupied from time to time by hunting parties (Larsen, 1968).

The Paleoarctic peoples clearly interacted with large ungulates, and the faunal remains indicate that large-mammal hunting was an important economic activity, probably the major one. There is evidence, however, of utilization of other resources. Ptarmigan, as we have noted, may have been utilized at Dry Creek, hare remains are abundant at Dyukhtai and Berelekh, and bird and fish remains have been recovered in small quantities at Dyukhtai Cave. Remains of fish, grouse and other birds, and small mammals are present at every level in the Bluefish Caves but are thought to have been introduced by perching raptors (Cinq-Mars, 1979). None of the Paleoarctic artifact inventories include objects identifiable as fish hooks, fish spears, net weights, or net floats, but the 10,300-year-old Ushki Level VI contains large quantities of salmon bones (and also cropstones attributed by Dikov, 1967, to ducks). In future excavations, more attention must be given to screen washing and water flotation in order to obtain a fuller representation of food remains in Paleoarctic sites. The effort has already begun in eastern Beringia with the excavation of the Bluefish Caves; now, these methods must also be applied in northeastern Siberia.

Only a few of the Paleoarctic sites are dated with adequate precision, but the record supports the inference that the span of Paleoarctic occupation included the interval during which much of the Pleistocene large-mammal fauna disappeared. Mammoth bones are numerous at deep levels in Dyukhtai Cave and, of course, at the 13,500-year-

[8] Quite different in outline but also startlingly reminiscent of North American stemmed points are the Ushki Points from Level VII at Ushki I (Dikov, 1977, 1978). Nothing resembling these apparently 14,000-year-old points has yet been found at early levels elsewhere in Beringia.

[9] Mainly on the basis of the presence of the thin, triangular points, W. R. Powers suggests in a manuscript cited by Morlan and Cinq-Mars that the inventory from Level I at Dry Creek (11,100 BP) differs from that from Level II (10,700 BP) in ways that suggest the presence of two different technological traditions, Level I being "broadly comparable to Paleo-Indian tool-kits from the plains of interior North America." Without necessarily wishing to dispute this suggested relationship, we note our belief that the differences between the two Dry Creek assemblages would probably appear less significant if a larger inventory were available from Level I.

[10] Mochanov's (1977) and Vereshchagin's (1977) descriptions of the stratigraphy of the Berelekh site strongly suggests to D.M.H. the deposits of a thermokarst lake whose collapsing banks undercut the house structures and associated occupation floor shortly after abandonment.

old Berelekh site. Mammoth remains in the Bluefish Caves are all older than the birch rise; thus older than 14,000 ± 500 years BP. Mammoth is lacking in other Beringian archaeological sites of late Pleistocene or early Holocene age. Horse, bison, and sheep are reported from Dyukhtai, and these are present along with wapiti in the Bluefish Caves. The 11,000 to 10,500-year-old Dry Creek site, however, contains only sheep, a few wapiti, and bison remains. Remains of a single bison have also been recovered from 10,300-year-old Level VI at Ushki (Dikov, 1977). Sites identified by West (1975) as part of the Denali Tradition—a late, East Beringian phase of the Paleoarctic tradition evidently dating from about 10,000 ± 1,000 BP—generally lack faunal remains but are located in situations such as river crossings and mountain passes that suggest exploitation of migrating caribou and perhaps also sheep.

The late Pleistocene archaeological record suggests that Beringia was occupied by fairly numerous human groups utilizing a diverse subsistence base that emphasized the hunting of large mammals. It is certain that man was present and vigorously interacting with the large mammal population at the end of the Duvanny Yar and throughout the time of the birch zone. All ungulate species present on the landscape at a given moment were utilized. Man was certainly a witness and in all probability a participant in the destruction of the late Pleistocene large-mammal fauna of Beringia.

Significance of Bifacial Projectile Points in Late Pleistocene Beringian Archaeological Sites

An interesting and possibly significant aspect of the late Pleistocene industries in Beringia is the presence in increasing abundance in increasingly younger sites of more or less symmetrical bifaces of the size and proportions of stone projectile points. Although use-wear analysis indicates that some of these are knives (R. A. Ackerman, personal communication 1982), we wish to pursue the implications that might follow if, in fact, many of these objects are, indeed, stone projectile points.

In a discussion defining the American stemmed-point tradition, Bryan (1980) argues that point form essentially reflects hafting methods: "(1) fluted and other thin-based points were presumably hafted onto split or beveled shafts; (2) notched points, hafted by wraparound ties; and (3) willow-leaf shaped and stemmed points with relatively thick, narrow bases, hafted onto socketed shafts" (p. 78). Although points using each of these three hafting styles have commonly been considered to be horizon markers in North America, stratigraphic distributions indicate that all three types were known and remained in use throughout much of Holocene time in Beringia. We see no reason to doubt that all three might have been known to the late Pleistocene inhabitants of Beringia as well. The presence of both stemmed and notched "points" attributed to the ca. 14,000-year-old Layer VII at Ushki and of both thick, contracting-based "points" and thin, triangular "points" in Paleoarctic sites suggests that indeed this may have been the case. The important question may not be "why do bifacial projectile points of a particular form appear at particular times and places in Beringia?" but rather "why do bifacial stone projectile points of any form become increasingly abundant in Beringia during late Duvanny Yar and birch-zone time?"

Bryan (1980) goes on to state that "Important for the adoption of stone projectile points was the introduction, perhaps by diffusion from the Old World, of the spearthrower and the discovery that this effective extension of the arm would work even more efficiently as a long distance propulsor if the spear were tipped with a heavier stone projectile point, rather than with the old lightweight wooden and bone tips" (p. 81). He states that the ballistics of stone points help carry *atlatl*-propelled spears farther and thus that stone points are especially useful in open, treeless country. This may be an oversimplification, and certainly the assertion needs experimental confirmation and documentation. Still, the idea is one that deserves serious consideration.

The fact is that slotted bone, antler, and ivory points armed with microblades must have continued in use in Beringia throughout the duration of the birch-pollen zone, judging from the consistent prominence of wedge-shaped cores and microblades in Paleoarctic industries. Nevertheless, the presence of increasingly well-made point-like pieces, side-by-side with the microblades, suggests that the *atlatl* may have been introduced into Beringia late in the Duvanny Yar interval and come into wide use by the time the birch-pollen zone was well underway. Perhaps the *atlatl* was the tool that allowed Beringian hunters to begin to have a drastic impact upon the ungulate fauna of the open landscape.

The nature of the Paleoarctic inventory permits us to suggest that *atlatls* and bifacial stone points had probably diffused into eastern Beringia earlier than 12,000 years ago. The diffusion route might then be traced southward through successively later appearances of bifacial points at progressively younger sites in the Americas. The so-called pre-projectile point industries in the Americas, if they exist at all, may simply have used perishable bone, antler, ivory, and wood projectile points prior to the introduction of the *atlatl*. Certainly, bone points continued in use in the Americas, as the evidence of the 12,000-year-old Sequim mastodon assures us (Gustafson and others, 1979).

The occasional presence of thin, triangular points in several Paleoarctic sites scattered across Beringia suggests that the split- or beveled-shaft method of hafting was widely known if infrequently used, as long ago as 12,000 years BP. Though unfluted, these points may well be technologically related to the thin-based Clovis, Folsom, and Plano Points of the American Great Plains. The final implication of our highly speculative discussion is that the sudden irruption of Clovis Points in North America, 11,500 years ago may simply reflect the rapid diffusion of a particularly satisfactory variant of an already well-known hunting technology rather than the arrival and rapid spread of a new human group.

Dissolution of the Ecosystem

We have seen that the Beringian ecosystem underwent abrupt and radical changes about 12,500 ± 500 years ago, and we have seen that these changes (rising sea level, submerging shelf, shrinking glaciers) were articulated by worldwide climatic and paleogeographic changes and ultimately by changes in the worldwide radiation balance. The interval that followed—the time of the birch pollen-assemblage zone—was a time of climatic and environmental instability, a lengthy transitional episode connecting the Beringia of the Pleistocene with Holocene Beringia.

Of the changes that took place during the time of the birch zone, the alteration and reduction in the large-mammal fauna was the most dramatic, but changes in vegetation, soils, and hydrology were hardly less significant. We have argued that the diverse ungulate fauna, by cropping, trampling, and manuring, helped to perpetuate open steppe-tundra vegetation during the Boutellier and Duvanny Yar intervals. Was this interaction so intense and so effective that steppe-tundra vegetation might have persisted, had the ungulate fauna survived? Our still-scanty knowledge of the paleoecology of Beringia during the previous Sangamon interglaciation suggests that the answer is no. Like the Holocene, the Sangamon seems to have been characterized by widespread peat accumulation and by vegetation consisting of tundra and taiga in about the same distribution and of about the same character as in present-day Beringia. To tell the truth, our knowledge is so scanty that we cannot be assured that mammoth, horse, bison, camel, and woolly rhinoceros actually persisted within Beringia during the Sangamon interglaciation, but if they did not, then population reservoirs were maintained elsewhere in close enough proximity to allow Beringia to be repopulated when environmental conditions again became favorable. But no matter where the ungulates persisted, their interaction with Beringian vegetation was not of itself sufficiently intense to maintain steppe-tundra in the face of Sangamon interglacial climatic change.

If the ungulate fauna was incapable of perpetuating steppe-tundra in the face of climatic change, was the converse true? Could the late Pleistocene changes in climate, vegetation, soils, and hydrology have been sufficiently drastic in themselves to eliminate the late Pleistocene large-mammal fauna? Vereshchagin and Baryshnikov think so, but we think not: the ungulate lineages that disappeared without issue between 14,000 and 10,000 years ago were lineages that had survived all previous interglacials. The climatic and environmental changes of the time of the birch zone are not a sufficient explanation for the destruction of the Beringian large-mammal fauna.

We have noted that the late Pleistocene transformation of the large-mammal fauna took approximately 4000 years to complete and that it involved the successive disappearances of mammoth and woolly rhinoceros, then horse, and finally bison. Meanwhile, Beringia saw a short-lived reappearance and then disappearance of wapiti/red deer, a reappearance of moose, and a rise to dominance of caribou/reindeer. Hibbert reminds us that it was the nature of the ungulate fauna that led to the first insights into the true nature of Pleistocene periglacial environments. So, too, can details of the late Pleistocene transformation of the ungulate fauna give us important clues to what was going on during the time of the birch zone.

It is significant that mammoth, woolly rhinoceros, and horse—unselective feeders and cecal digesters requiring large quantities of fibrous fodder—vanished first and that bison—the ruminant least selective in its feeding and most capable of coping with a high-fiber diet—ultimately disappeared as well. This sequence of ungulate extinctions coupled with the reappearance of wapiti/red deer underscores the significance of late Pleistocene vegetational change. In these events we discern the importance of an evidently drastic reduction in the extent of coverage by graminoids and ruderals coupled with the well-documented expansion of low shrubs that distinguishes the birch interval. However, the extermination of the large ungulates took some 4000 years to complete, while the expansion of dwarf-birch shrubs at the expense of graminoid and meadow vegetation was accomplished in less than 1000 years and may not have involved as drastic an increase in the shrub cover as the relative pollen data seem to suggest. Furthermore, the two ungulate species with the lowest reproductive potential—mammoth and woolly rhino—were the first to disappear. Intense predation is suggested. Since mammoth and rhino must have been only minimally vulnerable to predation by lions and wolves, it is hard to doubt that man was implicated in their early extermination.

Could man, then, be the agent solely responsible for the late Pleistocene extinctions? Paul Martin thinks so, but we disagree. We are convinced that man was on the Beringian scene as early as 35,000 years ago; furthermore, the traces that satisfy us concerning an early human presence also establish the fact of human predation upon the large-mammal population. Thus, man was interacting with Beringian ungulates more than 20,000 years earlier than the first extinction.[11] The simple presence of man, even as a predator, was not sufficient to transform the fauna.

Human technology no doubt progressed in Beringia as it progressed in other regions, and the human hunters of the Paleoarctic Tradition were no doubt better equipped and probably more specialized than the hunters whose shadowy presence we discern near the end of the Boutellier and early in the Duvanny Yar intervals. Still the extinctions did not begin until the initiation of the birch-zone transformation of the vegetation—and even as they began, an ungulate that had previously been absent for many millenia—wapiti/red deer—reoccupied Beringia.

Hopkins (1967b) stated that

"In Beringia, it seems likely that man not only contributed substantially to the demise of mammoth, horse, and bison, but also, in so doing, hastened a drastic change in the vegetation pattern that resulted ultimately in a considerable net reduction in his total meat supply . . . The rapid changes in climate and vegetation that took place at the end of the Wisconsin cold cycle must in themselves have resulted in severe reduction in the total area of suitable grazing land, and thus in a reduction in the total size of the large-mammal populations in Beringia—and the large mammal population reduction took place at a time when efficient hunters were present in numbers that had been nicely adjusted to larger game resources. During earlier Pleistocene interglacial cycles, remnant populations of grazing animals probably survived in local arctic grassland areas, grasslands perpetuated in part by active sand movement and loess deposition near active glacial-outwash streams and in part by the interaction of the herbivores themselves with the treeless vegetation of the time. But during late Wisconsin time, hunting pressures upon the diminishing game resource probably reduced many species to population levels where their simple presence no longer tended to perpetuate grasslands. The last small herds, restricted to those very few, very small areas where grasslands persist today, must have been . . . easy . . . quarry for stone age men. This interaction of changing climate, changing vegetation, and relentless hunting by man resulted in the extinction of major elements of the Beringian mammalian fauna and in a

[11] *Camelops*, of course, seems to have disappeared from Beringia as early as 30,000 years ago (Table 1). We don't wish to minimize the possible involvement of human hunters, but because North American camel is a rare fossil in Beringia and because it persisted south of the ice sheets for another ten or fifteen millenia, we assume for the moment that it was eliminated from our region simply by unfavorable climatic and environmental changes.

shift to a vegetation able to support only two herd animals—reindeer (or caribou) and musk ox" (pp. 478–9).

Subsequent studies have only strengthened our belief in Hopkins' simple scenario of 1967 while allowing us to refine and elaborate it. We believe that we have outlined the basic scenario for the transformation, during the birch interval, of the Beringian large-mammal fauna and for the dissolution of the Beringian steppe-tundra ecosystem.

—*C. E. Schweger, J. V. Matthews, Jr., D. M. Hopkins, and S. B. Young*

—THE END—

REFERENCES

Absolon, Karel, and Bohuslav Klima. 1977. *Predmostí, ein Mammut-Jägerplatz in Mähren.* Fomtes Archaeologiae Moravie 8. Prague.

Abramova, Z. A. 1979a. *Paleolit Eniseia: Afontovskaia kul'tura (The Paleolithic of the Yenisei: Afontova culture).* Novosibirsk: Nauka.

Abramova, Z. A. 1979b. *Paleolit Eniseia: Kokorevskaia kul'tura (The Paleolithic of the Yenisei: Kokorevo culture).* Novosibirsk: Nauka.

Ackerman, R. E., T. D. Hamilton, and R. Stuckenrath. 1979. Early culture complexes on the northern Northwest Coast. *Canadian Journal of Archaeology* 3:195–209.

Adam, K. D. 1951. Der Waldelefant von Lehringen, eine Jagdbeute des diluvialen Menschen. *Quartär* 5:79–92.

Adam, K. D. 1954. Die mittelpleistozänen Faunen von Steinheim an der Murr (Württemberg). *Quaternaria* 1:131–44.

Adam, K. D. 1961. Die Bedeutung per pleistozänen Saugetierfaunen Mitteleuropas fur die Geschichte des Eiszeitalters. *Stuttgarter Beitrage zur Naturkunde* 78.

Adovasio, J. M., J. D. Gunn, J. Donahue, R. Stuckenrath, J. Guilday, and K. Lord. 1978. "Meadowcroft Rockshelter," in *Early Man in America from a circum-Pacific perspective.* Edited by A. L. Bryan, pp. 140–80. University of Alberta Department of Anthropology Occasional Papers 1.

Afolayan, A. P. 1975. Effects of elephant activities on forest plantations in the Kilimanjaro forest game reserve in northern Tanzania. *Oikos* 26:405–10.

Ager, T. A.,1975. *Late Quaternary environmental history of the Tanana Valley, Alaska.* Ohio State University Institute of Polar Studies, Report 54.

Ager, T. A., and L. P. Ager. 1980. An ethnobotany of the Eskimos of Nelson Island, Alaska. *Arctic Anthropology* 17:26–48.

Ager, T. A., and J. D. Sims. 1981. Holocene pollen and sediment record from the Tangle Lakes area, central Alaska. *Palynology* 5:85–98.

Aguirre, E. 1969. Evolutionary history of the elephant. *Science* 164:1366–76.

Ahlgren, C. E. 1959. Some effects of fire on forest reproduction in northeastern Minnesota. *Journal of Forestry* 57:194–200.

Aigner, J. 1969. *The archaeology of Pleistocene China.* University of Wisconsin Ph.D. thesis.

Aigner, J. 1973. Relative dating of North Chinese faunal and cultural complexes. *Arctic Anthropology* 9:36–79.

Aigner, J. S. 1978a. The lithic remains from Anangula, an 8,500 year-old Aleut coastal village. *Urgeschichtliche Materialhefte* 3. Tübingen: Institut für Urgeschichte der Universität.

Aigner, J. S. 1978b. "The Paleolithic of China," in *Early Man in America from a circum-Pacific perspective.* Edited by A. L. Bryan, pp. 25–41. University of Alberta Department of Archaeology Occasional Papers 1.

Albrecht, G., J. Hahn, and W. G. Torke. 1972. *Merkmalanalyse von Geschoss-Spitzen des Mittleren Jungpleistozäns in Mittel und Osteuropa.* Stuttgart: Kohlhammer.

Alekseev, M. N., L. V. Golubeva, L. P. Karulova, O. M. Petrov, and E. V. Shantser. 1979. On the problem of the Neogene/Quaternary boundary at the Pacific coast of Asia. *14th Pacific Science Congress, Abstracts Committee* BIII:13–15. Khabarovsk.

Alexander, H. L. 1974. "The association of Aurignacoid elements with fluted point complexes in North America," in *International Conference on the Prehistory and Paleoecology of Western North American Arctic and Subarctic.* Edited by Scott Raymond and Peter Schlederman, pp. 21–31. Calgary: University of Calgary Archaeological Association.

Alexander, H. L. 1978. "The legalistic approach to early man studies," in *Early Man in America from a circum-Pacific perspective.* Edited by A. L. Bryan, pp. 20–22. University of Alberta Department of Archaeology Occasional Papers 1.

Aleksandrova, V. D. 1970. "The vegetation of the tundra zones in the USSR and data about its productivity," in *Productivity and conservation in northern circumpolar lands.* Edited by W. A. Fuller and P. J. Kevan. IUCN New Series 16:146–62. Morge, Switzerland.

Alexandrova, V. D. 1977. *Geobotanicheskoye rayonirovaniye Arktiki i Antarktiki (Geobotanical zonation of the Arctic and Antarctic).* Komarovskaye Chteniya 19, Leningrad (in Russian).

Anderson, D. D. 1968. A stone age campsite at the gateway to America. *Scientific American* 218(6):24–33.

Anderson, D. D. 1970. Akmak: An early archaeological assemblage from Onion Portage, northwest Alaska. *Acta Arctica* 16.

Anderson, D. D. 1972. An archaeological survey of the Noatak drainage, Alaska. *Arctic Anthropology* 9:66–117.

Anderson, D. D. 1978. "Western Arctic and Sub-arctic," in *Chronologies in New World Archaeology.* Edited by R. S. Taylor and C. W. Meighan, pp. 29–50. New York: Academic Press.

Anderson, D. D. 1980. Continuity and change in the prehistoric record from North Alaska. *Senri Ethnological Studies* 4:233–51.

Anderson, E. 1973. Ferret from the Pleistocene of central Alaska. *Journal of Mammalogy* 54:777–78.

Anderson, E. 1977. Pleistocene Mustelidae (Mammalia, Carnivora) from Fairbanks, Alaska. *Bulletin of the Museum of Comparative Zoology* 148:1–21.

Anderson, P. M. 1981. *Reconstructing the past: The synthesis of archaeological and palynological data, northern Alaska and northwestern Canada.* Brown University, Ph.D. thesis.

Andreev, V. N., and V. D. Aleksandrova. 1981. "Geobotanical division of the Soviet Arctic," in *Tundra ecosystems, a comparative analysis.* Edited by L. C. Bliss, O. W. Heal, and J. J. Moore. Cambridge: Cambridge University Press.

Andreev, V. N., and V. I. Perfilyev. 1975. Fragments of

steppe vegetation in northeastern Yakutia. *Biologicheskie problemy severa,* pp. 14–16. Yakutsk (in Russian).

Andrews, J. T., P. J. Webber, and H. Nichols, 1979. A late Holocene pollen diagram from Pangnirtung Pass, Baffin Island, N.W.T., Canada. *Review of Paleobotany and Palynology* 27:1–28.

Anonymous. 1934. The wild animals of the Indian Empire, Part II. Proboscidea: elephants. *Bombay Journal of Natural History Society* 37(1, suppl): 93–96.

Anthes, R. A., and T. T. Warner. 1978. Development of hydrodynamic models suitable for air pollution and other mesometeorological studies. *Monthly Weather Review* 106:1045–78.

Arkhipov, S. A. 1979. New ideas and trends in the investigation of Siberian glaciation. International Geological Correlation Programme, Project 73/1/24, *Quaternary glaciations in the northern hemisphere* 5:9–21.

Arkhipov, S. A., M. P. Votakh, A. V. Golbert, V. I. Gudina, L. A. Dovgal, and A. U. Yudkevich. 1977. *The last glaciation in the lower Ob' River region.* Novosibirsk: Izd. "Nauka" (cited in Arkhipov, 1979) (in Russian).

Armenta-Camacho, J. 1978. *Vestigios de labor humana en huesos de animales extintos de Valsequillo, Puebla, Mexico.* Puebla: Consejo Editorial el Gobierno del Estado Puebla.

Arslanov, Kh. A., and S. B. Chernov. 1977. "On the absolute age of the Selerikan horse," in *Fauni i flori Antropogena Severo-Vostok Sibiri.* Edited by O. A. Scarlato, pp. 76–78.

Arslanov, Kh. A., V. V. Lyubov, and T. V. Tertychnaya. 1981. "On the absolute age of the baby mammoth," in *Magadanskiy mamontenok.* Edited by N. K. Vereshchagin and V. M. Michel'son, pp. 50–51. Leningrad: Izd. "Nauka" (in Russian).

Avenarius, I. G., M. V. Muratova, and I. I. Spasskaya. 1978. Palaeogeography of northern Eurasia in the Late Pleistocene-Holocene and geographical reconstructions. Akad. Nauk SSSR, Geograficheskie Institut. Moscow (in Russian).

Axelrod, D. I. 1967. Quaternary extinction of large mammals. *University of California Publication in Geological Sciences* 74:1–42.

Bader, O. N. 1978. *Sungir': An upper Paleolithic site.* Akad. Nauk SSSR, Inst. Archaeol., Izd. "Nauka."

Bannikov, A. G. 1963. *Die saiga-antilope.* Wittenburg: A. Ziemsen.

Bannikov, A. G. 1967. "Some remarks concerning determination of biomass of wild ungulates in natural geographic zones of the USSR," in *Secondary productivity of terrestrial ecosystems: Principles and Methods.* Edited by K. Petrusevicz, pp. 255–60. Warsaw: Polish Academy of Sciences, 2 volumes.

Bannikov, A. G., L. V. Zhirnov, L. S. Lebedeva, and A. A. Fandeev. 1967. *Biology of the Saiga.* Jerusalem: Israel program for scientific translations. (Translation of *Biologiya Saigakal,* Izd. Sel'skokhozyaistvennoi Literatury, Zhurnalor i Platakov, 1961.)

Baranova, Yu. P., and S. F. Biske. 1964. *I storiya razvitiya rel'efa Sibiri i Dal'nego Vostoka: Severo-Vostok SSSR (History of the development of the relief of Siberia and the Far East).* Moscow: Izd. "Nauka" (in Russian).

Baranova, Yu. P., S. F. Biske, V. F. Goncharov, I. A. Kul'kova, and A. S. Titkov. 1968. *Kainozoi Severo-Vostoka SSSR (The Cenozoic of North-East USSR).* Moscow: Izd. "Nauka" (in Russian).

Barnes, R. F. W. 1979. *Elephant ecology in Ruaha National Park, Tanzania.* University of Cambridge, Ph.D. thesis.

Barney, R. J. 1971. "Wildfires in Alaska—some historical and projected effects and aspects," in *Fire in the northern environment, a symposium.* Pacific Northwest Forestry Range Experimentation Station, pp. 51–59. Portland: United States Department of Agriculture.

Barry, R. G. 1973. The conditions favoring glacierization and deglacierization in North America from a climatological viewpoint. *Arctic and Alpine Research* 5:171–84.

Barry, R. G. 1975. Climate models in palaeoclimatic reconstruction. *Palaeogeography, Palaeoclimatology, Palaeoecology* 17:123–37.

Barry, R. G. 1977. "Study of climatic effects on ice extent and its seasonal decay along the Beaufort and Chukchi Sea coasts," in *NOAA, Environmental Assessment of the Alaskan continental shelf. Quarterly Reports of Principal Investigators.* July-Sept.. 1977, III: 141–299.

Barry, R. G. 1978. "Study of climatic effects on ice extent and its seasonal decay along the Beaufort and Chukchi Sea coasts," in *NOAA, Environmental Assessment of the Alaskan Continental Shelf. Quarterly Reports of Principal Investigators.* October-December, 1977, II:141–71.

Barry, R. G., and F. K. Hare. 1974. "Arctic climate," in *Arctic and Alpine Environments.* Edited by J. D. Ives and R. G. Barry, pp. 17–54. London: Methuen.

Bartley, D. D. 1967. Pollen analysis of surface samples of vegetation from arctic Quebec. *Pollen et Spores* 9: 101–05.

Bartley, D. D., and B. Matthews. 1969. A paleobotanical investigation of postglacial deposits in the Sugluk area of northern Ungava (Quebec, Canada). *Review of Paleobotany and Palynology* 9:45–61.

Baryshnikov, G. F., I. E. Kuz'mina, and V. M. Khrabriy. 1977. "Results of measurements of tubular bones of mammoths from the Berelkh 'burial ground,'" in *Mamontovaya Fauna Russkoy Ravniny i Vostochnoy Sibiri.* Edited by O. A. Scarlato. Trudy Akad. Nauk SSSR, Zoologicheskie Instituta 72:58–67 (in Russian).

Batzli, G. O. 1975. "The role of small mammals in arctic ecosystems," in *Small mammals: Their productivity and population dynamics.* Edited by F. G. Golley, K. Petrusewicz, and L. Ryszkowski, pp. 243–68. Cambridge: Cambridge University Press.

Batzli, G. O. (editor). 1980. Patterns of vegetation and herbivory in arctic tundra. *Arctic and Alpine Research* 12:401–588.

Batzli, G. O. 1981. "Populations and energetics of small mammals in the tundra ecosystem," in *Tundra ecosystems, a comparative analysis.* Edited by L. C. Bliss, O. W. Heal, and J. J. Moore, pp. 377–96. Cambridge: Cambridge University Press.

Batzli, G. O., and H. G. J Juna. 1980. Nutritional geology

of microtine rodents: Resource utilization near Atkasook, Alaska. *Arctic and Alpine Research* 12:483–99.

Batzli, G. O., and S. T. Sobaski. 1980. Distribution, abundance, and foraging patterns of ground squirrels near Atkasook, Alaska. *Arctic and Alpine Research* 12: 501–10.

Bazé, W. 1955. *Just elephants.* London: Elek Books.

Bedwell, S. F. 1973. *Fort Rock Basin prehistory and environment.* Eugene: University of Oregon Books.

Behrensmeyer, A. K. 1975. The taphonomy and paleoecology of Plio-Pleistocene vertebrate assemblages east of Lake Rufolf, Kenya. Harvard University, *Bulletin of the Museum of Comparative Zoology* 146:473–578.

Behrensmeyer, A. K., and A. P. Hill, editors. 1980. *Fossils in the making: Vertebrate taphonomy and paleoecology.* Chicago: University of Chicago Press.

Beikman, H. M. 1974. Preliminary geological map, southwest quadrant of Alaska. *United States Geological Survey Miscellaneous Field Studies Map* MF-611.

Bell, F. G. 1969. The occurrence of southern, steppe and halophyte elements in Weichselian (last glacial) floras from southern Britain. *New Phytologist* 68:913–22.

Bell, K. L., and L. C. Bliss. 1980. Plant reproduction in a high arctic environment. *Arctic and Alpine Research* 12:1–10.

Bell, R. H. V. 1970. "The use of the herb layer by grazing ungulates in the Serengeti," in *Animal populations in relation to their food resources.* Edited by A. Watson, p. 2546. Oxford: Blackwell Scientific Publications.

Bell, R. H. V. 1971. A grazing system in the Serengeti. *Scientific American* 225:86–93.

Berger, A. L. 1978. "Insolation signatures of Quaternary climatic changes," in *Evolution of planetary atmospheres and climatology of the Earth,* pp. 233–48. Centre National d'Etudes Spatiales, Toulouse.

Berger, J. R. 1977. *Northern frontier, northern homeland: The report of the Mackenzie Valley pipeland inquiry, vol. 2. Terms and Conditions.* Ottawa: Supply and Services.

Berglund, B. E. (editor). 1979. Palaeohydrological changes in the temperate zone in the last 15,000 years. *International Geological Correlation Programme, Project 158,* Subproject B, Lake and Mire Environments. Lund, Sweden.

Berglund, B. E., Soren Hakanson, and Erik Lagerlund. 1976. Radiocarbon-dated mammoth (*Mammuthus primigenius* Blumenbach) finds in south Sweden. *Boreas* 5:177–91.

Berman, D. I., N. N. Vinokurov, and B. A. Korotyaev. 1979. On the Beringian connections of Curculionidae, Carabidae and Heteroptera faunas. *14th Pacific Science Congress, Abstracts, Committee CD:* 206–07. Khabarovsk.

Beschel, R. E. 1969. "Geobotanical studies on Axel Heiberg Island in 1962," in *Axel Heiberg Island; preliminary report 1961–1962,* pp. 199–215. McGill University.

Bhattacharya, D. K. 1977. *Paleolithic Europe.* Atlantic Highland, NJ: Humanities Press.

Bilello, M. A. 1974. Air masses, fronts and winter precipitation in central Alaska. *Cold Regions Research and Engineering Laboratory Research Report* 319.

Billings, W. D. 1974. "Arctic and alpine vegetation: plant adaptations to cold summer climates," in *Arctic and alpine environments.* Edited by R. G. Barry and J. D. Ives, pp. 403–43. London: Methuen.

Binford, L. R. 1978. *Nunamiut ethnoarchaeology.* New York: Academic Press.

Binford, L. R. 1981. *Bones: ancient man and modern myth.* New York: Academic Press.

Binford, L. R. In press. *Bones: ancient man and modern myths.* New York: Academic Press. (1982)

Bird, J. P. 1969. A comparison of South Chilean and Ecuadorian "fishtail" projectile points. *Kroeber Anthropological Society Papers* 40:52–71.

Birks, H. J. B. 1973. "Modern pollen rain studies in some arctic and alpine environments," in *Quaternary plant ecology.* Edited by H. J. B. Birks and R. G. West, pp. 143–68. New York: John Wiley.

Birks, H. J. B. 1977. Modern pollen rain and vegetation of St. Elias Mountains, Yukon Territory. *Canadian Journal of Botany* 55:2367–82.

Black, R. A., and L. C. Bliss. 1978. Recovery sequence of *Picea mariana-Vaccinium uliginosum* forests after burning near Inuvik, Northwest Territories, Canada. *Canadian Journal of Botany* 56:2020–30.

Black, R. A., and L. C. Bliss. 1980. Reproductive ecology of *Picea mariana* (Mill.) BSP., at treeline near Inuvik, Northwest Territories, Canada. *Ecology Monograph* 50:331–54.

Black, R. F. 1951. Eolian deposits of Alaska. *Arctic* 4:90–110.

Black, R. F. 1966. Late Pleistocene to recent history of Bering Sea—Alaska coast and man. *Arctic Anthropology* 3:7–22.

Bleich, K. E. 1975. Die Sedimente des Mauerner Tales. *Quartär* 26:100–106.

Bliss, L. C. 1962. Caloric and lipid content in alpine tundra plants. *Ecology* 43:753–57.

Bliss, L. C. 1975. Tundra grasslands, herblands and shrublands and the role of herbivores. *Geoscience and Man* 10:51–79.

Bliss, L. C. 1977. "General summary, Truelove Lowland ecosystem," in *Truelove Lowland, Devon Island, Canada: A high arctic ecosystem.* Edited by L. C. Bliss, pp. 657–75. Edmonton: University of Alberta Press.

Bliss, L. C. 1979. Vascular plant vegetation of the Southern Circumpolar Region in relation to antarctic, arctic and alpine vegetation. *Canadian Journal of Botany* 57:2167–78.

Bliss, L. C. 1981. "Productivity and nutrient content of tundra ecosystems," in *Handbook of nutrition and food.* West Palm Beach: CRC Press Inc.

Bliss, L. C., and J. E. Cantlon. 1957. Succession on river alluvium in northern Alaska. *American Midland Naturalist* 58:452–69.

Blood, D. A. 1973. Variation in reproduction and reproductivity of an enclosed herd of moose (*Alces alces*). *Transactions of the IX International Congress of Game Biology,* pp. 59–66

Bloom, A. L. 1980. The sea-level record of the structure of an ice age. *AMQUA Abstracts,* 6:29–30. Orono.

Bloom, A. L., W. S. Broecker, J. M. A. Chapell, R. K. Matthews, and K. J. Mesolella. 1974. Quaternary sea level fluctuations on a tectonic coast: New $^{230}Th/^{234}U$ dates from the Huon Peninsula, New Guinea. *Quaternary Research* 4:185–205.

Boaz, N. T. 1979. Early hominid population densities: New estimates. *Science* 206:592–94.

Bocher, T. W. 1949. Climate, soil, and lakes in continental West Greenland in relation to plant life. *Meddelelser om Grønland* 147:1–64.

Bocher, T. W. 1954. Oceanic and continental vegetational complexes in southwest Greenland. *Meddelelser om Grønland* 148: 1.

Bombin, Miguel. 1980. Early and mid-Wisconsin paleosols in the Old Crow Basin (Yukon Territory, Canada). *AMQUA Abstracts* 6:37–39. Orono.

Bonnichsen, Robson. 1973. "Some operational aspects of human and animal bone alteration," in *Mammalian osteo-archaeology: North America.* Edited by B. M. Gilbert, pp. 9–24. Columbia: Special Publication of the Missouri Archaeological Society.

Bonnichsen, Robson. 1978a. Clovis: Migration or *in situ* development?. *AMQUA Abstracts* 5:133–35. Edmonton.

Bonnichsen, Robson. 1978b. "Critical arguments for Pleistocene artifacts from the Old Crow Basin, Yukon: A preliminary statement," in *Early Man in America from a circum-Pacific perspective.* Edited by A. L. Bryan, pp. 102–18. University of Alberta Department of Archaeology Occasional Papers 1.

Bonnichsen, Robson. 1979. *Pleistocene bone technology in the Beringian refugium.* Canadian National Museum of Man Mercury Series (Archaeological Survey of Canada Paper No. 89).

Bordes, Francois and Cl. Thibault. 1977. Thoughts on the initial adaptation of Hominids to European glacial climates. *Quaternary Research* 8:115–27.

Bos, G. N. 1967. *Range types and their utilization by muskox on Nunivak Island: A reconnaissance study.* University of Alaska, M.S. thesis.

Bosinski, G. and Fischer, G. 1974. Die Menschendarstellungen von Gönnersdorf der Ausgrabung 1968. *Der Magdalénien-Fundplatz Gönnersdorf* 1. Wiesbaden: Steiner.

Bostock, H. S. 1966. Notes on glaciation in central Yukon Territory. *Geological Survey of Canada Paper* 65-36:18.

Bouchud, J. 1975. "Etude de la Faune de l'Abri Pataud," in *Abri Pataud.* Edited by H. L. Movius, pp. 67–153. Cambridge: Harvard University Press.

Bourliere, F. 1963. "Observations on the ecology of some large African mammals," in *African ecology and human evolution.* Edited by F. C. Howell, pp. 43–54. Chicago: Aldine.

Bourliere, F., and M. Hadley, 1970. The ecology of tropical savannas. *Annual Review of Ecology and Systematics* 1:125–52.

Bowling, S. A. 1979. Meteorological interpretations of ice age climate in Alaska (abstract). *Bulletin of the American Meteorological Society* 60:845.

Boyarskaya, T. D., P. A. Kaplin, O. B. Parunin, A. A. Svitoch, A. I. Shlyukov, and S. S. Faustov. 1971. "Absolute dating and paleomagnetic investigations in the Mammoth Mountain area," in *Problemy periodizatsii Pleistotsena,* pp. 253–57. Leningrad: Izd. "Nauka" (in Russian).

Boyce, M. S., and L. D. Hayden-Wing. 1979. North American elk: Ecology behavior and management. Laramie: University of Wyoming Press.

Braitseva, O. A., I. V. Melekestsev, I. S. Evteeva, and E. G. Lupikina. 1968. *Stratigrafiya chetvertichnikh otlozheniy i oledneniya Kamchatki (Stratigraphy of Quaternary deposits and Glaciation of Kamchatka).* Moscow: Izd. "Nauka" (in Russian).

Brande, A. 1975. Vegetationsgeschichtliche und pollenstratigraphische Untersuchungen zum Paläolithikum von Mauern und Meilenhofen. *Quartär* 26:73–100/103–106.

Brassard, G. R., and R. E. Longton. 1970. The flora and vegetation of Van Hauen Pass, northwestern Ellesmere Island. *Canadian Field-Naturalist* 84:357–64.

Bredkina, L. I., and I. I. Markova. 1978. "Desert-steppe species in the lichen flora of Yakutia and Middle Asia," in *Tezisy VI Konferentsii po sporovym rastitelniya sredney Azii i Kazakhstana.* Dushanbe, p. 280 (in Russian).

Breuil, Henri. 1939. Bone and antler industry of the Choukoutien *Sinanthropus* site. *Paleontologica Sinica,* n.s. D, No. 6 (whole series No. 117).

Briggs, N. D., and J. A. Westgate. 1978. Fission-track age of tephra marker beds in Beringia. *AMQUA Abstracts* 5:190. Edmonton.

Britton, M. E. 1966. "Vegetation of the arctic tundra," in *Arctic biology.* Edited by H. P. Hansen, pp. 67–113. Corvallis: Oregon State University Press.

Brown, J., P. E. Miller, L. L. Tiezen, and F. L. Bunnell, editors, 1980. *An Arctic ecosystem.* US/IBP Synthesis Series No. 13. Stroudburg, PA: Dowden, Hutchinson & Ross.

Bryan, A. L. 1969. Early man in America and the Late Pleistocene chronology of western Canada and Alaska. *Current Anthropology* 10:339–65.

Bryan, A. L., editor. 1978a. *Early man in America from a circum-Pacific perspective.* University of Alberta Department of Archaeology Occasional Papers 1.

Bryan, A. L. 1978b. "An overview of Paleo-American prehistory from a circum-Pacific perspective," in *Early Man in America from a circum-Pacific perspective.* Edited by A. L. Bryan, pp. 306–27. University of Alberta Department of Archaeology Occasional Papers 1.

Bryan, A. L. 1978c. An El Jobo mastodon kill at Taimiataima, Venezuela. *Science* 200:1275–77.

Bryan, A. L. 1980. "The stemmed point tradition: An early technological tradition in western North America," in *Anthropological papers in memory of Earl H. Swanson, Jr.* Edited by L. B. Harten, C. N. Warren, and D. R. Tuohy, pp. 77–107. Idaho State Museum Natural History Special Publication.

Bryant, J. P., and P. J. Kuropat. 1980. Selection of winter forage by subarctic browsing vertebrates. *Annual Review of Ecology and Systematics* 11:261–86.

Büdel, J. 1949. Die raumliche und zeitliche Gliederung des Eizeitklimas. *Die Naturwissenschaften* 36:105–12, 133–39.

Budyko, M. I. 1967. On the causes of the extinction of

some animals at the end of the Pleistocene. *Soviet Geographical Review and Translation* 8:783–93.

Bunnell, F. 1978. Horn growth and population quality in dall sheep. *Journal of Wildlife Management* 42:764–75.

Bunnel, F. L., S. F. Maclean, Jr., and J. Brown, 1975. "Barrow, Alaska, U.S.A.," in *Structure and function of tundra ecosystems.* Edited by T. Rosswall and O. W. Heal. *Swedish Natural Science Research Council Ecological Bulletin* 20:73–124.

Burt, W. H., and P. Grossenheider. 1976. *A field guide to the mammals,* 3rd edition. Boston: Houghton Mifflin Company.

Businger, J. A. 1973. "Turbulent transfer in the atmospheric surface layer," in *Workshop on Micrometeorology.* Edited by D. A. Haugen, pp. 67–100. Boston: American Meteorology Society.

Butler, B. R. 1963. An Early Man site at Big Camas Prairie, south-central Idaho. *Tebiwa* 6:22–33.

Butler, B. R., and R. J. Fitzwater. 1965. A further note on the Clovis Site at Big Camas Prairie, south-central Idaho. *Tebiwa* 8:38–40.

Butzer, K. W., and G. L. Isaac. 1975. "A delimitation of the geologic time term 'Middle Pleistocene' " Appendix 2, *After the Australopithecines.* Edited by K. W. Butzer and G. L. Isaac, pp. 901–903. The Hague: Mouton.

Calef, G. W. 1974. "The predicted effect of the Canadian Arctic Gas Pipeline Project on the Porcupine Caribou Herd," in *Environmental impact assessment of the portion of the Mackenzie gas pipeline from Alaska to Alberta* 4:101–20. Winnipeg: Environment Protection Board.

Campbell, J. B. 1977. *The Upper Palaeolithic of Britain. A study of man and nature in the late Ice Age.* 2 vols. Oxford: Clarendon Press.

Campbell, J. M. 1963. Ancient Alaska and Paleolithic Europe. *Anthropological Papers of the University of Alaska* 10(2):29–49.

Campbell, R. B. 1967. Glenlyon map-area, Yukon Territory, *Geological Survey of Canada Memoir* 352.

Carl, E. A. 1971. Population control in Arctic ground squirrels. *Ecology* 52:395–413.

Carrington, R. 1962. *Elephants.* Edinburgh: Penguin Books.

Carter, L. D. 1981a. A Pleistocene sand sea on the Alaskan arctic coastal plain. *Science* 211:381–83.

Carter, L. D. 1981b. Middle Wisconsinan through Holocene climate in the Ikpikpuk River region, Alaska. *University of Colorado, Institute of Arctic and Alpine Research, Arctic Workshop* 10:5–9.

Cates, R. G., and D. F. Rhoades. 1977. Patterns in the production of antiherbivore chemical defenses in plant communities. *Biochemical Systematics and Ecology* 5:185–93.

Chapin, F. S., III. 1980. The mineral nutrition of wild plants. *Annual Review of Ecology and Systematics* 11:233–60.

Chapin, F. S., K. van Cleve, and L. L. Tieszen. 1975. Seasonal nutrient dynamics of tundra vegetation at Barrow, Alaska, *Arctic and Alpine Research* 7:209–26.

Chapin, F. S., III, D. A. Johnson, and J. D. McKendrick. 1980. Seasonal movement of nutrients in plants of differing growth form in an Alaskan tundra ecosystem: Implications for herbivory. *Journal of Ecology* 68: 189–209.

Chapman, R. M., W. E. Yeend, W. P. Brosgé, and H. N. Reiser. 1975a. Preliminary geologic map of the Tanana and northeast part of the Kantishna River quadrangles, Alaska. *United States Geological Survey Open File Map* 75-337.

Chapman, R. M., W. E. Yeend, and W. W. Patton. 1975b. Preliminary reconnaissance geologic map of the western half of the Kantishna River quadrangle, Alaska. *United States Geological Survey Open File Report* 75-351.

Chappell, John. 1974. Geology of coral terraces, Huon Peninsula, New Guinea: A study of Quaternary tectonic movements and sea-level changes. *Geological Society of America Bulletin* 85:553–70.

Charney, J. 1975. Dynamics of deserts and drought in the Sahel. *Quarterly Journal of the Royal Meteorological Society* 101:193–202.

Chepurko, N. L. 1972. "The biological productivity and the cycle of nitrogen and ash elements in the dwarf shrub tundra ecosystems of the Khibini Mountains (Kola Peninsula)," in *Proceedings, 4th international meeting on the biological productivity of tundra.* Edited by F. E. Wielgolaski and T. Rosswall, pp. 236–47. Stockholm: Tundra Biome Steering Committee.

Cherskiy, I. D. 1879. Account of some fossil remains of mammalian stomachs excavated from the Nizhneudinsk caves. *Imp. Russ. Geograficheskie Obshchestva, Vostochny-Sibirsk Otdelnie, Izvestia* 10(1-2):7–35 (in Russian).

Chia Lan-po, W. C., and Li Chao-rong. 1979. Report on the excavation of Hsuchiayao man site in 1976. *Vertebrata Palasiatica* 17:277–98.

Chmielewski, Waldemar, and Henryk Kubiak. 1962. The find of mammoth bones at Skaratki in the Lowicz district. *Folia Quaternaria* 9

Christopherson, R. J. 1976. Effects of prolonged cold and the outdoor winter environment on apparent digestibility in sheep and cattle. *Canadian Journal of Animal Science* 56:201–12.

Churchill, E. D. 1955. Phytosociological and environmental characteristics of some plant communities in the Umiat region of Alaska. *Ecology* 36:606–27.

Churcher, C. S. 1978. Did the North American mammoth migrate? *AMQUA Abstracts* 5:111–12.

Cinq-Mars, Jacques. 1978. Northern Yukon Research Programme: Survey and excavations. *AMQUA Abstracts* 5:160–61. Edmonton.

Cinq-Mars, Jacques. 1979. Bluefish Cave I: A Late Pleistocene Eastern Beringian cave deposit in the northern Yukon. *Canadian Journal of Archaeology* 3:1–32.

Clark, D. W. 1972. Archaeology of the Batza Téna obsidian source, west-central Alaska. *Anthropological Papers of the University of Alaska* 15(2):1–21.

Clark, D. W. 1978. Discussion on "Clovis culture." *AMQUA Abstracts* 5:136–39. Edmonton.

Clark, D. W. In press. "Prehistory of the western Subarctic: Alaska, the Cordillera, and the Mackenzie Valley," in *Handbook of North American Indians,* Vol. 6, *The Subarctic.* Edited by June Helm.

Clark, D. W., and A. McFadyen Clark. 1975. Fluted points from the Batza Téna obsidian source of the Koyukuk River region, Alaska. *Anthropological Papers of the University of Alaska* 17(2):31–38.

Clark, D. W., and A. McFadyen Clark. 1980. "Fluted points from the Batza Téna obsidian source, northwestern Interior Alaska," in *Early Native Americans.* Edited by D. L. Brownman, pp. 141–59. The Hague: Mouton.

Clark, John, and K. K. Kietzke. 1967. Paleoecology of the lower nodular zone, Brule Formation, in the Big Badlands of South Dakota. *Fieldiana: Geology Memoirs* 5:111–37.

Clark, J. A., W. E. Farrell, and W. R. Peltier. 1978. Global changes in post-glacial sea level: A numerical calculation. *Quaternary Research* 9:265–87.

Clebsch, E. E. C., and R. E. Shanks. 1968. Summer climatic gradients and vegetation near Barrow, Alaska. *Arctic* 21:161–71.

CLIMAP project members. 1976. The surface of the ice-age earth. *Science* 191:1131–44.

Coe, M. J., D. H. Cumming, and J. Phillipson. 1976. Biomass and production of large African herbivores in relation to rainfall and primary production. *Oecologia* 22:341–54.

Colbaugh, P. R. 1968. *The environment of the Imuruk Lake area, Seward Peninsula, during Wisconsin time.* Ohio State University, M.S. thesis.

Coleman, D. C., R. Andrews, J. E. Ellis, and J. S. Singh. 1976. Energy flow and partitioning in selected man-managed and natural ecosystems. *Agro-Ecosystems* 3:45–54.

Colinvaux, P. A. 1964. The environment of the Bering Land Bridge. *Ecological Monographs* 34:297–329.

Colinvaux, P. A. 1967a. "Quaternary vegetation history of arctic Alaska," in *The Bering Land Bridge.* Edited by D. M. Hopkins, pp. 207–31. Stanford: Stanford University Press.

Colinvaux, P. A. 1967b. Bering Land Bridge: Evidence of spruce in late Wisconsin times. *Science* 156:380–83.

Colinvaux, P. A. 1967c. A long pollen record from St. Lawrence Island, Bering Sea, Alaska. *Palaeogeography, Palaeoclimatology, and Palaeoecology* 3:29–48.

Colinvaux, P. A. 1980. Vegetation of the Bering Land Bridge revisited. *Quarterly Review of Archaeology* 1:2–15.

Colinvaux, P. A. 1981. Historical ecology in Beringia: The south land bridge coast at St. Paul Island. *Quaternary Research* 16:18–36.

Colinvaux, P. A., and H. C. Noltimier. In press. Pleistocene chronostratigraphy in Beringia: A review of ^{14}C ages, paleosecular variation, and paleoecology of cores from Imuruk Lake. Manuscript.

Collins, D. 1969. Culture traditions and environment of early man. *Current Anthropology* 10:267–316.

Collins, F. R. 1958. "Vegetated sand dunes in central Alaska," in *Science in Alaska 1958,* Proceedings, Alaskan Science Conference, 9:72.

Cook, J. P. 1969. *The early prehistory of Healy Lake, Alaska.* University of Wisconsin, Ph.D. thesis.

Cook, J. P., editor. 1970. *Report of archaeological survey and excavations along the Alyeska Pipeline Service Company Haulroad and Pipeline Alignments.* College: University of Alaska.

Cook, J. P. 1975. Archaeology of interior Alaska. *Western Canadian Journal of Anthropology* 5:125–33.

Cook, J. P., editor. 1977. *Pipeline Archaeology.* College: University of Alaska, Institute of Arctic Biology.

Coonrad, W. L. 1957. Geological reconnaissance in the Yukon-Kuskokwim Delta region, Alaska. *United States Geological Survey Miscellaneous Geologic Investigations Map* I-223.

Coope, G. R. 1977. Fossil coleopteran assemblages as sensitive indicators of climatic changes during the Devensian (Last) cold stage. *Philosophical Transactions of Royal Society of London* B280:313–40.

Coope, G. R., and J. A. Brophy. 1972. Late Glacial environmental changes indicated by a coleopteran succession from North Wales. *Boreas* 1:97–142.

Coope, G. R., F. W. Shotton, and I. Strachen. 1961. A late Pleistocene fauna and flora from Upton Warren, Worchestershire. *Philosophical Transactions of the Royal Society of London,* Series B 244:379–421.

Corbett, P. S. 1972. The microclimate of arctic plants and animals on land and in fresh water. *Acta Arctica* 18: 1–43.

Corns, I. G. W. 1974. Arctic plant communities east of the Mackenzie Delta. *Canadian Journal of Botany* 52: 1730–45.

Cotter, J. C. 1962. The Paleo-Indian tradition in eastern North America, a reply. *Current Anthropology* 3: 250–52.

Coulter, H. W., D. M. Hopkins, T. N. V. Karlstrom, T. L. Péwé, C. Wahrhaftig, and J. R. Williams. 1965. Extent of glaciations in Alaska. *United States Geological Survey Miscellaneous Geological Investigations Map* I-415.

Coupland, R. T. 1974. *Producers: VI. Summary of studies of primary production by biomass and shoot observation methods.* Matador Project Technical Report No. 62. Saskatoon: University of Saskatchewan.

Courant, R., and D. Hilbert. 1962. *Methods of mathematical physics,* vol. 2. New York: Interscience.

Cox, A. V., D. M. Hopkins, and G. B. Dalrymple. 1966. Geomagnetic polarity epochs: Pribilof Islands. *Geological Society of American Bulletin* 77:883–909.

Cox, S. L. n.d. *A re-analysis of the Ghoop Site.* Smithsonian Institution unpublished manuscript.

Creager, J. S., and D. A. McManus. 1967. "Geology of the floor of Bering and Chukchi Seas—American studies," in *The Bering Land Bridge.* Edited by D. M. Hopkins, pp. 7–31. Stanford: Stanford University Press.

Cushing, E. J. 1964. Redeposited pollen in late-Wisconsin pollen spectra from east-central Minnesota. *American Journal of Science* 262:1075–88.

Cwynar, L. C. 1980. *A late-Quaternary vegetation history from Hanging Lake, northern Yukon.* University of Toronto, Ph.D. thesis.

Cwynar, L. C. In press. A late-Quaternary vegetation history from Hanging Lake, northern Yukon. *Ecological Monographs* 52:1–24 (1982).

Cwynar, L. C., and J. C. Ritchie. 1980. Arctic steppe-tundra: A Yukon perspective. *Science* 208:1375–77.

Dary, D. A. 1974. *The buffalo book.* Chicago: Swallow Press.

Daubenmire, R. 1968. Ecology of fire in grasslands. *Advances in Ecological Research* 5:209–66.

Davidovich, T. D., and V. F. Ivanov. 1976. "The climate of coastal areas of eastern Chukotka in the Late Pleistocene and Holocene," in *Geokryologicheskie usloviya formirovaniya verkhnepleistotsenovikh i golotsenovkh otlozheniy na Severo-Vostoke SSR,* pp. 22–23. Magadan (in Russian).

Davis, E. L. 1973. Paleo-Indian land use patterns at China Lake, California. *Pacific Coast Archaeological Society Quarterly* 10:3–16.

Davis, E. L., and Carol Panlaqui. 1978. "Stone tools, the action tools," in *The ancient Californians: Rancholabrean hunters of the Mojave Lakes country.* Edited by E. L. Davis, pp. 30–75. Natural History Museum of Los Angeles County Science Series 29.

Davis, M. B. 1967. "Late-glacial climate in northern United States: A comparison of New England and Great Lakes region," in *Quaternary Paleoecology.* Edited by E. J. Cushing and H. E. Wright, Jr. pp. 11–43. New Haven: Yale University Prrss.

Deardorff, J. W. 1972. Parameterization of the planetary boundary layer for use in general circulation models. *Monthly Weather Review* 100:93–106.

Delcourt, P. A., and H. R. Delcourt. 1980. Pollen preservation and Quaternary environmental history in the southeastern United States. *Palynology* 4:215–31.

Delcourt, P. A., H. R. Delcourt, R. C. Brister, and L. E. Lackey. 1980. Quaternary vegetation history of the Mississippi Embayment. *Quaternary Research* 13: 111–32.

Delorme, L. D., S. C. Zoltai, and L. L. Kalas. 1977. Freshwater shelled invertebrate indicators of paleoclimate in northwestern Canada during late glacial times. *Canadian Journal of Earth Sciences* 14:2029–46.

Denton, G. H. 1974. Quaternary glaciations of the White River Valley, Alaska, with a regional synthesis for the northern St. Elias Mountains, Alaska, and Yukon Territory. *Geological Society of America Bulletin* 85: 871–92.

Denton, G. H., and Minze Stuiver. 1967. Late Pleistocene glacial stratigraphy and chronology, northeastern St. Elias Mountains, Yukon Territory, Canada. *Geological Survey of American Bulletin* 78:485–510.

De Vos, A. 1969. Ecological conditions affecting the production of wild herbivorous mammals on grasslands. *Advancement in Ecological Research* 6:137–83.

Diamond, J. M. 1973. Distributional ecology of New Guinea birds. *Science* 179:759–69.

Dikov, N. N. 1977. *Arkheologicheskie pamyatniki Kamchatki, Chukotki i verkhey Kolymy (Archaeologic monuments in Kamchatka, Chukotka, and upper Kolyma).* Moscow: Izd. "Nauka" (in Russian).

Dikov, N. N. 1978. "Ancestors of Paleo-Indians and Proto-Eskimo-Aleuts in the Paleolithic of Kamchatka," in *Early Man in America from a circum-Pacific perspective.* Edited by A. L. Bryan, pp. 68–69. University of alberta Department of Anthropology Occasional Papers.

Dinter, D. A. In press. Holocene marine sediments on the middle and outer continental shelf of the Beaufort Sea, north of Alaska. *United States Geological Survey Miscellaneous Geology Investigations Map* I-1189-B.

Dixon, E. J. 1975. The Gallagher Flint station, an early man site on the north slope, Arctic Alaska, and its role in relation to the Bering Land Bridge. *Arctic Anthropology* 12:68–75.

Dixon, E. J. 1976. The Pleistocene prehistory of Arctic North America. *IXe Congrés, Union Internationale des Sciences Préhistoriques et Protohistoriques* 8:168–98.

Dolukhanov, P. M. 1978. Ecology *and economy in Neolithic eastern Europe.* New York: St. Martin's Press.

Donner, J. J. 1975. "Pollen composition," in *Abri Pataud.* Edited by H. L. Movius, pp. 160–70. Cambridge: Harvard University Press.

Dort, W., Jr., and S. Miller. 1977. Archaeological geology of Birch Creek Valley and the eastern Snake River Plain, Idaho. *Geological Society of America, Division of Archaeological Geology, Guidebook.* Idaho Falls.

Dosotovalov, B. N., and V. A. Kudryavtsev. 1967. *Obshchee merzlotovedenie (General geocryology),* pp. 391–98. Moscow: Izd. MGU (in Russian).

Dougall, H. W., and D. L. W. Sheldrick. 1964. The chemical composition of a day's diet of an elephant. *East African Wildlife Journal* 2:51–59.

Douglas, G. W. 1974. Montane zone vegetation of the Alsek River region, southwestern Yukon. *Canadian Journal of Botany* 52:2505–32.

Douglas-Hamilton, I. 1972. *On the ecology of behaviour of the African elephant.* University of Oxford, Ph.D. thesis.

Douglas-Hamilton, I. 1975. "Radio-elephants," in *Among the elephants,* chapter 7. London: Collins and Harvill Press.

Douglas-Hamilton, Ian, and Oria Douglas-Hamilton. 1975. *Among the elephants.* New York: Viking Press.

Dragoo, D. W. 1973. "The Wells Creek Site, Steward County, Tennessee," in *Archaeology of eastern North America.* Eastern States Archaeological Federation 1:1–56.

Dreimanis, A. 1967. Mastodons, their geologic age and extinction in Ontario, Canada. *Canadian Journal of Earth Sciences* 4:663–75.

Drew, J. V., and R. E. Shanks. 1965. Landscape relationships of soil and vegetation in the forest-tundra ecotone, Upper Firth River Valley, Alaska-Canada. *Ecological Monographs* 35:285–306.

Dumond, D. E. 1974. "Remarks on the prehistory of the North Pacific: to lump or not to lump," in *International Conference on the Prehistory and Paleoecology of Western North American Arctic and Subarctic.* Edited by Scott Raymond and Peter Schledermann, pp. 47–55. Calgary: University of Calgary Archaeological Association.

Dumond, D. E. 1977. *The Eskimos and Aleuts.* London: Thames and Hudson.

Dumond, D. E. 1979. Eskimo-Indian relationships: A review from prehistory. *Arctic Anthropology* 16:3–31.

Dumond, D. E. 1980a. A chronology of native Alaska subsistence systems. *Senri Ethnological Studies* 4:23–47.

Dumond, D. E. 1980b. The archaeology of Alaska and the peopling of America. *Science* 209:984–91.

Dumond, D. E., Winfield Henn, and Robert Stuckenrath.

1976. Archaeology and prehistory on the Alaska Peninsula. *Anthropological Papers of the University of Alaska* 18(1):17–29.

Dupré, W. R. 1977a. The Yukon Delta: Onshore impacts of offshore oil. *Geological Society of America Abstracts with Programs* 9:959–60.

Dupré, W. R. 1977b. Late Quaternary history of the Yukon-Kuskokwim Delta complex, Alaska. *Geological Society of America Abstracts with Programs* 9:413.

Dupré, W. R. 1978. Yukon Delta Coastal Processes Study. NOAA, *Environmental Assessment of the Alaskan Continental Shelf,* Annual Reports, Principal Investigators XI (Hazards): 384–446.

Dupré, W. R., and R. Thompson. 1979. The Yukon Delta: A model for deltaic sedimentation in an ice-dominated environment. *Proceedings of the 11th Annual Offshore Technology Conference,* Paper OTC 3434:657–64.

Duvall, J. G., and W. T. Venner. 1979. A statistical analysis of the lithics from the Calico site (SBCM 1500 A), California. *Journal of Field Archaeology* 6:455–62.

Dyck, W., and J. G. Fyles. 1964. Geological Survey of Canada Radiocarbon Dates III. *Geological Survey of Canada Paper* 64–40.

Eddy, J. A. 1976. The Maunder minimum. *Science* 192: 1189–1202.

Efremov, J. A. 1940. Taphonomy: A new branch of paleontology. *Pan-American Geologist* 74:81–93.

Eisenberg, J. F., and M. Lockhart. 1972. An ecological reconnaissance of Wilpattu National Park, Ceylon. *Smithsonian Contributions to Zoology* 101.

Eltringham, S. K. 1974. Changes in the large mammal community of Mweya Peninsula, Rwenzori National Park, Uganda, following removal of hippopotamus. *Journal of Applied Ecology* 11:855–65.

Eltringham, S. K. 1976. The frequency and extent of uncontrolled grass fires in the Rwenzori National Park, Uganda. *East African Wildlife Journal* 14:215–22.

Emiliani, C. 1955. Pleistocene temperatures. *Journal of Geology* 63:538–78.

England, R. E., and A. De Vos. 1969. Influence of animals on pristine conditions on the Canadian grasslands. *Journal of Range Management* 22:87–94.

Ermolova, N. M. 1978. *Theriofauna in the Angara Valley in the late Anthropogene.* Novosibirsk: Izd. "Nauka" (in Russian).

Estes, R. D. 1974. "Social organization of the African Bovidae," in *The behavior of ungulates and its relation to management.* Edited by V. Geist and F. F. Walther, pp. 166–205. International Union for Conservation of Nature and Natural Resources Publ. n. s. 24.

Everett, K. R. 1971. Composition and genesis of the organic soils of Amchitka Island, Alaska. *Journal of Arctic and Alpine Research* 3:1–16.

Ewer, R. F. 1968. *Ethology of mammals.* London: Logos Press.

Faegri, K., and J. Iversen. 1975. *Textbook of pollen analysis.* New York: Hafner Press.

Fahl, C. B. 1975. "Mean sea level pressure patterns relating to glacier activity in Alaska," in *Climate of the Arctic.* Edited by G. Weller and S. A. Bowling, pp. 339–46. Fairbanks: University of Alaska.

Fairbanks, R. G. 1980. The marine oxygen isotope record: A problematic ice-volume analogue. *AMQUA Abstracts* 6:72–73. Orono.

Farrand, W. R. 1961. Frozen mammoths and modern geology. *Science* 133:729–35.

Fernald, A. T. 1960. Geomorphology of the upper Kuskokwim region, Alaska. *United States Geological Survey Bulletin* 1071-G:191–279.

Fernald, A. T. 1964. Surficial geology of the central Kobuk River valley, northwestern Alaska. *United States Geological Survey Bulletin* 1181-K.

Fernald, A. T. 1965. Recent history of the upper Tanana River lowland, Alaska. *United States Geological Survey Professional Paper* 525-C:C124–C127.

Ferns, P. N. 1977. Muskox abundance in the southern part of the range in east Greenland. *Arctic* 30:52–60.

Field, C. R. 1971. Elephant ecology in the Queen Elizabeth National Park, Uganda. *East African Wildlife Journal* 9:99–123.

Field, C. R., and R. M. Laws. 1970. The distribution of the larger herbivores in the Queen Elizabeth National Park, Uganda. *Journal of Applied Ecology* 7:273–94.

Field, C. R., and I. C. Ross. 1976. The savanna ecology of the Kidepo Valley National Park. II: Feeding ecology of elephant and giraffe. *East African Wildlife Journal* 114:1–15.

Fischer, A. G. 1960. Latitudinal variation in organic diversity. *Evolution* 14:64–81.

Fitting, J. E., J. DeVisscher, and E. J. Wahla. 1966. The Paleoindian occupation of the Holcombe Beach, *Anthropological Papers of the University of Michigan Museum of Anthropology* 27:1–159.

Flerow, C. C. 1931. A trunk of mammoth, *Elephas primigenius* Blum., found in the Kolyma District, Siberia. *Izvestia Akad. Nauk SSSR,* pp. 863–70 (in Russian).

Flerow, C. C. 1935. "The reindeer from the Paleolithic of Siberia," in *Paleolit SSSR.* SSSR Gos. Akad. Instituta Materialny Kul'tury, Izvestia 118:271–73 (in Russian).

Flerow, C. C. 1950. "The morphology and ecology of the Cervidae in relation to their evolution," in *Materialy po Chetvertichnogo Period SSSR* 2:50–69. Moscow-Leningrad: Izd. ANSSR (in Russian).

Flerow, C. C. 1967. "On the origins of the mammalian fauna in Canada," in *The Bering Land Bridge.* Edited by D. M. Hopkins, pp. 271–80. Stanford: Stanford University Press.

Flerow, C. C. 1977. Bison of northeastern Asia. *Trudy Akad. Nauk SSSR, Zoologicheskie Instituta* 73:39–56 (in Russian).

Flerow, C. C. 1979. "Bison systematics and evolution," in *Zubr. M.* Edited by V. E. Sokolov, pp. 9–127. Izd. "Nauka" (in Russian).

Flohn, H. 1975. "History and intransivity of climate," in *The physical basis of climate and climate modeling.* World Meteorological Organization, Global Atmospheric Research Project Series 16:106–18. Geneva.

Flook, D. R. 1970. A study of sex differential in the survival of Wapiti. *Canadian Wildlife Service Report* 11.

Forbes, D. L. 1980. Late Quaternary sea levels in the southern Beaufort Sea. *Geological Survey of Canada Paper* 80-1B:75–87.

Formazov, A. N. 1946. Snow cover as an environmental

factor, and its significance in the life of mammals and birds in the U.S.S.R. *Materialy k Pozd. Fauni i Flori SSSR, n.s., Otdelnie Zoologicheskie* 5(20). Moscow.

Foscolos, A. E., N. W. Rutter, and O. L. Hughes. 1977. The use of pedological studies in interpreting the Quaternary history of central Yukon Territory. *Geological Survey of Canada Bulletin* 271, 48 pp.

Foster, H. L. 1969. Reconnaissance geology of the Eagle A-1 and A-2 quadrangles, Alaska. *U.S. Geological Survey Bulletin* 1271-G.

Fox, J. 1978. *Weather as a determinate factor in summer mountain goat activity and habitat use.* University of Alaska, M.S. thesis.

Fox, J. E. D. 1972. *The natural vegetation of Sabah and natural regeneration of the dipterocarp forests.* University of Wales, Ph.D. thesis.

Fredskild, B. 1973. Studies in the vegetational history of Greenland. *Meddelelser om Grønland* 198:4.

Freeland, W. J., and D. H. Janzen. 1974. Strategies in herbivory by mammals: The role of plant secondary compounds. *American Naturalist* 108:269–89.

Freeman, L. G. 1978. "The analysis of some occupation floor distributions from earlier and middle Palaeolithic sites in Spain," in *View of the past.* Edited by L. G. Freeman, pp. 57–116. The Hague: Mouton.

French, D. D. 1974. "Classification of IBP Tundra Biome sites based on climate and soil properties," in *Soil organisms and decomposition.* Edited by A. J. Holding et al., pp. 3–25. Stockholm: Tundra Biome Steering Committee.

Frenzel, B. 1967. *Die klimaschwangungen des Eiszeitalters.* Braunschweig: Vieweg.

Frenzel, B. 1968. *Grundzüge der Pleistozänen Vegetationsgeschichte Nord-Eurasiens.* Wiesbaden: Steiner.

Frenzel, B. 1968b. The Pleistocene vegetation of northern Eurasia. *Science* 161:637–49.

Frison, G. C. 1974. "Archaeology of the Casper site," in *The Casper site: A Hell Gap bison kill on the High Plains.* Edited by G. C. Frison, pp. 1–111. New York: Academic Press.

Frison, G. C. 1977. Bone butchering tools in prehistoric bison kill sites. *Abstracts of the 42nd Annual Meeting of the Society for American Archaeology,* p. 39.

Frison, G. C. 1978. *Prehistoric hunters of the High Plains.* New York: Academic Press.

Fuller, W. A. 1962. The Biology and management of the bison of Wood Buffalo National Park. *Canadian Wildlife Service, Wildlife Management Bulletin,* Series 1(16).

Fuller, W. A., A. M. Martell, R. F. C. Smith, and S. W. Speller, 1977. "Biology and secondary production of *Dicrostonyx groenlandicus* on Truelove Lowland," in *Truelove Lowland, Devon Island, Canada: A high arctic ecosystem.* Edited by L. C. Bliss, pp. 461–91. Edmonton: University of Alberta Press.

Funder, S. 1978. Holocene stratigraphy and vegetation history in the Scoresby Sund area, east Greenland. *Grønlands Geologiske Undersøgelse Bulletin* 129.

Furlow, J. J. 1979. The systematics of the American species of *Alnus* (Betulaceae). *Rhodora* 81:1–121, 151–248.

Fyles, J. G., J. A. Heginbottom, and V. N. Rampton. 1972. Quaternary geology and geomorphology, Mackenzie Delta to Hudson Bay. *24th International Geological Congress, Guidebook, Excursion A-30.* Montreal.

Gamitto, A. C. P. 1960. *King Kazemb* (translation by I. Cunnison) volumes 1 and 2. Lisbon: Estudios de ciencias politicas e socias.

Gardner, W. M. 1977. "Flint Run Paleoindian complex and its implications for eastern North American prehistory," in *Amerinds and their paleoenvironments in northeastern North America.* Edited by W. S. Newman and B. Salwen. *Annals of the New York Academy of Sciences* 288:257–63.

GARP (Global Atmospheric Research Program). 1975. *The physical basis of climate and climate modelling.* World meteorological Organization, Global Atmospheric Research Project Series 16:106–18. Geneva.

Garutt, V. E. 1964. *Das Mammut Die Neue Brehm-Bücherei.* Wittenberg Lutherstadt A. Ziemsen Verlag.

Garutt, V. E., Ye. P. Metel'tseva, and B. A. Tikhomirov. 1970. "New data on the food of the woolly rhinoceros in Siberia," in *Severo ledovityy okean i ego poberezh'e v Kainozoe,* pp. 113–25. Leningrad: Gidrometeoizdat (in Russian).

Garutt, V. E., and K. V. Yurtiyev. 1966. Mummified remains of a wild horse in permafrost in the Indigirka River basin. *Byulleten Kommissii Izucheniyu Chetvertichnogo Perioda* 31:86–92 (in Russian).

Gasanov, Sh. 1969. *Stroyeniya i istoriya formirovaniya merzlykh porod Vostochnoi Chukotki (Structure and history of the frozen rocks of East Chukotka).* Moscow: Izd. "Nauka" (in Russian).

Gasaway, U. C., and J. W. Coady. 1974. Review of energy requirements and rumen fermentation in moose and other ruminants. *Naturalist Canada* 101:227–62.

Gates, W. L. 1976a. Modeling the ice age climate. *Science* 191:1138–44.

Gates, W. L. 1976b. The numerical simulation of Ice-Age climate with a global general circulation model. *Journal of the Atmospheric Sciences* 33:1844–73.

van Geel, B., and E. Kolstrump. 1978. Tentative explanation of the late glacial and early Holocene climatic changes in northwestern Europe. *Geologie en Mijnbouw* 57:87–89.

Geist, V. 1971. *Mountain sheep: A study in behavior and evolution.* University of Chicago Press.

Geist, V. 1974. On the relationship of social evolution and ecology in ungulates. *American Zoology* 14:205–20.

Geist, V. 1977. A comparison of social adaptations in relation to ecology in gallinaceous birds and ungulate societies. *Annual Review of Ecology and Systematics* 8:193–208.

Geist, V. 1978. *Life strategies, human evolution, environmental designs; towards a biological theory of health.* New York: Springer-Verlag.

Geologiya SSR, Severo-Vostok SSSR. 1970. *Geologicheskoe Opisanie* Book 2, v. 30. Moscow: Izd. "Nedra" (in Russian).

Gerasimov, I. P. 1952. Contemporary vestiges of Late Glacial phenomena near the coldest region in the world. *Akad Nauk SSSR Izvestia, Seriya Geograficheskaia* 5:16–22 (In Russian).

Gerasimov, M. M. 1935. Excavations of the Palaeolithic site of Mal'ta (Preliminary report on the work of 1928–1932). Akad. Nauk SSSR, *Izvestia Gosudarstvennoy Akademii Istorii Materal'noy Kultury* 118:78–124 (In Russian).

Gerasimov, M. M. 1941. Worked bones from the Paleolithic site of Mal'ta. *Materialy i Issledovanii po Arkheologii SSSR* 2:65–85 (in Russian).

Getchinson, ? ?. 1900. Extinct monsters. *Obshchedotupnaya Nauchnaya Biblioteka* 16. St. Petersburg: Izd. "Znaie" (quoted by Vereshchagin "from Engineer Beckendorf's text, pp. 158–61") (in Russian).

Giddings, J. L. 1954. Early man in the Arctic. *Scientific American* 190(6):82–88.

Giddings, J. L. 1960. The archaeology of Bering Strait. *Current Anthropology* 1:121–38.

Giddings, J. L. 1964. *The archaeology of Cape Denbigh.* Providence: Brown University Press.

Gillespie, J. M. 1970. Mammoth hair: Stability of α-keratin structure and constituent proteins. *Science* 170:1100–02.

Giterman, R. E. 1963. *Etapy razvitiya chetvertichnoy rastitel'nosti Yakutii i ikh znachenie dlya stratigrafii (Stages of development of Quaternary vegetation of Yakutia and their importance for stratigraphy).* Akad. Nauk SSSR, Geologicheskie Institut Trudy 78 (in Russian).

Giterman, R. E. 1968. "Vegetation of the Early Pleistocene," in *Osnovnye Etapy Razvitiya Rastitel'nosti Severnoi Azii v Antropogene,* pp. 189–203. Izd. "Nauka" (in Russian).

Giterman, R. E. 1975. Periglacial flora and vegetation of the western part of Beringia in the Pleistocene. *Byulleten Kommissii po Izuchenyu Chetvertichnogo Perioda* 43:146–51 (in Russian).

Giterman, R. E. In press. The history of Pliocene and Pleistocene vegetation in the western part of Beringia. *4th International Palynological Conference, Proceedings,* pp. 134–39. Lucknow (1980).

Giterman, R. E., and L. V. Golubeva. 1965. "The history of development of East Siberian vegetation in the Anthropogene," in *Osnovnye Problemy Izucheniya Chetvertichnogo Perioda.* Moscow: Izd. "Nauka," pp. 365–75 (in Russian).

Giterman, R. E., and L. V. Golubeva. 1967. "Vegetation of eastern Siberia during the Anthropogene Period," in *The Bering Land Bridge.* Edited by D. M. Hopkins, pp. 232–44. Stanford: Stanford University Press.

Giterman, R. E., L. V. Golubeva, E. D. Zaklinskaya, E. V. Koreneva, O. V. Matveeva, and L. A. Skiba. 1968. *Osnovnye Etapy Razvitiya Rastitel'nosti Severnoi Azii v Antropogene (The main developmental stages of the vegetation of North Asia in the Anthropogene).* Akad. Nauk SSSR, Geo*logicheskie Institut Trudy* 177 (in Russian).

Glaser, P. H. 1978. *Recent plant macrofossils from the Alaska interior and their relationship to late-glacial landscapes in Minnesota.* University of Minnesota, Ph.D. thesis.

Golley, F. B. 1972. "Energy flux in ecosystems," in *Ecosystem structure and function.* Edited by J. A. Wiens, pp. 69–90. Corvallis: Oregon State University Press.

Golubeva, L. V. 1968. "Vegetation of the late Pleistocene," in *Osnovnye Etapy Razvitiya Rastitel'nosti Severnoi Azii v Antropogene,* pp. 203–30. Izd. "Nauka" (in Russian).

Gorchakovsky, P. L., and N. I. Andreyaskina. 1972. "Productivity of some shrub, dwarf shrub and herbaceous communities of forest-tundra," in *Proceedings 4th international meeting on the biological productivity of tundra.* Edited by F. E. Wielgolaski and T. Rosswall, pp. 113–16. Stockholm: Tundra Biome Steering Committee.

Gorchakovsky, P. L., and P. L. Shiyatov. 1978. The upper forest limit in the mountains of the boreal zone of the U.S.S.R. *Arctic and Alpine Research* 10:349–63.

Graham, R. W., and H. A. Semken. 1976. Paleoecological significance of the short-tailed shrew (*Blarina*), with a systematic discussion of *Blarina ozarkensis. Journal of Mammalogy* 57:433–49.

Grant, W. E., and E. C. Birney. 1979. Small mammal community structure in North American grasslands. *Journal of Mammalogy* 60:23–36.

Gray, J. M., and J. J. Lowe. 1977. *Studies in the Scottish Lateglacial environment.* Pergamon.

Grayson, D. K. 1977. Pleistocene avifaunas and the overkill hypothesis. *Science* 195:691–93.

Greene, F. E. 1963. The Clovis blades: An important addition to the Llano Complex. *American Antiquity* 29: 145–65.

Grichuk, M. P. 1960. Stratigraphic significance of the history of the middle reaches of the Yenisei and Ob' Rivers. *Mat-lov po Geol. Krasnoyarskogo Kraya.* Moscow: Gostoptekhizdat (in Russian).

Grichuk, M. P. 1967. The study of pollen spectra from recent and ancient alluvium. *Review of Palaeobotany and Palynology* 4:107–12.

Grichuk, M. P., and V. P. Grichuk. 1960. "On periglacial vegetation in the territory of the U.S.S.R.," in *Periglyatsial'nye Yavleniya na Territorii S.S.S.R.,* pp. 66–100. Moscow: Izd. MGU (in Russian).

Grichuk, V. P. 1946. "Zur Vegetationsgeschichte des Europaischen Russlands im Quartär," in *Problems of Quaternary,* pp. 105–12. Vortrage bei der Moskauer Quartärtagung im Feb. 1941. ersch. i. d. Arb. d. Geogr. Inst. d Akad., v. 37. Cited in Budel (1949). (In Russian.)

Grichuk, V. P. 1951. The post-glacial period of aridity in the European USSR. *Voprosy Geografii* 24 (in Russian).

Grichuk, V. P. 1971. The analysis of zonal structure of the Pleistocene vegetational cover across U.S.S.R. *Pollen et Spores* 13:101–16.

Grichuk, V. P. 1973. "Late Pleistocene natural phenomena in the periglacial regions of Europe. 4. Vegetation," in *Paleogeografiya Evropy v pozdnem Pleystotsene. Rekonstruktsii i Modeli. Opytnyy maket atlasa-monografii.* Akad. Nauk SSSR Instituta Geografiya. k IX INQUA Kongress, p. 182–219 (in Russian).

Griffin, J. B. 1979. "The origin and dispersion of American Indians in North America," in *The First Americans: Origins, affinities, and adaptations.* Edited by W. S. Laughlin and A. B. Harper, pp. 43–55. New York: Gustav Fischer.

Grigoryev, A. A., and M. I. Budyko. 1956. On the periodi-

cal law of geographic zonation. *Akad. Nauk SSSR Doklady* 110:129–32 (in Russian).
Grime, J. P. 1979. *Plant strategies and vegetation processes.* New York: Wiley.
Grimsdell, J. J. R. 1969. *Ecology of the buffalo,* Syncerus caffer, *in western Uganda.* University of Cambridge, Ph.D. thesis.
Gromov, I. M., and I. Ya. Polyakov. 1977. Voles (Microtinae). *Fauna SSSR Mlekopitayuie* 3(8). Leningrad: Izd. "Nauka."
Grootes, P. M. 1977. *Thermal diffusion, isotopic enrichment, and radiocarbon dating beyond 50,000 years BP.* Groningen University, Ph.D. thesis.
Grosswald, M. G. 1980. Late Weichselian ice sheet of northern Eurasia. *Quaternary Research* 13:1–32.
Gruhn, Ruth. 1961. *The archaeology of Wilson Butte Cave, south-central Idaho.* Idaho State College Museum Occasional Papers 6.
Gruhn, Ruth. 1965. Two early radiocarbon dates from the lower levels of Wilson Butte Cave, south-central Idaho. *Tebiwa* 8:57.
Guilday, J. E., H. W. Hamilton, and P. W. Parmalee. 1975. Caribou (*Rangifer tarandus* L.) from the Pleistocene of Tennessee. *Journal of Tennessee Academy of Sciences* 50:109–12.
Guilday, J. E., P. W. Parmalee, and H. W. Hamilton. 1977. The Clark's cave bone deposit and the late Pleistocene paleoecology of the central Appalachian Mountains of Virginia. *Carnegie Museum of Natural History Bulletin* 2:1–88.
Günther, E. W. 1975. Die Backenzähne der Elefanten von Ehringsdorf bei Weimar. *Abhandlungen des Zentralen Geologischen Instituts* 23:399–452.
Gustafson, C. E., Delbert Gilbow, and R. D. Daugherty. 1979. The Manis mastodon site: Early man on the Olympic Peninsula. *Canadian Journal of Archaeology* 3:157–64.
Guthrie, R. D. 1966. The extinct wapiti of Alaska and the Yukon Territory. *Canadian Journal of Zoology* 44: 45–57.
Guthrie, R. D. 1968a. Paleoecology of the large-mammal community in interior Alaska during the late Pleistocene. *American Midland Naturalist* 79:346–63.
Guthrie, R. D. 1968b. Paleoecology of a late Pleistocene small mammal community from interior Alaska. *Arctic* 21:223–44.
Guthrie, R. D. 1972. Recreating a vanished world. *National Geographic Magazine* 141(3):294–301.
Guthrie, R. D. 1976. "Environmental influences on body size, social organs, population parameters and extinction of Beringian animals," in *Beringiya v Kainozoe.* Edited by V. L. Kontrimavichus, pp. 296–322. Khabarovsk: Akad. Nauk SSSR, Dal'nyvostochnyy Nauchnyy Tsentr (in Russian, English abstract).
Guthrie, R. D. 1978. Bison and man in North America. *AMQUA Abstracts* 5:83–88. Tempe.
Guthrie, R. D. 1980. The first Americans? The elusive arctic bone technology. *Quarterly Review of Archaeology* 1(1):2.
Guthrie, R. D. 1981a. A preliminary report on the frozen bison mummy. *American Association for the Advancement of Science,* Alaska Science Conference 32:42.
Guthrie, R. D. 1981b. Mid-Wisconsin Beringian man: Elusive or illusive?. *Quarterly Review of Archaeology* 1(2):17–19.
Guthrie, R. D. In press. "A new ecological theory of the Late Pleistocene megafaunal extinctions," in *Pleistocene extinctions.* Edited by P. S. Martin and R. G. Klein. Tucson: University of Arizona Press.
Guthrie, R. D., and J. V. Matthews, Jr. 1971. The Cape Deceit Fauna—early Pleistocene mammalian assemblage from the Alaskan Arctic. *Quaternary Research* 1:474–510.
Guy, P. R. 1976. The feeding behavior of elephant (*Loxodonta africana*) in the Sengwa area, Rhodesia. *South African Journal of Wildlife Research* 6:55–63.
Haag, R. W. 1974. Nutrient limitations to plant production in two tundra communities. *Canadian Journal of Botany* 52:103–16.
Haberland, Wolfgang. 1969. Comment on Bryan, 1969. *Current Anthropology* 10:352.
Hahn, J. 1977. *Aurignacien. Das Ältere Jungpaläolithikum in Mittel- und Osteuropa.* Cologne: Böhlau.
Hall, E. S. 1969. Comment on Bryan, 1969. *Current Anthropology* 10:352–53.
Hall, E. S. 1975. An archaeological survey of interior Northwest Alaska. *Anthropological Papers of the University of Alaska* 17(2):13–30.
Hall, E. S., and R. A. McKennan. 1973. An archaeological survey of the Old John Lake area, northern Alaska. *Polar Notes* 13:1–31.
Hallam, S. J. 1977. The relevance of Old World archaeology to the first entry of man into New Worlds: Colonization seen from the Antipodes. *Quaternary Research* 8:128–48.
Hamilton, T. D. 1970. "Geologic relations of the Akmak assemblage, Onion Portage area," (appendix to "Akmak, an early archaeological assemblage from Onion Portage, northwest Alaska"). *Acta Arctica* 16:71–78.
Hamilton, T. D. 1979a. Quaternary stratigraphic sections with radiocarbon dates, Chandalar quadrangle, Alaska. *United States Geological Survey Open File Report* 79-751.
Hamilton, T. D. 1979b. Surficial geology, Wiseman quadrangle, Alaska. *United States Geological Survey Map* MF-1122.
Hamilton, T. D. 1980. Quaternary stratigraphic sections with radiocarbon dates, Wiseman quadrangle, Alaska. *United States Geological Survey Open File Report* 80-791.
Hamilton, T. D. 1981. Multiple moisture sources and the Brooks Range glacial record. University of Colorado, Institute of Arctic and Alpine Research, *Arctic Workshop* 10:16–18.
Hamilton, T. D. In press. A late Pleistocene glacial chronology for the southern Brooks Range: Stratigraphic record and regional significance. *Geological Society of America Bulletin.*
Hamilton, T. D., and D. M. Hopkins. 1982. Correlation of northern Alaskan glacial deposits—a provisional stratigraphic framework. *United States Geological Survey Circular* 844:15–17.

Hamilton, T. D., and S. C. Porter. 1975. Itkillik glaciation in the Brooks Range, northern Alaska. *Quaternary Research* 5:471–98.

Hamilton, T. D., R. Stuckenrath, and M. Stuiver. 1980. Itkillik Glaciation in the central Brooks Range: Radiocarbon dates and stratigraphic record. *Geological Society of America Abstracts with Programs* 12:109.

Hamilton, T. D., and R. M. Thorson. In press. "Cordilleran ice sheet in Alaska," in *Late Wisconsin Pleistocene environments of the United States.* Edited by S. C. Porter. Minneapolis: University of Minnesota Press.

Hanson, H. C. 1951. Characteristics of some grassland, marsh, and other plant communities in western Alaska. *Ecological Monographs* 21:317–78.

Hanson, H. C. 1953. Vegetation types in northwestern Alaska and comparisons with communities in other arctic regions. *Ecology* 343:111–40.

Hare, F. K., and J. C. Ritchie. 1972. The boreal bioclimates. *Geographical Review 62:* 333–65.

Harington, C. R. 1975. A bone tool found with Ice Age mammal remains near Dawson City, Yukon Territory. *The Arctic Circular* 23(1):2–5.

Harington, C. R. 1977. *Pleistocene mammals of the Yukon Territory.* University of Alberta, Ph.D. thesis.

Harington, C. R. 1978. *Quaternary vertebrate faunas of Canada and Alaska and their suggested chronological sequence.* Canadian National Museum of Natural Sciences, Syllogeus 15.

Harington, C. R. 1980a. Pleistocene mammals from Lost Chicken Creek, Alaska. *Canadian Journal of Earth Sciences* 17:168–98.

Harington, C. R. 1980b. Radiocarbon dates on some Quaternary mammals and artifacts from northern North America. *Arctic* 33:815–32.

Harington, C. R. 1981. "Pleistocene Saiga antelopes in North America and their paleoenvironmental implications," in *Quaternary Paleoclimate.* Edited by W. C. Mahaney, pp. 193–225. Norwich: Geoabstracts.

Harington, C. R., Robson Bonnichsen, and R. E. Morlan. 1975. Bones say man lived in Yukon 27,000 years ago. *Canadian Geographical Journal* 91:42–48.

Harington, C. R., and F. V. Clulow. 1973. Pleistocene mammals from Gold Run Creek, Yukon Territory. *Canadian Journal of Earth Sciences* 10:697–759.

Harington, C. R., and D. M. Shackleton. 1978. A tooth of *Mammuthus primigenius* from Chestermere Lake near Calgary, Alberta, and the distribution of mammoths in southwestern Canada. *Canadian Journal of Earth Sciences* 15:1272–83.

Harris, A. H., and P. Mundel. 1974. Size reduction in bighorn sheep (*Ovis canadensis*) at the close of the Pleistocene. *Journal of Mammalogy* 55:678–80.

Hartz, R. W., K. Holden, D. M. Hopkins, and G. Shearer. 1979. Location map and summary logs for the Geological Survey's 1979 Beaufort Sea over-the-ice drilling program. *United States Geological Survey Open File Report* 79-1303.

Hassan, A. A., and D. J. Ortner. 1977. Inclusions in bone material as a source of error in radiocarbon dating. *Archaeometry* 19:131–35.

Haugen, R. K., M. J. Lynch, and T. C. Roberts. 1971. *Summer temperatures in interior Alaska. Cold Regions Research and Engineering Laboratory Research Report* 244.

Haury, E. W., E. Antevs, and J. F. Lance. 1953. Artifacts with mammoth remains, Naco, Arizona. *American Antiquity* 19:1–14.

Haury, E. W., E. B. Sayles, and W. W. Wasley. 1959. The Lehner Mammoth site, southeastern Arizona. *American Antiquity* 15:2–30.

Hawley, A. W. L. 1978. *Comparison of forage utilization and blood composition of bison and Hereford cattle.* University of Saskatchewan, Ph.D. thesis.

Haynes, C. V. 1964. Fluted projectile points: Their age and dispersion. *Science* 145:1408–13.

Haynes, C. V. 1966. Elephant-hunting in North America. *Scientific American* 214:104–12.

Haynes, C. V. 1967. "Carbon-14 dates and early man in the New World," in *Pleistocene extinctions: The search for a cause.* Edited by P. S. Martin and H. E. Wright, Jr., pp. 267–86. New Haven: Yale University Press.

Haynes, C. V. 1968. "Geochronology of Late Quaternary alluvium," in *Means of correlation of Quaternary successions.* Edited by R. B. Morrison, pp. 591–631. Salt Lake: University of Utah Press.

Haynes, C. V. 1970. "Geochronology of man-mammoth sites and their bearing upon the origin of the Llano Complex," in *Pleistocene and recent environments of the Central Plains.* Edited by W. E. Dort and A. E. Johnson, pp. 77–92. Lawrence: University of Kansas Press.

Haynes, C. V. 1971. Time, environment, and early man. *Arctic Anthropology* 8:3–14.

Haynes, C. V. 1972. Stratigraphic investigations at the Williamson site, Virginia. *The Chesopiean* 10:107–13.

Haynes, C. V. 1973. Exploration of a mammoth-kill site in Arizona. *National Geographic Society Research Reports,* 1966 Projects: 125–26.

Haynes, C. V. 1976. "Mammoth hunters of the USA and USSR," in *Beringiya v Kainozoe.* Edited by V. L. Kontrimavichus, pp. 427–38. Khabarovsk: Akad. Nauk SSSR, Dal'nevostochnyy Nauchnyy Tsentr (in Russian, English abstract).

Haynes, C. V. 1980. The Clovis culture. *Canadian Journal of Anthropology* 1:115–21.

Haynes, C. V., and E. T. Hemmings. 1968. Mammoth-bone shaft wrench from Murray Springs, Arizona. *Science* 159:186–87.

Haynes, G. A. 1978. Morphological damage and alteration to bone: Laboratory experiments, field studies, and zoo studies. *AMQUA Abstracts* 5:210. Edmonton.

Haynes, G. A. 1980. Evidence of carnivore gnawing on Pleistocene and Recent mammalian bones. *Paleobiology* 6:341–51.

Heinlein, E. F. W., R. D. Holdren, and R. M. Yoon. 1966. Comparative responses of horse and sheep to different physical forms of alfalfa hay. *Journal of Animal Science* 25:740–43.

Hemming, J. E. 1971. The distribution and movement of caribou in Alaska. *Alaska Department of Fish and Game, Game Technical Bulletin* I.

Hemmings, E. T. 1970. *Early man in San Pedro Valley.* University of Arizona, Ph.D. dissertation.

Hemmings, E. T., and C. V. Haynes. 1969. The Escapule

mammoth and associated projectile points, San Pedro Valley, Arizona. *Journal of the Arizona Academy of Science* 5:184–88.

Herero, S. 1978. A comparison of some features of the evolution, ecology and behavior of black and grizzly/brown bears. *Carnivore* 1:7–17.

Herman, G. F., and W. T. Johnson. 1978. The sensitivity of the general circulation to arctic sea ice boundaries: A numerical experiment. *Monthly Weather Review* 106:1649–64.

Hernandez, H. 1974. "Possible effects on vegetation of the proposed gas pipeline from Prudhoe Bay, Alaska and the Mackenzie Delta to Alberta," in *Environmental impact assessment of the portion of the Mackenzie Gas Pipeline from Alaska to Alberta* 4:37–68. Winnipeg: Environment Protection Board.

Hertz, O. F. 1902. Official account of the expedition of the Imperial Academy of Science to Berezovka to excavate a mammoth carcass. *Imp. Akad. Nauk, Izvestia* 16(4) (in Russian).

Hester, J. J. 1966. Origins of the Clovis Culture. *26th Congreso Internacional de Americanistas* 1:129–42. Seville.

Hester, J. J. 1967. "The agency of man in animal extinctions," in *Pleistocene extinctions: the search for a cause.* Edited by P. S. Martin and H. E. Wright, Jr., pp. 169–92. New Haven: Yale University Press.

Hester, J. J. 1972. Blackwater Draw: A stratified Early Man site in eastern New Mexico. *Fort Burgwin Research Center Publication* 8.

Hettinger, L., A. Janz, and R. W. Wein. 1973. *Vegetation of the northern Yukon Territory.* Biological Report Series v. 1. Calgary: Canadian Arctic Gas Study Ltd.

Heusser, C. J. 1963. Postglacial palynology and archaeology in the Naknek River drainage area, Alaska. *American Antiquity* 29:74–81.

Heusser, C. J. 1973. Postglacial vegetation on Umnak Island, Aleutian Islands, Alaska. *Review of Palaeobotany and Palynology* 15:277–85.

Heusser, C. J. 1978. Postglacial vegetation on Adak Island, Aleutian Islands, Alaska. *Bulletin of the Torrey Botanical Club* 105:18–23.

Hibben, F. C. 1943. Evidences of early man in Alaska. *American Antiquity* 8:254–59.

Hibben, F. C. 1946. *The lost Americans.* New York: Crowell.

Hintz, H. F. 1969. Review article: Equine nutrition: Comparisons of digestion coefficients obtained with cattle, sheep, rabbits and horses. *The Veterinarian* 6:45 (cited by Robinson and Slade, 1974).

Hironaka, R. 1976. Feedlot finishing of cattle. *Agriculture Canada Publication* 1591.

Hirst, S. M. 1975. Ungulate-habitat relationships in a South African woodland/savanna ecosystem. *Wildlife Monograph* 44.

Hoare, J. M., and W. H. Condon. 1971a. Geologic map of the Marshall quadrangle, western Alaska. *United States Geological Survey Miscellaneous Geologic Investigations Map* I-668.

Hoare, J. M., and W. H. Condon. 1971b. Geologic map of the St. Michael quadrangle, Alaska. *United States Geological Survey Miscellaneous Geologic Investigations Map* I-682.

Hoefs, M. E. G. 1974. "Food selection by Dall's sheep, *Ovis dalli dalli*," in *The behavior of ungulates and its relation to management.* Edited by V. Geist and F. Walther, pp. 759–86. International Union for Conservation of Nature and Natural Resources Publ. n.s. 24.

Hoefs, M., I. M. Cowan, and V. J. Krajina. 1975. Phytosociological analysis and synthesis of Sheep Mountain, southwest Yukon Territory, Canada. *Syesis* 8 (Suppl. 1).

Holmen, K. 1957. The vascular plants of Peary Land, North Greenland. *Meddelelser om Grønland* 124, 9: 1–149.

Holmes. C. E. 1973. *The archaeology of Bonanza Creek valley, north-central Alaska.* University of Alaska, M.A. thesis.

Holmes, R. M., and A. N. Dingle. 1965. The relationship between macro- and micro-climate. *Agricultural Meteorology* 2:127–33.

Hopkins, D. M. 1959a. Cenozoic history of the Bering Land Bridge (Alaska). *Science* 129:1519–28.

Hopkins, D. M. 1959b. Some characteristics of the climate in forest and tundra regions in Alaska. *Arctic* 12: 215–20.

Hopkins, D. M., editor. 1967a. *The Bering Land Bridge.* Stanford: Stanford University Press.

Hopkins, D. M. 1967b. "The Cenozoic history of Beringia—a synthesis," in *The Bering Land Bridge.* Edited by D. M. Hopkins, pp. 451–84. Stanford: Stanford University Press.

Hopkins, D. M. 1967c. "Quaternary marine transgressions in Alaska," in *The Bering Land Bridge.* Edited by D. M. Hopkins, pp. 47–90. Stanford: Stanford University Press.

Hopkins, D. M. 1970. Paleoclimatic speculations suggested by new data on the location of the spruce refugium in Alaska during the last glaciation. *AMQUA Abstracts* 1:67. Bozeman.

Hopkins, D. M. 1972. The paleogeography and climatic history of Beringia during late Cenozoic time. *Internord* 12:121–50.

Hopkins, D. M. 1973. Sea level history in Beringia during the last 250,000 years. *Quaternary Research* 3:520–40.

Hopkins, D. M. 1978. A model for ancient population movements in and through Beringia. *AMQUA Abstracts* 5:5–7.

Hopkins, D. M. 1979a. "Landscape and climate of Beringia during Late Pleistocene and Holocene time," in *The First Americans: origins, affinities, and adaptations.* Edited by W. S. Laughlin and A. B. Harper, pp. 15–41. New York: Gustav Fischer.

Hopkins, D. M. 1979b. The Flaxman Formation of northern Alaska: Record of early Wisconsinan shelf glaciation in the high Arctic? *14th Pacific Science Congress, Abstracts, additional volume:* 15–16. Khabarovsk.

Hopkins, D. M., R. E. Giterman, and J. V. Matthews, Jr. 1976. Interstadial mammoth remains and associated pollen and insect fossils, Kotzebue Sound area, northwestern Alaska. *Geology* 4:169–72.

Hopkins, D. M., and R. W. Hartz. 1978a. Offshore permafrost studies, Beaufort Sea. NOAA, *Environmental Assessment of the Alaskan Continental Shelf,* Annual Reports, Principal Investigators, XI (Hazards):75–147.

Hopkins, D. M., and R. W. Hartz. 1978b. Coastal morphology, coastal erosion, and barrier islands of the Beaufort Sea, Alaska. *United States Geological Survey Open File Report* 78-1063.

Hopkins, D. M., F. S. MacNeil, and E. B. Leopold. 1960. The Coastal Plain at Nome, Alaska: A late Cenozoic type section for the Bering Strait region. *Report of the 11th International Geological Congress* 4:46–57. Copenhagen.

Hopkins, D. M., S. W. Robinson, and J. Buckley. 1981. "Radiocarbon dates from the Beaufort and Chukchi Sea coasts (1979–80)," in NOAA, *Environmental Assessment of the Alaskan continental shelf,* Annual Reports, Principal Investigators, IV (Hazards):203–20.

Hopkins, D. M., R. W. Rowland, R. E. Echols, and P. C. Valentine. 1974. An Anvilian (Early Pleistocene) marine fauna from western Seward Peninsula, Alaska. *Quaternary Research* 4:441–70.

Hopkins, D. M., P. V. Sellman, E. Chamberlain, R. E. Lewellen, and S. W. Robinson. 1979. "Buried valleys as a possible determinant of the distribution of deeply buried permafrost on the continental shelf of the Beaufort Sea," in NOAA, *Environmental Assessment of the Alaskan continental shelf,* Annual Reports, Principal Investigators, IX (Hazards):135–41.

Hopkins, D. M., P. A. Smith, and J. V. Matthews, Jr. 1981. Dated wood from Alaska and the Yukon: Implications for forest refugia in Beringia. *Quaternary Research* 15:217–49.

Howell, F. C. 1966. "Observations on the earlier phases of the European Lower Paleolithic," in *Recent studies in Paleoanthropology.* Edited by J. D. Clark and F. C. Howell. American Anthropologist Special Publication 68(2):88–201.

Howell, F. C. 1968. *Early Man.* New York: Time-Life Books.

Hoyer, H. 1914. Über die Haut und Behaarung des rhinoceros und mammuts von Starunia in Galizien. *Zeitschrift für Morphologie und Anthropologie* 18:207–26 (in German).

Hubert, B. A. 1977. "Estimated productivity of muskox on Truelove Lowland," in *Truelove Lowland, Devon Island, Canada: A high arctic ecosystem.* Edited by L. C. Bliss. Edmonton: University of Alberta Press.

Huckell, B. B. 1979. Of chipped stone tools, elephants, and Clovis hunters: An experiment. *Plains Anthropologist* 24:177–89.

Huckell, B. B. n.d. *Murray Springs debitage and Clovis technology.* Unpublished manuscript.

Hudson, R. J. 1976. Resource division within a community of large herbivores. *Le Naturaliste Canadian* 103: 153–67.

Hughes, O. L. 1969a. Pleistocene stratigraphy, Porcupine and Old Crow Rivers, Yukon Territory. *Geological Survey of Canada Paper* 69-1:209–12.

Hughes, O. L. 1972. Surficial geology of northern Yukon Territory and northwestern District of Mackenzie, Northwest Territories. *Geological Survey of Canada Paper* 69–36.

Hughes, O. L., C. R. Harington, J. A. Janssens, J. V. Matthews, Jr., R. E. Morlan, N. W. Rutter, and C. E. Schweger. 1981. Upper Pleistocene stratigraphy, paleoecology, and archaeology of the northern Yukon Interior, eastern Beringia. *Arctic* 34:329–65.

Hughes, O. L., V. N. Rampton, and N. W. Rutter. 1972. Quaternary geology and geomorphology, southern and central Yukon (northern Canada). *24th International Geological Congress, Guidebook, Excursion A-11.* Montreal.

Hultén, E. 1937. *Outline of the history of Arctic and Boreal biota during the Quaternary Period.* Stockholm: Bokforlagsaktiebolaget Thule.

Hultén, E. 1962. Flora and vegetation of Scammon Bay, Bering Sea Coast, Alaska. *Svensk Botanisk Tidskrift* 56:36–54.

Hultén, E. 1966. Contributions to the knowledge of flora and vegetation of the southwestern Alaskan mainland. *Svensk Botanisk Tidskrift* 60:175–89.

Hultén, E. 1968. *Flora of Alaska and neighboring territories.* Stanford University Press.

Humphrey, R. L. 1966. The prehistory of the Utukok River region, Arctic Alaska: Early fluted point tradition with Old World relationships. *Current Anthropology* 7:586–88.

Humphrey, R. L. 1970. *The prehistory* of *the Arctic Slope of Alaska: Pleistocene cultural relationships between Eurasia and North America.* University of New Mexico, Ph.D. thesis.

Hunt, B. G. 1977. A simulation of the possible consequences of a volcanic eruption on the general circulation of the atmosphere. *Monthly Weather Review* 105:247–60.

Hustich, I. 1953. The boreal limit of conifers. *Arctic* 6: 149–62.

Hustich, I. 1966. On the forest-tundra and the northern tree-lines. *Report of the Kevo Subarctic Station* 3: 7–47.

Huston, M. 1979. A general hypothesis of species diversity. *American Naturalist* 113:81–101.

Hyvarinen, H. 1976. Flandrian pollen deposition rates and tree-line history in northern Fennoscandia. *Boreas* 5:163–75.

Imbrie, J., and N. G. Kipp. 1971. "A new micropalaeontologic method for quantitative paleoclimatology: Application to a late Pleistocene Caribbean core," in *Late Cenozoic Glacial Ages.* Edited by K. K. Turekian, pp. 71–191. New Haven: Yale University Press.

Irving, W. N. 1971. Recent early man research in the North. *Arctic Anthropology* 8:68–82.

Irving, W. N. 1978. "Pleistocene archaeology in eastern Beringia," in *Early Man in America from a circum-Pacific perspective.* Edited by A. L. Bryan, pp. 96–101. University of Alberta Department of Archaeology Occasional Papers 1.

Irving, W. N., and Jacques Cinq-Mars. 1974. A tentative archaeological sequence for Old Crow Flats, Yukon Territory. *Arctic Anthropology* 11(supplement):65-81.

Irving, W. N., and C. R. Harington. 1973. Upper Pleistocene radiocarbon-dated artefacts from the northern Yukon. *Science* 179:335–40.

Irving, W. N., J. T. Mayhall, F. J. Melbye, and B. F.

Beebe. 1977. A human mandible in probable association with a Pleistocene faunal assemblage in eastern Beringia: A preliminary report. *Canadian Journal of Archaeology* 1:81–93.

Irwin-Williams, Cynthia. 1967. "Associations of early man with horse, camel, and mastodon at Hueyatlaco, Valsequillo (Puebla, Mexico)," in *Pleistocene extinctions—the search for a cause.* Edited by P. S. Martin and H. E. Wright, Jr., pp. 337–50. Yale University Press.

Irwin-Williams, Cynthia. 1978. "Summary of archaeological evidence from the Valsequillo region, Puebla, Mexico," in *Cultural continuity in Mesoamerica.* Edited by D. L. Browman, pp. 7–21. Mouton.

Ivanova, I. K. 1972. *Das geologische Alter des Fossilen Menschen.* Stuttgart: Kohlhammer.

Iversen, J. 1973. The development of Denmark's nature since the last Glacial. *Geological Survey of Denmark,* V, Series 7-C.

Jackson, L. E., Jr. 1980. Quaternary stratigraphy and history of the Alberta portion of the Kananaskis Lakes map area (82-J) and its implications for the existence of an ice-free corridor during Wisconsinan time. *Canadian Journal of Anthropology* 1:9–10.

Jacobs, J. D., and G. Sabo, III. 1978. Environments and adaptations of the Thule Culture on the Davis Strait coast off Baffin Island. *Arctic and Alpine Research* 10:595–615.

Jacobson, G. L., Jr., and R. H. W. Bradshaw. 1981. The selection of sites for paleovegetational studies. *Quaternary Research* 16:80–96.

Jaguttis-Emden, Martin. 1977. Zur Präzision archäologischer Datierungen. *Archaeologica Venatoria* 4.

Janis, C. 1976. The evolutionary strategy of the Equidae and the origins of rumen and cecal digestion. *Evolution* 30:757–74.

Janssen, C. R. 1966. Recent pollen spectra from the deciduous and coniferous-deciduous forests of northeastern Minnesota: A study in pollen dispersal. *Ecology* 47:804–25.

Janssens, J. A. P. 1981. *Subfossil bryophytes in Beringia, their paleoenvironmental and phytogeographic significance.* University of Alberta, Ph.D. thesis.

Janzen, D. H. 1975. *Ecology of plants in the tropics.* London: Arnold.

Jarman, P. J. 1974. The social organization of antelope in relation to their ecology. *Behavior* 48:215–67.

Jelinek, A. J. 1965. The Upper Paleolithic revolution and the peopling of the New World. *Michigan Archaeologist* 11:85–88.

Jelinek, A. J. 1967. "Man's role in the extinction of Pleistocene faunas," in *Pleistocene extinctions: The search for a cause.* Edited by P. S. Martin and H. E. Wright, Jr., pp. 193–200. New Haven: Yale University Press.

Jelinek, A. J. 1975. *The evolution of man.* London: Hamlyn Publication Group.

Jelinek, A. J. 1980. Human actions and reactions—the archaeological record. *AMQUA Abstracts* 6:108–09. Orono.

Jennings, J. D. 1974. *Prehistory of North America.* Second edition. New York: McGraw Hill.

Jingfors, K. T. 1980. *Habitat relationships and activity patterns of a reintroduced muskoxen population.* University of Alaska, M.S. thesis.

Johnson, A. W., L. A. Viereck, R. E. Johnson, and H. Melchior. 1966. "Vegetation and flora," in *Environment of the Cape Thompson region, Alaska.* Edited by N. J. Wilimovsky and J. N. Wolfe, pp. 277–354. Washington: U.S. Atomic Energy Commission, Division of Technical Information.

Johnson, D. L. 1970. The Bering Land Bridge: Not a prerequisite for Late Pleistocene peopling of the New World. *Abstracts of the 35th Annual Meeting of the Society for American Archaeology:*44.

Johnson, D. L. 1977. The California Ice-Age refugium and the Rancholabrean extinction problem. *Quaternary Research* 8:149–53.

Johnson, Eileen. 1977. Paleo-Indian bone expediency tools—Lubbock Lake and Bonfire Shelter. *Abstracts of the 42nd Annual Meeting of the Society for American Archaeology:*49–50.

Johnson, P. R., and C. W. Hartman. 1971. *Environmental atlas of Alaska.* Second edition. University of Alaska Institute of Arctic Environmental Engineering.

Jopling, A. V., W. N. Irving, and B. F. Beebe. 1981. Stratigraphic, sedimentological and faunal evidence for the occurrence of pre-Sangamonian artefacts in northern Yukon. *Arctic* 34:3–33.

Kalesnik, S. V. 1947. *Osnovy obshcego zemlevideniya (Fundamentals of general physical geography).* Moskva Gos. Universitet, Geogr. Facultet (in Russian).

Kallio, P. 1975. "Kevo, Finland," in *Structure and function of tundra ecosystems.* Edited by T. Rosswall and O. W. Heal, Swedish Natural Science Research Council, *Ecological Bulletin* 20:193–223.

Kaplina, T. N. 1979. Spore-pollen spectra of sediments of the "ice complex" in the maritime lowlands in Yakutia. Akad. Nauk SSSR, *Izvestia, Seriya Geograficheskaia* 2:85–93 (in Russian).

Kaplina, T. N., R. E. Giterman, O. V. Lakhtina, V. A. Avrashov, S. V. Kiselyov, and A. V. Sher. 1978. Duvannyy Yar, a key section of Upper Pleistocene sediments of the Kolyma Lowland. Akad. Nauk SSSR, *Byulleten'Kommissii po Izuchenyu Chetvertichnogo Perioda* 48:49–65 (in Russian).

Karavaev, M. N., and S. Z. Skryabin. 1971. *Rastitel'nyy mir Yakutii (The plant world of Yakutia).* Yakutsk (in Russian).

Karrow, P. F., T. W. Anderson, L. D. Delorme, and M. R. Skreenivasa. 1975. Stratigraphy, paleontology, and age of Lake Algonquin sediments in southwestern Ontario, Canada. *Quaternary Research* 5:49–87.

Kartashova, G. G. 1974. *Osnovnye Etapy Razvitiya Flory I Rastel'nostt Nizoviy Yani I Omoloya vo Vtoroy Polovine Paleogena I V Neogene (The main stages of flora and vegetation evolution in the lower courses of the Yana and Omoloy Rivers in the second half of the Paleogene and Neogene—palynological evidence).* Moskva Gos. Universitet, Avtroreferat dissertatsy kandidat geogr. nauk (in Russian).

Katz, N. Ya., S. V. Katz, A. V. Sher, and R. E. Giterman. 1970. "On the lower Pleistocene flora of the eastern part of the coastal plain," in *Severnyy Ledovityy Okean i ego poberezh'e v Kaynozoe.* Edited By A. I.

Tolmachev, pp. 483–93. Leningrad: Gidrometizdat (in Russian).

Kelsall, J. P. 1968. *The migrating barren-ground caribou of Canada.* Ottawa: Queen's Printer.

Kelsall, J. P., and W. Prescott. 1971. Moose and deer behavior in snow. *Canadian Wildlife Service Report* 15.

Kemp, W. B. 1971. The flow of energy in a hunting society. *Scientific American* 225:105–15.

Kennedy, P. M., and L. P. Milligan. 1978. Transfer of urea from the blood to the rumen of sheep. *British Journal of Nutrition* 40:149–54.

Kind, N. V. 1967. "Radiocarbon chronology in Siberia," in *The Bering Land Bridge.* Edited by D. M. Hopkins, pp. 172–92. Stanford: Stanford University Press.

Kind, N. V. 1972. Late Quaternary climatic changes and glacial events in the Old and New World—radiocarbon chronology. *24th International Geological Congress, Reports* 12:55–61. Montreal.

Kind, N. V. 1973. "Late Quaternary changes in climate, sea-level variations, and glacial events in the Old and New Worlds: Geology based on isotope data," in *Bering Land Bridge and its role for floras and faunas in the late Cenozoic,* pp. 101–02. Khabarovsk: Academy of Sciences of USSR, Far-Eastern Scientific Center.

Kind, N. V. 1975. Glaciations in the Verkhoyansk Mountains and their place in the radiocarbon geochronology of the Siberian late Anthropogene. *Biuletyn Peryglacjalny* 24:41–54.

Kindle, E. D. 1953. Dezadeash Map-area, Yukon Territory. *Geological Survey of Canada Memoir* 268.

Kiselyov, S. V. 1976. *Late Cenozoic Coleoptera of the Kolyma Lowland and their paleogeographical importance.* Moskva Gos. Universitet Avtoreferat dissertatsy kandidat biologicheskie nauk.: 3–20 (in Russian).

Kiselyov, S. V. 1979. Some paleogeographical aspects of the study of late Cenozoic insects of Beringia. *14th Pacific Science Congress Abstracts, Committee BIII:* 169–71. Khabarovsk.

Klein, D. R. 1965. Ecology of deer range in Alaska. *Ecological Monographs* 35:259–84.

Klein, D. R. 1970. Tundra ranges north of the boreal forest. *Journal of Range Management* 23:8–14.

Klein, R. G. 1969a. *Man and culture in the Late Pleistocene.* San Francisco: Chandler Publishing Company.

Klein, R. G. 1969b. Mousterian cultures in European Russia. *Science* 165:257–67.

Klein, R. G. 1971. The Pleistocene prehistory of Siberia. *Quaternary Research* 1:133–61.

Klein, R. G. 1973. *Ice-age hunters of the Ukraine.* Chicago: University of Chicago Press.

Klein, R. G. 1976. Cultural adaptations to Pleistocene "steppe tundras" in the Old World. *AMQUA Abstracts* 4:90–91. Tempe.

Klein, R. G. 1978. Stone Age predation on large African bovids. *Journal of Archaeological Science* 5:195–217.

Klein, R. G. 1979. Stone Age exploitation of animals in Southern Africa. *Scientific American* 67:151–60.

Klieber, M. 1961. *The fire of life: An introduction to animal energetics.* New York: John Wiley and Sons.

Klima, B. 1968. Das Pavlovien in den Weinbergöhlen von Mauern. *Quartär* 19:263–73.

Klima, Bohuslav. 1963. *Dolni Vestonice: vyzkum taboriste lovcu mamutu v letech 1947–1952 (Dolni Vestonice: investigation of a habitation site of mammoth hunters in 1947–1952).* Monumenta Archaeologica 11. Prague (in Czech with German abstract).

Klyukin, N. K., editor. 1960. *Applied climatological reference book for northeastern USSR.* Magadan (in Russian).

Knebel, H. J., and J. S. Creager. 1973. Yukon river: Evidence for extensive migration during the Holocene transgression. *Science* 179:1230–32.

von Koenigswald, W., and H. Müller-Beck. 1975. Das Pleistozän der Weingerghöhlen bei Mauern (Bayern), Nachtrag 1975. *Quartär* 26:107–18.

von Koenigswald, W., H. Müller-Beck, and E. Pressmar. 1974. *Die Archäologie und Paläontologie in den Weinberghöhlen bei Mauern (Bayern), Grabungen 1937–1967.* Tubingen: Archaeologica Venatoria.

Kolpakov, V. V. 1973. "Paleogeographic significance of the Quaternary eolian deposits of North East Siberia," in *Nekotorye voprosy regional'noy Geologii,* pp. 38–42. Moscow: Izd. MGU (in Russian).

Kolpakov, V. V. 1978. Early Quaternary deflation desert in Yakutia. *Akad. Nauk SSSR, Doklady,* Geologicheskie Sereya 241:174–76.

Komarkova, V., and P. J. Webber. 1980. Two low arctic vegetation maps near Atkasook, Alaska. *Arctic and Alpine Research* 12:447–72.

Komarkova, V., and P. J. Webber. 1981. Two low arctic vegetation maps along the Meade River at Atkasook, Alaska. *Arctic and Alpine Research* 12:447–71.

Konischev, V. N. 1970. "Conditions of formation of the Quaternary deposits with underground ice in the territory of the Yana-Indigirka lowland," in *Dokl. Vsesoyuzn, Sovesch. po Merzlotoved.* Moscow: Izd. MGU (in Russian).

Koring, P. 1977. *Canada North Almanac.* Yellowknife: Research Institute of Northern Canada.

Korotkov, I. A. 1976. Geographic regularities in the forest distribution in the Mongolian People's Republic. *Botanicheskie Zhurnal* 61:145–53 (in Russian).

Kostenko, V. A. 1979. Forming of the fauna of rodents in land and islands in the northwestern part of the Pacific region. *14th Pacific Science Congress, Abstract Committee CD:*64–65. Khabarovsk.

Kowalski, K. 1967. "The Pleistocene extinction of mammals in Europe," in *Pleistocene extinctions, the search for a cause.* Edited by P. S. Martin and H. E. Wright, Jr., pp. 349–64. New Haven: Yale University Press.

Kozhevnikov, Yu. P. 1977. Downstream by the Main and Anadyr Rivers from the Vaegi Village to the Ust-Belaya (Chukotka National Circuit). *Botanicheskie Zhurnal* 62:597–610 (in Russian).

Kozhevnikov, Yu. P. 1979. The historical interpretation of the modern peculiarities of the Chukotsk and Arctic-Alaskan floras (of vascular plants). *14th Pacific Science Congress, Abstract Committee CD:*217–18. Khabarovsk.

Kozlowski, J. K., B. van Vliet, E. Sachse-Kozlowska, H. Kubiak, and G. Zakrzewska. 1974. Upper Paleolithic site with dwellings of mammoth bones—Cracow, Spadzista Street B. *Folia Quaternaria* 44.

Kraev, V. F. 1975. On the periglacial character of Ukrainian loesses. *Biuletyn Peryglacjalny* 24:69–70.

Kreitzberg, C. W., and D. J. Perkey. 1976. Release of

potential instability: Part I. A sequential plume model within a hydrostatic primitive equation model. *Journal of Atmospheric Science* 33:456–75.

Kubiak, H. 1965. Examples of abnormalities in the dentition of fossil elephants. *Folia Quaternaria* 19:45–61 (In Polish with English summary).

Kubiak, H. 1969. Über die bedeutung der Kadaver des Wollhaarnashorns von Starunia. *Ber. deutsch. Ges. geol. Wiss. A, Geol. Paläont.* 14:345–47.

Kubiak, H. 1971. Radiocarbon datings of remains of the woolly rhinoceros from Starunia. *Wszechswiat* 10:267–68 (in Polish).

Kubiak, H. 1980a. The skulls of *Mammuthus primigenius* Blumenbach from Debica and Bzianka near Rzeszow, south Poland. *Folia Quaternaria* 51:31–45 (in Polish).

Kubiak, H. 1980b. The hyoid bone in the mammoth *Mammuthus primigenius* Blumenbach. *Folia Quaternaria* 51:47–56 (in Polish).

Kubiak, H., and B. Dziurdzik. 1973. Histological characters of hairs in extant and fossil rhinoceroses. *Acta Biologica Craccov., Series Zoologia* 16:55–61 (in Polish).

Kukla, G. J. 1975. Milankovitch and climate: The missing link? *Nature* 253:600–603.

Kurihara, Y., G. J. Tripoli, and M. A. Bender. 1979. Design of movable nested-nest primitive equation model. *Monthly Weather Review* 107:239–49.

Kurt, F. 1974a. Asiatic elephants: Shapers of the jungle. *Image Roche* 59:2–13.

Kurt, F. 1974b. "Remarks on the social structure and ecology of the Ceylon elephant in the Yala National Park," in *The behaviour of ungulates and its relation to management.* Edited by V. Geist and F. Walther, pp. 618–34. International Union for Conservation of Nature and Natural Resources Publ. n.s. 24.

Kurtén, B. 1968. *Pleistocene mammals of Europe.* Chicago: Aldine.

Kutzbach, J. E. 1981. Monsoon climate of the early Holocene: Climate experiments with the earth's orbital parameters for 9000 years ago. *Science* 214:59–61.

Kuzmina, I. Ye. 1971. Development of the theriofauna of the northern Urals in the late Anthropogene. *Akad. Nauk SSSR, Trudy Zoologicheskogo Instituta* 49:44–122.

Kuzmina, I. Ye. 1977. "On the origin and history of the mammalian fauna of the Siberian Arctic," in *Fauni i flori Antropogena Severo-Vostoka Sibiri.* Edited by O. A. Scarlato. Akad. Nauk SSSR, Trudy Zoologicheskogo Instituta 63:18–55 (in Russian).

Lackowitz, W., 1899. *Das Buch der Tierwelt.* Leipzig: Herlet.

Lahren, L., and R. Bonnichsen. 1974. Bone foreshafts from a Clovis burial in southwestern Montana. *Science* 186:147–50.

Lamb, H. H. 1977. *Climate: Present, past and future.* London: Methuen, 835 pp.

Lamb, H. H., and A. Woodroffe, 1970. Atmospheric circulation during the last ice age. *Quaternary Research* 1:29–58.

Lamb, H. H., R. P. W. Lewis, and A. Woodroffe. 1966. "Atmospheric circulation and the main climatic variables between 8000 and 0 B.C.: Meteorological evidence," in *World Climate 8000–0 B.C.* Edited by J. S. Sawyer, pp. 174–217. London: Royal Meteorological Society.

Landau, L. D., and E. M. Lifshitz. 1959. *Fluid mechanics.* London: Pergamon Press; Addison-Wesley.

Lang, H. 1925. Problems and facts about frozen Siberian mammoths (*Elephas primigenius*) and their ivory. *Zoologica* 4:25–53.

Lanning, Edward. 1970. Pleistocene man in South America. *World Archaeology* 2:90–111.

Larsen, Helge. 1968. Trail Creek: Final report on the excavation of two caves on Seward Peninsula, Alaska. *Acta Arctica* 15.

Lasker, G. W. 1976. *Physical anthropology.* 2nd edition. New York: Holt, Rinehart and Winston, Inc.

Laughlin, W. S. 1967. "Human migration and permanent occupation in the Bering Sea area," in *The Bering Land Bridge.* Edited by D. M. Hopkins, pp. 409–50. Stanford: Stanford University Press.

Laughlin, W. S. 1975. Aleuts: Ecosystem, Holocene history, and Siberian origin. *Science* 189:507–15.

Lavrenko, E. M. 1940. *The vegetation of the USSR.* 1. Moscow-Leningrad (in Russian).

Lavrushin, Yu. A. 1963. Alluvial plains of the rivers of the subarctic belt and periglacial regions of continental glaciations. *Akad. Nauk, Geologicheskii Institut Trudy* 87 (in Russian).

Laws, R. M. 1979. The Tsavo Research Project. *Journal of Reproductive Fertility*, suppl. 6:495–531.

Laws, R. M., and I. S. C. Parker. 1968. Recent studies on elephant popllations in East Africa. *Symposium of the Zoological Society of London* 21:319–59.

Laws, R. M., I. S. C. Parker, and R. C. B. Johnstone. 1970. Elephants and habitats in North Bunyoro, Uganda, *East African Wildlife Journal* 8:163–80.

Laws, R. M., I. S. C. Parker, and R. C. B. Johnstone. 1975. *Elephants and their habitats: The ecology of elephants in north Bunyoro, Uganda.* Oxford: Clarendon Press.

Lazarev, P. A. 1977a. "History of discovery of the Selerikan horse carcass and its study." in *Fauni i Flori Antropogena Severo-Vostoka Sibiri.* Edited by O. A. Scarlato. Akad. Nauk SSSR, Trudy Zoologicheskogo Instituta 63:56–69 (in Russian).

Lazarev, P. A. 1977b. "New discovery of the skeleton of a woolly rhinoceros in Yakutia," in *Fauni i Flori Antropogena Severo-Vostoka Sibiri.* Edited by O. A. Scarlato. Akad. Nauk SSSR, Trudy Zoologicheskogo Instituta 63:281–85 (in Russian).

Lazukov, G. I. 1957. The ecological conditions in the time of the Upper Paleolithic in the area of Kostienki-Borshevo. *Sovetskaya Arkheologiya* 1957:84–104 (in Russian).

Leith, C. E. 1975. "The design of a statistical-dynamical climate model and statistical constraints on the predictability of climate," in *The physical basis of climate and climate modelling.* World Meteorological Organization, Global Atmospheric Research Project Series 16:137–41, Geneva.

Leonhardy, F. C., editor. 1966. *Domebo: A paleoindian mammoth kill in the Prairie-Plains.* Lawton: Museum of the Great Plains.

Leopold, E. B. 1967. "Late Cenozoic patterns of plant extinction," in *Pleistocene extinctions: The search for a cause.* Edited by P. S. Martin and H. E. Wright, Jr., pp. 203–46. New Haven: Yale University Press.

Lettau, H. H., and K. Lettau. 1975. "Regional climatonomy of tundra and boreal forests in Canada," in *Climate of the Arctic.* Edited by G. Weller and S. A. Bowling, pp. 209–21. Fairbanks: University of Alaska.

Leuthold, W. 1972. Home range, movements and food of a buffalo herd in Tsavo National Park. *East African Wildlife Journal* 10:237–43.

Leuthold, W. 1977. *African ungulates: A comparative review of their ethology and behavioral ecology.* Berlin: Springer-Verlag.

Leuthold, W., and B. M. Leuthold. 1976. Density and biomass of ungulates in Tsavo East National Park, Kenya. *East African Wildlife Journal* 14:49–58.

Levenson, H. 1979. Sciurid growth rates: Some corrections and additions. *Journal of Mammalogy* 60:232–35.

Lichti-Federovich, S. 1973. Palynology of six sections of late Quaternary sediments from the Old Crow River, Yukon Territory. *Canadian Journal of Biology* 51: 553–64.

Lichti-Federovich, S. 1974. Palynology of two sections of late Quaternary sediments from the Porcupine River, Yukon Territory. *Geological Survey of Canada Paper* 74–23.

Lichti-Federovich, S., and J. C. Ritchie. 1968. Recent pollen assemblages from the western interior of Canada. *Review of Palaeobotany and Palynology* 7:297–344.

Lindroth, C. H. 1961–1969. The ground-beetles of Canada and Alaska, 6 pts. *Opuscula Entomologica.* Suppl. 20, 24, 29, 33, 35.

Livingstone, David. 1874. *The last journals of David Livingstone in central Africa from 1865 to his death.* Edited by H. Waller. 2 volumes. London: John Murray.

Livingstone, David, and Charles Livingstone. 1865. *Narrative of an expedition to the Zambezi and its tributaries.* London: John Murray.

Livingstone, D. A. 1955. Some pollen profiles from arctic Alaska. *Ecology* 36:587–600.

Livingstone, D. A. 1957. Pollen analysis of a valley fill near Umiat, Alaska. *American Journal of Science* 255:254–60.

Lodge, R. W., J. B. Campbell, S. Smoliak, and A. Johnston. 1971. Management of the western range. *Canada Department of Agriculture Publication* 1425.

Looman, J. 1964. The distribution of some lichen communities in the Prairie Province and adjacent parts of the Great Plains. *Bryologist* 67:209–23.

Lorenz, E. N. 1978. Climate Determinism. *Meteorological Monograph* 8:1–3.

Lorenz, E. N. 1969. The predictability of a flow which possesses many scales of motion. *Tellus* 21:289–307.

Lorenz, E. N. 1975. "Climatic predictability," in *The physical basis of climate and climate modelling.* World Meteorological Organization, Global Atmospheric Research Project Series 16:132–36. Geneva.

Lorenzo, J. L. 1978. "Early man research in the American Hemisphere: Appraisal and perspective," in *Early Man in America from a circum-Pacific perspective.* Edited by A. L. Bryan, pp. 1–9, University of Alberta Department of Archaeology Occasional Papers 1.

Löve, A., and D. Löve. 1974. "Origin and evolution of the arctic and alpine floras," in *Arctic and Alpine Environments.* Edited by J. D. Ives and R. G. Barry, pp. 571–603. London: Methuen.

Lozhkin, A. V. 1975. Absolute geochronology and occurrence of Pleistocene deposits in northeastern U.S.S.R. *Akad. Nauk SSSR, Severo-Vostochikh Kompleks Institut Trudy* 68:126–29.

Lozhkin, A. V. 1977. "The habitat of the Berekh population of mammoths," in *Mamontovaya fauna Russkoy ravninny i Vostochnoy Sibiri.* Edited by O. A. Scarlato. Akad. Nauk SSSR, Trudy Zoologicheskogo Instituta 72:119–20 (in Russian).

Lozhkin, A. V. 1979. Radiocarbon dating of upper Pleistocene sediments of the New Siberian Islands and age of the "Yedoma" Suite of northeastern U.S.S.R. *Akad. Nauk SSSR Doklady* 235:57–59 (in Russian).

de Lumley, H. 1966. Les Fouilles de Terra Amata a Nice. *Bulletin du Musée d'Anthropologie Préhistorique de Monaco* 13:29–51.

de Lumley, H. 1976. "Le Civilizations du Paléolithique inférieur en Provence," in *La Préhistoire Française.* Edited by H. de Lumley, pp. 819–29. Paris: Editions CNRS.

Lundelius, E. L. S. 1974. "The last 15,000 years of faunal change in North America," in *History and prehistory of the Lubbock Lake site.* Edited by C. C. Black. Museum Journal of Texas Technical University 15: 141–60.

Lydekker, R. 1924. *Game animals of India, Burma, Malaya, and Tibet.* London: Rowland Ward.

Lynch. T. F. 1974. The antiquity of man in South America. *Quaternary Research* 4:356–77.

MacArthur, R. H., and E. O. Wolson. 1967. *The theory of island biogeography.* Princeton: Princeton University Press.

MacDonald, G. F. 1968. Debert: A paleoindian site in central Nova Scotia. National Museum of Canada, *Anthropology Papers* 16:1–158.

Mackay, J. R. 1963. *The Mackenzie Delta area, N.W.T.* Canada Department of Mines and Technical Surveys, Geographic Branch, Memoir 8.

Mackay, J. R. 1970. Disturbances to the tundra and forest tundra environment of the western arctic. *Canadian Geotechnical Journal* 7:420–32.

Mackay, J. R. 1978. Freshwater shelled invertebrate indicators of paleoclimate in northwestern Canada during late glacial times: discussion. *Canadian Journal of Earth Sciences* 15:461–62.

MacNeish, R. S. 1976. Early man in the New World. *American Scientist* 64:316–27.

MacNeish, R. S. 1978. Late Pleistocene adaptations: A new look at early peopling of the New World as of 1976. *Journal of Anthropological Research* 34:475–96.

Maglio, V. J. 1973. Origin and evolution of the Elephantidae. *Transactions of the American Philosophical Society* 63(3):1–149.

Maher, L. J., Jr. 1963. Pollen analysis of surface materials from the southern San Juan Mountains, Colorado.

Geological Society of America Bulletin 74:1485–1504.
Maher, L. J., Jr. 1972. Absolute pollen diagram of Redrock Lake, Boulder County, Colorado. *Quaternary Research* 2:531–53.
Malpas, R. C. 1977. Diet and the condition and growth of elephants in Uganda. *Journal of Applied Ecology* 14:489–504.
Manabe, S., and D. G. Hahn. 1977. Simulation of the tropical climate of an Ice Age. *Journal of Geophysical Research* 82:3889–3911.
Mania, D., and Toepfer, V. 1973. *Königsaue: Gliederung, Ökologie und mittelpaläolithische Funde der letzten Eiszeit.* Berlin: Deutscher Verlag der Wissenschaften.
Markov, K. K. 1956. Physical-geographic conditions in the periglacial zone: Paleobotanical studies. *Biuletyn Peri-glacjalny* 3:145–50 (in Russian).
Marks, S. A. 1976. *Large mammals and a brave people: Subsistence hunters in Zambia.* Seattle: University of Washington Press.
Marks, S. A. 1977a. Hunting behavior and strategies of the Valley Bisa of Zambia. *Human Ecology* 5:1–36.
Marks, S. A. 1977b. Buffalo movements and accessibility to a community of hunters in Zambia. *East African Wildlife Journal* 15:151–261.
Marks, S. A. 1979a. Profile and process: Subsistence hunters in a Zambian community. *Africa* 49:53–67.
Marks, S. A. 1979b. *An integrated social-environmental analysis of the Luangwa Valley in Zambia.* Case Studies in Development. Washington: U.S. Agency for International Development.
Martin, P. S. 1974. *The discovery of America. Science* 179:969–74.
Martin, P. S. 1974. "Paleolithic players on the American stage," in *Arctic and Alpine Environments.* Edited by J. Ives and R. Barry, pp. 669–700. London: Methuen.
Martin, P. S., and H. E. Wright, Jr. Editors. 1967. *Pleistocene extinctions: Search for a cause.* New Haven: Yale University Press.
Mathews, W. H. 1980. Retreat of the last ice sheets in northeastern British Columbia and adjacent Alberta. *Geological Survey of Canada Bulletin* 331.
Matthews, J. V., Jr. 1968. A paleoenvironmental analysis of three late Pleistocene coleopterous assemblages from Fairbanks, Alaska. *Quaestiones Entomologicae* 4:202–24.
Matthews, J. V., Jr. 1970. Quaternary environmental history of interior Alaska: Pollen Samples from organic colluvium and peats. *Arctic and Alpine Research* 2:241–51.
Matthews, J. V., Jr. 1974a. Wisconsin environment of interior Alaska: Pollen and macrofossil analysis of a 27 m core from Isabella Basin (Fairbanks, Alaska). *Canadian Journal of Earth Sciences* 11:812–41.
Matthews, J. V., Jr. 1974b. Quaternary environments at Cape Deceit (Seward Peninsula, Alaska): Evolution of a tundra ecosystem. Geol*ogical Society of America Bulletin* 85:1353–85.
Matthews, J. V., Jr. 1974c. Fossil insects from the early Pleistocene Olyor Suite (Chukochya River: Kolymian Lowland, U.S.S.R.). *Geological Survey of Canada Paper* 74-1a:205–11.
Matthews, J. V., Jr. 1975. Insects and plant macrofossils from two Quaternary exposures in the Old Crow-Porcupine region, Yukon Territory, Canada. *Arctic and Alpine Research* 7:249–59.
Matthews, J. V., Jr. 1976a. Arctic-steppe—an extinct biome. *AMQUA Abstracts* 4:73–77. Tempe.
Matthews, J. V., Jr. 1976b. Insect fossils from the Beaufort Formation: Geological and biological significance. *Geological Survey of Canada Paper* 76-1B:217–27.
Matthews, J. V., Jr. 1979a. "Tertiary and Quaternary environments: Historical background for an analysis of the Canadian insect fauna," in *Canada and its insect fauna.* Edited by H. V. Danks. Memoirs of the Entomological Society of Canada no. 108.
Matthews, J. V., Jr. 1979b. Beringia during the late Pleistocene: Arctic-steppe or discontinuous herb-tundra? A review of the paleontological evidence. *Geological Survey of Canada Open File Report* 649.
Matthews, J. V., Jr. 1980. An early Wisconsinan warm interstadial in East Beringia? *AMQUA Abstracts* 6:130–31. Orono.
Matthews, J. V., Jr. In press. Quaternary vegetational history and the Arctic-steppe in Beringia. *4th International Palynological Conference, Proceedings* 3:82–91. Lucknow. (1980)
Mattson, W. J. 1980. Herbivory in relation to plant nitrogen content. *Annual Review of Ecology and Systematics* 11:119–62.
McCary, B. C. 1975. The Williamson paleoindian site, Dinwiddle County, Virginia. *The Chesopiean* 13:48–131.
McCulloch, D. S., and D. M. Hopkins. 1966. Evidence for an early Recent warm interval in northwestern Alaska. *Geological Society of America Bulletin* 77:1089–1108.
McKay, G. M. 1973. Behavior and ecology of the Asiatic elephant in southeastern Ceylon. *Smithsonian Contributions to Zoology* 125.
McKay, G. M., and J. F. Eisenberg. 1974. "Movement patterns and habitat utilization of ungulates in Ceylon," in *The behaviour of ungulates and its relation to management.* Edited by V. Geist and F. Walther, pp. 708–21. International Union for Conservation of Nature and Natural Resources Publ. n.s. 24(2).
McKendrick, J. D., G. O. Batzli, K. R. Everett, and J. C. Swanson. 1980. Some effects of mammalian herbivores and fertilization on tundra soils and vegetation. *Arctic and Alpine Research* 12:565–78.
McKennan, R. A., and J. P. Cook. 1968. Prehistory of Healy Lake, Alaska. *Proceedings of the 8th International Congress of Anthropological and Ethnological Sciences* 3:182–84.
McManus, D. A., and J. S. Creager. 1965. Bottom sediments from the continental shelf of the Chukchi and Bering Seas, 1959–1963. *University of Washington Department of Oceanography Technical Report 135* (2 v.).
McManus, D. A., J. S. Creager, R. J. Echols, and M. J. Holmes. In press. The Holocene transgression on the arctic flank of Beringia: Chukchi valley to Chukchi estuary to Chukchi Sea. *Symposium on Quaternary land-sea migration routes and human occupation of submerged coastlines,* La Jolla.
McNaughton, S. J. 1976. Serengeti migratory wildebeest:

Facilitation of energy flow by grazing. *Science* 191: 92–94.

McNaughton, S. J. 1979. Grazing as an optimization process; grass-ungulate relationships in the Serengeti. *The American Naturalist* 113:691–703.

Medway, Lord. 1972. Phenology of a tropical rainforest in Malaya. *Biological Journal of Linnean Society* 4: 117–46.

Mehl, M. G. 1966. The Grundel mastodon. *Report of Investigations of the Missouri Geological Survey and Water Resources* 35.

Mehringer, P. J., Jr., P. S. Martin, and C. V. Haynes, Jr. 1967. Murray Springs, a mid-postglacial pollen record from southern Arizona. *American Journal of Science* 265:786–97.

Mehringer, P. J., Jr., K. L. Peterson, and F. A. Hassan. 1979. A pollen record from Birket Qarun and the recent history of the Fayum, Egypt. *Quaternary Research* 11:238–56.

Merz, G. 1977. *Untersuchungen uber Ernahrungsbiologie und habitat preferenzen des Afrikanischen Waldenfonten,* Loxodonta africana cyclotis, *Matschie, 1900,* Heidelberg: Ruprecht Karl University, dissertation for zoology diploma.

Milankovitch, M. 1930. "Mathematische Klimalehre und Astronomische theorie der Klimaschwankungen I. Part A," in *Handbuch del Klimatologie,* pp. 1–176, Berlin: Verlag Borntroger (in German).

Miller, F. L., R. H. Russell, and A. Gunn. 1975. The recent decline of Peary Caribou on Western Queen Elizabeth Islands of Arctic Canada. *Polarforschung* 445: 17–21.

Miller, S. J., and Wakefield Dort Jr. 1978. "Early man at Owl Cave: Current investigations at the Wasden site, eastern Snake River Plain, Idaho," in *Early Man in America from a circum-Pacific perspective.* Edited by A. L. Bryan, pp. 129–39. University of Alberta Department of Archaeology Occasional Papers 1.

Mirabell, Lorena. 1978. "Tlapacoya: A late Pleistocene site in central Mexico," in *Early Man in America from a circum-Pacific perspective.* Edited by A. L. Bryan, pp. 221–30. University of Alberta Department of Anthropology Occasional Papers 1.

Mitchell, G. J., and S. Smoliak. 1971. Pronghorn antelope range characteristics and food habits in Alberta. *Journal of Wildlife Management* 35:238–50.

Mochanov, Yu. A. 1978. "The Paleolithic of NE Asia and the problem of the first peopling of America," in *Early Man in America from a circum-Pacific perspective.* Edited by A. L. Bryan. University of Alberta Department of Anthropology Occasional Papers 1:67.

Mochanov, Yu. A. 1976. "The Paleolithic of Siberia," in *Beringiya v Kainozoe.* Edited by V. L. Kontrimavichus, pp. 540–65. Khabarovsk: Akad. Nauk SSSR, Dal'nevostochnyy Nauchnyy Tsentr (in Russian, English abstract).

Mochanov, Yu. A. 1977. *Drevneishe etapy zaseleniia chelovekom Severo-Vostochnoi Azii (Most ancient stages of human settlement in Northeast Asia).* Novosibirsk: AN SSSR (in Russian).

Mochanov, Yu. A. 1978a. "Stratigraphy and absolute chronology of the Paleolithic of Northeast Asia," in *Early Man in America from a circum-Pacific perspective.* Edited by A. L. Bryan, pp. 54–66. University of Alberta Department of Archaeology Occasional Papers 1.

Mochanov, Yu. A. 1980. "Early migrations to America in the light of a study of the Dyuktai Paleolithic culture in Northeast Asia," in *Early Native Americans.* Edited by D. L. Brownman, pp. 119–31. The Hague: Mouton.

Moore, D. G. 1964. "Acoustic-reflection reconnaissance of continental shelves—eastern Bering and Chukchi Seas," in *Papers in Marine Geology, Shepard commemorative volume.* Edited by R. L. Miller, pp. 319–62. New York: MacMillan.

Morlan, R. E. 1970. Wedge-shaped core technology in northern North America. *Arctic Anthropology* 7:17–37.

Morlan, R. E. 1977a. Fluted point markers and the extinction of the Arctic-Steppe biome in eastern Beringia. *Canadian Journal of Archaeology* 1:95–108.

Morlan, R. E. 1977b. The Yukon Refugium Project. *National Museum of Man Canadian Studies Report* 1.

Morlan, R. E. 1978a. "Early man in northern Yukon Territory: Perspectives as of 1977," in *Early Man in America from a circum-Pacific perspective.* Edited by A. L. Bryan, pp. 78–95. University of Alberta Department of Archaeology Occasional Papers 1.

Morlan, R. E. 1978b. Technological characteristics of some wedge-shaped cores in northwestern North America and northeast Asia. *Asian Perspectives* 19:96–106.

Morlan, R. E. 1979. "A stratigraphic framework for Pleistocene artifacts from Old Crow River, northern Yukon Territory," in *Pre-Llano cultures in the Americas: Paradoxes and possibilities.* Edited by R. L. Humphrey and Dennis Stanford, pp. 125–45. Washington: Anthropological Society of Washington.

Morlan, R. E. 1980a. *Taphonomy and archaeology in the Upper Pleistocene of the northern Yukon Territory: A glimpse of the peopling of the New World.* Canadian National Museum of Man Mercury Series (Archaeological Survey of Canada Paper No. 94).

Morlan, R. E. 1980b. Taphonomy as an aid to archaeological reconnaissance and interpretation: An example from northern Yukon Territory. *AMQUA Abstracts* 6:143–44. Orono.

Morlan, R. E. 1980c. Evidence for stone artifacts in the Early Wisconsinan of northern Yukon Territory. *Abstracts of the 13th Annual Meeting of the Canadian Archaeological Association* p. 30.

Morlan, R. E. In press. Pleistocene archaeology in the Old Crow Basin: A critical reappraisal. *Quaternary Research.*

Morlan, R. E., and J. V. Matthews, Jr. 1978. New dates for early man. *GEOS,* Winter 1978:2–5.

Morlan, R. E., and W. B. Workman. 1980. "Prehistoric man in the southwest Yukon," in *Kluane, pinnacle of the Yukon.* Edited by J. B. Theberge, pp. 97–107. Toronto: Doubleday.

Morozova, T. D. 1971. Characterization of loess-like deposits of Central Yakutia. *Akad. Nauk SSSR Izvestia. Seriya Geograficheskaya* 5:95–104 (in Russian).

Mosimann, J. E., and P. S. Martin. 1975. Simulating

overkill by Paleoindians. *American Scientist* 63:304–13.

Mosquin, T. 1966. "Reproductive specialization as a factor in the evolution of the Canadian flora," in *The Evolution of Canada's flora.* Edited by R. L. Taylor and R. A. Ludwig, pp. 43–65. Toronto: University of Toronto Press.

Mozolevskaya, A. K. ed. 1963. *Agroclimatological reference book on the Yakutian ASSR.* Leningrad: Gidrometeoizdat (in Russian).

Müller-Beck, Hansjürgen. 1966. Paleohunters in America: Origins and diffusion. *Science* 152:1191–1210.

Müller-Beck, Hansjürgen. 1967. "On migrations of hunters across the Bering Land Bridge in the Upper Pleistocene," in *The Bering Land Bridge.* Edited by D. M. Hopkins, pp. 373–408. Stanford: Stanford University Press.

Müller-Beck, Hansjürgen. 1979. Das Ökosystem der Moschusochsenstation Umingmak. *Marburger Geographische Schriften* 79:97–112.

Murie, O. J. 1944. *The elk of North America.* Harrisburg: Stockpole.

Murray, D. F. 1980a. "Balsam poplar in Arctic Alaska," in *The ice-free corridor and peopling of the New World.* Edited by N. W. Rutter and C. E. Schweger. *Canadian Journal of Anthropology* 1:29–32.

Murray, D. F. 1981. "The role of Arctic refugia in the evolution of the Arctic vascular flora—a Beringian perspective," in *Evolution today: Proceedings of Second International Congress of Systematic and Evolutionary Biology.* Edited by G. G. Scudder and J. V. Reveal.

Murray, D. F., and A. R. Batten. 1977. A provisional *classification of Alaskan tundra.* Unpublished report, University of Alaska and Pacific Northwest Forest and Range Experiment Station.

Musil, R. 1969. Die Pferd der Pekárna-Höhle. Ein Beitrag zur Problematik der Evolution der Equiden. *Zeitschrift für Tierzüchtung und Züchtungsbiologie* 86:147–99.

Myers, Norman. 1980. The sinking ark. *Harrowsmith* 31:30–39.

National Academy of Sciences. 1975. *Understanding climatic change: A program for action.* Washington: National Academy of Sciences.

National Research Council. 1973. Nutrient requirements of horses. *Nutrient requirements of domestic animals* 6. Washington: National Academy of Science.

Naumov, E. M. 1973. "The basic types of genetic soil profiles and the specificity of the soil cover in the taiga zone of the extreme North-East," in *Pochvy i rasitelnost' merezlotnikh rayonov SSSR,* pp. 48–55. Magadan (in Russian).

Nehring, A. 1890. *Uber Tundren und Steppen der Jetzt- und Vorzeit, mit besonderer Berucksichtigung iberer Fauna.* Berlin: F. Dummler.

Nehring, A. 1895. Die kleinen Wirbeltiere vom Schweizersbild bei Schaffhaussen. *Neue Denkschriften der Allgemeinen Schweizerischen Naturforschende Gesellschaft* 35:39–78.

Neilson, J. E. 1968. New and important additions to the flora of the southwestern Yukon Territory, Canada. *Canadian Field-Naturalist* 81:114–19.

Nelson, C. H., D. M. Hopkins, and D. W. Scholl, 1974. "Cenozoic sedimentary and tectonic history of Bering Sea," in *Oceanography of the Bering Sea.* Edited by D. W. Hood and E. J. Kelley, pp. 485–516. Fairbanks, Alaska: Institute of Marine Science.

Nelson, R. E. 1979. *Quaternary environments of the arctic slope of Alaska.* University of Washington, M.Sc. thesis.

Nettasinghe, A. P. W. 1973. *The interrelationship of livestock and elephants at the Thamankaduwa farm, with special reference to feeding and environment.* University of Sri Lanka, M.Sc. thesis.

Neuville, H. 1917. On the extinction of the mammoth. *Bulletin of the Museum of Natural History* 23:374–87.

Niezabitowski, E. L. 1914a. "Slon mammu (*Elephas primigenius* Blum.)," in *Wykopaliska Starunskie,* pp. 95–190. Krakow (in Polish).

Niezabitowski, E. L. 1914b. "Nosorozec wlochaty (*Rhinoceros antiquitis* Blum., *tichorhuns* Fisch.)," in *Wykopaliska Starunskie,* pp. 181–339. Krakow (in Polish).

Norin, B. N., and I. V. Ignatenko. 1975. "Ary-Mas, U.S.S.R.," in *Structure and function of tundra ecosystem.* Edited by T. Rosswall and A. W. Heal. *Swedish Natural Science Research Council. Ecological Bulletin* 20:183–91.

North Slope Borough Contract Staff. 1979. *Native livelihood and dependence: A study of land use values through time.* Anchorage: U.S. Department of the Interior, National Petroleum Reserve in Alaska 105(C).

Nowak, J., E. Panow, J. Tokarski, W. Szafer, and J. Stach. 1930. The second woolly rhinoceros, *Coelodonta aniquitatis* Blum., from Starunia, Poland: Geology, mineralogy, flora, and fauna. *Bulletin Internationale de l'Academie Polonaise des Sciences et des Lettres.* Séries B, pp. 1–47, Krakow.

Ognev, S. I. 1948. *Animals of the U.S.S.R. and adjacent countries 4, Rodents.* Moscow-Leningrad: Izd. ANSSSR (in Russian).

Olivier, R. C. D. 1978. *On the ecology of the Asian elephant* (Elephas maximus *Linn), with particular reference to Malaya and Sri Lanka.* University of Cambidge, Ph.D. thesis.

Orlanski, I. A. 1976. A simple boundary condition for unbounded hyperbolic flows. *Journal of Computational Physics* 21:251–69.

Orlova, L. A. 1979. Radiocarbon age of fossil mammoths in the territory of the U.S.S.R. *Akad. Nauk SSSR, Izvestia Sibirskoe Otdelnie, Izvestia, Seriya Obshchestv. Nauk* 6(2):89–97 (in Russian).

Østbye, E. 1975. "Hardangervidda, Norway," in *Structure and function of tundra ecosystems.* Edited by T. Rosswall and O. W. Heal. *Ecological Bulletins* 20:255–62. Swedish Natural Science Research Council.

Oswald, E. T., and J. P. Senyk. 1977. *Ecoregions of Yukon Territory.* Canadian Forestry Service.

Painter, F. 1970. The Cattail Creek fluting tradition and its complex-determining lithic debris. *Southeastern Archaeological Conference Bulletin* 13:1–14.

Pakarinen, P., and D. H. Vitt. 1974. The major organic components and caloric contents of high arctic bryophytes. *Canadian Journal of Botany* 52:1151–61.

Parker, G. R. 1975. An investigation of caribou range in Southhampton Island, NWT. *Canadian Wildlife Service Report Series* 33.

Parrish, L. 1979. *A record of Holocene climatic changes from St. George Island, Pribilofs.* Ohio State University, M.S. thesis.

Patterson, A. 1974. *Subsistence harvests in five native regions.* Anchorage: Joint Federal-State Land Use Planning Commission of Alaska.

Peden, D. G., G. M. van Dyne, R. W. Rice, and R. M. Hansen. 1974. The trophic ecology of *Bison bison* L. on shortgrass plains. *Journal of Applied Ecology* 11: 489–98.

Peet, R. K. 1974. The measurement of species diversity. *Annual Review of Ecology and Systematics* 5:285–308.

Perkey, D. J., and C. W. Kreitzberg. 1976. A time-dependent lateral boundary scheme for limited-area primitive equation models. *Monthly Weather Review* 104:744–55.

Petersen, J. C. B., and R. L. Casebeer. 1971. A bibliography relating to the ecology and energetics of east African large mammals. *East Africa Wildlife Journal* 9:1–23.

Peterson, K. M., and W. D. Billings. 1980. Tundra vegetation patterns and succession in relation to microtopography near Atkasook, Alaska. *Arctic and Alpine Research* 12:473–82.

Petrides, G. A. 1956. Big game densities and range carrying capacities in East Africa. *Transactions North American Wildlife Conference* 21:525–37.

Petrides, G. A., and W. G. Swank. 1965. Estimating the productivity-energy relation of an African elephant population. *Proceedings, Ninth International Grassland Congress,* pp. 831–42. Sao Paulo.

Petrovsky, V. V. 1967. Essay on plant communities in the central part of Wrangel Island. *Botanicheskie Zhurnal* 52:332–43 (in Russian).

Petrovsky, V. V. 1978. Geographic connections of the flora of Wrangel Island with respect to the problem of the Bering Land Bridge. *Botanicheskie Zhurnal* 63: 637–48 (in Russian).

Péwé, T. L. 1955. Origin of the upland silt near Fairbanks, Alaska. *Geological Society of America Bulletin* 66: 699–724.

Péwé, T. L. 1965. Central and south-central Alaska. 7th INQUA Congress, *Guidebook, Field Conference F.* Boulder.

Péwé, T. L. 1968. Loess deposits of Alaska. *23rd International Geological Congress Proceedings* 8:297–309. Prague.

Péwé, T. L. 1975a. Quaternary geology of Alaska. *United States Geological Survey Professional Paper* 835.

Péwé, T. L. 1975b. Quaternary stratigraphic nomenclature in unglaciated central Alaska. *U.S. Geological Survey Professional Paper* 862.

Péwé, T. L., et al. 1953. Multiple glaciation in Alaska—a progress report. *United States Geological Survey Circular* 289.

Péwé, T. L., A. Journaux, and R. Stuckenrath. 1977. Radiocarbon dates and late-Quaternary stratigraphy from Mamontova Gora, unglaciated central Yakutia, Siberia, U.S.S.R. *Quaternary Research* 8:51–63.

Péwé, T. L., C. Wahrhaftig, and F. Weber. 1966. Geologic map of the Fairbanks quadrangle, Alaska. *United States Geological Survey Miscellaneous Geological Investigations Map* I-455.

Pfitzenmaier, E. W. 1939. *Siberian man and mammoth.* London: Blackie & Son.

Phillipson, J. 1975. Rainfall, primary production and carrying capacity of Tsavo National Park (East) Kenya. *East African Wildlife Journal* 13:171–201.

Pidoplichko, I. G. 1954. *On the glacial period,* volume 3. Kiev: Izd. AN Ukrain. SSR (in Russian).

Pidoplichko, I. G. 1956. *Materials on the study of the fossil fauna of the Ukraine,* no. 2. Kiev: Izd. Academy of Sciences, Ukrainian SSR (in Ukrainian).

Pidoplichko, I. G. 1976. *Mezhirichskie zhilishcha iz kostey mamonta (The Mezhirich mammoth-bone houses).* Akad. Nauk Ukraintskoy SSR Institut Zoologii. Kiev: Izd. "Naukova dumka."

Pielke, R. A. 1974. A three dimensional numerical model of the sea breezes over South Florida. *Monthly Weather Review* 102:115–39.

Pielke, R. A., and Y. Mahrer. 1978. Verification analysis of the University of Virginia three-dimensional mesoscale model prediction over South Florida for 1 July 1973. *Monthly Weather Review* 106:1568–89.

Pielou, E. C. 1975a. *Ecological diversity.* New York: John Wiley.

Pielou, E. C. 1975b. Production ecology of grassland plant communities in western North Dakota. *Ecological Monographs* 45:83–106.

Pienaar, de V. V. 1969. Observations on the developmental biology, growth and some aspects of the population ecology of African buffalo in the Kruger National Park. *Koedoe* 12:29–52.

Poplin, F. 1976. "Les grands vertébré de Gönnersdorf, Fouilles 1968," in *Der Magdalénien-Fundplats Gönnersdorf* 2. Mainz: Steiner.

Popov, A. I. 1965. "Subsurface ice in the Quaternary deposits of the Yana-Indigirka lowland as a genetic and stratigraphic marker," in *Osnovnye problemy izucheniya Chetvertichnogo Perioda,* 7th INQUA Congress, pp. 278–85. Moscow: Izd. "Nauka" (in Russian).

Porsild, A. E. 1964. *Illustrated flora of the Canadian Arctic Archipelago.* 2nd Edition. Ottawa.

Porter, Lee. 1979. *Ecology of a late Pleistocene (Wisconsin) ungulate community near Jack Wade, east-central Alaska.* University of Washington, M.S. thesis.

Porter, Lee, and D. M. Hopkins. In press. "Butchered caribou skulls, Pleistocene and Recent, from Beringia," in *Taphonomic analysis: Its role in early man archaeology in North America.* Edited by E. J. Dixon. Alaska Anthropological Society.

Porter, S. C., K. L. Pierce, and T. D. Hamilton. In press. "Mountain glaciation in western United States," in *Late Wisconsin Pleistocene environments of the United States.* Edited by S. C. Porter. Minneapolis: University of Minnesota Press.

Powers, W. R. 1978. Perspectives on early man. *AMQUA Abstracts* 5:114–22. Edmonton.

Powers, W. R., and T. D. Hamilton. 1978. "Dry Creek: A Late Pleistocene human occupation in central Alaska," in *Early Man in America from a circum-Pacific perspective.* Edited by A. L. Bryan, pp. 72–77. University of Alberta Department of Archaeology Occasional Papers 1.

Prest, V. K. 1969. Retreat of Wisconsin and Recent ice in North America. *Geological Survey of Canada Map* 1257A.

Protsch, R. R. 1978. *Catalog of fossil Hominids of North America.* New York: Gustav Fischer.

Pruitt, W. O., Jr. 1959. Snow as a factor in the winter ecology of the barren-ground caribou. *Arctic* 12:159–79.

Purdy, B. A., and H. K. Brooks. 1971. Thermal alteration of silica minerals: an archaeological approach. *Science* 1973:322–25.

Rainey, F. G. 1939. Archaeology in central Alaska. *Anthropological Papers of the American Museum of Natural History* 36(4):351–405.

Rainey, F. G. 1940. Archaeological investigations in central Alaska. *American Antiquity* 5:299–308.

Rampton, V. N. 1971. Late Quaternary vegetational and climatic history of the Snag-Klutlan area, southwestern Yukon Territory, Canada. *Geological Society of America Bulletin* 82:959–78.

Rampton, V. N. In press. Quaternary geology of the Yukon coastal plain. *Geological Survey of Canada Bulletin* 317.

Rampton, V. N., and M. Bouchard. 1975. Surficial geology of Tuktoyaktuk, District of Mackenzie. *Geological Survey of Canada Paper* 73-53.

Ravskii, Eh. I. 1961. Pleistocene periglacial phenomena and periglacial zones of eastern Siberia. *Voprosy Geologii Antropogena,* pp. 141–51. Izd. AN SSSR (in Russian).

Redmann, R. E. 1978. Plant and soil water potentials following fire in a northern mixed grassland. *Journal of Range Management* 31:443–45.

Reeburgh, W. S., and M. S. Young. 1976. University of Alaska radiocarbon dates I. *Radiocarbon* 18:1–15.

Reed, J. C., Jr. 1961. Geology of the Mount McKinley quadrangle, Alaska. *United States Geological Survey Bulletin* 1108-A.

Remizov, I. N., and P. V. Kovalev. 1965. "On the problem of the origin and stratigraphy of the loess rocks of the European part of the USSR," in *Osnovnye problemy izucheniya Chetvertichnogo Perioda,* 7th INQUA Congress, pp. 435–39. Moscow: Izd. "Nauka" (in Russian).

Repenning, C. A. 1967. "Palearctic-Nearctic mammalian dispersal in the late Cenozoic," in *The Bering Land Bridge.* Edited by D. M. Hopkins, pp. 288–311. Stanford University Press.

Repenning, C. A. 1980. Faunal exchange between Siberia and North America. *Canadian Journal of Anthropology* 1:37–44.

Reverdatto, V. V. 1940. Outline of the development of the post-Tertiary flora of central Siberia. *Sovetskaya Botanika* 2:48–64 (in Russian).

Reverdatto, V. V. 1960. "Glacial and steppe relicts in the flora of central Siberia and their bearing on the history of the flora," in *Nauchnye Chteniya Pamyati Mikhaila G. Popova.* Novosibirsk: Izd. SOAN SSSR (in Russian).

Rhoades, D. F., and R. G. Cates. 1976. "Toward a general theory of plant antiherbivory chemistry," in *Recent advances in phytochemistry.* 10: *Biochemical interaction between plants and insects.* Edited by J. W. Wallace and R. L. Mansell, pp. 168–213. New York: Plenum.

Richard, P. 1977. *Histoire Post-Wisconsinienne de la végétation due Québec meridional par l'analyse pollinique.* Service de la Recherche, Ministere des Terres et Forets, Québec. 2 volumes.

Richmond, R. J., R. J. Hudson, and R. J. Christopherson. 1977. Comparison of forage intake and digestibility by American bison, yak, and cattle. *Acta Theriologica* 22:225–30.

Riek, G. 1934. *Die Eiszeitjägerstation am Vogelherd.* Tübingen: Heine.

Riek, G. 1938. Ein Beitrag zur Kenntnis des süddeutschen Solutréen. *Germania* 22:147–50.

Riewe, R. R. 1977. "The utilization of wildlife in the Jones Sound region by the Grise Fiord Inuit," in *Truelove Lowland, Devon Island, Canada: A high arctic ecosystem.* Edited by L. C. Bliss, pp. 623–44. Edmonton: University of Alberta Press.

Ripley, E. A. 1976. Drought in the Sahara: Insufficient biogeophysical feedback? *Science* 191:100.

Ritchie, J. C. 1974. Modern pollen assemblages near the arctic treeline, Mackenzie Delta region, Northwest Territories. *Canadian Journal of Botany* 52:381–96.

Ritchie, J. C. 1976. The Late-Quaternary vegetational history of the western interior of Canada. *Canadian Journal of Botany* 54:1793–1818.

Ritchie, J. C. 1977. The modern and late Quaternary vegetation of the Campbell Dolomite Uplands, near Inuvik, N.W.T. *Ecological Monographs* 47:401–23.

Ritchie, J. C. 1980. "Towards a late-Quaternary paleoecology of the ice-free corridor," in *The ice-free corridor and peopling of the New World.* Edited by N. W. Rutter and C. E. Schweger. *Canadian Journal of Anthropology* 1:15–28.

Ritchie, J. C. 1982. The modern and Late-Quaternary vegetation of the Doll Creek area, northern Yukon, Canada. *New Phytologist* 81.

Ritchie, J. C., J. Cinq-Mars, and L. Cwynar. In press. L'environment tardiglaciaire du Yukon septentrional. *Geographie Physique et Quaternaire* no. 1–2. Montreal. (1982)

Ritchie, J. C., and S. Lichti-Federovich. 1965. Contemporary pollen spectra in Central Canada. II. The forest-grassland transition in Manitoba. *Pollen et Spores* 7:63–87.

Ritchie, J. C., and S. Lichti-Federovich. 1967. Pollen dispersal phenomena in Arctic-subarctic Canada. *Review of Paleobotany and Palynology* 3:255–66.

Robinson, D. W., and L. M. Slade. 1974. The current status of knowledge on the nutrition of equines. *Journal of Animal Science* 39:1045–66.

Roe, F. G. 1970. *The North American buffalo: A critical study of the species in its wild state,* 2nd edition. Toronto: University of Toronto Press.

Rogachev, A. N. 1957. The multilevel sites of the Kostienki-Borshevo region on the Don and the problem of the development of culture in the upper Paleolithic Epoch on the Russian Plain. *Materialy i Issledovanii po Arkheologii SSSR* 59:134 (in Russian).

Romanovskiy, N. N., and E. A. Shaposhnikova. 1971. "Study of the zonal character of frost cracking," in *Merzlotnye issledovaniya* 11:89–107. Moscow: Izd. MGU (in Russian).

Romer, A. 1966. *Vertebrate Paleontology.* 3rd edition. University of Chicago Press.

Rouse, Irving. 1976. Peopling of the Americas. *Quaternary Research* 6:567–612.

Rouse, W. R., and K. A. Kershaw. 1971. The effects of burning on the heat and water regimes of lichen-dominated subarctic subsurfaces. *Arctic and Alpine Research* 3:291–304.

Rowe, J. S., and G. W. Scotter. 1973. Fire in the boreal forest. *Quaternary Research* 3:444–64.

Rozenbaum, G. E. 1971. *Sovremennyi alluvi ravninnykh rek vostochnoi subarktiki (Recent alluvium of the plains river of the eastern subarctic).* Dissertation abstracts, MGU (in Russian).

Ruddiman, W. F. 1981. Oceanic mechanisms for amplification of the 23,000-year ice volume cycle. *Science* 212:617–27.

Ruddiman, W. F., and A. McIntyre. 1979. Warmth of the subpolar North Atlantic during Northern Hemisphere Ice-sheet growth. *Science* 204:173–75.

Rutter, N. W. 1978. Geology of the ice-free corridor. *AMQUA Abstracts* 5:2–12. Edmonton.

Rutter, N. W. 1980. Late Pleistocene history of the Western Canadian Ice-Free Corridor. *Canadian Journal of Anthropology* 1:1–8.

Ryder, M. L. 1974. Hair of the mammoth. *Nature* 249: 190–92.

Sachs, H. M., T. Webb, III, and D. R. Clark. 1977. Paleoecological transfer functions. *Annual Review of Earth and Planetary Science* 5:159–78.

Salensky, W. 1905. Cited in Lang (1925), without a title as Compt. Rend. *Sean. 6th International Zoological Congress.* Geneve (1904):72.

Salter, R. E., and R. J. Hudson. 1979. Feeding ecology of feral horses in western Alberta. *Journal of Range Management* 32:221–25.

Sancetta, Constance. 1981. Hydrography of the eastern Bering Sea during the last glacial event: evidence from fossil diatoms. *EOS* 62:297.

Sancetta, Constance. In press. "Fossil diatoms and the oceanography of the Bering Sea during the last glacial event," in *2nd International Conference on Siliceous Deposits of the Pacific Region.* Tokyo (1981).

Saunders, J. J. 1977. Lehner Ranch revisited. *The Museum Journal* 17:48–64.

Saunders, J. J. 1978. *Mammuthus* in the Clovis horizon of North America. *AMQUA Abstracts* 5:99–107. Emonton.

Schaeffer, A. L., B. A. Young, and A. M. Chimwano. 1978. Ration digestion and retention times of digesta in domestic cattle (*Bos taurus*), American bison (*Bison bison*) and Tibetan yak (*Bos grunniens*). *Canadian Journal of Zoology* 56:2355–58.

Schaller, G. B. 1972. *The Serengeti lion.* Chicago: University of Chicago Press.

Schaller, G. B. 1977. *Mountain monarchs: Wild sheep and goats in the Himalaya.* University of Chicago Press.

Schleisier, K. H. 1967. Sedna Creek: Report on an archaeological survey on the Arctic slope of the Brooks Range. *American Antiquity* 32:210–22.

Schneider, S. H. 1972. Cloudiness as a global climatic feedback mechanism: The effect on the radiation balance and surface temperature of variations in cloudiness. *Journal of Atmospheric Science* 29:1413–22.

Schoenwetter, J. 1966. A re-evaluation of the Navajo Reservoir pollen chronology. *El Palacio* 73:19–26.

Schweger, C. E. 1976. *Late Quaternary palaeoecology of the Onion Portage region, northwestern Alaska.* University of Alberta, Ph.D. thesis.

Schweger, C. E., and T. Habgood. 1976. The late Pleistocene steppe-tundra in Beringia—a critique. *AMQUA Abstracts* 4:80–81. Tempe.

Schweger, C. E., and J. A. P. Janssens. 1980. Paleoecology of the Boutellier nonglacial interval, St. Elias Mountains, Yukon Territory, Canada. *Arctic and Alpine Research* 12:309–17.

Schweger, C. E., and J. Martin. 1976. Grazing strategies of the Pleistocene steppe-tundra fauna. *AMQUA Abstracts* 4:157. Tempe.

Schweinfurth, U. 1957. Die horizontale and vertikale: verbreitung der vegetation im Himalaya. *Bonner Geographische Abhandlungen* 2:1–373.

Scotter, G. W. 1972. Reindeer ranching in Canada. *Journal of Range Management* 25:167–74.

Scudder, G. G. E., and S. G. Cannings. 1979. Entomological studies in the southern Yukon. *Information North:* 7.

Searby, H. W., and D. I. Branton. 1975. "Climatic conditions in agricultural areas in Alaska," in *Climate in the Arctic.* Edited by G. Weller and S. A. Bowling, pp. 281–92. Fairbanks: University of Alaska.

Sellmann, O. V. 1967. Geology of the USA CRREL permafrost tunnel, Fairbanks, Alaska. United States Army *Cold Regions Research and Engineering Laboratory Technical Report* 199.

Sellmann, P. V. 1972. Geology and properties of materials exposed in the USA CRREL permafrost tunnel. *United States Army Cold Regions Research and Engineering Laboratory Special Report* 177.

Sellmann, P. V., and J. Brown. 1973. "Stratigraphy and diagenesis of perennially frozen sediments in the Barrow, Alaska region," in *Permafrost, the North American Contribution to the Second International Conference.* pp. 171–81. National Academy of Sciences.

Sergin, S. Y., and Shcheglova, M. S. 1976. "The climate of Beringia during glacial epochs as a result of the influence of local and global factors," in *Beringiya v Kainozoe.* Edited by V. L. Kontrimavichus, pp. 171–76. Vladivostok (in Russian).

Shackleton, D. M. 1973. *Population quality and bighorn sheep.* University of Calgary, Ph.D. thesis.

Shackleton, J. 1979. *Paleoenvironmental histories from Whitefish and Imuruk Lakes, Seward Peninsula, Alaska.* Ohio State University, M.Sc. thesis.

Shackleton, N. J., and N. D. Opdyke. 1973. Oxygen isotope and paleomagnetic stratigraphy of equatorial Pacific core V28-238: Oxygen isotope temperatures and ice volumes on a 10^5 and 10^6 year scale. *Quaternary Research* 3:39–55.

Sharatchandra, H. C., and M. Gadgil. 1975. A year of Bandipur. *Journal of Bombay Natural Historical Society* 72:623–47.

Shebbeare, E. O. 1958. *Soondar Mooni: The life of an Indian elephant.* London: Victor Gollancz.

Sheludyakova, V. A. 1938. The vegetation of the Indigirka River drainage. *Sovetskaya Botanika* 4-5:43–79 (in Russian).

Sheludyakova, V. A. 1957. Steppe vegetation of Yakutia north of the Polar Circle. *Akad. Nauk SSSR, Yakutsk Bilian, Institut Biologichesie Trudy* 3:68–82 (in Russian).

Sher, A. V. 1967. Saiga remains mined in the northeast of Siberia and in Alaska. *Byulleten' Kommissii po Izucheniyu Chetvertichnogo Perioda* 33:97–112 (in Russian).

Sher, A. V. 1968. Fossil saiga in northeastern Siberia and Alaska. *International Geology Review* 10:1247–60.

Sher, A. V. 1974. Pleistocene mammals and stratigraphy of the far Northeast U.S.S.R. and North America. *International Geology Review* 16:1–284.

Sher, A. V., R. E. Giterman, V. S. Zazhigin, and S. V. Kiselyov. 1977. New data on the late Cenozoic of the Kolyma Lowland. *Akad. Nauk SSSR, Izvestia, Seriya Geol.* 5:69–83 (in Russian).

Sher, A. V., T. N. Kaplina, R. E. Giterman, A. V. Lozhkin, A. A. Arkhangelov, S. V. Kiselyov, Yu. V. Kouznetsov, E. I. Virina, and V. S. Zazhigin. 1979. Late Cenozoic of the Kolyma Lowland. *14th Pacific Science Congress, Guidebook, Tour XI.* Khabarovsk.

Shilo, N. A. 1964. On the evolution of the plains of the subarctic belt of Northeast Asia. Akad. Nauk SSSR, Sibirskoe Otdel., *Trudy Institut, Magadan* 11:154–69 (in Russian).

Shilo, N. A. 1971. Periglacial lithogenesis in the general scheme of the process of continental rock formation. Akad. Nauk SSSR, Sibirskoe Otdel., *Trudy Institut, Magadan* 38:3–56 (in Russian).

Shilo, N. A., and A. V. Lozhkin. 1981. "Radiocarbon-dating of the baby mammoth," in *Magadanskiy mamontenok.* Edited by N. K. Vereshchagin and V. M. Michel'son, pp. 48–49. Leningrad: Izd. "Nauka" (in Russian).

Shilo, N. A., A. V. Lozhkin, E. E. Titov, U. V. Shumilov. 1977. Unique new mammoth find. *Akad. Nauk Doklady* 236:1449–52 (in Russian).

Shvarts, S. S. 1963. *The adaptation of terrestrial vertebrates to the conditions of the subarctic:* vol. 1, *Mammals.* Akad. Nauk SSSR, Sibirskoe Otdelnie, Instituta Biologie 33.

Sikes, S. K. 1971. *The natural history of the African elephant.* New York: Elsevier.

Simberlof, D. F., and L. G. Abele. 1976. Island biogeography theory and conservation practice. *Science* 191: 285–86.

Simpson, G. G. 1950. *Horses.* Oxford: Oxford University Press.

Simpson, G. G. 1964. Species density of North American recent mammals. *Systematic Zoology* 13:57–73.

Simpson, R. D. 1978. "The Calico Mountains archaeological site," in *Early Man in America from a circum-Pacific perspective.* Edited by A. L. Bryan, pp. 218–20. University of Alberta Department of Archaeology Occasional Papers 1.

Sims, P. L., and J. S. Singh. 1978a. The structure and function of ten western North American grasslands. II. Intraseasonal dynamics in primary producer compartments. *Journal of Ecology* 66:547–72.

Sims, P. L., and J. S. Singh. 1978b. The structure and function of ten western North American grasslands. III. Net primary production, turnover and efficiencies of energy capture and water use. *Journal of Ecology* 66:573–98.

Sinclair, A. R. E. 1970. *Studies of the ecology of the east African buffalo.* Oxford University, Ph.D. dissertation.

Sinclair, A. R. E. 1974. "The social organization of the east African buffalo (*Syncerus caffer* Sparrman)," in *The behavior of ungulates and its relation to management.* Edited by V. Geist and F. Walther, pp. 676–89. Morges (Switzerland): International Union for Conservation of Nature and Natural Resources Publ. n.s. 24.

Sinclair, A. R. E. 1977. *The African buffalo—a study of resource limitation of populations.* University of Chicago Press.

Sinclair, A. R. E. 1975. The resource limitation of trophic levels in tropical grassland ecosystems. *Journal of Animal Ecology* 44:497–520.

Singh, V. B. 1969. The elephent (*Elephas maximus* Linn) in Utter Pradesh, India. *Journal of Bombay Natural Historical Society* 66:239–50.

Skarlato, O. A. (Editor). 1977. *Mamontovaya fauna Russkoy Ravniny i Vostochnoy Sibiri (Mammoth fauna of the Russian plain and eastern Siberia).* Akad. Nauk SSSR, Trudy Zoologicheskogo Instituta 72 (in Russian).

Skryabin, S. Z., and A. K. Konorovskiy. 1975. "The vegetation and soils of the steppes in the middle reaches of the Indigirka River," in *Botanicheskie Materialy po Yakutia,* pp. 38–47. Yakutsk (in Russian).

Slaughter, B. H., and B. R. Hoover. 1963. Sulphur river formation and the Pleistocene mammals of the Ben Franklin local fauna. *Southern Methodist University, Journal of Graduate Research Center* 31:132–48.

Slaughter, C. W., and K. P. Long. 1975. "Upland climatic parameters on subarctic slopes, central Alaska," in *Climate in the Arctic.* Edited by G. Weller and S. A. Bowling, pp. 276–80. Fairbanks: University of Alaska.

Smith, R. F. C., and L. C. H. Wong. 1977. "Arctic hares on Truelove Lowland," in *Truelove Lowland, Devon Island, Canada: A high arctic ecosystem.* Edited by L. C. Bliss, pp. 461–66. Edmonton: University of Alberta.

Sokolov, I. A., G. M. Bystryakov, and E. V. Kulinskaya. 1977. *K kharakteristike ul'trakontineltai'nogo kholodnogo aridnogo pochvoobrazonasiya (On the characterization of the ultra-continental cold arid soil-formation),* in *Spetsifika Pochvoorazoniya v Sibiri,* pp. 9–13. Novosibirsk (in Russian).

Solecki, R. S. 1951. Notes on two archaeological discoveries in northern Alaska, 1950. *American Antiquity* 17:55–57.

Spencer, R. F. 1959. The North Alaskan Eskimo, a study in ecology and society. *Smithsonian Institution Bureau of American Ethnology Bulletin* 171.

Spetzman, L. 1959. Vegetation of the Arctic Slope of Alaska. *United States Geological Survey Professional Paper* 302-B.

Spiker, Elliott, Lea Kelley, and Meyer Rubin. 1978. United States Geological Survey radiocarbon dates XIII. *Radiocarbon* 20:139–56.

Spinage, C. A. 1972. The evolution of the proboscidea. *Säugetierk Mitt.* 20:152–56.

Stanford, D. 1978. Some Clovis points. *AMQUA Abstracts* 5:140–41. Edmonton.

Stanford, Dennis. 1979a. "Afterword: Resolving the question of New World origin," in *Pre-Llano cultures of the Americas: Paradoxes and possibilities.* Edited by R. L. Humphrey and Dennis Stanford, pp. 147–50. Washington: Anthropological Society of Washington.

Stanford, Dennis. 1979b. "The Selby and Dutton sites: Evidence for a possible Pre-Clovis occupation of the High Plains," in *Pre-Llano cultures of the Americas: Paradoxes and possibilities.* Edited by R. L. Humphrey and Dennis Stanford, pp. 101–23. Washington: Anthropological Society of Washington.

Stanford, Dennis, Robson Bonnichsen, and R. E. Morlan. 1981. The Ginsberg experiment: Modern and prehistoric evidence of a bone flaking technology. *Science* 212:438–40.

Staniforth, A. N., and R. W. Daley, 1979. A baroclinic finite element model for regional forecasting with the primitive equations. *Monthly Weather Review* 107: 107–21.

Steen-McIntyre, V., R. Fryxell, and H. E. Malde. 1981. Geologic evidence for age of deposits at Hueyatlaco archaeological site, Valsequillo, Mexico. *Quaternary Research* 16:1–17.

Steiner, W. 1979. *Der Travertin von Ehringsdorf und siene Fossilien.* Wittenberg Lutherstadt: Ziemsen.

Stoddart, L. A., A. D. Smith, and T. W. Box. 1975. *Range Management,* 3rd edition. New York: McGraw Hill.

Strang, W. G., and G. J. Fix. 1973. *An analysis of the finite element method.* Englewood Cliffs, New Jersey: Prentice-Hall.

Streten, N. A. 1974. Some features of the summer climate of interior Alaska. *Arctic* 27:273–86.

Strugnell, R. G., and C. D. Pigott. 1978. Biomass, short production and grazing of two grasslands in the Rwenzori National Park, Uganda. *Journal of Ecology* 66: 73–96.

Stuart, A. J. 1977. The vertebrates of the last cold stage in Britain and Ireland. *Philosophical Transactions of the Royal Society of London* 289:295–312.

Stuckenrath, R., Jr., and J. E. Mielke. 1970. Smithsonian Institution radiocarbon measurements VI. *Radiocarbon* 12:203.

Sukachev, V. N. 1901. *A study of the vegetable remains of the mammoth found on Berezovka River in Yakutia.* Petrograd (in Russian).

Sutcliffe, R. C. 1947. A contribution to the problem of development. *Quaternary Journal of the Royal Meteorological Society* 73:370–83.

Svoboda, J. 1977. "Ecology and primary production of raised beach communities, Truelove Lowland," in *Truelove Lowland, Devon Island, Canada: A high arctic ecosystem,* pp. 185–216. Edited by L. C. Bliss. Edmonton: University of Alberta Press.

Swan, L. W. 1961. The ecology of the high Himalayas. *Scientific American* 205:68–78.

Tapp, M. C., and P. W. White. 1976. A non-hydrostatic mesoscale model. *Quarterly Journal of Royal Meteorological Society* 102:277–95.

Tauber, H. 1973. Copenhagen radiocarbon dates X. *Radiocarbon* 15:107.

Taylor, D. C. 1969. The Wilsall excavations: An exercise in frustration. *Proceedings of the Montana Academy of Science* 29:147–50.

Taylor, R. D., and B. H. Walker. 1978. Comparisons of vegetation use and herbivore biomass on a Rhodesian game and cattle ranch. *Journal of Applied Ecology* 15:565–81.

Tedford, R. H. 1970. Principles and practices of mammalian geochronology in North America. *Proceedings of the North American Paleontological Convention,* Sept. 1969. Part F:666–703.

Telfer, E. S., and G. W. Scotter. 1975. Potential for game ranching in boreal aspen forests of western Canada. *Journal of Range Management* 28:172–80.

Tener, J. S. 1965. *Muskoxen in Canada—a biological and taxonomic review.* Canadian Wildlife Service.

Terasmae, J., and A. Dreimanis. 1976. "Quaternary stratigraphy of southern Ontario," in *Quaternary stratigrapphy of North America.* Edited by W. C. Mahaney, pp. 51–64. Dowden, Hutchinson, and Ross.

Thompson, R. M. 1948. Notes on the archaeology of the Utukok River, northwestern Alaska. *American Antiquity* 14:62–65.

Thorson, R. M., E. J. Dixon, Jr., G. S. Smith, and A. R. Batten. 1981. Interstadial Proboscidian from south-central Alaska: Implications for biogeography, geology, and archaeology. *Quaternary Research* 16:404–17.

Thorson, R. M., and T. D. Hamilton. 1977. Geology of the Dry Creek site: A stratified early man site in interior Alaska. *Quaternary Research* 7:149–76.

Tikhomirov, B. A. 1958. Natural conditions and vegetation in the mammoth epoch in northern Siberia. *Problems of the North* 1:168–88.

Tikhomirov, B. A. 1970. "Forest limits as the important biogeographical boundary of the North," in *Ecology of the subarctic regions,* pp. 35–40. Paris: UNESCO.

Tikhomirov, B. A., and L. A. Kupriyanova. 1954. Pollen analysis of plant remains of the Beresovka Mammoth. *Doklady Akad. Nauk SSSR* 95:1313–15 (in Russian).

Tode, A. et al. 1953. Die Untersuchung der paläolithischen Freilandstation von Salzgitter-Lebendtedt. *Eiszeitalter und Gegenwart* 3:144–220.

Toepfer, V. 1970. "Stratigraphie und Ökologie des Paläolithikums," in *Periglazial-Löss-Paläolithickum in Jungpleistozän der Deutschen Demokratischen Republik.* Edited by W. Richter et al., pp. 329–422. Gotha: Haack.

Tomirdiaro, S. V. 1966. Lacustrine and thermokarst factors in Northern Plains and their geological engineering significance. Akad. Nauk SSSR, *Sibirskoe Inst. Merezlotovedeniia, Mater. VIII* 7:24–34 (in Russian).

Tomirdiaro, S. V. 1970a. "Permafrost of the Northeast SSSR," in *Prirodi uslovia i estestv. resursy SSSR*, pp. 133–49. Moscow: Izd. "Nauka" (in Russian).

Tomirdiaro, S. V. 1970b. Pleistocene glaciation of the arctic basin and its relation to the surface and subsurface glaciation of the land. *Arkticheskii i antarkticheskii nauchno-isslevovatel'skii Inst. Trudy* 294:172–82 (in Russian).

Tomirdiaro, S. V. 1972. *Vechnaya merzlota i osvdeie gornykh stran i Nizhmenostei (Permafrost and economic development of uplands and plains in the Magadan region and Yakutia).* Magadan: Izd. Magadanskoe Knizh (in Russian).

Tomirdiaro, S. V. 1978. *prirodn'e protsessy i osvoenie territoriy zony vehnoy merzloty (Natural processes of the ice-permafrost zone).* Moscow: Izd. "Nedra" (in Russian).

Tomirdiaro, S. V., and Z. V. Orlova. 1970. Nature and conditions of formation of the Pleistocene loess-ice plains in the Northeast. *Kolyma* 5 (in Russian).

Tomirdiaro, S. V., V. K. Ryabchun, Yu. V. Kuznetsov, and Z. V. Orlova. 1972. The loess-ice mantle of Northern Yakutia and the Novosibirsk Islands as the product of deflation of the shelf. *Kolyma* 11:38–42 (in Russian).

Touhy, D. R. 1968. "Some early lithic sites in western Nevada," in *Early Man in western North America.* Edited by C. Irwin-Williams, pp. 27–38. Portales: Eastern New Mexico University.

Tseitlin, S. M. 1979. *Geologiia Paleolita Severnoi Azii (Geology of the Paleolithic of northern Asia).* Moscow: Izd. "Nauka" (in Russian).

Tugarinov, A. Ya. 1929. Origin of the Arctic fauna. *Priroda* 7-8:653–80 (in Russian).

Ukraintseva, V. V. 1979. Plants of the warm periods in late Pleistocene and dying off of some large herbivorous mammals. *Botanicheskii Zhurnal* 64:318–28 (in Russian).

United States Forest Service. 1937. *Range plant handbook.* Washington: U.S. Government Printing Office.

Vail, Leroy. 1977. Ecology and history: The example of eastern Zambia. *Journal of Southern African Studies* 3:129–55.

Valoch, Karel. 1979. Paläolithische Artefakte aus früheren Grabungen in der Kulna-Höhle im Mährischen Karst. *Acta Musei Moraviae* 64:7–44.

Vancuylenberg, B. W. B. 1974. *The feeding behaviour of the Asiatic elephant in the southeastern Ceylon.* University of Sri Lanka, M.Sc. thesis.

Van Devender, T. R., and W. G. Spaulding. 1979. Development of vegetation and climate in the southwestern United States. *Science* 204:701–10.

Vangengeim, E. A. 1975. Sur la fauna periglaciaire du Pleistocene. *Biuletyn Peryglacjalny* 24:81–88 (in French).

Vangengeim, E. A. 1977. Paleontologicheskoe obosnovanie stratigrafii Antropogena severnoi Azii (Po Mlekopita yushchim) (The paleontological basis for Anthropogene stratigraphy in northern Asia: mammals). *10th INQUA Congress*, Birmingham.

Vangengeim, E. A., and Eh. I. Ravskii. 1965. "Inner-continental natural zonation in the Quaternary (Anthropogene)," in *Problemy Stratigrafii Kainozoya*, pp. 128–41. Moscow: Izd. "Nedra" (in Russian).

Vasari, T. Y. K. 1976. "Periglacial vegetation as an indicator of past natural conditions in eastern Fennoscaninavia," in *Geomorphology and Paleogeography* 1. *International Geography '76*, pp. 406–09. Moscow: XXII International Geographical Congress.

Vasil'evsky, R. S. 1975. "Problems of the origin of the ancient sea hunters' cultures in the northern Pacific," in *Prehistoric maritime adaptations of the circumpolar zone.* Edited by William Fitzhugh, pp. 113–21. The Hague: Mouton.

Vassiljevskaja, V. D., V. V. Ivanov, L. G. Botatyrev, E. B. Pospelova, N. M. Shalaeva, and L. A. Grishina. 1975. "Agapa, USSR," in *Structure and function of tundra ecosystems.* Edited by T. Rosswall and O. W. Heal. *Ecological Bulletins* 20:141–58. Swedish Natural Science Research Council.

Vas'kovskiy, A. P. 1957. "Spore-pollen spectra of the present-day plant associations of the extreme northeast of the U.S.S.R. and their importance for reconstructions of Quaternary vegetation," in *Materialy po Geologii i Poleznya Iskop. Severo-Vostoka S.S.S.R.* 2:130–78. Magadan (in Russian).

Velichko, A. A. 1973. *Prirodnyy Protsessy v Pleytotsene (Natural process in Pleistocene).* 9th INQUA Congress, New Zealand. Moscow: Izd. "Nauka" (in Russian).

Velichko, A. A. 1975. "Paragenesis of the cryogenic (periglacial) zone," in *Paleogeography and periglacial phenomena of the Pleistocene.* Edited by K. K. Markov and I. I. Spasskaya, pp. 89–100. Moscow: Izd. "Nauka" (in Russian).

Vereshchagin, N. K. 1967a. The geological history of moose and its utilization by primitive man. *Biologiya Promysellosya,* Moscow Sb., 3:1–36 (in Russian).

Vereshchagin, N. K. 1967b. "Primitive hunters and Pleistocene extinction in the Soviet Union," in *Pleistocene extinctions: The search for a cause.* Edited by P. S. Martin and H. E. Wright, Jr., pp. 365–98. New Haven: Yale University Press.

Vereshchagin, N. K. 1971. The cave lion and its history in the Holarctic and in the U.S.S.R. *Akad. Nauk SSSR, Trudy Zoologicheskogo Instituta* 49:123–99 (in Russian).

Vereshchagin, N. K. 1974. The mammoth "cemeteries" of North-East Siberia. *Polar Record* 17(106):3–12.

Vereshchagin, N. K. 1975. About a mammoth from the River Shandrina. *Vestnick Zoologii* 2:82–83 (in Russian).

Vereshchagin, N. K. 1979. Why *the mammoths became extinct.* Leningrad: Izd. "Nauka" (in Russian).

Vereshchagin, N. K., and G. F. Baryshnikov. 1980. "Range of the ungulate fauna of the U.S.S.R. in the Anthropogene," in *Mlekopitayushche Vostochnoy Evropy v Antropogene.* Edited by O. A. Scarlato. Akad. Nauk SSSR, Trudy Zoologicheskogo Instituta 93:3–20 (in Russian).

Vereshchagin, N. K., and I. E. Kuz'mina. 1979. "Mammal remains from Paleolithic sites on the Don and upper Desna," in *Mamontovaya Fauna Russkoy Ravniny i Vostochnoy Sibiri.* Edited by O. A. Scarlato. Trudy

Zoologicheskogo Instituta 72:138–78 (in Russian).

Vereshchagin, N. K., and P. A. Lazarev. 1977. "Description of soft parts and skeleton of the Selerikan horse," in *Fauni i Flori Antropogena Severo-Vostoka Sibiri.* Edited by O. A. Scarlato. Akad. Nauk SSSR, Trudy Zoologicheskogo Instituta 63:85–185 (in Russian).

Verneker, A. D. 1972. Long period global variation of incoming solar radiation. *Meteorological Monographs* 12:1–21.

Viereck, L. A. 1973. Wildfire in the taiga of Alaska. *Quaternary Research* 3:465–95.

Viereck, L. A. 1975. "Forest ecology of the Alaskan taiga," in *Proceedings of the circumpolar conference on northern ecology.* Ottawa: National Research Council of Canada.

Viereck, L. A., and E. L. Little. 1972. Alaska trees and shrubs. *United States Forest Service Handbook* No. 410.

Vincent, A. B. 1965. Black spruce: A review of its silvics, ecology, and silviculture. *Canada Department of Forestry Publication* 110.

Vinogradov, B. S. 1948. On the discovery of fossil sousliks in permafrost in the Indigirka Basin. *Akad. Nauk SSSR, Doklady* 62:553–56 (in Russian).

Vinokurov, N. N. 1975. A new species of Heteroptera, *Crophius bermani* sp. n. (Lygaeidae) from northeastern Yakutia. *Entomology Review* 54:110–12.

Virina, E. I. 1979. Paleomagnetism of late Cenozoic sediments of the northern and southern part of the Soviet Far East. *14th Pacific Science Congress, Abstracts, Committee BIII:* 150–52. Khabarovsk.

Vitt, D. H., and P. Pakarinen. 1977. "The bryophyte vegetation, production, and organic components of Truelove Lowland," in *Truelove Lowland, Devon Island, Canada: A high arctic ecosystem.* Edited by L. C. Bliss, pp. 225–44. Edmonton: University of Alberta Press.

Vogl, R. J. 1974. "Effects of fire on grasslands," in *Fire and ecosystems.* Edited by T. T. Kozlowski and C. E. Ahlgren, pp. 139–94. New York: Academic Press.

Volkov, I. A., M. G. Grosval'd, and S. L. Troitskiy. 1980. The discharge of proglacial waters during the last glaciation of western Siberia. *Polar Geography* 4:1–14.

Voorhies, M. R. 1969. Taphonomy and population dynamics of an Early Pliocene vertebrate fauna, Knox County, Nebraska. *University of Wyoming Contributions to Geology, Special Paper* 1.

Vtyurin, B. A. 1961. Particulars of the cryogenic structure of the unconsolidated deposits in the Tumara Basin. Akad. Nauk SSSR, *Trudy Inst. Merzlotoved* 17:34–38. Moscow (In Russian).

Walker, D. Ed. 1971. *Bridge and barrier: The natural and cultural Torres Strait.* Australian National University, Research School of Pacific Studies, Dept. of Biogeography and Geomorphology Publication BG/3. Canberra.

Walker, D. A., and P. J. Webber. 1979. Relationships of soil acidity and air temperature to the wind and vegetation at Prudhoe Bay, Alaska. *Arctic* 32:224–36.

Walker, D. A., K. R. Everett, P. J. Webber, and J. Brown. 1979. "Climate," in *Geobotanical Atlas of Prudhoe Bay, Alaska.* Hanover, NH: U.S. Army Cold Regions Research Engineering Laboratory.

Warren-Wilson, J. 1959. Notes on wind and its effect in arctic-alpine vegetation. *Journal of Ecology* 47: 415–27.

Watts, W. A. 1973. "Rates of change and stability in vegetation in the perspective of long period of time," in *Quaternary Plant Ecology.* Edited by H. J. B. Birks and R. G. West. Oxford: Oxford University Press.

Watts, W. A. 1977. The late Devensian vegetation of Ireland. *Philosophical Transactions of the Royal Society of London* B280:273–93.

Webb, T. III, and R. A. Bryson. 1972. Late- and postglacial climatic change in the northern Midwest, U.S.A.: quantitative estimates derived from fossil pollen spectra by multivariate statistical analysis. *Quaternary Research* 2:70–115.

Webb, T., III, and D. R. Clark. 1977. Calibrating micropaleontological data in climatic terms: A critical review. *Annals of the New York Academy of Science* 288:93–118.

Webber, P. J. 1978. "Spatial and temporal variation of the vegetation and its production, Barrow, Alaska," in *Vegetation and production ecology of an Alaskan arctic tundra.* Edited by L. L. Tieszen, pp. 37–112. Ecological Studies 29. New York: Springer-Verlag.

Webber, P. J., and D. A. Walker. 1975. "Vegetation and landscape analysis at Prudhoe Bay, Alaska: A vegetation map of the tundra biome study area," in *Ecological investigations of the tundra biome in the Prudhoe Bay region, Alaska.* Edited by J. Brown, pp. 81–91. Biological Papers of the University of Alaska Special Report 2.

Weber, F. R., and T. L. Péwé. 1970. Surficial and engineering geology of the central part of the Yukon-Koyukuk lowland, Alaska. *United States Geological Survey Miscellaneous Geological Investigations Map* I-590.

Weber, F. R., H. L. Foster, T. E. C. Keith, and C. Dusel-Bacon. 1978. Preliminary geologic map of the Big Delta quadrangle, Alaska. *United States Geological Survey Open File Report* 78-529A.

Weber, F. R., T. D. Hamilton, D. M. Hopkins, C. A. Repenning, and Herbert Haas. 1981. Canyon Creek: A late Pleistocene vertebrate locality in interior Alaska. *Quaternary Research* 16:167–80.

Webster, P. J., and N. A. Streten. 1978. Late Quaternary ice age climates of tropical Australasia: Interpretations and reconstructions. *Quaternary Research* 10:279–309.

Wei Chi. 1979. Ricerca dei discendentii dell'uomo di Pechino. *Minerva Medica* 70:2723–28.

Weigert, R. G., and D. F. Owen. 1971. Trophic structure, available resources and population density in terrestrial vs. aquatic ecosystems. *Journal of Theoretical Biology* 30:69–81.

Wein, R. W. 1976. Frequency and characteristics of arctic tundra fires. *Arctic* 9:213–22.

Wein, R. W., and L. C. Bliss. 1973. Changes in Arctic *Eriophorum* tussock communities following fire. *Ecology* 54:845–52.

Weller, G., and G. Holmgren. 1974. The microclimates of the arctic tundra. *Journal of Applied Meteorology* 13:854–62.

West, F. H. 1967. The Donnelly Ridge site and the definition of an early core and blade complex in central Alaska. *American Antiquity* 32:360–82.

West, F. H. 1975. Dating the Denali complex. *Arctic Anthropology* 12:6–81.

West, F. H. 1976. "Old World affinity of archaeological complexes from Tangle Lakes, Alaska," in *Beringiya v Kainozoe.* Edited by V. L. Kontrimavichus, pp. 439–58. Khabarovsk: Akad. Nauk SSSR, Dal'neovostochnyy Nauchnyy Tsentr (in Russian, English abstract).

West, F. H. 1980. "Late Paleolithic cultures in Alaska," in *Early Native Americans.* Edited by D. L. Browman, pp. 161–87. The Hague: Mouton.

West, R. G. 1964. Inter-relations of ecology and Quaternary palaeobotany. *Journal of Ecology* 52 (Suppl.): 47–57.

Westgate, J. A., N. D. Briggs, and G. W. Pearce. 1978. *The Old Crow tephra and its stratigraphic significance: A progress report.* Paper presented to the 11th Annual Meeting of the Canadian Archaeological Association. Quebec City.

Westgate, J. A., J. V. Matthews, Jr., and T. D. Hamilton. 1981. Old Crow tephra: A new Late Pleistocene stratigraphic marker across Alaska and the Yukon Territory. Geological Society of America, *Abstracts with Programs,* p. 579.

Westra, R., and R. J. Christopherson. 1976. Effects of cold on digestibility, retention time of digesta, reticulum motility and thyroid hormone in sheep. *Canadian Journal of Animal Science* 56:699–708.

Wethereald, R. T., and S. Manabe. 1975. The effects of changing the solar constant on the climate of a general circulation model. *Journal of Atmospheric Science* 32:2044–59.

Wheeler, J. O. 1961. Whitehorse Map-area, Yukon Territory. *Geological Survey of Canada Memoir* 312.

White, R. G., F. L. Bunnell, E. Gaare, T. Skogland, and B. Hubert. 1981. "Ungulates on arctic ranges," in *Tundra ecosystems: A comparative analysis.* Edited by L. C. Bliss, O. W. Heal, and J. J. Moore, pp. 397–483. Cambridge: Cambridge University Press.

White, R. G., B. R. Thompson, T. Skogladd, S. J. Person, D. E. Russell, D. F. Holleman, and J. R. Luick. 1975. "Ecology of caribou at Prudhoe Bay, Alaska," in *Ecological investigations of the tundra biome in the Prudhoe Bay region, Alaska.* Edited by J. Brown, pp. 155–201. Biological Papers of the University of Alaska, Special Report 2.

White, R. G., and J. Trudell. 1980. Habitat preference and forage consumption by reindeer and caribou near Atkasook, Alaska. *Arctic and Alpine Research* 12:511–29.

Whitmore, F. C., and H. L. Foster. 1967. *Panthera atrox* (Mammalia: Felidae) from central Alaska. *Journal of Paleontology* 41:247–51.

Whitmore, T. C., editor. 1972. *Tree flora of Malaya: A manual for foresters,* v. 1. London: Longmans Group.

Whittaker, R. H. 1975. *Communities and ecosystems,* 3rd edition. New York: Macmillan.

Whittaker, R. H., and P. P. Feeny. 1971. Allelochemics: Chemical interactions between species. *Science* 171: 757–70.

Witthoft, J. 1952. A Paleoindian site in eastern Pennsylvania: An early hunting culture. *Proceedings of the American Philosophical Society* 96:464–95.

Williams, J. 1978. "The use of numerical models in studying climatic change," in *Climatic change.* Edited by J. Gribben, pp. 178-90. Cambridge: Cambridge University Press.

Williams, J., and R. G. Barry. 1975. "Ice Age experiments with the NCAR General Circulation Model: Conditions in the vicinity of the northern continental ice sheets," in *Climate of the Arctic.* Edited by G. Weller and S. A. Bowling, pp. 143–49. Fairbanks: University of Alaska.

Williams, J., R. G. Barry, and W. M. Washington. 1974. Simulation of the atmospheric circulation using the NCAR global circulation model with Ice Age boundary conditions. *Journal of Applied Meteorology* 13:305–17.

Williams, J. R. 1962. Geologic reconnaissance of the Yukon Flats district, Alaska. *United States Geological Survey Bulletin* 1111-H:289–331.

Williams, J. R., L. D. Carter, W. E. Yeend, and T. D. Hamilton. 1977. Preliminary surficial deposits map of National Petroleum Reserve-Alaska. *United States Geological Survey Open File Report* 77-868.

Williamson, B. R. 1975. The condition and nutrition of elephant in Wankie National Park. *Arnoldia (Rhod)* 7(12):1–20.

Wilmsen, E. N. 1974. *Lindenmeir: A Pleistocene hunting society.* New York: Harper & Row.

Wing, L. D., and I. O. Buss. 1970. Elephants and forests. *Wildlife Monographs* 19:1–92.

Woillard, G. M. 1978a. Grande Pile peat bog: A continuous pollen record for the last 140,000 years. *Quaternary Research 9:1–21.*

Woillard, G. M. 1978b. The last interglacial-glacial cycle at Grande Pile in Northeastern France. *Travaux du Laboratoire de Palynologie et Phytosociologie.* Université du Louvain.

Wolfe, J. A. 1972. "An interpretation of Alaskan Tertiary floras," in *Floristics and Paleofloristics of Asia and Eastern North America.* Edited by A. Graham, pp. 201–33. Elsevier.

Wood, A. J., I. M. Cowan, and H. C. Nordan. 1962. Periodicity of growth in ungulates as shown by deer of the genus *Odocoileus. Canadian Journal of Zoology* 40:593–603.

Workman, W. B. 1978. *Prehistory of the Aishihik-Kluane area, southwest Yukon Territory.* Canadian National Museum of Man Mercury Series (Archaeological Survey of Canada Paper No. 74).

Workman, W. B. 1980a. Holocene peopling of the New World: Implications of the Arctic and Subarctic area. *Canadian Journal of Anthropology* 1:129–39.

Workman, W. B. 1980b. Continuity and change in the prehistoric record from southern Alaska. *Senri Ethnological Series* 4:49–101.

Wright, H. E., Jr. 1970. "Vegetational history of the central plains," in *Pleistocene and Recent environments of the central Great Plains.* Edited by W. Dort, Jr., and J. K. Jones, pp. 157–72. University Press of Kansas.

Wright, H. E., Jr. 1977. Quaternary vegetation history—

some comparisons between Europe and America. *Annual Review of Earth and Planetary Sciences* 5: 123–58.

Wyatt, J. R., and S. K. Eltringham. 1974. The daily activity of the elephant in the Ruwenzori National Park, Uganda. *East African Wildlife Journal* 12:273–89.

Yarovoi, M. I. 1939. The vegetation of the Yana River drainage and of the Verkhoyansk Range. *Sovetskaya Botanika* 1:21–40 (in Russian).

Yermoleva, I. M. 1978. *The theriofauna of the Angara Valley during the late Anthropogene.* Leningrad: Izd. "Nauka" (in Russian).

Young, B. A. 1977. One degree equals one percent. *56th Annual Feeders' Day Report,* pp. 38–42. Edmonton: University of Alberta.

Young, B. A. 1978. Cold stress as it affects animal production. *A.D.S.A.-A.S.A.S. Symposium,* East Lansing, Mich.

Young, S. B. 1971. The vascular flora of St. Lawrence Island with special reference to floristic zonation in the Arctic regions. Cambridge: Harvard University *Contributions from the Grey Herbarium* 201:11–115.

Young, S. B. 1974. The environment of the Noatak River basin, Alaska. *Contributions of the Center for Northern Studies* 1:1–584. Wolcott.

Young, S. B. 1976. Is steppe tundra alive and well in Alaska? *AMQUA Abstracts* 4:84–88. Tempe.

Young, S. B., and C. H. Racine. 1977. *Vegetational and floristic analysis and discussion of the Quaternary environment of the Kobuk Valley.* Wolcott: Center for Northern Studies.

Yurtsev, B. A. 1962. On the floristic interrelations between steppes of Siberia and prairies of North America. *Botanicheskie Zhurnal* 47:317–36 (in Russian).

Yurtsev, B. A. 1963. On the floristic relations between steppes and prairies. *Botaniska Notiser* 116:396–408 (in Russian).

Yurtsev, B. A. 1966. *The hypoarctic phytogeographic zone and the origin of its flora.* Moscow-Leningrad: Izd. "Nauka" (in Russian).

Yurtsev, B. A. 1968. "Steppe communities in the Chukotka tundra and the question of the Pleistocene 'tundra-steppe,'" in *Problemy izucheniya chetvertichnogo perioda,* pp. 133–35. Khabarovsk (in Russian).

Yurtsev, B. A. 1972a. "Phytogeography of Northeastern Asia and the problem of trans-Beringian floristic interrelations," in *Floristics and Paleofloristics of Asia and eastern North America,* pp. 19–54. Amsterdam: Elsevier.

Yurtsev, B. A. 1972b. "Steppe communities in the tundra of Chukotka and the problem of Pleistocene 'tundra-steppe,'" in *Problemy Izucheinya Chetvertichnogo Perioda,* pp. 471–79. Izd. "Nauka" (in Russian).

Yurtsev, B. A. 1974a. Steppe communities in the Chukotka tundra and the Pleistocene "tundra-steppe." *Botanicheskie Zhurnal* 59:484–501.

Yurtsev, B. A. 1974b. *Problemy Botanicheskoy Geografii Severo-Vostochnoy Azii (Problems of phytogeography of the Northeastern Asia).* Leningrad: Izd. "Nauka" (in Russian).

Yurtsev, B. A. 1978. Some questions of typology of steppe communities in northeastern Asia. *Botanicheskie Zhurnal* 63:1566–78 (in Russian).

Yurtsev, B. A., V. N. Andreev, V. I. Perfilyeva, and Z. P. Savkina. 1975. *Botanical excursion guide: North East Yakutia.* Leningrad.

Yurtsev, B. A., and A. A. Korobkov. 1979. Floristic finds in the Anadyr drainage in 1977. *Botanicheskie Zhurnal* 64,5 (in Russian).

Yurtsev, B. A., A. I. Tolmatchev, and O. V. Rebristaya. 1978. "The floristic delimitation and subdivision of the Arctic," in *Arkicheskaya floristicheskaya oblast',* Edited by B. A. Yurtsev, pp. 9–104. Leningrad: Izd. "Nauka."

Zaklinskaya, J. D. 1954. On the question of plants during the life and death of the Taimyr mammoth. *Doklady Akad. Nauk SSSR* 98:571–74 (in Russian).

Zamyatin, S. N. 1961. The Paleolithic station of Stalingrad. *Kratkie Soobshcheniya Instituta Istorii Material'noy Kul'tury* 82:5–36 (in Russian).

GENERAL INDEX

INDEX TO FAUNAL AND FLORAL TAXA